A Companion to Cognitive Science

Blackwell Companions to Philosophy

This outstanding student reference series offers a comprehensive survey of philosophy as a whole. Written by today's leading philosophers, each volume provides lucid and engaging coverage of the key figures, terms, topics, and problems of the field. Taken together, they provide the ideal basis for course use, representing an unparalleled work of reference for students and specialists alike.

Published

1 A Companion to Ethics
Edited by Peter Singer

2 A Companion to Aesthetics
Edited by David Cooper

3 A Companion to Epistemology
Edited by Jonathan Dancy and Ernest Sosa

4 A Companion to Contemporary Political Philosophy
Edited by Robert E. Goodin and Philip Pettit

5 A Companion to the Philosophy of Mind
Edited by Samuel Guttenplan

6 A Companion to Metaphysics
Edited by Jaegwon Kim and Ernest Sosa

7 A Companion to Philosophy of Law and Legal Theory
Edited by Dennis Patterson

8 A Companion to Philosophy of Religion
Edited by Philip Quinn and Charles Taliaferro

9 A Companion to the Philosophy of Language
Edited by Bob Hale and Crispin Wright

10 A Companion to World Philosophies
Edited by Eliot Deutsch and Ron Bontekoe

11 A Companion to Continental Philosophy
Edited by Simon Critchley and William Schroeder

12 A Companion to Feminist Philosophy
Edited by Alison Jaggar and Iris Young

13 A Companion to Cognitive Science
Edited by William Bechtel and George Graham

Blackwell Companions to Philosophy

A Companion to Cognitive Science

Edited by

WILLIAM BECHTEL and GEORGE GRAHAM

Advisory editors

David A. Balota
Paul G. Chapin
Michael J. Friedlander
Janet L. Kolodner

Blackwell Publishers Ltd
108 Cowley Road
Oxford OX4 1JF
UK

Library of Congress Cataloging-in-Publication Data

A companion to cognitive science / edited by William Bechtel and
George Graham; advisory editors, David A. Balota . . . [et al.].
p. cm. — (Blackwell companions to philosophy; 13)
Includes bibliographical references and indexes.
ISBN 1–55786–542–6 (hardcover: alk. paper)
1. Cognitive science. I. Bechtel, William. II. Graham, George,
1945– . III. Balota, D. A. IV. Series.
BF311.C578 1998
153—dc21 97–38757
CIP

British Library Cataloguing in Publication Data

A CIP catalogue record for this book is available from the British Library.

Typeset in 10 on 12 pt Photina
by Graphicraft Typesetters Limited, Hong Kong
Printed in Great Britain by T.J. International, Padstow, Cornwall

This book is printed on acid-free paper

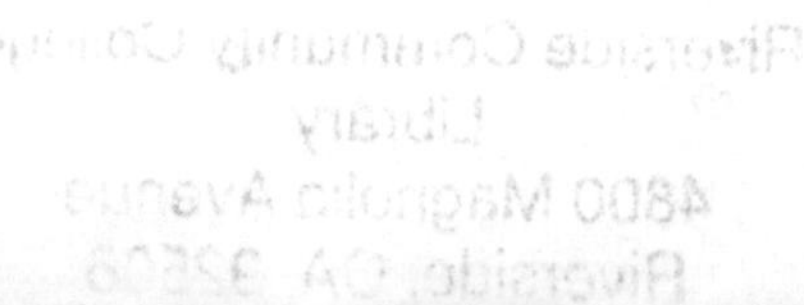

Contents

Contributors

Adele Abrahamsen, Linguistic Studies Program and Department of Psychology, Washington University in St Louis

Rhianon Allen, Department of Psychology, Long Island University

Rita E. Anderson, Department of Psychology, Memorial University of Newfoundland

Vivek Anumolu, CompuWare, Inc., Milwaukee, Wisconsin

Irene Appelbaum, Department of Philosophy, University of Montana

Elizabeth Bates, Center for Research in Language, University of California at San Diego

William Bechtel, Philosophy–Neuroscience–Psychology Program and Department of Philosophy, Washington University in St Louis

Marc Bekoff, Department of Environmental, Population, and Organismic Biology, University of Colorado

Dorrit Billman, School of Psychology, Georgia Institute of Technology

Kathryn Bock, Department of Psychology, University of Illinois at Urbana–Champaign

Norman W. Bray, Department of Psychology, University of Alabama at Birmingham

William F. Brewer, Department of Psychology, University of Illinois at Urbana–Champaign

John T. Bruer, James S. McDonnell Foundation, St Louis

Randy L. Buckner, Department of Psychology, Washington University in St Louis

Alison L. Chasteen, Department of Psychology, Washington University in St Louis

Andy Clark, Philosophy–Neuroscience–Psychology Program and Department of Philosophy, Washington University in St Louis

Austen Clark, Department of Philosophy, University of Connecticut

Peter Dayan, Department of Brain and Cognitive Science, Massachusetts Institute of Technology

Terrence W. Deacon, Department of Anthropology, Boston University

Kevin Dunbar, Department of Psychology, McGill University

Jeffrey L. Elman, Department of Cognitive Science, University of California at San Diego

K. Anders Ericsson, Department of Psychology, Florida State University

Paul A. Estin, Department of Psychology, University of Michigan

Owen Flanagan, Department of Philosophy, Duke University

Kathryn L. Fletcher, Department of Psychology, University of Miami

Robert Frank, Department of Cognitive Science, Johns Hopkins University

Christopher D. Frith, Wellcome Department of Cognitive Neurology, Institute of Neurology, London

Susan M. Garnsey, Department of Psychology, University of Illinois at Urbana–Champaign

Dedre Gentner, Department of Psychology, Northwestern University

Lyn M. Goff, Department of Psychology, Washington University in St Louis

Art Graesser, Department of Psychology, University of Memphis

George Graham, Department of Philosophy, University of Alabama at Birmingham

Paul E. Griffiths, Department of Philosophy, Otago University

Lisa A. Grupe, Department of Psychology, University of Alabama at Birmingham

Valerie Gray Hardcastle, Department of Philosophy, Virginia Polytechnic Institute and State University

Gilbert Harman, Department of Philosophy, Princeton University

Terence Horgan, Department of Philosophy, University of Memphis

Lisa F. Huffman, Department of Psychology, University of Alabama at Birmingham

Mark H. Johnson, MRC Cognitive Development Unit, London

Mark L. Johnson, Department of Philosophy, University of Oregon

Annette Karmiloff-Smith, MRC Cognitive Development Unit, London

Alex Kirlik, Center for Human–Machine Systems Research, Georgia Institute of Technology

David Klahr, Department of Psychology, Carnegie–Mellon University

Alan J. Lambert, Department of Psychology, Washington University in St Louis

Barbara Landau, Department of Psychology, University of Delaware

D. Terence Langendoen, Department of Linguistics, University of Arizona

David B. Leake, Computer Science Department, Indiana University

Donald G. MacKay, Department of Psychology, University of California at Los Angeles

Brian MacWhinney, Department of Psychology, Carnegie–Mellon University

Robert N. McCauley, Department of Philosophy, Emory University

Barbara C. Malt, Department of Psychology, Lehigh University

Douglas Medin, Department of Psychology, Northwestern University

Punyashloke Mishra, Department of Psychology, University of Illinois at Urbana–Champaign

P. Read Montague, Division of Neuroscience, Baylor College of Medicine

Jennifer Mundale, Department of Philosophy, Hartwick College

Nancy J. Nersessian, Cognitive Science Program, Georgia Institute of Technology

Douglass C. North, Department of Economics, Washington University in St Louis

Charles W. Nuckolls, Department of Anthropology, Emory University

Domenico Parisi, Institute of Psychology, National Research Council, Rome, Italy

Steven E. Petersen, Department of Neurology, Washington University Medical School

Kim Plunkett, Department of Experimental Psychology, Oxford University

Alexander Pollatsek, Department of Psychology, University of Massachusetts at Amherst

Keith Rayner, Department of Psychology, University of Massachusetts at Amherst

Arthur S. Reber, Department of Psychology, Brooklyn College, City University of New York

Kevin D. Reilly, Department of Computer and Information Science, University of Alabama at Birmingham

Robert C. Richardson, Department of Philosophy, University of Cincinnati

Lance J. Rips, Department of Psychology, Northwestern University

Edwina L. Rissland, Department of Computer Science, University of Massachusetts at Amherst

Henry L. Roediger III, Department of Psychology, Washington University in St Louis

Herbert L. Roitblat, Department of Psychology, University of Hawaii

Barry Saferstein, Communication Program, California State University, San Marcos

Christian D. Schunn, Department of Psychology, Carnegie–Mellon University

T. R. Stanford, Department of Neurobiology and Anatomy, Bowman Gray School of Medicine of Wake Forest University

B. E. Stein, Department of Neurobiology and Anatomy, Bowman Gray School of Medicine of Wake Forest University

Robert S. Stufflebeam, Department of Philosophy and Religion, University of Tulsa

Ron Sun, Department of Computer Science, University of Alabama at Tuscaloosa

Paul Thagard, Departments of Philosophy and Computer Science, University of Waterloo

John Tienson, Department of Philosophy, University of Memphis

Pam Tipping, Department of Psychology, University of Memphis

Michael Tomasello, Department of Psychology, Emory University

A. H. C. van der Heijden, Department of Psychology, Leiden University

Cees van Leeuwen, Faculty of Psychology, University of Amsterdam

Mark F. Villa, Department of Computer and Information Science, University of Alabama at Birmingham

M. T. Wallace, Department of Neurobiology and Anatomy, Bowman Gray School of Medicine of Wake Forest University

Sandra R. Waxman, Department of Psychology, Northwestern University

James V. Wertsch, Department of Education, Washington University in St Louis

J. Frank Yates, Department of Psychology, University of Michigan

Tadeusz Zawidzki, Philosophy–Neuroscience–Psychology Program, Washington University in St Louis

Website notice

A website has been established for this volume at http://www.artsci.wustl.edu/~wbechtel/companion/. At the time of publication it will include a description of the volume, the address of each contributor, and a supplement to the biographical section at the end of the volume. Comments, discussion, and errata received by the editors may also be posted; their addresses are available on the website.

Preface

You have a companion in your hands – a companion of a special sort. It is a guide to one of the most important scientific developments of the end of the twentieth century: multidisciplinary cognitive science.

The expression *cognitive science* is used to describe a broadly integrated class of approaches to the study of mental activities and processes and of cognition in particular. Cognitive science is broad not just in the sense of encompassing disciplines as varied as neuroscience, cognitive psychology, philosophy, linguistics, computer science, and anthropology, but also in the sense that cognitive scientists tend to adopt certain basic, general assumptions about mind and intelligent thought and behavior. These include assumptions that the mind is (1) an information processing system, (2) a representational device, and (3) (in some sense) a computer.

As the Companion reveals, various interpretations of, as well as relations among, the above assumptions exist within cognitive science. Indeed, the entire set is not shared by all who dub themselves cognitive scientists. Partly because of diverse interpretations and other differences, cognitive science has generated vigorous dialogues concerning the nature of mental activities and processes, as well as over the nature of science and the structure of disciplines.

What makes this a Companion to cognitive science?

What makes the book a *companion* to cognitive science is that it presents everything needed to acquire working familiarity with cognitive science: its origins, central research areas and methodologies, main achievements, intellectual stances and controversies, and likely future developments. It should serve as a reference book, classroom text, and resource guide. It should be readable by nonacademics and nonspecialists, graduate and undergraduate students taking first courses in cognitive science, and also specialists in disciplines which are part of cognitive science but who wish an overview of topics outside their own specialty.

Although cognitive science has existed as a multidisciplinary research endeavor for a couple of decades, its character and content have not been static, and are indeed undergoing fundamental changes at present. This volume is organized so as to describe not only the past and present of cognitive science, but future problems for inquiry and new approaches to conceptualizing cognitive phenomena, including perspectives from neuroscience and from social and ecological studies. Moreover, the volume includes articles that examine not just the central inquiries of cognitive science but real-world applications of work that has been done in cognitive science.

Following each article is a list of references. Some (indicated by asterisks) are recommended readings that were selected to take the reader to the next level of understanding of the topics covered in the article. The others are works cited in the article. The number of citations has been minimized so as to allocate the greatest amount of space to exposition.

The organization of the Companion

The companion is organized in six parts. What are those parts? And what are they designed to do?

Part I is an **overview** and anticipation of the whole of cognitive science. It describes the origins of contemporary cognitive science, depicts the contributions of different disciplines to cognitive science, explains why and how cognitive science is transforming the understanding of mind and behavior, discusses institutional structures that have developed to facilitate cognitive science research, and attempts to provide a clear, readable introduction to the other parts and issues of the Companion. Coupling metaphor to exposition, Part I conceives of cognitive science as a developing organism with a biography of its own. Yesterday's birth becomes today's development and tomorrow's remembered achievement; there are family gatherings, social tensions, lines of ancestral influence, anxieties over self-definition, and aspirations for the future. Part I is entitled "The Life of Cognitive Science."

Part II is devoted to **areas of study** within cognitive science. A number of different phenomena comprise mental activities and processes. These provide areas of study for investigators within cognitive science – to name just a few: attention, consciousness, imagery, language, memory, perception, and reasoning. It is largely as a result of focusing on these common phenomena that cognitive scientists, though coming from different disciplines and using different research methods, interact with each other. The articles in Part II attempt to characterize the problems that arise in the various areas of study and some of the outstanding discoveries that have been made. The articles are intended to offer grounding in actual research accomplishments in cognitive science that will be useful in taking up more theoretical matters or attempts to relate work in cognitive science to more real-world human endeavors.

One reason why cognitive science is such a dynamic research field is that researchers bring a broad range of research methodologies to bear on phenomena of common interest. Typically these research methods develop primarily in one cognitive science discipline, but they are borrowed and often modified by those in other disciplines. Thus there is not a rigid connection between a specific research methodology and a given discipline. The articles in Part III, on **methodologies of cognitive science**, indicate the range of methodologies within cognitive science.

Also adding to the vitality of cognitive science is the fact that different cognitive science researchers adopt different stances on cognitive phenomena. These stances shape researchers' inquiries by directing their attention to particular questions and conceptions of what count as answers to those questions. A stance is an overall perspective on what should be studied and by what methods, and how explanations should be framed. As with methodologies, frequently a stance is embraced first in one of the participant disciplines of cognitive science and then migrates to influence researchers in other contributing disciplines. The articles in Part IV characterize the different

stances and exhibit how advocates of each go about the practice of cognitive science. Although those who adopt different stances may frequently engage in dialogue, the differences between stances do not readily or commonly lend themselves to empirical resolution or inquiry, and thus the discussions tend to settle into ongoing debates. Frequently those who adopt one stance are led to investigate particular problems that can be answered within the perspective of that stance, while those who adopt a different stance will be directed to different problems.

In addition to broad theoretical stances, cognitive science inquiry is characterized by a number of controversies that reach across various areas of study. These are examined in Part V, on **controversies**. They concern particular features of the cognitive system or ways of examining it. Unlike stances, the controversies are often objects of empirical investigation. Empirical evidence to date has not resolved these controversies, but has regularly forced changes in positions.

While cognitive scientists typically have construed their work as part of basic science, some have anticipated the consequences which their investigations may have for other aspects of human life. Increasingly, cognitive scientists devote themselves more explicitly to relating their inquiries to those other areas. The focus on real-world problems is, in turn, transforming some of the basic science inquiries. The articles in Part VI on **cognitive science in the real world** discuss both current endeavors relating cognitive science to other human pursuits and the potential for further developments in these directions.

What is different about this book?

Many books describe limited aspects of cognitive science or take approaches to cognitive science which reflect emphases on certain disciplines or assumptions rather than others. Readers familiar with other books on cognitive science may want to know what is different about this Companion.

The most obvious difference is its forward-looking organization. The excitement – and the anxiety – of cognitive science is that it reaches into the twenty-first century with unsettled self-definition. Much editorial attention has been paid, therefore, into making this book anticipatory of future developments of cognitive science in neuroscience, socioculturally embedded cognition, the emotions, and animal modeling, and into showing that cognitive science is not made of disciplinary steel. Its character is open to theoretical refinement and empirical revision.

Much attention has also been paid to designing a book that is as comprehensive as possible in its depiction of cognitive science, but also as reader-accessible and learner-friendly as a guide to a science under construction can be. It is meant to provide a coherent view of a broad scientific terrain, offering different points of entry for different sorts of readers to aspects of cognitive science.

How to read this book

The Companion's six parts serve distinct if related purposes. They are assembled so that they may be read independently of one another. They contain, in effect, self-contained essays on a variety of topics in cognitive science, arranged alphabetically by topic and with frequent cross-references. Thus someone who wants to learn about connectionism, artificial life, and dynamical systems can read chapter 38 in Part IV (Stances) as a self-enclosed account of those topics. When read together with the other essays in Part IV, the reader may compare and contrast different stances on cognitive phenomena which have shaped cognitive scientific inquiry. Similar reading and learning opportunities apply to the other five parts.

When all six parts of the Companion are read in order, however, they tell a more or less unified story of cognitive science. The importance of becoming familiar with cognitive science through a unified story should not be underestimated. There is, understandably, among students and nonpractitioners of cognitive science, a widespread anxiety about learning cognitive science. There is *so* much literature, *so* much activity, *so* many conferences, *so* many people calling themselves cognitive scientists, that initiation into the field can feel more like a lurching or leaping than an acquiring – and it is easy to slip into a shadow unilluminated by what is truly important. A good companion to cognitive science tells a directed tale; it keeps light aimed on main themes, ideas, and issues.

Suggestions for how to teach from this book

Some readers will become acquainted with this book in the classroom; some will teach from it. Teachers may find the organization of the Companion useful in class or seminar instruction.

In particular, suppose you are teaching an introduction to cognitive science. How may this book be used as a text? You might use Part I, The Life of Cognitive Science, to organize the course across most of the semester, periodically taking excursions into other parts of the book by following some of the cross-references to areas, methodologies, and stances and examining various controversies and applications. Some of the suggested readings at the ends of articles might be selected for class use as well.

Alternatively, you may be teaching a thematically focused course from a broad, cognitive science perspective. In that case you would pick and choose segments of the book that address your theme, and you might also make heavy use of the suggested readings. A course on cognitive neuroscience or mind/brain, for example, may begin with the overview of cognitive science, paying special attention to the role of neuroscience within the evolution of cognitive science, and then consider such topics as brain mapping and language evolution and neuromechanisms in Part II, neuroimaging, single neuron electrophysiology, and deficits and pathologies in Part III, neurobiological modeling and perhaps one or two other articles in Part IV. The course might end with the controversies on binding, innateness, modularity, and levels and cognitive architectures in Part V.

A third route through the book is to emphasize a particular cognitive science discipline. A philosophy class, for example, might begin with the historical perspective provided in section 2.6 of Part I, Getting a Philosophy. Philosophical concerns are prominent in Part V (Controversies), and also make some appearances in Part IV (Stances). There are articles on such topics as conceptual change, consciousness, and reasoning in Part II and ethics in Part VI. Finally, an instructor might use some of the articles, especially in Part III, as material for addressing issues in philosophy of science.

William Bechtel and George Graham

Acknowledgments

Our debts of gratitude and appreciation are deep and many. Preparations for the Companion began in 1993, prompted and generously reinforced by an invitation from Stephan Chambers of Blackwell Publishers. He asked us to conceive of an ambitious volume the likes of which had yet to appear in the service of cognitive science. We would like to thank him for his interest and support. We would also like to thank our editor, Steve Smith, and his staff at Blackwell, especially Mary Riso, Margaret Aherne, and Jean van Altena. Thanks are owed to our home universities of Washington University in St Louis and the University of Alabama at Birmingham for a variety of forms of assistance, including a grant in one case and teaching leave in another. Beth Stufflebeam helped to produce the final manuscript. Suggestions, information, and help of other sorts came not just from our distinguished advisory editors, David A. Balota, Paul G. Chapin, Michael J. Friedlander, and Janet L. Kolodner, but from a host of friends, colleagues, and students, including (among others) John Bruer, Chris Eliasmith, Pete Mandik, Donald Norman, Martin Ringle, Betty Stanton, and Joseph Young. We also wish to thank Adele Abrahamsen, whose assistance with the volume went well beyond her own overt contributions to the book. She became a kind of virtual Companion editor.

We dedicate the book to Adele for her assistance and to George's wife, Patricia, for her encouragement and support over the several years of composition and construction.

PART I

THE LIFE OF COGNITIVE SCIENCE

WILLIAM BECHTEL, ADELE ABRAHAMSEN, AND GEORGE GRAHAM

Contents

Preliminaries

Let's begin prematurely. Let's try to characterize cognitive science:

> Cognitive science is the multidisciplinary scientific study of cognition and its role in intelligent agency. It examines what cognition is, what it does, and how it works.

That proposition may appear more definitive than it truly is. Which creatures or sorts of things count as intelligent agents? Insofar as cognitive science seeks to be multidisciplinary, which scientific disciplines are included? Do they interact substantively – share theses, methods, views – or do they simply converse? Finally, how does one discover what cognition is, what it does, and how it works? Cognitive scientists answer these questions in a variety of ways. No answer is without dissent. Each inspires controversy: everyone likes some answer, but no one likes every answer.

Shall we chart the answers? Only a conceptual botanist would delight in that task; besides which, it would be a premature and unhelpfully abstruse way in which to introduce both cognitive science and the content of this Companion. To those two related ends we prefer a short anecdote, then a long story – a very long story. We shall revisit the above characterization at the very end of the story, for by then the abstruse will have metamorphosed into the familiar, and any sources of controversy will be intelligible if not eliminable.

An anecdote: Building 20

Though all three of us objected to the Vietnam War, one of us (GG) was formally classified as a conscientious objector and, during the early 1970s, performed civilian alternative work service for New England Deaconess Hospital in Boston. One day – his day off – on a rather aimless walk through the campus of Massachusetts Institute of Technology in Cambridge, he came upon some stoically wooden buildings set unobtrusively in the middle of the campus. One was marked simply "Building 20." Looking for a telephone, the future co-editor asked a student standing in front of the building, "Is there a public phone in 20?" "I don't know," replied the student. "All I know about 20 is that Noam Chomsky works here."

"Noam Chomsky?" One hates to admit such ignorance, but being new to Cambridge and unfamiliar with *Syntactic Structures*, perhaps one can be forgiven.

"What?!" Befuddled, but trying to be polite: "Why, he's the world's leading linguist." In retrospect, I had stumbled into the domain of one of the prime movers of modern cognitive science. Chomsky was both icon of the Cambridge anti-war movement and hero of the battle against anti-cognitive psychology – behaviorism.

"Without Chomsky," added the student, "you would be left with B. F. Skinner and his rats up at Harvard."

It was the early 1970s. Talk of cognition thickened the air; cognitive science was growing up. So how *did* cognitive science form? How did it self-conceive and mature? Certainly Chomsky played a key role. Others did too. Time for the long story.

A predecessor: behaviorism

In North America something dramatic happened in psychological science in the 1950s, something often referred to, in retrospect, as the *cognitive revolution*, something Howard Gardner characterized as "the unofficial launching of cognitive science" (Gardner, 1985, p. 7). The revolt was against behaviorism, which was heralded in John Watson's 1913 manifesto and quickly came to largely dominate psychology and linguistics, and influence other disciplines in North America. Behaviorism turned away from earlier, mentalistic attempts to analyze the mind; instead it focused on overt behavior and the discovery of regularities involving observable events and behaviors. "Psychology," wrote Watson, "as the behaviorist views it is a purely objective experimental branch of natural science" (1913, p. 158). Behaviorism was a blend of Darwinism, functionalism in psychology, and anti-introspectionism. It was a normative meta-psychology; it tried, from its own platform, to legislate psychologists into being good empirical scientists. Here, very quickly, most roughly, and simplified stepwise, is how behaviorism said psychology should be done:

Step One: Observe behavior.

Step Two: Select descriptions of behavior which are nonmentalistic – that is, which do not presuppose theorizing about the internal psychology of the organism or agent in question.

Step Three: Select descriptions of the environment (in which the observed behavior takes place) which themselves are nonmental in that they do not presuppose theorizing about how the organism or agent represents its environment.

Step Four: Note that certain nonmental aspects of behavior (such as its frequency of occurrence, physical direction, and so forth) seem to be correlated with certain nonmental aspects of the environment (physical stimuli which are present).

Step Five: Judiciously vary – in a laboratory model and experimental setting – the environmental aspects; thereby determine the class of environmental events and the class of behaviors covered by the correlation.

Step Six: Speak of the behavior (response) as a function of the environment (stimuli); refer to environmental stimuli and behavioral responses as existing in a functional relationship.

A compressed example illustrates:

A rat scurries across the alley. It turns left towards a tipped garbage can and ingests food. Remove the rat from the alley. Place it in a laboratory maze. Vary the location of food pellets with the direction of its turning (whether it turns left or right). Notc that under certain conditions the behavior of turning left or right is correlated with its immediate history of ingesting food. The history is "responsible" for the direction. Left turning is a function of a food-left history; right turning is a function of a food-right history.

The specification of functionally related stimuli and responses posed a number of problems for behavioristically oriented psychology, itself sometimes called "the experimental analysis of behavior." Often, for example, stimuli and responses selected for a functional class cannot be usefully characterized in an *apsychological* (nonmental) vocabulary. Consider, for example, the temptation to classify the rat's responses as *seeking* food and *remembering* whether it was found to the left or right. Mentalistic attribution is a tough temptation to resist. In some cases – human verbal behavior, for instance – it is impossible to resist. However, let's return to the chronology.

In North America behaviorism reigned for decades as a remarkably resilient, influential, and in many ways laudable doctrine that resonated through a number of disciplines beyond psychology. In linguistics it helped to displace philology (the study of the histories of particular languages) with empirical studies of language use. Under the leadership of Leonard Bloomfield, linguistic behaviorism aspired to carry out a program in which linguists would collect speakers' utterances into a corpus and produce a grammar that described it. Explicitly excluded were any mentalistic assumptions, inferences, or explanations.

In philosophy, the logical positivism of Rudolf Carnap and Carl Hempel was congenial to behaviorism. Each tried to develop behavioristic canons for the meaningfulness and empirical grounding of scientific hypotheses. Hempel himself eventually abandoned this effort: "In order to characterize the behavioral patterns, propensities, or capacities . . . we need not only a suitable behavioristic vocabulary, but psychological terms as well" (Hempel, 1966, p. 110). Others maintained a thoroughgoing empiricism. Willard van Orman Quine imposed behavioristic standards on the task of interpreting the speech of another person (or oneself) and argued that the only evidence available was the sensory input from the environment. He argued that from this evidence alone the meaning of a sentence would always be indeterminate, and therefore concluded that the notion of meaning was vacuous. He made an exception only for those statements most firmly rooted in sensory experience (observation statements).

The story to be told

Not everyone agreed with behaviorist strictures. To such critics as the aforementioned resident of Building 20, behaviorism was a severely truncated, virtually atheoretical stance. The historical events to be discussed in the next section clearly represent a rebellion against behaviorism and the birth of a new approach. The first stirrings of life of a cognitive science revolution occurred at the end of World War II. The concept of *information* came to center stage in cybernetics, information theory, and early neural networks. This enabled cognitive researchers to cast off their fear of mentalism and attempt to understand the processing of information in the head – in the mind – that underlies behavior. By the mid-1970s the conceptual and methodological frameworks of linguistics, psychology, and philosophy were fundamentally altered in ways characteristic of what Thomas Kuhn (1962/1970) has referred to as a "scientific revolution." A generation of new thinkers, including Chomsky, George Miller, and Hilary Putnam, had created a new *paradigm*, and a new generation of researchers took up the banner and gave birth to a radically different set of research agendas. In addition, a brand new discipline – artificial intelligence – emerged, and such leaders as Allen Newell and Herbert Simon linked its approach to those of the other disciplines. Neuroscience also made major advances, but within its own paradigm.

The story that follows is about the development of cognitive science as an intellectual enterprise *and* as an institution. The choice of conjunction is intentional. The intellectual enterprise of cognitive science did not develop independently of its institutions; so if we are interested in the enterprise of cognitive science, we need to mention the kinds of social mechanisms, ranging from journals to graduate programs, that often both reflected and helped to support intellectual changes.

The story that we tell, like all stories, is selective. Without filling the entire volume with historical narrative, there is no way we could cover all the plot lines of research and theory that contributed to what is now known as cognitive science. We have chosen to emphasize work that is interdisciplinary in its nature or impact, with the result that a number of major researchers doing core work within each discipline have been left out. Our other constraint is our goal of providing a context for the contemporary scene, especially as it is portrayed in the rest of the volume. Accordingly, this essay is skewed toward earlier developments, becoming briefer as we get closer to the current scene. The reader is encouraged to follow the links to the other contributions, which we have marked by setting their titles in small capitals. These links become ever more numerous as we go along, so as to fill out the story.

1 Gestation and birth of the cognitive revolution

Cognitive science did not emerge suddenly. Like a person, it went through a long period of gestation. Unlike a person, it has no official and unambiguous birthdate. Following one of its pioneers, George Miller, we have selected 1956 as a plausible year of birth. Most of the preparatory developments occurred in computer science, psychology, and neuroscience. Only in the last stages of this gestation did Chomsky arrive on the scene to begin transforming linguistics.

1.1 The seeds of computation

The attempt to design intelligent machines played a critical role in the development of cognitive science. Three different research traditions contributed to the development of such machines: cybernetics, artificial neural networks, and symbolic artificial intelligence (AI). While symbolic AI garnered attention as a central contributor when cognitive science took formal, institutional shape in the 1970s, both cybernetics and artificial neural network research played a major historical role as well, providing many of the ideas that allowed for characterization of events inside a person's head in cognitive or information processing terms. Early developments in artificial neural networks will be discussed in the context of neuroscience in section 1.3.5. For now, let's focus on cybernetics and artificial intelligence.

1.1.1 Cybernetics

How do living organisms maintain themselves in the face of changing and often threatening external environments? Here, roughly, is the answer of Claude Bernard, a mid-nineteenth-century physiologist: "Each living organism is composed of different component sub-systems; these respond when particular features of the organism's internal environment exceed – under pressure from its external environment or malfunctioning of a sub-system – their normal range. They act so as to restore that feature to the normal range." Bernard's notion of internal componential adjustment to external change contained the germ of the notion of feedback, which is central to cybernetics.

The idea of feedback was developed more thoroughly by Norbert Wiener, who had interests and training in biology before getting a degree in mathematical logic. He conceptualized feedback as consisting in the feeding of information generated by a system back into the system, thereby enabling it to adjust its behavior. Wiener (1948) coined the term *cybernetics*, which derived from the Greek for *helmsperson*, for this idea, and he proposed that natural and artificial systems could steer themselves by using feedback.

At MIT, where he was professor of mathematics, Wiener collaborated with Vannevar Bush, who in the 1930s had begun to develop an analog computer. At the start of World War II, they set out to design a system for improving anti-aircraft fire in which feedback would play a critical role. Information from radar would be employed to calculate adjustments to gun controls; after new shots were fired, information about the results would be used to readjust the gun controls. If this rather Bernardian procedure were automated, one would have a self-steering device; even if humans were part of the loop, the overall activity would count as one of self-steering by means of feedback.

As the war continued, Wiener collaborated with two other researchers, Julian Bigelow (an engineer) and Arturo Rosenblueth (a physiologist). The three scientists offered a cybernetic theory of "control and communication in the animal and machine." Rosenblueth gave the first public presentation in 1942 at a conference on Cerebral Inhibition sponsored by the Josiah Macy Foundation, which was soon to play a critical role in developing the cybernetic framework. Together Rosenblueth, Wiener, and Bigelow published a paper entitled "Behavior, Purpose and Teleology" in *Philosophy of Science* in 1943, in which they ventured to use the concept of feedback to legitimize the notion of *teleology* (goal direction); in their view, feedback enabled systems, both living and artificial, to be goal-directed.

In January 1945, Wiener, together with Howard Aiken and John von Neumann, brought together a group of theorists from a broad range of disciplines for a meeting on the notion of feedback. Among the participants were the neurophysiologists Warren McCulloch and Rafael Lorente de Nó, the logician Walter Pitts, and Samuel Wilkes (a statistician), Ernest Vestine (a geophysicist), and Walter E. Deming (a quality control theorist). On January 24 Wiener wrote to Rosenblueth (who had returned to Mexico), emphasizing the potential for integrating the study of the brain with engineering work on artificial systems:

> The first day von Neumann spoke on computing machines and I spoke on communication engineering. The second day Lorente de Nó and McCulloch joined forces for a very convincing presentation on the present status of the problem of organization of the brain. In the end we were all convinced that the subject embracing both the engineering and the neurology aspects is essentially one, and we should go ahead with plans to embody these ideas in a permanent program of research. (Quoted in Heims 1980, p. 186)

In addition, the participants talked of creating a journal and a scientific society after the war. These plans did not come to fruition, but Heinz von Foerster and McCulloch, with support from the Macy Foundation, did organize twice-yearly meetings of the group and invited investigators from an even broader array of backgrounds, such as psychologist Kurt Lewin and anthropologists Gregory Bateson and Margaret Mead. Originally, the conference series was called the Conference for Circular Causal and Feedback Mechanisms in Biological and Social Systems, but in 1949 it adopted Wiener's term *cybernetics* and changed its name to the Conference on Cybernetics.

The last meeting was in 1953, and the cybernetics movement waned. Some of its key ideas, such as the notion of feedback of information from environment to behaving system, were further developed later by cognitive scientists. Cybernetics also represented a first attempt at a broad, multidisciplinary endeavor to explain mental phenomena. An especially noteworthy difference from the cognitive science of the 1970s was the central role of neuroscience in cybernetics. Some of its products, such as W. Ross Ashby's *Design for a Brain*, put forth ideas that were in a sense ahead of their time and are only now bearing fruition in cognitive science.

1.1.2 Computers and artificial intelligence

Of all the research fields that would come to play a major role in cognitive science, ARTIFICIAL INTELLIGENCE, usually classified as a branch of computer science, was the newest, having to await the invention of the computer itself. The digital computer, as we know it, was another product of World War II, though the idea of automated computing goes back much further. One key element of computing is the idea of a set of instructions that can be applied mechanically. An early version of this idea was found in an 1805 device of Joseph-Marie Jacquard which used removable punch cards to determine the pattern which a loom would weave. In the 1840s, Charles Babbage made use of this idea in his design for an *analytical engine*, which was to have been a steam-driven computational device. Babbage never succeeded in actually building the engine. He did, however, engage in a fruitful collaboration with Lady Lovelace (Ada Augusta Byron), who worked out ideas for programming Babbage's machine. Ada, the modern programming language, was named in her honor.

A major hurdle faced by Babbage in the nineteenth century was the lack of sufficiently precise manufacturing for the components of his engine. However, by the start of the twentieth century, precision had improved to the point where mechanical calculators could be manufactured by companies such as Tabulating Machine Company, which later merged into IBM. These machines were purely mechanical – without electrical components – but in the late 1930s Claude Shannon showed that electric switches could be arranged to turn one another on and off in such a way as to perform arithmetic operations. The idea of using electronic circuits to carry out calculations was put into practical use during World War II in England by Alan Turing and his collaborators at Bletchley Park in the effort to decipher German military communications. The German cipher machine Enigma was a particular challenge, since it was built out of a set of rotors which permuted the letters of the alphabet; the rotors were mechanically coupled so as to constantly change the alphabetic substitutions employed in the cipher. The challenge to Turing and his colleagues was to examine all combinations of encoding assignments in the machine to find the one used in the cipher – a huge computational task. The result was Bombe, which employed a single electronic valve for fast switching. For highest-level communications, Germany employed an even more sophisticated cipher, which produced what researchers at Bletchley Park referred to as "Fish" cipher text. To decipher these messages, Turing and his colleagues designed a vacuum tube-based special-purpose machine, Colossus, which employed thousands of electronic valves.

Another World War II era computer, ENIAC (Electronic Numerical Integrator and Calculator), was developed by J. Presper Eckert and John Mauchly at the Moore School of the University of Pennsylvania. It was designed to calculate artillery tables, which would specify how to aim artillery on various terrains so as to hit desired targets. Despite massive effort, ENIAC remained incomplete until 1946. John von Neumann designed the basic architecture for ENIAC – the "von Neumann architecture." It was, however, only fully realized in ENIAC's successor, EDVAC (Electronic Discrete Variable Computer), and has continued to play a central role in computing to the present.

At the heart of von Neumann architecture is a distinction between a computer's memory and its central processing unit (CPU). One of von Neumann's innovations was to recognize that the instructions comprising a program could be stored in memory in the same manner as the data being operated upon. Computer operations are carried out in cycles in the CPU; in each cycle both data and instructions are read from memory into the CPU, which carries out the instructions and returns the results to memory.

We now come closer to the role of the computer in the birth of cognitive science, but we need to make another brief digression. After the war, computers became increasingly powerful. And with such power a possibility began to be realized that had first been envisaged by Gottfried Wilhelm Leibniz, the famous seventeenth-century philosopher at the University of Leipzig. He had proposed that numbers could be assigned to concepts, and that formal rules for manipulating those numbers would in effect also manipulate the concepts to which they were assigned. In 1854, the English mathematician George Boole had taken a major step in developing this idea in a book called *The Laws of Thought*. Boole formulated several operations that could be performed on sets. He also showed that these operations correspond to logical operators (*and*, *or*, *not*) which could be applied to propositions. He suggested that the laws governing these operations could serve as laws of thought. The switches that Shannon had devised in the late 1930s performed these basic Boolean operations, with the resulting state of the switches (*on* or *off*) corresponding to the truth values of the proposition (*true* or *false*).

Boole's system was limited to operations on complete propositions (e.g., "The woman is a lawyer") and could not deal with structure internal to the proposition (e.g., the fact that the predicate "is a lawyer" is being predicated of "the woman"). Gottlob

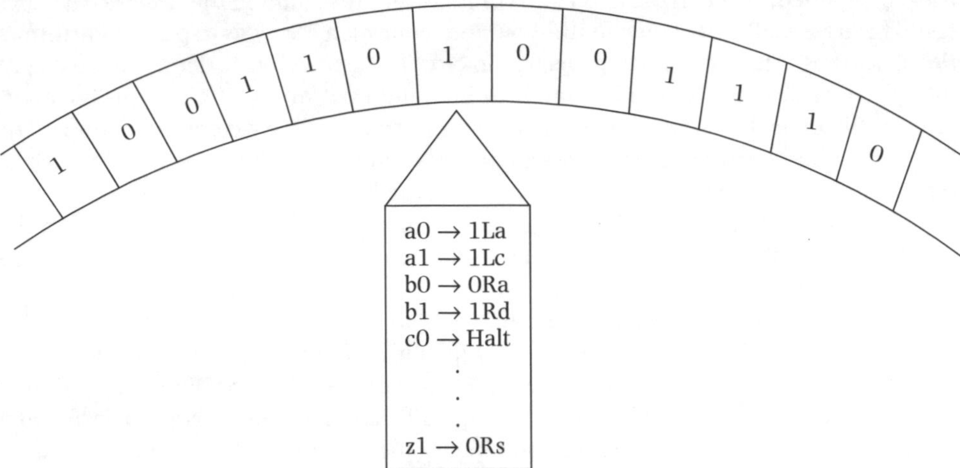

Figure I.1 A Turing machine.

Frege, though, expanded the system in 1879 to deal with such predications (permitting representations of arguments from premises such as "All lawyers have passed the bar exam" and "The woman is a lawyer" to "The woman has passed the bar exam"); the resulting system of predicate calculus provided a way of formalizing inferences that has been extremely influential. The idea of formally representing information in symbolic notation and using formal operations to transform this information provided a critical entrée to the use of computers to simulate reasoning.

Two additional ideas that have guided the use of computers to model thinking were contributed by Alan Turing. Even before the first digital computer was built, Turing (1936) proposed a simple machine for performing computations. The Turing machine (figure I.1) consisted of a read/write head and an infinite tape. The tape consisted of squares, each containing either a 0 or 1. The head could move one square to the left or right along the tape, read the numeral off that square, and write a replacement numeral. The head could be in any one of a finite number of states, each of which would specify what to do in response to a given number on the tape (e.g., if the square contains a 0, write a 1, move one square to the left, and remain in state *c*). Turing showed that for any well-defined series of formal operations (such as those of arithmetic), one could design a Turing machine which could carry it out. Also in the 1930s Alonzo Church independently proposed that any process for which there is a decision procedure could be carried out through such a series of operations; accordingly, the Church–Turing thesis proposes that any decidable process can be implemented in a Turing machine. Turing also proved that it was possible to construct a universal Turing machine that could simulate the operation of any given Turing machine; a universal Turing machine would thereby be capable of carrying out any well-defined series of operations.

If it were possible to provide it with infinite memory, a properly programmed von Neumann computer would be a universal Turing machine. The challenge would then be to provide it with the right program for carrying out all of the necessary formal operations. But is carrying out formal operations sufficient for thinking – for conceptual thought? Part of the difficulty in answering this question is that we lack a notion of thinking that is sufficiently clear for us to decide whether it could be accomplished in formal operations. What is needed is a way of specifying when something is thinking.

Turing had an ingenious proposal here as well. He offered a test – not the sole test, but a test – for thinking (Turing, 1950). His suggestion was to approach the question in terms of the behavior of the machine: could its behavior pass for that of a thinking person? If yes, it thinks. In what is now known as the Turing test, one decides whether a machine is thinking by arranging for a human interrogator at a keyboard to ask questions of both the machine and another human – both of whom are unseen but whose answers are displayed. If the interrogator, even after sophisticated questioning, cannot differentiate the computer from the human, then the computer's activity counts as thinking. Turing recognized that it would require a very complicated machine to engage in any protracted dialogue with humans and not be detected, but he believed that a computer would eventually pass this test.

By the early 1950s the theoretical foundations for artificial intelligence had been established; what remained was to actually build systems that exemplified aspects of human thinking. This task fell to a younger generation of investigators who were just then launching their careers. The team that set the prototype for the new enterprise of modeling intelligence – of producing an artificially intelligent system – consisted of Herbert Simon and Allen Newell. Neither Simon nor Newell was initially oriented towards computer science. Simon's background was in political science, and his appointment was in the Graduate School of Industrial Administration at Carnegie Tech (now Carnegie–Mellon University); he first made his reputation, and later won a Nobel Prize, in economics for his analysis of the functioning of human organizations. This work led him to challenge one of the tenets of modern economics, the assumption that agents are perfectly rational in the choices which they make (see Article 57, INSTITUTIONS AND ECONOMICS). Simon, to the contrary, emphasized that rationality was *bounded* and that, rather than examining all possibilities they face and then choosing one, humans generally accept the first option which meets a predetermined standard. Simon called this approach to decision making *satisficing* (see Article 44, HEURISTICS AND SATISFICING). He also drew from his work on human organizations the recognition that humans often rely on stock recipes, or *heuristics*, rather than seeking optimal solution procedures that guarantee correct answers. Lastly, Simon noted how lower divisions of a corporation typically pursue subgoals of the corporation's overall goal, thus suggesting the strategy of subgoaling in computer programs.

Starting in 1952, Simon became a consultant for the RAND Corporation, and on a visit to the RAND offices was intrigued by a printer he saw producing maps using characters other than numerals. He found the idea of a computer manipulating nonnumeric symbols attractive. At RAND he also met Allen Newell, who was developing such maps as part of a project of modeling an air defense center. The prospect that intrigued Newell came from Oliver Selfridge, who was using digital computers to simulate a neural net-like model for pattern recognition which he dubbed "Pandemonium." Such a model organizes a number of specialized agents (*demons*) into layers, with those in each layer competing in parallel to recognize a pattern. The output of lower layers serves as input to higher layers, as illustrated in figure I.2 for a model of letter recognition. Selfridge's work led Newell to the notion of a complex process being achieved through the interaction of simpler subprocesses.

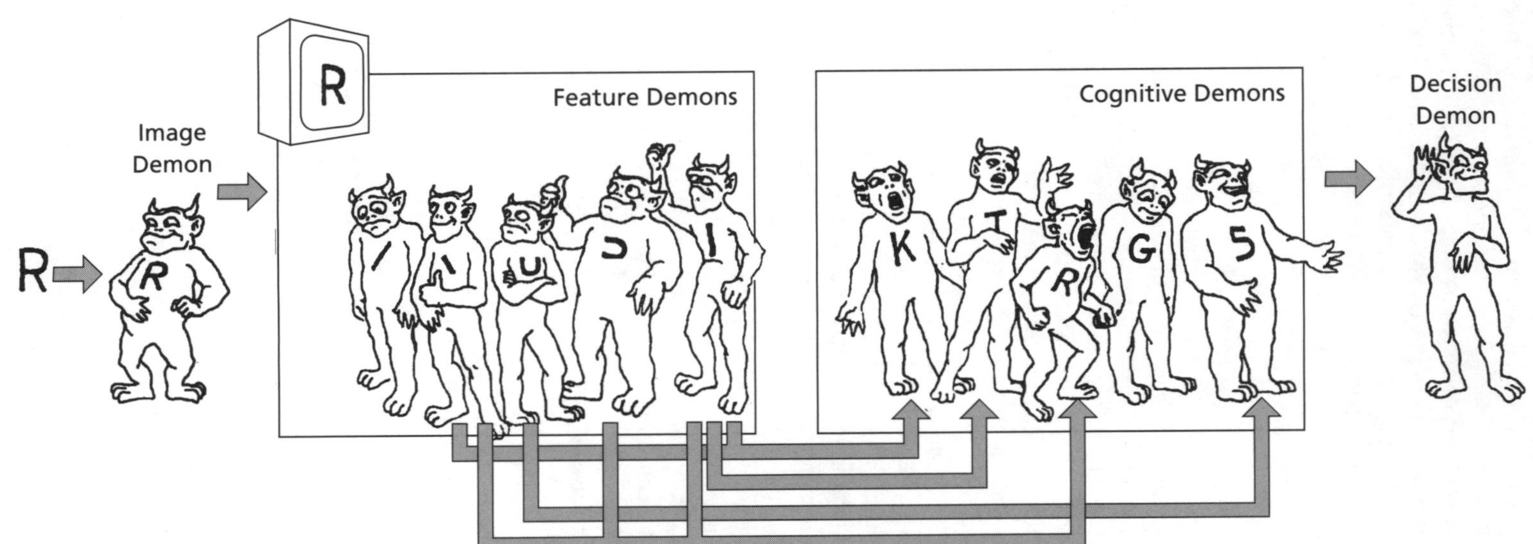

Figure I.2 Rendition of Oliver Selfridge's Pandemonium model. The *demons* at each level beyond the image demon (which merely records the incoming image of a letter) respond to those demons at the preceding level with which they have links (indicated by arrows). Figure drawn by Jesse Prinz.

Having both been drawn to ideas about problem solving through means–goal (means–end) reasoning via heuristics, Simon and Newell set out with computer programmer J. Clifford Shaw to develop a system that could prove theorems in formal logic. They first implemented the system with human agents (including Simon's children) playing the roles of the various parts of the program; each agent carried out rules written on index cards. When actually programmed in 1956, their so-called Logic Theorist proved 38 theorems from Russell and Whitehead's *Principia Mathematica*, one more elegantly than had Russell and Whitehead. The actual implementation of Logic Theorist represented more than the first apparent success of a computer program in performing a task requiring intelligence; it also brought the development of a list processing language, IPL (Information Processing Language). The symbols in a list could be stored at arbitrary memory addresses in the computer, and links could be added between one item and another simply by specifying at the first site the address of the related item.

Two other important pioneers in artificial intelligence were Marvin Minsky and John McCarthy. Minsky, after completing a dissertation on neural networks at Princeton in 1954, was drawn to modeling intelligence by writing programs for von Neumann-style computers. McCarthy also did his doctoral work at Princeton, first doing research on finite automata (systems like the read/write head of a Turing machine which could, by following rules, progress through a finite number of different states). During a summer at IBM in 1955, though, he too became attracted to modeling intelligent processes on digital computers. Minsky and McCarthy joined forces to organize a pivotal two-month workshop, the Dartmouth Summer Research Project on Artificial Intelligence. They secured funding from the Rockefeller Foundation, which enabled eight key researchers to join them at Dartmouth College during the summer of 1956. Nathaniel Rochester and Oliver Selfridge were using digital computers to simulate neural networks. The organizers and most of the other participants (Ray Solomonoff, Trenchard More, and Arthur Samuel, as well as Simon and Newell) were taking advantage of the computer's ability to manipulate symbols to simulate thinking more directly. The research of the eighth participant, Claude Shannon, involved information theory (see below) and automata theory. Most of the discussion during the conference was programmatic, taking off from the proposition initially put forward in the application to the Rockefeller Foundation that "every aspect of learning or any other feature of intelligence can in principle be so precisely described that a machine can be made to simulate it." In some respects, the highlight of the conference was the presentation of Logic Theorist, the prototype of an intelligent system. However, this also became the source of some social tension; because Newell and Simon were busy programming Logic Theorist during the summer, they attended the conference for only one week.

At the Dartmouth Conference the new enterprise was christened *artificial intelligence (AI)*. However, as so often happens at christenings, the name served both to unify and to divide. *Artificial* suggested that the form of intelligence exhibited by computers might differ from that of humans, and indeed, at the time, Minsky and McCarthy were not strongly committed to the idea that artificial intelligence would be particularly revealing about human cognition. Newell and Simon, who had the only working program, were more concerned with human cognition and were more directed toward psychology. For a number of years they resisted the label "artificial intelligence" and referred to "complex information processing." The very idea of information processing has an ecumenical conceptual feel to it, suggesting that just as humans process information, so, too, could computers, and perhaps even in the same way.

1.2 *Quickening: psychology makes its moves*

Many of the developments prior to the actual birth of cognitive science – developments discussed above – involved mathematicians and engineers laying the foundations for artificial intelligence. But AI was not the natural disciplinary locus for the study of thinking and reasoning. That, obviously, had to be psychology. And indeed, as cognitive science took life, psychology was one of its most important disciplinary contributors.

1.2.1 *Origins of psychology*

There is a very long history of inquiry into psychological questions within philosophy, and this left a substantial endowment to the newer discipline of psychology that emerged in the late nineteenth century; the best-known example is the idea that mental life can be understood in terms of elementary sensations that are combined by association (*associationism*). Nonetheless, it was developments in nineteenth-century German physiology that gave the immediate impetus to the emergence of psychology as a distinct, experimental science of the mind. Between 1850 and 1870, for example, Hermann Helmholtz devised a way to measure the speed of nerve transmission, proposed the notion of *unconscious inference* for operations occurring in the mind outside the reach of consciousness, and made major contributions to understanding how we see and hear (including his component theory of color vision and his studies of how subjects reorganize their perceptual experience after wearing distorting lenses for some time).

Helmholtz preferred not to label his work as psychology, but his laboratory assistant from 1858 to 1864 in Heidelberg, Wilhelm Wundt, was calling his own course "Physiological Psychology" by 1867 and had completed a major book on this topic by 1874. Moving to the University of Leipzig in 1875, Wundt established a demonstration laboratory which he upgraded to a research laboratory in 1879 – the event most often cited as the birth of psychology. Wundt's reach was extraordinary: he pursued more topics using a greater diversity of methods, taught more students (28,000), and published more pages (almost 60,000) than any of his friends or foes. During the last 20 years of his career, he focused on the use of nonexperimental methods to understand social life and language use. In the *Völkerpsychologie* that resulted, Wundt stretched beyond his inherited tradition of associationism to emphasize sentences as structured wholes that could not be reduced to a succession of words. Prior to this, Wundt was an experimentalist. In one line of research, he refined the Helmholtz–Donders technique of mental chronometry to produce surprisingly modern reaction time studies (see Article 27, BEHAVIORAL EXPERIMENTATION). He obtained time estimates for processes beyond simple perception, including *apperception* (achieving a synthesis or awareness of structure by means of conscious attention to a particular stimulus), cognition (required to discriminate one stimulus from others), and association (required when different stimuli have different responses). Wundt is better known, however, for his use of *introspection*, in which highly trained observers systematically analyzed their own mental experiences in order to identify the elements. Edward Titchener, the leading expositor of Wundt in the English-speaking world, focused almost exclusively on this technique.

Introspection was a major target of behaviorists; accordingly, Wundt's legacy of opening up a variety of scientific approaches to the study of mental activity was largely overlooked in North America during and after the ascendancy of behaviorism.

Whereas Wundt developed experimental foundations for the study of mental experience, William James emphasized the theoretical. At Harvard he too created a demonstration laboratory for psychology in 1874–5, but his genius lay in synthesizing and interpreting others' research rather than carrying out detailed experimental manipulations and measurements himself. Influenced by Darwin, James emphasized the adaptive function of behavior. His functionalism and his concern with individual differences contrasted with Wundt's structuralism and Kantian emphasis on intrapsychic universals – a theoretical divide that is still with us today in the camps of language acquisition researchers who emphasize language function and conversation versus those who follow Chomsky, for example. James developed his ideas in a monumental and engagingly written work, *Principles of Psychology*, ten years in the writing and finally published in 1890. His phenomenological descriptions of how mental activity is consciously experienced and how it figures in ordinary life are unparalleled and still frequently quoted. For James, habit functions as "the enormous fly-wheel of society, its most precious conservative agent" (vol. 1, p. 121), and consciousness is a continuously changing "stream of thought" (ibid., p. 224) rather than a construction of elements. James also contributed an influential theory of EMOTIONS, in which he claimed that "we feel sorry because we cry" (vol. 2, p. 450) rather than the reverse. By contrast, Wundt painstakingly gathered and analyzed introspections, arriving at three dimensions of variation in emotion: pleasantness/unpleasantness, activity/passivity, and tension/relaxation – which are strikingly similar to the results of Osgood, Suci, and Tannenbaum's (1957) factor analysis of rating scale data in the 1950s (section 1.2.2).

From its beginnings with Wundt and James, psychology developed quickly as a discipline. Many universities created chairs or departments of psychology, and the American Psychological Association was founded in 1892. Psychology was thus well established prior to the emergence of behaviorism, which not only repudiated the early tool of introspection but also, in the radical form advocated by John Watson in his manifesto noted above, prohibited any appeal to mental processes in explaining behavior.

1.2.2 The era of behaviorism

Watson can be credited with founding the behaviorist movement in the United States, but there are many other portraits with many other names in the behaviorist gallery: among others, Ivan Sechenev, Ivan Pavlov, Edwin Guthrie, Edward Thorndike, Edward Tolman, Clark Hull, Kenneth Spence, and, of course, B. F. Skinner. Some of these behaviorists were actually interested in cognition. Tolman, in particular, studied the ability of rats to navigate their environments and proposed that they did so by constructing *cognitive maps*; he also demonstrated learning without reward (*latent learning*) and posited a role for expectations and other *intervening variables*. He tried to show, in the words of one observer, "that a sophisticated behaviorism can be cognizant of all the richness and variety of psychological events" (cited in Bry, 1975, p. 59).

Tolman's primary rival, neobehaviorist Clark Hull at Yale University, pursued a less cognitive theory of stimulus–response learning. In his enormously influential 1943 book, *Principles of Behavior*, Hull systematized many of the research findings on instrumental conditioning in a *mathematico-deductive* theory. For Hull, how quickly a rat responded by pressing a lever (a measure of response strength) depended on such input variables as hours of food deprivation and number of trials reinforced by a food pellet. These are all observable variables; the tricky part was arriving at equations that would link them to posited intervening variables. *Drive* was a function of number of hours of food deprivation; *habit strength* was a negatively accelerated function of number of reinforcements and of drive reduction; and the *excitatory potential* that would lead to the actual response was a multiplicative function of drive, habit strength, and other variables. Hull's theory set much of the research agenda on learning in the 1940s. All too often, research results failed to support the theory, and it would be rescued with revisions until the next challenge.

Hull's camp (including Kenneth W. Spence at Iowa) continued sparring with Tolman's camp, but by the time of Hull's death in 1952 it was becoming increasingly obvious that neither theory had won and that no new grand theory would emerge. Tolman's cognition-friendly approach would not be appreciated again until the late 1960s. For those students who still wished to study animal learning, B. F. Skinner's (1938) *operant conditioning* paradigm was waiting in the wings at Harvard and was dominant by the late 1950s. Skinner designed elaborate schedules of reinforcement that produced pleasing regularities in the timing of naturally emitted behaviors (*operants*) or shaped them into more elaborate behaviors. Intervening variables and formal theory had no place in Skinner's *radical behaviorism*, which was more akin to Watson's than to Hull's thinking. A completely different, but still respectable, direction was to go inside the organism and work on *physiological psychology*. At Yale, for example, Neal Miller turned to studying the physiological underpinnings of learning, Frank Beach worked on neural and hormonal control of sexual and maternal behavior, and Karl Pribram did innovative work in neuroscience.

Yale's graduate students took various parts of Hull's legacy off in different directions; we will note three. First, some who were students in the 1940s continued the work on *verbal learning* that had been one of Hull's concerns, developing increasingly complex *mediation theories* (e.g., Osgood, 1953). The idea was to account for human language phenomena such as semantic generalization and transfer of training by positing chains of internal stimuli and responses that mediated between the observable ones. James J. Jenkins and colleagues (section 2.4) disconfirmed a key prediction in 1963, but optimism reigned during the 1950s. There was also an increased emphasis on methodology and measurement (e.g., it was shown that nonsense syllables varied in their meaningfulness, as indicated by the number of word associations they could produce; this became a factor to control in the design of experiments).

Second, some key students in the next generation, the early 1950s, gradually moved away from their Hullian roots in verbal learning to study memory, language, or visual imagery in the 1960s and beyond. George Mandler, for example, looking back on his Yale days (in Baars, 1986, p. 254), noted that "'Cognition' was a dirty word for us . . . because cognitive psychologists were seen as fuzzy, hand-waving, imprecise people who never really did anything that was testable." Nonetheless, Mandler became interested in the idea that memory was organized and made that one focus of his research at Harvard, Toronto, and then the new psychology department at the University of California, San Diego. Gordon H. Bower (a Neal Miller student) moved right from his

Ph.D. at Yale to an influential career at Stanford, and Roger Shepard (a student of Hull's student, Carl Hovland) went to Bell Laboratories and Harvard University before settling at Stanford in 1968. All three were leaders in creating the cognitive psychology of the 1960s and beyond (see section 2.4).

Third, a set of researchers that overlapped somewhat with the erstwhile verbal learners revamped Hull's *mathematical modeling* strategy. The key contribution came from an unexpected person and place: William K. Estes, who had obtained his Ph.D. under Skinner at Minnesota and followed him to Indiana but was strongly influenced by Hull and Guthrie. His 1950 stimulus sampling theory was a less global but better motivated learning model that helped kick off two vigorous decades of work in mathematical psychology. Estes moved to Stanford in 1962, where he joined Richard C. Atkinson (who had been his student at Indiana), Patrick Suppes, Bower, and a new generation of students. Although Estes left in 1968, this critical mass of researchers at Stanford would play a major role in producing improved learning models. Other centers of activity included Indiana University (where Atkinson's student Richard M. Shiffrin made his career), Harvard and MIT in Cambridge (including, at various times, C. F. Mosteller, Robert R. Bush, R. Duncan Luce, and the ubiquitous George Miller), the University of Pennsylvania (where Eugene Galanter collaborated with Luce and Bush in the 1960s), and Bell Laboratories. It would take us too far from our main story to meaningfully describe the work produced by these mathematical psychologists in the 1950s. However, the same individuals played an important role in the transition to cognitive psychology in the 1960s, and we will meet some of them again in sections 2.1–2.4.

1.2.3 *Alternatives during the era of behaviorism*

In addition to diversity within the behaviorist camp, a variety of psychological endeavors thrived beyond the bounds of behaviorism. Many of these were situated in Europe, where behaviorism actually had little impact. We will briefly explore a few examples of nonbehaviorist research in the first half of the twentieth century that later contributed to the development of cognitive psychology.

We start in Britain with Sir Frederic Bartlett (1932), an experimental psychologist who studied the role of subjective construction in MEMORY. Memories, he claimed, are not simple recordings of experienced events, but are filled in by their subjects and embellished with details not present in the original context. For example, when asked to recall a Native American folktale, "The War of the Ghosts" from the Kwakiutl people, his subjects made changes in the plot of the story which tended to Westernize it. To explain this, Bartlett proposed that they employed their existing *schemata* to organize events in the story. As we will see, the notion of a schema as a structure for organizing information in memory has played a major role in subsequent cognitive psychology and in cognitive science generally. Bartlett also trained a number of influential British psychologists, including David Broadbent, who pioneered ATTENTION research using multi-channel listening techniques.

Beginning in the 1920s in Switzerland, Jean Piaget produced a huge and impressive body of work in the field he called *genetic epistemology*. Piaget's route into psychology was via his precocious studies of biology – he published at age 10 and received a doctorate at 21 – and his early and long-standing desire to integrate scientific and epistemological concerns. Pursuing postdoctoral studies in psychology and philosophy, he worked for a time in the Binet Laboratory in Paris and became intrigued by children's errors on standardized reasoning tests. He proceeded to devise ingenious methods for uncovering children's changing competencies, and over the years worked out an elaborate and unabashedly mentalistic theory that laid out stages of development and the internal processes responsible for children's movement through these stages. The theory was an edifice of twentieth-century thought that spawned hundreds of disciples, critics, and revisionists throughout the world, including many in North America (see Article 6, COGNITIVE AND LINGUISTIC DEVELOPMENT).

During the 1920s in the (then) Soviet Union yet another tradition with a cognitive or psychological flavor emerged from the group at the Institute of Psychology in Moscow. Lev Vygotsky developed a cultural-historical approach to psychology which guided his empirical work on children's cognitive and linguistic development; Alexander Luria maintained a like degree of theoretical breadth while focusing much of his empirical work on language disorders and functions of the frontal cortex. Particularly influential was their proposal that cognitive abilities emerge in interpersonal interactions (e.g., talking to others) before they assume a central role in private mental life (e.g., thinking to oneself in language). Today's inquiries into MEDIATED ACTION are rooted in this Soviet tradition.

An especially important counterpoint to America's behaviorism was the emergence in Germany and Austria of Gestalt psychology in 1912. Gestalt psychologists primarily studied PERCEPTION, especially perceptual and cognitive organization. They observed that the global properties of a whole object, such as its overall contour, are often more salient in perception than are component parts. A foundational study of the Gestalt movement was Max Wertheimer's (1912) examination of the so-called *Phi* phenomenon: the apparent motion when one light flashes on-off a split second after a nearby light has flashed on-off. Wertheimer argued against accounts of the Phi phenomenon according to which it was built out of separate recognition of the two lights flashing and offered an alternative account in terms of so-called field properties of the brain.

The notion that persons and animals often see or perceive things whole was a central conviction of Gestalt psychology (rather than a secondary theme, as it was for Wundt). In some cases the whole is spatial, as when perceiving the roundness of an object; in other cases it is temporal, as when an individual imagines goals and organizes behavior as a means to those goals. An example of perceiving temporal wholes is Wolfgang Köhler's research from 1913 to 1917 with chimpanzees in the Canary Islands. Köhler posed the problem of securing a banana that was out of reach, for example, and noted that chimpanzees solved the problem not by random trial and error, but rather by an overall reorganization of the parts of the situation that enabled an intelligent solution. This reorganization seemed to be discovered and implemented after a period of quiet planning. Hence, a chimpanzee would observe the situation and then go to a tree, tear off a branch, and use the branch to get the bananas. Otto Selz argued that such PROBLEM SOLVING required organizing the problem in a stepwise manner, with each step involving an operation on a representation of the problem (his work was influential in the development of Simon's thinking in AI). Köhler himself characterized such solutions as exhibiting insight (the *aha!* experience).

Returning to North America, the extent of behaviorism's reach varied with location and research area. In particular, the research area of sensation and perception felt little real impact from behaviorism – due in part to deep historical roots in a line of empirical inquiry going back to the mid-nineteenth century. Most researchers proceeded along lines laid down long before the emergence of behaviorism. Nevertheless, individual circumstances created niches outside the mainstream for a few novel approaches. First, the major proponents of Gestalt psychology relocated to such colleges as Swarthmore and Smith in the 1930s due to Hitler's rise to power in Europe. Some of their ideas and phenomena got picked up, but the theory as a whole gained few converts. Second, J. J. Gibson developed an ecological approach that Eleanor J. Gibson extended to perceptual development. Both were exposed to Gestalt psychology during their early careers at Smith College, and Eleanor Gibson did her Ph.D. work with Hull and Hovland at Yale; but they turned their backs on these approaches to pursue the idea that information and invariances in the environment are available for *direct perception*. The Gibsonian sphere of influence extended from their base at Cornell University to such places as Haskins Laboratory and the University of Connecticut, but in recent years it has broadened (see section 3.3).

Returning to mainstream research on sensation and perception, we said this research had deep historical roots. How deep? In one area, color vision, a major achievement in 1968 was Jameson and Hurvich's integration of two opposing theories first proposed by Hering and Helmholtz in the nineteenth century (see Article 19, PERCEPTION: COLOR). In another area, psychophysics, researchers seek to establish relationships between physical stimuli and subjects' subjective experiences of them. In the nineteenth century, Gustav Fechner (building on work by Ernst Weber) demonstrated that with respect to such variables as brightness or loudness, the perceived intensity of a stimulus is proportional to the logarithm of its physical intensity. Although perceived intensities are subjective, psychophysicists were able to develop methods for obtaining behavioral reports that lent themselves to systematic analysis. For example, a standard stimulus would be paired with a succession of comparison stimuli, with the subject reporting for each pair whether they appeared to be the same or different; from these reports Fechner calculated the *just noticeable difference (JND)* at various intensities and derived *Fechner's law* (the logarithmic scale noted above). With some refinements, methods of this kind continued to dominate psychophysics for a century. Then in 1956 at his Psychoacoustic Laboratory at Harvard, S. S. Stevens introduced the new, more direct method of magnitude estimation, in which subjects assigned a number to one stimulus at a time. Using this method he obtained *Stevens' law* (in which the scale is based on a power function). This type of scientific method and progressive refinement of laws was considered respectable, even to most behaviorists. It is not merely coincidental that several of the researchers most responsible for the early development of cognitive psychology, including George Miller, Ulric Neisser, and Donald A. Norman, received their Ph.D. training in psychoacoustics.

Next we briefly note three research areas that arose more recently and are more diverse than sensation and perception. First, developmental psychology in North America was pluralistic enough to serve as a seedbed and safe haven for the study of mental functioning at the same time that some developmental researchers pursued the implications of behaviorism and others limited themselves to descriptive studies. Piaget's influence on developmental psychology in North America was delayed until the 1960s, when researchers replaced his flexible *revised clinical method* with standardized procedures that confirmed his empirical findings. But even prior to this, Arnold Gesell's maturational approach and Heinz Werner's organismic-developmental psychology were available as alternatives to behaviorist studies of learning and conditioning in children.

Second, social psychology was even more pluralistic. One relatively cognitive line of researchers began with Kurt Lewin, a nonorthodox Gestalt psychologist who was among those relocating to North America in the 1930s. He was mentor to such major figures as Leon Festinger, who proposed his famous theory of cognitive dissonance in 1957. The claim was that subjects would modify their beliefs so as to reduce the inconsistency or dissonance between their beliefs and their behaviors. When they were unable to eliminate the dissonance, they would exhibit psychological discomfort. Just one of the many kinds of evidence which Festinger obtained involved the dissonance that smokers experienced as information began to appear in the 1950s that smoking causes lung cancer. He showed that heavier smokers who did not succeed in stopping were more reluctant to accept the evidence than were more moderate smokers who also could not stop.

Finally, the antecedents of today's clinical psychology should not be forgotten. There was considerably less specialization in the first half of the twentieth century than in the second half, and a surprising number of America's leading experimental psychologists devoted part of their careers to investigating such topics as personality and aptitude assessment, hypnosis, emotion, and psychodynamic theory (some had even been psychoanalyzed).

1.2.4 Happenings at Harvard

Although we have identified a number of alternatives to behaviorism, its influence, at least in North America, was powerful and widespread. Accordingly, a major transformation was required before psychology in the United States could become a contributor to cognitive science: it had to become cognitive again. Mediation theory had already prepared the way for rejecting the behaviorist proscription on appealing to mental events in explaining behavior; the cognitivists completed the job by also rejecting the stimulus–response framework for conceptualizing internal events. However, cognitivism retained other aspects of the behaviorist legacy: (1) its principle that behavior provided psychology's objective evidence, and (2) its methods for systematically gathering and analyzing that evidence (especially statistical significance testing and mathematical modeling). Cognitivists would posit mental events without apology, but only after predicting and confirming their effects on observable behaviors.

Many of the first stirrings of a cognitive psychology that would eventually overthrow behaviorist strictures originated with two Harvard psychologists, Jerome Bruner and George Miller. Harvard's administration split its psychologists between two departments in 1946 (and rejoined them in 1972–3): a new interdisciplinary Department of Social Relations and a reconstituted Department of Psychology with Edwin G. Boring as chair. In addition to the personality conflicts that occasioned the reorganization, the two departments provided homes to two very different theoretical orientations – Boring called them *sociotropy* and *biotropy*. Bruner and Miller found themselves, despite converging interests, in different departments. Bruner was in the Department of Social Relations. One of his first contributions was the development of the New Look movement in the psychology of PERCEPTION, which (a) emphasized the contribution of internal mental states (partly determined by social factors) of the perceiver to what is perceived and (b) denied that the external stimulus is *the* determining factor in perception. In the psychophysics that dominated the study of perception at the time, variability in perceptual judgments was regarded as an impurity, but Bruner brought variability to center stage.

In 1947 Bruner and Cecile Goodman, a Harvard undergraduate, performed a study showing that children's judgments of the sizes of coins varied with their value: the size of lower-valued coins was underestimated, while that of higher-valued coins was overestimated. This contradicted the general principle of the psychophysical Law of Central Tendency, according to which judgments of smaller and larger items should err in the direction of the mean or central tendency in a series. Further, the overestimates of higher-valued coins were even larger for poorer children, revealing a social effect on perception.

Most of the major studies that defined the New Look were done collaboratively by Bruner and Leo Postman. Together they began to study the ability of subjects to read words flashed quickly through a tachistoscope. They discovered that the time required to read a word varied with a number of factors, including whether the word was closely associated with values that were strongly held by the subject. What was surprising about this and related findings was that in some way the semantic significance of a word could affect processes prior to the actual recognition of the word itself. Equally surprising were the results of their 1949 study of tachistoscopic perception of playing cards, some of which were anomalous in that the color was reversed (e.g., a red 10 of clubs). For each card, very brief exposure times were lengthened until it was recognized. With the anomalous cards, subjects would initially respond with a suit appropriate to the color displayed (e.g., a red 10 of clubs would be seen as a 10 of diamonds). Only at very long durations were they able to recognize the anomalous combination of suit and color. After experience with some of the anomalous cards, though, subjects learned to recognize them as rapidly as the normal cards. This set of experiments revealed both the role of expectations in perception and the possibility of learning to see new things or old things in new ways. (At the time of the Bruner–Postman experiments, Thomas Kuhn was a Harvard Fellow. In his 1962 book, *The Structure of Scientific Revolutions*, he employed the New Look results as evidence against the claim that scientists are purely objective reporters; rather, they may fail to see certain phenomena until a new paradigm changes their expectations.)

Bruner soon turned from showing the role of thinking in perception to a more direct exploration of thinking, which culminated in his 1956 book with Jacqueline Goodnow and George Austin, *A Study of Thinking*. Bruner and his colleagues regarded the learning and use of categories as central to thought. To investigate category learning, they built on a procedure from Lev Vygotsky using arrays of cards with geometrical patterns (e.g., two black circles surrounded by a single border). The investigator mentally chose a rule defining a category (e.g., all cards with two circles), and subjects tried to discover the rule by picking one card at a time for the investigator to identify as an exemplar or nonexemplar of the category. Each subject's sequence of selections could then be examined to determine what strategies were being used. A common, generally successful strategy was for the subject to find one positive instance of the category, and then to systematically pick other cards that differed in one attribute: if the new card was also a member of the category, then the dissimilar attribute should not be relevant to the category assignment. This program of exploring thought through the window of category reasoning was expanded in a variety of important and interesting directions in subsequent years. For example, in the 1970s Peter Wason and Philip Johnson-Laird conducted studies which were interpreted as showing that subjects manifested a *confirmation bias* in category-reasoning tasks, continuing to examine instances that would support a hypothesized rule rather than seeking out cases that might falsify it. Also in the 1970s Eleanor Rosch, one of Bruner's graduate students, began a program of study of natural categories which revealed that generally they lacked defining rules such as Bruner and his colleagues employed in their studies of contrived categories (see Article 8, CONCEPTUAL ORGANIZATION, and Article 25, WORD MEANING). The foundation for these and many other developments, however, was the seminal work reported in *A Study of Thinking*.

While Bruner carried out his work on perception and categorization in Harvard's Department of Social Relations, George Miller was at work in the Department of Psychology. That department was dominated by two personalities: B. F. Skinner, who returned to Harvard from Indiana in 1948, and S. S. Stevens. During World War II Miller worked in Stevens's laboratory on optimal signals for spot jamming of speech, which was classified military research. At his Ph.D. oral, Miller had to present his results very discreetly; accordingly, he spoke not of jamming, but rather of the effects of noise on intelligibility of speech. One of his observations was that certain messages were harder to jam (or easier to understand in noisy environments) than others, a finding that did not seem explicable in terms of the physical acoustical data alone but was soon to receive a novel interpretation. After receiving his Ph.D. in 1946, Miller remained at Harvard until 1968 (except for the early 1950s, when he spent a year in Princeton and then four years at MIT). Miller's interest in mathematical analysis led him to actively follow developments in statistics and information theory, especially Shannon's (1948) paper on information theory.

Claude Shannon, whom we have already encountered as a designer of switching circuits that implemented Boolean operations, was a mathematician at Bell Telephone Laboratories. This organization had a natural interest in understanding the laws governing transmission of information and especially in determining the maximum quantity of information that could be transmitted over an acoustic channel. Shannon's theorizing about information started from the observation that the quantity of information transmitted over a channel depends on variation in a signal. In the simplest case, the variation would involve just two equally likely alternatives: *on* or *off*. Then the basic unit of information, a *bit* (binary unit), could be defined as the amount of information transmitted by selecting one of these alternatives rather than the other (a binary decision). A single bit can distinguish between just two alternatives (1 corresponds to *on*, 0 to *off*). Adding a second bit (an additional *on/off* signal) doubles the number of alternatives that can be distinguished to four (00, 01, 10, 11). A third bit doubles the information again – a three-bit sequence distinguishes among eight alternatives. (Wendell Garner, who played a major role in applying Shannon's ideas to psychology, pointed out in 1988 that this approach defines information in terms of all the possible events that could have occurred, not just the actual event. For example, the informativeness of the event 10 – *on* then *off* – depends on the fact that it excludes exactly three other events.)

This general approach can be extended to more complex situations in which there are more than two alternatives or alternatives have unequal probabilities – for example, any message in English – and can be used to measure the amount of redundancy in such messages. Shannon (1948, 1951) presented a text one letter at a time to subjects whose task was to predict the next letter. There were 26 alternatives at each point, and they had unequal probabilities due in part to context. For example, *x* has a low probability overall, but is highly probable following *mailbo*. Shannon defined redundancy as the reciprocal of the average number of guesses needed to generate the correct letter. Averaging across the entire text, subjects required an average of two guesses per letter, yielding a redundancy estimate of about 50 percent for printed English. Shannon's information theory provided the key to interpreting Miller's dissertation result that messages differed in how easily they could be understood in noisy environments. Miller and Selfridge (1950) found further application for information theory in a list learning experiment: the closer the word lists came to resembling English sentences (i.e., the greater their redundancy), the more words a subject could remember.

In one of the most influential papers of this period, Miller (1956) addressed more extensively the question of the cognitive structure of MEMORY. The study of human learning and memory had long moved along the path laid down by Hermann Ebbinghaus (1885/1913), who served as his own subject in a prolonged series of experiments in order to bring higher mental processes under experimental control and quantitative analysis. In his attempts to eliminate extraneous influences, Ebbinghaus arrived at the idea of using pronounceable nonsense syllables such as DAX and PAF as his stimuli rather than words. He studied lists of these nonsense syllables daily, and then tested himself to determine rates of learning and forgetting. Ebbinghaus uncovered important functional relations (e.g., repetition yields better retention, especially if distributed across several days; the amount retained is a logarithmic function of time), but the down side was his neglect of the cognitive structures and processes that meaningful stimuli so readily engage. Frederick Bartlett's (1932) previously described idea that schemata help organize memory offered a corrective to the limitations of the Ebbinghaus tradition, but Bartlett's early impact was felt primarily in England. Verbal learning in North America (as discussed in section 1.2.2) pursued updated variations on the Ebbinghaus tradition by asking, for example, which particular model of stimulus–response conditioning might best account for the accumulated data on paired-associate learning. Retention was an indicator of learning, not a clue to the nature of the memory system within.

Miller made an exploratory claim about the underlying memory system by pointing in his 1956 paper to an interesting limitation. Over a short period of time, humans can retain only about seven items in memory ("the magical number seven, plus or minus two"). This limit could be overcome if the items formed coherent units, or chunks (as do the letters of a familiar word or acronym). Thus, the sequence of letters I, B, M, C, I, A, B, B, C, U, S, A exceeds the limit, and so is very difficult to remember. When it is chunked as IBM, CIA, BBC, USA, it falls well within the limit and is easy to remember. The limit reemerges, however, in that humans can retain only about seven chunks unless those can themselves be rechunked. For Miller, the limitation to seven items was not an isolated, odd fact. He began his paper: "My problem is that I have been persecuted by an integer. For seven years this number has followed me around, has intruded in my most private data, and has assaulted me from the pages of our most public journals" (1956, p. 81). For a variety of activities, such as distinguishing phonemes from one another, making absolute distinctions amongst items, as well as remembering distinct items, when the number of items reached seven, significant changes arose. Miller concluded: "There seems to be some limitation built into us either by learning or by design of our nervous system, a limit that keeps our channel capacities in this general range" (p. 86). Miller's essay became a classic of cognitive psychology, in part because it suggested that there was structure to the internal processing system, whose character could partly be discovered via BEHAVIORAL EXPERIMENTATION.

In the research of Bruner and Miller, an approach to psychology was stirring which retained the emphasis on behavioral data but rejected behaviorism's suspicion of mental machinery. The enterprise would not be named *cognitive psychology* until 1967, but its *modus operandi* – seeking structure in the mind which could explain features of behavior – was already apparent.

1.3 The brain develops

After cognitive science left its conceptual womb, as it were, and began to mature, it went through a long period in which contributions from neuroscience were either ignored or actively dismissed as irrelevant to the pursuits of cognitive science. But while it was in the womb during the 1940s and 1950s, advances in understanding the brain contributed to researchers' thinking about how concepts such as information and computation might provide a basis for understanding mental processes. We have already observed links between psychology and work on computation in the cybernetics movement. The idea that psychology could benefit from collaborations with neuroscience as well was articulated by psychologist Donald Hebb of McGill University in the preface to his classic 1949 book, *The Organization of Behavior*:

> There is a considerable overlap between the problems of psychology and those of neurophysiology, hence the possibility (or necessity) of reciprocal assistance. . . . Psychologist and neurophysiologist thus chart the same bay – working perhaps from opposite shores, sometimes overlapping and duplicating one another, but using some of the same fixed points and continually with the opportunity of contributing to each other's results. (pp. xii and xiv)

The spirit of Hebb's book was clearly evident in the September 1948 conference on "Cerebral Mechanisms in Behavior" sponsored by the Hixon Fund. Speakers included neurophysiologists Warren McCulloch and Rafael Lorente de Nó; biologically oriented psychologists Ward Halstead, Heinrich Klüver, and Karl Lashley; Gestalt psychologist Wolfgang Köhler; and computer scientist John von Neumann. The papers covered a wide range of topics, including von Neumann's analysis of the similarity of computers to the brain, Köhler's study of evoked potentials during pattern perception, Klüver's comparison of functional contributions of the occipital and temporal lobes, and Halstead's attempt to relate intelligence to the brain. As the discussion recorded in Jeffress (1951) makes evident, the divides between psychologists and neuroscientists were not large, and discussion flowed easily from psychological phenomena to neural mechanisms.

While disciplines such as psychology and linguistics were in a position to contribute to the birth of cognitive science only after undergoing internal revolution, and artificial intelligence had first to be created, neuroscience had a much longer and continuous history. The idea that the brain was not merely the organ of mental processes but might be decomposed into component systems which perform different specific functions in mental life was a product of the nineteenth century. The challenge to neuroscience, then and now, is to parse the brain into its functional components and (a more difficult task) to figure out how they work together as a system. The implication for cognitive science is that information about the distinct functions performed in the brain could be used to corroborate or guide the development of psychological models of cognitive activities. Exploring this functional decomposition and localization depended in part on the development of appropriate tools. What follows is a very selective review of research from the nineteenth century through 1960 that began to determine the structure of the brain and the relation of its components to mental life.

1.3.1 Neural architecture

Before scientists could make claims about the functional organization of the brain, they needed to learn something about its general architecture. At the end of the nineteenth century major advances were made at both the micro and the macro level in understanding the brain. At the micro level the crucial breakthrough was the discovery that nerve tissue is made up of discrete cells – neurons – and that there are tiny gaps between the axons that carry impulses away from one neuron and the dendrites of other neurons that pick up those impulses. In the 1880s Camilio Golgi introduced silver nitrate to stain brain slices for microscopic examination. Silver nitrate had the unusual and useful feature of staining only certain cells in the specimen, thereby making it possible to see individual cells, with their associated axons and dendrites, clearly. Using this stain, Santiago Ramón y Cajal argued that the nervous system was comprised of distinct cells (a view that Golgi, however, never accepted). Sir Charles Scott Sherrington then characterized the points of communication at the gaps between neurons as *synapses* and proposed that this communication was ultimately chemical in nature.

While processes at the micro level of the neuronal substrate would figure prominently in understanding cognitive processes such as learning (which is widely thought to involve changes at synapses that alter the ability of one neuron to excite or inhibit another) and became the inspiration for computational modeling using neural networks (see section 1.3.5), another kind of advance involved linking different macro-level brain areas with specific cognitive functions. This required overcoming the view widely shared in the eighteenth century that the brain, especially the cerebral cortex, operated holistically, without any localized differentiation of function.

The major credit for promoting the idea that the macrostructure of the brain was divided into distinct functional areas is due to Franz Joseph Gall. Working in the early nineteenth century, he proposed that protrusions or indentations in the skull indicated the size of underlying brain areas. He further thought that it was the size of brain areas that determined how great a contribution they made to behavior. Accordingly, he proposed that by correlating protrusions and indentations in individuals' skulls with their excesses and deficiencies in particular mental and character traits, he could determine which brain areas were responsible for each mental or character trait. *Phrenology*, the name given to Gall's views by his one-time collaborator Johann Spurzheim, has been much derided as quackery. Nonetheless, Gall's fundamental claim that differentiation of structure corresponds to differentiation of function in the cortex came to be widely accepted, so much so that those espousing localization of function in the latter part of the nineteenth century were often referred to as neophrenologists.

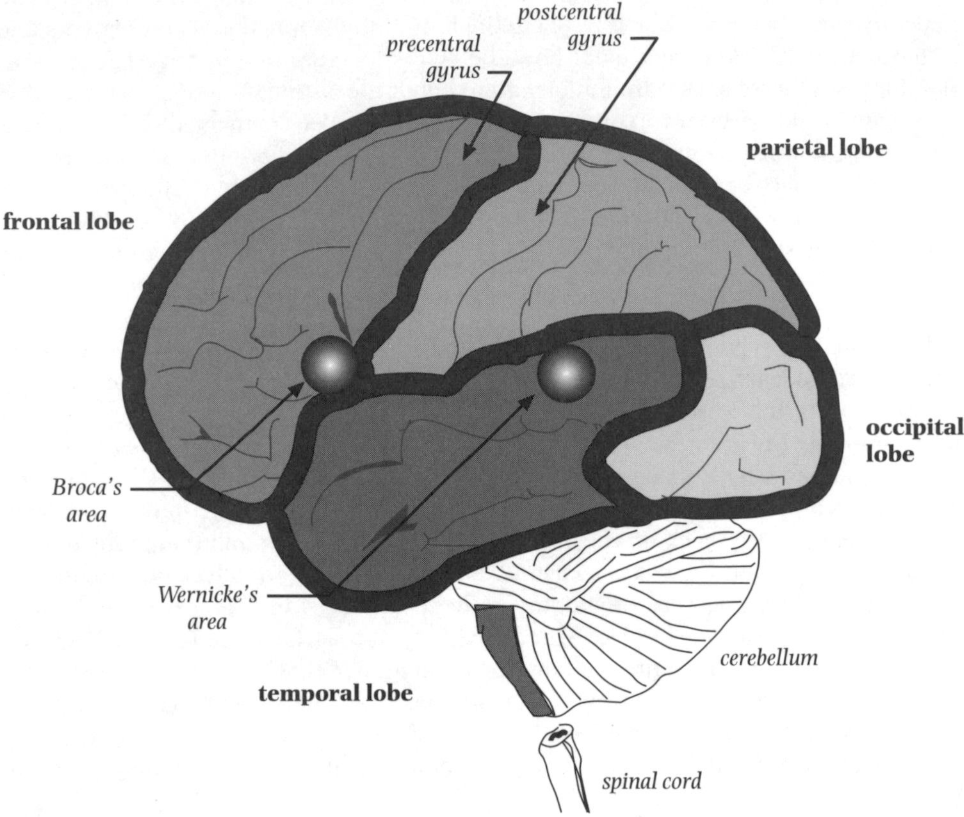

Figure I.3 Major features of the brain's architecture. Shown are the cerebellum and the four lobes of the cortex: frontal, parietal, temporal, and occipital. Also indicated are the approximate locations of Broca's and Wernicke's areas and the precentral gyrus (primary motor cortex) and the postcentral gyrus (primary sensory cortex). Figure by Robert S. Stufflebeam.

One problem that researchers faced in attempting to localize mental functions in the brain was the lack of any standardized way of designating parts of the brain. The folding of the cortex creates *gyri* (hills) and *sulci* (valleys); anatomists have named some of them and used the most prominent sulci to divide the brain into different lobes, as shown in figure I.3. But each lobe itself contains a number of anatomically distinct regions. Using such criteria as responses to various stains and the distribution of cells between cortical layers, a number of researchers at the end of the nineteenth century produced more detailed atlases of the brain. Of these, that by Korbinian Brodmann (1909) became the most widely adopted, and his numbering of brain regions is still widely employed today (see Article 4, BRAIN MAPPING).

A guiding principle for Brodmann and others who tried to map the brain was that different brain regions would perform different functions. The principle of localization of function has had vocal opponents. In the 1820s, Marie-Jean-Pierre Flourens voiced objections to Gall, while a century later neuropsychologist Karl Lashley was the most prominent critic of Brodmann. Arguing that higher cognitive processes (ones involved in memory and learning) were not localized but distributed, Lashley (1929) introduced two alternative principles: equipotentiality and mass action. *Equipotentiality* refers to the ability of brain regions to take on different functions as needed (e.g., if the region that previously performed a function were damaged), while *mass action* refers to the idea that the ability to perform higher functions relates to the total available cortex, not to any one part of it. In 1950 Lashley presented his arguments against localization of specific memory traces in the very often cited paper "In search of the engram," in which he recounted the repeated failures to localize such major functions as habitual behavior.

Despite the doubts of Flourens, Lashley, and others, most researchers have assumed that – at some level of detail in the analysis of function – functions are localized in the brain. To obtain evidence for particular localizations, researchers have had to develop a number of research techniques. We briefly review a number of them and some of the more prominent results obtained by using them.

1.3.2 *Deficit studies*

One of the earliest and most fruitful sources of information about the function of brain areas is the study of deficits that ensue when a neurostructure is damaged (see Article 29, DEFICITS AND PATHOLOGIES). The path from damage to conclusion is difficult to traverse, however: anatomical and functional brain areas vary across individuals, and it is also tricky to determine precisely what contribution the damaged part makes to normal function. Some of these challenges are well illustrated in the *locus classicus* of deficit studies, Paul Broca's (1861/1960) famous case of Monsieur Leborgne (more commonly known as "Tan" for the one syllable he could utter). When Tan was brought to Broca, he suffered not only loss of articulate speech, but also epilepsy and right hemiplegia. Tan died six days after Broca first saw him. Broca performed an autopsy, which revealed massive damage to the left frontal lobe. The central region of this lesion came to be known as *Broca's area* (see Article 13, LANGUAGE EVOLUTION AND NEUROMECHANISMS).

Broca's research led to a much more positive response to claims that mental functions are localized in the cortex. But not everyone agreed that mental functions should be identified with the regions in which damage could lead to loss of function. The variability in relations between structures in the brain and mental functions was emphasized by Charles-Edouard Brown-Séquard who, in response to Broca, presented atypical cases in which aphasia developed after damage to the right frontal cortex, cases in which lesions outside Broca's area affected speech, and cases in which damage to Broca's area did not result in speech deficits. Even some of those who accepted the claim that damage to specific areas would result in specific functional deficits rejected the claim that the function itself was performed in that area. This view is exemplified by Carl Wernicke (1874), who, in the decade after Broca, presented cases in which damage to an area in the temporal lobe that came to be known as *Wernicke's area* resulted in a loss of comprehension of speech while leaving the ability to speak intact. Wernicke, however, operating out of an associationist framework, viewed the site of damage as a locus of association between simple ideas, not the locus where comprehension was achieved.

Wernicke's associationism was largely overlooked. The standard view, one espoused especially by Norman Geschwind (1974), was that Wernicke's area is the locus of speech comprehension and Broca's area the locus of speech production. More recently, this decomposition of function into comprehension and production was challenged by researchers influenced by Chomsky. In a Chomskian LINGUISTIC THEORY there are separate components for phonology, syntax, and semantics. The theory is intended to be neutral with respect to comprehension and production. In 1980 Bradley, Garrett, and Zurif produced evidence that lesions in Broca's area produce deficits of syntactic processing that, although more obvious in production, can be observed in comprehension as well. This gave rise to a new standard view, which localized semantics in Wernicke's area and syntax in Broca's area.

The deficits discussed so far stemmed from natural lesions. But in the first half of the twentieth century surgeons sometimes induced lesions in the brain to alleviate effects of diseases such as epilepsy. Some of these patients subsequently experienced unanticipated effects, which provided suggestive evidence of the functional significance of the structure that had been lesioned. One prominent example of such surgery in the 1950s involved a patient who has become known in the literature by his initials, H.M. (Milner, 1965). In 1953, while in his twenties, H.M. had his hippocampus removed. This resulted in *anterograde amnesia*, loss of memory for events that happened after his surgery. H.M. could learn new skills, but could not learn new facts or events (e.g., he could not recall ever meeting physicians who saw him on a daily basis). This indicates that the hippocampus plays a crucial role in the storage of new information in memory; however, since H.M. retained previous memories, it is presumably not the site of the memories themselves. Also, the specificity of the deficit (H.M.'s intelligence was unaffected) constituted strong evidence against Lashley's doctrine of mass action.

1.3.3 Stimulation studies

From the fact that a lesion is accompanied by the loss of a function, one cannot conclude that the site of the lesion is itself the locus of the function; it may merely be the site of some necessary but relatively minor component of the function. One way of ameliorating this limitation of deficit studies is to augment them with information obtained by stimulating a specific part of the brain and observing the response. The classical example of this approach occurred just a few years after Broca's observations. In 1870 Gustav Fritsch, an anatomist, and Eduard Hitzig, a psychiatrist, collaborated on a study in which they applied low levels of electrical stimulation to different cortical areas of a dog. (Studies of electrical stimulation had been common earlier, but generally suffered from employing too strong an electrical current.) They found that stimulating specific sites on the cortex resulted in muscular responses on the opposite side of the body in the forepaw, hindpaw, face, and neck. They combined this stimulation study with a lesion study in which they removed the portion of the cortex that activated the forepaw; the result was impairment (although not total loss) of movement in the forepaw. Fritsch's and Hitzig's stimulation studies were followed up by David Ferrier, who performed comparative studies on a wide range of species and developed techniques for eliciting relatively fine movements such as the twitch of an eye.

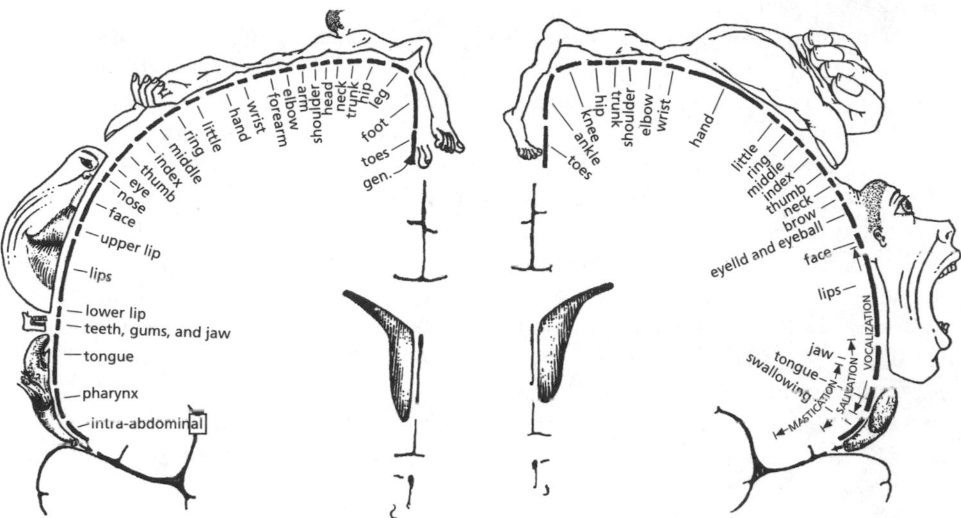

Figure I.4 Representations of the primary sensory cortex (left) and primary motor cortex (right), using homunculi to show the relative amounts of brain tissue devoted to different portions of the body. Redrawn from Penfield and Rasmussen (1950).

Stimulation studies gained practical applications as neurosurgery became more common in the twentieth century. The Canadian neurosurgeon Wilder Penfield, for example, developed a procedure of removing the portion of the brain in which epileptic attacks seemed to originate in severe epileptics. To reduce the risk of causing cognitive or behavioral deficits as a result of these surgeries, Penfield would insert tiny electrodes and induce a small current into the site of surgery to see what responses the patients would make. Employing patients' verbal reports of sensation as well as their own observations of motor responses, Penfield and Rasmussen (1950) were able to develop far more detailed maps than had been possible previously for the primary motor cortex and primary sensory cortex (located on the precentral gyrus and postcentral gyrus respectively). Figure I.4 illustrates the most striking feature of these maps: different parts of the body are represented by disproportionate amounts of brain area. The hand and face are represented in great detail, while representations of the trunk and neck are localized in much smaller areas. In other research Penfield stimulated areas of the temporal lobe; this often triggered highly detailed memories – memories more detailed than subjects would recall voluntarily. Although critics often emphasized the variability of these results and the fact that reports of supposed memories seemed to be highly related to current topics of conversation (see Valenstein, 1973), Penfield's observations excited many researchers with the prospect of localizing specific cognitive states in the brain. The stimulation technique also produced controversial claims of centers for hunger, aggression, pleasure, and so forth.

1.3.4 Single neuron electrophysiology

One of the major advances around 1950 was the development of SINGLE NEURON ELECTROPHYSIOLOGY – the technique of using microelectrodes to record the activity of individual nerve cells. Some of this research took place with simple organisms, as in Ratliff and Hartline's (1959) research on the retina of the horseshoe crab. Stimulating photoreceptor cells with bars of light and measuring activity in the ganglion nerve cells to which they were connected, the researchers found that the ganglion cells were more responsive to the edge between light and dark than to the center of the light bar. They explained this by positing that the *excitatory* projection from an aligned receptor was only one determinant of a ganglion cell's activity. There were also *inhibitory* connections from the neighboring receptors to the same ganglion cell (*lateral inhibition*), and the ganglion cell's response depended on the difference between the excitatory and inhibitory inputs. Ganglion cells connected to receptors near the edge of the light bar would receive as much excitation as those connected to receptors at the center of the light bar, but less lateral inhibition. Making the maximum response, they functioned as *edge detectors*.

Working with cats rather than crabs during the same period, but also using light bars as stimuli, David Hubel and Torsten Wiesel (1962) recorded activity further along the neural pathways involved in vision. They found cells in the visual cortex that were maximally responsive to light bars of particular width and orientation in specific locations in the visual field. They also identified cells that responded to light bars of particular width and orientation but were much less restrictive as to location. Hubel and Wiesel suggested that there might be a hierarchy of processing cells, such that cells detecting lines at various specific locations would feed into higher-level cells that detected lines at any of those locations. One of the best-known exemplars of this line of research was the demonstration by Jerome Lettvin, Humberto Maturana, Warren S. McCulloch, and Walter Pitts (1959) of cells in the frog's brain that responded to small dark roundish shapes in motion, and hence seemed to serve as *bug detectors*. Findings of this kind led to speculative discussions of the extent of localization and specialization of encoding in the human brain, with so-called *grandmother detectors* at one end of the spectrum and Lashley's mass action principle at the other.

1.3.5 Computational modeling: neural networks

A major new development in the 1940s and 1950s was the emergence of computational analyses of neural systems and the beginnings of brainlike computational modeling (an approach which, via the mediation of Donald Hebb, took over the term *connectionism* from earlier, associationist approaches to conceptualizing the brain such as Wernicke's). A key figure in this development was Warren McCulloch, a neurophysiologist who began his career at the University of Chicago. He collaborated with Walter Pitts, then an 18-year-old logician, in a widely cited 1943 paper that analyzed networks of neuron-like units. McCulloch and Pitts showed that these networks could evaluate any compound logical function and claimed that, if supplemented with a tape and means for altering symbols on the tape, they were equivalent in computing power to a universal Turing machine. The units of the network were intended as simplified model neurons and have been referred to ever since as *McCulloch–Pitts neurons*. Each unit is a binary device (i.e., it can be in one of two states: *on* or *off*) that receives excitatory and inhibitory inputs from other units or from outside the network. The state of a network of these units emerges over a number of cycles. On a given cycle, if a unit receives any inhibitory input, it is blocked from firing. If it receives no inhibitory input, it fires if the sum of equally weighted excitatory inputs exceeds a specified threshold. A unit with this design is appropriate not only as a model of a simplified neuron but also as a model of an electrical relay – a basic component of a computer – and hence McCulloch–Pitts neurons helped draw the connection between the brain and computers that was emphasized by others, including John von Neumann and Marvin Minsky. McCulloch and Pitts also made a link to logic: the neurons could be associated with propositions, and because of the binary nature of these units, their activation states could be associated with truth values.

As attractive as some theorists found the comparison of the brain to a computer at the architectural level, many others moved beyond the logic-gate level of focus and began trying to analyze how nervous systems carried out more complex psychological tasks, such as those of perception. These ambitious researchers included Pitts and McCulloch themselves, who, in a 1947 paper, tackled two knotty problems: how someone can recognize an object as the same when it appears in different parts of the visual field and how the superior colliculus is able to transform spatial maps of sensory inputs into motor maps that direct such activities as eye movements. Here they abandoned the earlier paper's focus on propositional logic in favor of spatial representations and analog computations. A further departure from the earlier paper is an emphasis on networks that rely on statistical order and operate appropriately despite small perturbations. Moreover, as part of their evidence for specific computational models, they compared diagrams of these computational systems with diagrams of specific neural structures.

The focus on perception continued in the central parts of Donald Hebb's 1949 book, *The Organization of Behavior*. The subtitle, "Stimulus and response – and what occurs in the brain in the interval between them," points to one of the main emphases of Hebb's analysis: the development of internal structures that mediate stimulus and response. Hebb sought to overcome the opposition between the more localizationist switchboard theories emphasizing sensory–motor connections and the anti-localization approaches of the Gestalt theorists and his own mentor, Lashley. The key to his alternative was the notion of neuronal cell assemblies, which consisted of interconnected, and hence self-reinforcing, sets of neurons which represent and transform information in the brain.

> Any frequently repeated, particular stimulation will lead to the slow development of a "cell-assembly," a diffuse structure comprising cells in the cortex and diencephalon (and also, perhaps, in the basal ganglia of the cerebrum), capable of acting briefly as a closed system, delivering facilitation to other such systems and usually having a specific motor facilitation. A series of such events constitutes a "phase sequence" – the thought process. Each assembly action may be aroused by a preceding assembly, by a sensory event, or – normally – by both. The central facilitation from one of these activities on the next is the prototype of "attention." (1949, p. xix)

Hebb proposed that these subassemblies were created by an interaction between cells whereby every time one cell figured in the firing of another, their connection was strengthened: "When an axon of cell A is near enough to excite a cell B and repeatedly or persistently takes part in firing it, some growth process or metabolic change takes place in one or both cells such that A's efficiency, as one of the cells firing B, is increased" (p. 62).

One of the first attempts to model neural processes on a digital computer was Rochester, Holland, Haibt, and Duda's (1956) study of Hebb's proposal for cell assemblies, carried out at IBM. They discovered the need for several additions and modifications to Hebb's proposal, including (a) inhibitory connections within the net and (b) a mechanism for normalizing connection weights so that they would not grow without bound. Perhaps most importantly, they modified the learning rule so that connection weights would be reduced when one unit was active but the other was not. The modified Hebb rule has become one of the standard *learning rules* employed in subsequent connectionist modeling (see Article 16, MACHINE LEARNING).

The culmination of this early work on modeling the brain was psychologist Frank Rosenblatt's work on perceptrons at Cornell (Rosenblatt, 1962). Figure I.5 shows a simplified perceptron (omitting backward and lateral connections). It had an input device or retina consisting of a number of binary units. Each of these had weighted connections to a number of other units (associator units), which would become active whenever the combined activation from the input units exceeded a threshold. These associator units were connected in turn to response units. The perceptron's task was to activate the appropriate response units when a pattern was activated on the input units. The perceptron's ability to do this depended on its having appropriate weights on its connections – those that would produce the intended classification of the input patterns.

One of Rosenblatt's major contributions was to develop a learning procedure for adjusting the strength of the connections from the associator units to the response units. The procedure began by supplying a pattern to the input units and allowing the network to generate a response. The activity of each response unit was then compared with the desired response. If the unit was on when it should have been off, then each connection from an active associator unit was weakened; if the unit was off when it should have been on, each connection from an active associator unit was strengthened. Rosenblatt proved the Perceptron Convergence Theorem, which established that this procedure would succeed in finding connection weights that would permit the network to produce correct responses unless no such set of connection weights existed. Rosenblatt discovered that there were some sets of input–output pairings which no set of connection weights could generate. He explored a number of variations, including networks with feedback from response units to associator units and networks with multiple layers of associator units. He also explored procedures through which the network would self-organize as a result of inhibitory connections between units, hence learning response patterns without a trainer. These projects were never completed; Rosenblatt died in his early forties in 1971.

Many of Rosenblatt's ideas have been taken up by more recent neural network researchers (see Article 38, CONNECTIONISM, ARTIFICIAL LIFE, AND DYNAMICAL SYSTEMS), but one thing that was distinctive about Rosenblatt's work was that he built actual physical devices, rather than limiting himself to mathematical analyses of hypothetical devices or simulations on digital computers. One of these devices was the Mark I

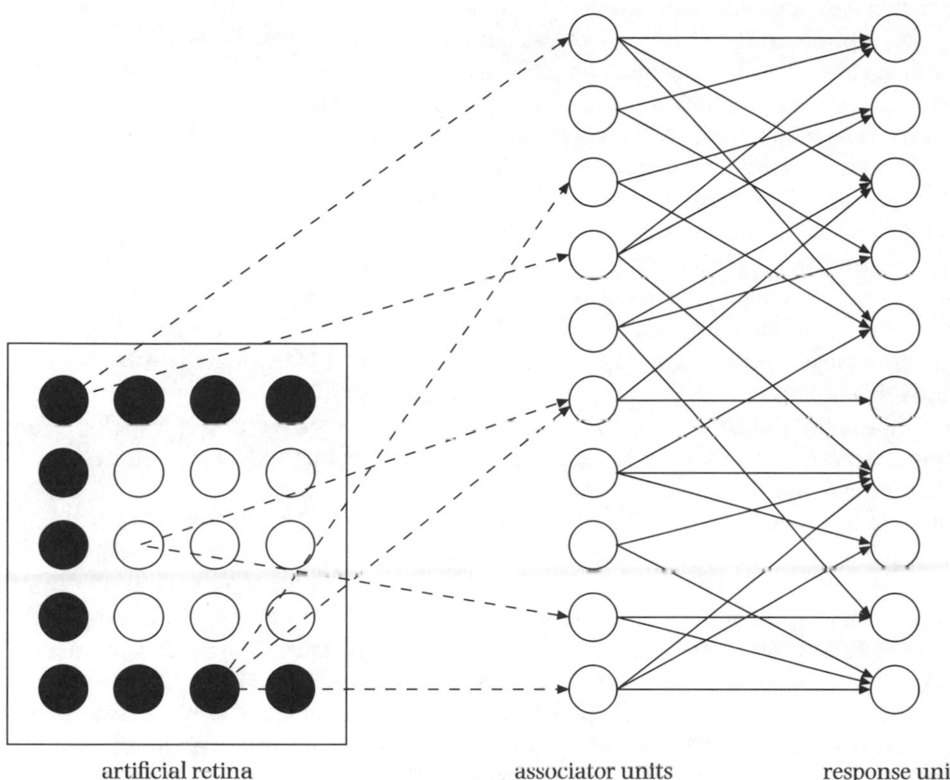

Figure I.5 A simplified perceptron. Each unit in the artificial retina is connected to a randomly selected set of associator units. Each associator unit is in turn connected to a randomly selected set of response units. Rosenblatt (1962) fixed the weights of the first set of connections (dashed lines) but developed learning procedures for the second set of connections (solid lines).

Perceptron, which had a 20 by 20 input grid of photoelectric cells, 512 associator units, and 8 response units. The variable connections between associator and response units were implemented by motor-driven potentiometers. Some of the experiments conducted on the Mark I included studies with noisy target patterns or with damaged networks (accomplished by removing actual wire connections). In a subsequent model, named "Tobermory" after his own cat, Rosenblatt ambitiously attempted to model the visual system of a cat.

1.4 Viability: the transformation of linguistics

Like psychology, linguistics had to undergo a transformation in order to make its contribution to cognitive science. The central figure in this transformation was Noam Chomsky; but before turning to his innovations, we should consider briefly the nature of linguistics prior to Chomsky. In a landmark achievement that began in the late eighteenth century, *historical linguists* reconstructed the long extinct Proto-Indo-European language and traced its divergence into contemporary languages of Europe, Iran, and northern India. In the late nineteenth century, largely inspired by the Swiss linguist Ferdinand de Saussure, the focus shifted: *structural linguists* began to describe the current structure of particular languages rather than their history. This involved analyzing a corpus to identify units at several levels. For example, the word *boats* is composed of two morphemes, one composed of three phonemes, /bot/, and one composed of one phoneme, /s/. European structuralists tended to take an explanatory stance and to emphasize the role of language in conveying meaning. North American structuralists, by contrast, became descriptivists with a vengeance. Franz Boas (a contemporary of Saussure, but educated in anthropology), his student Edward Sapir (who combined Boas's descriptivism with broader, almost European concerns), and Leonard Bloomfield (the previously mentioned behaviorist linguist of the 1920s through 1940s) put considerable emphasis on developing appropriate methods for identifying the phonemes and morphemes of any language. With Native American languages rapidly losing their speakers, they felt some urgency in applying these methods to the analysis of as many indigenous languages as possible.

Bloomfield's distinctive contribution was his positivist stance, reflected in his affinity for psychological behaviorism and his insistence on a rigorously empiricist linguistic methodology: "Accept everything a native speaker says in his language and nothing he says about it." Bloomfield's influence was enhanced by his energy for organization: he played a key role (with Boas and Sapir) in creating the Linguistic Society of America in 1924 and sought to focus it on the "science of language" (versus such pursuits as literary analysis) in its journal, *Language*, and in the courses offered in its summer Linguistic Institute. These are still the premier institutions in American linguistics, but they played an especially critical role in defining the field during an era in which most linguists were employed in language or anthropology departments. By the 1940s and 1950s, Bloomfield's students and followers – the *post-Bloomfieldians* – dominated linguistics. They prided themselves (overly optimistically) on their near completion of a system of *discovery procedures* – an objective sequence of operations sufficient to arrive at a phonological and morphological analysis of any language. New technologies even created the dream that these procedures might be mechanized. The sound spectrograph was invented at Bell Laboratories and made public in 1945; its visual display of speech frequencies across time fostered a surge of new knowledge concerning acoustic phonetics that had implications for linguistics. Digital computers gave rise to machine translation projects in the early 1950s (including Victor H. Yngve's group at MIT) and to early parsers in the late 1950s (including one programmed under the direction of Zellig Harris; see Joshi & Hopely, 1997, for an account).

Meanwhile, new ideas about language were flowing from two adjacent fields. First, the counterparts of the post-Bloomfieldians in psychology – Hull's successors – were pursuing mediation theories of language (section 1.2.2). Second, information theory was suggesting ways to approach speech statistically (section 1.2.4). In retrospect, both these approaches are notable for their transitional character. Mediation theory was constructed in an S–R framework, but drew ever closer to mentalism. Information theory contributed the idea of information, but as applied in cognitive science, its statistical approach to information was largely superseded.

Hull died in 1952, and Bloomfield in 1949. Their successors, enjoying optimistic times in their respective disciplines of psychology and linguistics, were ready to rediscover the advantages of interdisciplinary cooperation. At an eight-week summer seminar sponsored by the Social Science Research Council in 1953 (following a smaller, preliminary meeting in 1951), *psycholinguistics* as a term and an endeavor was rediscovered, and an ambitious agenda for cooperative research was published (Osgood & Sebeok, 1954). The specific linguistic and psychological theories that the participants brought to the table look dated now, as do many of the information theory concepts they embraced. By contrast, the research goals and methods were carried forward with surprisingly little change into the next era of psycholinguistic research (section 2.2). For example, the goal of establishing the *psychological reality* of linguistic constructs, such as the phoneme, was to be pursued by such methods as the analysis of speech errors. A more ambitious idea was to "suppose that speech sounds can be regarded as occupying positions in a multidimensional space" (p. 78) and to discover that space by applying an advanced psychological scaling technique, *factor analysis*, to subjects' judgments of the similarity between particular sounds. One participant adapted this method to semantics, uncovering three dimensions of connotative meaning (evaluation, potency, and activity), and obtained cross-cultural evidence of their universality (Osgood et al., 1957). The title of this book, *The Measurement of Meaning*, says it all: for latter-day Hullian psychologists, it was OK to study meaning if you could come up with a good way to measure it.

The idea of a space of speech sounds was linked (in ways too complex to pursue here) to Roman Jakobson's analysis of the world's phonemes in terms of a small set of binary *distinctive features* (e.g., *voiced* vs *unvoiced, nasal* vs *oral, front* vs *back*). A variety of the distinctive feature approach called *componential analysis* also found application within anthropological linguistics. For example, seminar participant Floyd G. Lounsbury (1956) analyzed kinship terms using semantic features like *female/male kinsman, female/male ego, generation 1/generation 2*, as well as some less obvious ones.

No one suspected it at the time, but the rug was about to pulled out from under these structural linguists and their dance partners from psychology; the youngest and most nimble would learn to dance to a new linguistic tune, and the others would end up on the sidelines. The change began with a post-Bloomfieldian who was unusually adept at discovering discovery procedures – Zellig Harris at the University of Pennsylvania. He and other linguists had achieved considerable success with phonology and morphology, but in the early 1950s syntactic analysis was still underdeveloped. In the quest to make syntax tractable, Harris had the idea of normalizing complex sentences by using *transformations* to relate them to simpler *kernel sentences*; further analysis could then focus on the kernels. For example, the passive sentence "Titchener was defeated by the behaviorists" and the cleft sentence "It was the behaviorists who defeated Titchener" can both be transformationally related to the same kernel sentence: "The behaviorists defeated Titchener."

The notion of a transformation blossomed into a revolution in the hands of Harris's student Noam Chomsky. After receiving his M.A. in 1951, Chomsky moved to Harvard as a Junior Fellow of the Society of Fellows from 1951 to 1955. During this period he wrote a large, difficult work entitled *The Logical Structure of Linguistic Theory*, which remained unpublished until 1975. He submitted part of it for his Ph.D. from the University of Pennsylvania in 1955 and joined the faculty at MIT (home of the aforementioned Building 20). His first book, *Syntactic Structures* (1957), made Chomsky's ideas more accessible. Taking a combative stance, Chomsky won only a few converts from the established generation of post-Bloomfieldian linguists. But, more important, he and his collaborator Morris Halle attracted the best of the new generation of graduate students to MIT, and it was these students who were hired as linguistics departments were begun or expanded in the 1960s.

One of Chomsky's key departures from Bloomfieldian linguistics was that he construed a grammar as a generative system – a set of rules that would generate all and only members of the infinite set of grammatically well-formed sentences of a language. Chomsky took up the question of what sort of computational system (automaton) was needed to realize a generative grammar for natural language (cf. Harris's search for a computational system that could discover a grammar from a corpus of sentences). Chomsky argued that two sorts of systems then being considered were inadequate. One was a finite state automaton (Markov process), which consists of a finite number of states and probabilistic transitions between states. As applied to generating sentences, in an initial state there would be a choice of words with which to begin the sentence (e.g., *Noam* or *Linguists*); selecting one of them determines the next state, where again there is a choice of words (e.g., the choices might include *proposes, disposes*, and *sleeps* following *Noam* versus *propound, disagree*, and *sleep* following *Linguists*). An important limitation is that such a device has no memory of the path by which it reached its current state; it has access only to the next set of choices (the possible transitions from that state to the next). Chomsky argued persuasively that no finite state automaton could be adequate to the task, because English (as an example of a natural language) is not a finite state language. The complete argument cannot be easily summarized (it includes the assumption that natural languages have an infinite number of sentences, at which some readers balked), but the flavor is conveyed by considering the difficulty of designing a finite state automaton to generate sentences with embedded clauses ("The woman who was talking to the astronauts is the President") while maintaining the correct dependencies between nonadjacent words (e.g., subject–verb agreement between *woman* and *is*).

Next, Chomsky argued for the inadequacy of phrase structure grammars, which generate phrase structure trees by successively applying rewrite rules such as $S \rightarrow NP\ VP$ and $NP \rightarrow Adj\ N$. (These rules state that a sentence can be composed of a noun phrase followed by a verb phrase; in turn, the noun phrase can be composed of an adjective followed by a noun. This STRUCTURAL ANALYSIS is often displayed in an inverted tree structure diagram, with *S* at the top.) He pointed out that such grammars would have to be unnecessarily complex due to their inability to take advantage of such regularities as (a) the underlying similarity between a kernel sentence and sentences related to it transformationally, and (b) the dissimilarity between different readings of certain ambiguous sentences, such as "Flying planes can be dangerous."

Accordingly, Chomsky advocated transformational grammars, in which rewrite rules generate an underlying tree structure, and transformational rules then apply to obtain a derived phrase structure. In our example, an active sentence and a corresponding passive sentence would have the same initial phrase structure (called *deep structure*) but different derived structures (called *surface structure*); the passive sentence would have the passive transformation included as part of its derivation. Chomsky worked out a fragment of a transformational grammar of English in *Syntactic Structures* and added revisions and elaborations in what came to be known as his Standard Theory in *Aspects of the Theory of Syntax* (1965). Over a long career, he has repeatedly revisited the question of what kind of grammar best captures generalizations about language (see Article 15, LINGUISTIC THEORY).

In many respects, Chomsky's transformational grammar was a natural extension of the structuralist program of Bloomfield and Harris. Like the structuralists, he focused on the formal structures of a language. His claim of the autonomy of syntax – the claim that we can model syntactic knowledge independently of concerns for meaning (semantics) or the pragmatics of communication – comported well with structuralist principles. But Chomsky incorporated his grammar within an ambitious theoretical framework that repudiated the behaviorism of the structuralists by emphasizing the creative open-endedness of syntax and especially by moving to a Cartesian mentalism in the 1960s (see section 2.2).

1.5 *Inside the delivery room: the events of 1956*

So far we have emphasized the intellectual developments which provided the foundation for cognitive science. We have also seen that figures in the history of cognitive science often did not work alone. One of the important forums in which they communicated were conferences.

Within established fields of inquiry, such as cognitive science has now become, the major conferences are highly formalized. They occur on a regular basis, and although each year a different program committee places its own stamp on the gathering by emphasizing some topics and de-emphasizing others, the programs usually adhere to an established script. At the outset of a new field of inquiry, however, the meetings have a much different character. Investigators are putting forward programs, not incremental advances. The discussion from the audience focuses on the wisdom of the new program and is not consumed with challenges directed at details. Informal conversations in the hallways or over drinks are even livelier than in more settled times. Furthermore, a new scientific inquiry does not draw upon an already established lineage of investigators and may even rely on the ability of investigators from different disciplines to communicate across differences in terminology, techniques, and criteria for success. Sometimes the salience of what investigators in other disciplines are doing is so tangible that it is possible, at least momentarily, to glimpse the value of an integrated pursuit and transcend the differences.

We have noted a number of conferences already in which such crossing of disciplinary boundaries occurred, but a meeting at MIT on September 10–12, 1956, seems to have taken this a step further, so much so that George Miller fixes on the second day of the conference, September 11, as the birthdate of cognitive science ("the day that cognitive science burst from the womb of cybernetics and became a recognizable, interdisciplinary adventure in its own right" (Miller, 1979, p. 4)). Miller reports going away from the 1956 Symposium on Information Theory "with a strong conviction, more intuitive than rational, that human experimental psychology, theoretical linguistics, and the computer simulation of cognitive processes were all pieces from a larger whole, and that the future would see a progressive elaboration and coordination of their shared concerns" (p. 9).

The papers in this symposium brought together some of the major contributions discussed earlier. The first day was devoted to coding theory and included papers by Shannon among others. A symposium on automata started the second morning, and the first paper, "The logic theory machine: a complex information processing system" by Newell and Simon presented their Logic Theorist's proof of theorem 2.01 of Whitehead and Russell's *Principia Mathematica*. The following paper, by Rochester, Holland, Haibt, and Duda, presented their computer implementation of Hebb's neurophysiological theory of cell assemblies. The next symposium, on information sources, included a paper by the young Chomsky entitled "Three models of language," in which he presented his arguments for transformational grammar. The third symposium of the day, on information users, included Miller's paper on the magic number 7.

One reason for focusing on this conference as representing the birth of cognitive science is that it did more than bring together some of the most important accomplishments of cognitive science's gestation period. It also made clear that the basic theme linking the various disciplines of the mind was a common conception of the mind as engaged in processing information. Shannon had formulated the idea of information as a measure of what could be transmitted over a channel like a telephone wire, but the different fields began to focus more on how information could be represented and operated upon. From linguistics, Chomsky presented the idea that the representation of linguistic knowledge involves rewrite rules and transformational rules. From psychology, Miller's discussion of the magic number 7 introduced the notion of memory as a limited-capacity information storage system that forces us to use hierarchical encodings of information. Rochester and colleagues' implementation of Hebb's model of cell assemblies suggested how information processing might be accomplished in the brain as well as in computers. Thus, the various speakers found that they shared the conception of the mind as an information processing system and saw that different disciplines could each contribute to understanding it.

The year 1956 stands out as seminal beyond these important papers – we have encountered it a number of times. New twists were added to some traditional fields, including Stevens's magnitude estimation technique in psychophysics. Other fields became more explicitly cognitive: 1956 was the year that Bruner, Goodnow, and Austin published *A Study of Thinking* and Festinger published his cognitive dissonance theory. The Dartmouth Conference got artificial intelligence off to a start. Anthropologists Goodenough and Lounsbury each published a componential analysis of kinship terms. Ulric Neisser received his Ph.D. from Harvard, and within a few years so did many others of the new generation of researchers who would bring cognitive science to initial maturation. We now turn to that part of the story.

2 Maturation, 1960–1985

A birth represents a transition between a time of gestation and a period of maturation. If we assume that cognitive science was born in the late 1950s (Miller's Tale of 1956 plus or minus), the next 25 years, roughly from 1960 to 1985, represents its initial maturation. During this period cognitive science began to be refined and to bear fruit, both intellectually and institutionally.

In this section we will examine a number of these intellectual accomplishments, including the building of new institutional frameworks in cognitive science. One consequence for refinement, largely unintended, was that some elements of the original broad perspective of cognitive science were initially left out. These included, most notably, neuroscience and artificial neural networks. Kintsch, Miller, and Polson, in their 1984 collection *Method and Tactics in Cognitive Science*, for example, included only AI, linguistics, and psychology as cognitive science disciplines. The reasons for this shrinking in perspective will be discussed both in this section and in section 3 (where we consider how fields like neuroscience were reincorporated into cognitive science after 1985). But we shall also see that another discipline, philosophy, started to play a more important role during the phase of initial maturation.

2.1 Early development: a distinctively cognitive model of mind

During the gestation period the different disciplines of cognitive science each developed proposals for thinking about cognitive processes without managing to produce a cohesive model of the mechanisms that might realize cognition. But in 1960 George Miller, together with Eugene Galanter and Karl Pribram, developed just such a proposal for a basic cognitive mechanism, which they called a *TOTE* unit. They described these units and argued that they could provide the basis for intelligent behavior in *Plans and the Structure of Behavior*, perhaps the most influential of the early books in cognitive psychology. *Plans* was composed during the 1959–60 academic year, which Miller, Galanter, and Pribram spent at the Center for Advanced Studies in the Behavioral Sciences at Stanford. Like Miller, Galanter had been trained in psychophysics by Stevens; however, he also had some background in mathematical psychology, engineering, and philosophy (especially logic) and so was attracted to the idea that algorithmic procedures might provide an account of mental activities. Pribram, a student of Karl Lashley, was then a young neuroscientist.

During the year at the Center, Miller organized a conference at which there was a lively debate between experimentalists and mathematical modelers. According to Galanter, the experimentalists criticized the mathematical modelers for failing to provide an explanatory mechanism for the functional relationships between variables and behavior which their equations described. For example, in a simple linear learning model, the probability of a correct response on the next trial equals the probability of a correct response on the current trial plus an increment that is proportional to the amount still to be learned. How high a proportion may depend upon such factors as the size of the reward. But what internal mechanism would work in such a way as to produce a proportional increment? And beyond that, what mechanisms might be in place to produce more complex and interesting behaviors? *Plans* was envisaged as a speculative book that would help chart a new course: "The aim, we all agreed, was to replace behaviorism (really our name for associationism) and trivial experimentation (the T-maze and the Skinner box) with the new insights and maybe even the new technology from computational logic, linguistics, and neurophysiology" (Galanter, 1988, p. 38).

Miller, Galanter, and Pribram took as their focus purposive action (executed *Plans*) and proposed that Plans took the form of TOTE units: Test–Operate–Test–Exit. Borrowed from the feedback loops of cybernetics, TOTE was advanced as an alternative to the classical reflex arc as a model for the basic unit of mental activity. Galanter relates the discussion which gave rise to the TOTE unit:

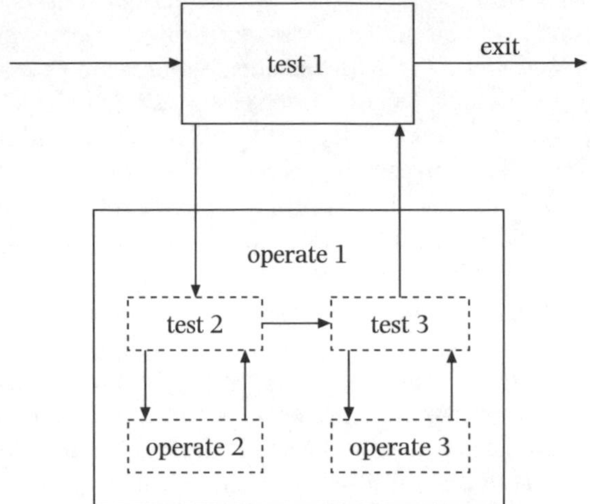

Figure I.6 A recursively embedded set of TOTE units as proposed by Miller, Galanter, and Pribram (1960). If test 1 reveals that its goal was not satisfied, operation 1 is invoked, which itself imposes tests and operations. The operations specified by any test would be repeated until the test showed that the goal was satisfied, at which point that TOTE unit would be exited.

> At one point, George proposed that we examine some intentional human act.
> "Flying a plane," I suggested.
> "No – too much. How about crossing a street. An equally dangerous act in the Bay area," Karl responded. I went to the blackboard and started a flow chart. The boxes, lines, and arrows snaked around the board as step after step was drawn.
> "No," George said, "all that stuff on the board is only a string of reentrant reflexes. Let a whole piece of the action be repeated until it's finished."
> "How will it know?" from Karl.
> "With a cybernetic test," replied George.
> "But how do I draw it?" I asked.
> "Like this," said George, and the TOTE replacement for the reflex was designed.
> (p. 40)

The key idea was that within an agent an initial test would evaluate a current situation against a goal; if they did not match, an operation was carried out to reduce the difference. The cycle of test and operate would be repeated until the goal was matched, at which point the TOTE unit would be exited. Miller, Galanter, and Pribram also proposed that TOTE units could be recursively organized by inserting TOTEs within the operation cycle of another TOTE, as shown in figure I.6. One of the applications they envisioned (but did not work out in detail) was a realization of Chomsky's phrase structure trees and transformations.

Historically, the TOTE unit can be viewed as realizing Lashley's (1951) call for some sort of central, hierarchically structured control over sequential behavior (he had argued persuasively that fluent sequences such as those in typing and speaking occurred too rapidly to be produced by simple associative chains from one element of the sequence to the next). The realization was only schematic, but traces of some of the ideas (considerably modified) can be seen in Newell and Simon's production systems architecture (see Article 42, PRODUCTION SYSTEMS).

2.2 *Learning to talk: Chomsky's impact reaches psycholinguistics*

One of the first sustained collaborations in cognitive science was that between psychology and linguistics. As we saw in section 1.4, the collaboration began in the early 1950s, prior to the cognitive turn in both disciplines. In 1957 Chomsky proposed his transformational grammar in *Syntactic Structures*. The impact began to be absorbed within linguistics, then psycholinguistics, and then helped to shape the new cognitive psychology as well.

In the meantime, Chomsky himself was elaborating his initial proposals. He made a number of changes to the grammar itself, but perhaps more important was his movement towards a Cartesian rationalism and bold claims about *Language and Mind* (Chomsky, 1968). Adopting a strongly mentalistic perspective, Chomsky proposed that (a) generative grammar was a formalized expression of people's *tacit knowledge* of their language, and (b) the primary data for linguistic analysis should be speakers' judgments as to which sentences are grammatical. Chomsky was well aware that people sometimes produce ungrammatical sentences. However, he introduced a distinction between *competence* and *performance*, which enabled him to attribute to people correct tacit knowledge of their language in the face of flawed performance (e.g., a speaker might forget that the subject of a sentence in progress was singular, and hence fail to inflect the verb with the agreement affix, *-s*).

Early in the development of these ideas, Chomsky (1959) repudiated behaviorism in a review in *Language* of B. F. Skinner's 1957 behaviorist account of language, *Verbal Behavior*. In part of the review Chomsky focused on the vacuity of Skinner's major explanatory concepts – stimulus, response, and reinforcement – in explaining language use. For example, the stimulus control for a given behavior might be defined well enough in animal conditioning contexts, but not in the context of an adult human using someone's name (we do not use a name only in the presence of the person whose name it is). Chomsky also emphasized that language use is a creative activity, in the sense that there is no bound to the novel but grammatically well-formed sentences one might produce or hear. Finally, Chomsky introduced a nascent form of what later came to be known as his *poverty of the stimulus* argument for the innateness of basic linguistic competence: "The fact that all normal children acquire essentially comparable grammars of great complexity with remarkable rapidity suggests that human beings are somehow specially designed to do this" (p. 58). As he later developed the idea (e.g., Chomsky, 1965, 1968), the sentences in a child's environment provide too impoverished a database to make it credible that ordinary learning processes can account for the child's competence; an innate *Universal Grammar (UG)* must guide the child's inductions from input (see Article 45, INNATE KNOWLEDGE).

Chomsky's incursions into psychology generated considerable controversy, even to this day. But his formalizations of generative grammar have probably had a broader practical impact in psychology, because they suggested how mental representations might look at a time when psychologists were ready to ask this question. George Miller was ready earlier than most. His interest in information theory had led him to explore the potential of statistical approaches for understanding human information processing.

Although his goal had not been to provide a grammar for English, Miller quickly appreciated the importance of Chomsky's contention about the kind of automaton needed to generate natural human languages and saw its relevance to his own project: if finite state automata were inadequate, Miller's own statistical approach was doomed. He employed Chomsky as his assistant for a summer seminar on mathematical psychology in 1957, and together they wrote a 1958 paper on finite state languages.

In a more practical vein, this led to a research program for Miller in which cognitive activity was construed in terms of operations on symbolic representations. Chomsky's 1957 grammar had transformations derive one tree structure from another. While it is certainly not necessary to use a linguistic grammar as a direct model for positing a series of mental operations performed by a speaker or hearer of a language (it could be viewed as simply a compact representation of the grammatically well-structured sentences of a language), the idea was attractive enough to launch an exciting era of psycholinguistic research. Using Chomsky's transformational grammar to suggest experiments (rather than speculations about the uses of Plans as in the 1960 book), Miller gathered collaborators and presented the preliminary results of several studies in the *American Psychologist* (Miller, 1962). Both memory for sentences (Mehler) and response times (Miller, McKean and Slobin) revealed that the more transformations in a sentence's derivation, the more difficult it was to process. This was taken as evidence for transformational grammar's *psychological reality.* (The phrase was borrowed from Edward Sapir, but one goal was to supersede the psychological reality claims of the psycholinguistics of the 1950s.)

The specific form of this research program in psycholinguistics was short-lived, as it soon became apparent that not all transformations added processing time. The new generation of psycholinguistic researchers (particularly Jerry Fodor, Merrill Garrett, and Thomas Bever at MIT) concluded that the relation between competence and performance must be more abstract than originally thought. Also favoring this more nuanced approach was the fact that Chomsky changed his theory over the years so as to de-emphasize transformations, removing the explanation for the data that Miller and others had gathered. The focus of psycholinguistic research shifted from transformation counting to explorations of the psychological reality of deep structure and of the processes involved in parsing surface structures, and became broader still as the field matured (see Article 14, LANGUAGE PROCESSING).

The heady psycholinguistics of the 1960s eventually settled into a more sedate *normal science* mode, but cognitive psychology and cognitive science more generally had been permanently transformed by Chomsky's ideas about how to describe and explain the linguistic competence of speakers. To get some feel for the extent of the change, consider the titles of articles published in the *Journal of Verbal Learning and Verbal Behavior*. The first volume appeared in the same year as Miller's paper on transformations (1962) and did not yet reflect Chomsky's influence; among the titles were "Verbal mediation in paired-associate and serial learning," "Aural paired-associate learning: pronunciability and the interval between stimulus and response," and "Associative indices as measures of word relatedness: a summary and comparison of ten methods." A few papers reflecting Chomsky's influence appeared in 1963, and by 1968 there were a number of papers with such titles as "The role of syntactic structure in the recall of English nominalizations," "The perception of grammatical relations in sentences: a methodological exploration," and "Semantic distinctions and memory for complex sentences." Finally the title of the journal caught up with the content in 1985, when it was renamed *Journal of Memory and Language.*

Chomsky's impact was not limited to psycholinguistic studies of adult subjects in laboratories. People investigating children's language began calling their field *developmental psycholinguistics* and using Chomskian grammars as a framework. Eric Lenneberg published his landmark *Biological Foundations of Language* in 1967, helping to make *neurolinguistics* an exception to the general neglect of neuroscience in cognitive psychology during the 1960–85 period. *Computational linguistics* became a distinct research area, in which the design of parsing programs was one focus. Even *sociolinguistics* and *anthropological linguistics* made more room than usual for formal approaches to language. These fields are discussed in several of the chapters that follow; here we must return to developments in psychology.

2.3 *A first home: the Center for Cognitive Studies at Harvard*

As noted earlier, two of the founders of today's cognitive psychology, Bruner and Miller, were housed in separate departments at Harvard. The difference in orientation between these two departments was considerable. Yet, Bruner and Miller clearly recognized their intellectual affinities and had been teaching a course called "Cognitive Processes" together for several years. In 1960, after Miller's return from the Center for Advanced Studies in the Behavioral Sciences, they put together a proposal for a Center for Cognitive Studies. It was funded for ten years by the Carnegie Corporation. Although Bruner and Miller were both psychologists, the Center was designed from the outset to have an interdisciplinary character.

One of the things the Center did was to provide support for long-term research fellows, many of whom were scholars at the beginning of their careers who would go on to make their own mark in such cognitive science fields as COGNITIVE AND LINGUISTIC DEVELOPMENT, CONCEPTUAL ORGANIZATION, IMAGERY AND SPATIAL REPRESENTATION, LANGUAGE PROCESSING, MEMORY, and ATTENTION. Among them were Arthur Blumenthal, Janellen Huttenlocher, Paul Kolers, David McNeill, Donald Norman, and Nancy Waugh. These fellows developed a large number of research projects, many of them in collaboration with one another, with Miller or Bruner, or with one of the annual visitors. There were also a number of graduate students in the Harvard departments who worked with the research fellows and visitors and later had influential careers in cognitive science, such as Ursula Bellugi, Susan Carey, Patricia Greenfield, Jacques Mehler, and Dan Slobin.

One of the major justifications for the Center was to bring in outside visitors to spend a year doing research and contributing to the community there. As Miller relates, "The way this worked was that he [Bruner] and I got together at least once each year over a bottle of madeira (or was it a bottle of port?) and discussed people whose ideas we found exciting. Anyone whose ideas appealed to both of us was invited to join us for a year" (Miller, 1979, p. 11). A number of scientists, both from the Cambridge area and the rest of the world, spent a year at the Center. Among these were researchers focusing on language, such as Chomsky, Roger Brown, Roman Jakobson, Jerrold Katz, and Willem J. M. Levelt. Roger Brown's longitudinal study of the early language of three children provided the basis for much subsequent work in developmental psycholinguistics (see Article 6, COGNITIVE AND LINGUISTIC DEVELOPMENT). Levelt went on to establish the Max-Planck-Institut für Psycholinguistik at Nijmegen, one of the premier centers in the world for psycholinguistic research (see Articles 14, 24, and 25: LANGUAGE PROCESSING, UNDERSTANDING TEXTS, and WORD MEANING). Several other visitors, such as Daniel Kahneman, Amos Tversky, and Peter Wason, made REASONING and PROBLEM SOLVING a focus of their research. Each of them produced evidence that human reasoning does not conform to the norms of proper reasoning advanced in logic (see Article 44, HEURISTICS AND SATISFICING).

A major activity of the Center was its Thursday afternoon colloquia, a series which was announced to the broader public and drew audiences from a number of other institutions. This series continued a previous series that Bruner had established in 1952 as part of his Cognition Project (the Center's predecessor). To give the flavor of the breadth of topics covered in the series, the schedule for the first year is reproduced in box I.1.

While formal activities such as colloquia are perhaps the easiest to document, participants at the Center tended to focus on the informal activities. Thus Norman and Levelt relate:

> Bruner states that "the intellectual life of the Center revolved around the seminars, the Thursday lunches, and the weekly colloquia." Perhaps. But that is not our memory. For us, the intellectual life was in the routine daily activities, in the offices and halls, in the labs late at night, and in the social interactions. The excitement was in the personal interaction and the private discussions and arguments. The formal seminars and lunches and colloquia were, well, formalities: the public display of the refinements. . . .
>
> The prototypical lunchtime seminar – or at least, prototypical in our memory – is of everyone assembled around the large wooden seminar table with an active, young cast of protagonists (perhaps Mehler, Bever, Fodor, and Katz), each paraphrasing and explaining to the lunchtime audience what the one had tried to explain to us what another had just said what yet another had just previously said that Noam would have said in retort to whatever the issue was at the time. All the time, Chomsky sitting and listening to the others explaining his mind. (Norman and Levelt, 1988, pp. 100–1).

The Harvard Center provided a prototype for many cognitive science centers developed later and left a lasting imprint, but after a decade it was dissolved. Its core consisted of two individuals, but in 1967 Miller left for Rockefeller University. Bruner directed the Center by himself for a couple of years, but it was closed in 1970, and Bruner moved to Oxford University in 1972.

2.4 *Cognitive psychology learns to walk and travels to other institutions*

Although Bruner, Miller, and those connected with the Center at Harvard had played an important role in giving a cognitive focus to psychology, the endeavor quickly became mobile, and cognitive psychology became part of the program in experimental psychology at numerous institutions.

The new endeavors became known as *cognitive psychology,* largely due to Ulric Neisser's influential book in 1967 by that name. In addition to a name, this book gave cognitive psychology a broad vision. Neisser, then a very junior professor at Brandeis and frequently associated with the Harvard Center, had entered psychology just as the crucial ideas of information theory and cybernetics were being adopted by Miller and

Box I.1 Thursday afternoon colloquia, Harvard Center for Cognitive Studies, 1960–1

November 10: Jerome S. Bruner, Harvard University, "Similarity and Difference: Two Approaches to Knowing"

November 17: Ernst Mayr, Harvard University, "A Discussion of the Phylogenetics of Information Processing"

December 1: Daniel Berlyne, Boston University, "Epistemic Curiosity"

December 8: Roger Brown, Massachusetts Institute of Technology, "The Acquisition of Grammar"

December 13: Roman Jakobson, Harvard University, "Infants' and Aphasics' Testimonies on Language and Thought"

January 12: Walter A. Rosenblith, Massachusetts Institute of Technology, "Computer-aided Electrophysiological Studies in Sensory Communication"

January 19: Richard Held, Brandeis University, "Exposure to Strange Environments and an Ordering Principle in Perception"

January 26: Raymond A. Bauer, Harvard University, "A New Look at an Old Problem: Rational Versus Emotional Appeals"

February 2: Eric H. Lenneberg, Harvard Medical School, "The Biological Matrix of Speech"

February 9: Zoltan Dienes, University of Adelaide, "A Theory of Learning Mathematics"

February 16: Frederick C. Frick, Massachusetts Institute of Technology, "Pattern Recognition"

February 23: Ulric Neisser, Brandeis University, "A Theory of Intuitive Thinking"

March 2: Gordon Allport, Harvard University, "Intuition Revisited"

March 9: David Page, University of Illinois, "On Teaching Mathematics"

March 16: Walter Mischel, Harvard University, "Cognitive Activity and Delay of Gratification"

March 23: Ivor A. Richards, Harvard University, "Film Sequences in the Investigation of Learning"

March 30 and April 13: Noam Chomsky, Massachusetts Institute of Technology, "Grammatical Factors in the Perception of Sentences"

April 20: Harold Conklin, Columbia University, "The Cultural Relevance of Cognitive Contrast"

April 27: Jules Henry, Washington University, "Cultural Factors in Elementary School Readers"

May 4: Gerard Salton, Harvard University, "Some Problems in Automatic Information Retrieval"

May 11: David French, Reed College, "Anthropology, Ethnoscience, and Cognition"

his associates. He describes taking Miller's course, "The Psychology of Speech and Communication," in 1949: "George taught us about decibels and filters, phonemes and morphemes, Shannon's theorem and Zipf's law, coding principles and the redundancy of English. Naturally there was no standard textbook for such a course, so he was writing one himself. In 1949, we worked from mimeographed copies of the first draft. We didn't know then that it was a first draft of the future" (Neisser, 1988, p. 82).

Even then, however, Neisser's interests were broader than those Miller was pursuing; he had sympathies with the Gestalt psychologists, many of whom had just emigrated to the USA, and he accordingly began graduate school at Swarthmore with Köhler. But after a couple of years Neisser returned to work with Miller and then, like several other pioneers, completed his Ph.D. in psychophysics under Stevens. While

Neisser found aspects of information theory attractive, he was restless for experiments that could genuinely reveal the nature of cognition. At an MIT symposium on information theory he met Oliver Selfridge, then working on his Pandemonium model of pattern recognition (see section 1.1.2). Neisser and Selfridge collaborated on visual scanning experiments and on further development of the Pandemonium model. From these and other experiences, Neisser developed a vision of what cognition might be like:

> By 1964, it had come together in my head. In principle, I thought, one could follow the information inward from its first encounter with the sense organ all the way to its storage and eventual reconstruction in memory. The early stages of processing were necessarily wholistic (an idea I borrowed from Gestalt psychology) and the later ones were based on repeated recoding (an idea borrowed, even more obviously, from George Miller). But the processing sequence was by no means fixed; at every point there was room for choice, strategy, executive routines, individual constructive activity. Noam Chomsky's linguistic arguments had shown that an activity could be rule governed and yet indefinitely free and creative. People were not much like computers (I had already sketched out some of the differences in a 1963 *Science* paper), but nevertheless the computer had made a crucial contribution to psychology: It had given us a new definition of our subject matter, a new set of metaphors, and a new assurance. (Neisser, 1988, p. 86)

With this vision in mind, Neisser took a leave from Brandeis and wrote *Cognitive Psychology*. Published in 1967, it served to both introduce and synthesize the work on information processing that was beginning to burgeon, particularly emphasizing attention and pattern recognition, and it quickly became the bible for a new generation of students.

One set of topics was rooted in Broadbent's (1958) argument that there are three types of MEMORY stores. Oversimplifying and using a later set of names for the stores, here are some highlights. (1) Sperling (1960) found evidence for the first type, a *sensory register* in which a visual or auditory stimulus briefly persists after it has been turned off. Sperling devised a way to actually measure the rapid decay of information from this register. Presented with 3 by 3 array of letters for 50 msec, subjects can generally report four or five letters (i.e., one or two per row). But if one row is cued by a tone exactly when the array is turned off, subjects can still "see" it briefly and, by concentrating just on their image of that row, report all three letters. If the cue is delayed for a split second, the image in the sensory store has partly decayed, and the advantage of cueing is less. At a delay of one second, decay is complete. (2) The second type of memory discussed by Neisser is the *short-term store* that was the focus of Miller's seven-plus-or-minus-two paper. Peterson and Peterson (1959) traced decay from this store by presenting three consonants to subjects (well within the memory span) and then requiring them to count backwards by threes for periods varying from 3 to 18 seconds before recall. Ability to recall the consonants dropped sharply between 3 and 9 seconds' delay. In an alternative procedure, Waugh and Norman (1965) used a sequential probe technique to prevent rehearsal. Subjects heard 15 digits and then, probed with one of the digits, tried to report the digit that had followed it. The further back the probe had been in the sequence, the poorer the performance. Waugh and Norman attributed this to later digits replacing earlier digits in the limited number of slots in the short-term store. (3) The third type of memory discussed by Neisser is a

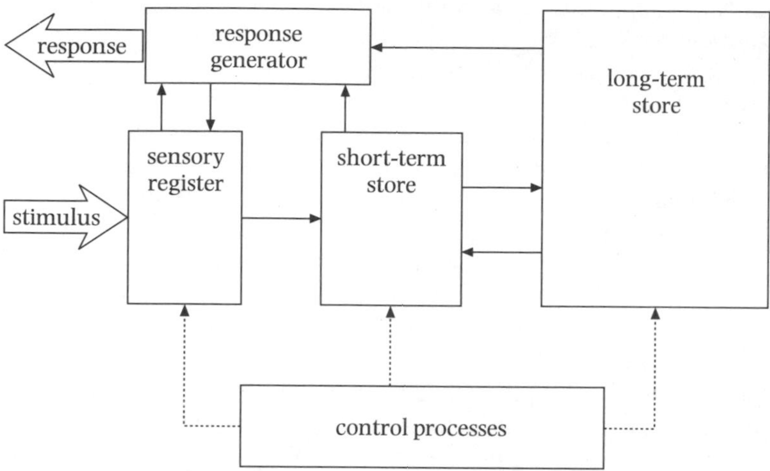

Figure I.7 Atkinson and Shiffrin's (1968) memory model.

long-term store. Waugh and Norman argued that when subjects have time to rehearse an item, it thereby retains its place in the short-term store and also gains a chance of being added to the long-term store. In general, the short-term store has an auditory character (the inner voice), whereas the long-term store is meaning-based. (Later theories, such as that of Alan D. Baddeley (1976), include a visual short-term store as well.)

This idea of a short-term versus a long-term store was not new. Waugh and Norman cited William James on the two memory systems and used his terms of *primary* and *secondary memory* to refer to them. What they added were an experimental technique, a theory about rehearsal, and a mathematical model in which the probabilities of retrieval from the two stores were statistically independent. These contributions played an important role in the many information processing models of memory that emerged around the same time that Neisser's book was published. The most influential integrative model (shown in figure I.7) was developed at Stanford University by Richard Atkinson and Richard Shiffrin (1968). Key to their model was a distinction between fixed and flexible structures. The sensory register and short-term and long-term stores are fixed. The flexible structures are control processes which operate on the fixed structures. For example, ATTENTION is a control process that determines which information gets transferred from the sensory register to the short-term store (in a new format), and rehearsal is a control process that determines which information is retained there long enough to be copied into the long-term store. Items could be lost from the short-term store (bumped by new items or simply through decay), but Atkinson and Shiffrin claimed unlimited capacity for the long-term store. Forgetting would be due to retrieval failure, not loss. An important part of their work was to develop mathematical models of the workings of this system that would specify, for example, the probability that an item would be retrieved from at least one store during a memory test.

These are just a few pieces of work from a very active period. One way to view the spread of the cognitive approach is to focus on how it developed at different universities. We have picked three institutions to get a representative range of developments – each gained strength in cognitive psychology by a different route. A portrait of each of these exemplars is suggestive of the important developments occurring at a number of other institutions as well.

2.4.1 Stanford University

Stanford had a well-established and well-regarded psychology department that was nimble enough to quickly gain strength in cognitive psychology in the 1960s. It attracted a stellar faculty, including Atkinson, Bower, and Shepard, and has probably been the premier university in producing graduate students who went on to shape the field. Two younger faculty members whose influential work helped to shape the department in the 1970s were Herbert Clark and Edward Smith. Clark's work on LANGUAGE PROCESSING was strongly interdisciplinary; he often collaborated with Eve Clark in the linguistics department. The Clarks were proponents of semantic feature approaches (descended from Jakobson's phonemic features) and produced models of sentence processing which specified, for example, the extra time taken to process negative sentences or to deal with mismatches between word order and referent order. Smith's research was on CONCEPTUAL ORGANIZATION. With graduate students Lance Rips and Edward Shoben, he predicted reaction times from a model of semantic memory that used multidimensional scaling to represent basic concepts like *robin* and superordinates like *bird*. Multidimensional scaling, developed more recently than the factor analysis used by Osgood, is a family of methods for using data on pairwise item comparisons to uncover a particular representation of those items – a multidimensional psychological space. For example, Rips, Shoben and Smith (1973) found birds to be organized in a space with dimensions they labeled as size and predacity. In general, greater distances between the point representing a prototypical bird and the point representing a particular bird were associated with longer reaction times: it took longer to verify "A duck is a bird" than "A robin is a bird."

Shepard's forte has been the use of mathematical models to understand human thought. His core concern has been to show that psychological laws can be obtained by formulating them in relation to a psychological space. His formal work in this vein (e.g., on nonparametric multidimensional scaling in the 1960s and on generalization later) is highly respected and influential, but the work that has received the broadest attention is on mental imagery and mental rotation. In a study with his colleague Jacqueline Metzler, he presented subjects with pairs of geometrical forms. Subjects were asked whether they were images of the same object (differing only in rotation, versus mirror images). Reaction times increased linearly with the degree of rotation, which Shepard and Metzler (1971) interpreted as evidence that subjects were mentally rotating one of the figures at a constant rate. These initial results inspired further studies by Shepard and his students, especially Lynn Cooper (1975). After so many years of anti-mentalism, and the recent emergence of a proposition-based mentalism, these results persuaded many that analog images and processes not only existed but could be brought under experimental control and measurement (see Article 12, IMAGERY AND SPATIAL REPRESENTATION).

Gordon Bower's early work was on mathematical models of memory; like Shepard, he moved towards more cognitively oriented research on the nature of mental representation. En route he did work in the 1960s on the role of organization, mental imagery, and mnemonics in memory. A researcher of unusual breadth and legendary as a mentor, he directed the dissertations of numerous students in the 1960s through 1980s who would become major contributors to almost every major area of cognitive science, including investigations of both analog and propositional mental representations. To provide just a few examples from the early 1970s: on the analog side, Stephen Kosslyn designed ingenious experiments to support his claim that the mind employs image-like REPRESENTATIONS; he is now a major figure in cognitive neuroscience (see section 3.2). Preferring a propositional approach, John Anderson constructed a semantic network-based model of associative memory (HAM) and performed numerous experiments to test its predictions. The earliest work is described in Anderson and Bower's 1973 book, *Human Associative Memory*. Much of Anderson's subsequent career, based at Carnegie–Mellon University after a few years at Yale, has been devoted to developing a computational framework, ACT* (Adaptive Control of Thought), which incorporates both PRODUCTION SYSTEMS and semantic networks that allow spreading activation between nodes in memory (Anderson, 1983). Another influential Bower student is Robert Sternberg, whose studies of mental representation and processing as a graduate student launched him into a career focused on new approaches to intelligence. Sternberg's concerns have included decomposing intelligence test tasks into their cognitive components, broadening the conception of intelligence beyond academic intelligence, and understanding individual differences.

Students and faculty in psychology at Stanford had ample opportunity to benefit from a talented faculty in areas of psychology besides cognition and also from groundbreaking work outside their own department. Particularly noteworthy were the AI researchers in the computer science department (John McCarthy, Edward Feigenbaum, and Roger Schank) and nearby research centers (e.g., Daniel Bobrow at Xerox Palo Alto Research Center and Bertram Raphael at Stanford Research Institute). Also, the Center for Advanced Study in the Behavioral Sciences brought many of the world's leading cognitive psychologists to Stanford as one-year visitors.

2.4.2 *University of California, San Diego (UCSD)*

In contrast to Stanford, UCSD was a new university that admitted its first undergraduates in 1964 and initially emphasized the sciences. It attracted a vigorous faculty that welcomed the opportunity to design programs from the ground up, and hence provided a relatively open niche to be colonized by cognitivists. The first three faculty arrived in 1965, including George Mandler as chair. Like Bower, he is renowned as a mentor and contributed some of the early studies that persuaded psychologists that memory involves active organizational processes. He also encouraged the adaptation of Signal Detection Theory to the analysis of recognition memory data; this is a statistical procedure adopted earlier in psychophysics (Green & Swets, 1966) for separating sensitivity (memory strength) from bias. Mandler also had long-standing interests in emotion and in the history and philosophy of psychology as a science, and eventually he was one of the first cognitive psychologists to bring CONSCIOUSNESS within the scope of inquiry. In 1966 Mandler recruited Donald Norman from Harvard and hired a new Ph.D. from the University of Toronto, Peter Lindsay. They were soon joined by David Rumelhart, a mathematical psychologist who worked with Estes, receiving his Ph.D. from Stanford in 1967. These three created the LNR (Lindsay–Norman–Rumelhart, or ELINOR) research group, which included as graduate students and outside visitors several researchers who later played important roles in cognitive science. The group's 1975 book, *Explorations in Cognition*, ended with a suggestion that the "concerted efforts of a number of people from . . . linguistics, artificial intelligence, and psychology may be creating a new field: *cognitive science*" (Norman and Rumelhart, 1975, p. 409). This prescient statement, plus the subtitle of a 1975 book edited by Daniel Bobrow and Allan Collins, are the earliest published uses of the term *cognitive science* that we have identified.

The LNR group pursued research across a broad sweep of topics and methodologies. There were mathematical models of word recognition, mathematical and computational models of analogy, and studies of memory, perception, imagery, sentence processing, story understanding, and more. One of the main joint concerns of the group was the organization of information in long-term memory, and they developed a semantic network format for representing information that was a descendant of Ross Quillian's work in artificial intelligence (section 2.3.3). In a typical semantic network, nodes representing concepts are connected by labeled, directed relations. To this the LNR group added procedural information specifying how tasks should be performed (in the same format as was used for knowledge representation) and a notation for schemata such as those inspired by Bartlett (see above) and Piaget, yielding what they called *active structural networks*. The active semantic network was not just a theoretical proposal: a version named MEMOD was partially implemented on a computer: "This was a major project in itself, and the resulting program became one of the largest and most complex interactive programs operating on the campus computer center computer (a Burroughs 6700)" (p. xiv).

The 1975 book reporting this work juxtaposes primarily experimental chapters with primarily theoretical chapters in several areas. Topics included the active structural network, MEMORY, PERCEPTION, REPRESENTATIONS of knowledge, PROBLEM SOLVING, LINGUISTIC THEORY, LANGUAGE PROCESSING, and augmented transition network parsers. In the middle of the book were three chapters from the "verb group," which had developed semantic representations for verbs composed of primitive predicates that were linked to their arguments via labeled relations inspired by Charles Fillmore's case grammar (e.g., agent, object, source, goal). Rumelhart and James Levin used these representations in developing a computer-implemented language comprehension system. Dedre Gentner focused on fractionating verbs of possession into their primitive predicates, and demonstrated that children's learning of those verbs progressed from those involving fewer components (e.g., *give* involves an agent **doing** something that **causes** a **transfer** of goods) to those including more (e.g., *sell* adds a **contract** for **transfer** of money). The relatively simple representation of *give* is illustrated in figure I.8. Adele Abrahamsen proposed a method for constraining the analysis of verbs of motion into component predicates: she asked subjects to recall a story containing these verbs and used recall errors to suggest components that had been added or deleted from the original verbs.

Other research groups were also part of the cognitive science community at UCSD. The psychology department itself had several strong experimental and social psychologists. At the nearby Salk Institute, Ursula Bellugi, also a product of the Harvard Center, began psycholinguistic studies of American Sign Language in the early 1970s. With such collaborators as Edward Klima and Susan Fischer (MIT-educated linguists) and Patricia Siple (a recent UCSD Ph.D. in psychology), she did much of the pioneering work showing that American Sign Language was a true language both in structure and processing. (Later this topic was addressed in the psychology department by

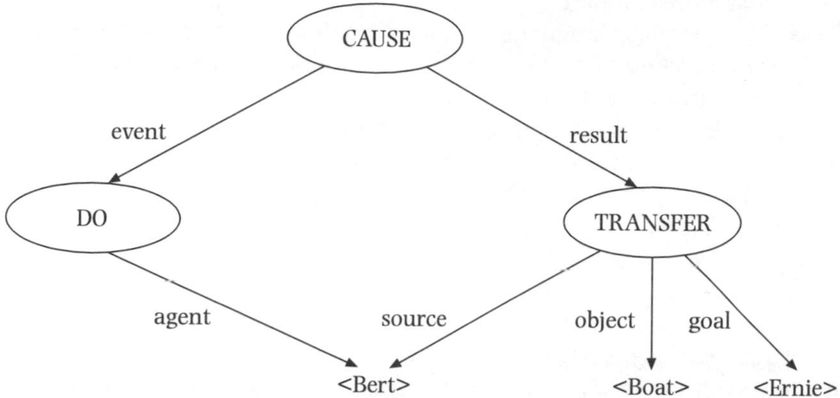

Figure I.8 The LNR verb group's representation of the meaning of the verb in the sentence "Bert gives a boat to Ernie." Each entity (angle-bracketed nodes) has a labeled case relation to a primitive predicate (oval nodes).

Elissa Newport and Ted Supalla, in the medical school by Helen Neville, and at Salk by Howard Poizner.) In the linguistics department, there was strong influence from Chomsky and also ongoing discussion of *generative semantics*. Finally, Roy D'Andrade was a pioneer in COGNITIVE ANTHROPOLOGY, and Aaron Cicourel pursued a cognitive approach in sociology (see Article 30, ETHNOMETHODOLOGY).

The creation of an Institute for Cognitive Science at UCSD was not to occur until the late 1970s (becoming a department in 1987), but the psychology graduate program had been designed from its beginning in the mid-1960s to encourage collaborative work with faculty, interaction across disciplinary boundaries, and even interaction across institutions: the LNR group had fruitful contacts with researchers on both coasts during the exciting period of the early 1970s. On the East Coast, this included cognitive scientists at Bolt Beranek and Newman outside Boston (especially Allan Collins, William Woods, Joseph Becker, and the late Jaime Carbonell), Ronald Kaplan at Harvard University, and more sporadic contacts with the very active AI lab at MIT. On the West Coast, there were three working visits with Stanford's cognitive psychologists and also fruitful interactions with Stanford-area AI researchers (fostered especially by Norman, who coauthored publications with Bobrow and with Schank).

2.4.3 University of Minnesota

The University of Minnesota provides an interesting alternative model of the development of cognitive psychology. It was a department with strong roots in learning theory and long-time home of two major centers: the Center for Research in Human Learning and the Institute for Child Development. Not as strongly identified with cognitive science as Stanford or UCSD, it nonetheless generated some important research in this field.

One of the leaders of the move towards cognitivism at Minnesota was James J. Jenkins, who began his career by trying to develop a mediational theory of learning in the behaviorist tradition that would be sufficient to explain the acquisition of language. The mediators were envisioned as internal stimuli and responses, a series of which was thought to facilitate the ability to make associations needed for language. But Jenkins discovered that more complex mediations failed to develop automatically, in the manner necessary to explain language, and as a result began to abandon behaviorism. He also began to recognize that the sort of grammar for which mediation theory was designed, a slot grammar of the form advocated in structural linguistics, was being superseded by Chomskian transformational grammars, which required rules, not associations. Jenkins spent the 1959–60 academic year at the Center for Advanced Studies in the Behavioral Sciences (as did George Mandler), the same year in which Miller, Galanter, and Pribram were writing *Plans and the Structure of Behavior*. Then in 1964–5 Jenkins spent another year at the Center, this time in the company of Jerry Fodor and Sol Saporta. Fodor (section 2.6) was a philosopher who had by this time become a major expositor of Chomsky. At the Center he and Saporta directed a tutorial reading of *Syntactic Structures*.

Returning to Minnesota, Jenkins was appointed director of the new Center for Research in Human Learning. One of the first activities of the Center was a summer program which involved a number of recent Minnesota Ph.D.s. The outside instructors were Fodor, Walter Reitman, and David Premack. Reitman's role was to introduce new developments in artificial intelligence, using his then new textbook *Cognition and Thought*. Fodor advanced the new cognitive views, especially those of Chomsky, while Premack defended somewhat more traditional behaviorist positions. Fodor and Premack sparred throughout the summer. Walter Weimer, then a graduate student at Minnesota, describes the interactions this way:

> Premack represented one direction – trying to resuscitate behaviorism and be "sophisticated" by saying we can address all those higher mental processes, too. Premack was playing catch-up ball, but without knowing he was playing catch-up against Dallas in the last minutes of the Super Bowl. Another direction was Fodor, who just stole the show. He was already so involved in the linguistic argument that he was doing normal-science transformational work, that hot-off-the-mimeo-presses, "This is the latest grammar from MIT." He was already doing normal science in the new linguistics, and saying to us "All right, this is what you must do in psychology to catch up." (Interview in Baars, 1986, p. 300)

Jenkins has directed more than 70 dissertations, in almost every area of cognitive psychology. Two of them were by John Bransford and Jeffrey Franks, who carried out a series of influential experiments that provided evidence for internal cognitive processes. Drawing upon Michael Posner and Steven Keele's demonstration that subjects could abstract a prototype they had never seen from random dot patterns that were generated from the prototype, Bransford and Franks (1971) developed their own stimuli – geometrical patterns that were derived from a prototype in a rule-governed way – and showed that when subjects were asked to draw the typical figure they had just seen, they drew the prototype they had not seen. From this starting point in visual perception, Bransford and Franks moved to language. They created complex sentences such as "The ants in the kitchen ate the sweet jelly which was on the table" to function as prototypes. They then extracted the four component propositions (e.g., "The jelly was sweet") and recombined these into derived sentences expressing one, two, or three of the propositions. Subjects were presented with subsets of the derived sentences from more than one prototype sentence in scrambled order. Then they were given old and new sentences to recognize, in which the new sentences were combinations of the four component propositions that had not yet been presented. Not only were subjects poor at distinguishing between old and new sentences; they actually gave their highest confidence ratings to the never presented prototypes. In a later study, Bransford, Barclay, and Franks (1972) showed that never presented inferences also were prone to false recognition. These studies were taken as evidence that memory is constructive.

2.5 Learning to think: artificial intelligence (AI)

At the time of the 1956 Dartmouth Conference, the Newell–Simon–Shaw Logic Theorist was the only functioning AI program. The 1960s and 1970s, though, saw rapid expansion. During this period, three centers for AI research assumed prominence: Allen Newell joined Simon at Carnegie–Mellon in the Graduate School of Industrial Administration; Minsky and McCarthy joined forces to create the Artificial Intelligence Group at MIT; and McCarthy left MIT in 1962 to join the computer science department at Stanford and founded an AI laboratory there. Artificial intelligence research during this time was expensive, since running the programs was computationally very demanding, and the existing computers were considerably less powerful than contemporary notebook computers. One of the major sources of funding was the Advanced Research Projects Agency (ARPA) of the Department of Defense, created after Sputnik in an attempt to ensure US competitiveness in science and technology. Unlike other granting agencies, ARPA did not rely on peer review and often focused on funding individuals thought to be promising rather than specific projects. One of its first major grants was a $2,220,000 grant to MIT for a project known as MAC (for both Multiple Access Computer and Machine-Aided Cognition). A third of this money went to the Artificial Intelligence Group; given McCarthy's departure, this money went to support Minsky and his new collaborator, Seymour Papert. We will return to Minsky's work below, but first we need to capture some of the atmosphere of early AI.

As soon as AI began to develop, it attracted attention in the popular press. One of the most visible domains into which AI ventured was game playing. Checkers and chess, in particular, are games of strategy that seem to require intelligence; thus, if AI systems could succeed in these games, this might be taken as evidence that artificial intelligence is possible. One of the early leaders in these efforts was Arthur Samuels, who in 1946 left Bell Laboratories for the University of Illinois. Aware that computers were on the horizon, he sought to have the university either buy or build one for him. As part of a scheme to raise money, his research group came up with the idea of building a computer to play checkers and then, by defeating the world checkers champion, raise additional money. Although they failed to develop either the hardware or the program, Samuels was bitten by the bug to build a checkers-playing computer and continued pursuing the idea when he went to IBM in 1949. When in 1951 IBM produced the 701, its first commercial computer, one of the programs prepared to run on it was a crude checkers-playing program by Samuels. Samuels kept improving on the program, partly by providing it with a capacity to learn from previous games, and by 1961 it was playing at master's level.

As challenging as checkers turned out to be, it was chess that emerged as the holy grail for AI. Early on it was realized that chess, unlike checkers, could not be played successfully by simply considering all possible sequences of moves that constituted games and choosing only sequences that led to winning (or at least not losing) – for the simple reason that there are too many such sequences (on the order of 10^{120}). In principle, one could represent all possible games by constructing a tree structure in which chess positions are nodes and each possible move from a position is a branch from its node. Proceeding from the root of the tree (initial chessboard position), a path down a series of branches to a terminal node would represent a single game. To make search of such a tree manageable, a procedure was required to prune it – that is, to rule out certain branches, and all branches that branch off from them, as unhelpful and thus not meriting further consideration. In a *Scientific American* article in 1950 Claude Shannon had proposed a strategy of following out all paths to a certain depth in the tree (e.g., four moves), evaluate the resulting board positions, and then use a minimax strategy to choose the best next move. (A minimax procedure is one which selects the move whose worst outcome is better than the worst outcome of all competitors.) Variants of Shannon's strategy permitted continued search of those branches which either seemed to involve potential risk or held the most promise. Other AI researchers quickly joined in the effort to develop a championship-level chess program; for example, even before they had developed Logic Theorist, Newell, Simon, and Shaw had embarked on developing a chess-playing computer. In 1957 they predicted that a computer would be world chess champion (if the rules allowed it) within a decade. But by 1967, chess-playing programs were still so inadequate that international master David Levy confidently bet that no computer would beat him within a decade. He not only won that bet, but won a renewed bet. Finally (shortly after the penultimate draft of this introduction was written), an IBM system dubbed "Big Blue" won a rematch with world champion Garry Kasparov on May 11, 1997 – four decades after the original optimistic prediction. Ironically, it appears that Big Blue's victory was clinched by its ability to unnerve Kasparov, rather than to outthink him; the opponents were tied when Kasparov fell apart in the sixth and last game.

2.5.1 Simulating human performance

Despite the crowd-pleasing appeal of such contests, game-playing programs have lived at the periphery of AI. Far more central was the goal of simulating human REASONING in a variety of domains. Throughout the 1960s and beyond, Newell and Simon at Carnegie–Mellon were a major presence at the intellectual center of this effort. As a strategy for developing computer programs that simulated human performance, they adopted a strategy developed by two psychologists, O. K. Moore and S. B. Anderson (1954), of asking subjects to *think aloud* while solving puzzles (see Article 33, PROTOCOL ANALYSIS). From the resulting reports they extracted reasoning strategies or heuristics to be incorporated into their next program, General Problem Solver (GPS). As its name suggests, GPS was designed to employ general procedures that could be widely applied in PROBLEM SOLVING. GPS was designed as a PRODUCTION SYSTEM, which consists of a working memory and a set of production rules. The working memory provides for the temporary storage of symbolic expressions. The heuristics were programmed in the form of production rules that paired a condition with an action: "If circumstances *X* obtain, do *Y*." *X* is an expression that might appear in working memory. If it did, then the production rule would *fire*, directing the system to perform action *Y*, which might involve adding or removing content from working memory or sending output from the system.

One kind of problem to which Newell and Simon applied GPS was cryptarithmetic. The challenge in cryptarithmetic is to replace the letters with numbers in an equation such as

$$\begin{array}{rl} & DONALD \\ + & \underline{GERALD} \quad D = 5, \\ & ROBERT \end{array}$$

so that the result is a valid addition problem. The key to their approach was *means–end* reasoning wherein the machine compares a description of the goal state with a description of its current state; if there is a difference, it employs a variety of operators to reduce this difference. One important aspect of the strategy involves working backwards: if no operator will take the system directly to the goal state, then operators are identified which would take it to the goal state from another state. Then a goal is established of reaching this new state, and again operators are sought to reduce the difference. Once GPS was up and running, Newell and Simon devoted much of their effort to comparing its performance to that of humans and revising the program to better simulate human performance. This project culminated in the 1972 publication of *Human Problem Solving*.

GPS was grounded on the assumption that intelligent behavior stemmed largely from general reasoning principles, not detailed knowledge of particular circumstances. Increasingly during the 1970s, though, AI researchers began to recognize the relevance of specific knowledge in solving problems. One direction of this effort led to the development of expert systems, programs that would incorporate knowledge gained from interviewing human experts in a particular domain. The designers would represent the expert knowledge in the form of rules and then apply general reasoning principles to arrive at inferences and answers to queries. Often a significant amount of additional manipulation of the system was required to get it to perform well, but some of the systems ultimately equaled or exceeded the performance of human experts. An exemplar of such research is DENDRAL, developed at Stanford University and described in 1971 by three researchers: Simon's student Edward Feigenbaum, philosopher-turned-computer scientist Bruce Buchanan, and Nobel Laureate geneticist Joshua Lederberg. DENDRAL performed at an expert level in analyzing data from mass spectrographs to determine the molecular structure of the organic compound being analyzed. In another project that relied on expert knowledge, Buchanan's graduate student Edward Shortliffe (1976) developed MYCIN, a program for diagnosis of infectious blood diseases.

Many of the early pioneering efforts in AI occurred at MIT during the period when McCarthy and Minsky were both there. John McCarthy developed what was to become the standard language for AI programming, LISP (LISt Processing language). It took from Newell and Simon's language IPL the idea of working with lists, each item of which would index the next item. LISP advanced beyond IPL in part by utilizing the lambda calculus, which allows functions to be treated as objects and hence as arguments in other functions. Another advance was to allow the items to represent not just simple imperatives (do this) but conditional statements like those in production systems (if specified conditions are met, do this).

In 1968 Minsky edited an influential book, *Semantic Information Processing*, in which McCarthy gave an overview of LISP, and a number of doctoral students published abbreviated versions of their dissertations. In general, each dissertation involved a computer program written to simulate some aspect of human cognition. For example, Thomas Evans's program ANALOGY was designed to solve visual geometric analogy problems in which it had to pick one of five possible solutions to problems of the form "*A is to B as C is to?*" Descriptions of the geometrical forms were represented in propositional expressions in LISP. The program then identified the difference between the representations of A and B and applied this difference operation to the representation of C to arrive at its answer.

While adequate programming languages were important to the development of AI, it was also necessary, if AI systems were to appear intelligent to humans, that users be able to interact with them in ordinary language. Daniel Bobrow's MIT dissertation, also included in the 1968 book, described a program called STUDENT which solved algebra story problems such as: "Bill's father's uncle is twice as old as Bill's father. Two years from now Bill's father will be three times as old as Bill. The sum of their ages is 92. Find Bill's age." In order to solve the problem, STUDENT needed first to transform the story problems into equations, which it did by matching the sentences against stored templates (e.g., ____ times ____) and from that extracting the equations.

Programs like STUDENT had to *kluge* (a term of art popularized at MIT for ad hoc computational solutions) serious problems, such as really understanding English. Matching sentences against stored templates would work sometimes, but sentences were easily constructed on which the strategy failed. To move from these *toy* cognitive activities to cognitive activities on the scale of real life required providing AI systems with vision and a more principled ability to understand natural language. Some early AI researchers thought vision, at least, would be easy, since even simple organisms could sense features of their environment, and modestly more complex organisms could detect visual layouts and recognize objects. Simplicity in organism was thought to betoken simplicity in task. Minsky, who had received ARPA funding for vision research, therefore assigned it as a summer project in 1966 to a precocious undergraduate, Gerald Sussman.

Soon realizing the challenge of the task, Minsky and Papert made the development of an AI system for vision a major part of their mission. To make the project more tractable, they developed the idea of using a simplified visual world as the target: the blocks micro world. Blocks were chosen because of their straight edges and relatively smooth appearance. The blocks world also became a target for work in robotics (getting robots to move blocks around) and natural language processing (with the goal of eventually using English to direct robots' interactions with the blocks). One successful program, written by Patrick Winston for his dissertation, used semantic networks of the sort developed by Quillian (see below) to acquire information about the blocks world linguistically. For example, Winston's program would be presented with a symbolic description of a configuration such as an arch and would be told that it was an arch. It would also be presented with some non-arches. After being fed several examples, the program was able to form a general description that would include the arches and exclude the non-arches.

Developing AI systems capable of navigating real environments was the objective of a major AI research group located at the Stanford Research Institute (which subsequently dropped its official affiliation with Stanford and was renamed SRI International). Also heavily supported by ARPA funding, Charles Rosen headed a team that included Bertram Raphael (who received his Ph.D. with Minsky), Richard Fikes, and Nils Nilsson. In 1969 they built a robot, Shakey, which propelled itself on wheels and used a TV camera for visual input. Although mobile, Shakey was restricted to a suite of seven rooms, some of which contained boxes that Shakey could push around or stack. Fikes and Nilsson developed a control system for Shakey called STRIPS (Stanford Research Institute Problem Solver), which consisted of three-part rules: one part would specify preconditions (x is on the table, nothing else is on x, and Shakey's hand is empty), a second part would delete conditions (x is on the table and Shakey's hand is empty), and a third part would add conditions (Shakey is holding x). Goals would also be specified in simple predicate structures, and a goal would invoke a rule if the rule contained as one of its add conditions a state specified in a goal. STRIPS could reason backwards from goals to preconditions, set the preconditions as goals, etc., until it reached preconditions that could be executed. Further, once a plan was formed in STRIPS for reaching a goal, that plan could be stored, with variables replacing names, so as to be employed in similar circumstances in the future. STRIPS thus contained many of the features of PRODUCTION SYSTEMS.

Perhaps the most impressive research with a blocks world, however, used a simulated rather than real blocks world and robot in order to focus on how a program could communicate in English about acting in a blocks world. For his dissertation under Papert, Terry Winograd (1972) used a computational representation of a blocks world which his program, SHRDLU, could also display on a computer monitor. SHRDLU would process sentences supplied to it by applying both syntactic and semantic rules of unprecedented sophistication. SHRDLU was able to answer a broad range of questions about the blocks world and could also carry out actions in response to commands. Some of the commands could be satisfied by a single action such as grasping a block; others required a sequence of preliminary actions. An important aspect of SHRDLU was its underlying mode of operation. It did not employ the theorem-proving approach of such investigators as Newell, Simon, and McCarthy. Instead, using PLANNER, a LISP-based language, it operated with various subprograms, each of which pursued an independent goal; a subprogram could, in appropriate circumstances, take control of the operation of the program until its goal was realized or failed.

2.5.2 *AI aims to get real*

With programs such as GPS, STUDENT, and SHRDLU, AI had clearly achieved some success. But the hype for early AI had been much greater: the promise of a computer becoming world chess champion in the 1960s and of demonstrating humanlike intelligence by the 1970s. When these aspirations were not realized, funding reductions and critical assessment began. In the mid-1970s ARPA reduced its funding of AI research. But perhaps as important was critical assessment within and outside the AI community. One of the major problems facing AI was identified by John McCarthy in collaboration with philosopher Patrick Hayes (McCarthy and Hayes, 1969). McCarthy's approach to AI – using formal logical derivations to arrive at actions – required a complete internal representation of all relevant features of the world, a frame of reference. Difficulties ensued because activities in the world, including those initiated by the AI system, could change this frame. McCarthy and Hayes labeled the problem of how to update the frame of reference the *frame problem*. Solving it seemed to require providing the computer with complete knowledge about what would change and what would remain the same in the world as a result of an activity. Providing such knowledge would be, at best, a mammoth undertaking. What the frame problem indicated was that successes like SHRDLU depended critically on the limitations imposed on the micro-world in which it operated. Since the behavior of that world was itself controlled by deterministic laws in the computer, it was possible for the system to update itself on all changes. But this would not carry over when an AI system was operating in the real world – as the brain of a robot, for example.

Almost from its inception, critics of AI emerged who questioned whether problems like the frame problem could ever be solved. The most influential of these was the philosopher Hubert Dreyfus. Dreyfus had an early clash with Herbert Simon after a colloquium at MIT in 1961, where Dreyfus was teaching. Dreyfus was then hired as a consultant by the RAND Corporation for the summer of 1964 and produced a report entitled "Alchemy and AI," which, as its title indicates, challenged the legitimacy of the whole AI enterprise. (Seymour Papert wrote a response, entitled "The artificial intelligence of Hubert L. Dreyfus: a budget of fallacies," which RAND decided not to publish for fear of a libel suit, but which later appeared as an MIT technical report.) Dreyfus's report became the basis for his 1972 book *What Computers Can't Do.*

Dreyfus's objection went to the very foundation of AI. He questioned whether "processing data representing facts about the world using logical operations" was sufficient to account for what our cognitive systems do. Part of Dreyfus's critique involved noting that for an AI system working on logical operations, no meaning is attached to the data structures or symbolic representations. To deal with meaning, he contended, one had to get beyond formal rules and deal with the world and body as experienced – what philosophers call *phenomenological experience.* One of the features of our phenomenological experience is what Dreyfus (in the manner of William James) calls *fringe consciousness,* our awareness of features of the world that are not focal (i.e., features to which we are not explicitly attending but which nonetheless influence our focal consciousness). He contended that the effects of fringe consciousness could not be simulated by simply providing more symbolic representations and rules; to have fringe consciousness, one must have a body operating in a world. A long-distance runner, for example, may respond to the twists and turns of the road on which she runs without consciously representing those twists and turns to herself. For humans, cues and constraints from the world shape our understanding, and these are unavailable to the AI system – or so claimed Dreyfus.

While the objections raised by Dreyfus sparked bitter controversy, some of his concerns began to be addressed in research during the 1970s. One of the limitations of early AI programs was that they essentially treated each piece of information as a separate proposition, which would then be related to other propositions through associations or logical relations. One remedy was to develop more complex data structures that represented information in relation to other information. An idea of how to develop relational data structures – Ross Quillian's semantic network approach – was actually developed quite early. Quillian developed his basic framework as a graduate student under Herbert Simon in the mid-1960s, and a modified version of his dissertation was published in Minsky's (1968) book *Semantic Information Processing.* Quillian thought in terms of a network of nodes. For each sense of a particular word (a *word concept*) there was one type node (similar to a dictionary entry) and numerous token nodes (for each particular use of the word concept). The node for one word concept

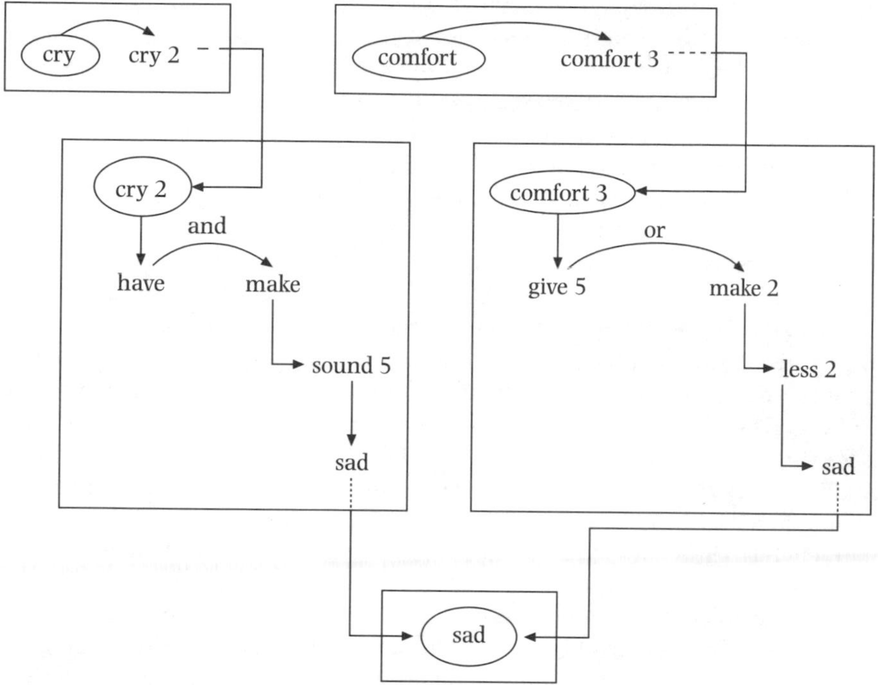

Figure I.9 A portion of a semantic network as explored by Quillian (1968). Two concepts, *cry* and *comfort*, are shown to be related in that paths lead from both of them to *sad*.

was connected to nodes for other word concepts that figured in its definitions, and these to nodes that figured in their definitions, creating a configuration of links. One method used by Quillian to explore the behavior of such networks was to write a computer program to compare any two word concepts. Starting at the two type nodes, the program worked its way outwards and placed activation tags on each node encountered. The overlaps in meaning between the two word concepts were determined in this way, and Quillian even wrote another program to give a crudely worded English report of the comparison. He also wrote a program that did a fair job of disambiguating words in sentences. Figure I.9 provides a relatively simple example of what happens when two related concepts are compared: the two paths traversed by moving out from *cry* and *comfort* converge at *sad*.

Quillian's networks were even more computationally demanding than other early AI programs; given the limitations of the available computers, he never got beyond the demonstration phase for this particular network design. But his ideas had enormous influence. We have already encountered two later projects that used modified network designs to simulate human performance: the LNR group's ELINOR model at UCSD and Anderson and Bower's SAM model at Stanford. Additionally, Quillian collaborated with psychologist Allan Collins to predict human reaction time patterns from a simplified version of Quillian's network (Collins and Quillian, 1969). This spawned the spreading activation research paradigm, which has had a long, vigorous life within mainstream experimental psychology.

One of the prime movers in developing larger-scale knowledge structures was Roger Schank. Originally trained at the University of Texas as a linguist, he broke fundamentally with the Chomskian focus on autonomy of syntax and reinvented himself as an AI researcher. Radically de-emphasizing syntax, Schank argued that meaning representations could be extracted directly from sentences. As an assistant professor in linguistics and computer science at Stanford, Schank tried to write computer programs capable of understanding natural language. Analyzing the meaning of verbs, he proposed eleven primitive actions that could be used to obtain semantic representations for a large number of verbs. Examples included PTRANS (transferring the physical location of an object), MTRANS (transferring mental information within or between subjects), and ATRANS (transferring the abstract location of an object; i.e., possession). Each primitive action involved roles such as actor, object, source, and goal. When the program needed to interpret a particular verb in a sentence, it started constructing a meaning representation containing the appropriate primitive actions; the specified roles for those actions would then need to be filled. If the program could not fill them, it would reassess the actions that had been assigned to the verb. For example, *sold* frequently involves an ATRANS, but in the phrase *sold out* it does not. Schank therefore attached other general rules to the various predicates, which enabled the system to go beyond the literal meaning of a sentence to make plausible inferences.

Schank's first effort to employ such rules in sentence interpretation, MARGIE (Memory, Analysis, Response Generalization in English), suffered from the fact that typically there were too many plausible inferences that might be licensed. After moving to Yale, Schank began a fruitful collaboration with psychologist Robert Abelson. They overcame these limitations by developing higher-order knowledge structures, *scripts*, which characterized the general structure of events in a common experience from the point of view of a specific participant (Schank and Abelson, 1977). Their best-known script was for a diner going to a restaurant. Since the events that transpire in a restaurant depend upon the kind of restaurant, the restaurant script contained tracks for fast food restaurant, cafeteria, coffee shop, etc. Within each track were a number of roles and a number of scenes (e.g., entering, ordering, eating, and exiting), as well as a typical sequence of primitive actions. The following is the portion of the coffee shop track of the restaurant script in which the customer (S) requests a menu from the waiter (W) and the waiter begins to respond:

S MTRANS signal to W
W PTRANS W to table
S MTRANS "need menu" to W
W PTRANS W to menu

Although only a few of the primitive actions might be mentioned specifically in a story, Schank and Abelson proposed that readers supply the others by using scripts to understand the story. Hence, readers should be able to answer questions about primitive actions not mentioned in a story and include some of these in a paraphrase of the story. (See Article 51, REPRESENTATIONS.) To deal with nonstereotyped circumstances, Schank and Abelson added the recognition that humans generally have goals, as well as plans for meeting them.

Schank's graduate students proceeded to develop a number of programs that used scripts to understand stories, including Gerry DeJong's FRUMP, which analyzed stories on the UPI news wire, and Janet Kolodner's CYRUS, which emulated government official Cyrus Vance. Employing FRUMP to learn from the UPI about Vance, CYRUS could then answer questions and make inferences about Vance. (Subsequently Schank's research led him to recognize the importance of even larger structures encoding knowledge of previous experiences of a kind, such as previous meals one has cooked. These cases provide the basis for CASE-BASED REASONING models. In addition, since moving to Northwestern in 1989 to direct the Institute for the Learning Sciences, Schank has become increasingly concerned with applications of AI to EDUCATION.)

Another proposal for higher-level knowledge structures was found in Minsky's (1975) work on frames and frame systems. A frame is a data structure representing stereotypical situations in terms of features that are always true of such situations, as well as terminal slots for features which may take on a variety of different values but must be assigned some value in a given situation. For example, in looking at a room from a given viewpoint, there will always be walls in the scene, but several options for the color of the walls. Slots will generally have default values associated with them, but these can be dislodged if alternative information is presented. The slots themselves may be filled by frames, providing for a recursive representational system. Another source of systematicity is that frames are related to one another by transformations (e.g., the transformation of moving from one viewpoint in a room to another). Some transformations will result in changes in the slot-fillers, while others will retain the same values. Minsky proposed that when a person encounters a situation, what the person tries to do is to match the information about the situation to a frame in a stored frame system; once a possible frame is proposed, it generates expectations that guide further search for information. If such a search produces information inconsistent with the frame, then a new proposal for a frame must be advanced. Minsky linked his notion of a frame to *schemata*, which figured in Bartlett's (1932) account of memory, and to *paradigms*, which figured in Thomas Kuhn's (1962/1970) account of normal science.

The introduction of large-scale knowledge structures such as semantic networks, scripts, and frames went part way toward addressing Dreyfus's concerns, as he acknowledged in an extended introduction to the second edition of his book *What Computers Can't Do* in 1979. But they still did not meet his objective of having systems engage the real world through real bodies. Without that, Dreyfus contended that such systems would never achieve real knowledge. An extreme example of this can be found in the program ELIZA, designed by Joseph Weizenbaum to carry on conversations with human interlocutors without any understanding of the topic of the conversation. Weizenbaum pulled this off by modeling ELIZA after Rogerian psychotherapists, who are known for their technique of nondirectively reflecting back to the client what the client has said. One strategy was to insert "Why do you think ____" before repeating what the interlocutor said. Another would monitor for key words such as *mother* in the interlocutor's statements and respond "Tell me about your ____." The program would also make appropriate substitutions, such as *your* for *my*, so as to mimic a real conversation.

Many people who interacted with ELIZA were seduced into thinking that it understood what they were saying and would engage in elaborate conversations. In those cases, ELIZA seemed to be passing the Turing test (interacting indistinguishably from a human), and this was impressing people far more than Weizenbaum thought it should. In *Computer Power and Human Reason* (1976) he strongly criticized the tendency in AI to overinterpret performances achieved using symbolic representations and rules. Because AI systems lacked any understanding of the symbols they were manipulating, they might imitate humans (to a degree) but could never really replicate human intelligence. The moral risk, then, was to be taken in by mimicry and turn over to machines decisions that required true human intelligence.

2.6 Getting a philosophy

Philosophers have theorized about the mind for 2,500 years; thus, it may be a bit surprising that we have reached this point in describing the emergence of cognitive science and have yet to focus on the contributions of philosophers. One major reason for this is that the endeavors that motivated the development of modern cognitive science were mostly empirical efforts: experiments and other data-based studies in psychology and neuroscience, analysis of sentences judged to be well formed in linguistics, and construction of machines (or programs for machines) that could carry out cognitive activities in AI. Thus, it might seem that cognitive science, like so many other sciences, would leave philosophy behind after developing its empirical side. But this was not to be: during this period of initial maturation of cognitive science, philosophical inquiry (sometimes by nonphilosophers) came to play an increasingly important role.

Perhaps the oldest philosophical problem relevant to cognitive science is the mind–body problem, which took its modern form with the seventeenth-century philosopher René Descartes' contention that the mind is distinct from the body. Descartes thus defended a dual substance ontology, which raised as a central question how nonmaterial minds could interact with physical bodies, including brains. The primary opposition to dualism has come from theorists labeled collectively as *materialists*; materialism, however, is such a broad camp that it includes a number of very different views about how mental states relate to physical ones, including how psychological states relate to states of the brain. One very prominent position, advanced by Gilbert Ryle in the 1940s, seemed to comport well with the behaviorism then current in psychology. Ryle proposed that mental predicates (such as *believes)* do not designate internal states of agents but rather describe propensities of agents to behave in certain ways. This approach, however, seemed incapable of accounting for other mental states, such as sensations. To handle sensations, in the 1950s philosopher J. C. Smart and philosophically minded psychologist U. T. Place advanced the mind–brain *identity theory*, which identified sensations with brain states; for example, pain was simply a particular brain state. The identity theory eventually became generalized to other mental states such as beliefs. At the time of the birth of cognitive science, the identity theory was the dominant version of materialism advanced by philosophers.

But the identity theory did not seem to fit well with a major feature of the new cognitive approach: if mental states were identical with brain states, then particular mental states were limited to organisms with similar brains. Many cognitivists were attracted to the idea that the same sort of cognitive state or condition could occur in diverse brains or central nervous systems. Different sorts of animals may perceive the rainfall, taste a lemon, attend to a red triangle. Those attracted to AI were further enticed by the fact that computers too might perceive, attend, and reason. Reflecting on animal minds and the possibility of artificial minds, as well as minds in other organisms, Hilary Putnam, a philosopher who began his career at Princeton and then moved to

Harvard, argued against the identity of mental states with brain states (Putnam, 1960). He claimed that states such as pain and hunger could occur in organisms whose brains are as different from ours as those of octopuses. He thus introduced the idea that mental states might be *multiply realizable*, and consequently cannot be identified with any of their realizations.

Rejecting the identity between psychological states and brain states might seem to be part of an argument for dualism, but Putnam certainly did not construe himself as abandoning materialism. What, then, are mental states if they are not to be identified with brain states? Putnam (1967) proposed that mental states be identified in terms of their causal or functional role in mediating between sensations and behavior. That role – rather than its neural realization – determines the nature or identity of the state. For example, what determines whether a particular cognitive process or state is a preference for an ice-cold beer, as opposed to the perception that a flower is red, is that the state is caused by thirst and gives rise to other mental states such as planning to move towards the refrigerator or requesting that someone bring over a beer. The state causes other mental states and then beer-seeking behavior, because, roughly, it is a preference for beer. The brain must somehow implement the causal role, but it is the role – not the brain – that makes the state a beer preference.

Putnam dubbed his theory *functionalism*. His use of the term is confusing in some respects. Within psychology, for example, we have seen that *functionalism* is used to describe an approach to psychological theory construction stemming from James that has an evolutionary orientation. What Putnam had in mind was not Darwin's work on adaptivity, but rather Turing's work on machine tables. However, the essence of Putnamian functionalism is strikingly clear, and the term aptly captures that essence. For Putnam cognitive states are defined by their typical causes and effects. (Other defenders of functionalism, such as Daniel Dennett and William Lycan, maintain Putnam's emphasis on causal role, but, as a result of allowing for multiple iterations of functional decomposition, allow for a greater connection between psychological functional roles and the neural processes that realize them.)

Functionalism fits well into the cognitive agenda. By drawing links between mental processes and operations in computers, functionalists saw that one could avoid the chauvinism of limiting cognitive states to systems with our kind of brain while seemingly also avoiding some of the problems of behaviorism (see, however, Block, 1978, for an influential criticism that denies the potential for finding such a middle ground). Functionalism (in one form or another) was rapidly endorsed by many philosophers, who acclaimed it as providing the cognitivist solution to the mind–body problem. But in the hands of one of Putnam's students, Jerry Fodor, it became just one of several contentious claims in an ambitious theory that attempted to answer a key question for cognitive science: "What would a satisfactory theory of cognition look like?" Neisser and others had asked the question and sketched suggestive answers. In *The Language of Thought* (1975) Fodor, then at MIT, offered one of the most detailed answers to that question, one that has continued to play an influential, often divisive role in cognitive science. The heart of his answer was something he called the *Language of Thought* (LOT) hypothesis. Here is a capsule characterization:

> LOT: To be a cognizer is to possess a system of syntactically structured symbols-in-the-head (mind/brain) which undergo processing that is sensitive to that structure. Cognition,

> in all of its forms, from the simplest perception of a physical stimulus to the most complex judgment concerning the grammaticality of an utterance, consists of manipulating symbols-in-the-head in accord with that syntax. The system of primitive, innate symbols-in-the-head and their syntactic combination in sentence-like structures is sometimes called "mentalese."

Fodor sometimes referred to the syntactically structured symbols-in-the-head as *representations* (see Article 50, REPRESENTATION AND COMPUTATION). By this he meant that these symbols represent the world by referring to things and by predicating or ascribing properties to them. "Water is wet" (or its mentalese equivalent), for example, refers to water and attributes the property of being wet to water. This is the meaning of the sequence of symbols (its semantics). An important claim, however, is that semantics need not be accessed when operating on the symbols: operations are licensed solely on the basis of form (the "shape" of each symbol and the syntax by which they are combined).

Since 1975 Fodor has put forward a number of arguments for LOT. Initially, many of his arguments focused on learning. He argued that in order for a cognitive system to learn a language like English, for example, it had to be able to advance hypotheses (e.g., *water* means water) and then test these. Thus, the cognitive system needed a mode of representation capable of expressing any hypothesis it might want to test. One consequence of this kind of reasoning was that LOT could not be a natural language like English, but had to be available to the system before it could acquire language. Fodor therefore followed another of his mentors, Chomsky, in arguing for a strong nativism. While Chomsky had argued that knowledge of basic grammatical rules must be innate, Fodor argued that a powerful representational system had to be innate (see Article 45, INNATE KNOWLEDGE).

More recently Fodor (1987) has employed LOT to explain other features of cognitive systems, particularly their productivity and systematicity. Productivity refers to the idea that cognitive systems are not bounded: one can always generate new thoughts. Systematicity refers to the idea that having a given cognitive capacity guarantees having certain related capacities. To adapt one of Fodor's remarks, you don't find people who can think Morgan loves Shannon but cannot think Shannon loves Morgan ("mLs but not sLm" in an abbreviated notation). Productivity and systematicity are no accidents, according to Fodor. They fall out of the syntactic structure of the language of thought. Productivity follows from the fact that the rules for building up syntactic structures are recursive, so that one can repeatedly combine composed structures into still larger structures. Systematicity is accounted for by the LOT hypothesis that the brain state that encodes one mental representation (say, "mLs") is a syntactic rearrangement of the brain state that encodes another representation (say, "sLm"). The patterns of possible thoughts and preferences depend on the structure of the language of thought. It is the nature of that structure that regulates the manner in which one thought may lead to or include another.

LOT fits together well with the multiple realizability arguments advanced for functionalism. According to LOT, cognition has nothing directly to do with its species-specific neurobiological embodiment or implementation, but rather concerns processes operating on the common language of thought shared by all entities capable of being in the same cognitive state. Cognition *per se* is not neural; cognition is computation in mentalese. Being in the same cognitive state (say, preferring water) is characterized in LOT as employing the same set of internal symbols and syntactic rules.

Over time, it has been recognized that neither LOT nor functionalism is free of liability. First, we have no inkling as to how or where the syntactically structured mental representations of LOT are stored. In the minds of some critics, that was one huge drawback of Fodor's hypothesis. Second, LOT and functionalism both fail to connect psychology to neuroscience in any experimentally tractable way. For Fodor and Putnam this dissociation between the study of cognition and the study of the brain was not a shortcoming but a strength. Moreover, it fitted well with the *zeitgeist* of cognitive science during the 1970s, which preferred independence from neuroscience. On a practical note, there seemed to be little work in neuroscience that could address cognitive questions such as how people solve problems or learn languages. Through the multiple realizability argument and LOT, Putnam and Fodor seemed to give principled legitimacy to the then current approaches in cognitive science. But by the late 1970s, some philosophers were objecting to the divorce of cognitive science from neuroscience, Paul M. and Patricia S. Churchland foremost amongst them. They tended to continue to endorse a version of the identity theory and to reject LOT (P. S. Churchland, 1986) (see Article 48, LEVELS OF EXPLANATION AND COGNITIVE ARCHITECTURES).

Not all meta-theorizing about cognitive science was done by professional philosophers. We will examine two extremely influential analyses of the project of cognitive science, both of which cohered in many respects with the views of Putnam and Fodor. The first was advanced by neurophysiologist-turned-AI-researcher David Marr in his influential 1982 book *Vision*. In the first chapter he relates his growing disillusionment with the prospects of figuring out how the brain performs cognitive tasks by starting with the response patterns of individual neurons, such as the edge detectors identified by Hubel and Wiesel. He himself had contributed to this project in an important study of the cerebellar cortex (Marr, 1969). Marr's disillusionment was due principally to his coming to realize that discovering the response patterns and wiring diagrams of individual brain parts could only *describe* what was happening in the brain, not *explain* how it accomplished its tasks. He concluded:

> There must exist an additional level of understanding at which the character of the information-processing tasks carried out during perception are [sic] analyzed and understood in a way that is independent of the particular mechanisms and structures that implement them in our heads. This was what was missing – the analysis of the problem as an information-processing task. (1982, p. 19)

Marr went on to argue that in understanding any computational system, such as the brain, two additional levels were needed above the level at which the details of the physical device (the brain) were analyzed. He called his highest level *computational theory* (a label that many have found misleading; it is somewhat akin to Chomsky's notion of competence and might best be called task analysis). Computational theory characterizes the task to be performed, thus answering the questions: What is the goal of this computation? Why is it appropriate? In building an adding machine, for example, the abstractly characterized function of addition is the task we want the machine to carry out. Marr emphasized that the answers to these questions would constrain work at lower levels. The next level, that of *representation and algorithm*, specifies (a) a system of representations (e.g., arabic and roman numerals provide two different representational systems for numbers) and (b) the operations (algorithms) to be performed on them so as to satisfy the function specified in the computational theory (e.g., arithmetic). Marr, like Putnam, emphasized the multiple realizability of computational goals using different representational systems and algorithms. Only at the lowest level, the level of *hardware implementation*, did one turn to the actual physical device and show how it implemented the representations and algorithms from level 2.

It is important to recognize that Marr's levels identify three different ways to analyze the same system; they do not characterize levels of organization in nature, which typically are related in a part–whole fashion. Moreover, while he emphasized the differences between levels and noted the multiple realizability of systems satisfying higher levels in the analysis, he also recognized a loose coupling of the levels: the analysis advanced at any given level constrained analyses at the others. Finally, while Marr is sometimes portrayed as proposing that we work solely from the highest level down, he clearly emphasized constraints coming from the bottom up as well and employed them in his attempt to explain visual processing. His concern was ultimately to figure out what an adequate explanation would look like, which brought him to the claim that it would have to provide analyses at each of these three levels.

While Marr's analysis of levels is in some respects compatible with Putnamian functionalism (but recognizes more constraints between levels than Putnam indicated), the other case of meta-theorizing by nonphilosophers, Newell and Simon's *Physical Symbol System Hypothesis*, shared Fodor's emphasis on symbolic processing as a key to understanding cognition. The hypothesis (as presented in Newell, 1980) states that a physical symbol system satisfies the necessary and sufficient conditions for exhibiting "general intelligent action" (p. 170). The kind of system envisaged by Newell and Simon is a universal system that can carry out any mapping from input states to output states and does this by operating on symbols. What makes something a symbol is its designation or reference:

> The most fundamental concept for a symbol system is that which gives symbols their symbolic character, i.e., which lets them stand for some entity. We call this concept designation, though we might have used any of several other terms, e.g., reference, denotation, naming, standing for, aboutness, or even symbolization or meaning. . . .
>
> Let us have a definition:
>
> > *Designation*: An entity X designates an entity Y relative to a process P, if, when P takes X as input, its behavior depends on Y.
>
> There are two keys to this definition: First, the concept is grounded in the behavior of a process. Thus, the implications of designation will depend on the nature of this process. Second, there is action at a distance . . . This is the symbolic aspect, that having X (the symbol) is tantamount to having Y (the thing designated) for the purposes of process P. (p. 156)

Consider, by way of analogy, a map that represents Boston. If the map has the right resources for designation, and someone who knows how to read maps applies that process appropriately, then Boston-on-the-map can stand for – is tantamount to – Boston in Massachusetts.

What philosophers give, they also can take away. Dreyfus and his fellow Berkeley philosopher John Searle were especially critical of approaches to cognition like those of Putnam and Fodor as well as Newell and Simon. Both these philosophers denied that cognition consists in syntactic processing of symbols alone. Dreyfus (1972/1979) argued that much natural cognition is not representable in a symbolic code, but is, in some sense, out in the environment. To do cognitive science, according to Dreyfus, one must work with embodied systems, not syntactic processors – a theme some cognitive scientists embraced in the 1990s (see Article 39, EMBODIED, SITUATED, AND DISTRIBUTED COGNITION).

Searle was a skeptic of his own kind. Symbolic codes, Searle claimed, are not even cognitive; they lack genuine *cognitivity* (what Searle calls *intrinsic intentionality*, the property of being about something). Real mental states possess INTENTIONALITY in themselves; your belief that there is a cold beer in the refrigerator has that content in itself and does not depend on an external interpreter to assign it that content. Codes possess only derived intentionality (based on outside interpretation). Searle (1980) argued for these claims by constructing a thought-experiment which he dubbed the "Chinese Room." In the thought-experiment he plays the role of a computer that is programmed to answer in Chinese questions asked in Chinese, much in the manner of trying to pass a Turing test. Searle imagines himself in possession of a book of syntactic rules which tells him how to replace strings of Chinese characters (in questions) with other strings of Chinese characters (in answers). Outside observers may believe that the computer/person in the room understands Chinese; the *room* will pass the Turing test. But the rule book is a grammatical manual, not a bilingual dictionary. It does not tell the denizen of the room what the characters mean. So Searle contends that he could execute all the syntactic manipulation proposed by a theory like LOT, yet not understand – not cognize in – Chinese. Therefore computation over a system of meaningless symbols is insufficient to account for cognition. Such symbolic codes are not cognitive.

The objections of Dreyfus and Searle, to the extent that they are valid, may threaten not cognitive science in general, but rather those cognitive science research programs that are wedded to an *exclusively* computational view of cognition. One may concede that contemporary computers are purely syntactic engines that lack intentionality. Hence, they are not cognitive. But what this means is that to answer Searle, one must develop some way to *ground* the symbols of a system so as to give them genuine content. Some theorists have suggested that it is the way in which the symbols are acquired by the system (e.g., through a connectionist learning process in which the structure of the representation is acquired through interaction with the objects to which they refer) that makes them intentional (Harnad, 1990). Another strategy is to deny that our mental states enjoy intrinsic intentionality – they only seem to do so because we have the ability to use yet other symbols, such as words in our natural language, to specify their referents. To accommodate Dreyfus, cognitive science may have to take seriously that our cognitive systems are embodied and situated in the world, a theme that, as we have noted, a variety of contemporary researchers are pursuing.

2.7 *Getting an identity*

In the modern world, intellectual developments such as those we have been discussing require institutions which frame and support their activities. These institutions include departments or other administrative units in universities, as well as journals and professional societies. We turn now to the process by which cognitive science developed its institutional identity towards the end of its period of initial maturation.

2.7.1 *Cognitive science centers: the legacy of the Sloan Foundation*

Academic researchers are generally hired into departments, and the structure of departments plays a critical role in their ongoing life. Departments allocate the important commodities, such as space, money, and teaching assignments. They control the curriculum for both undergraduate and graduate education, and thus determine the intellectual frameworks and tools that new investigators will be prepared to employ. Even more important, they make important personnel decisions about whom to hire and whom to tenure.

As cognitive scientists from different disciplines began to explore their common interests, they sought an institutional structure in their home universities. Sometimes new academic pursuits result in new departments, and this has happened at a few locations for cognitive science (e.g., at UCSD and Johns Hopkins University). But creating new departments is not always the best path of development for new areas of inquiry to take, especially when they bridge existing disciplines rather than offer new methodologies for exploring new domains. First, departments cost money, and universities are often reluctant to use scarce resources in this way. Second, moving into a new department can isolate a researcher from others who share a common disciplinary foundation (e.g., it could isolate a cognitive psychologist from other psychologists, many of whom employ similar research tools and theoretical frameworks).

A solution to the desire of cognitive scientists both to retain affiliation with their home disciplines and to affiliate with other cognitive scientists from different departments is to create a center. The Harvard Center for Cognitive Studies, discussed above, was a prototype. The creation of such centers is not unique to cognitive science; universities have created centers (sometimes called *committees* or *programs*) in a variety of interdisciplinary areas, including women's studies, African-American studies, history and philosophy of science, evolutionary biology, and materials science. In some instances, these programs or centers become virtual departments, running their own degree programs, tenuring their own faculty, and so forth. But generally they serve more as meeting grounds, by sponsoring colloquia and postdoctoral fellowships; sometimes they also develop a curriculum comprised of courses from various departments as well as new, explicitly interdisciplinary courses. Instead of offering their own degrees, at least at the Ph.D. level, they may offer certificates which complement a graduate degree in one of the existing departments.

The center model was quickly adopted in cognitive science. As of 1983, the following North American universities had all developed centers for cognitive science: Brown University, Carnegie–Mellon University, the University of Chicago, the University of Colorado, Cornell Medical School, the University of Illinois at Urbana, the University of Massachusetts at Amherst, Massachusetts Institute of Technology, the University of Michigan, the University of Pennsylvania, the University of Pittsburgh, Princeton University, the University of Rochester, Rutgers University, Stanford University, the University of Texas at Austin, the University of Western Ontario, Yale University, and four campuses of the University of California (Berkeley, Irvine, Santa Barbara, and San Diego). The initiative for creating these centers in part arose from faculty members who realized that many of their closest intellectual alliances involved faculty from other departments. But another powerful incentive behind the creation of many of these centers was financial. Starting in 1977, the Alfred P. Sloan Foundation provided substantial grants to universities to help them establish institutional structures supporting research in cognitive science. In the Foundation's annual report for 1977, this initiative was characterized as follows:

> Scientists in several disciplines, including psychology, linguistics, computer science, and neuroscience, believe that by pooling their diverse knowledge they can achieve a more advanced understanding of human mental processes such as memory, perception, and language. The Particular Program in Cognitive Sciences seeks to accelerate this cross-fertilization at some of the most promising centers for such research. (p. 48)

During its first two years of funding cognitive science, the Sloan Foundation focused on activities that were *exploratory*: "intended to enable institutions with varying disciplinary strengths to consider the kinds of coordinated research and training programs which they might develop in the future" (report for 1978, pp. 55–6). In the initial year of support, grants primarily supported workshops and visiting scientists. The largest of these, to UCSD, supported a number of postdoctoral fellowships and a major conference in summer 1979 which was to become the first conference of the Cognitive Science Society (see below).

In the course of identifying its mission in cognitive science, the Sloan Foundation commissioned a report on the state of the art in cognitive science from a committee of advisers from the cognitive sciences. In its report of October 1, 1978 (excerpted in Pylyshyn, 1983, appendix, p. 75), the committee argued that an "autonomous science of cognition has arisen in the past decade" on the basis of "the richly articulated pattern of interconnection among [the] subdomains" of neuroscience, computer science, psychology, philosophy, linguistics, and anthropology. They presented the pattern of connections in a diagram, in which six nodes represent the contributing disciplines, and connections between them represent interdisciplinary endeavors (figure I.10). The labeled connections represent well-defined areas of cross-disciplinary inquiry. For example, psycholinguistics joins the research methods and conceptual underpinnings of experimental psychology to those of linguistics. The four dotted lines represent interdisciplinary connections that had not yet taken form. The committee also noted that one can identify some three-way connections, such as the cooperation among psychology, linguistics, and philosophy in the study of language. The authors then offered a statement of the common research objective that links all the disciplines of cognitive science: "to discover the representational and computational capacities of the mind and their structural and functional representation in the brain" (p. 76).

With the vision of what cognitive science should be, and on the basis of the initial exploratory grants, the Sloan Foundation moved into a second phase, "the establishment of formal training programs in the emerging discipline of cognitive science" (report for 1979, p. 49). In that year they made six grants in the $400,000 to $500,000 range to UCSD, MIT, Carnegie–Mellon University, the University of Pennsylvania, the University of Texas, and Yale University. A brief sketch of the initiative at two of these institutions gives a sense of the scope of the cognitive science centers that Sloan was supporting. Sloan characterized the interests of the members of the Center for Human

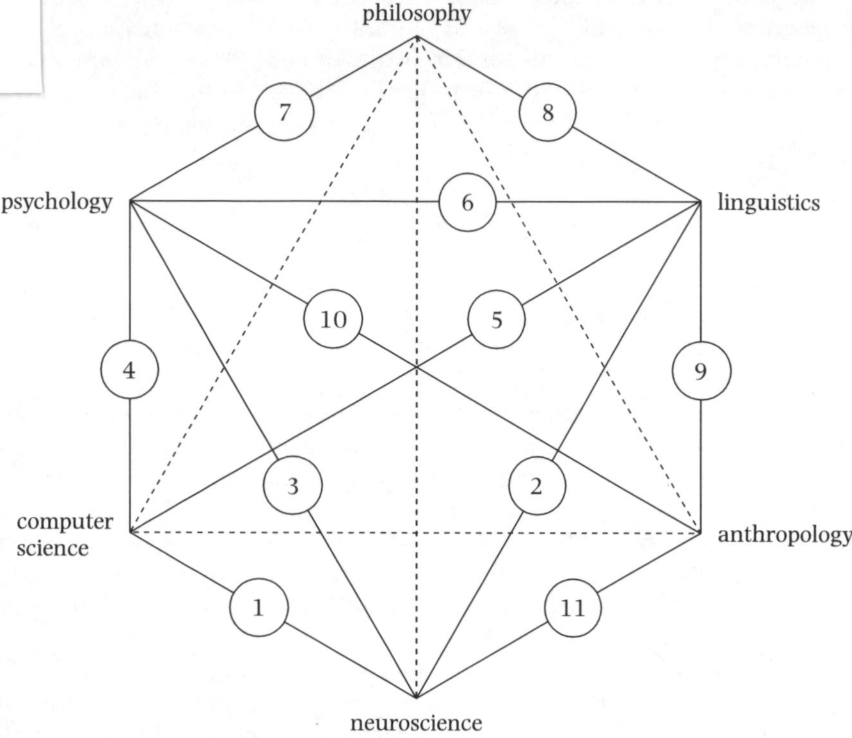

Figure I.10 Representation of the domains and subdomains of cognitive science according to the 1978 Sloan Report on Cognitive Science. Each of the six contributing disciplines is represented by a node. The labeled connections between nodes represent formally organized interdisciplinary collaborations: (1) cybernetics, (2) neurolinguistics, (3) neuropsychology, (4) simulation of cognitive processes, (5) computational linguistics, (6) psycholinguistics, (7) philosophy of psychology, (8) philosophy of language, (9) anthropological linguistics, (10) cognitive anthropology, (11) evolution of the brain. The dotted lines represent collaborations which had not yet formally developed.

Information Processing at UCSD as including "formal analyses of linguistic systems, neural mechanisms of cognitive functioning, and anthropological investigation of human belief systems." With the new grant, "UCSD is now instituting a postdoctoral training program drawing upon insights gained from computational methods, theoretical psychology, neuroscience, and related experimental techniques" (pp. 49–50). MIT, by comparison, is characterized as having "rich resources in linguistics, philosophy, artificial intelligence, speech research, cognitive psychology, and neuroscience. Its administration has supported the creation of a Center for Cognitive Science which will draw upon all of those resources and provide some new ones" (p. 50). The following year Sloan made seven additional grants of similar size and focus to the University of Chicago and University of Michigan (in a inter-university collaboration), Stanford University, Brown University, the University of Massachusetts at Amherst, Cornell Medical College (in conjunction with Rockefeller University), and UC–Irvine. The grant to Cornell Medical College is noteworthy, since it was the only grant to place primary emphasis on neurophysiological information (by seeking to use analyses of brain damage to make inferences about cognitive functions in non-brain-damaged individuals).

The following year, 1981, Sloan took its most ambitious step by committing amounts ranging from $500,000 to $2.5 million to the "establishment at each participating institution of an identifiable, self-sustaining center, institute, department, or other administrative entity where a continuing program of research and training will be conducted in cognitive science" (report for 1981, p. 12). The following list identifies the institutions selected and the amounts they received:

Massachusetts Institute of Technology	$2.5 million
University of California, Berkeley	$2.5 million
Carnegie–Mellon University	$1.0 million
Stanford University	$1.0 million
University of Pennsylvania	$1.0 million
Cognitive Neuroscience Institute	$0.5 million
University of California, Irvine	$0.5 million
University of Rochester	$0.5 million
University of Texas, Austin	$0.5 million

While most of these schools and institutes were ones previously supported by Sloan (the Cognitive Neuroscience Institute was a continuation of the program at Cornell Medical College under Gazzaniga), there were some interesting changes. UCSD, which had been heavily funded in the two previous phases, was not included (perhaps because it already had a relatively stable institutional base in place), whereas the University of Rochester, in a program under the direction of philosopher Patrick Hayes, was newly selected (perhaps in part as a result of the University's express commitment to cognitive science in the form of an established chair).

Another relative newcomer to Sloan support was the University of California at Berkeley, which received an exploratory grant only after other institutions were already being selected for training grants. The Sloan Foundation explained its grant to Berkeley in the following way (it helps to keep in mind that Searle and Dreyfus taught at Berkeley):

> The unorthodox approach researchers at Berkeley take to cognitive science – sometimes called the "Berkeley approach" – is characterized by a vigorous skepticism toward some of the basic assumptions underlying research at other cognitive science centers. Berkeley philosophers, for example, have been extremely critical of some prominent claims made for artificial intelligence. There is no reason, they assert, to believe that machine intelligence can ever approximate human intelligence or achieve anything like human consciousness. Similarly, the Berkeley linguistics department has become a center of opposition to transformational theory as developed at M.I.T. A major attempt has been made at Berkeley to develop a theory of language that is part of a general theory of action, an approach that differs sharply from the more analytic attempts of transformational linguists to segregate linguistic competence from the other psychological mechanisms involved in language use. Because the field of cognitive science is so new and research is still so rudimentary, it seems to us and to our outside advisory committee only prudent for the Foundation to place some bets on those who dissent from the majority views. (Report for 1983, p. 16)

After ten years, in which it provided approximately $17.4 million for direct costs, the Sloan Foundation brought its initiative in cognitive science to a close. During part of this period (1982–4) a second foundation, the System Development Foundation, was also pursuing an initiative in areas related to cognitive science. Focusing on computational linguistics and speech, it invested a total of $26 million. The largest single grant went to support the Center for the Study of Language and Information (CSLI) at Stanford University. CSLI was created in 1983, as a research institute connecting faculty at Stanford with researchers at SRI and Xerox-Parc in collaborative inquiry into human–computer interaction, language processing, and reasoning. The participants were drawn primarily from linguistics, philosophy, and computer science. One of the best-known projects of CSLI was initiated by philosophers Jon Barwise and John Perry: they developed a model of situation semantics which analyzed the meaning of mental states and linguistic utterances in terms of their relations to situations in the actual world.

The funding by both the Sloan Foundation and the System Development Foundation helped to build the institutional base for cognitive science, but since both foundations committed support for only a relatively short period, a serious question arose as to sources of future funding. In anticipation of this problem, on March 30 and April 1, 1985, a workshop was held under the sponsorship of the National Science Foundation and the System Development Foundation. Reflecting the orientation of the System Development Foundation, this workshop did not adopt the term *cognitive science* but rather referred to the *study of information, computation, and cognition*. It nonetheless identified the same contributing disciplines as had the Sloan Foundation, with the notable omission of neuroscience. The workshop assessed the current state of funding in the cognitive sciences. While some cognitive science research was supported by various agencies within the Department of Defense, especially the Office of Naval Research, the most dependable source of funding was the National Science Foundation. The report advocated a significant increase in funding through NSF and other federal agencies.

2.7.2 The journal and the society

While it might have seemed easy for researchers sharing an interest in cognition to communicate with one another, this actually became one of the most serious challenges for cognitive science. Researchers trained in different fields conceived of their inquiries differently, used different research tools, spoke in different vocabularies, and read different literatures.

A study that Zenon Pylyshyn conducted in the late 1970s, the period when cognitive science was already taking form, revealed a serious communication gap (Pylyshyn, 1983). He surveyed citation patterns within artificial intelligence and cognitive psychology publications, sampling a total of 528 references from the proceedings of the International Joint Conference on Artificial Intelligence (IJCAI) in 1977 and from two years of the journal *Artificial Intelligence*. Of the references that could be categorized according to the discipline of origin, 300 of these were to other papers and books in artificial intelligence; only 35 were to books or journals in psychology and 7 to journals in linguistics. An important feature of the citations is that many were to unpublished technical reports, which often would not be easily available to those outside the field. The situation in psychology was similar. Of the 1,200 citations that Pylyshyn sampled in *Cognitive Psychology*, *Cognition*, and *Memory and Cognition*, nearly 1,000 were to books or journals in psychology, with a majority of the journal references to journals in general experimental psychology that did not emphasize a cognitive or information processing perspective. Only 50 citations were to AI sources, 70 to linguistics articles, and 16 to papers in philosophy. The majority of the references in the latter two categories appeared in *Cognition*, which explicitly solicited interdisciplinary work.

An obvious reason for the relative paucity of references across cognitive science disciplines was that, except in specially arranged meetings, researchers in one area would not easily encounter the work of those in related areas. Generally a cognitive psychologist or linguist would not subscribe to an AI journal, and vice versa. They would also not attend each others' professional meetings, unless explicitly invited to give a talk. For one thing, there was a significant cost to subscribing to journals or registering for meetings outside one's field. In addition, each profession had its own network of people who gave talks and wrote papers intended for each other. It was not easy for an outsider to gain admittance. If there were to be real interdisciplinary communication, cognitive science clearly needed to establish forums devoted to this goal.

Accordingly, in 1977 a new journal called *Cognitive Science* was founded under the editorship of Roger Schank, Allan Collins, and Eugene Charniak. With their diverse backgrounds, the three editors seemed to represent the interdisciplinary mix that was beginning to characterize cognitive science. As noted above, Schank's Ph.D. was in linguistics, but his subsequent research led him into computer science and AI. Allan Collins received his Ph.D. in psychology from Michigan in 1970 and then held a research position at Bolt Beranek and Newman (a Boston-area high technology firm which employed a number of cognitive science researchers). Following his collaboration with Quillian (discussed above), at BBN Collins developed research approaches that belonged to cognitive science itself rather than any traditional discipline, and then joined Schank in moving towards educational applications (especially through the Institute for the Learning Sciences). Eugene Charniak's Ph.D. was in Computer Science from MIT, and at the time he was at the Institut pour les études semantiques et cognitives in Geneva, Switzerland. But the mix was not quite as eclectic as it might seem. While Schank was a linguist by training, his interest was in systems for natural language understanding and informal reasoning. Moreover, he was already an outspoken critic of Chomsky; accordingly, the MIT orientation to linguistics, which had played a significant role in the genesis of cognitive science, was not represented. Indeed, the banner underneath the title of the journal read "A multidisciplinary journal of artificial intelligence, psychology, and language," referring to language rather than linguistics.

The first issue began with an editorial by Allan Collins entitled "Why Cognitive Science" in which he described the converging interest in cognition involving a number of disciplines and characterized cognitive science as a new discipline with a distinctive view of natural and artificial intelligence: "This view has recently begun to produce a spate of books and conferences, which are the first trappings of an emerging discipline. This discipline might have been called applied epistemology or intelligence theory, but someone on high declared it should be cognitive science and so it shall. In starting the journal we are just adding another trapping in the formation of a new discipline" (p. 1).

A fuller indication of the orientation of the new journal is provided in box I.2, which shows the table of contents of the first volume, together with the names of the authors and their institutional affiliations. The authorship of papers reveals a significant mix

Box I.2 Table of Contents from first volume of *Cognitive Science*

Volume 1, Number 1

Daniel G. Bobrow (Xerox Palo Alto Research Center) and Terry Winograd (Computer Science, Stanford University), "An Overview of KRL, a Knowledge Representation Language"

Wendy Lehnert (Computer Science, Yale University), "Human and Computational Question Answering"

Andrew Ortony and Richard C. Anderson (Psychology, University of Illinois at Urbana–Champaign), "Definite Descriptions and Semantic Memory"

Ira Goldstein and Seymour Papert (MIT AI Labs), "Artificial Intelligence, Language, and the Study of Knowledge"

Volume 1, Number 2

John R. Anderson (Psychology, Yale), "Induction of Augmented Transition Networks"

Jerome A. Feldman (Computer Science, University of Rochester) and Robert F. Sproull (Xerox Palo Alto Research Center), "Decision Theory and Artificial Intelligence II: The Hungry Monkey"

B. Bhaskar and Herbert A. Simon (Psychology, Carnegie–Mellon University), "Problem Solving in Semantically Rich Domains: An Example from Engineering Thermodynamics"

M. J. Steedman (Psychology, University of Sussex), "Verbs, Time, and Modality"

Volume 1, Number 3

Yorick Wilks (Artificial Intelligence, Edinburgh), "What Sort of Taxonomy of Causation Do We Need for Language Understanding?"

Stephen M. Kosslyn and Steven P. Shwartz (Psychology, The Johns Hopkins University), "A Simulation of Visual Imagery"

J. R. Hayes, D. A. Waterman, and C. S. Robinson (Psychology, Carnegie–Mellon University), "Identifying the Relevant Aspects of a Problem Text"

Chuck Rieger (Computer Science, University of Maryland), "Spontaneous Computation in Cognitive Models"

Volume 1, Number 4

Eugene Charniak (Institut pour les études semantiques et cognitives, University of Geneva), "A Framed Painting: The Representation of a Common Sense Knowledge Fragment"

James A. Levin and James A. Moore (USC Information Sciences Institute), "Dialogue-Games: Metacommunication Structures for Natural Language Interaction"

Roger C. Schank (Computer Science, Yale University), "Rules and Topics in Conversation"

of psychologists and AI researchers. In terms of citation statistics, Pylyshyn (1983) reported that *Cognitive Science* represented a far greater amount of cross-disciplinary integration than the other journals he examined: "Out of 331 citations of the last two years, 110 were judged to be clearly psychological (31 books and 79 journal articles in psychology, with 29 of these in cognitive psychology); 55 were artificial intelligence papers; 14 were articles in computer science journals; 50 were citations of other articles in *Cognitive Science* itself; 40 were journal articles in philosophy and logic; 26 were linguistics papers; 7 were neuropsychological papers; and the remaining 36 citations were distributed among a variety of areas, including library science, education, business, anthropology, and book reviews" (p. 72). What these statistics do not reveal is whether individual authors were citing papers outside their own disciplines, revealing integration at the level of individual research, or whether *Cognitive Science* was simply providing a common forum.

The original conception of *Cognitive Science* emphasized psychology and AI. Such disciplines as neuroscience and philosophy were not expressly included in its scope, and linguistics of the Chomskian sort was not represented. But *Cognitive Science* was not the only new journal trying to reach an interdisciplinary community in cognitive science. In 1977, the same year that *Cognitive Science* was first published, the Society for the Interdisciplinary Study of the Mind began to publish its own newsletter, the *SISTM Quarterly*. In 1979–80 *SISTM Quarterly* incorporated the *Brain Theory Newsletter*, and took the name *Cognition and Brain Theory: The Newsletter of Philosophy, Psychology, Linguistics, Artificial Intelligence, and Neuroscience.* As the term *newsletter* suggests, it was initially published as a stapled, 8.5 by 11 inch pamphlet and was distributed directly by Vassar College where Martin Ringle was the editor. In addition to news notices, it published abstracts, bibliographies, and articles. Publication of *Cognition and Brain Theory* was transferred in 1981 to Lawrence Erlbaum Associates, where it took on the format of a standard journal. Both as a newsletter and as a journal, *Cognition and Brain Theory* tended to give greater coverage to philosophy and neuroscience than did *Cognitive Science.* In its first two issues in journal format, *Cognition and Brain Theory* included papers by linguist Noam Chomsky, neuroscientists Michael Arbib and Karl Pribram, neuropsychologists Daniel Bub and Harry Whitaker, psychologist Dedre Gentner, AI researchers Nils Nilsson and Roger Schank, and philosopher William Lycan. After 1984, *Cognition and Brain Theory* merged with *Cognitive Science*, but with a commitment from *Cognitive Science* to broaden its scope and publish more neuroscience and philosophy.

In 1978, one year after *Cognitive Science* and *Cognition and Brain Theory* were established, Stevan Harnad, then a graduate student at Princeton, created yet another new journal, *Behavioral and Brain Sciences* (*BBS*), published by Cambridge University Press. *BBS* adopted a distinctive format of primary papers (called *target articles*), followed by a dozen or more short commentaries. It invited participation from the entire cognitive science community in writing both target articles and commentaries. Target articles were selected with an eye to their potential for stimulating multidisciplinary dialogue. For example, in 1980 *BBS* published John Searle's "Minds, brains, and programs," which advanced his Chinese room critique of computational accounts of mind (see above), together with responses by psychologists such as Robert Abelson and Bruce Bridgeman, philosophers such as Ned Block, Daniel Dennett, Jerry Fodor, and William Lycan, neuroscientists such as John Eccles and Benjamin Libet, and AI researchers such as Zenon Pylyshyn, John McCarthy, Marvin Minsky, and Roger Schank.

At this point (the late 1970s), there were at least three journals supporting the emerging interdisciplinary field of cognitive science. Two new academic presses soon created similar opportunities for publishing book-length works. First, Lawrence Erlbaum established Lawrence Erlbaum and Associates, which quickly amassed a strong list of books in cognitive psychology and gradually expanded into other cognitive science fields such as artificial intelligence and linguistics. Second, Harry and Betty Stanton established a small press, Bradford Books, which was subsequently acquired by MIT Press but retained its separate editorial process and imprimatur. Bradford's initial focus was on philosophy texts such as *Brainstorms* by Daniel Dennett and *Knowledge and the Flow of Information* by Fred Dretske, and it gradually expanded into the cognitive sciences more generally.

But there was not yet any professional organization. One of the most important functions performed by professional organizations is to sponsor annual meetings at which researchers can report on work they are doing and learn about new theoretical ideas and empirical results obtained by others. Discussions, both formal ones after presentations and informal ones in hallways and lounges, serve as a powerful stimulus to new thoughts and help to ratify promising ideas. Conferences were even more important before the era of rapid electronic communication than they are today, since circulation of early reports of research was generally limited to small groups of fellow researchers. As we have noted, conferences were arranged in cognitive science without the sponsorship of a professional organization, but they depended upon the interests and resources of the particular researchers who organized them. Once a professional organization is in place, it raises funds, both through dues and grants, to cover the costs of planning a meeting, and deputizes people to carry out the necessary work (arrange meeting locations, plan a program, and so on).

It was in the course of planning what was to be just one more individually arranged conference that the idea of the Cognitive Science Society took form. The newly developed interdisciplinary program in cognitive science at UCSD, in the context of its grant from the Sloan Foundation (see above), set the La Jolla Conference on Cognitive Science for August 13–16, 1979. The goal of the conference was to assess the state of the art in cognitive science by having ten leading figures in the profession each address "some of the hopes, aspirations, and critical issues that face the development of a cognitive science" (Norman, 1981, p. v). As Norman relates, the goal of the conference was nothing less than to define *cognitive science*:

> It was to be the "defining meeting," the meeting where many of those concerned with the birth of Cognitive Science could record its origins, speak of its hopes, and chart its course. We knew these aspirations to be unrealistic, but did not let that knowledge deter us. The speakers at the conference – the contributors to this volume – all work within the sibling disciplines that comprise Cognitive Science. All were charged with the task of presenting broad, overview statements of their views, statements that would last beyond the year of the conference and that would help set the definition of the field, statements that would prove useful in the initial stages of the discipline and that would provide examples of what we are, what we wish to become, and even what we should not be. (Ibid.)

Five of the eleven speakers came from AI (Herbert Simon, Allen Newell, Marvin Minsky, Roger Schank, and Terry Winograd), two from psychology (Donald Norman and Philip Johnson-Laird), one from neuroscience (Norman Geschwind), one from linguistics (George Lakoff), and two from philosophy (John Searle and Mark Johnson, coauthor of the paper with Lakoff). Norman notes that there was concern about the omission of "what might be called the 'MIT school of linguistics'," but that none of the three investigators who were invited were able to attend. One feature of defining activities is that they are selective: they emphasize certain perspectives at the expense of others. This was true of the La Jolla meeting. AI was clearly the dominant discipline, and it has continued to play a fundamental role in the Cognitive Science Society, whereas Chomskian approaches received less attention. Nonetheless, the Cognitive Science Society grew rapidly. Its meetings now consist of a large number of paper sessions scheduled in parallel, as well as large poster sessions, and attract about a thousand attendees. In addition to publishing a large volume of papers from its annual meeting, in 1980 it became the sponsoring organization for the journal *Cognitive Science*.

3 Identity crises: 1985–1999

The focus of cognitive science had narrowed in the period 1960–1985 from its earlier breadth. Beginning in the 1980s, it regained that breadth and more by expanding in two directions: vertically into the brain and horizontally into the environment. These expansions, however, have induced an identity crisis for the developing cognitive science enterprise. During its phase of initial maturation the scope of cognitive science was reasonably well demarcated, with an accepted framework of representations and computations over them defining the accepted explanatory approach. The movement outwards into the environment and downwards into the brain has given rise to other models for cognitive science. Many of these developments are covered in detail elsewhere in the Companion; accordingly, our interest here is simply to place them in historical context.

3.1 Rediscovering neural networks

After a flourish of research in the initial, revolutionary period of cognitive science (see section 1.3.5), research into artificial neural networks largely disappeared from artificial intelligence and cognitive science after 1970. One reason for this was that, like traditional AI researchers, neural network researchers such as Rosenblatt often made claims that they could not substantiate. A second reason was that, while neural networks' forte is pattern recognition, an activity that seems most suited to analyzing sensory input, much of the interest in cognition was focused on higher cognitive processes such as reasoning and problem solving. A third reason for the decline of neural network research was a serious limitation found in neural networks of the sort Rosenblatt was developing. As discussed in section 1.3.5, his greatest achievement was to develop and prove the effectiveness of a learning procedure for modifying connection weights in networks with one layer of connections. But there were a large number of problems which could not be solved by such a simple perceptron.

Rosenblatt was aware of this limitation, and was in the process of exploring networks with multiple layers of connections when Minsky and Papert published an extremely influential book entitled *Perceptrons* in 1969. The single-layer perceptron was made the object of detailed mathematical study. One of the results the authors presented was a proof that such networks could not perform certain computations, such as *exclusive or* (*XOR*). (A XOR B is true if and only if just one of the two propositions, A or B, is true. It contrasts with the usual logical connective *or*, which is true when at least one of the two propositions is true.) Although Minsky had himself initially worked on neural network models, he became disaffected from that approach and adopted the symbolic framework instead. *Perceptrons* was broadly perceived as a fundamental attack on the whole neural network approach. It is less clear, however, whether it was responsible for driving away funding for neural networks generally, or was rather only one sign that the approach had fallen into disfavor.

Whatever the causes, neural network research did not show the same growth pattern during the 1970s as did the symbolic approach. Nonetheless, some important research continued to be carried out by researchers such as James Anderson, Teuvo Kohonen, Christoph von der Malsburg, and Stephen Grossberg. At the beginning of the 1980s interest began to revive. In part this was due to emerging dissatisfaction with some of the perceived limits of symbolic models, such as their failure to degrade gracefully as the systems were partially damaged, restricted ability to generalize to new cases, and inefficiency as models grew more complex. An important testimony to renewed interest in the new approach (now often referred to by the terms *connectionism* and *parallel distributed processing* as well as *neural networks*) was the publication of *Parallel Models of Associative Memory* in 1981, edited by Geoffrey Hinton and James Anderson. It contained papers from a conference held at UCSD in June, 1979, which included neuroscientists, cognitive psychologists, AI researchers, mathematicians, and electrical engineers, all exploring the neural network approach. Many of the participants were recent Ph.D. recipients whose involvement breathed new life into the neural network program.

Two important papers in the 1980s further fueled the rejuvenation of interest in neural networks. The first, by John Hopfield (1982), provided a very clear exposition of how computation was possible in neural networks; it was influential, though, largely because of Hopfield's status as a leading physicist:

> John Hopfield is a distinguished physicist. When he talks, people listen. Theory in his hands becomes respectable. Neural networks became instantly legitimate, whereas before, most developments in networks had been the province of somewhat suspect psychologists and neurobiologists, or by those removed from the hot centers of scientific activity. (Anderson and Rosenfeld, 1988, p. 457)

The second paper, by David Rumelhart, Geoffrey Hinton, and Ronald Williams (1986), cracked the problem of how to train multilayered networks. The extra layer(s) of connections together with the new learning procedure enabled these networks to solve the XOR problem and others that had figured in Minsky and Papert's critique of Rosenblatt. The learning procedure, known as *back-propagation*, took Rosenblatt's procedure for one-layer networks as a starting point. That is, the network would use its current weights to respond to an input pattern, and the difference between the actual output and the target output for each unit would constitute an error signal which would be passed back through the network. Weights were adjusted in a direction that would reduce the error on that input pattern. Rumelhart et al. found a way to repeat this process for additional layers of connections. (See Article 16, MACHINE LEARNING.) Subsequent research established that networks with two layers of connections and sufficient numbers of hidden units could be trained in this way to compute any computable function, thereby rendering them equivalent in computational power to universal Turing machines or digital computers. In 1986 Rumelhart, James McClelland, and the PDP Research Group (based at UCSD) published a two-volume collection entitled *Parallel Distributed Processing: Explorations in the Microstructure of Cognition*, which quickly became the bible of the connectionist movement.

Neural networks provided an answer to several of the sources of dissatisfaction that were emerging with symbolic models. When networks are damaged, they do not simply crash; rather, their performance generally degrades gradually (e.g., responses will become increasingly less accurate). The same weights that enable the networks to master training cases enable them to generalize to new cases. In addition, research on neural networks introduced different ways of thinking about mental representations. Representations in networks are often distributed; that is, the representation of a single item will involve more than one unit, and a given unit will participate in the representation of more than one item. Individual units then encode information at a *sub-symbolic* level (Smolensky, 1988). For example, a set of words might be represented by distinctive but overlapping patterns of activation values across ten units. The values could be arbitrary, or the units could be engineered to correspond to phonemic or semantic features. Distributed representations turn out to be robust not only against damage but to permit interesting generalizations to new cases.

The re-emergence of neural networks in the 1980s resulted in considerable rancor in the cognitive science community. To theorists such as Jerry Fodor and Zenon Pylyshyn, the return of neural networks represented a relapse into associationism, which they contended had already been shown to be inadequate by such proofs as Chomsky's arguments about the insufficiency of statistical models to account for grammatical structure in language. Fodor and Pylyshyn (1988) presented a broad theoretical critique of connectionism, whose centerpiece was the objection that since connectionist networks did not employ a representational system with combinatorial syntax and semantics, they could not account for the productivity and systematicity of thought. We encountered these notions above in discussing Fodor's arguments for a language of thought. Far from surrendering in the face of such objections, however, connectionists have issued a number of vigorous responses (see Article 48, LEVELS OF EXPLANATION AND COGNITIVE ARCHITECTURES, and Article 52, RULES). Neural networks have become increasingly influential with cognitive researchers, many of whom regard them as simply part of the tool-kit for modeling cognition, rather than a call to arms in a theoretical battle.

What made neural network models radical for some cognitive scientists was that they seemed to do away with syntactically structured representations as the currency within a cognitive system. An even more radical approach has emerged more recently under the banner *dynamical systems theory* (*DST*). The DST approach has sought to apply to cognitive phenomena the same type of dynamical equations that have proved successful in physics, beginning with Newton's formulation of the force laws. The key to the approach is to develop mathematical relations, often nonlinear, among parameters characterizing features of a cognitive system (sometimes relating these to parameters characterizing features of the environment). The nonlinearities in the equations often lead to complex patterns of change, but methods for obtaining geometric representations make it easier to track the behavior of such systems (Port and van Gelder, 1995). Among the more radical claims of some DST advocates has been a repudiation of (a) the framework of REPRESENTATION AND COMPUTATION that has been basic to most cognitive science and (b) the general strategy of explaining cognitive processes in terms of different components performing different sub-tasks (see section 4).

Another theoretical perspective that emerged at the end of the 1980s and derived in part from neural network research is research on *Artificial Life* (Langton, 1995). This rubric incorporates a wide range of approaches which generally share the perspective of taking living systems, not just cognitive systems, as their focus. One inspiration for it was the discovery that one could develop computer models that exhibited a broad range of behaviors characteristic of living organisms. Another was the discovery by investigators such as Rodney Brooks that one could develop robots which could negotiate environments without a centralized system for planning movement. A final inspiration was the discovery by John Holland of computer algorithms (*genetic algorithms*) which used simulated evolution to develop new programs that were better adapted to the tasks they faced than were their ancestors. While often themselves critical of cognitive science, Artificial Life researchers have developed a range of new ideas that have percolated through cognitive science. (For more on the approaches introduced in this section, see Article 38, CONNECTIONISM, ARTIFICIAL LIFE, AND DYNAMICAL SYSTEMS).

3.2 Rediscovering the brain: cognitive neuroscience

During 1960–80, most cognitive scientists only occasionally considered neuroscience and the brain as they went about the business of developing and testing theories and models. Reasons for ignoring neuroscience were both pragmatic and principled. On the pragmatic side was the fact that the questions asked and the tools used in much neuroscience research were remote from the inquiries being conducted in cognitive science, especially in cognitive psychology. (An exception is that many psychologists and linguists kept an eye on the intriguing flow of results from neurolinguistics.) On the principled side, the very metaphor that enabled artificial intelligence to play such a major role in cognitive science, the computer model, also provided a framework for minimizing the relevance of neuroscience. The idea was that the relation between psychology and neuroscience was like the relation between hardware and software. Hardware is required for any software to run – to adapt the language of Putnamian functionalism, the hardware implements the functional roles constitutive of the software – but the details of the hardware generally do not matter. One can run the same high-level program on different hardware systems, with no differences in what the software does. (This requires the appropriate lower-level software – a compiler or interpreter – for each hardware system. There may be differences in how long the program takes to run, what memory registers get used, and so on, but these can be dismissed as *implementation* details.) If mental processes constitute something like the software responsible for behavior, while the brain is the hardware, then it would seem that cognitive science could proceed without concern for neuroscience.

These reasons for neglecting the brain began to be challenged in the 1980s. On the one hand, the computer metaphor began to lose its hegemony. In part this was because connectionist (neural network) models of cognitive function re-emerged, thereby undercutting the principled arguments for dismissing the relevance of neuroscience. On the other hand, there was increasing emphasis in neuroscience itself at the *systems* level, where researchers focused on the organization of the visual system or memory systems rather than the cellular or molecular level. A major factor in this move to the systems level was work on mapping the brain. Beginning in the 1960s, neuroanatomists were increasingly able to use patterns of connectivity between cells to infer the existence of specialized areas (those that were highly interconnected). This strategy produced a finer level of resolution than Brodmann's areas, and in many cases the newly identified areas could be analyzed in terms of their contribution to cognitive functions in the way Brodmann had envisaged (see Article 4, BRAIN MAPPING).

To take the best-developed example, the mapping of visual processing pathways, Cragg (1969) and others identified five different pathways from V1 (for visual area 1, located at the rear of the brain in the occipital lobe) to nearby areas labeled V2, V3, superior temporal sulcus, V4, and V4a. By providing various visual stimuli and recording from cells in each of these areas, researchers such as Semir Zeki (1976) discovered that cells in different regions responded to different types of information –

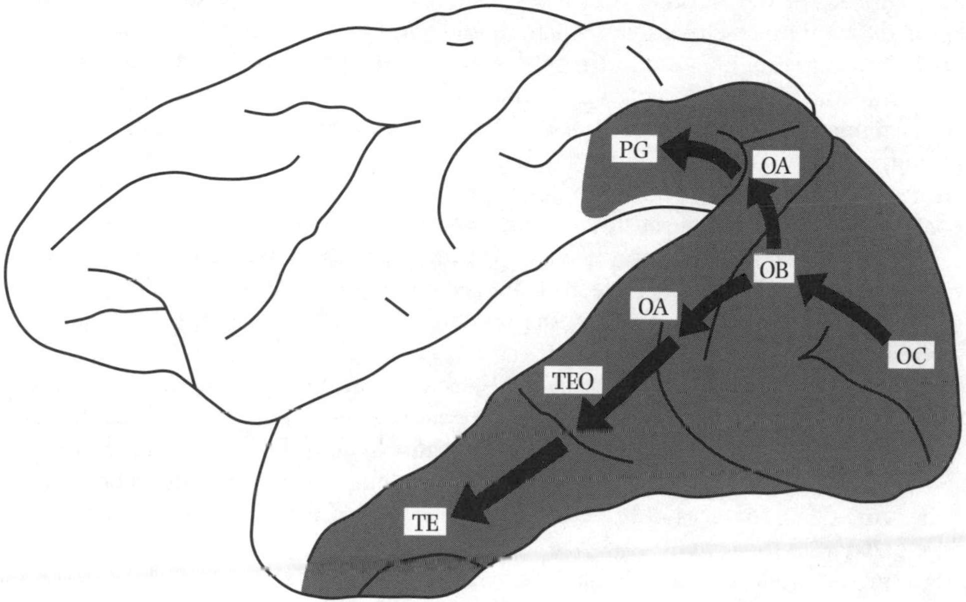

Figure I.11 Two pathways of visual processing in the rhesus monkey proposed by Mishkin, Ungerleider, and Macko (1983). Each begins in area OC (primary visual cortex, also called V1) and projects into prestriate areas OB (V2) and OA (V3, V4, and MT). The *what* pathway then projects ventrally into inferior temporal cortex (areas TEO and TE), whereas the *where* pathway projects dorsally into inferior parietal cortex (area PG).

e.g., those in V2 to binocular disparity, those in V4 to colors, and those on the superior temporal sulcus to motion (see Article 34, SINGLE NEURON ELECTROPHYSIOLOGY). Then Mortimer Mishkin and Leslie Ungerleider, relying primarily on lesion studies in monkeys, differentiated two main routes for processing visual information, as shown in figure I.11 (Mishkin et al., 1983). One route proceeds ventrally into the posterior temporal cortex along the inferior longitudinal fascilus to areas TEO and TE. Based on the fact that lesions in the posterior temporal area result in loss of pattern discrimination and that lesions in TE in particular result in failure to recognize previously presented objects, Mishkin and Ungerleider proposed that this pathway analyzed the physical properties of visually perceived objects, such as size, color, texture, and shape. One feature of this ventral pathway is that neurons further along the pathway have increasingly large receptive fields, suggesting that the more distant neurons take responsibility for recognizing objects independently of where they appear in the visual field. The other route proceeds dorsally into the posterior parietal cortex. Lesions in the posterior parietal cortex in monkeys produce an inability to select the one of two food wells that is closer to a visual landmark, suggesting that this pathway figures in perception of spatial relations.

The two systems differentiated by Mishkin and Ungerleider quickly became known popularly as the *what* and the *where* systems. Research on these pathways has continued. One discovery is that the two pathways diverge even before V1 and employ different cells in V1. Another is that the visual system has a surprisingly large number of different processing areas (Felleman and van Essen, 1991, distinguish 32 areas in the Macaque monkey), and that there are numerous connections between them (Felleman and van Essen identify more than 300). Of particular interest, many of these connections run between the what and the where pathways. Although the story is proving to be ever more complex, it is possible, especially via SINGLE NEURON ELECTROPHYSIOLOGY and NEUROIMAGING, to determine what kinds of information the cells in particular areas are most sensitive to, and to begin to model how information is processed in the visual cortex (van Essen and Anderson, 1990). (For another example of how discoveries of neural architecture are providing guidance in discovering the processing involved in performing cognitive functions, see Article 13, LANGUAGE EVOLUTION AND NEUROMECHANISMS.)

With recognition growing in the 1980s that a variety of research endeavors in neuroscience were relevant to cognitive modeling, neuroscientists and cognitive scientists increased their communication and began to actively collaborate. The name *cognitive neuroscience* has been adopted for this growing enterprise. A number of major researchers on both sides of the intellectual aisle played a key role. We will focus on three: two from the cognitive science side, Steve Kosslyn and Michael Posner, and one from the neuroscience side, Michael Gazzaniga.

From his days as a psychology graduate student at Stanford (see section 2.4), Kosslyn's research focused on the use people make of mental imagery in solving problems. (See Article 12, IMAGERY AND SPATIAL REPRESENTATION.) He soon became embroiled in a vigorous debate as to whether the underlying mental processes operate on representations employing a depictive (quasi-pictorial) format or on ones employing a propositional format. Kosslyn (1980) argued for a depictive format and supported his view with data from a variety of ingenious tasks. He found, for example, that scanning a mental image for a small property (e.g., a cat's claws) took longer than scanning it for a large property (e.g., a cat's head). The case against a depictive format was argued most forcefully by Zenon Pylyshyn (1973, 1981). From his key premise that minds (like computers) store information in structures with a sentential syntax, Pylyshyn concluded that mental representations must be propositional. Examining the available data from experiments, he saw no incompatibility with his claim that subjects perform imagery tasks by operating on propositional representations.

The issue between Pylyshyn and Kosslyn is not whether something that *looks* like a picture, rather than a sentence, is before the mind's eye; it is conceded by propositionalists that mental images may seem introspectively to be pictorial. The issue is whether the actual image, as an underlying representation or mental code, is picture-like or proposition-like. The way pictures represent referents is different in various respects from the way in which propositions or sentences represent them. One of these differences (emphasized by Pylyshyn) is that pictures have spatial properties, whereas propositions do not. Compare, for example, a map which tells how far Paris is from Madrid with a sentence that provides that information. The sentence may simply report the number of miles; from the map we will also be able to tell that Madrid is south of Paris, that there are highways connecting the two cities, and so on.

In 1978, John Anderson contended that the imagery debate could not be resolved by behavioral data, since pictorial accounts could always be mimicked by propositional accounts, and vice versa. Accordingly, in more recent work Kosslyn (1994) and other psychologists, including his former student Martha Farah (1989), have employed a variety of neuroscience tools, including studies of DEFICITS AND PATHOLOGIES and NEUROIMAGING, to show that the same areas of the cortex are involved in both vision and visual imagery, and that these areas use topographical (hence depictive) representations.

Michael Posner's research, like Kosslyn's, began firmly rooted in psychology, and he too gradually came to regard links with neuroscience as crucial. In his early research, Posner was a major developer of chronometric methods for identifying elementary cognitive operations (see Article 27, BEHAVIORAL EXPERIMENTATION). In so doing, he was a modern-day counterpart to the Dutch psychologist Frans Cornelis Donders (1868/1969), who pioneered the use of reaction time patterns to analyze cognitive activities into component operations. Donders developed what is known as the *subtractive method*, in which a researcher has a subject perform two tasks (e.g., discriminate between two signals) which are thought to differ in that one involves an additional mental operation not required for the other (e.g., simply react to a signal). The difference in reaction times yields an estimate of how much time the additional operation requires. An objection to the subtractive technique is that it assumes that the additional processing is a pure insertion of a new elementary operation that does not alter the execution of those operations that are common to the two tasks. Saul Sternberg (1969) extended Donders's method into an *additive factors* approach: the researcher uses multiple manipulations, each of which should have an independent effect on the time required for a particular component operation. Sternberg had subjects memorize a short list of items and then say whether or not a test item was on the memorized list. Time to encode the test item should be affected by the clarity of its visual display, and time to mentally scan the memorized list should be affected by its length. If the slope of the list length effect is the same for both clear and degraded test items, then these factors are additive, and one has evidence of separate processing components. A problematic feature of both the subtractive and the additive factors approaches is that they assume that mental operations are performed sequentially, an assumption that is sometimes false (see Article 53, STAGE THEORIES REFUTED). Partly to overcome this objection, Posner often based his conclusions on interference effects (the extent to which performing an additional activity would increase the time required to perform a primary task). Interference indicates the use of a shared resource, whereas no interference indicates independent operations.

Many of Posner's studies of mental chronometry involved mechanisms of ATTENTION while performing two tasks. In one of Posner's experiments subjects were asked to indicate by a key press with one hand whether two letters were the same, while with the other hand they press a different key whenever they hear a tone. Following a visual warning, the two letters were presented sequentially on a visual display. Reaction times to the tone were not affected during the interval between the warning signal and the first letter, but were lengthened during the interval between the two letter presentations. This indicated that an attentional resource was consumed during the interval between letters but not during the earlier interval. While chronometric techniques work from the behavioral level to identify elementary mental operations, Posner foresaw that they might provide the links to underlying neural systems (Posner and McLeod, 1982, p. 478). In 1982 Posner, Roy Pea, and Bruce Volpe published a prescient paper entitled "Cognitive-neuroscience: developments toward a science of synthesis." They focused on a number of techniques by which one might relate neuroprocesses to the elementary mental operations isolated by means of chronometric methods. These included studies of individuals with brain lesions, changes in electrical potentials recorded on the scalp (event-related potentials or ERPs), and measurements of blood flow by means of *positron emission tomography* (*PET*), then under development at Washington University in St Louis (where Marcus Raichle was exploring its potential for studying mental processes).

From 1985 to 1988 Posner left the University of Oregon to collaborate with Raichle and Steven Petersen at Washington University in developing the use of PET as a tool for imaging brain processes during the performance of cognitive activities (see Article 32, NEUROIMAGING). A key to this development was a transformation of Donders's subtractive method from a temporal context to a spatial one. Again, subjects were asked to perform two tasks thought to differ in just one component. Instead of seeking differences in reaction times, these investigators looked for differences in blood flow as indicative of which brain regions were most involved in that component. The collaboration between Posner, Raichle, and Petersen produced a series of influential studies. In a 1988 *Science* paper with Peter T. Fox, they constructed a series of word tasks that involved progressively more cognitive operations and analyzed PET data subtractively in order to identify brain regions that were distinctively active for particular operations. For example, subjects showed more brain activity in an area in the anterior left frontal lobe when they were required to generate and say aloud a verb semantically associated with a visually presented noun than when they only had to read the noun aloud. Even in this research Posner has continued to pursue his interest in mechanisms of attention, and his more recent work integrates chronometric measures, neuroimaging, and ERP results in the attempt to distinguish different attentional mechanisms. (See Posner and Raichle, 1994, which also discusses *functional magnetic resonance imaging* (*fMRI*), a new technique for neuroimaging that has advantages and disadvantages compared to PET.)

On the neuroscience side, Michael Gazzaniga was one of the chief initiators of greater collaboration between neuroscience and cognitive science. His early work focused on the study of split-brain patients: that is, patients whose corpus callosum (the major neural pathway between the two hemispheres) was severed. Developed by Rochester, New York, neurosurgeon William Van Wagenen, this procedure was used to relieve symptoms in a small proportion of epileptic patients. Initially it seemed not to result in any cognitive deficits. But as a graduate student and postdoctoral fellow with Roger Sperry at the California Institute of Technology, Gazzaniga demonstrated that when communication between the hemispheres was impaired, patients indeed suffered cognitive deficits. Gazzaniga uncovered these deficits by providing stimuli to sensory receptors that passed information to one hemisphere and requiring responses to be made by motor systems controlled by the other hemisphere. This involved, for example, presenting visual stimuli briefly to the left or right of a fixation point. Information from stimuli presented to the left of the fixation point would reach only the right hemisphere, and split-brain patients in whom language production was lateralized in the left hemisphere would be unable to name them. One of Gazzaniga's major findings came from a split-brain patient, P.S., who had suffered considerable damage to the left hemisphere early in his life. One consequence of early damage to language areas in the left hemisphere is greater involvement of the right hemisphere in language tasks. As a result, P.S. was able initially to respond to a question, a crucial part of which was visually presented to the right hemisphere, by spelling out the answer with Scrabble letters; subsequently, he was able to respond orally as well (Gazzaniga and LeDoux, 1978). At this later stage, if the word *cupcake* was presented so that *cup* was seen by the right hemisphere and *cake* by the left hemisphere, he would report seeing the words *cup* and *cake*, but not *cupcake*, indicating that the two hemispheres had each processed part of the word, but were not able to integrate their results.

In addition to his own attempt to integrate cognitive and neuroscience techniques in analyzing split-brain patients, Gazzaniga became a major force in developing the institutional structures required for cognitive neuroscience. At Dartmouth and, more recently, at the University of California, Davis, Gazzaniga has developed academic programs emphasizing cognitive neuroscience (he recently returned to Dartmouth). Since 1989 he has directed annual summer institutes in cognitive neuroscience which have been designed to bring together researchers with different primary research orientations but sufficient background to enter into a common dialogue. In 1988 he created a new journal, *Journal of Cognitive Neuroscience*, which he continues to edit, and in 1993 he founded the Society for Cognitive Neuroscience. For MIT Press he edited a hefty tome, *The Cognitive Neurosciences*; published in 1995, this volume offers a comprehensive review of work at the interface of neuroscience and cognitive science.

One factor in the renewed interest in the relation of the brain to cognitive science was foundation support. In 1984, in the context of its existing cognitive science initiative, the Sloan Foundation instigated another initiative devoted to linking cognitive science and neuroscience. As with the original cognitive science initiative, this support initially emphasized conferences: "One arrangement that seems especially well suited to fostering productive interaction between these two groups of researchers is indeed simple – regular meetings of working groups in which no more than a dozen scientists come together two or three times a year to discuss research topics and strategies" (report for 1984, p. 20). One of these grants went to Johns Hopkins University for the purpose of supporting a series of meetings between neuroscientists and cognitive scientists concerned with memory:

> The study of memory is a central concern in both neuroscience and cognitive science. In neuroscience, memory is studied by neurophysiologists to determine the synaptic changes that underlie neural plasticity, by molecular biologists to determine the molecular processes governing synaptic behavior, and by neuroanatomists to locate the major brain centers that mediate memory. In cognitive science, memory is studied by computer scientists interested in building electronic learning systems and by cognitive psychologists to understand the performance of human memory. These lines of research have developed independently of each other in the past. (Ibid., pp. 20–1)

The following year this project became refined into a sub-initiative targeting computational neuroscience, an endeavor in which tools of mathematics and computer science are employed to model neural processes (extending from the level of single neurons up to the level of coordinated systems involved in vision or action). In 1986 and 1987, 11 computational neuroscience grants ranging from $100,000 to $300,000 were made to collaborative teams, often from more than one institution. The largest of these went to David Sparks, Frank Amthor, and Michael Friedlander at the University of Alabama, Birmingham, for developing and testing computational models of perception of direction of motion, control of visual attention, and control of saccadic eye movements. Their investigations involved collaboration with theorists at MIT, California Institute of Technology and the National Institutes of Health (NIH).

In 1987 the System Development Corporation sponsored a symposium that led to publication of *Computational Neuroscience*, a large collection of papers which attempted to define the field (Schwartz, 1990). Research in this field has expanded dramatically in recent years. Starting with CNS*92 in 1992 in San Francisco, there has been a Computational and Neural Systems meeting annually. The *Journal of Computational Neuroscience* commenced publication in 1994.

More recently, two other foundations joined forces in supporting the building of bridges between cognitive science and neuroscience: the James S. McDonnell Foundation and the Pew Charitable Trusts. In 1989 they awarded substantial grants to establish centers for cognitive neuroscience at eight institutions: the University of Arizona, UCSD, Dartmouth College (later moved to the University of California, Davis), Johns Hopkins University, MIT, the Montreal Neurological Institute, the University of Oregon, and Oxford University. In addition, they have provided training grants and individual research grants as well as support for the summer institutes directed by Gazzaniga.

3.3 Rediscovering the environment: ecological validity and situated action

In its early period cognitive science tended to limit its focus to events presumed to be taking place within the mind/brain. While all researchers would acknowledge that minds exist within bodies and that these bodies have to deal with the external world (both physical and social), most researchers assumed that they could disregard these considerations when studying cognition. Cognition focused on the processing of information inside the head of the person. In order for this to happen, information had to be represented mentally; cognitive processes could then operate on these representations. Subsequently, represented information had to be translated into commands to the motor system, but this took place after cognitive processing as such was finished. Jerry Fodor (1980) articulated such theoretical justification for ignoring both the external world and the body in cognitive science, labeling the resulting framework *methodological solipsism*, but opposition was already gathering in a number of quarters.

One of the major inspirations for challenging methodological solipsism was the work of J. J. Gibson, a psychologist working at Cornell contemporaneous with the early period of cognitive science but whose impact fell elsewhere. Gibson studied visual perception, but instead of concentrating on the information processing going on within individuals as they see, he examined the information that was available to the organism from its environment. His major contention was that there was much more information available *in the light* than psychologists recognized, and that organisms had only to *pick up* this information (Gibson, 1966). They did not need to construct the visual world through a process of inference or hypothesis formation. He argued, for example, that people do not need to construct a three-dimensional representation of the world; rather, there is information specifying the three-dimensional nature of the visual scene in the gradients of texture density, changes in occlusion of objects as the perceiver moves about in the environment, and so forth. One of Gibson's major contentions was that the perceiver must be understood as an active agent using its own motion to sample information about the environment. Gibson also stressed that not all organisms pick up the same information from the environment, but rather would *resonate with* information that is coordinated with their own potential for action. Accordingly, he introduced the notion of an *affordance*; different objects afford different actions to different agents (e.g., a baseball affords throwing to us, but not to frogs), and it is these affordances which organisms are tuned to pick up. (See Article 18, PERCEPTION.)

Gibson launched what has come to be known as the *ecological* approach to perception, one that has been pursued by a number of researchers who have made important discoveries about the sorts of information available to cognitive agents. A well-known example is Gunnar Johansson's (1973) use of motion pictures of people walking in a dark room with lights attached to their ankles, knees, shoulders, elbows, and wrists; he demonstrated that observers were able to see the form and motion of a walking human in a brief film clip. In another example, from studying plovers (shorebirds) diving into water to catch fish, David Lee (1976) identified a variable τ, defined as the inverse of the relative rate of expansion of the image of the object in the visual field, which predicts the time remaining before impact with the object. Plovers apparently track the value of τ as they are diving and use that to determine the moment to close their wings. Finally, developmental psychologist Eleanor Gibson applied the ecological approach to understanding the development of perceptual competence in infants (e.g., her well-known *visual cliff* experiments) and in older children (e.g., her studies of how children learn to differentiate the critical features in a set of meaningless shapes). She produced a theory of perceptual learning and development that emphasized detection rather than construction of information.

Although Gibson was generally either ignored or severely attacked by cognitive scientists (see Fodor and Pylyshyn, 1981, and Ullman, 1980), his emphasis on the ecological components of perception strongly influenced Ulric Neisser after Neisser joined the Cornell faculty in 1967. In his 1976 book, *Cognition and Reality*, which appeared less than a decade after his *Cognitive Psychology*, Neisser attempted an integration of the Gibsonian approach with information processing. Unlike Gibson, Neisser did not completely reject the idea of internal information processing, but he departed from the highly mechanistic conception which had become common in computational models. Instead, he adopted Bartlett's notion of a schema, a highly structured internal state acquired in the course of experience which partially determines what one perceives and remembers. But, like Gibson, Neisser emphasized the complex information available to the organism and the role of the organism in navigating its environment. He criticized much experimental work in psychology for failing to achieve *ecological validity* and relying too exclusively on artificial laboratory tasks. Such research failed to reveal the types of information that cognizers gain from their environments and the ways in which they explore their environments to gain information. Instead of Fodor's solipsistic cognizer, Neisser advocated the perceptual cycle (figure I.12), in which perceivers apply schemata to information received from the environment, which then leads to exploration in the environment and the pickup of new information, and so forth.

In subsequent work Neisser has continued to emphasize the importance of an ecological perspective, but he has carried it beyond perception to more cognitive domains such as CONCEPTUAL ORGANIZATION and MEMORY. For example, he has emphasized the importance of studying memory as it functions in real life. Thus, he analyzed John

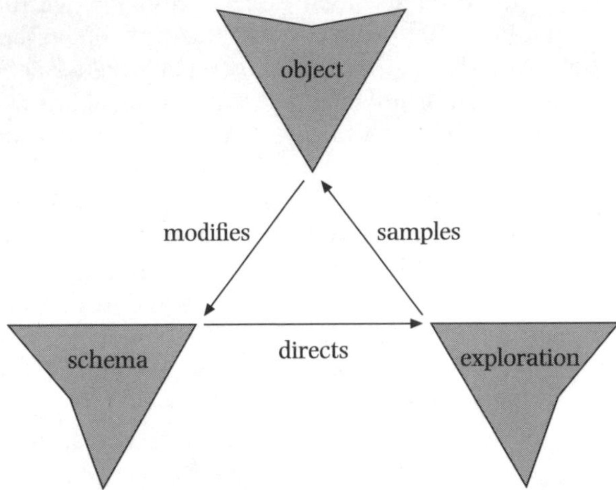

Figure I.12 The perceptual cycle according to Neisser (1975). Perceptual input from an object modifies a schematic representation, which then directs further exploration, leading to further sampling of information from the object.

Dean's testimony on events surrounding Watergate as well as flashbulb memories (memories of when one first learned of major events such as the Challenger crash and the San Francisco earthquake that seem etched in one's mind). He showed in both cases how recall is often erroneous due to the processes by which the subject reconstructs the event. While Neisser attempted to integrate Gibson's perspective with a more traditional cognitive one, others influenced by Gibson such as Michael Turvey, Robert Shaw, and Scott Kelso have pursued a more radical approach, one that is now leading to links with the dynamical systems perspective discussed in section 3.1.

Although the influences stemming from Gibson have been a major factor in the move toward a more ecologically valid cognitive science, there are other major figures who have moved in an ecological direction independently of Gibson. One of these is Donald Norman, whose career has traversed a path from the University of Pennsylvania to Harvard's Center for Cognitive Studies to UCSD (where he helped to establish the LNR group and later the cognitive science program) to Apple Computer to Hewlett-Packard. In his later research he began to focus on how cognizers operate in real world contexts, and especially on ways in which artifacts produced by them favor or impede their performance. (See Article 56, EVERYDAY LIFE ENVIRONMENTS). The focus on how cognition occurs in relation to artifacts in the environment has also been pursued by Norman's UCSD colleague, Ed Hutchins, who has made remarkable studies of how cognition is distributed between agents and artifacts in such activities as ship navigation (see Article 39, EMBODIED, SITUATED, AND DISTRIBUTED COGNITION, and Article 40, MEDIATED ACTION).

A recent development in PERCEPTION has also played a role in reconnecting cognitive systems to their environments. A traditional view of perception, advanced by David Marr among others, is that the task of the perceptual system is to build up a comprehensive representation of the visual scene. But a number of investigators from computer science (Dana Ballard), neuroscience (Terrence Sejnowski), neuropsychology (V. S.

Ramachandran), and philosophy (Patricia Churchland) have begun to argue that this badly mischaracterizes what the visual system does. Rather, they argue that it constantly interacts with visual scenes, extracting information relevant to the organism's motoric goals. Because the visual system is capable of rapid eye movements (*saccades*), we can, by using eye movements, gain information about any part of the visual field. This helps generate the illusion that we encode the whole visual field. In addition to eye movements, these researchers emphasize how cognitive agents move their heads and their bodies to gain even more information about the layout of their environments. Churchland, Ramachandran, and Sejnowski (1994), draw heavily upon both behavioral studies of perceptual behavior and findings from neuroscience (in particular, the prevalence of recurrent pathways from higher brain centers down to basic visual areas and the physiological effects of later processing stages on the response patterns of neurons involved in early visual processing).

The recognition that cognitive systems are in constant interaction with their environment has resulted in an infusion into cognitive science of research and methodologies in such social sciences as anthropology and sociology (see Article 30, ETHNOMETHODOLOGY). There has been tension between mainstream cognitive scientists (whose tools are primarily directed at understanding processes inside the head) and those who have embraced social science perspectives (whose tools are concerned with the social contexts of such processes); see, for example, the conflicts in the 1993 special issue of *Cognitive Science* devoted to situated action. However, constructive interactions have increasingly been achieved. One consequence has been the reintroduction of work from the Soviet tradition in psychology that developed from the research of Lev Vygotsky (see Article 40, MEDIATED ACTION).

3.4 *Rediscovering function: cognitive linguistics*

During the middle period in the history of cognitive science, Chomsky's radical new approach had gradually become dominant in linguistics and influential in other disciplines. This is not to say, however, that Chomsky continued to affirm the same grammatical models as he had advanced in the 1950s and 1960s. In his initial generative grammars the emphasis was on transformational rules that could be used to derive surface structures from underlying deep structures. Increasingly, Chomsky has deemphasized transformational rules and the distinction between deep structure and surface structure. In X-bar theory, government and binding theory, and most recently minimalism, the transformational rules have been reduced to the point that minimalism posits just one rule, *move alpha* (move any category anywhere). It generates far more structures than the old transformational rules, but constraints filter out the ungrammatical ones. Partly as a result of the different grammars that Chomsky has advanced over the decades, which some researchers have continued to pursue even after Chomsky himself has abandoned them, partly due to splits from Chomsky and alternative theories within linguistics, and partly due to investigators who have developed grammars for different purposes (such as natural language processing in AI), there is now a rich variety of approaches in LINGUISTIC THEORY. *Optimality theory* (Prince and Smolensky, in press) is particularly distinctive and important. More computational and statistical than Chomskian linguistics, it retains the separation between such components as semantics, syntax, and phonology and has been especially influential in the 1990s in phonology.

A number of linguists led by John R. Ross (a former student of Chomsky), George Lakoff, Paul Postal, and James McCawley broke from Chomsky as early as the 1960s over his insistence on the autonomy of syntax. Chomsky held that syntactic principles are not the product of other linguistic or cognitive processes, hence that one can characterize syntax independently of other aspects of language (e.g., its semantics or the pragmatics involved in its use). Adopting the label *generative semantics*, these early critics sought to extend Chomsky's generative program into the domain of semantics by developing rules that would generate syntactic structures from semantic representations without using a privileged intermediate level, deep structure, to segregate semantic from syntactic parts of the derivation. Chomsky rejected these extensions of his endeavors, and the disputes between these generative semanticists and Chomsky resulted in the *linguistic wars* of the late 1960s and early 1970s (Harris, 1993). After a period of heated controversy, generative semantics morphed into a less cohesive but innovative variety of alternative approaches. Chomskian linguistics retained its cohesiveness but also underwent considerable change.

One generative semanticist, George Lakoff, began in the 1980s at UC–Berkeley to develop a new approach eventually known as *cognitive linguistics*. (There is now a professional society and a journal that employ that name.) Lakoff especially emphasized the structure of concepts and (with philosopher Mark Johnson) did influential work on metaphor. At UCSD, Ronald Langacker and his former student, Gilles Fauconnier, developed a highly systematized theory of cognitive grammar that emphasized the grounding of language in highly abstract spatial representations. In general, cognitive linguists shared a conviction that syntax, far from being autonomous, was the product of more basic cognitive operations. Loosely affiliated with cognitive linguistics are other linguists and psycholinguists who focus more on the pragmatics or function of language use and who try to account for syntactic structures in terms of such functions (see Article 37, COGNITIVE LINGUISTICS, and Article 31, FUNCTIONAL ANALYSIS).

4 Coming of age: downwards and outwards

We have explored the origins and early development of cognitive science, but where is it going? Just as it is hard to know how adolescents will resolve their identity crises and choose from the many paths that lie before them, it would be presumptuous for us to predict the precise form that cognitive science will take as it confronts the choices posed in the previous section. The one thing that seems certain is change: cognitive science is being pulled vertically *down* into the brain and horizontally *out* into the environment.

Where might these lines of growth take the field? As a starting point, let us return to the characterization of cognitive science that we put forward at the beginning:

> Cognitive science is the multidisciplinary scientific study of cognition and its role in intelligent agency. It examines what cognition is, what it does, and how it works.

The changes will involve multiple disciplines; how they develop and interact will determine the shape of cognitive science as we enter the next millennium. We will return to this topic shortly, but first we consider answers to the questions of what cognition is, what it does, and how it works.

The cognitive science of the 1970s answered these questions from an information processing perspective. *What cognition is*: the processing of information in the head. *What cognition does*: it enables an agent to exhibit intelligent behavior, which is prototypically manifested in such activities as solving the Tower of Hanoi problem or understanding sentences. Psychologists confined such activities to the laboratory for study, and AI researchers modeled them in programs which might be judged adequate to the extent that they passed the Turing test (by generating behavior indistinguishable from that of a human). *How cognition works*: like a computer. Information is encoded in a symbolic representational format upon which rules operate – much like algorithms in a programming language.

Through the identity crisis that we explored in section 3, cognitive scientists began to reassess some of these answers. We can look at the new thinking on the three questions in reverse order. *How cognition works*: it proved difficult to accommodate some of the data about how humans actually behave without going beyond the computer metaphor. Even rich data structures like schemas and frames were inadequate to capture the fluid character of human cognition. To some cognitive scientists the neurally inspired approach of connectionism offered a way to realize previously neglected characteristics of human intelligent activity, such as graceful degradation and soft constraint satisfaction. The return to the brain, though, has brought more than a new computational framework. Minimally, many cognitive scientists would insist that any answer to how cognition works must be compatible with emerging knowledge of how the brain works. Others go much further, maintaining that the best way to gain clues as to how cognition works is to study the brain. This does not mean giving up a computational perspective, since many researchers in cognitive neuroscience take as their focus developing computational models of neural activity. It does mean, however, that the inspiration for developing accounts of how cognition works is no longer the digital computer; instead, knowledge about how the brain works increasingly provides the foundation for theoretical modeling.

Another element in the identity crisis involved reassessment of *what cognition does*. The return to the environment and the body refocused attention on how cognition facilitates life in the real world. Although many cognitive psychologists have continued to emphasize laboratory studies of problem solving, reasoning, memory, and language processing, the new concern for *ecological validity* has redirected others to study skills and abilities as exercised in the real world. These have included the old abilities studied in new contexts (a server in a restaurant remembering orders), as well as abilities newly of interest (a navigator guiding a large ship into a harbor). In AI, the Turing test began to lose its status as a sufficient indicator of intelligent agency in artificial systems, and increasing numbers of researchers turned from modeling rational thought to such projects as building robots that could operate in real environments. It became recognized that what cognition does is to provide an open-ended capacity to respond appropriately and flexibly to whatever may come along or appear. It provides a capacity to respond reasonably, as Descartes remarked in the *Discourse on Method*, to "the contingencies of life."

Neither of these developments led to a reassessment of *what cognition is* – the processing of information within the head of the agent. Very recently, however, even this answer has begun to be questioned. We consider three aspects of the challenge in turn.

(i) On one front, investigators emphasizing MEDIATED ACTION and EMBODIED, SITUATED, AND DISTRIBUTED COGNITION are questioning whether analyses that separate and focus on activities *within* the brain are adequate. The alternative view is that interactions between the brain and the environment within which it functions are so intricate and pervasive that the primary unit of analysis must be the system formed by the interacting brain and the environment.

(ii) Expanding the boundaries of the cognitive system so as to incorporate parts of the world is supported by some of the advocates of a move beyond connectionism to dynamical systems theory. In dynamical models, researchers seek to identify a variety of parameters that affect the performance of a system and to develop mathematical laws, frequently in the form of equations employing first or second derivatives, that describe the changes in the system over time. Such mathematical accounts do not impose a boundary at the skin. Even if one does develop a dynamical model limited to processes occurring inside the skin, the fact that one can always couple dynamical systems that share parameters into a compound system ensures the potential for linking models of activities within the brain to those in the environment in a single theoretical model (see Article 38, CONNECTIONISM, ARTIFICIAL LIFE, AND DYNAMICAL SYSTEMS).

(iii) On a third front, as researchers attend more to the brain, their attention is drawn to aspects of mental life, such as EMOTIONS and CONSCIOUSNESS, which may not be best described in terms of information processing, at least as it has been understood so far. Emotional responses are largely under the control of midbrain structures that comprise the limbic system, rather than cortical structures, and these systems do not seem to work by encoding and processing information. While it has been acknowledged that emotions and consciousness modulate cognitive activity, cognitive scientists have frequently assumed that they could disregard their impact and study cognition in isolation. But some cognitive scientists are coming to believe that responsiveness to the contingencies of life relies in significant part on these other responses of the brain. A domain such as chess is highly restricted and may be mastered by computers employing sophisticated inference strategies. A rook is just a rook. But let our eyes sweep across a great landscape, and the richness of sensory detail seems informationally overwhelming. This is because, claim some cognitive scientists, visually experienced detail *is* beyond information processing. The conscious and emotional world has a character, a subjective quality, an identity, which cannot be captured on a purely information processing account. Meanwhile, that subjective quality helps us to respond appropriately to life's contingencies (see Lahav, 1993).

A comprehensive answer to the question of what cognition is, if it is not limited to information processing within the brain, has not yet been developed. Accordingly, it is not yet clear whether information processing will turn out to be a component of a more comprehensive characterization of cognition, or whether it will have been a false step. Given that what cognition does is to enable agents to interact intelligently with their environments, it seems plausible that in some way information processing will be a part of a more adequate conception of cognition. However, the recent developments we have touched upon suggest that it will not exhaust cognition.

Given these views about what cognition is, what it does, and how it works, let us return to the question of what disciplines figure in cognitive science's attempt to understand it. In the phase of gestation, the three principal disciplines from which researchers began to interact and formulate a plan of study were neuroscience, psychology (especially cognitive), and computer science (AI). Drawing from sources such as cybernetics and information theory, researchers from all three disciplines proposed that cognition should be understood computationally, with computational models being constrained by neural and behavioral data. This interaction is represented graphically as a triangle in figure I.13a, with the three disciplines represented by the three vertices. The fact that the names of the disciplines are all in the same font size indicates that they were roughly equal contributors to the new enterprise. Following the format of figure I.10 above, the lines connecting the nodes represent the interdisciplinary interactions. The fact that they are of the same width indicates that the interactions were equally potent.

Around the crucial year 1956, however, major changes in the collaborations occurred. Figure I.13b illustrates the new mix of fields and interdisciplinary connections that characterized cognitive science during the period of its initial maturation (roughly until 1985, as discussed in section 2). A new program in linguistics, advanced by Noam Chomsky, boosted its influence to about the same level as psychology. And at nearly the same moment, neuroscience began to play a much less significant role. Three other disciplines – sociology, anthropology, and philosophy – began to play an ancillary role. Finally, computer science (AI) became the preeminent cognitive science discipline during this period of initial maturation.

As in figure I.13a, relative prominence and influence are indicated by font size, but with the additional disciplines, the initial triangle has become a hexagon, and many more interdisciplinary interactions are possible. Some of these are better developed than others, and the thickness of the lines connecting the various disciplines are intended to be indicative of the extent to which these connections were pursued during the period of initial maturation.

Recently, the relative importance of the various contributing disciplines has changed yet again, as illustrated in figure I.13c. As we have noted, neuroscience is playing an increasingly important role in cognitive science. Techniques of functional investigation in neuroscience are becoming increasingly important in guiding thinking about cognition. Social and cultural studies originating in sociology and anthropology are also coming to a play a more influential role, whereas both computer science (AI) and linguistics have become less influential. The diminished role of computer science (AI) is due to at least two factors. On the one hand, internal pressures in computer science to focus on the development of such basic computer tools as computer operating systems and compilers have provided a less hospitable home for AI. On the other hand, computational modeling tools, once primarily the property of computer scientists interested in artificial intelligence, are becoming tools of practitioners in other cognitive science disciplines. Thus, these disciplines are all represented in the same size font as philosophy, whose importance remains stable.

As important as the transitions in the significance of individual disciplines, however, has been the increased importance of some of the interdisciplinary interactions. One thriving area of inquiry is cognitive neuroscience, which combines the behavioral tools of cognitive psychology with the functional methods of neuroscience in identifying the brain areas involved in performing cognitive tasks. The field became established as different sources of information about the contribution of different brain areas to cognition became available: first deficits and pathologies, then single neuron electrophysiology, then event-related potentials in electroencephalograms. It is thriving

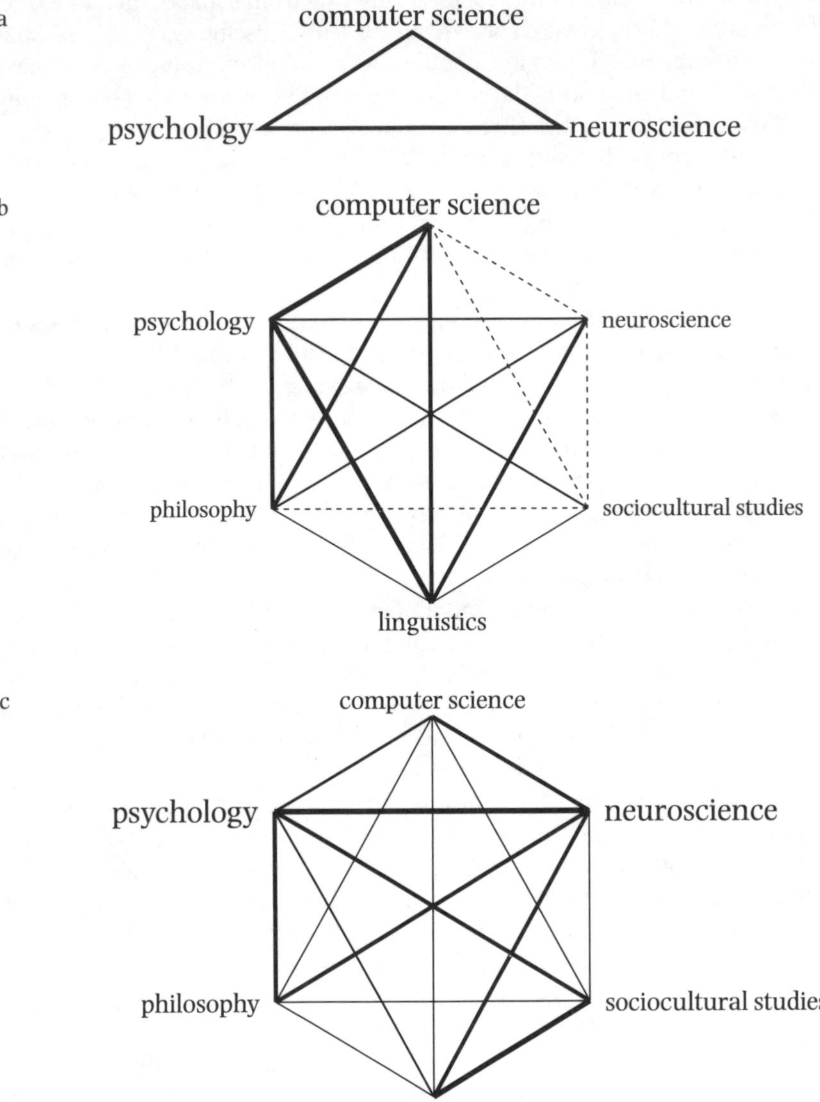

Figure I.13 Contributing disciplines and interdisciplinary connections during three different stages in the development of cognitive science. The font size in which the name of the discipline is printed reflects the relative importance of that discipline to cognitive science during the stage in question. The lines between disciplines represent interdisciplinary connections. The thickness of the line represents the activity level of that interdisciplinary connection; a dotted line represents an essentially undeveloped connection. a. The three disciplines that were central to cognitive science during most of its gestation (linguistics became an important contributor only at the very end of this period). b. The six disciplines contributing to cognitive science during the period of initial maturation. c. The six disciplines currently contributing to cognitive science.

now, due to the advent of neuroimaging, which provides relatively direct information from intact human brains at good temporal and spatial resolutions. Also newly important is the impact of sociology and anthropology on both cognitive psychology and computer science (AI), as evidenced in the role of mediated and situated action in some psychological models and robotic simulations.

Some of the already established interdisciplinary connections have changed their character. For example, the interaction between psychology and linguistics took the form of psycholinguistic inquiries into the psychological reality of grammar during cognitive science's initial maturation. Interactions are more pluralistic in the current era; for example, cognitive linguistics denies the autonomy of syntax and instead appeals to cognitive processes to try to explain the link between meanings and phonological forms. Optimality theory can retain an autonomous syntax but proposes to account for syntax in terms of a system of soft constraints. Meanwhile, Chomskian language development researchers study parameter setting within an autonomous syntax module that retains a classic architecture.

Figure I.13c provides a picture of the disciplines of cognitive science and their interrelations that are likely to figure in the immediate future. But this picture is no more likely to remain static than the previous ones. At different times different disciplines will be better positioned than others to advance our understanding of cognition. Indeed, we can identify one field which is just beginning to make contributions and may soon become a highly important player: behavioral psychopharmacology (see Willner, 1991). For the most part, researchers have focused on electrical properties of the brain in seeking to understand its relation to cognition. But a fundamental discovery of early twentieth-century neuroscience was that the nervous system is not reticular: neurons are separated by synapses. Communication across synapses is mediated by chemicals. Starting from a few key neurotransmitters, researchers have identified a large class of chemicals that serve this function. These have been shown to be critical to normal cognition, but a detailed understanding of how they figure in cognition remains to be developed.

If the nature of chemical processes in the brain turns out to be critical to cognition, this will certainly strengthen the tendency in cognitive science to pursue a downwards direction of inquiry. But the development of behavioral psychopharmacology may, as the name suggests, also direct researchers outwards to the environment. The effectiveness of some psychopharmacological agents is linked with, or dependent upon, changes in the environment of subjects using the particular drug (see Whybrow, 1996). Behavioral psychopharmacology may thus encourage further efforts by cognitive scientists both in searching for underlying mechanisms and in examining relations to environments.

Thus, cognitive science has been, and promises to remain, broadly interdisciplinary. Interdisciplinary research always involves a tension, since practitioners of different disciplines generally bring to the interaction different agendas, different research tools, and different models of satisfactory answers. Successful interdisciplinary research requires rendering these differences compatible. Sometimes, despite the loftiest goals, interfield collaboration founders. Cognitive science has experienced its share of tensions. Conflicts between symbolic modeling (inspired by the digital computer) and neural network modeling (inspired by the brain) are a current case in point. Another has been the conflict over the autonomy of syntax advocated by Chomskian linguistics and rejected by the cognitive and functional grammarians. So far, the institutions of cognitive science, such as the journal *Cognitive Science* and the Cognitive Science Society, have proved capable of remaining fairly inclusive.

But will cognitive science remain viable as a field of interdisciplinary cooperation? The simultaneous pulls *downwards* into the brain and *outwards* into the world may prove to be too much pulling, and lead to the disintegration of cognitive science. On the other hand, awareness that brain processes and events in the external world interact in crucial ways may suffice to hold the inquiry together. The attempt to combine the two in this volume represents the conviction that, to at least a significant degree, cognitive science will continue as a robust interdisciplinary endeavor.

There is a feature of the competing pulls downwards and outwards, though, that increases the risk of a serious rift. One important aspect of the information processing perspective that was adopted in cognitive science was the attempt to specify mechanisms underlying cognition. Each operation upon information represented a process occurring within the cognitive system. Behaviorists and mathematical psychologists had not attempted to identify such mechanisms; thus, the cognitivists' rebellion against behaviorism and mathematical models of learning involved embracing a different conception of explanation – one in which it was not sufficient to identify laws or mathematical regularities in behavior, but actual mechanisms responsible for it. The current turn downwards into the brain is sometimes represented as *reductionistic*. Since the word *reduction*, however, is understood in a host of different ways, we will not employ it here. What it clearly represents is a further step in the continuing quest to identify the underlying mechanisms responsible for intelligent behavior.

In treating the quest for explanation as a quest for mechanisms, cognitive scientists are adopting a perspective on explanation that has been widely shared in the life sciences. For example, biologists seek to explain such processes as energy liberation and reproduction in animals by characterizing the mechanisms which make these possible. Discovery of these mechanisms has generally involved identifying a function that a system performs (providing energy for work, or comprehending and producing linguistic utterances), *decomposing* that function into component functions, and *localizing* those component functions in the physical system (Bechtel and Richardson, 1993). Sometimes the decomposition may produce a linear sequence of component functions, but it need not (see Article 53, STAGE THEORIES REFUTED). The performance of different tasks may be highly integrated (e.g., through backwards or recurrent connections). Moreover, sometimes localization will identify one discrete area of the system responsible for a component task, but other times the component function may be distributed throughout the system. Further, while researchers may aspire ultimately to identify the actual physical locus, often they must settle for indirect evidence that such a locus exists (e.g., by demonstrating that each of two different functions can be preserved while the other is incapacitated: a double dissociation). Cognitive science, especially cognitive psychology, can be seen as proposing decompositions of the cognitive system. Without invoking neuroscience, the evidence for the underlying mechanisms performing the different functions remains indirect. The support for these mechanisms increases when one combines behavioral and neural sources of evidence.

While the downward pull has not challenged the emphasis on mechanism in cognitive science, some researchers pursuing the outward pull have questioned it and have advocated a return to an explanatory framework that in some respects resembles that of behaviorists and mathematical psychologists. Thus, some advocates of dynamical analyses (van Gelder, 1995) suggest that it is sufficient to identify critical variables characterizing the state of systems and to construct mathematical laws to account for the ways in which the values of these variables change over time. These theorists reject the idea that information is represented in the system, and that it is representations that are operated on by different components of the system (see Article 50, REPRESENTATION AND COMPUTATION). That is, they reject the information processing perspective.

If there is incommensurability between mechanistic and dynamical models, and if it tends to correspond to the difference between efforts to go downwards into the brain and outwards into the environment, then we may have the seeds of a significant fracture in cognitive science. The developmental psychologist Erik Erikson (1968, p. 136) coined the appositely awkward expression *distantiation* to identify what happens in the life of an individual who repudiates elements in her or his personality which seem incompatible or threatening. As the areas of cognitive neuroscience and dynamical systems theory are progressively delineated, they may fortify their gains by overvaluing their differences and hence increasing the distance.

A distantiated future will not satisfy many committed to cognitive science. But those committed to an integrated cognitive science may discover that the potential for fracture is not as serious as it seems. At present the dynamicists' challenge is not fully formed. Central to the challenge are the notions of information processing and representation, but these notions are currently vague and must be theoretically regimented. It may well be that a mature dynamical account will posit genuine information processing and representation, although the representations employed will not be syntactically structured or sentence-like. The model of syntactically structured or sentence-like representations (Fodor's language of thought) is, in any case, under severe attack from a number of quarters in contemporary cognitive science. Other models of representations have come to the fore: graphs, maps, holograms, house plans, and other nonsentential schemes, and many investigators are exploring the idea that the brain may process information using one or more of these other kinds of representation. The path has already been cleared by theorists of perception (see Article 18, PERCEPTION, also Article 19, PERCEPTION: COLOR), who have advocated understanding perception in terms of transformations of spatial or quasi-pictorial representations.

There is further reason to doubt the seriousness of the dynamicist/mechanist rift. On the one hand, neuroscientists often emphasize that recurrent or backward projections in the brain outnumber forward connections, suggesting that the brain itself is a highly interconnected system. Accordingly, they expect that a computation carried out by any given brain part may be highly influenced by activities elsewhere in the brain. In such a complex system, those attempting to develop mechanistic accounts may require the dynamicists' tools. On the other hand, successful dynamical accounts relating different brain and environmental parameters themselves call out for explanation. What underlying mechanisms produce the behavior described at an abstract level by the dynamical equations? Unless one accepts action at a distance, one is back to the search for intermediate processes (in the head), and it is natural to characterize these processes in terms of how they carry information and represent self and world. Accordingly, it may be possible for dynamicists and mechanists to coexist within cognitive science, and even to collaborate with each other.

In the life of cognitive science time is of the essence. Cognitive science is not a static entity. Nor are the disciplines that comprise it. In its gestation period it lacked institutional organization; integration resulted from the power of the idea of information processing that suggested a way to view the activities of the nervous system and release psychology from behaviorism. Between 1960 and 1985 it matured, developing its identity in terms of computational models of a variety of cognitive activities. With the support of generous benefactors, it developed its institutional base. But having reached adulthood, it now recognizes some of the advantages of approaches overlooked in its development and has been drawn back downwards into the brain and outwards into the world. As it pursues its adult career, these and other factors will create tensions. But we are optimistic that cognitive science will not only endure but will develop into an even more interesting domain of science as it confronts challenges not yet recognized, devises theories not yet anticipated, and encompasses stances and deploys methods not yet imagined.

5 References and recommended reading

Alfred P. Sloan Foundation 1975–87: *Annual Report*. New York: Alfred P. Sloan Foundation.

*Anderson, J. A. and Rosenfeld, E. 1988: *Neurocomputing: Foundations of Research*. Cambridge, Mass.: MIT Press.

Anderson, J. R. 1978: Arguments concerning representations for mental imagery. *Psychological Review*, 85, 249–77.

—— 1983: *The Architecture of Cognition*. Cambridge, Mass.: Harvard University Press.

Anderson, J. R. and Bower, G. H. 1973: *Human Associative Memory*. New York: Winston.

Ashby, W. R. 1952: *Design for a Brain*. New York: John Wiley and Sons.

Atkinson, R. C. and Shiffrin, R. M. 1968: Human memory: a proposed system and its control processes. In K. W. Spence and J. T. Spence (eds), *The Psychology of Learning and Motivation: Advances in Research and Theory*, vol. 2, New York: Academic Press, 89–195.

*Baars, B. J. 1986: *The Cognitive Revolution in Psychology*. New York: Guilford Press.

Baddeley, A. D. 1976: *The Psychology of Memory*. New York: Basic Books.

Bartlett, F. C. 1932: *Remembering*. Cambridge: Cambridge University Press.

*Bechtel, W. 1988: *Philosophy of Mind: An Overview for Cognitive Science*. Hillsdale, NJ: Erlbaum.

Bechtel, W. and Richardson, R. C. 1993: *Discovering Complexity: Decomposition and Localization as Scientific Research Strategies*. Princeton: Princeton University Press.

Block, N. 1978: Troubles with functionalism. In C. W. Savage (ed.), *Perception and Cognition. Issues in the Foundations of Psychology*. Minnesota Studies in the Philosophy of Science, vol. 9, Minneapolis: University of Minnesota Press, 261–325.

*Blumenthal, A. L. 1970: *Language and Psychology: Historical Aspects of Psycholinguistics*. New York: Wiley.

Bobrow, D. G. and Collins, A. (eds) 1975: *Representation and Understanding: Studies in Cognitive Science*. New York: Academic Press.

Bradley, D., Garrett, M. and Zurif, E. 1980: Syntactic deficits in Broca's aphasia. In D. Caplan (ed.), *Biological Studies of Mental Processes*, Cambridge, Mass.: MIT Press, 269–86.

Bransford, J. D. and Franks, J. J. 1971: The abstraction of linguistic ideas. *Cognitive Psychology*, 2, 331–50.

Bransford, J. D., Barclay, J. R. and Franks, J. J. 1972: Sentence memory: a constructive versus an interpretive approach. *Cognitive Psychology*, 3, 193–209.

Broadbent, D. E. 1958: *Perception and Communication*. New York: Pergamon Press.

Broca, P. 1861: Remarques sur le siège de la faculté suivies d'une observation d'aphéme. *Bulletin de la Société Anatomique de Paris*, 6, 343–57. Trans. by G. von Bonin in *idem* (ed.), *Some Papers on the Cerebral Cortex*, Springfield, Ill.: Charles C. Thomas, 49–72.

Brodmann, K. 1909: *Vergleichende Lokalisationslehre der Grosshirnrinde*. Leipzig: J. A. Barth. English translation by L. J. Garey (1994), *Brodmann's "Localisation in the Cerebral Cortex."* London: Smith-Gordon.

*Bruner, J. S. 1983: *In Search of Mind: Essays in Autobiography*. New York: Harper and Row.

Bruner, J. S. and Goodman, C. 1947: Value and need as organizing factors in perception. *Journal of Abnormal and Social Psychology*, 42, 33–44.

Bruner, J. S. and Postman, L. 1949: On the perception of incongruity: a paradigm. *Journal of Personality*, 18, 206–23.

Bruner, J. S., Goodnow, J. and Austin, G. 1956: *A Study of Thinking*. New York: Wiley.

Bry, A. 1975: *A Primer of Behavioral Psychology*. New York: Mentor/New American Library.

Chomsky, N. 1957: *Syntactic Structures*. The Hague: Mouton.

—— 1959: Review of Skinner's *Verbal Behavior*. *Language*, 35, 26–58.

—— 1965: *Aspects of the Theory of Syntax*. Cambridge, Mass.: MIT Press.

—— 1968: *Language and Mind*, 1st edn. New York: Harcourt.

—— 1975: *The Logical Structure of Linguistic Theory*. New York: Plenum.

Chomsky, N. and Miller, G. A. 1958: Finite-state languages. *Information and Control*, 1, 91–112.

Churchland, P. S. 1986: *Neurophilosophy*. Cambridge, Mass.: MIT Press.

Churchland, P. S., Ramachandran, V. S. and Sejnowski, T. 1994: A critique of pure vision. In C. Koch and J. L. Davis (eds), *Large-scale Neuronal Theories of the Brain*, Cambridge, Mass.: MIT Press, 23–61.

Collins, A. 1977: Why cognitive science. *Cognitive Science*, 1, 1–2.

Collins, A. and Quillian, M. R. 1969: Retrieval time from semantic memory. *Journal of Verbal Learning and Verbal Behavior*, 8, 240–7.

Cooper, L. A. 1975: Mental rotation of random two-dimensional shapes. *Cognitive Psychology*, 7, 20–43.

*Corsi, P. 1991: *The Enchanted Loom. Chapters in the History of Neuroscience*. New York: Oxford University Press.

*Corsini, R. J. 1994: *Encyclopedia of Psychology*. New York: Wiley.

Cragg, B. G. 1969: The topography of the afferent projections in the circumstriate visual cortex (C. V. C.) of the monkey studied by the Nauta method. *Vision Research*, 9, 733–47.

*Crevier, D. 1993: *AI: The Tumultuous History of the Search for Artificial Intelligence*. New York: Basic Books.

Donders, F. C. 1868: Over de snelheid van psychische processen. Onderzoekingen gedaan in het Psyiologish Laboratorium der Utrechtsche Hoogeschool: 1868–1869. *Tweede Reeks*, II, 92–120. English translation by W. G. Koster (1969), On the speed of mental processes. *Acta Psychologica*, 30 (1969), 412–31.

Dreyfus, H. 1972/1979: *What Computers Can't Do*. New York: Harper and Row.

Ebbinghaus, H. 1885/1913: *Memory: A Contribution to Experimental Psychology*. New York: Columbia Teachers' College.

Erikson, E. H. 1968: *Identity, Youth and Crisis*. New York: W. W. Norton.

Farah, M. J. 1989: Mechanisms of imagery–perception interaction. *Journal of Experimental Psychology: Human Perception and Performance*, 15, 203–11.

Feigenbaum, E. A., Buchanan, B. G. and Lederberg, J. 1971: On generality and problem solving: a case study using the DENDRAL program. In B. Melzer and D. Michie (eds), *Machine Intelligence*, vol. 6, Edinburgh: Edinburgh University Press, 165–90.

Felleman, D. J. and van Essen, D. C. 1991: Distributed hierarchical processing in the primate cerebral cortex. *Cerebral Cortex*, 1, 1–47.

Festinger, L. 1957: *A Theory of Cognitive Dissonance.* Evanston, Ill.: Row, Peterson.
*Finger, S. (1994). *Origins of Neuroscience.* New York: Oxford University Press.
Fodor, J. A. 1975: *The Language of Thought.* Cambridge, Mass.: MIT Press/Bradford.
—— 1980: Methodological solipsism considered as a research strategy in cognitive psychology. *Behavioral and Brain Sciences*, 3, 63–109.
—— 1987: *Psychosemantics: The Problem of Meaning in the Philosophy of Mind.* Cambridge, Mass.: MIT Press.
Fodor, J. A. and Pylyshyn, Z. W. 1981: How direct is visual perception: some reflections on Gibson's "ecological approach." *Cognition*, 9, 139–96.
—— —— 1988: Connectionism and cognitive architecture: a critical analysis. *Cognition*, 28, 3–71.
*Fodor, J. A., Bever, T. G. and Garrett, M. F. 1974: *The Psychology of Language.* New York: McGraw-Hill.
Fritsch, G. and Hitzig, E. 1870/1960: On the electrical excitability of the cerebrum. In G. von Bonin (ed.), *Some Papers on the Cerebral Cortex*, Springfield, Ill.: Charles C. Thomas, 73–96.
Galanter, E. 1988: Writing *Plans* . . . In W. Hirst (ed.), *The Making of Cognitive Science: Essays in Honor of George A. Miller*, New York: Cambridge University Press, 36–44.
*Gardner, H. 1985: *The Mind's New Science.* New York: Basic Books.
Garner, W. R. 1988: The contribution of information theory to psychology. In W. Hirst (ed.), *The Making of Cognitive Science: Essays in Honor of George A. Miller*, New York: Cambridge University Press, 19–35.
Gazzaniga, M. S. (ed.) 1995: *The Cognitive Neurosciences.* Cambridge, Mass.: MIT Press.
Gazzaniga, M. S. and LeDoux, J. E. 1978: *The Integrated Mind.* New York: Plenum.
Geschwind, N. 1974: *Some Selected Papers on Language and Brain.* Dordrecht: Reidel.
Gibson, J. J. 1966: *The Senses Considered as Perceptual Systems.* Boston: Houghton-Mifflin.
Green, D. M. and Swets, J. A. 1966: *Signal Detection Theory and Psychophysics.* New York: Wiley.
Harnad, S. 1990: The symbol grounding problem. *Physica D*, 42, 335–46.
*Harris, R. A. 1993: *The Linguistics Wars.* New York: Oxford University Press.
Harvard University 1961–9: *Annual Report of the Center for Cognitive Studies.* Cambridge, Mass.: Harvard University Press.
*Hearst, E. 1979: *The First Century of Experimental Psychology.* Hillsdale, NJ: Erlbaum.
Hebb, D. O. 1949: *The Organization of Behavior.* New York: Wiley.
Heims, S. J. 1980: *John von Neumann and Norbert Wiener: From Mathematics to the Technologies of Life and Death.* Cambridge, Mass.: MIT Press.
Hempel, C. 1966: *Philosophy of Natural Science.* Englewood Cliffs, NJ: Prentice-Hall.
*Hilgard, E. R. 1987: *Psychology in America: A Historical Survey.* New York: Harcourt Brace Jovanovich.
Hinton, G. and Anderson, J. A. (eds) 1981: *Parallel Models of Associative Memory.* Hillsdale, NJ: Erlbaum.
*Hirst, W. (ed.) 1988: *The Making of Cognitive Science: Essays in Honor of George A. Miller.* New York: Cambridge University Press.
Hopfield, J. J. 1982: Neural networks and physical systems with emergent collective computational abilities. *Proceedings of the National Academy of Sciences*, 79, 2554–8.
Hubel, D. H. and Wiesel, T. N. 1962: Receptive fields, binocular interaction and functional architecture in the cat's visual cortex. *Journal of Physiology*, 160, 106–54.
James, W. 1890: *Principles of Psychology*, 2 vols. New York: Holt.
Jameson, D. and Hurvich, L. M. 1968: Opponent-response functions related to measured cone photopigments. *Journal of the Optical Society of America*, 58, 429–30.
Jeffress, L. A. (ed.) 1951: *Cerebral Mechanisms in Behavior: The Hixon Symposium.* New York: Wiley.
Johansson, G. 1973: Visual perception of biological motion and a model for its analysis. *Perception and Psychophysics*, 14, 201–11.

Joshi, A. K. and Hopely, P. 1997: A parser from antiquity. *Natural Language Engineering*, 2, 291–4.

*Kessen, W., Ortony, A. and Craik, F. 1991: *Memories, Thoughts, and Emotions: Essays in Honor of George Mandler*. Hillsdale, NJ: Erlbaum.

Kintsch, W., Miller, J. R. and Polson, P. G. 1984: *Method and Tactics in Cognitive Science*. Hillsdale, NJ: Erlbaum.

Kosslyn, S. M. 1980: *Image and Mind*. Cambridge, Mass.: Harvard University Press.

—— 1994: *Image and Brain: The Resolution of the Imagery Debate*. Cambridge, Mass.: MIT Press.

Kuhn, T. 1962/1970: *The Structure of Scientific Revolutions*. Chicago: University of Chicago Press.

*Lachman, R., Lachman, J. L. and Butterfield, E. C. 1979: *Cognitive Psychology and Information Processing: An Introduction*. Hillsdale, NJ: Erlbaum.

Lahav, R. 1993: What neuropsychology tells us about consciousness. *Philosophy of Science*, 60, 67–85.

Langton, C. G. 1995: *Artificial Life*. Cambridge, Mass.: MIT Press.

Lashley, K. S. 1929: *Brain Mechanisms and Intelligence*. Chicago: University of Chicago Press.

—— 1950: In search of the engram. *Symposia of the Society for Experimental Biology*, 4, 454–82.

—— 1951: The problem of serial order in behavior. In L. A. Jeffries (ed.), *Cerebral Mechanisms in Behavior: The Hixon Symposium*, New York: Wiley, 112–36.

Lee, D. N. 1976: A theory of the visual control of breaking based on information about time to collision. *Perception*, 5, 437–59.

Lenneberg, E. H. 1967: *Biological Foundations of Language*. New York: Wiley.

Lettvin, J. Y., Maturana, H. R., McCulloch, W. S. and Pitts, W. H. 1959: What the frog's eye tells the frog's brain. *Proceedings of the IRE*, 47, 1940–51.

Lindsay, P. and Norman, D. A. 1972: *Human Information Processing*. New York: Academic Press.

Lounsbury, F. G. 1956: A semantic analysis of the Pawnee kinship usage. *Language*, 32, 158–94.

Marr, D. 1969: A theory of the cerebral cortex. *Journal of Physiology*, 202, 437–70.

—— 1982: *Vision*. San Francisco: Freeman.

McCarthy, J. and Hayes, P. J. 1969: Some philosophical problems from the standpoint of artificial intelligence. In D. Michie (ed.), *Machine Intelligence*, vol. 4, Edinburgh: Edinburgh University Press, 463–503.

*McCorduck, P. 1979: *Machines Who Think: A Personal Inquiry into the History and Prospects of Artificial Intelligence*. San Francisco: Freeman.

McCulloch, W. S. and Pitts, W. H. 1943: A logical calculus of the ideas immanent in nervous activity. *Bulletin of Mathematical Biophysics*, 5, 115–33.

Miller, G. A. 1956: The magic number seven, plus or minus two: some limits on our capacity for processing information. *Psychological Review*, 63, 81–97.

—— 1962: Some psychological studies of grammar. *American Psychologist*, 17, 748–62.

—— 1979: *A Very Personal History*. Cambridge, Mass.: MIT Center for Cognitive Science, Occasional Paper no. 1.

Miller, G. A. and Selfridge, J. A. 1950: Verbal context and the recall of meaningful material. *American Journal of Psychology*, 63, 176–85.

Miller, G. A., Galanter, E. and Pribram, K. 1960: *Plans and the Structure of Behavior*. New York: Holt.

Milner, B. 1965: Memory disturbance after bilateral hippocampal lesions. In P. M. Milner and S. E. Glickman (eds), *Cognitive Processes and the Brain*, Patterson, NJ: Van Nostrand, 97–111.

Minsky, M. (ed.) 1968: *Semantic Information Processing*. Cambridge, Mass.: MIT Press.

—— 1975: A framework for representing knowledge. In P. H. Winston (ed.), *The Psychology of Computer Vision*, New York: McGraw-Hill, 211–79.

Minsky, M. and Papert, S. 1969: *Perceptrons*. Cambridge, Mass.: MIT Press.

Mishkin, M., Ungerleider, L. G. and Macko, K. A. 1983: Object vision and spatial vision: two cortical pathways. *Trends in Neuroscience*, 6, 414–17.

Moore, O. K. and Anderson, S. B. 1954: Modern logic and tasks for experiments on problem solving behavior. *Journal of Psychology*, 38, 151–60.
Neisser, U. 1967: *Cognitive Psychology*. New York: Appleton-Century-Crofts.
—— 1976: *Cognition and Reality*. San Francisco: Freeman.
—— 1988: Cognitive recollections. In W. Hirst (ed.), *The Making of Cognitive Science: Essays in Honor of George A. Miller*, New York: Cambridge University Press, 81–8.
Newell, A. 1980: Physical symbol systems. *Cognitive Science*, 4, 135–83.
Newell, A. and Simon, H. 1956: The logic theory machine. *IRE Transactions on Information Theory*, 3, 61–79.
—— —— 1972: *Human Problem Solving*. Englewood Cliffs, NJ: Prentice-Hall.
*Newmeyer, F. J. 1986: *Linguistic Theory in America*, 2nd edn. New York: Academic Press.
Norman, D. A. 1981: Preface to D. A. Norman (ed.), *Perspectives on Cognitive Science*, Norwood, NJ: Ablex.
Norman, D. A. and Levelt, W. J. 1988: Life at the Center. In W. Hirst (ed.), *The Making of Cognitive Science: Essays in Honor of George A. Miller*, New York: Cambridge University Press, 100–9.
Norman, D. A., Rumelhart, D. E. and the LNR Research Group 1975: *Explorations in Cognition*. San Francisco: Freeman.
Osgood, C. E. 1953: *Method and Theory in Experimental Psychology*. New York: Oxford University Press.
Osgood, C. E. and Sebeok, T. A. (eds) 1954: *Psycholinguistics: A Survey of Theory and Research Problems*, Indiana University Publications in Anthropology and Linguistics, Memoir 10 of *International Journal of American Linguistics*. Bloomington: Indiana University Press.
Osgood, C. E., Suci, G. J. and Tannenbaum, P. H. 1957: *The Measurement of Meaning*. Urbana: University of Illinois Press.
*Osherson, D. N. (ed.) 1995–1997: *An Invitation to Cognitive Science*, 4 vols. Cambridge, Mass.: MIT Press.
Penfield, W. and Rasmussen, T. 1950: *The Cerebral Cortex of Man: A Clinical Study of Localization of Function*. New York: Macmillan.
Peterson, L. R. and Peterson, M. J. 1959: Short-term retention of individual verbal items. *Journal of Experimental Psychology*, 58, 193–8.
Pitts, W. and McCulloch, W. S. 1947: How we know universals: the perception of auditory and visual forms. *Bulletin of Mathematical Biophysics*, 9, 127–47.
Port, R. and van Gelder, T. 1995: *Mind as Motion*. Cambridge, Mass.: MIT Press.
Posner, M. I. and McLeod, P. 1982: Information processing models: in search of elementary operations. *Annual Review of Psychology*, 33, 477–514.
Posner, M. I. and Raichle, M. E. 1994: *Images of Mind*. San Francisco: Freeman.
Posner, M. I., Pea, R. and Volpe, B. 1982: Cognitive-neuroscience: developments toward a science of synthesis. In J. Mehler, E. C. T. Walker, and M. Garrett (eds), *Perspectives on Mental Representation*, Hillsdale, NJ: Erlbaum, 251–76.
Posner, M. I., Petersen, S. E., Fox, P. T. and Raichle, M. E. 1988: Localization of cognitive operations in the human brain. *Science*, 240, 1627–31.
Prince, A. and Smolensky, P., in press: *Optimality Theory: Constraint Interaction in Generative Grammar*. Cambridge, Mass.: MIT Press.
Putnam, H. 1960: Minds and machines. In S. Hook (ed.), *Dimensions of Mind*, New York: New York University Press, 148–79.
—— 1967: Psychological predicates. In W. Capitan and D. Merrill (eds), *Art, Mind, and Religion*, Pittsburgh: University of Pittsburgh Press, 37–48.
Pylyshyn, Z. W. 1973: What the mind's eye tells the mind's brain: a critique of mental imagery. *Psychological Bulletin*, 80, 1–24.
—— 1981: The imagery debate: analogue media versus tacit knowledge. *Psychological Review*, 87, 16–45.

—— 1983: Information science: its roots and relations as viewed from the perspective of cognitive science. In F. Machlup and U. Mansfield (eds), *The Study of Information: Interdisciplinary Messages*, New York: John Wiley and Sons, 63–80.

Quillian, M. R. 1968: Semantic memory. In M. Minsky (ed.), *Semantic Information Processing*, Cambridge, Mass.: MIT Press, 227–70.

Ratliff, F. and Hartline, H. K. 1959: The response of limulus optic nerve fibers to patterns of illumination on the receptor mosaic. *Journal of General Physiology*, 42, 1241–55.

Rips, L. J., Shoben, E. J. and Smith, E. E. 1973: Semantic distance and the verification of semantic relations. *Journal of Verbal Learning and Verbal Behavior*, 12, 1–20.

Rochester, N., Holland, J. H., Haibt, L. H. and Duda, W. L. 1956: Tests on a cell assembly theory of the action of the brain, using a large digital computer. *IRE Transactions on Information Theory*, 2, 80–93.

Rosenblatt, F. 1962: *Principles of Neurodynamics: Perceptrons and the Theory of Brain Mechanisms*. Washington, DC: Spartan Books.

Rosenblueth, A., Wiener, N. and Bigelow, J. 1943: Behavior, purpose, and teleology. *Philosophy of Science*, 10, 18–24.

Rumelhart, D. E., Hinton, G. E. and Williams, R. J. 1986: Learning representations by back-propagating errors. *Nature*, 323, 533–6.

Rumelhart, D. E., McClelland, J. L. and the PDP Research Group 1986: *Parallel Distributed Processing: Explorations in the Microstructure of Cognition*, vol. 1: *Foundations*. Cambridge, Mass.: MIT Press.

Schank, R. C. and Abelson, R. P. 1977: *Scripts, Plans, Goals and Understanding*. Hillsdale, NJ: Erlbaum.

Schwartz, E. L. (ed.) 1990: *Computational Neuroscience*. Cambridge, Mass.: MIT Press.

Searle, J. 1980: Minds, brains, and programs. *Behavioral and Brain Sciences*, 3, 417–57.

Shannon, C. E. 1948: A mathematical theory of communication. *Bell System Technical Journal*, 27, 379–423, 623–56.

—— 1950: A chess-playing machine. *Scientific American*, 182(2), 48–51.

—— 1951: Prediction and entropy of printed English. *Bell System Technical Journal*, 30, 50–64.

Shepard, R. N. and Metzler, J. 1971: Mental rotation of three-dimensional objects. *Science*, 171, 701–3.

Shortliffe, E. H. 1976: *Mycin: Computer Based Medical Consultations*. New York: Elsevier.

Skinner, B. F. 1938: *The Behavior of Organisms: An Experimental Analysis*. New York: Appleton-Century.

—— 1957: *Verbal Behavior*. Englewood Cliffs, NJ: Prentice-Hall.

Smolensky, P. 1988: On the proper treatment of connectionism. *Behavior and Brain Sciences*, 11, 1–74.

Sperling, G. 1960: *The Information Available in Brief Visual Presentations*, Psychological Monographs, 74 (whole no. 498).

Sternberg, S. 1969: The discovery of processing stages: extension of Donders' method. *Acta Psychologica*, 30, 276–315.

Turing, A. M. 1936: On computable numbers, with an application to the Entscheidungs-Problem. *Proceedings of the London Mathematical Society*, 2nd ser., 42, 230–65.

—— 1950: Computing machinery and intelligence. *Mind*, 59, 433–60.

Ullman, S. 1980: Against direct perception. *Behavioral and Brain Sciences*, 3, 373–416.

Valenstein, E. S. 1973: *Brain Control: A Critical Examination of Brain Stimulation and Psychosurgery*. New York: Wiley Interscience.

van Essen, D. C. and Anderson, C. H. 1990: Information processing strategies and pathways in the primate retina and visual cortex. In S. F. Zornetzer, J. L. Davis, and C. Lau (eds), *An Introduction to Neural and Electronic Networks*, New York: Academic Press, 43–72.

van Gelder, T. 1995: What might cognition be, if not computation? *Journal of Philosophy*, 92, 345–81.

Watson, J. 1913: Psychology as a behaviorist views it. *Psychological Review*, 20, 158–77.

Waugh, N. C. and Norman, D. A. 1965: Primary memory. *Psychological Review*, 72, 89–104.

Weizenbaum, J. 1976: *Computer Power and Human Reason: From Judgment to Calculation*. San Francisco: Freeman.

Wernicke, C. 1874: *Der aphasiche Symptomcomplex: Eine psychologische Studie auf anatomischer Basis*. Breslau: Cohen and Weigert.

Wertheimer, M. 1912: Experimentelle Studien über das Sehen von Bewegungen. *Zeitschrift für Psychologie*, 61, 161–265.

Whybrow, P. 1996: *A Mood Apart: Depression, Manic Depression, and Other Afflictions of the Self*. New York: Basic Books.

Wiener, N. 1948: *Cybernetics: Or, Control and Communication in the Animal Machine*. New York: Wiley.

Willner, P. (ed.) 1991: *Behavioral Models in Psychopharmacology: Theoretical, Industrial, and Clinical Perspectives*. Cambridge: Cambridge University Press.

Winograd, T. 1972: *Understanding Natural Languages*. New York: Academic Press.

Zeki, S. M. 1976: The functional organization of projections from striate to prestriate visual cortex in rhesus monkey. *Cold Spring Harbor Symposium on Quantitative Biology*, 40, 591–600.

PART II

AREAS OF STUDY IN COGNITIVE SCIENCE

1

Analogy

DEDRE GENTNER

Analogies are partial similarities between different situations that support further inferences. Specifically, analogy is a kind of similarity in which the same system of relations holds across different objects. Analogies thus capture parallels across different situations.

Analogy is ubiquitous in cognitive science. First, in the study of learning, analogies are important in the transfer of knowledge and inferences across different concepts, situations, or domains. They are often used in instruction to explain new concepts, such as electricity or evaporation (see Article 54, EDUCATION). Second, analogies are often used in PROBLEM SOLVING and REASONING. Third, analogies can serve as mental models for understanding a new domain. For example, novices in electricity often reason about electric current using mental models based on analogies with water flow or with crowds of moving entities. These analogical mental models can be misleading as well as helpful. The cognitive anthropologist Willet Kempton interviewed home-owners about how their furnaces worked and found that many of them applied incorrect analogies, such as a gas pedal model whereby the higher the thermostat is set, the faster the furnace heats up the house. Fourth, analogy is important in creativity. Studies in history of science show that analogy was a frequent mode of thought for such great scientists as Faraday, Maxwell, and Kepler. More direct evidence comes from studies by Kevin Dunbar, who traced scientists' day-to-day activities and discussions in four different microbiology laboratories. He found that frequent use of analogy was one of the chief predictors of research productivity. Fifth, analogy is used in communication and persuasion. For example, environmentalists may compare the Earth to Easter Island, where overpopulation and exploitation of the island's bountiful ecology led to massive loss of species, famine, and societal collapse. The invited inference is that the point of no return may pass unnoticed. A final reason to study analogy is that analogy and its cousin, similarity, underlie many other cognitive processes. For example, most theories of conceptual structure assume that items are categorized in part on the basis of similarity between the current situation and the prior exemplars or prototype (see Article 25, WORD MEANING). Much of human categorization and reasoning may be based on implicit or explicit analogies between the current situation and prior situations. As another example, analogical processes are involved in using conceptual metaphors, such as "Love is a journey" or "Time is a commodity." Such metaphors have been claimed to perform a structuring role across different domains (see Article 37, COGNITIVE LINGUISTICS).

History

The study of analogy has been characterized by fruitful interdisciplinary convergence between psychology and artificial intelligence, with significant influences from history of science, philosophy, and linguistics. Important early work came out of philosophy, notably Mary Hesse's (1966) analysis of analogical models in science. However, early psychological research on analogy mostly focused on simple four-term analogies of the kind used in intelligence testing. David Rumelhart and Adele Abrahamsen modeled analogy as a mapping from one mental space to another and found that respondents given analogies like "Horse is to zebra as dog is to——?" would choose the answer (e.g., *fox*) whose position relative to *dog* was the same as that of *zebra* relative to *horse*. Robert Sternberg measured solution times to solve such analogies as a way of studying component processes – *encoding, inference, mapping, application*, and *response* – and individual differences in their use.

In the early 1980s, a new breed of analogical models appeared that assumed complex representations and processes. Artificial intelligence researchers like Patrick Winston and Jaime Carbonell suggested computational principles applicable to human processing and inspired psychologists to create explicit models of representation and process. In cognitive science, a multidisciplinary approach grew up in which analogy was viewed as a mapping between structured representations, such as propositional representations or schemata.

Processes of analogical use

To model the use of analogy in learning and reasoning, current accounts distinguish the following subprocesses: (1) *retrieval*: given some current situation in working memory, the person accesses a prior similar or analogous example from long-term memory; (2) *mapping*: given two cases in working memory, *mapping* consists of *aligning* their representational structures to derive the commonalities and *projecting inferences* from one to the other. Mapping is followed by (3) *evaluation* of the analogy and its inferences and often by (4) *abstraction* of the structure common to both analogs. A further process that may occur is (5) *re-representation*: *adaptation* of one or both representations to improve the match. We begin with the processes of mapping through evaluation, reserving retrieval for later.

Analogical mapping

The core process in analogy is *mapping*: the process by which one case is used to explain and predict another. In mapping, a familiar situation – the *base* or *source* analog – provides a kind of model for making inferences about an unfamiliar situation – the *target* analog. One of the first theories to focus on this process was *structure-mapping* theory (Gentner, 1983). According to this theory, an analogy conveys that a system of relations that holds in the base domain also holds in the target domain, whether or not the actual objects in the two domains are similar. The alignment must be *structurally consistent*: there is *one-to-one correspondence* between elements in the base and elements in the target, and the arguments of corresponding predicates must also correspond (*parallel connectivity*). A further assumption is the *systematicity principle*: systems of relations connected by higher-order constraining relations such as *cause* contribute more to analogy than do isolated matches or an equal number of independent matches. The information highlighted by the comparison forms a connected relational system, and commonalities connected to the matching system gain in importance. For example,

Clement and I found that people given analogous stories judged that corresponding assertions were more important to the analogy when they were connected to other matching information than when they were not. A parallel result was found for inference projection: people were more likely to import a fact from the base to the target when it was connected to other predicates shared with the target. Thus there is a kind of "no match is an island" phenomenon. In analogical matching, people are not interested in isolated coincidental matches; rather, they seek causal and logical connections that give the analogy its inferential force.

Another important approach to analogy that grew up in the 1980s was Holyoak's pragmatic account. Focusing on the use of analogy in problem solving, this approach emphasized the role of pragmatics in analogy – how current goals and context guide the interpretation of an analogy. Holyoak defined analogy as similarity with respect to a goal and suggested that mapping processes are oriented towards attainment of goal states. Holyoak and Paul Thagard (1989) combined this pragmatic focus with the assumption of structural consistency and developed a multi-constraint approach to analogy in which similarity, structural parallelism, and pragmatic factors interact to produce an interpretation.

Evaluation

Evaluating an analogy involves at least three kinds of judgment. One criterion is structural soundness: whether the alignment and the projected inferences are structurally consistent. Another is the factual validity of the projected inferences in the target. Because analogy is not a deductive mechanism, these candidate inferences are only hypotheses; their factual correctness is not guaranteed by their structural consistency and must be checked separately. Brian Falkenhainer's Phineas program operationalized this by first attempting to prove the inferences true or false in the target domain. If this failed, an empirical test was derived. A third criterion, which applies in a problem-solving situation, is whether the analogical inferences are relevant to the current goals. An analogy may be structurally sound and yield true inferences, but still fail the relevance criterion if it does not bear on the problem at hand. A related issue discussed by Mark Keane is the *adaptability* of the inferences to the target problem.

Schema abstraction

In *schema abstraction*, the common system that represents the interpretation of an analogy is retained for later use. For example, Mary Gick and Holyoak's (1983) research on problem solving provided evidence that people can abstract the relational correspondences between examples into a schema. Comparing structurally similar problems leads to improved performance on further parallel problems and promotes transfer from concrete comparisons to abstract analogies.

Computational models of analogical mapping

Computing an analogy typically involves both matching the representations and projecting new inferences. Computational models can be classified into *projection-first* and *alignment-first* models, according to which process occurs first. In *projection-first* models, the analogical abstraction is derived initially from the base alone and projected onto the target, after which it is aligned (or matched) with the target's representation. In this kind of model, the first step in processing an analogy is to find a projectable schema or derive an abstraction in the base and project it onto the target. Then the system attempts to verify the projected structure in the target: to discover for each assertion either (a) that it can be proved correct in the target on the basis of existing knowledge (or is already present), or (b) that it can be proved false in the target (in which case the analogy must be rejected), or (c) that it can neither be proved nor disproved, in which case it stands as a possible new inference. Two recent simulations that fit loosely into the projection-first mode are Keane's IAM (Incremental Analogy Machine) and Hummel and Holyoak's LISA model.

In *alignment-first* models, the common system arises interactively, via processes of alignment, with the projection of inferences as a second step. For example, the Structure-Mapping Engine (SME) of Falkenhainer, Kenneth Forbus, and myself begins by aligning two representations and then carries over further predicates from the base to the target. When given two potential analogs, it first finds all possible local matches between elements of the base and the target. Then it combines these into kernels – little clusters of connected correspondences – and finally it merges the kernels into the two or three maximal structurally consistent systems of matches, which represent possible interpretations of the analogy. It performs a structural evaluation of the analogy, which reflects the size and depth of the matching system, and draws spontaneous *candidate inferences* using a process of structural completion from base to target. (See Forbus, Gentner, and Law, 1995, for a description.)

Another alignment-first model is Holyoak and Thagard's (1989) Analogical Constraint Mapping Engine (ACME), which uses a local-to-global algorithm similar to SME's. However, ACME differs from SME in some important ways. First, it is a multiconstraint connectionist system. In ACME, structural consistency is only a soft constraint, along with semantic similarity and pragmatic bindings. This allows more flexibility in the mapping process, but with the cost that structurally inconsistent mappings can easily occur. Second, whereas SME typically produces two or more winning interpretations, ACME uses a winner-take-all algorithm, producing one interpretation that is the best compromise among the three constraints. Third, candidate inferences are requested by the user rather than being generated automatically as in SME.

Projection-first models are particularly apt when there is one main schema associated with the base that can be projected onto the target. However, they encounter difficulties when more than one schema is associated with the base. A further drawback is that they do not readily capture emergent processing, in which the juxtaposition of two cases leads to insights not initially obvious in either representation. Alignment-first models seem apt for processing new comparisons, in which the common abstraction is not already a salient schema. However, in cases where the base domain possesses a conventionalized abstract schema, it seems likely that the learner will simply project this schema rather than deriving a new match from scratch. Thus a complete account of analogical processing will probably involve both kinds of algorithms, at different stages of knowledge. For example, given the analogy "The atom is like a solar system," an advanced learner might process it in projection-first mode, by projecting the abstraction *central force system* from the base (the solar system). However, a novice learner who has not yet explicitly stored the central-force abstraction will need to derive the common *central force* abstraction via an alignment process. As discussed above, one result of such an alignment is that the common system becomes more salient, thus promoting the development of a relational abstraction. Thus, with increasing domain expertise, projection-first processes may supplant alignment-first processes.

Retrieval

A striking and robust finding is that people often fail to retrieve potentially useful analogs, even when it is clear that they have retained the material in memory. For example, Gick and Holyoak (1983) gave subjects a classic thought-problem: How can a surgeon cure an inoperable tumor without using so much radiation that the surrounding flesh will be killed? Only about 10 percent of the participants came up with the ideal solution, which is to converge on the tumor with several weak beams of radiation. If given a prior analogous story in which soldiers converged on a fort, three times as many people (about 30 percent) produced convergence solutions. Surprisingly, the majority of participants still failed to think of the convergence solution. Yet when these people were simply given a hint to think about the story they had heard, the percentage of correct convergence solutions again nearly tripled, to about 80 percent. We can infer that the fortress story was stored in memory and was potentially useful, but it was not retrieved by the analogous tumor problem. This failure to access potentially useful analogs is a major cause of the *inert knowledge* problem in EDUCATION.

Evidence further suggests that memory not only fails to produce analogous items, but often produces superficially similar items instead. Gentner, Rattermann, and Forbus gave subjects a set of stories to remember and later showed them probe stories that were either surface-similar to their memory item (e.g., similar objects and characters) or structurally similar (i.e., analogous, with similar higher-order causal structure). Subjects were told to write out any of the prior stories that they were reminded of. Surface commonalities were the best predictor of memory access: Recall rates were two to five times higher when the probes had surface commonalities than when the probes had structural commonalities. However, in a separate rating task, structurally similar pairs were rated much higher in inferential soundness and even in rated similarity than surface-similar pairs. Participants rated their own surface-similar remindings as low in inferential value and in similarity. The good news here is that although surface similarity has a large say in initial memory retrieval, people often reject purely superficial matches quickly, retaining matches with structural commonalities for further processing. (See Forbus, Gentner, and Law, 1995, for further details.)

Similar results have been found in problem-solving tasks. (See Reeves and Weisberg, 1994, for a comprehensive review.) Research by Brian Ross, Miriam Bassok, Laura Novick, and others bears out the finding that remindings of prior problems are strongly influenced by surface similarity, although structural similarity better predicts success in solving current problems. This gap between access and use may be less pronounced for experts in a domain. For example, Novick found that people with mathematics training retrieved fewer surface-similar lures in a problem-solving task and rejected them more quickly than did novice mathematicians. This may stem in part from experts' encoding patterns. There is evidence that relational remindings increase when the same relational terminology is used in the memory item and the probe item, suggesting that uniform encoding of the items is important in promoting retrieval.

Computational models of similarity-based retrieval

There are two main approaches to similarity-based retrieval. The CASE-BASED REASONING approach is founded on the view that much human reasoning is based on retrieval and on use of specific episodes in memory. This research focuses on how memory can be organized such that relevant cases are retrieved when needed. The second approach, more prevalent among psychologists, aims to capture the phenomena of human memory retrieval, including errors based on surface similarity. Two models in this spirit are Analog Retrieval by Constraint Satisfaction (ARCS), by Thagard, Holyoak, Nelson, and Gochfeld, and Many Are Called/but Few Are Chosen (MAC/FAC), by Forbus, Gentner, and Law. For example, MAC/FAC utilizes a two-stage process. An initial content-similarity stage is followed by a structural alignment process (that of SME) that filters the initial pool of potential retrievals. ARCS uses a competitive retrieval algorithm combination of content similarity, structural similarity, and pragmatic relevance. (See Holyoak and Thagard, 1995, for a discussion of ARCS, and Forbus, Gentner, and Law, 1995, for a comparison of the two models.)

Development of analogy

Even preschool children appear to engage in metaphor and analogy, as Ann Brown, Usha Goswami, and others have shown. However, there are marked developmental changes. Young children are likely to interpret analogies and metaphors in terms of thematic connections or common object properties rather than common relational systems. For example, when I asked children "Why is a cloud like a sponge?", five-year-olds respond that both are round and fluffy, rather than responding that both can hold and release water (the adult response). Some theorists explain this relational shift as due to children's increasing domain knowledge, with mapping processes remaining roughly constant across age. Others suggest that the shift is due to a change in processing capacity. (See Goswami, 1992; Halford, 1993; and Gentner and Rattermann, 1991, for further discussion.)

Extensions of analogy theory

Theories of analogy have been extended to ordinary (literal or overall) similarity. The basic idea is that overall similarity can be thought of as an especially rich analogy – one that shares both structural and surface commonalities. (See Gentner and Markman, 1997, and Medin, Goldstone, and Gentner, 1993, for reviews.) This perspective suggests a continuum of similarity types. At one end lies abstract analogy, in which the two terms share only a common relational system, as in the match between [*hen and chick*] and [*mare and colt*]. As more commonalities are added, the comparison becomes one of overall literal similarity, as in the match between [*hen and chick*] and [*duck and duckling*]. There is evidence that many of the same processes of structural alignment and mapping described for analogy also apply to overall similarity comparisons. As in analogy, common systems of connected information become more salient in literal comparison, and inferences are projected using a kind of structural-completion process. Analogical alignment processes have also been extended to metaphor, decision making, and categorization.

References and recommended reading

Forbus, K. D., Gentner, D. and Law, K. 1995: MAC/FAC: a model of similarity-based retrieval. *Cognitive Science*, 19, 141–205.

Gentner, D. 1983: Structure-mapping: a theoretical framework for analogy. *Cognitive Science*, 7, 155–70.

Gentner, D. and Markman, A. B. 1997: Structure-mapping in analogy and similarity. *American Psychologist*, 52, 45–56.

Gentner, D. and Rattermann, M. J. 1991: Language and the career of similarity. In S. A. Gelman and J. P. Brynes (eds), *Perspective on Thought and Language: Interrelations in Development*, London: Cambridge University Press, 225–77.

Gick, M. L. and Holyoak, K. J. 1983: Schema induction and analogical transfer. *Cognitive Psychology*, 15, 1–38.

*Goswami, U. 1992: *Analogical Reasoning in Children*. Hillsdale, NJ: Erlbaum.

Halford, G. S. 1993: *Children's Understanding: The Development of Mental Models*. Hillsdale, NJ: Erlbaum.

*Hesse, M. B. 1966: *Models and Analogies in Science*. Notre Dame, Ind.: University of Notre Dame Press.

Holyoak, K. J. and Thagard, P. R. 1989: Analogical mapping by constraint satisfaction. *Cognitive Science*, 13, 295–355.

*—— —— 1995: *Mental Leaps: Analogy in Creative Thought*. Cambridge, Mass.: MIT Press.

Medin, D. L., Goldstone, R. L. and Gentner, D. 1993: Respects for similarity. *Psychological Review*, 100, 254–78.

Reeves, L. M. and Weisberg, R. W. 1994: The role of content and abstract information in analogical transfer. *Psychological Bulletin*, 115, 381–400.

2

Animal cognition

HERBERT L. ROITBLAT

Animal cognition is the study of the minds of animals and the mechanisms by which those minds operate. It touches on and illuminates a wide variety of issues at the foundation of cognition science. The methods developed for its study have broad application, and its theories provide essential links between brain and behavior and between evolution and cognition. Among the foundational issues it addresses are: (1) What do we mean by mind? (2) What role does language play in the mind? (3) What are the cognitive processes that operate during perception and recognition? (4) What is the nature of memory? (5) What is the relation between brain and behavior? (6) How does experience affect behavior?

A major focus in cognitive science has been on modeling the performance of tasks that characterize the achievements of intelligent, educated, language-using humans. These tasks include planning, problem solving, scientific creativity, and the like. They depend on conscious and deliberate effort; and they are tasks that people perform badly and relatively slowly. In contrast, humans also perform many tasks apparently automatically and without effort. These tasks include recognition, spatial navigation, sensorimotor coordination and balance, recognition of objects, path planning, obstacle avoidance, learning from experience, anticipation, and effective response to changing environmental conditions. These tasks are technologically rather mundane, but they appear nonetheless to require rather sophisticated processes (as shown, for example, by the difficulty of performing these tasks with computers). For the most part, cognitive scientists have not yet developed fully effective theories of their performance, and they tend to be tasks that are performed by both human and nonhuman species.

Given that the functionality appears in both humans and nonhumans, there are advantages to investigating these processes in nonhumans. Three factors make it difficult to study these processes in humans. First, because all the investigators studying these phenomena are human, they are so familiar with the products of these mechanisms that they easily confuse what is familiar with what is easy. We have evolved specific mechanisms that make these processes automatic and thereby *transparent*. We may thus fail to recognize the difficulty in performing these tasks, albeit by automatic mechanisms. Second, introspection is an unreliable instrument, so much of the evidence on which our own mental analysis is based is derived from faulty instruments. Introspection does not have even the appearance of validity when considering animals; so we avoid the self-deception that introspection entails. Third, language is such a ubiquitous feature of human cognitive experience that it is difficult to disentangle our verbal representations of objects and events from the perceptual and conceptual features of those objects and events. It is difficult to tell how much of our intelligence is due to language and how much is due to the processes that provide the foundation for that language.

The study of animal cognition allows us to view human cognition as one among alternative forms of cognition. Because we generally do not have strong intuitions about the nature of animal cognitive processing, as we do about our own processing, we can approach animal cognition from a more skeptical and analytic perspective. With animals it is more difficult to miss assumptions about the processes that are hidden by our own familiarity with our own cognitive processes.

The meaning of mind

Investigating the minds of animals forces us to face the problem of what it means to have a mind. There are at least two definitions of mind that people have used, without being particularly clear about the distinction between them. Confusing the two definitions leads to unnecessary difficulties. The first definition emphasizes the phenomenology of mind, the second emphasizes the functionality of mind. The phenomenological issue is exemplified by Thomas Nagel's (1974) question about what it is like to be a bat. Because we are so familiar with our own minds, we know quite well what it is like to be ourselves; but we have little direct idea of what it is like to be another person, let alone an animal of another species. Things do not just happen to us; rather, we experience them. The mind is the mechanism that does the experiencing. What is it like to be another person or a member of another species? The phenomenological definition of mind argues that to have a mind is to have the experience of self-conscious existence. This kind of definition, as Nagel pointed out, is extremely problematic when applied to animals. Because we can have the experience of only one mind, we cannot know the mind of any other individuals and must infer its presence from their behavior. We have no principled way to judge whether the people around us are merely simulating individuals with minds or whether they too have minds just like our own.

At the end of the last century, George John Romanes (1848–94) attempted to infer the phenomenological presence of mind in animals on the same basis that he inferred the presence of mind in others. He observed their behavior, and by thinking about what processes would be going on in his own mind were he to undertake the same behavior, he sought to infer what was happening in the animal's mind. Most modern investigators of animal behavior reject Romanes's methods as unwarranted anthropomorphism. Nevertheless, a small group of investigators, most notably Gordon Gallup at SUNY Albany, have attempted to determine whether animals have anything resembling personal experience and self-awareness.

Their main test for the presence of self-awareness is the mirror test. An animal such as a chimpanzee or a gorilla is anesthetized. While it is asleep, a mark is painted on the animal's head with some dye that presumably it cannot feel. The only way for the animal to detect its presence is visually. When the animal wakes up, it is given a mirror. Looking at its reflection in the mirror the animal can see the mark on its forehead. If it reaches up and touches its head where the mark has been placed, Gallup and his colleagues infer that the animal is self-aware – it can recognize its own reflection in the mirror. If the animal does not reach for its own head, but tries to touch the reflection in the mirror, this is taken as evidence that the animal detects the mark, but does not recognize its reflection as an image of itself. If Gallup and his colleagues are correct, the mirror test can determine that chimpanzees and the few other species that *pass* the mirror test have a phenomenological mind, but it cannot tell us very much about the nature of that mind.

Most investigators of animal cognition have focused instead on the functional characteristics of mind. On this approach, an animal's mind is the set of cognitive structures, processes, representations, and skills that allow it to process, represent, store, and manipulate information. The question is not whether an animal has a mind, but rather what kind of mind it has. How does it obtain, store, and use information to guide its behavior? What computational processes constitute its information processing capacities?

One problem that some investigators have with the functional definition of mind is its implication that mere computation is sufficient to constitute a mind. On this definition, not only do animals have minds, but computers must also be said to have minds, and perhaps even microwave ovens and wristwatches. According to some investigators, such as Stevan Harnad (1990), more than computation is needed to constitute a mind. The missing ingredient is some way to escape from pure syntax. Syntactic processing is mindless; minds inextricably require semantics. Syntactic processing is just the use of rules that transform one set of symbols into another set of symbols. For example, the string of symbols 2 + 2 is transformed to the symbol 4 by the addition rule. A computer can be programmed to perform this transformation without knowing anything about arithmetic or about what the symbols stand for.

Semantics, according to Harnad, can be provided by properly grounding the symbols: that is, by insuring that the symbols do actually stand for something and are not just meaningless squiggles and squoggles. Harnad argues that sensorimotor transduction (i.e., the mechanisms that translate symbolic activity into movement and those that translate physical characteristics, such as light/dark patterns, into neural symbols) provides this grounding for biological entities, but that computers, which lack such processes, are incapable of having minds. Animals, on the other hand, are equipped with sensory systems and so, on this definition, are clearly capable of having grounded symbols and minds. We may not be able to know what it is like to be a bat, but we can know something about the cognitive processes that mediate between its sensory system and its behavior.

Mechanisms of perception and recognition

We are able to gain perspective on sensory and perceptual processes by investigating animals with unique sensory systems. For example, both bats and dolphins use a kind of biological sonar (called *echolocation*, though it is used for much more than just location). Dolphins, for example, can identify many characteristics of submerged objects, including their size, structure, shape, and material composition (Nachtigall, 1980; Au, 1993). Dolphins can detect the presence of small (7.6 cm diameter) stainless-steel spheres at distances up to 113 m; can discriminate between aluminum, copper, and brass circular targets; and can discriminate between circles, squares, and triangular targets.

Dolphins echolocate by generating an extremely brief, broadband click deep within their nasal apparatus, below the blow hole. The sound emerges in a narrow beam from the dolphin's forehead and reflects off objects in its path. The dolphin detects the returning echo through its jaw, which connects via a sound-conducting fat channel to its inner ear, and uses the echo characteristics to identify the features of the object that returned the echo. The dolphin's hearing is exquisitely sensitive, with a range up to about 150 kHz. Humans, by contrast, tend to be limited to the hearing range (up to 15–20 kHz).

Because the dolphin uses distinct clicks to sense the properties of objects that it echolocates, each echo returns a discrete packet of information about the object. The distribution of clicks can be used as an indicator of the animal's attention. The discreteness of the echolocation clicks raises a number of questions about how the animal combines information over these discrete packets. How, for example, does it generate a coherent representation of an object that it senses only in relatively widely spaced acoustic glimpses (analogous to a strobe light)? During the time between clicks, the object and the dolphin may both be moving. This movement makes it difficult to determine how to combine information that may originate at different angles and different distances from the dolphin. Detailed computational models of the dolphin's echolocation performance have been developed by the group at the Hawaii Institute of Marine Biology.

Because the dolphin's echolocation system is so different from our own sensory systems, this research allows us to examine questions about the nature of perceptual representations. Vision is such a primary sense for us that we tend to conceive of these representations in vision-like terms and to think of vision as being somehow more primary than other senses. For example, if we want to know what something really is, we need (or at least think we need) to see it. The information that a dolphin gets through echolocation is nearly as complex as the information we receive through vision – object structure and material composition – but it comes through a very different sensory modality. This raises the possibility that dolphins have images of the objects they echolocate, but these images are unlikely to use the same brain mechanisms or processes as visual images. By understanding dolphin images or whatever representations they have of echolocated objects in their environment, we should be able to gain powerful insights into the nature of these cognitive systems and our own.

Language and thought

Investigations of animal cognition also help to illuminate the relationship between language and thought. No animal is known to use anything like a full-blown human language system (i.e., with the language capacity of an average adult human). They nevertheless share many cognitive processing mechanisms with humans. Understanding these mechanisms in animals can help us to understand the role that language plays in human thought, by identifying the nature of mechanisms that do not require language. For example, Herbert Terrace at Columbia University and Karyl Swartz at Hunter College (e.g., Swartz et al., 1991) have been investigating the ability of pigeons and monkeys, neither of which show any evidence of a language capacity, to process and remember sequentially structured stimuli. They show an animal a set of several pictures, and the animal's task is to touch the pictures in a specific order. By using different types of pictures and different combinations of previously learned sequences, they can identify the processes that the animals use to represent the serial order of the pictures without the use of language.

Similarly, Anthony Wright, at the University of Texas in Houston, has examined the ability of pigeons and monkeys to remember long lists of pictures (Wright et al., 1985). Humans appear to be able to remember hundreds of items, based partly, at least, on verbal descriptions of those items. When Wright tested humans with kaleidoscope pictures (i.e., photographs of the patterns generated by a kaleidoscope), which are difficult to describe, their performance approximated that of the monkeys. Wright thus demonstrated the important role that language plays in facilitating memory, but he was also able to demonstrate what people and animals could do without the benefit of language.

Language training

Another way in which animal cognition addresses fundamental issues in cognitive science concerns the investigation of the language abilities of animals. No animal, so far as we can determine, has a communication system with the flexibility and power of human language (as seen in ordinary adult language capabilities). Language use is so central to our conceptualization of what it means to be human that, following Descartes, language has been taken by many to be the skill that separates humans from other animals. Furthermore, some investigators, most notably Noam Chomsky (1959), have argued strenuously that human language is the result of a *language organ* that evolved uniquely in humans. Hence, any demonstration of linguistic capacity in nonhuman animals would be a serious challenge to the view that humans are unique.

Limited linguistic capacities have been demonstrated in a number of species, including dolphins (Louis Herman at the University of Hawaii), sea lions (Ronald Schusterman at the University of California, Santa Cruz), and a parrot (Irene Pepperberg at the University of Arizona). The most sophisticated language capacity has been demonstrated in the pygmy chimpanzee, or bonobo, by Sue Savage-Rumbaugh at Georgia State University (see chapters in Roitblat et al., 1993). These studies generally show that animals can learn to use symbols to stand for things. When a dolphin is given the *sentence* SURFBOARD BALL FETCH, for example, it can take a ball floating in the tank and carry it to a surfboard floating in the tank among a large number of other objects.

Similarly, these studies show that the animals can learn at least rudimentary syntax. For example, the same dolphin can respond differently to the sentence, BALL SURFBOARD FETCH, by taking the surfboard to the ball. The dolphin has clearly learned that changing the order of the words changes the meaning of the sentence. The animals also demonstrate other uses of word order: for example, by attaching modifiers (adjectives) to the appropriate nouns.

The bonobo, Kanzi, at Georgia State University, has demonstrated evidence of somewhat more complex syntax. Even more striking, however, is his apparent ability to understand spoken sentences, even when tested under conditions that prevent him from using alternative cues. For example, the speaker can be in another room and present sentences to Kanzi over head phones or from behind a one-way mirror (Savage-Rumbaugh et al., 1993).

An interesting feature of Kanzi's performance is that he was not trained explicitly to perform these tasks, but rather acquired his protolinguistic skills through observation of his mother being trained and through natural interactions with his human companions. Although tested under very stringent formal conditions, he was never explicitly trained either to produce language or to respond to it.

Kanzi has not – or at least not yet – demonstrated the full range of complexity seen in adult human language. He has not, for example, demonstrated any capacity to produce complex syntactic constructions, such as "The boy who liked the girl left." In fact, he has shown little inclination to use much syntax in his productions at all. The incompleteness of his language mastery has led some investigators, such as the University of Hawaii's Derek Bickerton (1990), to deny that Kanzi has language. Bickerton allows that Kanzi has perhaps demonstrated protolanguage, but demonstration of full language requires evidence of the use of complex recursive syntax. Nonetheless, Kanzi's skills are impressive and do demonstrate at least rudimentary forms of language capacity that might be analogous to those possessed by our ancestors during the course of human evolution.

Conclusion

Investigation of the cognitive processes of animals helps us to gain a clear perspective on fundamental issues in cognitive science. It helps us to see that there are ways of doing cognition other than those with which we are familiar. When combined with computational models of this performance, studies of animal cognition help to highlight the sophisticated mechanisms that are hidden beneath our familiarity with our own processes. Studies of animal cognition also lead us to question our assumptions about the nature of mind and help to clarify potential confusions. Furthermore, because much of cognitive neuroscience depends on our ability to manipulate brain functions in ways that cannot ethically be done with humans, we need an adequate understanding of cognition in animals if we are to have any hope of identifying the cognitive features mediated by these brain sites. Finally, understanding our own relationship with nature depends on discovering what makes us unique relative to other animals and on identifying the evolutionary pathways by which we became human. Although no current animal species is ancestral to humans (or any other species), understanding the different evolutionary pathways taken by different species can help us in our attempts to understand human evolution and existence.

References and recommended reading

Au, W. W. L. 1993: *The Sonar of Dolphins*. New York: Springer Verlag.

*Bekoff, M. and Jamieson, D. 1996: *Readings in Animal Cognition*. Cambridge, Mass.: MIT Press.

*Bickerton, D. 1990: *Language and Species*. Chicago: University of Chicago Press.

Chomsky, N. 1959: Review of *Verbal Behavior*, by B. F. Skinner. *Language*, 35, 26–58.

*Gallistel, C. R. 1992: *Animal Cognition*, Cambridge, Mass.: MIT Press.

Harnad, S. 1990: The symbol grounding problem. *Physica D*, 42, 335–46.

Nachtigall, P. E. 1980: Odontocete echolocation performance on object size, shape, and material. In R. G. Busnel and J. F. Fish (eds), *Animal Sonar Systems*, New York: Plenum Press, 71–95.

Nagel, T. 1974: What is it like to be a bat? *Philosophical Review*, 83, 435–57.

*Roitblat, H. L. 1987: *Introduction to Comparative Cognition*. New York: W. H. Freeman.

*Roitblat, H. L. and Meyer, J.-A. (eds) 1995: *Comparative Approaches to Cognitive Science*. Cambridge, Mass.: MIT Press.

Roitblat, H. L., Herman, L. M. and Nachtigall, P. E. (eds) 1993: *Language and Communication: Comparative Perspectives*. Mahwah, NJ: Erlbaum.

Savage-Rumbaugh, S., Murphy, J., Sevcik, R. A., Brakke, K. E., Williams, S. L. and Rumbaugh, D. M. 1993: Language comprehension in Ape and Child. *Monographs of the Society for Research in Child Development*, 58, 1–221.

Swartz, K. B., Chen, S. and Terrace, H. S. 1991: Serial learning by rhesus monkeys: I. Acquisition and retention of multiple four-item lists. *Journal of Experimental Psychology: Animal Behavior Processes*, 17, 396–410.

Wright, A. A., Santiago, H. C., Sands, S. F., Kendrick, D. F. and Cook, R. G. 1985: Memory processing of serial lists by pigeons, monkeys, and people. *Science*, 229, 287–9.

3

Attention

A. H. C. VAN DER HEIJDEN

Introduction

The phenomena referred to by the term *attention* were not discovered by scientific psychology. They were discovered and described within philosophy and gently handed over to the emerging academic psychology of the nineteenth century. The main contributors and contributions to the delineation and construction of attention as an empirical phenomenon and a topic for theorizing were Aristotle, who noticed that not all that reaches the senses is clearly perceived; Augustine, who interpreted attention as an effort of the soul; Descartes, who distinguished active attention from passive attention; and Leibniz, who discussed the relation between attention, perception, and consciousness. (For further information on attention's philosophical past the reader is referred to Neumann's (in press) brief history of this philosophical past that is badly in need of an English translation.)

These features or aspects of the empirical concept of attention formed, in varying combinations, the topics of the two main approaches to attention that subsequently appeared in the two flourishing periods of experimental psychology: the approach of early experimental psychology, with its centers of gravity in Germany and the USA, and the approach of contemporary information processing psychology, with its centers of gravity in the USA and the UK. These two psychologies – separated in time by about 40 years of behaviorism – have to be clearly distinguished, because they are completely different and, in a sense, complement each other. The early experimental psychology was a *mind* psychology. Systematic introspection in well-controlled settings was its method, the descriptions of experiences formed its data, and empirical generalizations were its theoretical aim. The information processing psychology is a *behavior* psychology. The registration of aspects of behavior in well-controlled settings is its method, the quantitative summaries of observed behavior form its data, and the construction of a hypothetical information processor that can produce that behavior is its theoretical goal.

Nearly all ideas on attention in contemporary information processing psychology were initially shaped by work on auditory attention. Especially during the last 20 years, however, attention research has been concerned almost exclusively with visual PERCEPTION. Therefore, in this chapter, I will mainly confine myself to a description of attention in vision. Within this field of visual attention research I will deal only with the phenomena of covert attention: that is, phenomena that are observed in situations where overt eye movements are excluded. William James (1890/1950, p. 437) aptly described the phenomenology of this form of attending:

> It has been said . . . that we may attend to an object on the periphery of the visual field and yet not accommodate the eye for it. Teachers thus notice the acts of children in the school-room at whom they appear not to be looking . . . The object under these circumstances never becomes perfectly distinct – the place of its image on the retina makes distinctness impossible – but (as anyone can satisfy himself by trying) we become more vividly conscious of it than we were before the effort was made.

In the next three sections I treat the main features of the empirical concept of attention under the headings "Limited capacity and unlimited capacity," "Active attending and passive attending," and "Early selection and late selection." Each section starts with a brief characterization of the pertinent empirical phenomena by means of an example taken from early experimental psychology and continues with a description of the further elaboration of, and the theoretical state of affairs with regard to, that feature of attention in contemporary information processing psychology. In the last section I present a principled summary.

Limited capacity and unlimited capacity

Something like Aristotle's observation that not all that reaches the senses is clearly perceived was often confirmed in early experimental psychology. Hermann Helmholtz, for instance, was very explicit about this observation and its relation to attention. He describes experiments starting with an observer sitting in a dark room, looking at a small, slightly illuminated pinhole in a page with large printed letters. Then the page was illuminated by a flash of light. The flash was of such short duration that the observer only saw the letters as an afterimage. Under these exposure conditions, directed eye movements were not possible. Helmholtz found that he was able to decide in advance which part of the field was going to be perceived; in that region of the field the letters were recognized, while letters in other parts of the field, even in the vicinity of the fixation point, were not perceived. He describes how, in subsequent trials, he could direct his perception to different regions of the visual field while his eyes always fixated the central pinhole. According to Helmholtz, these observations demonstrate that one can concentrate attention through a *voluntary kind of intention* or *conscious and voluntary effort* on any part of a dark field of view and at the same time exclude attention from all other parts. The part attended to is then perceived, and the part not attended to is not perceived.

At the start of information processing psychology in the 1950s, Aristotle's limits on perception and something like Helmholtz's absence of perception in the absence of attention were abundantly confirmed by behavioral experiments. A basic observation was that people show all kinds of performance limitations: that is, they cannot do everything and certainly cannot do everything at the same time. Because the theoretical aim of cognitive science in general, and of information processing psychology in particular, is to explain observed phenomena in terms of internal structure and function of a hypothetical information processing system, a model was required that explained this phenomenon. Donald Broadbent (1958) presented an information processing model that accounted for the performance limitations observed in auditory information processing tasks. Central to that model is the assumption that the nervous system can be regarded as a single communication channel with a limited capacity.

So, in Broadbent's model, the observed behavioral limitations are explained in terms of postulated internal limitations. In their influential works, Daniel Kahneman (1973) and Ulric Neisser (1967) presented highly similar variants of this central limited capacity view for auditory and visual information processing.

Today this limited central capacity assumption is generally accepted. Nearly all (visual) information processing theories are two-stage, limited capacity – limited capacity theories. They postulate an initial stage of information processing in which all the information available is processed, but only partly. That stage is followed by a stage in which only part of the information is further processed, but that part completely. The theories do not completely agree about the amount and kind of processing that goes on in the first stage and the amount and kind of work that is consequently left for the second stage. The theories completely agree, however, that the first stage is *preattentive*, and that only in the second stage is attention involved in the processing of the information. The operations performed in the first stage are often characterized by terms like *automatic, parallel*, and, incorrectly, *unlimited*. The operations performed in the second stage are often characterized by terms like *controlled, serial*, and *limited*.

Nevertheless, despite this almost universal acceptance of the limited central capacity assumption, some theorists have pointed out that an explanation of observed limits in terms of internal limits is a *virtus dormitiva* explanation – an explanation of the sort ridiculed by Molière with a candidate doctor who explained the soporific working of opium in terms of its power to put one asleep (see especially Neumann, in press). These theorists have further pointed out that, to begin with, it is important to distinguish between auditory information processing and visual information processing (see Neumann et al., 1986). They agree that for auditory information processing something like limited central capacity might be real, and therefore a valid explanatory feature in the information processing model. For visual information processing, however, they doubt the validity and hence the explanatory power of this assumption. They point out that there is virtually no independent behavioral or other (e.g., neurobiological) evidence indicating that the visual information processing system suffers from this kind of capacity limitation. They emphasize that the visual information processing system is a massive, highly diverging, complexly connected, parallel operating, modular system that has to function in consistent goal-directed internal and external vision-based actions (see Allport, 1987). To perform these actions, selection of visual information is required; to prevent behavioral chaos, the information processing system must limit itself in its use of information.

These starting points lead to a radically different point of view with regard to the capacity issue. In the dominant models there is attentional selection, because there is a limited central capacity. In the theoretical alternatives there are observable external limitations, because there is internal attentional selection. In other words, in the alternatives, the widely held assumption of a limited central capacity is simply irrelevant.

Active attending and passive attending

Something like Descartes' distinction between passive attention and active attention – Augustine's "attention as an effort of the soul" – reappeared in early experimental psychology in various forms. James (1890/1950), for instance, was explicit in this regard, and this for an obvious reason. Like Helmholtz, he recognized a form of attention that is guided by *a voluntary intention* or *a voluntary effort*, but for him that attention was not an internal force, but some kind of internal *ideational preparation, anticipatory imagination*, or *preperception* – in general, a form of *thought*. With only this form of attending, however – a form of attending through which, according to James, each of us *chooses* the world in which she/he lives – the external world has limited opportunity to make itself known. James therefore recognized a second form of attention: passive, immediate sensory attention. This form of attention can be captured by very intense, voluminous, or sudden sense impressions, and then the nature of the impression is irrelevant. It can also be captured by instinctive stimuli that appeal to some one of our normal congenital impulses; but then, of course, the nature of the stimulus is what matters.

At the start of information processing psychology, Descartes' admiration or James's passive immediate sensory attention was not a topic of active investigation. This information processing psychology was primarily interested in determining how subjects process information, not how information guides subjects. The construction of an active hypothetical information processor was the goal. Only during the last ten years has research returned to *involuntary* attention.

With regard to active or *voluntary* attention, nearly all theorists assume that the preattentive stage of information processing delivers a crude, global picture of the world and that *the subject* directs his attention to an *object, element*, or *region* in that world, and that attention then finishes the perceptual work. In most theories it is assumed that *the subject* directs *his/her attention* at a single, contiguous, geometrically defined region of visual space – e.g., a circular or elliptical region of greater or lesser size. To characterize the spatial and temporal properties of attention, advocates of these views use metaphors such as *spotlights* that move and *zoom lenses* that change the field of view. What attention does exactly is mostly left unspecified; attention simply *processes* the information. The theoretical construct *the subject*, which does most of the theoretical work in these theories of active, voluntary, attention, is almost never seen as problematic.

In contrast to most theories, one influential line of theorizing, initiated by Anne Treisman, has something to say about the kind of *processing* that attention does (see Treisman, 1988, for an overview). In this *feature integration theory* it is assumed that in the first, preattentive, stage of information processing, sensory features such as color, shape, and position are abstracted and represented in the information processing system within separate specialized modules: a shape module, a color module, and a position module. In the second, attentive stage, a serial spotlight of attention is directed at a position in the position module. This directing of attention to one of the positions results in separated features, present in the attended location, being made available for further perceptual processing and being combined into a representation of the attended object. So attention serves to integrate the initially separated features. In this theory, the control of attention is still a highly problematic issue: In this general framework for perceptual processing in object perception, attention comes from nowhere and is controlled by nothing. Attention is not directed *by* a location but is called *to* a location and has to *home* in on that location. In other words, somewhere hidden in this theory of active attention there is still *the subject* as a theoretical construct that does most of the difficult explanatory work; *the subject* actively directs attention.

Recently a number of attempts have been made to replace *the subject* as a component in the information processing theories and to account for the active, voluntary control of attention in a less dualistic way (see Duncan, 1996). In these attempts the obsession to answer the question "At what is attention directed?" is suppressed in favor of the question "Where does attention come from?" Just as in Treisman's theory, these proposals start with the important constraining information, provided by neurobiology, that the visual information processing system is a modular information processing system (see Article 49, MODULARITY). It is assumed that within each module a gain in activation of the representation of (an attribute of) one object is bought at a loss of activation of the representations of (that attribute of) all other objects. Such a gain in activation in one module is transmitted through orderly connections from one module to corresponding positions in all other modules, so that there too the representation of (another attribute of) the object gains activation and suppresses other activations.

In such a highly connected modular system, attention consists of additional activation stemming from task-dependent external or internal advance priming of units representing task-relevant attributes in task-relevant modules. That attention is directed by internal activations which stand for the information processor's goals and intentions. In such a conceptualization there is no principled distinction between active and passive attending; these forms of attending differ only in size and in complexity of the set of connected modules involved.

Early selection and late selection

The elaboration of the relation between attention, perception, and consciousness proposed by Leibniz continued in early experimental psychology. We have already seen that Helmholtz held that what is attended to is perceived and what is not attended to is not perceived. Wundt and Titchener, however, saw this differently. For them the difference between what is attended to and what is not was not that absolute. For Wilhelm Wundt, the processes within the range of consciousness consisted of two sets: those within the field of consciousness (*Blickfeld*) and those within the focus of consciousness (*Blickpunkt*). The processes in the *Blickpunkt* were the attended processes seen against a background of unattended processes in the *Blickfeld*. In Titchener's psychology of observables, attention was that which changes in sensory experience when attention is said to change. In his view, a change in attention consisted in a change of degree of clearness of the sensory processes. So, in his view, sensory processes had, besides the attributes of quality and intensity, the attribute of clearness. Later Titchener used the terms *vividness* and *attensity* for this attribute.

At the start of information processing psychology, when most research was still concerned with auditory information processing, the relation between attention and perception became a central topic. Basically there were two points of view: the early selection view and the late selection view. After Broadbent (1958), proponents of the early selection view held that the information processing system has a limited central capacity for processing information and that a selection mechanism, early in a linear stream of information processing, had to protect this system against overload. In this view, attentional selection precedes, and is required for, perception and recognition. After Deutsch and Deutsch (1963), proponents of the late selection view rejected the assumption that the information processing system suffers from capacity limitations and suggested that a selection mechanism, late in a linear stream of information processing, selects parts of the processed information for task performance or storage in memory. In this view, attentional selection follows perception and recognition and can be based on significance and meaning.

Today, it is generally accepted that the early selection view is the correct one. This conviction that in vision selection is early is mainly based, however, upon irrelevant evidence. The conviction should be based on evidence regarding selection, and, indeed, there is abundant evidence that attentional selection in vision intervenes at a relatively early stage of information processing: that is, in vision, and not in some kind of semantic memory. Selection on the basis of visual attributes – for example, color or position – is easy, and selection on the basis of meaning – for example, letter or digit – is difficult. But the conviction is mainly based on the questionable a priori theoretical assumption of limited central capacity. The generally accepted reasoning is that the information processing system suffers central capacity limitations; therefore, proper operation requires early selection. There are at least two problems with this line of argument. The first is that there is a flaw in the reasoning. While limited central capacity requires early selection, early selection does not require limited central capacity. And early selection is the given; limited central capacity is only an a priori assumption. The second problem is, of course, that this a priori theoretical stand prevents serious investigation of an alternative theoretical possibility: namely, of early selection for external or internal action in a system with unlimited capacity. This is certainly a viable alternative. In a modular network with spreading activation, selection effected through enhanced activation in no sense logically entails rejection or exclusion of nonselected information for further processing (see Allport, 1987).

One possible reason for maintaining the limited central capacity assumption is the widespread conviction that only the combination of assumptions of limited central capacity and early selection can account for subjective conscious experience in visual perception. Unfortunately, neither the limited central capacity assumption in general, nor the more specific two-stage theories derived from that assumption in particular, clarify very much in this regard. A critical evaluation of the limited capacity, two-stage literature reveals that within information processing psychology the descriptions of visual perception as a subjective experience are extremely shallow, and the explanations wholly inadequate. With the first, preattentive stage of processing, nothing is explained, because the required mental phenomena are either introduced through a simple identification or borrowed from an alien approach. With the second, attentive stage, nothing is explained, either because there is no need for any further explanation or because what has to be explained in fact remains fully unexplained (see Van der Heijden, 1996).

In short, information processing psychology has not produced the concepts that would allow us to explain visual perception as a subjective conscious experience. Of course, such an explanation should not be expected from an approach that is based purely on behavior.

A constructive overview

The approach of information processing psychology and the state of the art with regard to the study of (visual) attention can be summarized as follows:

- The aim of information processing psychology, concerned with perception, cognition, and attention, is to infer aspects of the internal structure and functioning of a behaving organism from the overt behavior of that organism, so that it becomes possible to explain the organism's behavior in terms of its internal structure and functioning. The overt behavior is observed in laboratory experiments. The inferred structure and functioning are expressed in terms of a model of a hypothetical information processor.
- In the theoretical endeavor of information processing psychology, constraining information about the structure and functioning of the central nervous system is of fundamental importance. The neurobiological evidence with regard to structure, which indicates that the (human) information processing system is a complex, highly connected, modular system, with different components performing different specialized tasks, is such a piece of constraining information. The psychological evidence with regard to function, which indicates that the control of attention through goals and intentions requires massive interactions between modules, is another piece of constraining information.
- In theorizing about information processing and the role of (selective) attention therein, a clear distinction has to be made between visual, auditory, and other kinds of information processing. Just because different kinds of information, different peripheral senses, and different devoted central information processing components are involved, there is no justification whatsoever for starting with the assumption that these different modalities exhibit the same or even similar modes of processing and selecting operations.
- Neither the behavioral evidence obtained in information processing experiments nor other complementary (e.g., neurobiological) evidence indicates that it is theoretically fruitful to regard the visual information processing system as a system with severe central capacity limitations. The visual information processing system is a massive, highly diverging, complexly connected, parallel operating, modular system, and there is no reason whatsoever to assume that it is not perfectly capable of processing all the information that is delivered by the eyes.
- Selective attention is best regarded as an additional activation process in regions of greater or lesser size in subsets of modules and corresponding connections between modules in the visual information processing system. That activation interacts with other activations in the modular information processing system – for example, with peripherally originating activations as a result of visual stimulation and with centrally originating activations representing goals, intentions, and expectations.
- Selective attention in vision performs early selection in the sense that it consists of additional activation in modules concerned with aspects of visual perception, not in higher-order modules concerned with meaning and significance or with goals and intentions. It is highly likely that primarily the modules concerned with the analysis of three-dimensional spatial properties of the visual world are essential in generating and regulating this additional activation, and thereby produce the spatiotemporal *spotlight* or *zoom lens*-like properties reported in the literature.
- There is passive, sensory (bottom-up) selective attention when events in the world (e.g., sudden onsets or changes) are solely responsible for the spatiotemporal pattern of additional attentional activation in the visual system. There is active, *voluntary* (top-down) selective attention when higher-order modules with patterns of

activation representing expectations, goals, and/or intentions codetermine the spatiotemporal pattern of additional activation in the visual system.

- This whole visual information processing system with its interacting activations in action can be identified with visual perception as a subjective experience in action. When information processing psychology, a behavior-based psychology, claims that it sheds more light on visual perception as a conscious subjective experience, it simply boasts.

References and recommended reading

Allport, D. A. 1987: Selection for action: some behavioral and neurophysiological considerations of attention and action. In H. Heuer and A. F. Sanders (eds), *Perspectives on Perception and Action*, Hillsdale, N. J.: Erlbaum, 395–419.

—— 1989: Visual attention. In M. I. Posner (ed.), *Foundations of Cognitive Science*, Cambridge, Mass.: MIT Press, 631–82.

Broadbent, D. E. 1958: *Perception and Communication*. London: Pergamon Press.

Deutsch, J. A. and Deutsch, D. 1963: Attention: some theoretical considerations. *Psychological Review*, 70, 80–90.

Duncan, J. 1996: Coordinated brain systems in selective perception and action. In T. Inni and J. L. McClelland (eds), *Attention and Performance XVI*, Cambridge, Mass.: MIT Press, 549–78.

James, W. 1890/1950: *The Principles of Psychology*, Vol. 1. Authorized edn, New York: Dover Publications, Inc.

Kahneman, D. 1973: *Attention and Effort*. Englewood Cliffs, NJ: Prentice-Hall.

Neisser, U. 1967: *Cognitive Psychology*. Englewood Cliffs, NJ: Prentice-Hall.

Neumann, O., in press: *Konzepte der Aufmerksamkeit. Bielefeld University*. Göttingen: Hogrefe Verlag.

*Neumann, O. and Sanders, A. F. 1986: *Handbook of Perception and Action*. Vol. 3: *Attention*. London: Academic Press.

Neumann, O., Van der Heijden, A. H. C. and Allport, D. A. 1986: Visual selective attention: introductory remarks. *Psychological Research*, 48, 185–8.

*Ten Hoopen, G. 1996: Auditory attention. In Neumann and Sanders (eds), 79–112.

Treisman, A. M. 1988: Features and objects: the fourteenth Bartlett memorial lecture. *Quarterly Journal of Experimental Psychology*, 40A, 201–37.

*Van der Heijden, A. H. C. 1992: *Selective Attention in Vision*. London and New York: Routledge.

—— 1996: Selective attention as a computational function. In A. F. Kramer, M. G. H. Coles, and G. D. Logan (eds), *Converging Operations in the Study of Visual Selective Attention*, Washington, DC: American Psychological Association, 459–82.

4

Brain mapping

JENNIFER MUNDALE

Introduction

One important way in which neuroscience, particularly neuroanatomy, contributes to cognitive science is by providing a model of the brain's architecture, which, in turn, can be utilized as a guide to the architecture of cognition. This project assumes commitment to a view, now well established, that different mental processes, such as perceiving and remembering, employ different parts of the brain (where *part* is loosely construed so as not to exclude entities which may themselves be composite). These parts may differ according to their internal organizational structure, information processing capacities, and degree of informational connectivity with other parts. Later I will present an example of research in visual processing where researchers are currently attempting to realize the heuristic potential of the brain's architecture in order to understand cognitive organization. Of prior importance, however, is an understanding of how neuroscientists decipher the architecture of the brain itself. This task of mapping the brain, of *carving it up* into distinct areas, is not as straightforward as one might think. There are, for example, several different criteria by which one might fix a given boundary between brain regions. Below, I not only highlight some of the more interesting facets of brain mapping research but also elaborate on why such research is of interest to cognitive scientists.

In order to better understand the nature and complexity of brain cartography, consider an example from another type of cartography – delineating the various *parts* of a country. First, with regard to the borders of the country itself, in some instances gross natural landmarks make for obvious boundary decisions. In the case of the United States, for example, the Atlantic and Pacific Oceans demarcate the eastern and western boundaries; but with respect to its northern and southern boundaries, natural landmarks play a relatively minor role. Thus the Great Lakes divide Canada from the USA for only a small fraction of the border, and even then, portions of some lakes are shared by both countries. Most of the northern border determinations were the result of factors of a quite different kind, including war, treaty negotiation, and settlement patterns of Native Americans, English, and French.

There is also division into states, cities, counties, etc., where natural boundaries may or may not play an important role (for that matter, the decision to divide at the state level is itself an active choice, largely driven perhaps by practical, political utility, though not naturally dictated by any inherent properties of the larger whole). The eastern states of the USA, for example, have very irregular borders when compared with most of the western states, because the western states were affected by the Northwest Ordinance, which imposed a grid pattern of survey lines, one square mile sector at a time. These survey lines are a more abstract, less obvious means of dividing territory and require rather sophisticated instruments to fix their location. Even with the grid in place, however, it often turned out to be more convenient to follow natural landmarks rather than grid lines. The Mississippi River, for example, which follows a curved course, forms either the eastern or western border for all the states it runs through, even though the resulting borders cross-cut the survey grid. Islands are also a clear example of areas whose peripheral boundaries, at least, are entirely determined by one gross natural landmark – water. On the other hand, local population patterns (human and otherwise) may also play a role in establishing boundaries. Certainly, there are countless factors which figure in the establishment of the complete set of boundaries, throughout all levels from global to local, and some divide the territory differently from others.

In the brain too, there are multiple means by which a given region may be carved out from the rest and set aside as distinct. As in the case of countries, some of the boundaries are natural and obvious, such as the interhemispheric fissure, which, as its name implies, divides the two hemispheres of the brain, much as the Mississippi River divides the United States into two large regions. The cerebellum is like a peninsula, physically separate from the rest of the brain except at its points of attachment, or peduncles. The metaphor only takes us so far, of course; but in a preliminary way, it helps to illustrate many important aspects of delineating boundaries in the brain.

Before discussing modern technologies and techniques of brain mapping, an excursion into its historical evolution will afford us a better understanding of its nature and importance to cognitive science. Schemes for dividing the brain into various regions of special significance appeared concomitantly with coming to view the brain as the locus of mentation. Prior to this, other candidates, such as the heart and ventricles of the brain (open cavities within the brain), were also divided into various specialized regions (see Finger, 1994). This long, conceptually tortuous story, however, is not of immediate concern. From the modern standpoint, it is more informative to begin the historical narrative with Franz Josef Gall (1757–1828). Widely credited with putting into scientific play the concept of the cortical (referring to the surface of the brain) localization of function, his reputation as a charlatan is not entirely deserved.

Franz Josef Gall

Gall's theory of organology, or *phrenology* (a term introduced by his colleague Spurzheim), was that the cortex was divided into discrete regional *organs*, each with specific functions and capacities (see figure 4.1). Gall thought, erroneously, that the size of a given brain region corresponded directly to the degree to which a given person possessed the function with which it was associated. Further, he thought that the skull conformed exactly to the cortex, and thus that any regional enlargements of the underlying brain would be detectable indirectly as bumps on the person's skull. The functions which Gall thought himself to have localized are sometimes quite whimsical (e.g., the sense of metaphysics, comparative sagacity, the feeling of property), and this doubtless detracted from his credibility. Others, however, have a quite contemporary ring to them (e.g., the sense of language, memory of things, memory of words, and sense of colors).

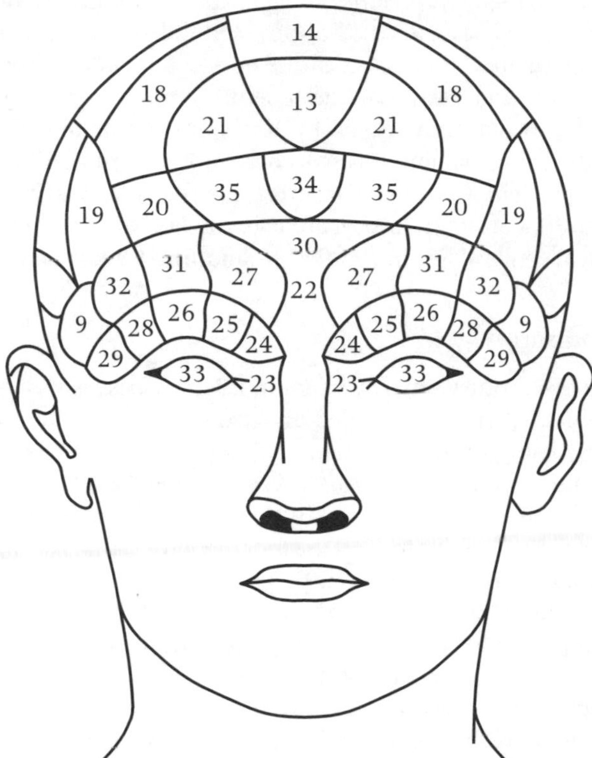

Figure 4.1 Phrenological map of the human skull. Each region shown on the skull localizes a distinct mental function to that part of the brain directly underlying it. Gall thought that an enlargement of a given brain region signified an increased capacity for whatever function that particular part of the brain was supposed to carry out, and further, that the skull conformed directly to the contour of the brain, thus allowing him to *read* someone's mental makeup by feeling the bumps on their skull. From G. Spurzheim, *Phrenology or the Doctrine of the Human Mind*, 3rd American edn (Boston: Marsh, Capen and Lyon, 1834).

For all his errors, Gall's impact on the scientific community was enormous: both negatively, in terms of the reactions against him, and positively, for the conceptual advance his work represented. On the negative side, when Gall fell into disrepute, he temporarily took localization theory down with him. The controversy ignited the debate between holists and localizationists which continues to smolder to the present day. At the time, the major factor behind Gall's decline was the anti-phrenological campaign of Marie-Jean-Pierre Flourens, who opposed Gall's theory both on holistic grounds and because he was concerned to defend such Cartesian tenets as the immateriality of the soul and the unity of mind. Localizationist thinking did not begin to appear scientifically viable again until Paul Broca produced compelling evidence, based on a case of brain damage, of the localization of a specific language area in 1861, and Gustav Fritsch and Eduard Hitzig in 1870 induced muscular movement through the stimulation of cortex with electric current.

On the positive side, many historians of science have commented on the theoretical fecundity of phrenology's basic tenet, that physical differentiation denotes functional differentiation. In particular, there is a definite intellectual link between Gall's concept of cortical localization and early twentieth-century brain mapping research. Gall's contribution is explicitly acknowledged by Korbinian Brodmann (1868–1918), for example, one of the most influential researchers of early modern brain cartography. He writes: "I look upon it as a pious duty to stress that the much maligned and . . . highly meritorious research neuroanatomist, Gall, was the first to introduce a practical system of physiological cerebral localization" (Brodmann, 1994, p. 250).

Korbinian Brodmann

In the late nineteenth century, improvement in microscopes, as well as techniques for staining neural tissue, permitted detailed analyses of cortical structure. Early in the twentieth century, using these and other techniques, several researchers produced detailed maps of the cerebral cortex. Alfred Walter Campbell, Oskar and Cécile Vogt, Constantin von Economo, and Gerhardt von Bonin all produced candidate maps of the cortex, but the one which has endured as the touchstone for all subsequent parcellations of the cortex was that of Brodmann (1909).

The guiding assumption of Brodmann's parcellation of the cortex was that areas of the brain which appeared to be physically distinct from surrounding tissue were likely to be functionally distinct as well. He writes, for example: "It is a basic biological principle that the function of an organ is correlated with its elementary histological structure" (1994, p. 243). Here, he is making the additional claim that not only is functional differentiation correlated with physical differentiation, but the level of *histological*, or *cellular*, differentia is significant in this regard.

The delineations in Brodmann's map (figure 4.2) are based primarily on regional differences in the types of cells and their density and distribution within a given area of cortex, otherwise known as the cytoarchitectonics (or cytoarchitecture) of a region. One particularly clear example of how Brodmann employed cytoarchitectonic criteria was in establishing the boundary between areas 3 and 4, which are recognized as the primary somatosensory area and the primary motor area, respectively. Brodmann observed a striking difference between these two areas with respect to the appearance of their cortical layers (figure 4.3), particularly the fourth layer. Today we recognize the fourth layer as a heavy input layer, so it is not surprising that one should see such a marked transition between the primary sensory cortex, where it is large, and the primary motor cortex, where it is thin to nondiscernible. These areas lie right next to each other; yet, at the cytoarchitectonic level, in appropriately stained and prepared tissues, one can easily discern a clear boundary between them.

Unfortunately, Brodmann was not always specific about how he applied such variations in cell types and densities, and he is often charged with making highly subjective judgments about cytoarchitectonic boundaries. His staunchest critic, who opposed not only Brodmann, but most attempts at the localization of function (cytoarchitectonically based or otherwise), was Karl Lashley (1890–1958). In 1946 he, along with George Clark, published a lengthy study of cytoarchitectonic methods. As part of the study, each of them undertook to generate his own cortical map of the spider monkey (*Ateles geoffroyi*), on the basis of work with a different, single specimen. While there were

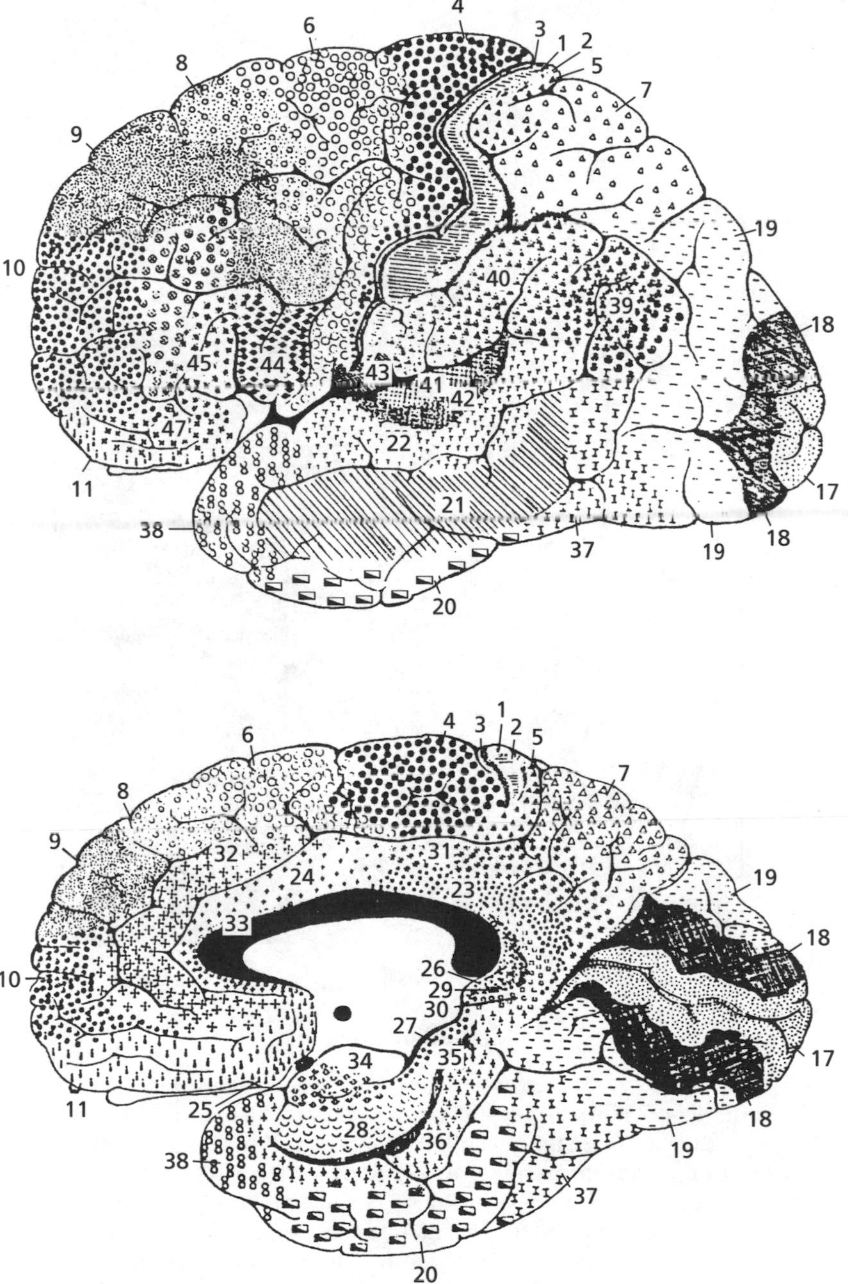

Figure 4.2 Brodmann's (1909) cytoarchitectonic map of the human cortex. Relying primarily on regional differences in cell type, density, and distribution – a method broadly referred to as *cytoarchitectonics* – Brodmann identified over forty distinct areas of human cortex. As he was among the first to recognize, cytoarchitectonic boundaries tend to demarcate functional boundaries as well, so the map, still in use today, was intended to be a guide to the localization of specific functional regions across the cortex.

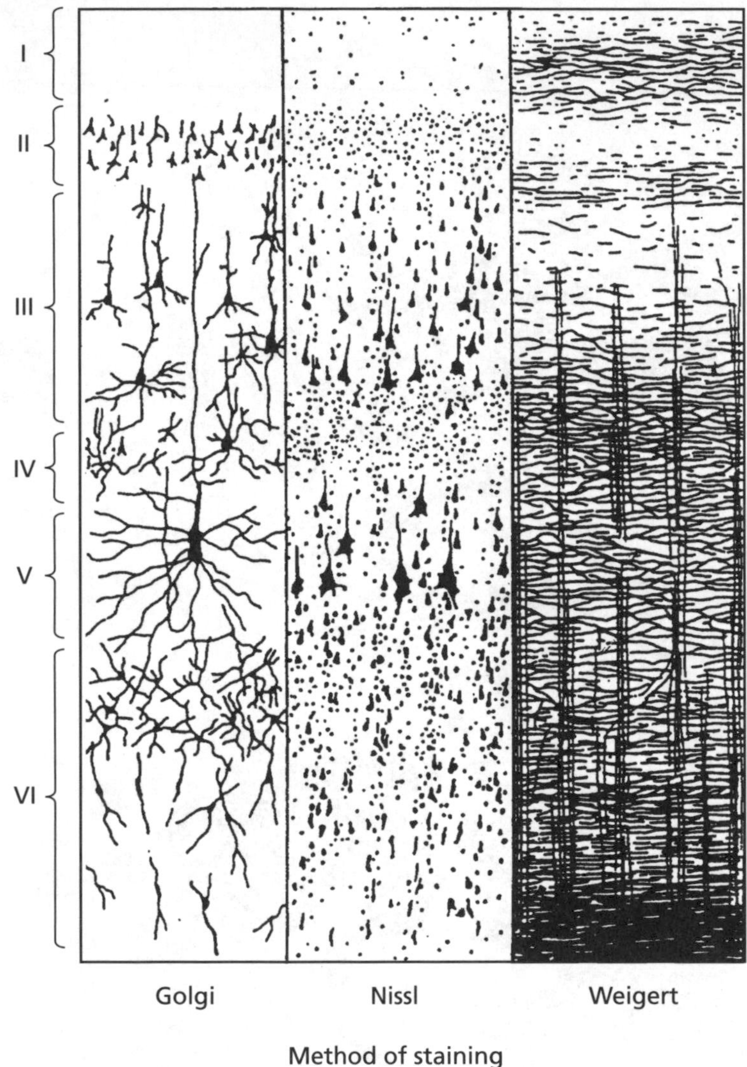

Figure 4.3 Brodmann's (1910) representation of the laminar structure of the cortex, differentially stained. This figure shows the layered, or laminar, structure of the cortex, as revealed by three different staining methods. Based partly on the work of Brodmann and others, it is generally accepted today that the human cortex, as well as the cortex of several other species, has six separate layers (a number once hotly disputed). In moving from one Brodmann area to another, one observes cytoarchitectonic changes in these layers. A given layer, for example, may be thinner from one area to the next, or have slightly different cellular densities, or populations. This figure also shows how what one observes in a given area of cortex depends partly on how the tissue has been prepared, and how different methods may produce different results.

similarities between their cortical maps, there were significant differences as well. In attempting to account for these differences, they discovered that each had employed different delineating criteria. They also discovered significant differences between the brains each had studied, even though the two specimens were from organisms of the same species. Moreover, they noted that each had set out to identify areas found in the published maps of the brains of other primates, rather than approaching the task *de novo*. This should have resulted in greater agreement; however, their failure to generate such agreement prompted the charge that the cortical maps of Brodmann and others were essentially useless, highly unreliable indicators of functional division.

While some of Lashley's criticisms concerning the lack of critical analysis and standardization of cytoarchitectonic methods continue to be regarded as serious (see e.g., Carmichael and Price, 1994), and researchers now recognize that there are differences between brains of members of the same species, subsequent research has tended overall to confirm the legitimacy of Brodmann's work in demarcating different cortical regions. His original map has undergone some minor revisions, many involving the further subdivision of his areas. Overall, though, his map remains the standard by which researchers identify brain regions.

Modern brain cartography

Today, methods similar to those employed by Brodmann continue to play an important role, but they have been supplemented by an additional array of powerful tools and sophisticated methods for demarcating parts of the brain. In addition to *cyto*architectonics (the common term for the types of tools Brodmann used), *chemo*architectonics, for example, serves to mark differential patterns of particular chemical substances in brain tissue. This method may reveal the presence of specific enzymes, neurotransmitters, or metabolic by-products in different parts of the brain; since these substances all play a role in how the brain functions, differential distributions of them may indicate functional divisions of a very specialized nature.

Connectivity analysis – examination of how a region is neurally connected to other regions of the brain – is presently one of the most heavily weighted criteria in distinguishing parts of the brain. Basic neuronal tracing methods were developed as early as 1850, by Augustus Waller, but present methods are vastly improved. Aided by the electron microscope and by modern dyes and tracing agents, researchers can plot entire neuronal pathways. By injecting specialized tracers in a given region, researchers can visualize both the incoming pathways to a region as well as the outgoing pathways. Knowing the neuronal connections between different regions also helps to establish the flow of information through the brain. If two nearby areas send projections in different directions, that is a good indication that one is dealing with two functionally distinct areas.

Topographic considerations also enter into delineating brain areas. Topography refers to the way some parts of the brain contain maps or congruent representations of corresponding sensory regions (see figure I.4 on p. 29). The visual cortex (Brodmann's area 17), for example, has a map of the entire visual field. The work of Tootel and others shows that neural activation patterns in this early visual processing area in the macaque, correspond, point for point, to the shape of the object in the macaque's visual field (Tootel et al., 1988). These activation patterns are so congruent with the object being viewed that, based on these patterns alone, researchers can accurately tell which object the monkey is observing. Similarly, the motor cortex (Brodmann's area 4) and somatosensory cortex (Brodmann's areas 1, 2, and 3) also contain maps of the body, with point-for-point correspondences for each bodily region represented (see e.g., Kandel et al., 1991, ch. 26). The fact that over a region of cortex a representation of a whole part of the body or of the world is laid out provides a good reason for thinking that that region of cortex constitutes one functional area. In addition to these factors, there are also other purely structural features, including broader anatomical or organizational characteristics of a region, such as the strikingly distinctive appearance of the layers in the visual cortex, that researchers may appeal to in decisions delimiting a region of the brain.

The methods described above fall primarily within the province of neuroanatomy, and they have generally played a primary role in brain mapping. But anatomy does not directly reveal the physiological processes occurring in a brain region, and, since Gall, a wide range of more physiological techniques have evolved. They are sometimes thought to provide a more direct indication of boundaries of regions in the brain that are relevant to function. At the micro level, for example, SINGLE NEURON ELECTROPHYSIOLOGY can reveal the sorts of stimuli to which a given cell responds. When used in such applications as mapping the visual system, recording from a single cell in the visual pathway can indicate something about the receptive field properties of that cell, or what sorts of visually presented stimuli will activate (or deactivate) the cell. When researchers collect this information over a whole region of cells, they are able to mark out the boundaries of regions in which cells will respond to a particular kind of information (e.g., about color or motion).

At a grosser level, clinical cases have often provided clues about the functional significance of a particular region. One of the more famous cases is Paul Broca's aphasic patient Leborgne, otherwise known as "Tan" for the one utterance he was able to make. By the time Broca was able to conduct an autopsy, much of Tan's frontal cortex had been destroyed, but in trying to determine the likely origin of the destruction, he argued that a region in the left third front lobe was the locus of articulate speech. Phineas Gage is another well-known example of a clinical case in which a person's mental deficits were correlated post mortem with a lesion in a certain region of the brain. In this case, the locus of damage was in the ventral medial prefrontal cortex, and theorists such as Damasio (1994) have attempted to identify a specific area where rewards are linked with plans for action. Sometimes, as in the patient known as "EVR," in whom surgical excision of part of his brain (to alleviate epileptic seizures) resulted in profound anterograde amnesia, the location of the lesion is known prior to postmortem examination. In all these cases of DEFICITS AND PATHOLOGIES, investigators have tried to use the extent of an area in which a particular deficit arises to determine a specialized area of the brain. One challenge faced by this approach, however, is that damage in human patients often occurs over a range of cortical regions which, while in the general vicinity of what is taken to be Broca's area, nonetheless results in a variety of deficits. Thus, there is a considerable amount of variability in the literature as to the precise locus of Broca's region.

More recently, NEUROIMAGING tools such as PET and functional MRI (fMRI) have been developed, which allow researchers to correlate specific tasks with distinct patterns of cortical activation in the brains of subjects actually performing a cognitive task. These correlations, combined with sophisticated subtraction techniques, contribute compelling information about functional, regional specificity. At this stage of development, however, the spatial resolution of these techniques is not yet sufficient to enable researchers to draw sharp boundaries around functionally relevant areas. When combined with evidence from the other methods described above, however, this liability is minimized, and the combination of techniques can make for very robust confirmations of the functional specificity of an area. Other difficulties in neuroimaging involve variability in brain size across test subjects. This difficulty is magnified by the fact that images of activation patterns are usually averaged across several subjects in order to minimize image *noise*. In order to accommodate this variation, there have been some attempts at standardization through the use of numerical algorithms. Recently, it has become common practice to adopt the Talairach–Tournoux grid, or coordinate system. This makes it possible to refer to activation patterns in terms of numerical coordinates, rather than, for example, Brodmann areas, and thus, to normalize data from several individuals to a standard scale.

Functional methods, architectonics, connectivity analysis, topography, and structural considerations are only some of the more commonly employed means of delineating brain areas. Together they are popularly known in the neuroscientific community by the acronym FACTS. One example of how these techniques are applied is in the work of Felleman and David van Essen (1991) in mapping the visual system of the macaque monkey. By first developing a procedure for flattening the monkey's cortex and then applying several of the methods described above, they have developed a detailed map of 32 distinct visual processing areas. They relied particularly heavily on connectivity analysis, architectonics, topographic organization, and functional considerations (such as the receptivity of a cell to specific kinds of visual stimuli).

Although this powerful array of cartographic tools can be useful in providing multiple means of confirmation, this variety itself presents problems. They are not equally applicable to every situation and, even when applicable, may produce indeterminate results, or in some cases, even incompatible results. This forces researchers to make decisions as to what criteria they will employ in a particular case, and how they will respond if there is incompatibility between results obtained with different tools. This can be recognized in the work of Felleman and van Essen, for there was not full convergence among the defining criteria designated by the acronym FACTS. That is, with respect to the 32 areas they identified, most were picked out by only two or three of the five available techniques. In other brain mapping research, one can find instances where different criteria seem to give different boundaries, and researchers have had to make decisions as to which criteria should be trusted in a given case.

Felleman and van Essen's research told us something not only about how many different visual processing areas there are, but also about how interconnected they are. Among the 32 areas they identified, there were 305 neuronal connections, most of which were reciprocal: that is, if there are sets of neurons which carry information from A to B, there are also neurons which carry information from B to A. Such reciprocity seems to challenge a simple feed-forward view of cognitive processing (see Article 53, STAGE THEORIES REFUTED). Moreover, the degree of interconnectivity, to the extent that it represents informational exchange, also challenges claims of informational encapsulation.

Applications to cognitive science

This research into brain cartography has implications for a number of broader issues in cognitive science. I will briefly discuss implications for the following issues: (1) organization and possible modularity of cognitive functions, (2) the relation between psychology and neuroscience, and (3) localization of psychological function.

As noted at the outset, brain mapping research, in the determination of both what the functional processing areas are and how they are connected relative to each other, has obvious potential significance for debates within cognitive science as to the nature of the mental processing architecture. One clear example concerns the MODULARITY hypothesis. Jerry Fodor (1983), for instance, characterizes modules as domain (content)-specific, innately specified, not assembled (directly mapped onto neural implementation), hard-wired, computationally autonomous, and informationally encapsulated. These features of the modularity hypothesis invite empirical evaluation, first for the identification and testing of putative cases of modularity and second, depending on these results, for the broader hypothesis itself. The program of mapping the brain would seem to fit well with a modular approach, since, as I have noted throughout, a common motivation for mapping the brain is to identify brain regions responsible for performing discrete functions. But the discovery of the degree of connectivity in the brain, especially the prevalence of collateral and feedback connections between brain regions, casts some doubt on whether processing in individual brain regions is as encapsulated as Fodor suggests. Paul Churchland (1992) draws attention to the prevalence of recurrent connections, for example, in his negative assessment of the modularity hypothesis.

The project of mapping the brain also has consequences for how we construe the relation between psychology and neuroscience (see Article 48, LEVELS OF EXPLANATION AND COGNITIVE ARCHITECTURES). First, since a major goal of mapping the brain is the delineation of distinct functional areas, and a subset of functions which the brain subserves are psychological functions, there has traditionally been interdisciplinary involvement in this research. Moreover, as I have noted, sometimes functional considerations figure in the very enterprise of determining what is a brain region. Conversely, the understanding of psychological functions has continually been illuminated by neuroscientific data, so the relationship between neuroscience and psychology, particularly with regard to taxonomic issues, has often been mutually informative. Thus, the project of mapping the brain seems incompatible with any proposal to isolate psychology and neuroscience but, rather, takes its place in the context of a strongly interactive inquiry.

Recognizing the interactive nature of the project of mapping the brain has a further consequence for the philosophical account of mental states known as "functionalism." Functionalism was proposed by philosophers such as Hilary Putnam as an alternative to proposals to identify mental states with brain states. Putnam and others argued that, since instances of the same mental state (e.g., hunger) can be physically realized by different brain states, and instances of the same neurological state can give rise to multiple mental states, mental states could not be identical with brain states. This, however, requires us to have some sense of just what counts as the *same* psychological and neuroscientific states. But, as I have shown, brain parts, let alone states, are not just given – they are identified through a variety of criteria that are still evolving. Until we have a fuller understanding of the functional and physical boundaries within the brain (and the working taxonomies which will evolve therefrom), arguments premised upon the multiple realizability of mental states will retain a *smoke and mirrors* quality about them. As brain cartography progresses, it will certainly help to mediate such discussions. But, given the use of functional, including psychological, criteria in demarcating brain parts, there is reason to be pessimistic about the strength and importance of the conclusions that multiple realizability can be made to support (see also Mundale, 1997).

Lastly, as noted throughout my discussion, brain mapping endeavors are closely linked to the debate over whether psychological functions are localized in the brain. The prime motivation for identifying brain parts is to localize different functions in different parts. If brain operation is not determined by local differential contributions of parts, but by more holistic principles, the enterprise of brain mapping will be much less significant for cognitive science. The debate between localizationists and holists continues to this day, but generally is less polarized than it was in the days of Gall and Flourens. The major issues now concern (a) *which* functions are taken to be localizable and (b) the *degree* of localization (see Finger, 1994, ch. 4). Almost no one now denies the high degree of localizability for somatosensory functions, for example; but the localizability of such higher functions as MEMORY, CONSCIOUSNESS, and ATTENTION still provokes debate.

References and recommended reading

Brodmann, K. 1910: Feinere Anatomie des Grosshirns. In Lewandowsky's *Handbuch der Neurologie*, Berlin, Bd. v, 206–307.

Brodmann, K. 1994: *Brodmann's 'Localisation in the Cerebral Cortex*,' trans. and ed. Laurence J. Garey. London: Smith-Gordon, 1994. Originally published as *Vergleichende Lokalisationslehre der Grosshirnrinde in ihren Prinzipien dargestellt auf Grund des Zellenbaues*. Leipzig: Barth, 1909.

Carmichael, S. T. and Price, J. L. 1994: Architectonic subdivision of the medial prefrontal cortex in the macaque monkey. *Journal of Comparative Neurology*, 346, 366–402.

Churchland, P. M. 1992: *A Neurocomputational Perspective*. Cambridge, Mass.: MIT Press.

Damasio, A. R. 1994: *Descartes' Error*. New York: G. P. Putnam's Sons.

*Felleman, D. J. and Van Essen, D. C. 1991: Distributed hierarchical processing in the primate cerebral cortex. *Cerebral Cortex*, 1, 1–47.

*Finger, S. 1994: *The Origins of Neuroscience*. New York: Oxford University Press.

Fodor, J. A. 1983: *The Modularity of Mind*. Cambridge, Mass.: MIT Press.

Kandel, E. R., Schwartz, J. H. and Jessell, T. M. 1991: *Principles of Neural Science*. New York: Elsevier.

Lashley, K. S. and Clark, G. 1946: The cytoarchitecture of the cerebral cortex of Ateles: a critical examination of architectonic studies. *Journal of Comparative Neurology*, 85, 223–305.

Mundale, J. 1997: *How do you Know a Brain Area when you "See" One?: A Philosophical Approach to the Problem of Mapping the Brain and its Implications for the Philosophy of Mind and Cognitive Science*. St Louis, Mo.: Washington University Press.

Tootel, R. B. H., Switkes, E., Silverman, M. S. and Hamilton, S. L. 1988: Functional anatomy of macaque striate cortex. *Journal of Neuroscience*, 8, 1531–68.

5

Cognitive anthropology

CHARLES W. NUCKOLLS

The study of the relationship between culture and mind is cognitive anthropology. Its primary objects of study are knowledge and thinking, mostly as these appear in naturally occurring settings. Cognitive anthropology's main contribution has been to show that there are important cultural differences in perception, memory, and inference. Recently, the field has begun to consider emotions and their power to motivate cognition, historically among the most neglected subjects in cognitive science. This is leading to an interesting *rapprochement* with psychoanalysis and to the partial eclipse of information processing models and their computer analogs.

Perception

Objects in the world are not grouped the same way by everybody. For example, when presented with pictures that could be grouped in terms of function, shape, or color, Western children tend to group by color at younger ages; but as they grow up, they group by shape and then by function. That is why Western adults generally group by function (Bruner et al., 1966). Given the same sorting task, however, adult Africans tend to group objects by color. Cognitive anthropologists believe that such differences reflect cultural, not psychological, differences, and that thinking is always socially situated.

The same is true of color perception, as Kay and Kempton (1984) demonstrated using native speakers of English and Tarahumara, a Uto-Aztecan language of northern Mexico. In Tarahumara, there is only one term for both green and blue. Using a series of color chips graded from pure green to pure blue, Tarahumara and English speakers were asked to judge which chip was the most different in color. The English terms *green* and *blue* passed between two chips, which for Tarahumara speakers were known by the same term. The English speakers therefore perceived the two chips as different, whereas the Tarahumaras saw them as the same. The experiment shows that how people label things affects the perception of differences between these things, and since labels vary culturally, the study of human perception must begin by attending to cultural differences.

Memory

Cultural differences are also found in MEMORY. For instance, reliance on an oral tradition apparently makes people better at remembering. Ross and Millson (1970) compared the memories of American and Ghanaian college students and found that, generally, the Ghanaian students were better. On the other hand, Cole and his colleagues found that nonliterate Africans did not perform better on lists of words, as opposed to stories, suggesting that cultural differences in memory as a function of oral tradition may be limited to meaningful narratives (Cole et al., 1971).

Inference and judgment

Inference is another aspect of reasoning that differs cross-culturally. In a classic study, the Soviet neuropsychologist Luria (1971) showed that East and Central Asian peasants were generally unable to provide answers to syllogisms that contained unfamiliar information. "In the Far North," said Luria to his informants, "where there is snow, all bears are white. Novaya Zemyla is in the Far North and there is always snow there. What color are the bears there?"

(*Informant*):	"We always speak only of what we see; we don't talk about what we haven't seen."
(*Ethnographer*):	But what do my words imply? (The syllogism is repeated.)
(*Informant*):	"Well, it's like this: our tsar isn't like yours, and yours isn't like ours. Your words can be answered only by someone who was there, and if a person wasn't there he can't say anything on the basis of your words."
(*Ethnographer*):	. . . But on the basis of my words – in the North, where there is always snow, the bears are white, can you gather what kind of bears there are in Novaya Zemyla?
(*Informant*):	"If a man was sixty or eighty and had seen a white bear and had told about it, he could be believed, but I've never seen one and hence I can't say. That's my last word. Those who saw can tell, and those who didn't see can't say anything!" (Luria, 1971, pp. 108–9)

Individuals from the same culture who had received a single year of schooling could respond *correctly*, however, demonstrating that the difference was not developmental, but related to exposure to Western education and styles of reasoning (see Article 21, REASONING).

In a study of personality judgment, Shweder found that Americans tend to remember what is *true* about someone as an enduring set of personality traits, and thus to say "She is good" or "He is a jerk." But Oriya Brahmins (in South India) tend to tell stories about what someone has done, without emphasizing abstract character features. Shweder demonstrates that the difference is due not to an Oriya failure to develop abstract reasoning, or to poor education, or to some other kind of problem. Rather, it is to be explained by a cultural difference in world view. Oriyas are *context-dependent* thinkers and understand persons situationally, whereas Americans are *context-independent* thinkers and understand persons more in terms of *traits* they assume do not change much with time (Shweder and Bourne 1984).

Schemas and cultural models

It will be seen that the cognitive anthropology of perception, memory, and inference involves the study of culturally constructed categories (see Article 8, CONCEPTUAL ORGANIZATION). Research has shown that many categories are organized around prototypes. The prototype is an organized bundle of default features, so that, for example, when one thinks of a *bird*, one tends to think of something that looks like a robin (the prototype) rather than an ostrich. Most of this work has been done by linguists, while cognitive anthropologists have focused on a related knowledge structure, the *schema*. The difference between a prototype and a schema is that while both are stereotypic, a prototype consists of a specified set of expectations, whereas a schema is an organized framework of relations which must be filled in with concrete detail. Schemas are highly generalized knowledge structures which help generate appropriate inferences. They fill in the gaps, supplying the information that is usually taken for granted, thus enabling individuals to identify actions and events based on brief exposure to only partial information.

Suppose we want to understand the following two sentences: "John went to a party." "The next morning he woke up with a headache." We immediately recall that people who go to parties often smoke and drink too much. They wake up the next morning feeling hung over. Notice that our causal explanation goes considerably beyond the information given. For all we know, the subject may have been run down by a bus or food-poisoned. The chosen explanation is an inference produced from our organized cultural knowledge concerning what people do at parties. The arrangement of this knowledge is almost story-like; we seem to know what should happen next. Thus, by having access to this culturally constituted, story-like arrangement, we are able to make sense out of the ambiguous two sentences above.

Cognitive anthropologists have found that much of everyday, applied social knowledge exists in schematic arrangements, known also as *scripts* or *event scenarios*. An even more fashionable term at present is *cultural model*, defined by Holland and Quinn as "presupposed, taken-for-granted models of the world that are widely shared (although not necessarily to the exclusion of other, alternative models) by the members of a society and that play an enormous role in their understanding of that world and their behavior in it" (1987, p. 4). Holland and Quinn find, for example, that a small set of highly salient cultural models govern college women's assessments of men as potential romantic partners. According to the central model, the two parties are equally attractive and equally attracted to one another. The male demonstrates his interest by concerning himself with the female's needs and desires, while the female acts on her assessment by permitting greater intimacy (p. 101). At the same time, there are variations on this model; typically they differ according to the type of male behavior they foreground: Don Juans, turkeys, gays, etc.

Another example of the power of a cultural model is the American folk model of the mind. According to this model, intentions and actions should be related, and acts are to be judged by whether or not they follow from an intention. This is quite different from the South Asian model of the mind, according to which no meaningful distinction can be made between the two. But in America, the implications are serious, as in this exchange between a seven-year-old child and her mother:

Mother: Rachel, you're making me mad!
Rachel: I didn't mean to make you mad.
Mother: Well, you sure seem to be trying.
Rachel: But I didn't mean to. If I didn't mean to, how could I be trying?

Rachel employs the cultural model according to which there should be a connection between intentions and actions. *Trying* is an action that is supposed to bring about an intention, but if there was no intention, then Rachel could not have been *trying* to anger her mother. What this example shows is that acculturated individuals, including young children, reason effectively from implicit models.

Another demonstration of the prevalence of cultural models in everyday thinking is home heat control. Most people in the United States, who live in centrally cooled/heated houses, have arguments with each other over proper temperature adjustment. According to Kempton, the reason is that Americans have two cultural models: the *feedback theory* and the *valve theory*. According to the first, the thermostat turns the furnace on or off according to room temperature. Left at one setting, the thermostat will switch the furnace on and off to maintain room temperature at this setting. According to the second theory, however, the thermostat does not maintain the temperature; it is the human operator, who adjusts the thermostat as he or she would a valve, releasing differing amounts of cool or warm air. It is like the accelerator pedal in an automobile and requires constant intervention. Since people who adhere to different models sometimes live in the same house – often as husband and wife – the thermostat has been found to be a center of considerable domestic discord (Kempton, 1987).

New directions: cognition and motivation in cultural context

In cognitive anthropology, at least, the limitations of the cognitive perspective have been recognized, mainly with reference to the problem of motivation. If people have a lot of scripts and schemas in their heads, what makes them emotionally compelling (see Article 11, EMOTIONS)? In a study of Trobriand myth, Hutchins (1987) shows that cultural models function as a disguised representation of repressed thoughts and fears concerning relations between the self and a deceased relative. He discusses mythic schemas as a series of episodic scenarios, much like information processing specialists do; but instead of ignoring motives, he uses the language of psychoanalysis to show that schema selectively distort repressed motivational schemas. The value of Hutchins's analysis is in demonstrating that cognitive schemas and psychodynamic processes are intimately linked, and may be conceptualized thus:

1 Cognitive schemas summarize past experiences into composite forms, allowing incoming information to be measured against the existing composite for *goodness of fit*. Incoming information may be distorted or partially deleted, in order to achieve this fit, thus explaining some of the *errors* or *gaps* in remembered accounts. Schemas enable rapid perception, as information is assimilated to the existing composite, but they also lead to patterned and recurrent distortions.
2 Multiple schemas may be applied simultaneously and unconsciously to the interpretation of information. Multiple parallel channels operate in unconscious information processing, but conscious thought tends to proceed in one or only a few of these. As a result, there may be competition for priority among these multiple schemata.
3 Conscious reflection on schemas is possible, especially when schemas are given symbolic representation in words, signs, or gestures. This may facilitate changes in how the schemas are used in appraisals, decision, and negotiations of meaning, leading to changes that can construct new schemas or that integrate old ones.

Following Hutchins (1987) and also Horowitz (1988), some cognitive anthropologists now posit the existence of a repertoire of relational schemas which contains archaic components, many formed in childhood, that can never be erased. Such early schemas of self and others are constrained by mature concepts of self which contain and integrate immature self schemas. Nevertheless, in a kind of parallel processing, earlier forms continue to unconsciously appraise current events, possibly following primitive association and the logic of the *primary process*. The link between the dynamic unconscious and the cognitive schematic could then be stated as follows: Access to conscious symbol systems, including cultural images, words, and action scenarios, might occur only through the information organized by unconscious schemas, such that to instantiate one is to instantiate the other.

For example, children on the South Pacific atoll of Ifaluk experience an intensely close, nurturing dependence which they lose, suddenly and dramatically, when younger siblings usurp the position of favored offspring. Following dethronement, children become aggressive and competitive with each other, but soon they are socialized to the positive value of nonaggressiveness and cooperation. They give up fighting and begin fearing violence, the supreme negative value. It is as if the desire for a return to the dependency of early childhood remains, now realized symbolically in the value of food. That is why the fear of aggression continues, because competition over possession of this valued substance, and what it represents, is still possible.

Significantly, in Ifaluk culture there are two main emotion scripts or cultural models, one focused on *fago* (compassion/love/sadness), the other on *song* (justifiable anger). These are the cultural models people use to interpret the emotions of themselves and others. The *fago* script is represented in episodic form thus, following the cognitivist convention of listing such things serially as an information processing sequence:

> (1) Someone is in need → (2) It is pointed out by someone → (3) That person reacts by bestowing compassion.

The second one, the *song* script, is said to look like this:

> (1) There is a rule violation → (2) It is pointed out by someone → (3) That person simultaneously calls for the condemnation of the act → (4) The perpetrator reacts in fear to that anger → (5) And he mends his ways.
> (Lutz, 1988, p. 157)

What gives these models the motivational salience which they need to become effective? Surely it is not the cognitive sequence itself. Notice that the social roles that underpin the division of the Ifaluk emotional landscape into *fago* and *song* have two conspicuous features, both traceable to childhood. The first is the nurturing bond that binds child and parent. It is realized in the giving of food and represented in the act of sharing. *Fago* (compassion/love/sadness) is the value whose correlate in development is associated with this episode in family history. It refers to an experience that gave one pleasure. One seeks to repeat it by extending the circle of dependency as widely as possible. The second is the sudden break that takes place when parents withdraw their nurturance of one child and bestow it on another, usually a younger sibling. This, too, is realized in the transaction of food – not in giving it but in taking it away –

and represented in the competition among children for its possession. *Song* (justifiable anger) is the value whose correlate in development is associated with this episode in family history. It refers to an experience that produced suffering. One seeks to restrict it, by morally impugning those who refuse to share.

Cultural models like these are what Luborsky and his colleagues (1991) refer to as "core conflictual relationship themes." Higher-order knowledge structures may be accessible only by activating lower ones, in the dynamic unconscious. If so, then it is possible that higher-order structures reproduce in their form and function the conflicts of the lower-order relational structures (Nuckolls, 1996).

For most of its history, cognitive anthropology has conformed to a distinction first articulated by Aristotle, differentiating the *dianoetic* (intellectual) from the *orectic* (emotional) capacities. Its chief contribution to cognitive anthropology was as a challenge to the universalistic assumptions of psychology. Perception, memory, and inference – the focal subjects of experimental cognitive psychology – are understood not just as influenced by cultural context, but as powerfully constructed by it, or, rather, in relationship with it. Now the field is discovering new opportunities for synthesis and cross-disciplinary fertilization, especially where motivation is concerned. This may lead, at long last, to the breakdown of the Aristotelian division, and to a cultural cognitive science that is truly holistic.

References and recommended reading

Bruner, J., Oliver, R. and Greenfield, P. 1966: *Studies in Cognitive Growth*. New York: Wiley.

Cole, M., Gay, J., Glick, J. and Sharp, D. 1971: *The Cultural Context of Learning and Thinking*. New York: Basic Books.

*D'Andrade, R. 1995: *Cognitive Anthropology*. Cambridge: Cambridge University Press.

*Dougherty, Janet (ed.) 1985: *Directions in Cognitive Anthropology*. Urbana, Ill.: University of Illinois Press.

Holland, D. and Quinn, N. (eds) 1987: *Cultural Models in Language and Thought*. Cambridge: Cambridge University Press.

Horowitz, M. 1988: *Cognition and Psychoanalysis*. Chicago: University of Chicago Press.

Hutchins, E. 1987: Myth and experience in the Trobriand Islands. In Holland and Quinn (eds), 269–89.

Kay, P. and Kempton, W. 1984: What is the Sapir–Whorf hypothesis? *American Anthropologist*, 86, 65–79.

Kempton, W. 1987: Two theories of home heat control. In Holland and Quinn (eds), 222–42.

Luborsky, L., Crits-Christoph, P., Friedman, S. H., Mark, D. and Scheffler, P. 1991: Freud's transference template compared with the core conflictual theme (CCRT): illustrations by the two specimen cases. In M. Horowitz (ed.), *Person Schemas and Maladaptive Interpersonal Patterns*, Chicago: University of Chicago Press, 167–96.

Luria, A. 1971: Towards the problem of the historical nature of psychological processes. *International Journal of Psychology*, 6, 259–72.

Lutz, C. 1988: *Unnatural Emotions*. Chicago: University of Chicago Press.

*Nuckolls, C. W. 1996: *The Cultural Dialectics of Knowledge and Desire*. Madison, Wis.: University of Wisconsin Press.

Ross, B. and Millson, C. 1970: Repeated memory of oral prose in Ghana and New York. *International Journal of Psychology*, 5, 173–81.

Shweder, R. and Bourne, E. 1984: Does the concept of the person vary cross-culturally? In R. Shweder and R. LeVine (eds), *Culture Theory*, Chicago: University of Chicago Press, 158–99.

6

Cognitive and linguistic development

ADELE ABRAHAMSEN

Aisha, age 24 months, sees her older sister trip over a toy and squeals "Kiki fell!" At 30 months her doll falls behind the couch, and she asks "Where dolly falled?" At 48 months her computer mouse falls behind a stack of computer manuals under the desk, and she asks "Where did that silly mouse fall?" Both her world and her sentences are getting more complex as she gets older, but there is one oddity: the past-tense verb *fell* shows a U-shaped developmental pattern. Aisha gets it right at 24 months, wrong at 30 months, and right again at 48 months. Why?

It is not uncommon for scholars who start out wanting to understand adult cognition and language to wander into developmental psychology seeking the origins of these functions and become so intrigued by puzzles like this one that they stay. Some (like Jerome Bruner and Jean Piaget) settle into developmental psychology as their home base for decades. Others breeze back and forth like emissaries between developmental psychology and the generally adult-centered field of cognitive science. These two centers of inquiry into the nature and workings of the mind overlap less than they should, but the emissaries ensure a two-way flow of ideas and findings.

When you explore the field of developmental psychology, you encounter a paradox. Cognition (including language) is both simpler and more complex in children than in adults. One way it is simpler is that various adult abilities can be traced back to their nascent forms, before the vagaries of subsequent experience and the inexorable hardware changes of biological maturation have elaborated upon their basic design. One way it is more complex is that the child's abilities are not a stable, finished product; various change-producing processes are intertwined with whatever you wish to study. You can try taking snapshots, but you will lose the dynamics that are such a crucial part of the story.

Here is an example: Elissa Newport and her colleagues showed that children exposed to American Sign Language (ASL) or to a second spoken language before age six or seven extract and correctly use the smallest meaningful components of the language (called *morphemes* in LINGUISTIC THEORY), whereas later learners do insufficient analysis and end up with "frozen forms" from which they cannot extract individual morphemes for productive recombination. The clearest examples are from languages that emphasize the design principle of building complex words by combining several morphemes, such as ASL, Hungarian, Swahili, and Michoacan Aztec. The idea is that a fluent speaker of Aztec could construct the word *nokalimes* (my houses) by combining the morphemes *no* (my), *kali* (house), and *mes* (plural); a late learner might produce the word only as a single, unanalyzed unit.

Newport (1988) proposed an ingenious explanation for her findings concerning the critical period for language acquisition: the "Less is More" hypothesis. It is known that short-term memory improves with maturation. Newport's idea is that a learner of limited capacity will naturally store only a few components of form and meaning at a time, and therefore will be able to match up forms (morphemes) with their meanings frequently enough to end up with a good analysis of the language. Older learners suffer from an embarrassment of riches: they can – and therefore do – store all the morphemes in several words and a good deal of the nonlinguistic context. When all that material is stored at once, the learner must try to line up all the morphemes against all the components of meaning and somehow get the right matches. Often the result is failure, and the only thing that will be learned is that the word or phrase as a whole corresponds to the meaning as a whole. From such a frozen form, appropriate recombination of the elements in new situations will not be possible.

Newport's proposal displays an interesting mix of the simple and the complex. A simpler (smaller capacity) short-term memory, due to a simpler (less mature) brain, yields a more complex analysis of the language heard. Maturation is one of several sources of change that work together to make development itself complex and difficult to study. The idea that limited memory might yield better initial learning has also inspired a successful connectionist network model: Jeffrey Elman's "starting small" approach to learning to deal with syntactically complex sentences (see Article 46, INNATENESS AND EMERGENTISM).

The above example illustrates how fruitful it can be when mainstream cognitive science and developmental research merge seamlessly. More often, though, researchers within each approach do their own thing and make occasional forays into the other camp to get or give something. To convey the flavor, a few examples of research influences in each direction follow.

Cognitive scientists go to developmental psychology

There are some big issues that cognitive scientists sometimes try to settle by looking at children (usually without complete success: some of these issues have bedeviled scholars for centuries and just get harder and more complicated as the stacks of evidence get higher). To begin with one that Chomsky made famous: Is there an innate Universal Grammar (UG) underlying the apparent diversity of human languages? In the fifth century BC Herodotus reported that the Egyptian king Psammetichus had posed a less subtle question about the innateness of language: Which of the many languages in the world was spoken by the most ancient race of humans? Though confident that this first and best tongue was Egyptian, the king was enough of a scientist that he sought empirical confirmation by arranging for two children to be raised hearing no language spoken. The shepherd entrusted with their care made his report when he noticed both children consistently saying "becos" – not Egyptian, but rather the Phrygian word for bread.

King Psammetichus's basic premise sounds odd to the modern mind, so we would make much less of the outcome than he did. For today's developmental psycholinguists, it is Chomsky's idea of an innate Universal Grammar that invites investigation. Chomsky proposed that there is a basic design framework underlying all human languages that is part of a baby's genetically transmitted biological endowment. Every language links sounds (or manual signs) to meanings by means of a layered system that generates phonemes, morphemes, words, phrases, and sentences. There is more

Table 6.1 Major semantic relations in two-word utterances (Brown's stage I)

Semantic relation	*Spoken example*	*Gestured example*
Agent + Action	daddy sit	MARCH *soldier*
Action + Object	throw stick	*drum* BEAT
Agent + Object	mommy sock	*duck mother*
Action + Location	sit chair	EAT *kitchen*
Entity + Location	toy floor	*bandaid finger*
Entity + Attribute	crayon big	*bottle* LITTLE
Possessor + Possession	my teddy	*picture Abe*

Note: In the gestured examples, italics are used for referents indicated by pointing, and upper case is used for invented gestures.

than one linguistic theory for describing this system. Chomsky's own theory of government-binding (GB; also principles and parameters) is usually regarded as providing the best current sketch of the design of Universal Grammar.

A direct inference from Chomsky's proposed biological origins for Universal Grammar is that this basic framework should be available to a baby even in the absence of exposure to any particular language. King Psammetichus's experiment would no longer be considered remotely ethical, but three researchers at the University of Pennsylvania realized that deaf children with hearing, nonsigning parents lack exposure to language long enough that they present a unique opportunity for testing the inference. For their doctoral dissertations, Heidi Feldman and Susan Goldin-Meadow videotaped and then painstakingly analyzed the gestural communications of six deaf preschoolers. In 1978, joined by Lila Gleitman, they published "Beyond Herodotus: the creation of language by linguistically deprived deaf children." It turned out that each child used a large number of gestures that were equivalent to words: for instance, fingers fluttering downwards for "snow," arms flapping for "bird," and pointing to a nearby person or object to indicate it. Hearing children use a few such gestures along with their earliest words, but then speech takes over. But Feldman et al.'s deaf children continued to use their earliest gestures, added many more, and began combining them into short utterances. These utterances had rudimentary grammar (e.g., preferred word orders) and expressed the same semantic relations as Roger Brown (1973) had found in his landmark analysis of the linguistic progress of Adam, Eve, and Sarah. Brown selected five levels of grammatical complexity, which he called *stages* (though he meant less by this term than did Piaget, as described below). The age of entry into stage I varies considerably, but 18–24 months is typical. Table 6.1 shows how it looks.

Feldman et al.'s six deaf children varied in how far they progressed beyond stage I. The parents never became as fluent as their children in gestural language, and the lack of a fully elaborated model language clearly impeded the children's progress. Nonetheless, the children's invented languages showed enough of the rudiments of Universal Grammar that Feldman et al. saw in them an unmistakable manifestation of our species' innate capacity for language. Many cognitive scientists have found this a compelling argument; others (such as Elizabeth Bates) have proposed alternative interpretations emphasizing an intricate interaction among cognition, language, and the environment. The same split develops over many other kinds of evidence. (For a more extensive discussion, see Article 45, INNATE KNOWLEDGE, and Article 46, INNATENESS AND EMERGENTISM.)

A further proposal about Universal Grammar is that it includes constrained sets of choices called *parameters*. For example, the subject omission parameter has two values: each language either permits leaving out the subject of a sentence or forbids it. Babies come with the switch set to permit subject omission (SO). A baby learning a language like Spanish need do no more; sentences like "Tiro pelota" (throw ball) are fine. A baby learning English will produce sentences like "Throw ball" in stage I, but keeps hearing sentences like "I threw the ball." Eventually the English baby will switch the parameter setting to forbid subject omission (No SO). The baby is biologically prepared to use either setting but to prefer SO by virtue of the innate Universal Grammar.

Child data have also been brought to bear on a related issue: Do humans mentally represent their knowledge in the form of rules operating on strings of symbols? Behaviorists did not think in terms of rules, but Chomsky did. Generative grammar blew in like a fresh wind, and there was considerable interest in demonstrating the relevance of rules during the 1960s. Some of the most persuasive evidence came from children acquiring language. Consider the sentence that Aisha used at 30 months: "Where dolly falled?" It provides not just one but two kinds of evidence that the child is acquiring rules. First, she is overgeneralizing the regular past-tense morpheme (*-ed*): the rule is so salient that it displaces the correct form (*fell*) which she already knows. (Learning to use grammatical morphemes like *the*, *be*, *-s*, and *-ed* is the major task of Roger Brown's stage II, which begins around 24–30 months.) Second, as Roger Brown and his colleagues described for children at stage III (beginning around 30–36 months), Aisha has learned one of the transformational rules needed to generate wh-questions as proposed in Chomsky's 1965 Standard Theory of grammar: she moved the appropriate wh-word for the locative, *where*, from the end of the sentence to the beginning. She has not yet learned the rest of the rules, which introduce the auxiliary verb *did*. (The transformational rules involved in forming questions are further explained in Article 35, STRUCTURAL ANALYSIS.) The piecemeal learning suggested to researchers in the 1960s that children were actually discovering or constructing rules rather than simply memorizing and imitating what was given directly in the input. It was further claimed, following Chomsky, that children did this by submitting the sentences they heard to an innate language acquisition device (LAD) that combined its knowledge of Universal Grammar with the language-specific input. Initially contentious, this account had gained widespread acceptance by the 1970s – which meant that it was time for a new generation to put forward yet another account.

The connectionist alternative to RULES and symbolic REPRESENTATIONS was first developed in the early 1980s. On this account, language input is used to gradually adjust weights in a connectionist network so as to produce the rule-like patterns of acquisition. One of the first phenomena modeled was preschoolers' overregularization of the past tense. The initial 1986 paper on this by Rumelhart and McClelland spawned a whole cottage industry that has not yet closed up shop; still in 1997 there are conferences that annually include the latest paper on the past tense and (often noisily) someone's objections to it. (See Article 38, CONNECTIONISM, ARTIFICIAL LIFE, AND DYNAMICAL SYSTEMS.)

To summarize, in the 1960s language acquisition data were used to argue that children induce abstract rules rather than imitate what they hear. In the 1980s more detailed data on the same acquisitions were used to argue that children are adjusting weights in connectionist networks that generate rule-like performance without using mental rules.

Developmental psychologists go to cognitive science

The other kind of interaction between mainstream cognitive science and developmental psychology involves influence in the opposite direction: developmental psychologists discover a phenomenon, and either it suggests new avenues for thinking about adult cognition, or the developmental phenomenon seems so important on its own that it needs to be included in any account of what it is to be a human cognizer. Some of the most interesting examples are from the acquisition of WORD MEANING in the first few years of life. As Lev Vygotsky pointed out, word meanings are the intersection between concepts and the sounds of language. The concepts underlying adult words are fairly restricted; for example, there are no words with meanings like "two or blue." Within the first 75 or so words, however, around ages 12–24 months, about one-third are overextended beyond their adult meaning, some in such a way that they do not correspond to any possible adult meaning if taken at face value. For example, in his classic diary study of his daughter Hildegard, Leopold noted the following sequence of referents for *sch*: noise of train, music, noise of any movement, wheels, balls. These suggest a nonadult-like complexive concept based on loose or shifting associations, something like "noise of movement or round," joined by the fact that trains both make noise and have round wheels. More common are overextensions that do correspond to an adult meaning, but for a different word. In particular, Eve Clark proposed in 1973 that many overextensions are based on linking the word to just one of several perceptual features in its meaning (a ball is anything that is round). Katherine Nelson pointed out that these might be functional or event-based instead (a ball is anything you can roll), and she built an influential theory around the idea that event-based knowledge representations such as Schank-style scripts are crucial in development (see Nelson, 1996, for a summary and recent extensions). Other researchers have cautioned that overextensions may give a distorted view of children's knowledge. Words that are overextended in production tend to be correctly comprehended in forced-choice tasks. Perhaps, then, children are simply using the closest words at their disposal to communicate. Where does this leave the idea that some early words have nonadult meanings? Somewhat in limbo, because it is hard to do laboratory comprehension tasks with children younger than 18 months. Newer techniques, such as selective looking, might give a clearer answer. This technique has been used by Kathy Hirsh-Pasek and Roberta Golinkoff (1996) to show, for example, that children understand the subject–verb–object order in simple sentences before they have begun combining words themselves (based on the fact that they look longer at a video that matches a sentence like "Big Bird is tickling Cookie Monster!" than at a mismatched video with the reverse meaning). More recently, the selective looking technique has been used by Douglas Medin and Sandra Waxman to compare basic and superordinate levels of CONCEPTUAL ORGANIZATION as they relate to word acquisition at 12–14 months.

With most children, vocabulary grows slowly during the second year of life, when the earliest words emerge and the meanings of some are overextended. Although there is considerable variation across individuals, a typical child produces 3 words at 12 months and 50 words by 18–21 months. Then something happens; children begin combining words and also become voracious consumers of new words ("word magnets" in Susan Carey's turn of phrase). They increase their rate of acquisition to more than 30 new words per month by 21 months. By age three to five years they are adding two to four new words per day, as well as adding twice that number to the words they can understand but not produce. That is, preschoolers are learning the meaning of a new word every one or two hours they are awake, and this pace continues into the school years (one estimate is 3,750 words per year). Fascinated by this, Carey arranged for children to hear an unfamiliar color term, *chromium*, used for suitable referents. She found that as little as one exposure to a teacher's use of the term led to comprehension of the term weeks later by some three-year-olds and dubbed this phenomenon *fast mapping*. It is even more impressive when you realize that during the same preschool years the child is acquiring most of the grammar and phonology of one or more languages, gaining conversational skill, and successfully traversing a number of nonlinguistic paths of development (including, but not limited to, gross and fine motor control, symbolic play, PERCEPTION, MEMORY, PROBLEM SOLVING, SOCIAL COGNITION, social interaction, personal identity, ETHICS, and EMOTIONS). Often these domains interact with language development; for example, Lois Bloom (1993) has examined the relation between early language and emotion.

Words continue to be added at a rapid clip during the school years, and the lexicon gradually attains its adult form. Hierarchical structure has long been recognized as an important arena for development. More recent approaches include pursuit of (a) Eleanor Rosch's ideas that one level of hierarchy is *basic* and that category membership is a graduated function of typicality rather than yes/no; (b) the claim that children's word meanings are constrained by assumptions, some of which will later be abandoned (e.g., the mutual exclusivity assumption that each object has exactly one label); (c) the view that concepts are not isolated sets of features but, rather, are organized within larger frameworks of knowledge (schemata); (d) Gopnik and Meltzoff's (1996) "theory theory" that children construct word meanings by forming and revising theories. This last point can be applied to developmental psychology itself: one of our most important tasks is to put the myriad specific findings together into a big picture. We now turn to that task.

Theories and the big picture

We begin, like many baby books, with a straightforward chart of easily noticed changes organized by age and domain (table 6.2). Charts like this have their uses, but they leave a lot unexplained. Developmental theorists have tried to dig beneath and between the lines to answer two big questions:

- What develops? (What is the design of the underlying system that controls the observable behaviors?)
- How does it develop? (What are the mechanisms or processes by which that system changes?)

Table 6.2 Developmental milestones in two domains

Age(mo)	*Eye–hand coordination*	*Expressive language*
4	Grasps cube on contact	Cooing, vowel sounds, laughter
6	Grasps cube on sight	Simple babbling (ma, di)
8	One cube per hand	Reduplicated babbling (mama, didi)
10	Removes cube from cup	Tries to imitate familiar words
12	Releases cube in cup	Produces 1–3 words, imitates a few
14	Builds tower of 2 cubes	Uses gestures to make requests
15	Puts 6 cubes in and out of cup	Jargon, tries to imitate novel words
18	Tower of 3–4 cubes; 10 in cup	Produces 3–50 words, makes verbal requests
21	Tower of 5–6 cubes	Two- or three-word utterances
24	Tower of 6–7 cubes; leaves cup full	More complex utterances, e.g. pronouns

The best-known theory that addresses both these questions is that of Jean Piaget, the Swiss psychologist who preferred to call his life's work "genetic epistemology," because he was studying developmental change to get at fundamental questions about knowing. A keen observer, ingenious experimenter, and bold theorist, he built an edifice between the 1920s and 1960s that still cannot be ignored. For Piaget, what developed were mental structures (often called *schemes*) that were dynamically attuned to the child's ongoing experience and became increasingly organized into coherent systems. For example, he posited a looking scheme and a grasping scheme that initially developed separately (the four-month-old can see a cube and can grasp a cube that she happens to touch) and later become coordinated (for the six-month-old, seeing a cube can elicit and guide grasping the cube). What drove the changes was a tendency to adapt by means of two complementary processes. The child (and adult as well) *assimilates* incoming events and perceptions to the most suitable schemes for dealing with them and simultaneously *accommodates* the schemes to the particular input. Seeing a cube, a child might assimilate the perceptual experience by electing to exercise the coordinated looking–grasping scheme and accommodate to the size of the cube by adjusting the grasping hand.

These are just Piaget's initial answers to the What and How questions; the theory gets much more complex. In another part of his What theory, he posits stages of development. Specifically, three major stages of development are each marked by concurrent (yoked) changes across a number of domains, which culminate in a qualitatively different, unified mental structure. Each stage is further divided into substages. The sensorimotor stage (birth to about 24 months) culminates in sensorimotor stage VI, in which sensory and motor schemes have achieved the same kind of structure as a mathematical group. For example, spatial relations exhibit the group properties of composition (two paths combined make a third path) and inversion (a path can be reversed to return to the starting point). The next stage is lengthy and has three main substages. (1) It begins with the flowering of the symbolic function in play, language, and imitation between ages two and four years. (2) This preoperational intelligence becomes increasingly systematic (four to seven years). (3) These changes culminate in the attainment of concrete operations (7 to 11 years). Least studied is the third major

Table 6.3 Piaget's six sensorimotor stages for the object concept in infancy

Stage	*Age (mo)*	*Behavior indicating stage of object permanence*
I	0–2	Out of sight is out of mind: no search.
II	2–4	Passively looks where an object disappeared.
III	4–8	Searches for an object that is partly covered.
IV	8–12	Searches for an object that is completely covered.
V	12–18	Searches for an object after visible displacements.
VI	18–24	Searches for an object after invisible displacements.

stage, in which formal operations enable hypothetico-deductive reasoning (12 years through adolescence).

In studying each stage, Piaget developed nuanced tasks and analyses that went far deeper than the chart of behavioral milestones in table 6.2. For example, he took what would seem to be a simple concept – the object – and deconstructed it in order to observe how the infant *constructed* it across the six sensorimotor stages. By the end, infants seemed to have robustly attained the notion of an object as something that occupies its own position within a common space and maintains its characteristics and its very existence regardless of whether it is in the infant's perceptual field (*object permanence*). Piaget tried hiding objects fully or partially under a cloth. For older infants he complicated the procedure by carrying out a series of visible displacements (letting the infant watch as he moved the object under cloth A, then under cloth B, and then under cloth C, where he left it). These displacements were visible, because the child could see the object in Piaget's hand as he moved it between cloths. Enclosing the object in his fist converted this into a series of invisible displacements (and here the infant would need to understand that the object could have been left at any of the three locations). By observing how infants behaved in eliciting situations like these, Piaget identified six stages that are summarized in table 6.3.

Piaget did not stop here. He linked these six stages of object permanence to six stages of development regarding space, time, and causality; six stages for means–end relations; six stages for imitation; and so forth. What was important was the underlying character of each stage. Development in each domain had to fall into line for the whole stage theory to maintain its integrity. This theory-based approach has a great deal more to say than does a simple listing of landmarks, and it is correspondingly more vulnerable to critiques or falsification.

Skipping to the stage of concrete operations (7–11 years), patterns of performance on an array of tasks again link up in a way that can only, for Piaget, mean an underlying common trajectory through stages. Here is where Piaget was most ingenious as an experimenter. One of his best-known tasks is conservation of liquid. The child agrees that two identical glasses each contain the same amount of water. Then the water from one glass is poured into a taller container. A preoperational child says that there is more water in the tall container than in the remaining original container. A child with concrete operations will say that the amounts are still the same and will justify that judgment verbally. Similarly, in a number conservation task the child must cope with one of two rows of small items like buttons being stretched out or compressed while the other row remains as it was. The child must explain why the number of buttons in the two rows is still the same.

Piaget was at the height of his influence in much of the world, including North America, during the 1970s. Other researchers had confirmed Piaget's findings on "ages and stages," and some were pursuing the idea that language development depended upon cognitive development more generally, as captured in Piaget's stages. Simultaneously, though, challenges were arising from several directions, and Piaget increasingly became less the visionary leader and more the foil against which new ideas and data were directed.

The least contentious of the new ideas focused on the second question: how does development occur – what are the mechanisms of change? If you read any of Piaget's numerous long books, you cannot help but be impressed by the detailed attention he gave to the question of how children progressed from one substage to the next, assimilating and accommodating as they went. But you also cannot help but think that a satisfying account had eluded him. In the 1970s, Robbie Case (1985) and other investigators proposed neo-Piagetian stage theories in which concepts from information processing psychology were called in to play an explanatory role. Automatization of processing and improvements in short-term memory were emphasized, for example. Contrary to Newport's "less is more" twist, the original idea was that "more is more." In related work that included computer implementation, David Klahr and his colleagues at Carnegie–Mellon University designed PRODUCTION SYSTEMS that simulated stages of development on conservation and other tasks. More recently, ideas about how to make these models self-modifying have been partially implemented. Also at CMU, Robert Siegler has sought to identify the rules or strategies that children use in producing different patterns of performance. Siegler emphasizes that more than one rule or strategy may be available at a time, producing a kind of Darwinian competition and soft (probabilistic or varying) transitions between stages.

Successful as they have been at providing an updated understanding of Piaget's stages (Siegler, 1991), these initial mechanistic accounts have yielded only preliminary answers to the How question. In the opinion of some, it is the connectionist models that arose in the 1980s that have provided the best mechanistic interpretation of Piaget's rather mysterious processes of assimilation and accommodation. A network takes in (assimilates) an input pattern and adjusts (accommodates) to that pattern by propagating activation appropriately across its layers. In learning mode, a more permanent accommodation is made by readjusting the weights on connections. A disadvantage of currently implemented networks, however, is that they are too passive to capture the constructive character of development. Dynamical systems theory, as applied by Esther Thelen and Linda Smith (1994) to such developmental domains as motor development, may not share this disadvantage and has the advantage that temporal dynamics and change are built into its basic architecture. Alternatively, Karmiloff-Smith suggests that connectionist models can be designed to exhibit a constructionist tendency which she calls "representational redescription" (see Article 49, MODULARITY, for an explanation of this specific idea, and Article 38, CONNECTIONISM, ARTIFICIAL LIFE, AND DYNAMICAL SYSTEMS, for an introduction to these general types of models).

To summarize: Researchers grappling with the How question have adopted some kind of stage theory, though evidence has pushed them to softer versions than Piaget's.

Switching now to researchers who have focused on the What question, their predominant move has been explicit rejection of Piaget's account. The usual tactic has been to raise a When question by designing new tasks and methodologies that yield substantially younger ages at which children seem to exhibit each competency. The "ages and stages" part of Piaget's account hence collapses into an unstructured heap.

The first What/When inquiries took Piaget's tasks as a starting point and modified them, sometimes in seemingly minor ways, with the result that three- to five-year-olds (depending on the task) could display knowledge that had been thought not to arise until age eight or so. For example, in her 1972 "magic" experiments Rochelle Gelman got children to call a plate with three toy mice attached the "winner" and a similar plate with two toy mice the "loser." Then a new plate with certain changes was surreptitiously substituted for one of the original plates. If the three-mouse plate was replaced by a two-mouse plate, for example, the children were surprised and searched for the missing mouse. But if it was replaced by another three-mouse plate in which the mice were spread further apart, children as young as age three were untroubled and still called it the "winner." Gelman concluded that they could conserve number in small arrays. (See Gelman and Gallistel, 1978/1986.)

In the 1980s, even more dramatic claims of early competence arose from new procedures that could be used to assess knowledge in infants without relying on motoric responses. Regarding the concept of number, for example, Karen Wynn obtained evidence of early or even innate knowledge of counting. Regarding the object concept, Renée Baillargeon, Elizabeth Spelke, and others have obtained evidence not only that infants in the range two to eight months infer the continued existence of a hidden object (object permanence) but also that they expect a rigid object moving behind a screen to continue on a connected, unobstructed path and retain its shape. In their basic procedure, they show certain familiarization events and then a test event in which an object moves behind a screen and the screen is then raised to reveal the object in a state and location that is either consistent with a mature knowledge of objects or violates some aspect of that knowledge. Since infants tend to look longer at novel events, they should (and do) look longer if a violation has occurred. Spelke suggests that these experiments are tapping a core of knowledge that is a foundation for later developments but does not itself change. Not everyone is so ready to draw this kind of conclusion: the rich array of early competence findings have elicited lots of interest but little agreement on how to incorporate them into developmental theory.

If Piaget's theory is at best incomplete and at worst wrong, what can take its place? Within standard inside-the-head psychology, the most obvious contenders are (1) descendants of theories that emphasize innate knowledge or constraints; (2) connectionism, artificial life, dynamical systems, and other accounts that emphasize adaptive functioning and the interaction of multiple factors, or *soft constraints*. Increasingly, though, researchers are going outside the head to incorporate external tools and social interactions as crucial aspects of thought and language. To note just a few examples, James Wertsch, Katherine Nelson, and others have taken Vygotsky as a touchstone in developing a MEDIATED ACTION or sociocultural approach to development. Jerome Bruner and his colleagues undertook influential explorations of the functions of language and the role of joint attention that have continued to be pursued by many others. Eleanor Gibson pioneered the ecological approach to development that now flourishes in several variations. A few investigators have tried to integrate EMBODIED, SITUATED, AND

DISTRIBUTED COGNITION with the mainstream orientation towards computation (e.g., Frawley, 1997); this more integrated approach has the potential to influence developmental researchers.

In our post-Piagetian era, then, developmental theory has fragmented into a number of cliques, but there has also been a good deal of cross-talk. Will one of these tendencies prevail? Arriving at an answer will require audience participation.

References and recommended reading

Bloom, L. 1993: *The Transition from Infancy to Language: Acquiring the Power of Expression.* New York: Cambridge University Press.

Brown, R. 1973: *A First Language: The Early Stages.* Cambridge, Mass.: Harvard University Press.

Case, R. 1985: *Intellectual Development: A Systematic Reinterpretation.* New York: Academic Press.

Clark, E. V. 1973: What's in a word? On the child's acquisition of semantics in his first language. In T. E. Moore (ed.), *Cognitive Development and the Acquisition of Language,* New York: Academic Press, 65–110.

*Damon, W. (ed.) 1988: *Handbook of Child Psychology,* 5th edn, 4 vols; vol. 2, *Cognition, Perception, and Language* (eds D. Kuhn and R. S. Siegler). New York: Wiley.

*Feldman, H. H., Goldin-Meadow, S. and Gleitman, L. R. 1978: Beyond Herodotus: the creation of language by linguistically deprived deaf children. In A. Locke (ed.), *Action, Gesture, and Symbol: The Emergence of Language,* London: Academic Press, 351–414.

*Flavell, John H. 1993: *Cognitive Development,* 3rd edn. Englewood Cliffs, NJ: Prentice-Hall.

*Fletcher, P. and MacWhinney, B. (eds) 1995: *The Handbook of Child Language.* Oxford: Blackwell.

Frawley, W. 1997: *Vygotsky and Cognitive Science: Language and the Unification of the Social and Computational Mind.* Cambridge, Mass.: Harvard University Press.

*Gelman, R. and Au, T. K.-F. (eds) 1996: *Perceptual and Cognitive Development.* New York: Academic Press.

Gelman, R. and Gallistel, C. R. 1978/1986: *The Child's Understanding of Number.* Cambridge, Mass.: Harvard University Press.

*Gleason, J. B. (ed.) 1997: *The Development of Language,* 4th edn. Boston: Allyn and Bacon.

Gopnik, A. and Meltzoff, A. N. 1996: *Words, Thoughts, and Theories.* Cambridge, Mass.: MIT Press.

Hirsh-Pasek, K. and Golinkoff, R. M. 1996: *The Origins of Grammar: Evidence from Early Language Comprehension.* Cambridge, Mass.: MIT Press.

Nelson, K. 1996: *Language in Cognitive Development: The Emergence of the Mediated Mind.* New York: Cambridge University Press.

Newport, E. L. 1988: Constraints on learning and their role in language acquisition: studies of the acquisition of American Sign Language. *Language Sciences,* 10, 147–72.

—— 1990: Maturational constraints on language learning. *Cognitive Science,* 14, 11–28.

Piaget, J. and Inhelder, B. 1969: *The Psychology of the Child,* trans. H. Weaver. New York: Basic Books.

*Pinker, S. 1994: *The Language Instinct: How the Mind Creates Language.* New York: William Morrow.

*Siegler, R. S. 1991: *Children's Thinking,* 2nd edn. Englewood Cliffs, NJ: Prentice-Hall.

Thelen, E. and Smith, L. 1994: *A Dynamic Systems Approach to the Development of Cognition and Action.* Cambridge, Mass.: MIT Press.

7

Conceptual change

NANCY J. NERSESSIAN

Introduction

Much of the attention of philosophy of science, history of science, and psychology in the twentieth century has focused on the nature of conceptual change. Conceptual change in science has occupied pride of place in these disciplines, as either the subject of inquiry or the source of ideas about the nature of conceptual change in other domains. There have been numerous conceptual changes in the history of science, some more radical than others. One of the most radical was *the chemical revolution*. In the seventeenth century, chemists believed that the processes of combustion and calcination involved the absorption or release of a substance called *phlogiston*. On this theory, when an ore is heated with charcoal, it absorbs phlogiston to produce a metal; when a metal is burned, it releases phlogiston and leaves behind a residue, or *calx*. The concept of phlogiston derived from a quite complex Aristotelian/medieval structure that included three concepts central to chemical theory: *sulphur*, the principle of inflammability; *mercury*, the principle of fluidity; and *salt*, the principle of inertness. All material substances were believed to contain these three principles in the form of *earths*. The phlogiston theory held that in combustion, the sulphurous earth (phlogiston) returns to the substance from which it escaped during some earlier burning process in its history, and that in calcination the process is reversed. However, chemists also knew that a calx is heavier than the metal from which it was derived. So, the theory implies that phlogiston has a negative weight, or a *positive lightness*. This did not present a problem, though, because it was compatible with the Aristotelian *elements* of fire and air (the others being earth and water), which were not attracted towards the center of the earth. The development of the oxygen theory of combustion and calcination by Lavoisier in the late eighteenth century has been called *the chemical revolution* because it required replacing the whole conceptual structure with, for example, different concepts of *substance* and *element* and new concepts of *oxygen* and *caloric*. In the new system, it was no longer possible to believe in the existence of substances with negative weight. According to the oxygen theory, oxygen *gas* is released in combustion and absorbed in calcination. Thus *calx* is metal (substance) plus oxygen, rather than metal minus phlogiston. The concept of phlogiston was eliminated from the chemical lexicon. The reconceptualization of chemical phenomena that took place in the chemical revolution made possible the atomic theory of matter, which, as we know, posits quite different constituents of material substances from the *principles* central to the earlier conceptual structure. Just what constitutes *conceptual change*, how it relates to theory change, and how it relates to changes in belief continues to be a subject of much debate. Clearly, though, as the preceding example demonstrates, the three are significantly interrelated.

There has been considerable mutual influence among the disciplines of philosophy of science, history of science, and psychology as regards research on the topic of conceptual change. With the development of the interdisciplinary field of cognitive science, that influence has been transformed into a deliberate use of research across a subset of practitioners of these disciplines. The future in this area, as in the field of cognitive science generally, lies in the synthesis of research in the different disciplines and in collaborations among them. In this discussion we will focus on the three most interactive areas that address conceptual change: cognitive development, science learning, and scientific change. In each of these areas there is considerable debate as to what constitutes conceptual change and how significant it is to understanding development, learning, and science, respectively. The objective of this chapter is to provide some insight into the issues and research of those who contend that conceptual change plays a major role in human cognition and to weave some connecting threads among them.

Briefly, how does the problem of conceptual change arise in each of these areas? In the area of cognitive development, the pioneering work of Jean Piaget found that children have concepts that are significantly different from those of adults and argued that a child's concepts change over the developmental process. Much contemporary research into children's intuitive understandings across a wide range of phenomena, including the shape of the earth, the day/night cycle, living things, and mental states, supports and extends at least the broad outlines of Piaget's findings. From this perspective, then, understanding cognitive development and its relation to maturation requires addressing the nature and processes of conceptual change (see Article 6, COGNITIVE AND LINGUISTIC DEVELOPMENT).

In the area of science learning, especially with the development of the field of *cognition and instruction* in the late 1970s, considerable research has established that students are not blank slates on which teachers can imprint scientific knowledge. Students come to school with intuitive conceptualizations of physical phenomena in several domains which differ from those of science. Further, these intuitive conceptualizations prove to be highly resistant to instruction. Developing pedagogical strategies to facilitate the shift from intuitive to scientific understanding requires insight into the nature and processes of conceptual change (see Article 54, EDUCATION).

Finally, in the area of scientific change, the histories of the various sciences exhibit considerable conceptual innovation and change. In some cases, such as the advent of relativity and quantum mechanics early in this century, the change in the conceptualization of nature has been so radical as to warrant the designation of a "revolution." Even though revolutionary conceptual change is infrequent, creating concepts through which to understand, structure, and communicate about physical phenomena occupies a central position in the scientific enterprise. Explaining scientific change requires examining how new conceptual structures emerge in science, what relation they bear to existing structures, and how they come to replace these. Most of our discussion will focus on scientific change, because research by philosophers and historians of science, to date, has had more impact on the areas of cognitive development and science learning than the reverse (see Article 60, SCIENCE). However, thinking about conceptual change in contemporary philosophy and history of science has been influenced considerably by other areas of cognitive science described in this volume.

Scientific change

Addressing the problem of conceptual change is essential to understanding the nature and development of scientific knowledge. The major changes in physical theory at the turn of the twentieth century thrust the problem of conceptual change into the spotlight for philosophers and historians of science. For nearly 300 years Newtonian mechanics had been held to be a true theory, and change within physics was seen largely as elaborating and extending the Newtonian world view. Characterizing conceptual change in such a way as to reconcile the seemingly radical reconceptualizations offered by relativity and quantum mechanics with an understanding of scientific change as by accretion shaped the initial perception of the problem in this century.

Until the advent of cognitive science, the two most influential views on conceptual change in science were the *received* view associated with logical positivist philosophers, chief among them Rudolph Carnap, Hans Reichenbach, and Carl Hempel, and the *radical* view of certain historicist philosophers, Norwood Hanson, Paul Feyerabend, and Thomas Kuhn. For the first half of the century, the characterization of the logical positivists held sway. They were called *positivists* because they identified with the tradition, originating with August Comte in the early nineteenth century, that held science to be the paradigm of empirical knowledge. In Comte's own view, science was the final, *positive* stage in the history of humankind's attempt to understand nature. The *logical* came from their advocacy of the newly developed symbolic logic as the primary methodological tool for analyzing scientific knowledge. The logical positivists approached the problem of conceptual change as follows. First, they characterized conceptual change as *continuous and cumulative*, holding the new conceptual structures to be logical extensions of previous ones. Second, they viewed scientific conceptual structures as languages and explored the relation of the terms (scientific concepts) in these languages to empirical phenomena and to one another using the methodological tools afforded by logic.

Beginning in the 1960s, the historicist philosophers offered critiques of positivism from the perspective that understanding scientific change requires reference to the actual history of science. The *radical* contingent argued that science history shows that major changes are best characterized as revolutions, in that they involve overthrow and replacement of the reigning conceptual system with one that is, at the very least, logically inconsistent with it, and at worst, what they called *incommensurable* with it. To take one of the stock examples, *mass* in Newtonian mechanics is an invariant quantity, whereas *mass* in relativity theory varies with motion. Though the same term is used in both cases, the meanings are so different that the relativistic conceptual structure cannot simply incorporate or *translate* the Newtonian structure within its bounds but replaces it. Thus, scientists maintaining the Newtonian world view should be unable to communicate with scientists maintaining the relativistic world view. They see and understand the world through radically different lenses. These philosophers were called "radical" because they characterized conceptual change as abrupt and discontinuous, which seemed to imply that scientific change is not a rational process. Significantly, though, what both the positivists and their historicist critics have in common is that the *problem of conceptual change* centers on the nature of the relations between the old and the new linguistic structures.

After much philosophical ink had been spilt in responding to the various conundra which the radical criticisms had raised, such as the problem of incommensurability mentioned above, the problem of conceptual change largely faded from the literature. This occurred not from a sense of the problem having been solved, but rather from a sense of the increasing sterility of the analyses presented. In particular, it seemed that many of the problems might be artifacts of the methodological tools and the implicit presuppositions with which the problem of scientific change, more generally, was being approached. From the perspective of a *cognitive* history and philosophy of science (HPS) that has been developing over the last 15 years, the major stumbling block is in fact how to determine the proper tools of analysis. Cognitive HPS maintains not only that the problem of conceptual change in science lies at the intersection of philosophy, history, and psychology, but that progress towards a solution requires combining the analytical resources and investigative findings of all three in *cognitive-historical* analyses.

First, understanding change requires engaging not simply in endpoints analysis, where the components of the completed conceptual structures are compared, such as in the *mass* example. Fine-structure analyses of scientific practices during the periods of emergence of, and transition between, conceptual structures must be constructed. Second, these examinations of scientific practices need to capture science as conducted in social contexts. These contexts provide material, conceptual, analytical, and cultural resources that facilitate and constrain concept formation and change. Third, scientists bring ordinary human cognitive resources and limitations to bear on their scientific representational, reasoning, and decision-making practices. What we are learning about these in the sciences of cognition need to be incorporated into investigations of scientific practice. The reverse also holds. Current theories of cognitive processes can be evaluated in light of how well they fit the scientific cognition evidenced in the historical cases.

Thus far, no single comprehensive model of conceptual change has emerged through cognitive-historical analysis. Significantly, though, it has recast the problem and set the agenda for contemporary investigations. Cognitive HPS focuses on the practices of scientists in creating conceptual change, not on the conceptual structures per se. Investigations of these practices have led to the view that much conceptual change can be characterized as continuous but noncumulative. For example, it is possible to trace a pattern of descent for the concept of a field from Faraday to Einstein, yet features of Einstein's concept of field are so different from those of any previous concept, such as Faraday's or Maxwell's or Lorentz's, that it cannot be viewed as simply an extension of any of these (see Nersessian, 1984). Research towards an explanatory account that might accommodate this characterization is proceeding in two directions. The first provides a new twist to the traditional problems concerning the nature of the structure of scientific conceptual systems and of the relations among old and new systems. The second addresses a problem traditionally thought to be intractable: the nature of the processes through which new conceptual structures are created.

The first line of research focuses on a central metatheoretical problem: What is the form of the representation of a concept? Underlying both the positivist and the radical historicist accounts is at least tacit acceptance of the *classical* notion that a concept is represented by a set of necessary and sufficient defining conditions. Attempting to fit science concepts and conceptual change into this notion has proved to be notoriously difficult. In science, new concepts are created, such as *spin* in quantum mechanics, and existing ones disappear, such as *phlogiston* from chemistry. However, a significant proportion of concepts in the new system seem to be *conceptual descendants* of existing ones. In the history of science we seem to be able to trace a distinct line of descent and a pattern of progress over time in a conceptual domain using concepts that are not consistent with one another, such as the concept of *inertia* from Galileo to Newton or the concept of *field* from Faraday to Einstein, or even the concept of *mass* from Newton to Einstein. Further, in the case of some, such as *ether*, which appears to have been eliminated, significant aspects have been absorbed by other concepts, in this case *field* and *space-time*. If situations of creation and disappearance were all that one encountered, conceptual change could be characterized fully in terms of the replacement of one structure by another. But, the existence of descendants and of absorption show the need to account for change in individual concepts, as well as entire systems, and to account for it in such a way as to accommodate continuous, noncumulative change at that level. The classical form of representation cannot provide such an account; for, if the defining conditions of a concept were to change, that would simply create a different concept.

Cognitive-historical analysis places the problems that arise with respect to scientific concepts within the context of the problem of human concept representation generally. There is extensive research in cognitive psychology and in psycholinguistics on categorization that provides evidence from ordinary human representation against the classical notion of concepts as well (see Article 25, WORD MEANING, and Article 8, CONCEPTUAL ORGANIZATION). This research was inspired by, and lends empirical support to, an early challenge to the classical notion offered by the philosopher Ludwig Wittgenstein, who argued that it is possible for various instances of a category to be related, even though some of the instances have no features in common (e.g., as in the series AB, BC, CD). He proposed that a better form of concept representation than a definition would be a set of overlapping features, or *family resemblances*.

Although there is consensus that the classical notion of concept representation is inadequate, there is none on which of the several alternative accounts that interpret these data is most satisfactory. Such data and interpretations have been utilized to examine scientific concepts in several different ways. I have adapted a *prototype* notion of a concept, associated with the work of Eleanor Rosch, to develop a *schema* representation of a scientific concept as an overlapping set of features. This analysis enables one to articulate the structure of individual concepts and their interrelations as they have developed over time. Recently Peter Barker and his collaborators (Anderson et al., 1996) have been drawing on the *frame* representation of a concept developed by Lawrence Barsalou (1992) to capture hierarchical and other relations among the features associated with a concept. Their analysis demonstrates how piecemeal transformations in concept frames can end in revolutionary change, making for continuous but noncumulative change. Paul Thagard has drawn on the *WordNet* model of lexical memory developed by George Miller to address the problem of how concepts within a structure are related to one another. This analysis treats concepts as nodes and articulates the structure of conceptual systems in terms of kind and part–whole relations and rules. Further, Thagard (1992) has used this model to argue that *explanatory coherence* of a conceptual system is a primary factor in the choice of one system over another.

The second line of research centers on the practices through which scientific concepts are constructed. Although still in the early stages, this research promises to move beyond description to explanatory accounts of the *mechanisms* or processes of conceptual change. The problem of how new conceptual structures arise was ruled out of philosophical analysis by the logical positivists. They equated scientific method with logic and held that there could be no *logic of discovery*. However, historical analyses across the sciences establish that conceptual innovation and change occur in a problem-solving process. Further, in numerous instances, extensive use is made of heuristics such as analogies, visual representations, and thought-experiments. Are these mere aids to thinking, as construed traditionally, or are they significant mechanisms for generating conceptual changes? The main problem which philosophers have had in even countenancing these as methods is that they are nonalgorithmic in application and even if used *correctly* may lead to the wrong solution or to no solution at all. This very feature, however, makes them more realistic from a cognitive-historical perspective. Scientists often use the same kinds of reasoning to go down fruitless paths as they do in successful outcomes. Viewing these reasoning practices in the light of cognitive research on ordinary reasoning practices provides support for their salience and insight into how they function in scientific reasoning (see Articles 1, 12, 20 and 21: ANALOGY, IMAGERY AND SPATIAL REPRESENTATION, PROBLEM SOLVING, and REASONING).

Cognitive-historical analyses lead to the interpretation that the problem-solving practices exhibited in historical records of concept formation and change are forms of *model-based reasoning*, specifically analogical and imagistic modeling and thought-experimenting and other forms of mental model simulation. Conceptual change often results from a model construction process involving different forms of abstraction (limiting case, idealization, generalization, generic abstraction), constraint satisfaction, adaptation, simulation, and evaluation. To engage in this practice, a scientist needs to know the generative principles and constraints for physical models in one or more domains. New representations are created from existing ones through processes that abstract and integrate source and target constraints into new models. Thus analogy plays a central role in accounting for the continuous, noncumulative nature of conceptual change. But the nature of creative uses of analogy in scientific conceptual changes still needs to be understood better. For example, some cases have involved not simply retrieving ready-to-hand analogies and mapping their salient structures to the new domain, but rather constructing, modifying, and merging imaginary analogs in interaction with constraints of the problem domain under investigation. Further, some reasoning in conceptual change exhibits visual modeling used in conjunction with analogy and dynamical mental simulation. Within cognitive science, analogy, visual representation and reasoning, and mental modeling have been functioning largely as separate areas of investigation. The case of scientific conceptual change shows that a unified account is necessary if we are to construct rich accounts of complex problem solving.

Cognitive development and science learning

Although each of these areas has produced a vast literature on conceptual change, we will discuss both together here, because parallel views in relation to scientific change have arisen in each area. Further, because much cognitive development takes place during the student years, the problems of conceptual change in learning and in development, though often addressed separately, are interrelated. The line of research we will consider in these areas has made considerable use of characterizations of conceptual change developed by philosophers and historians of science. As in the case of scientific change, conceptual change and theory change are thought to be intertwined in development and learning.

Piaget continues to have a profound impact on the current problem situation in these areas. This is especially so with respect to our topic, because the claim that scientific change is relevant to development and learning at all bears his imprint. Piaget's emphasis was on the domain-independent logical structures which he hypothesized were acquired at specific ages. He constructed a *stage theory*, in which cognitive development consists in the unfolding of biologically determined stages, and conceptual change is a process of *assimilation* and *accommodation* through which innate conceptions are restructured in the light of experience. In contrast, the focus of contemporary research has been on the nature of the domain-specific content of children's concepts, such as Susan Carey's (1985) seminal study of the concept of *living thing*. Studies such as those by Carey, Stella Vosniadou and William Brewer (1992), and Alison Gopnik (1996), among others, have led to what has been called the "theory theory account." These psychologists have proposed a theory that the processes of conceptual change in cognitive development take the form of theory formation and change (thus, the *theory theory*). That children converge on specific concepts and theories at roughly the same age is a function of having roughly similar biological theory formation capacities, social supports for learning, and evidential inputs. For many of these researchers, the history of scientific change provides a basis from which to argue that the conceptual changes which children undergo in development are similar in specific characteristics to the kinds of changes in conceptual structure that have taken place in scientific *revolutions*. A similar, related view, now also called the *theory theory*, was proposed earlier in science learning by Michael McClosky (1983) and John Clement (1983) among others. In this domain, the *theory theory* account proposes that students (of all ages) construct intuitive theories of phenomena encountered in their experience. The nature of the ontological commitments and explanatory structure of intuitive theories are thought to be sufficiently different from those of scientific theories to account for the robustness of intuitive concepts in the face of science instruction. Most proponents agree, though, that intuitive theories are rudimentary, and not as systematic and articulated as scientific theories. One implication of the view is that if learning is akin to theory change, difficulty learning abstract scientific concepts may be less a function of a student not having the requisite cognitive resources due to level of development than a function of our not having developed appropriate teaching methods for facilitating change.

Again, these researchers' use of the history of scientific change contrasts with Piaget's. Piaget noted that there are intriguing parallels between conceptualizations of nature produced during the course of cognitive development and those developed over the course of the history of science. He took the parallels to indicate a conceptual form of "ontogeny recapitulates phylogeny" and attempted to fit the entire unfolding of the history of science into the framework of his theory of cognitive development. Contemporary psychologists have continued to find and elaborate intriguing parallels, but their strategy is to use the historical cases to assist in characterizing the kinds of transformations that need to take place in learning and development. The interdisciplinary research area of *cognition and instruction* has made considerable use of the history of scientific change in considering the problem of how to help students learn the conceptual structure of a science. This problem has generated numerous investigations into novice (*naive* or *untutored*) representations of specific domains, including object motion, electricity and magnetism, and human physiology. These investigations comprise a vast and persuasive literature demonstrating that the conceptualizations which students have of many phenomena prior to and often post instruction differ significantly from those of the sciences. Thus, students are thought to have to undergo a major conceptual *restructuring* in order to learn a physical theory.

Despite controversies over the nature of the structures which students may be calling upon and the nature of the *restructuring*, there is widespread agreement that students are not empty vessels into which teachers pour knowledge. This raises the issue of how existing concepts affect learning new material. Take, for example, the case of learning Newtonian mechanics. A nearly universal finding is that students enter (and often exit!) a beginning physics class with a belief that can be characterized as "all motion implies force" (see McDermott, 1984, for a survey). Newtonian mechanics holds that "accelerated motion implies force." Further complicating this learning task, student explanations of motion reveal that their concepts of *motion* and *force* are quite different from the Newtonian concepts. In Newtonian mechanics *motion* is a state in which bodies remain unless acted upon by a force. *Rest* and *motion* have the same ontological status: they are both states. Like rest, motion per se does not need to be explained, only changes of motion. *Force* is a functional quantity that explains changes in motion. Newtonian forces are relations between two or more bodies. Students, however, conceive of *motion* as a process that bodies undergo, and they believe that *all* motion needs an explanation. They conceive of *force* as some kind of power imparted to a body by an agent or another body. This makes *force* ontologically a property or perhaps even an entity, but not a relationship. Thus, learning that "accelerated motion implies force" requires constructing new representations for *motion* and *force* (Nersessian, 1989). That difference in ontological categories is a significant marker of conceptual change figures prominently in Michelene Chi's (1992) recent research on learning. She argues that whole complexes of new ontological categories need to be constructed in order to learn science.

The main support for the psychological hypothesis that conceptual changes in learning and in development are similar to that in scientific change comes from research that describes the initial state of a child's or student's representation of a domain and compares that state with the desired final state. Susan Carey has argued that the *kinds* of change necessary to get from one state to the other resemble those that have taken place in scientific revolutions as characterized by Kuhn (1970). She uses the notion of *incommensurability* between child and adult concepts as evidence of conceptual change in cognitive development. The end-state comparisons, such as in the *motion* and *force* example, do give a sense that the kinds of change necessary for students to learn may be like those in what philosophers and historians of science have characterized as *scientific revolutions*. Significantly, though, even if the *kinds* of change required are strikingly similar in learning, development, and science, this does not mean that the *processes* of change will in any way be alike. It is an open question in need of investigation whether historical conceptual changes can provide models for either learning or cognitive development at the level of processes or *mechanisms* of change.

Implications for cognitive science

Precisely because the advocates of the *theory theory* are not Piagetian with respect to the processes of cognitive development, learning, and conceptual change, they owe us an account of the nature of the processes of conceptual change and of what the activity of *theorizing* comprises, and some are beginning to address these issues. However, cognitive psychologists are still working with accounts of theory and concept formation and change that are influenced by a mixture of earlier empiricist and Kuhnian philosophical accounts, and this presents an obstacle to their undertaking. As we have seen, these accounts provide no insight into the processes of conceptual change. The objectives of those working in development and learning would be better served by collaborating with researchers in cognitive HPS who are addressing the problem of the processes of conceptual change in science. From a cognitive-historical perspective, the problem-solving strategies which scientists have invented and the representational practices they have developed over the course of the history of science are sophisticated, refined outgrowths of ordinary reasoning and representational processes. From childhood through adulthood, ordinary and scientific reasoning and representation processes lie on a continuum, because they originate in modes of learning which evolution has equipped the human child with for survival into adulthood. On this *continuum hypothesis* the cognitive activity of scientists is potentially quite relevant to learning, and vice versa. But again, there are quite difficult questions that need to be addressed in exploring and testing this hypothesis, such as how maturation might complicate learning. Further, in addition to the similarities, there are significant differences that need to be addressed. For example, in cognitive development, change seems just to happen, but in education it requires explicit teaching, and in science it requires explicit and active investigation. Further, in both psychological cases, there are culturally available concepts which the child must acquire in order to mature properly or which we desire them to learn in the educational context, but in the scientific case conceptual change is truly creative. Exploring these and other fundamental questions about the nature and processes of conceptual change in science, cognitive development, and learning in a collaborative undertaking promises to yield richer, possibly unified accounts of this aspect of human cognition.

References and recommended reading

Anderson, H., Barker, P. and Chen, X. 1996: Kuhn's mature philosophy of science and cognitive science. *Philosophical Psychology*, 9, 347–63.

Barsalou, L. 1992: Frames, concepts, and conceptual fields. In A. Lehrer and E. Kittay (eds), *Frames, Fields, and Contrasts: New Essays in Semantical and Lexical Organization*, Hillsdale, NJ: Erlbaum, 21–74.

Carey, S. 1985: *Conceptual Change in Childhood*. Cambridge, Mass.: MIT Press.

Chi, M. 1992: Conceptual change within and across ontological categories: examples from learning and discovery in science. In R. Giere (ed.), *Cognitive Models of Science*, Minneapolis: University of Minnesota Press, 129–86.

Clement, J. 1983: A conceptual model discussed by Galileo and intuitively used by physics students. In D. Gentner and L. Stevens (eds), *Mental Models*, Hillsdale, NJ: Erlbaum, 325–40.

*Gooding, D. 1990: *Experiment and the Making of Meaning*. Dordrecht: Kluwer Academic Publishers.
Gopnik, A. 1996: The scientist as child. *Philosophy of Science*, 63, 485–514.
*Hempel, C. G. 1952: *Fundamentals of Concept Formation in Empirical Science*. Chicago: University of Chicago Press.
*Hoyningen-Huene, P. 1993: *Reconstructing Scientific Revolutions: Thomas S. Kuhn's Philosophy of Science*. Chicago: University of Chicago Press.
*Kuhn, T. S. 1970: *The Structure of Scientific Revolutions*, 2nd edn, enlarged. Chicago: University of Chicago Press.
McClosky, M. 1983: Naive theories of motion. In D. Gentner and A. L. Stevens (eds), *Mental Models*, Hillsdale, NJ: Erlbaum, 229–334.
McDermott, L. 1984: Research on conceptual understanding in mechanics. *Physics Today*, 37, 24–32.
Nersessian, N. J. 1984: *Faraday to Einstein: Constructing Meaning in Scientific Theories*. Dordrecht: Kluwer Academic Publishers.
—— 1989: Conceptual change in science and in science education. *Synthese*, 80, 163–84.
—— 1992: How do scientists think? Capturing the dynamics of conceptual change in science. In R. Giere (ed.), *Cognitive Models of Science*, Minneapolis: University of Minnesota Press, 3–44.
Thagard, P. 1992: *Conceptual Revolutions*. Princeton: Princeton University Press.
*Tweney, R. D. 1985: Faraday's discovery of induction: a cognitive approach. In D. Gooding and F. A. J. L. James (eds), *Faraday Rediscovered*, New York: Stockton Press, 189–210.
Vosniadou, S. and Brewer, W. 1992: Mental models of the earth: a study of conceptual change in childhood. *Cognitive Psychology*, 24, 535–85.

8

Conceptual organization

DOUGLAS MEDIN AND SANDRA R. WAXMAN

Introduction

Questions about concepts bring into play all the cognitive science disciplines. For many centuries, concepts belonged to philosophy; but more recently, these original caretakers have shared responsibility for this domain with cognitive and developmental psychology, linguistics, artificial intelligence, anthropology, and neuroscience. Each of these fields has offered insights into these building blocks of thought, and each has contributed a unique perspective on fundamental questions about the nature of minds. However, the integrative approach of cognitive science holds the promise of providing new vantage points from a range of disciplines on this core issue.

Our goal here is to consider the nature of the interplay between culture, language, and thought in the development and modification of conceptual systems. Two questions serve as a unifying theme: How do peoples across the world and across development organize their knowledge about objects? How can we best capture the similarities and differences in these systems across cultures?

We have selected this topic because the past several years have witnessed a burgeoning appreciation of the influence of culture and its artifacts (especially language) in shaping human conceptual systems. Cognitive psychologists have developed a virtual army of experimental techniques to examine the acquisition of concepts and taxonomies and to probe the psychological consequences of these structures in reasoning. Developmental psychologists have introduced powerful, innovative techniques that enable us to tap into the early acquisition of conceptual and linguistic systems of organization and to trace their development over time. But as a rule, cognitive and developmental psychologists have carried out their research on limited cultural populations and with an unsystematic sampling of biological kinds. By contrast, anthropologists have devoted considerable attention to documenting taxonomies of the biological world from many distinct geographical and cultural regions; they have also provided detailed accounts of convergences between patterns of object classification and nomenclatural patterns (see Article 5, COGNITIVE ANTHROPOLOGY). But as a rule, these anthropological accounts have left issues concerning the perceptual and conceptual processing mechanisms that support these capacities largely unexplored. As these diverse contributions to the study of human conceptual organization are shared across disciplines, it becomes clear that the potential synergy among these interdisciplinary contributions makes the cognitive science enterprise especially promising.

We will describe two programs of research that are cross-cultural and cross-disciplinary in character. One program is focused on cross-cultural similarities and differences in systems of biological categorization. A central focus in this program is the notion of a privileged level within a hierarchical system. Noting the considerable cross-cultural agreement in categorization, we go on to consider whether this agreement reflects a pattern in nature, where certain categories stand out as "beacons on the landscape of biological reality" (Berlin, 1992), a position advanced by anthropologist Brent Berlin at the University of California, Berkeley. Alternatively, cross-cultural agreement may reflect universal properties of the human mind, a position advanced by anthropologist Scott Atran at the University of Michigan and CNRS in Pris, France (Atran, 1990). The second research program traces the early establishment of conceptual systems and asks two interrelated questions: What initial cuts do infants and young children make in categorizing objects in the world? How do they go beyond these initial cuts to form the complex, flexible conceptual systems that characterize adult cognition? Central to this enterprise is the role of language in directing children's attention to object categories at various hierarchical levels.

It will become clear that these research programs have a great deal in common. Each is concerned with the acquisition and consequences of hierarchical systems of knowledge; each utilizes a multidisciplinary approach; each signals the importance of the interplay between processes inherent in the human mind and learning from the environment; and each points to the need for additional research from a cross-cultural, developmental perspective.

Privilege in taxonomic hierarchies: folk-biological classification and reasoning

One important aspect of categorization is that any individual may belong to multiple, hierarchically organized categories. For example, a furry creature may be categorized as a grey squirrel, a squirrel, a mammal, a vertebrate, an animal, a living thing, and so on. One of the major observations over the past two decades of research on concepts is that these levels are not equally salient psychologically. Instead a single level, called the "basic level" in psychology, appears to be privileged. What do we mean by *privileged*? Informally, the basic level constitutes the *best name* for something, the one that adults prefer to use in naming and, perhaps not coincidentally, the one that word-learners master first. Actually, there are a number of criteria which one could use for basicness, and the remarkable thing is that some pioneering observations by Eleanor Rosch at Berkeley, and her co-workers suggest that these various measures all mark the same level as privileged. Let us begin with her studies.

Rosch, Mervis, Gray, Johnson, and Boyes-Braem (1976) argued that the correlational structure of entities in the world creates natural clusters, and that concepts correspond to these clusters. For example, things with feathers are likely to have beaks, wings, and two legs and to fly. Creatures without feathers are less likely to have these other properties. This means not only that knowledge of some features can be used to predict other properties, but also that entities are distributed as clusters, or groups of similar things. Entities near the center of such clusters are said to be *better* or more typical examples of the category than more peripheral, atypical examples. One important theoretical idea is that the mental representation of these categories takes the form of a prototype which summarizes the central tendency of the category (cluster). The closer an example is to the prototype, the more typical it is. A contrasting view is that the category representation is simply the disjunction of the representations of individual category examples (the so-called exemplar view). Data from experiments tend to favor the latter, exemplar-based view over the prototype view. For purposes of the present essay, this distinction is less important than the idea that categories reflect chunks or clusters of similar entities.

Rosch and her associates (1976) argued that there was one level of granularity at which these clusters stood out the most. They called this level the "basic level." Basic-level categories such as chair, hammer, and dog may be contrasted with more general superordinate categories (furniture, tool, animal) and more specific, subordinate categories (recliner, hammer, poodle). Rosch et al. evaluated a number of criteria for category use, and they all pointed to a single level as privileged. Basic-level categories are the most inclusive categories that (a) possess numerous common attributes, (b) have similar shapes and can be identified from averaged shapes of members of the class, and (c) involve the same movements when handled or interacted with. In addition, basic-level category labels are preferred in adult naming and are learned first by children. Moreover, adults can identify entities at the basic level more rapidly than at more subordinate and superordinate levels. Finally, across languages and cultures, these basic level clusters are the ones that tend to be named. It is as if the structure of nature imposes itself on the human mind in the case of basic-level categories.

Observations from anthropology also point to one level that is psychologically privileged. Indeed, Berlin (1992) uses the same structure-in-the-world framework in arguing that one level in a taxonomic hierarchy is "crying out to be named."

Here is where it starts to get interesting, for what seems like a convergence between these two fields of research is actually a deep puzzle. The level that ethnobiological studies suggest is basic corresponds more or less to the genus level in scientific taxonomy. However, Rosch et al. found that the genus level was not basic: rather than *robin*, *trout*, and *maple* being privileged, Rosch et al. found that *bird*, *fish*, and *tree* met their criteria for basicness.

Why do anthropological and psychological measures of the basic level disagree? One possibility is related to expertise. Perhaps the Berkeley undergraduates in Rosch's studies knew little about biological categories, especially relative to people of the agricultural societies investigated in most ethnobiological studies. It seems plausible that people in cultures that are organized around technology may display less understanding of the natural world than people in traditional agrarian societies.

But this line of reasoning raises a developmental question: Would young children from technological and traditional societies show more agreement in the basic level than adults? If so, what factors might be responsible for this divergence with development? Does it stem from (a) expertise (in traditional societies) leading to the acceptance a more specific basic level or from (b) deterioration of knowledge in technologically oriented cultures leading to the acceptance of a more general basic level?

Answers to these questions will depend upon cross-cultural developmental programs of research. To the best of our knowledge, there have been no systematic examinations of this issue. However, Brian Stross's observations of children in Chiapas, Mexico (Stross, 1973), and Janet Dougherty's observations of children in Berkeley, California (Dougherty, 1979), suggest that there may be cultural differences in the privileged level, as indexed by children's preferred level of naming plants. Their results are preliminary, but they indicate a clear course for future research. What is required is cross-cultural developmental research on language and conceptual organization from infancy, throughout acquisition, across cultures.

A second potential reason for the Rosch–Berlin disparity is that the distributional patterns of flora and fauna differ across traditional and technological cultures. Biodiversity is greater in Central American rain forests than in more temperate climates. Furthermore, technological advances, sadly, are coupled with pollution, acid rain, and destruction of habitat, which trigger a further loss of diversity. If there are fewer distinct natural kinds in technological cultures, then coarser conceptual cuts may suffice. In the limiting case of one kind of fish and one kind of bird, labels like *bird* versus *fish* serve us as well as *downy woodpecker* and *steelhead trout*. Of course we are not (yet) at the limiting case. However, distributional patterns need to be taken seriously. This is an area in which anthropologists' tools have been honed more sharply than those of psychologists. For example, ethnobiological research typically is based upon botanical and zoological surveys of local plants and animals. Psychologists are too often guilty (the authors are no exception) of compiling stimuli without sufficiently addressing their representativeness.

A third difference between the work of psychologists and anthropologists is that different measures are used in the two types of investigations. For example, Berlin's observations relied heavily on linguistic measures like naming, whereas Rosch et al. focused more on perceptual tasks. Perhaps if the same measures were used, the differences in the apparent basic level would disappear. Clearly needed is a systematic comparison with a common set of measures.

Having raised these questions, it would be nice if we could answer them. But at best, we have only a piece or two of what promises to be a large (and fascinating) puzzle. Our first pass at comparability actually involves a measure used by neither Rosch nor Berlin – category-based induction. In this task, participants are told that some property is true of some category and are then asked to evaluate how likely that property is to be true of some other category. For example, if downy woodpeckers have sesamoid bones, how likely is it that ringneck pheasants also have sesamoid bones? Alternatively, one might ask questions involving different taxonomic levels: for example, if downy woodpeckers have sesamoid bones, how likely is it that all woodpeckers (all birds, all animals) have sesamoid bones?

Our experiments were designed and conducted by John Coley, Douglas Medin, and Elizabeth Lynch from Northwestern University in collaboration with Scott Atran (Coley et al., 1997). We used a range of plants and animals and abstract properties (e.g., has enzyme *x*) projected to different taxonomic levels (e.g., woodpecker, bird, animal). The participants in our first few studies were Northwestern University undergraduates whose knowledge of the natural world was, to say the least, limited. If the levels that Rosch found to be basic are privileged in induction, then anytime we ask our participants to project a property above that level, the rating or confidence should show a sharp drop. For example, from the statement "Downy woodpeckers have enzyme *x*," they should be fairly sure that all woodpeckers have the enzyme, somewhat less sure that all birds have it, and not at all sure that all animals have it. Berlin's observations would lead to the expectation that the biggest drop would come earlier, in this case in going from woodpeckers to all birds. In each of our studies (they involved ruling out alternative explanations and other methodological issues that we won't bore you with) we found that, consistent with Berlin's ideas about basicness, the folk-generic level (corresponding closely with the genus level in scientific taxonomy) acted as privileged in induction. We then ran more or less the same study with Itzaj Maya men and women from a community in Guatemala. The Itzaj have managed to live in the rain forest without destroying it, and both men and women are highly knowledgeable about the biological world. You will not be surprised to learn that we had to make a few adjustments in our procedure (e.g., different plants and animals, a verbal measure of confidence rather than a numerical rating scale). Nonetheless, our results were virtually identical – the genus or folk-generic level was clearly privileged relative to higher levels.

But how likely is it that our students and the Itzaj have the same privileged level on other measures of basicness, such as speeded categorization or names first learned by children? This, of course, is an empirical question, and we are far from sure that these parallels will continue. In fact, we have evidence from the same sort of reasoning tasks that points to some differences. Typically, undergraduates show strong similarity effects in category-based reasoning. For example, if told that sparrows have some disease, they are more sure that robins (which are similar birds) can also get this disease than that, for example, pheasants can get the disease. Although the Itzaj Maya sometimes give the same answers, they provide very different justifications for them. In particular, the judgments of the Itzaj are heavily based on specific ecological knowledge rather than similarity. To give a hypothetical example, they might explain that sparrows and robins both eat some insect that could give them the disease. Undergraduates cannot do this sort of reasoning very well, because typically they do not have the ecological knowledge to support it.

These investigations of categorization and reasoning among peoples from diverse cultures, language groups, and natural environments have identified points of universality and of difference. But it is also important to go further in this interdisciplinary endeavor, to pinpoint the mechanisms responsible for the similarities and differences we have observed across cultures. For example, to ascertain whether these cultural differences are related primarily to differences in expertise, we are currently studying the categorization and reasoning of selected subpopulations of North Americans (e.g., bird-watchers, tree experts, etc.) to see how novices and experts differ. Another approach is to chart the emergence and modification of these systems over time. This is where the developmental component of this multidisciplinary endeavor becomes essential. In the next program of research it will become clear that developmental work can reveal the *initial cuts* that infants make in categorizing the objects they encounter; it can also illustrate the powerful role of language in shaping the acquisition of hierarchical systems.

Language and the acquisition of hierarchical systems of knowledge: developmental and cross-linguistic considerations

One of the most robust findings in the developmental literature is that infants and young children first succeed in labeling and categorizing objects at a mid-level position within a hierarchical system, well before they do so at other hierarchical levels. We note that developmentalists face the same puzzles that we identified earlier concerning this preferred level. For example, there has been some debate as to the precise scope of children's first categories (e.g., duck versus bird); it has also been difficult to provide a formal account of this privileged level. These uncertainties notwithstanding, the notion that mid-level basic object categories are privileged in development is well established.

But how do children progress beyond these initial, privileged mid-level categories to build hierarchical systems of organization? Developmental research has revealed that language serves as a catalyst for the acquisition of concepts, particularly those at *nonprivileged* levels. To ascertain whether children direct their ATTENTION differently in the context of word learning than in nonlinguistic contexts, researchers have introduced children to novel words and have observed the effects of these labels in object classification at various hierarchical levels. Several different laboratories have revealed that children direct their attention differently in the context of learning a novel word than in neutral situations that include no novel words (see Waxman, 1994, for a review). Sandra Waxman at Northwestern University and her colleagues have shown that by two or three years of age, children interpret novel count nouns as referring to object categories and interpret novel modifiers (e.g., adjectives) as referring to properties of objects and subordinate-level categories. Thus, when children hear an object labeled, the linguistic form of the label directs their attention to particular aspects of the object.

These correlations, or linkages, between linguistic form (e.g., noun, adjective) and hierarchical level reveal one way in which language influences conceptual organization. These linkages have been invoked to help explain how toddlers so rapidly map words to their meanings and so successfully establish hierarchical systems of categories. Notice that these linkages, which have also been noted in the ethnobiological literature (Berlin, 1992), insure that the labeling practices of the adult community will shape the lexical and conceptual systems of the young.

But how do these linkages unfold? To answer this question, Waxman and her students pursued two complementary lines of research, examining the influence of language on categorization in two distinct populations: 12- to 14-month-old infants from English-speaking families, who have just begun to produce their first words, and preschool-aged children acquiring either English, French, or Spanish as their native language.

In the infancy studies, Waxman employed a novelty-preference task to examine the influence of novel words on 12-month-olds' object categorization. In the familiarization phase, an experimenter offered an infant four different toys from a given category (e.g., four animals), one at a time. In the test phase, the experimenter presented both (a) a new member of the given category (e.g., another animal) and (b) an object from a novel contrasting category (e.g., a tool). Infants were tested on both basic (e.g., cats versus horses) and superordinate (e.g., animals versus vehicles) level categories. The logic of this paradigm for examining infant categorization is as follows. If the infant notices the commonalities among the familiarization stimuli, then the infant's attention during familiarization should decrease; at test, the infant should show a preference for the novel, over the familiar, test object. Further, if novel words direct infants' attention toward object categories, then infants who hear novel words in conjunction with the objects presented during familiarization should be more likely to categorize in this task than should control subjects who hear no category labels. The data revealed a consistent effect of novel words in these infants on the brink of producing language.

Infants who heard novel words were more likely to form object categories than were those in a no word control condition.

Three points are especially germane here. First, infants in all conditions formed basic-level categories (e.g., cats versus horses) successfully; novel words did not influence infants' successful performance at this level. This accords well with assertions regarding the developmental primacy of these mid-level categories. Second, at nonbasic levels (e.g., animals versus vehicles), words exerted a clear influence. Only infants hearing novel words successfully formed object categories; those in the no word control condition exhibited no such pattern. This indicates that labels serve as a catalyst in conceptual development, particularly when the perceptual support for a category is not as compelling as it is at the privileged basic level. Third, the linkage between language and conceptual organization is relatively general during infancy: both nouns and adjectives highlight object categories, particularly at superordinate levels. This general linkage becomes more specific over development: by at least three years of age, children distinguish between nouns and adjectives, assigning to each particular types of meaning. Therefore, between infancy and the preschool years, there is a growing sensitivity to using linguistic form as a cue to meaning.

Waxman next asked whether and how these linkages are influenced by the language being acquired. In collaboration with Anne Senghas, an MIT-trained linguist, Luis and Susana Benveniste, from Buenos Aires, and Danielle Ross, a native of Montreal, she conducted a series of cross-linguistic, developmental experiments with young monolingual children in the process of acquiring either English, French, or Spanish.

A comprehensive review of the cross-linguistic literature highlighted the cross-linguistic stability of the grammatical category *noun*, as compared to *adjective*. We therefore predicted that the linkage between count nouns and object categories, which emerges early in development, would be evident across human languages. We also predicted that there would be cross-linguistic variation in the interpretation of adjectives (see Waxman et al., 1997, for a more complete account).

Despite the similarities among English, French, and Spanish, there is an important difference in the grammatical use and referential status of adjectives in these three languages. In Spanish (but not English or French), adjectives commonly appear in many of the same syntactic contexts as nouns and (like nouns) often refer to object categories. To examine the consequence of this syntactic and semantic overlap in children's expectations concerning word meanings, we adopted a forced-choice procedure. Children were introduced to a target object (e.g., a dog) and four alternatives: two members of the same superordinate-level category as the target (e.g., bear, fox) and two thematically related alternatives (e.g., dog bone, dog's bowl). In each language, children were assigned to one of three conditions, depending upon how the target objects were introduced: either with a *novel noun* (e.g., "Look at the *dax*"), a *novel adjective* (e.g., "Look at the *dak-ish* one"), or *no word* ("Look at this").

Children's interpretations of novel nouns were uniform across the languages examined. The expectation that a novel noun can be extended to include the target object and other members of its superordinate-level kind was evident in French- and Spanish-speaking children, just as it has been in English-speaking children and in infants in an English-speaking environment. This is consistent with the prediction that the noun-category linkage would be stable across development and across languages.

Children's interpretations of novel adjectives varied across the languages, clearly implying an important role for language-specific learning. In English and French, children revealed no preference for the taxonomic (or thematic) alternatives in the *novel adjective* conditions in our categorization task. By contrast, in Spanish, children exhibited a strong tendency to extend novel adjectives, like novel nouns, to the taxonomic alternatives. In Spanish, then, where adjectives are habitually permitted to adopt some of the syntactic and semantic features associated with count nouns, children have learned that adjectives, like nouns, may be used in a categorical sense. Thus, children acquiring different languages revealed different tacit expectations regarding the range of meanings associated with the grammatical category *adjective*.

In sum, children acquiring English, French, or Spanish share an expectation that a count noun applied to an individual will refer to that individual and can be extended to include other members of the superordinate-level kind. However, experience with these different native languages leads to different outcomes in children's expectations concerning the range of meaning associated with novel adjectives. This is consistent with the hypothesis that the linkage between nouns and object categories emerges early and is a candidate for universality, and that the meanings associated with adjectives may be more language-specific.

These cross-linguistic findings dovetail with those from infants. Basic-level categories appear to be salient to infants and young children. Categorization at this level is successful under a variety of circumstances. Yet, at nonbasic levels, labels play a powerful role, guiding the formation of categories beyond the privileged mid-level. Infants share with preschool-aged English-, French-, and Spanish-speaking children an expectation that count nouns can be extended to categories of objects. This expectation facilitates the formation of superordinate-level categories. By contrast, the mappings between adjectives and their meanings appear to emerge later in development, and to vary systematically according to the particular language under acquisition. Thus, early in acquisition, infants share a common set of expectations concerning the linkages between word meaning and conceptual organization, and these initial expectations become more entrained with age and language experience.

Implications for cognitive science

There are several unifying themes in the research described in this essay. Both the adult and the developmental programs draw upon multidisciplinary findings to address fundamental questions in the acquisition and modification of conceptual systems of organization. Both underscore the importance of the interplay between processes inherent in the human mind and learning based upon the input from the environment. Both expose the power and complexity of language as a force in the establishment of hierarchical systems of knowledge. A number of other common concerns and themes have been implicitly interwoven: conceptual acquisition and change; ranks and levels in hierarchically organized categories; and the complex and powerful role of language in shaping both. But perhaps most important, these programs illustrate vividly the possibilities that arise in cognitive science once interdisciplinary borders become permeable. We are optimistic about further integrating cross-cultural, cross-disciplinary, and developmental programs of research to address fundamental issues in cognitive science.

References and recommended reading

Atran, S. 1990: *Cognitive Foundations of Natural History: Towards an Anthropology of Science.* Cambridge: Cambridge University Press.

Berlin, B. 1992: *Ethnobiological Classification: Principles of Categorization of Plants and Animals in Traditional Societies.* Princeton: Princeton University Press.

Coley, J. D., Medin, D. L. and Atran, S. 1997: Does rank have its privilege? Inductive inferences within folkbiological taxonomies. *Cognition*, 64, 73–112.

Dougherty, J. W. D. 1979: Learning names for plants and plants for names. *Anthropological Linguistics*, 21, 298–315.

*Goldstone, R. L. 1994: The role of similarity in categorization: providing a groundwork. *Cognition*, 52, 125–57.

*Hirschfeld, L. A. and Gelman, S. A. 1994: *Mapping the Mind.* Cambridge: Cambridge University Press.

*Markman, E. M. 1989: *Categorization and Naming in Children: Problems of Induction.* Cambridge, Mass.: MIT Press.

*Medin, D. L. and Heit, E. 1995: Categorization. In D. E. Rumelhart and B. O. Martin (eds), *Handbook of Cognition and Perception: Cognitive Science*, San Diego, Calif.: Academic Press.

Rosch, E., Mervis, C. B., Gray, W. D., Johnson, D. M. and Boyes-Braem, P. 1976: Basic objects in natural categories. *Cognitive Psychology*, 8, 382–439.

Stross, B. 1973: Acquisition of botanical terminology by Tzeltal children. In M. S. Edmonson (ed.), *Meanings in Mayan Languages*, The Hague: Mouton, 107–42.

Waxman, S. R. 1994: The development of an appreciation of specific linkages between linguistic and conceptual organization. *Lingua*, 92, 229–57.

*Waxman, S. R., Senghas, A. and Benveniste, S. 1997: A cross-linguistic examination of the noun-category bias: its existence and specificity in French- and Spanish-speaking preschool-aged children. *Cognitive Psychology*, 43, 183–218.

9

Consciousness

OWEN FLANAGAN

What is consciousness? What role, if any, does consciousness play in the explanation of cognition? Can consciousness be studied empirically? These are the questions. Here are the answers.

Consciousness and intelligence

In 1997, what had been predicted for almost half a century happened. The reigning world chess champion, Garry Kasparov, possibly the best chess-player ever, was defeated by Deep Blue, an IBM computer, in a six-game match. The computer does not know chess from checkers. Furthermore, Deep Blue's owners, agents, seconds, trainers – it is hard to know what to call them – could not beat Kasparov individually if their lives depended on it and if Kasparov had ingested more than his fair share of vodka. Deep Blue *knows* about the history of great chess games and can compute possible moves more quickly and deeply than Kasparov. Most think that this is why Deep Blue won the match. However Deep Blue did it, it beat Kasparov. In one sense, Deep Blue is very intelligent. But Deep Blue is totally unconscious.

This example shows that problems of intelligence or rationality can, in principle, be prized apart from questions of conscious experience. David Chalmers calls the problems associated with intelligent information processing and action guidance "the easy problems" in the sense that we can picture how the natural sciences can solve them. Consciousness is "the hard problem."

> [E]ven when we have explained the performance of all the cognitive and behavioral functions in the vicinity of experience – perceptual discrimination, categorization, internal access, verbal report – there may still remain a further unanswered question: Why is the performance of these functions accompanied by experience? . . . This further question is the key question in the problem of consciousness. Why doesn't all this information-processing go on "in the dark," free of any inner feel? (Chalmers, 1995, p. 203)

If cognitive science defines its job as explaining the capacities that govern intelligence, information processing, logical inference, categorization, and verbal reports, then perhaps it can do its job without invoking consciousness. Such a strategy might be considered particularly wise if consciousness is considered, as it has been historically, the essence of mentality but not part of the natural order. I'll explain.

Consciousness and causation

Consciousness is the most familiar thing of all, but it is hard to say what it is. Thomas Nagel (1974) puts it this way: to be conscious, or in a conscious mental state, is to be in a state that seems a certain way to you, the subject of experience. There is something *it is like* to be in a conscious mental state, whereas there is nothing it is like to be in an unconscious mental state. Unconscious mental states, such as the states that my mind must be in to type these words – getting thoughts (some of which I have not articulated) to my fingers – don't seem any way at all. Still, such states are, for the reasons just given, causally efficacious. My desire to finish this essay, on the other hand, is vivid. The editors are after me to finish. I have an intense fear that they may even have hired thugs to harm my knees if I don't meet their final deadline. I also have vivid thoughts, such as that this topic really interests me. I am consciously aware of these thoughts and feelings, and they seem causally efficacious, part of the causal chain leading to the production of this essay. But are they really important players in the causation of my writing? How could one ask such a silly question? But consider this analogy.

My computer printer reveals my thoughts in tangible form. But my printer is not in any important sense involved in the process of composing what I write. Despite the perspicacity, noisiness, and fulfilling role it plays in the display of my written words, my printer is really, truth-be-told, just a public relations device. It gives voice to, or makes public, what has already been produced. Just as one might overrate the role of one's printer in the process of writing, one might overrate the causal efficacy of the conscious mind in the generation of action (Dennett, 1978). It is possible, after all, that conscious mental states lie at the end of causal chains. Perhaps they are broadcasts of what has happened, or is happening, in some largely unconsciously driven system, and are not, as they seem, critical initiators of action. I don't believe this is true, but it is not an incoherent idea. The point is that if consciousness is epiphenomenal, a noisy side effect of what the cognitive system is doing, then there is some justification for cognitive science to take a wait-and-see approach to consciousness. See how far one can get in explaining cognition, intelligence, rationality, categorization, and verbal reports without invoking consciousness. If and when consciousness reveals itself as an important player in information processing, then and only then invoke it.

Consider again the case of Deep Blue and Kasparov. We know that Deep Blue was doing something inner as it computed its moves. Call this Deep Blue's psychology. Kasparov was also doing inner things. He was trying to concentrate, to figure out what Deep Blue was thinking. He was also doing things unconsciously – computing moves and the like – that he can't report very clearly or reliably. Call this Kasparov's psychology. So both Deep Blue and Kasparov have psychologies – inner lives, as it were. Furthermore, it was a good match. Perhaps this is because their psychologies match up pretty well. But this can't be right. Kasparov, after all, cannot look ahead a gazillion moves, and Deep Blue doesn't care how it performs – the game doesn't matter to Deep Blue one bit. Deep Blue plays autonomously, without motivation, mindless deep chess. Kasparov plays autonomously, but with enormous motivation, mindful chess.

Once the match was over and Deep Blue had won, Kasparov was upset, embarrassed, and frustrated. At a news conference he said many things – things about his feelings, things about being duped by Deep Blue's trainers – in particular, about the trainers not sharing information about what drove or *motivated* Deep Blue, or about the extent of Deep Blue's historical knowledge of chess.

As I've said, one difference between Deep Blue and Kasparov is that the computer, but not the person, can do massively deep look-aheads. The difference that I'm emphasizing now, however, is not this one regarding computational ability. Rather, it is this:

although both Deep Blue and Kasparov have a psychology, there is nothing that it is like to be Deep Blue, whereas there is something that it is like to be Kasparov. Kasparov is conscious; Deep Blue is not. But the question remains: Is being conscious a difference that makes a difference? Or, to put it another way, does it make any other difference than that one system has experiences and the other doesn't?

Why are chess-playing computer programs important? Why are thermostats important, or guided missiles, or satellites? Practically, they are important because they do things we care about, things that we would like to have done or that need to be done. Philosophically, they are important because they do these things without consciousness. This raises the possibility that intelligence, cognition, and information processing do not require consciousness.

One might take a different but related line. Sciences such as thermodynamics factor out messy features of the world that make the explanations and predictions that their laws propose appear inaccurate. Consider the ideal gas laws, and what makes them ideal. They assume closed systems – no energy escaping or entering – and that molecular collisions will be more elegant geometrically than we have reason to think they could be. Imagine the following proposal for an ideal cognitive science: "Yes, consciousness does play a causal role in chess, in love, in life, etc., but its role is more one of interfering with cognitive processes that are designed to function well and generally do so. We factor consciousness out of our ideal cognitive science explanations because conscious mental events cause perturbations similar to those caused by trains passing the laboratory of a scientist who works on gases. To be sure, Kasparov was playing excellent chess against Deep Blue until he got upset after the second game. This caused a perturbation in his cognitive processing. For this he needs therapy, not cognitive science."

Cognitive science and consciousness

Cognitive science has always had an ambivalent relationship with the phenomenon of consciousness. On the one hand, as the successor doctrine to behaviorism, cognitive science is friendly to the *inside story* – friendly, that is, to drawing inferences about processes mediating stimuli and responses. On the other hand, as a research program inspired primarily, at least in its early stages, by computer science, the possibility that inner phenomena such as intelligence, thought, cognition, and the like, could occur – as they do in computers – without consciousness was rightly taken very seriously. A computer chess champion was no real surprise to those who had worked to create intelligent systems. Computer science and artificial intelligence have been very successful, but they are premised on the idea that cognition does not require consciousness. The very idea of a mind, especially a high-performing mind, lacking consciousness is hard to swallow, however, for those who assume that consciousness must play an essential role in intelligent information processing.

Cognitive science, it seems to me, was proposed on the basis of three main assumptions. First, the concepts of *information* and *information processing* are key to the science of the mind. Mentality is information processing. Second, the laws and generalizations of mind science will be information processing generalizations and laws. Third, intelligent information processing is what cognitive science studies. What does this have to do with consciousness? The answer is *nothing*, unless consciousness is thought or found to be involved in information processing in the systems, such as humans, that are conscious.

A student recently said to me that it was obvious what consciousness was for – happiness and love, that sort of thing. True, consciousness enables happiness and love. And no doubt Deep Blue can't experience these things: Kasparov felt terrible after losing, Deep Blue felt nothing. To be sure, most of us would not like to live life without happiness and love. But that aside, the philosophical question remains: can intelligence, cognition, and, most importantly, reproduction occur without conscious thoughts and feelings? The answer seems to be *yes*: Deep Blue is the possibility proof of intelligence without consciousness; the work of cognitive scientists on unconscious information processing (see Article 23, UNCONSCIOUS INTELLIGENCE) is the possibility proof that much cognition, probably most (if we knew how to measure), takes place unconsciously; and insects – for example, luna moths – that copulate on pheremonal signals without feeling sexy are the possibility proof (if we needed it) that mindless sex is possible (and actual).

The return of consciousness: philosophical therapy

The way in which I have been telling the story so far has rehearsed some of the reasons that the cognitive revolution did not, and still has not, returned consciousness to pride of place. I have mentioned the importance of the computer metaphor as one reason for thinking that cognition could be studied without invoking consciousness. What I haven't mentioned so far are the metaphysical fears that made many cognitive scientists shy away from talk about consciousness. Consciousness is metaphysically spooky, the proverbial ghost in the machine. Many prominent contemporary philosophers claim not to understand how to begin to explain consciousness or to make space for it in our scientific picture of the world (Nagel, 1974; McGinn, 1989; Levine, 1983).

Others, however, see a way to think about consciousness and its causal role that reduces its mysteriousness. The following four principles are not shared by all naturalistically inclined philosophers and cognitive scientists, but they provide a sense of the sort of commitments that engender confidence that the problem of consciousness can be made to yield a solution and that conscious mental states can be, as they seem to be, causally efficacious and not merely epiphenomenal:

1 *Principle of supervenience*: There exist emergent properties such as liquidity or solidity. Consciousness is in all likelihood an emergent property of complex states of the nervous system.
2 *Principle of macro causation*: Emergent, macro level properties, can causally interact with other emergent, macro-level events or processes, as well as with (often because of interactions with) micro-level events and processes. For example, water can put out fire (due to what large amounts of H_2O do to oxygen (make CO_2)). So too, emergent conscious events and processes can interact causally with conscious and nonconscious mental events (understood now as emergent neural events).
3 *Principle of organismic integrity*: That consciousness exists is amazing. But to paraphrase Dewey, "once we acknowledge that consciousness exists, there is no mystery in its being connected with what it is connected with." The basic motive behind this principle is to soothe, then remove, certain troublesome intuitions about subjectivity. Given that emergent properties are possible, and that consciousness is

likely to be such a property, then there should be no surprise in the fact that each person has her own, and only her own, experiences. It is because of the design of the nervous system. We are uniquely hooked up to ourselves. Given that there are experiences at all, it makes perfect evolutionary and physiological sense that I have my experiences and that you have yours.

4 *Principle of biological naturalism*: "Consciousness . . . is a biological feature of human and certain animal brains. It is caused by neurobiological processes and is as much a part of the natural biological order as any other biological features such as photosynthesis, digestion, or mitosis" (Searle, 1992, p. 90). Stated this way, the principle does not deny that consciousness could conceivably occur in systems with alternative biologies (non-carbon-based ones, for example) or even in robots made of inorganic parts. It simply provides a rationale for anyone who wants to understand consciousness to study the systems that exist in our vicinity that are known to be conscious.

The return of consciousness: cognitive neuroscience

The return of consciousness to respectability and to actual empirical investigation was not of course the straightforward outcome of having certain metaphysical fears allayed. It probably, and more importantly, had to do with the emergence of cognitive neuroscience as – depending on one's point of view – an expansion of the program of cognitive science or as the successor doctrine to traditional cognitive science. The emergence of cognitive neuroscience in the last two decades is linked not only to the view that the study of mind generally, and the conscious mind in particular, should be more realistic biologically – that is, it should involve the study of the brain – but also to the emergence of NEUROIMAGING techniques that allow us to link the phenomenological, psychological, and neural levels of explanation. I close by discussing two examples where profitable insight into the nature of both conscious and nonconscious mental processes has occurred.

Neural correlates of subjective visual attention

What happens if we try to get information about what an animal is experiencing, on the one hand, and information about, say, what is going on at the same time in the brain, and look for correlations? One fascinating set of studies linking particular types of awareness with particular types of neural activity has recently been made on rhesus macaques and exploits the well-known phenomenon of perceptual rivalry. Gestalt images, like the Necker cube or the duck–rabbit illusion, are rivalrous. The stimulus pattern (the visual information that actually reaches the eye) remains unchanged, but perception (what we experience, what we say we see) flip-flops. Binocular rivalry is a particular type of perceptual rivalry that exploits the fact that the visual system tries to come up with a single percept even though each of the two eyes may be receiving different visual information. In experiments involving binocular rivalry the visual input presented to the two eyes is incompatible. For example, a line moving upward is presented to the left eye, and a line moving downward is presented at the same time to the right eye. Because the two incompatible visual stimuli cannot be fused by the visual system to give a single, unified percept, humans report that what they see is, first, a line moving downward, then a line moving upward. The perception (how things appear to the experiencing subject) alternates rapidly between what the left eye sees (the stimulus pattern reaching the left eye) and what the right eye sees (see Logothetis and Schall, 1989, p. 761).

Suppose we wanted to perform a similar experiment with a monkey. How could we find out what the monkey is experiencing? The answer is that we can train the monkey prior to the experiment to be a reliable reporter of whether it perceives a line moving up or a line moving down. This can be done by training monkeys to give bar-press reports or, more surprisingly, by training the monkeys to execute a saccade (a quick eye movement) to a spot on the left if a downward movement is perceived and to a spot on the right if an upward movement is perceived. A monkey's report of how things appear to it at any moment (the monkey's phenomenology) tells us about the rate of perceptual shifting (what researchers call *psychophysical data*) and raises interesting questions about why perception shifts rapidly back and forth instead of locking on one of the two rival interpretations. But the phenomenological data ("The line is now moving upward") and the psychophysical data about the time between perceptual switches tell us nothing about what is going on in the nervous system as perception alternates between rival interpretations. Can we learn anything by looking more directly at events in the brain?

It is well known that monkeys have as many as 32 visual areas in the neocortex (see Article 4, BRAIN MAPPING). The bulk of the output from the retina to the brain projects through the lateral geniculate nucleus of the thalamus to the primary visual cortex at the back of the brain that computes edges. Other areas are interested in color and shape, position relative to the organism itself, facial features, and motion and depth. Motion detection occurs primarily in the middle temporal and medial superior temporal areas in the superior temporal sulcus (STS). Researchers have monitored the individual activity of a large number of neurons in the STS as the monkeys are reporting their perceptions of upward or downward motion. The activity of many of the neurons appears to be dictated by the retinal stimulation. Of these neurons, some show one pattern of activity throughout the experiment, apparently in response to the visual input to the left eye, while others show a different pattern in response to the visual stimuli reaching the right eye. There are other neurons, however, which show a change in activity as the monkeys report shifts in their perception of direction of motion (Logothetis and Schall, 1989, p. 761). The principle of supervenience says that every difference at the level of subjective experience must be subserved by a difference at some lower level (although the converse need not hold). This experiment indicates how the robust phenomenological difference between an upward- or downward-moving image might be subserved (although undoubtedly not exclusively) by small but detectable changes in activity in the cortical areas subserving motion detection.

The experiment is an excellent example of how subjective awareness – which we may sometimes consider to be beyond the reach of scientific understanding – can be studied by drawing together information gathered at different levels of analysis and by distinctive techniques. First, there is the assumption that there is something it is like for a monkey to have visual experience. Second, good old-fashioned psychological techniques of operant conditioning are used to train the monkeys to provide reports about what they see. Finally, these reports are linked with detailed observations of the activity of a large number of individual neurons to yield information about distinct brain processes that are correlated with the monkey's shifting perceptual experience of upward- and downward-moving lines. We might look at these results as a kind of *knowledge by triangulation*. While no one approach taken alone can tell us what we want to know about conscious experience, several different approaches taken together seem to give us a firmer fix on an otherwise elusive phenomenon. Let's look at another example.

Conscious event memory

Many years ago a famous neurological patient H.M. had medial temporal lobe surgery to cure his epilepsy. H.M.'s ability to remember new events and facts died with the excision. Still, H.M. remembered how to perform standard motor tasks and how to use language. Indeed, to this day H.M. is something of a crossword puzzle aficionado. His intact semantic memory is tapped by questions directly before his eyes, and his place in the puzzle is visually available to him. When H.M. is away from a puzzle he is working on, he cannot remember how far along on it he is, or even that he is working on a puzzle. There are other sorts of games or puzzles that H.M. was not familiar with before surgery but that he has become good at. The Tower of Hanoi puzzle requires that one move a set of rings down a three-post line-up (one ring and one post at a time) so that at the end they are in order of size on the last post (see Article 20, PROBLEM SOLVING). When the game was first taught to H.M., he played as poorly as any first-timer. But gradually he caught on. Each time the puzzle is presented to him, he claims never to have seen it before. But even after long periods of not working on the puzzle, he shows the clear effects of prior experience. He is better than any novice and invariably starts at a level of proficiency close to where he left off. H.M. has no conscious declarative memories about the Tower of Hanoi puzzle, but quite clearly information has sunk in about how to do it.

H.M.'s good general intelligence and his semantic memory can sustain him during a short period of newspaper reading. But minutes after he has read the newspaper, he is blank about what he read. It is tempting to say that H.M. has an intact short-term memory but no long-term memory. But this is too simplistic, for a variety of reasons. First, he retains information over the long term about how to do the Tower of Hanoi puzzle. The information is just not retained consciously. Second, when he does a crossword puzzle or reads a newspaper, he depends on something longer than short-term memory, which on standard views is very short. As long as H.M. is paying attention to a task, like a crossword puzzle or a newspaper article, he can retain information about it. When his attention is removed from the overall task, his memory goes blank (Flanagan, 1992).

H.M. can give us a phenomenology, a set of reports about what he remembers and what he doesn't. He also reveals in his behavior that he remembers certain things that he can't access consciously. Other human amnesiacs provide further phenomenological and behavioral data. Putting together the phenomenological and behavioral data with the knowledge that there is damage or excision to the medial temporal lobe leads to an initial hypothesis that this area plays some important role in fixation and retrieval of memories of conscious events.

The medial temporal lobe is a large region that includes the hippocampus and associated areas, as well as the amygdala and related areas. Magnetic resonance imaging (MRI) has allowed very precise specification of the damaged areas in living human amnesiacs and in monkeys. This research reveals that the hippocampus is the crucial component. When there is serious damage or removal of the hippocampal formation, the entorhinal cortex, and the adjacent anatomically related structures, the perirhinal and parahippocampal cortices, the ability to consciously remember novel facts or events is lost. Removal or serious lesions of the amygdala profoundly affect EMOTIONS, but not memory. It is not as if memories were created and set down in the hippocampal formation. The hippocampal formation is necessary to lay down memories, but it is not remotely sufficient for the conscious memory of facts and events. For habit and skill learning, it is not even necessary.

The complex network involved in fixing such memories has now been mapped in some detail by using monkey models and by drawing inferences about human memory function based on comparative deficit and normal data. Larry Squire and Stuart Zola-Morgan (1991) report an elegant experiment in which monkeys were given a simple recall task. Each monkey was presented with a single object. Then after a delay (of 15 seconds, 60 seconds, 10 minutes), it was presented with a pair consisting of the original object and a new one. The monkeys were trained to pick the novel object to get a reward. The task is trivial for monkeys with intact hippocampal formations, but there is severe memory impairment (increasing with latency) in monkeys with destruction to the hippocampal formation. The procedure is a reliable test of perceptual memory.

A skeptic could admit this but deny that there is any evidence that monkeys consciously remember what they saw. To be sure, the information about the original stimulus is processed and stored. The monkeys' behavior shows that these memories are laid down. But for all we know, these memories might be completely unconscious, like H.M.'s memories vis-à-vis the Tower of Hanoi puzzle. They might be. However, on the basis of anatomical similarities between monkey and human brains, the similarity of memory function, and evolutionary considerations, it is credible that the monkeys' selection of the appropriate stimulus indicates what they consciously remember, and thus that it can be read as providing us with a phenomenology, with a set of reports of how things seem to them.

The next step is to put the phenomenological and behavioral data from humans and monkeys together with the data about specific types of lesions suffered and to join both to our best overall theories of the neural functions subserving memory and perception. Coordination of these theories and data yield the following general hypothesis. When a stimulus is observed, the neocortical area known to be sensitive to different aspects of a visual stimulus are active. For example, in the simple matching task, the areas responsible for shape and color detection are active. In a task where the object is moving, the areas involved in motion detection become involved. Activity of the relevant cortical areas is sufficient for perception and immediate memory.

> Coordinated and distributed activity in neocortex is thought to underlie perception and immediate (short-term) memory. These capacities are unaffected by medial temporal lobe damage. . . . As long as a percept is in view or in mind, its representation remains coherent in short-term memory by virtue of mechanisms intrinsic to neocortex. (Squire and Zola-Morgan, 1991, p. 1384)

A memory is set down only if the information at the distributed cortical sites is passed to three different areas close to the hippocampus and then into the hippocampus itself.

The hippocampus then passes the message back through the medial temporal lobe out to the various originating sites in the neocortex that processed the relevant aspects of the stimulus in the first place. The memory is laid down in a distributed fashion. It is activated when the connections between the hippocampus and the various areas it projects to are strengthened by, for example, the request for recall. Once the memory is laid down, especially after it is strengthened, *Proust's principle* comes into play. The memory can be reactivated by the activation of any important node, such as the one subserving object shape or color, even smell or sound, without having to be turned on by the hippocampus directly. In this way the hippocampal formation passes the "burden of long-term (permanent) memory storage" to the neocortex and is freed to make new memories (ibid., p. 1385).

This theory is credible and powerful. It explains how damage to the hippocampal formation can destroy the capacity to form new memories and why old memories are retained despite hippocampal destruction. This research is also an elegant example of how work in cognitive neuroscience integrates several different lines of analysis to gain a purchase on the nature of mind in general and of conscious and unconscious mental states in particular. There are the phenomenological and behavioral data coming from H.M. and other human patients. There are the behavioral data coming from the monkeys, which we claim ought to be read as informing us about how things seem to monkeys. There is the prior work linking visual processing with certain brain processes. And there is the eventual theory that explains how conscious memories are fixed, how they are recalled, what they are doing when they are not active (they are dispositions laid down in neural nets), and so on.

The examples I have reviewed show the power of drawing together information obtained from what at first sight might appear to be conflicting or incompatible approaches, in order to learn something about the nature of consciousness, or about the nature of one kind of consciousness, or, if you like, about the similarities and differences between two kinds of consciousness. Specifically, they show that phenomenological facts, cognitive models, and neurophysiological analyses can be brought into reflective equilibrium in certain cases. They also show how we are beginning to fit conscious mental states into the natural fabric of the mind.

Conclusion

At the start I posed three questions: What is consciousness? What role, if any, does consciousness play in the explanation of cognition? Can consciousness be studied empirically? I promised answers. Here they are. Phenomenologically, consciousness is experience. What and whether consciousness is one thing at the neural level, we do not yet know. As DNA is to life, there may be some oscillatory frequency or aspect of neural architecture that is the key, the necessary marker, the sufficient condition, for consciousness. But most certainly consciousness plays a role in the explanation of cognition and action. Both the way things seem and the experiments I have discussed support the idea that consciousness is not epiphenomenal; it plays a causal role. Can consciousness be studied empirically? Yes. Consciousness is part of the natural order, and we are beginning to understand better what it is and what it does. The interdisciplinary effort behind cognitive neuroscience seems to me to hold the key to unlocking the mystery of consciousness.

References and recommended reading

*Block, N., Flanagan, O. and Ÿzeldere, G. 1997: *The Nature of Consciousness*. Cambridge, Mass.: MIT Press.

Chalmers, D. 1995: Facing up to the problem of consciousness. *Journal of Consciousness Studies*, 2, 200–19.

*Crick, F. and Koch, C. 1990: Towards a neurobiological theory of consciousness. *Seminars in the Neurosciences*, 2, 263–75.

Dennett, D. 1978: *Brainstorms*, Cambridge, Mass.: Bradford Books, MIT Press.

—— 1991: *Consciousness Explained*. Boston: Little, Brown.

*Flanagan, O. 1992: *Consciousness Reconsidered*. Cambridge, Mass.: MIT Press.

Levine, J. 1983: Materialism and qualia: the explanatory gap. *Pacific Philosophical Quarterly*, 64, 257–72.

Logothetis, N. and Schall, J. D. 1989: Neuronal correlates of subjective visual perception. *Science*, 245, 761–3.

McGinn, C. 1989: Can we solve the mind–body problem? *Mind*, 98, 349–66.

Nagel, T. 1974: What is it like to be a bat? Repr. in *Mortal Questions*, Cambridge: Cambridge University Press, 1979, 000–00.

Searle, J. 1992: *The Rediscovery of Mind*. Cambridge, Mass.: MIT Press.

Squire, L. and Zola-Morgan, S. 1991: The medial temporal lobe memory system. *Science*, 253, 1380–6.

*Turing, A. M. 1950: Computer machinery and intelligence. *Mind*, 54, 433–60. Repr. in A. R. Anderson (ed.), *Minds and Machines*, Englewood Cliffs, NJ: Prentice-Hall, 1964.

10

Decision making

J. FRANK YATES AND PAUL A. ESTIN

Modern scholarship on decision behavior dates from the late 1940s. But that scholarship has been preoccupied with two ideas that are much older. One is the notion of expected utility, first articulated in the scholarly literature by Daniel Bernoulli in 1738. In its simplest form, the expected utility concept applies to monetary gambles. Imagine you are asked to choose between two gifts, either gamble G, which you would then play and get either \$9 or nothing, or else a simple, direct payment of \$3. As shown in the decision tree in figure 10.1, you receive gamble G's \$9 payoff if the toss of a fair die yields one or two pips, but you receive nothing otherwise. (In decision trees, choice points are represented by square nodes, chance points by circular nodes.) The expected value or expectation of a gamble is the sum of its payoffs, each weighted (i.e., multiplied) by its probability. Thus, the expected value of G is EV(G) = (1/3)(\$9) + (2/3)(\$0) = \$3, the same as the direct payment. So, if you generally make decisions like this one so as to maximize expected value, you would be indifferent between the present options. However, if you are like most people, you are *not* indifferent. Instead, you are risk-averse, preferring the guaranteed \$3. Why?

According to Bernoulli's proposal, in situations like the one we have described, people do not attempt to maximize expected value. Rather, they decide as if they were trying to maximize expected utility – essentially, expected *subjective* value. Moreover, within that framework, people's risk aversion is a consequence of their utility functions for money exhibiting diminishing marginal value (implying, for instance, that the additional subjective value provided by an extra \$5 is less when one already has \$100 than when one starts with only \$10). In the case of our gamble G, this would mean that the expected utility of the \$3 gift would exceed that of gamble G: i.e., u(\$3) > (1/3)u(\$9) + (2/3)u(\$0), where u(\$X) represents the utility of \$X, and hence the choice of the \$3 for sure.

Can *expected utility (EU) theory* successfully explain all similar situations involving choice under uncertainty? No. Over the years, researchers have identified numerous circumstances in which people's decisions are not in line with EU maximization. An illustration, depicted by the decision trees in figure 10.2, is based on a problem described by Kahneman and Tversky (1979). In situation A, the decision-maker is given a choice between two prizes. Prize A1 is a lottery ticket promising a 50 percent chance of a three-week tour of England, France, and Italy and a 50 percent chance of nothing. Prize A2 is simply a one-week tour of England. In situation B, prize B1 is a 5 percent chance of a three-week tour of England, France, and Italy and a 95 percent chance of nothing. Prize B2 offers a 10 percent chance of a one-week tour of England and a 90 percent chance of nothing. Most people prefer prize A2 in situation A but prize B1 in situation B. It is straightforward to show that a person exhibiting such preferences

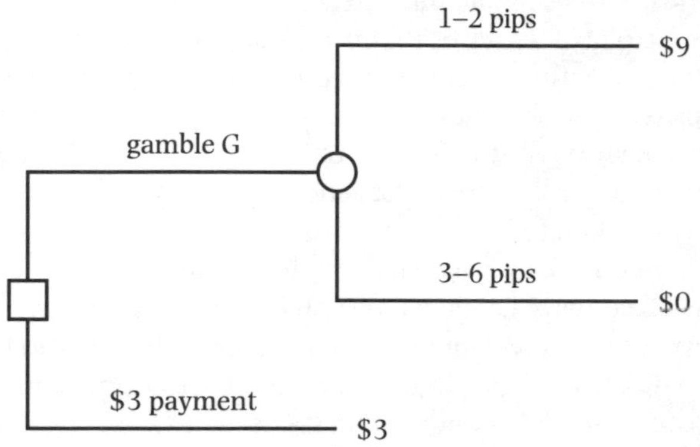

Figure 10.1 Some gift options.

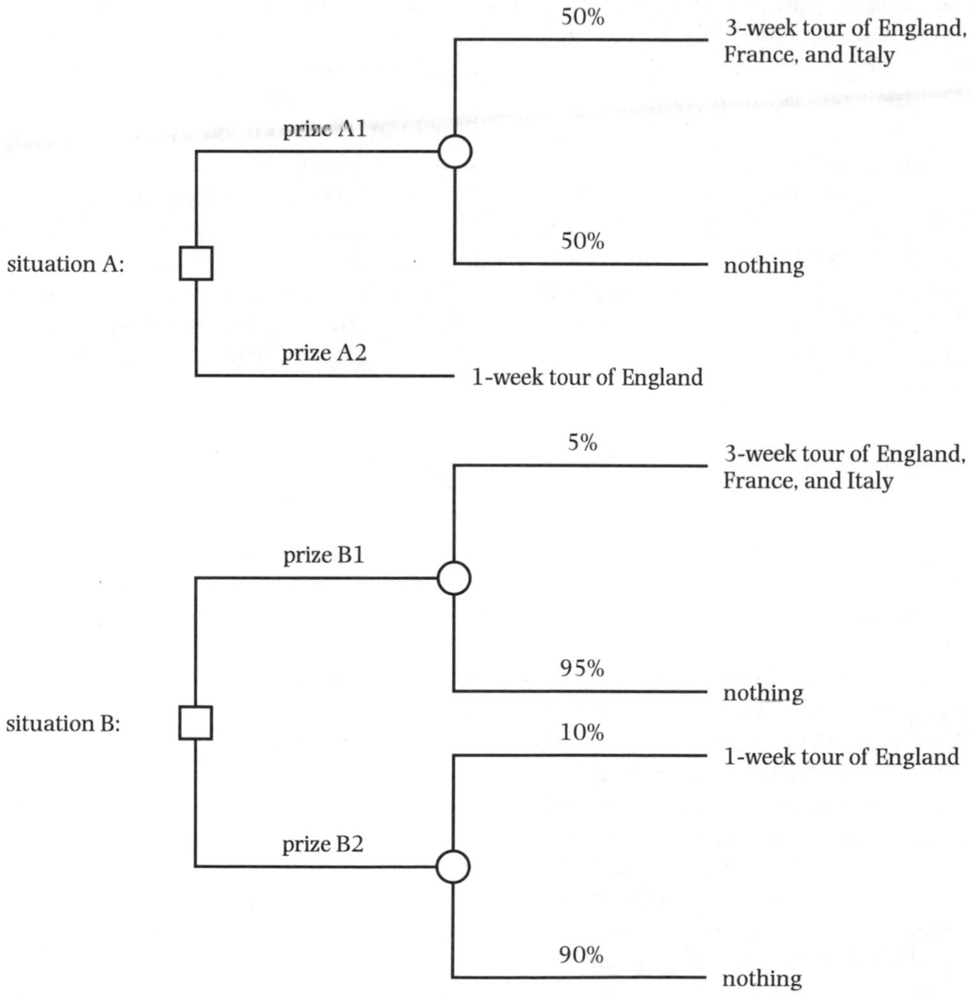

situation A:
prize A1
50%
3-week tour of England, France, and Italy
50%
nothing
prize A2
1-week tour of England

situation B:
prize B1
5%
3-week tour of England, France, and Italy
95%
nothing
prize B2
10%
1-week tour of England
90%
nothing

Figure 10.2 Prize options in two situations.

could not have chosen so as to maximize expected utility. (If you represent the utilities of the prizes by any plausible set of numbers – e.g., u(three-week tour) = 5, u(one-week tour) = 2, and u(nothing) = 0, and substitute them into expectations, you will observe a contradiction.) As we suggest above, much early modern decision scholarship sought to determine why what people actually do in various situations – contrived ones like those here, as well as more realistic ones – disagrees with what EU theory predicts that they will do (cf. Yates, 1990, chs 8–11).

The second old idea that has, directly and indirectly, dominated modern decision psychology is identified with Thomas Bayes, who, like Daniel Bernoulli, wrote in the eighteenth century. In its most common contemporary interpretation, Bayes's theorem applies to a situation like the following. Suppose you are entertaining some hypothesis H – for example, that candidate Smith would succeed at a certain job you must fill. The problem is to arrive at a probability assessment that Smith will indeed perform the job well. Let us say that 30 percent of all candidates in fact are successful. That ratio is taken as the *prior probability* for the hypothesis: that is, P(H) = 30 percent. Assume you know nothing about Smith other than that he or she is a candidate for the job. Then, if you are like the typical person, you will take 30 percent as your own judgment of Smith's chances of success.

Now let us change the situation a bit and imagine that you discover datum D, that Smith passed a certain screening test. We can then ask about P(H|D), the *posterior probability* for Smith's success on the job – *posterior* to your learning Smith's test score. (The expression P(H|D) is read "the probability that H is true given knowledge that D is true.") In particular, we might ask about the relation between P(H) and P(H|D), whether and how you revise your initial opinion about Smith's chances of success in light of the test results. Common sense says that the answer ought to depend on the connection that generally exists between D and H. Bayes's theorem formalizes this intuition and makes it precise:

$$P(H \mid D) = \frac{P(D \mid H)}{P(D \mid H)P(H) + P(D \mid \text{not } H)P(\text{not } H)} P(H)$$

Suppose that P(D|H) = P(D|not H). That is, the probability that a successful job candidate passes the screening test is the same as that of an unsuccessful candidate doing the same. Then it is easy to see that Bayes's theorem would yield P(H|D) = P(H). In other words, as seems eminently reasonable, the results of a patently useless test should leave your initial opinion unchanged. But suppose that P(D|H) is much larger than P(D|not H) – for example, 80 percent versus 15 percent, respectively – what you would hope for with something called a "screening test." An inspection of Bayes's theorem implies that the posterior probability of Smith's chances of succeeding would be substantially higher than the prior probability – approximately 70 percent rather than 30 percent for the specific figures given.

Do real people – like yourself – generally revise their opinions about chances in agreement with the prescriptions of Bayes's theorem? There have been numerous demonstrations in which they regularly do otherwise. And again, as we implied above, a good deal of the attention of behavioral decision specialists has been consumed by efforts to understand why this is so. Consider Bayes's theorem as not only a prescription for how people ought to change their minds about chances in the light of new evidence, but as a descriptive model as well – that is, as a proposal for how they in fact change their minds. Then we could view such scholars' efforts as attempts to develop more accurate characterizations of people's real beliefs (cf. Yates, 1990, chs 5–7).

Behavioral decision researchers have had dual aims. We have already suggested their purely scholarly aspirations: to achieve a deep understanding of various puzzles concerning how people actually decide. But the ever present companions to those aims have been practical ones: How can people be assisted in making *better* decisions? In recent years, pursuit of these practical goals has intensified in such diverse arenas as business, health care, human factors, and counseling. Significantly, these pragmatic concerns have led to at least an implicit recognition of numerous fundamental issues that were neglected in the past. Arguably, this neglect occurred in part because investigators were preoccupied with notions like EU maximization and Bayes's theorem. The main purpose of this chapter is to sketch some of these previously overlooked questions, presaging what we envision to be major emphases of future scholarship on decision behavior (cf. Yates, forthcoming).

There is a community of scholars who refer to themselves as "cognitive scientists." There is another community of self-described "behavioral decision scientists," researchers who pursue the kinds of ideas described above. The intersection of these communities is surprisingly small. In the past, *mainstream* cognitive scientists seldom read or cited the work of behavioral decision scientists, and vice versa. Thus, some observers, such as Simon and Kaplan (1989), were forced to conclude that, for instance, work on decision making under uncertainty has been "rather peripheral to the cognitive science endeavor" (p. 36). Hence our second set of objectives: to indicate some of the key differences between traditional scholarship on decision behavior and other topics in cognitive science and to explain why these traditions are likely to converge in the future.

Decision quality

What is good decision making? In much behavioral decision research, this difficult question has been deftly avoided. In other instances, good choices have sometimes been defined or assumed to be ones that maximize expected utility, or even expected value. And the analogous notion of good judgment has often been equated with opinions about chance being consistent with the prescriptions of Bayes's theorem and other probability theory rules, a concept sometimes referred to as *coherence*. Significantly, a major impetus for the very birth of modern decision behavior scholarship was von Neumann and Morgenstern's (1944) proof of a so-called *representation theorem* for EU theory. That theorem is often considered a watershed development because, in a particular way, it has sometimes been taken as a strong argument for the sensibility – some say *rationality* – of maximizing expected utility.

Such conceptions of decision quality are not actually wrong, but they *are* inadequate. And it has taken attempts to apply the results of decision scholarship beyond the laboratory to make that inadequacy apparent (Klein et al., 1993). For concreteness, consider the problem of deciding what field of work to enter. Developers have built decision aids that do things like, essentially, help users correctly determine the expected utilities of their career options. By all indications, such aids have been virtually ignored in practice. Our interpretation of this experience is that, although these aids assure high-quality decision making in the expected utility sense, they neglect broader and more practically significant senses of decision quality. And potential users eventually – actually, quickly – come to recognize this fact.

Before we describe our synthesis of the quality conception that is evolving in contemporary efforts to assist decision-makers, we must back up a moment and confront some questions that are even more fundamental, such as: Exactly what is a *decision?* How does decision making differ from other cognitive activities? We take a decision to be the selection of an alternative with the aim of producing outcomes that are favorable from the decision-maker's perspective (cf. Yates, 1990). There are several subtle but important features of this definition. The first is that, at some level, decision making is deliberate (cf. *aim*). Second, the decision-maker must acknowledge at least two different courses of action that could be taken (cf. *alternative*). And, finally, value – *subjective* value – is entailed (cf. *favorable . . . decision maker's perspective*).

The value element of decision making is especially critical – its hallmark in fact. That is what makes decision making so different from the other forms of PROBLEM SOLVING which cognitive scientists are accustomed to examining, and perhaps why it might not make sense to consider decision making to be purely *cognitive*; after all, value implies affect – that is, feeling (see Article 11, EMOTIONS). It is also one reason why decisions are so hard to make and to study. Contrast the task of responding to an item on a multiple choice calculus exam and deciding what career to pursue. Why would the typical decision scientist immediately acknowledge the career problem as a *decision* task yet hesitate to characterize the calculus problem that way? The main reason is that virtually everyone would agree on what is the most preferable outcome of the calculus problem, but that is hardly the case for the career problem. Preferences for various career experiences – the *values* attached to them – differ markedly from one person to the next.

This example also makes apparent some of the complications of aiding decision making. The correct answer to the calculus problem is the correct answer for everybody. But clearly, the best career for Jane is unlikely to be the best career for even her best friend Sue. The underlying individual differences in people's preferences pose formidable hurdles for applied decision scientists. Thus, a system for supporting career decisions must somehow tailor its recommendations to every user's personal tastes. This is difficult indeed.

Back to the original question: What is good decision making? Our synthesis of the quality concept implicit in current applied behavioral decision scholarship is most easily described in reverse: *A good decision is one that has few serious decision deficiencies.* The following are deficiencies that are commonly acknowledged, even if only implicitly and sometimes in different terms.

Type 1: aim deficiency When people appraise their own decisions, they often refer to their objectives or aims. Thus, we often hear decision-makers say things like "That was a pretty good choice, since things worked out as I had hoped." On the other hand, when a decision fails to achieve the decision-maker's aim(s) to some degree (e.g., a job eventually pays much less than the person sought when choosing it), we say that the decision exhibits an *aim deficiency*.

Type 2: instigating need deficiency Every decision episode begins when the decision-maker experiences some sort of discomfort. The discomfort can arise from either current circumstances that are less than ideal or from the decision-maker anticipating that such conditions are impending. The decision-maker is spurred to action in the hope that a good decision would ease the discomfort. The ultimate source of the unease – which may well be unrecognized and whose relief is thus not acknowledged as an aim – is the decision episode's *instigating need*. Once again, our career choice scenario permits an illustration. Whether they realize it or not, some people choose a particular field of work because they crave the respect of others who are important to them (e.g., their friends). When our decisions fail to satisfy our instigating needs, we say that the decision exhibits an *instigating need deficiency*.

Type 3: aggregate outcomes deficiency In most real-world situations, decisions do more than fulfill or fail to fulfill the decision-maker's aims or instigating needs; they affect other aspects of the decision-maker's life too, for better or worse. Suppose that, in the aggregate, taking all the decision's consequences into account, the decision-maker is distinctly worse off than some reference, say, if no active decision had been made at all. Then we say that the decision suffers from an *aggregate outcomes deficiency*. Return to our previous example and picture a career decision-maker seeking a comfortable living. In a not impossible scenario, she might unwittingly choose a field of work that pays well yet does irreparable damage to her health. Although her choice meets her aim and perhaps her instigating need, its *excess baggage* more than offsets those benefits.

Type 4: competitors deficiency Consider a decision-maker (e.g., a student picking a career) who selects an alternative that satisfies her aims and instigating need and whose entire ensemble of consequences is, on the whole, quite satisfying (e.g., the student picks a field that pays well and actually improves her health). But the decision-maker then discovers another option that is all that and more (e.g., a career that pays well, is good for one's health, and also very visibly benefits society). The decision-maker will almost certainly experience some measure of disappointment, and properly so in the present view. The decision that was made suffers from what we describe as a *competitors deficiency*: namely, there exists at least one alternative that is superior to it in terms of producing favorable aggregate outcomes.

Type 5: process cost deficiency Making decisions requires the expenditure of resources, including (among others) time, money, and tolerance for *aggravation* – for example, anxiety. When such costs are excessive for a given decision, we say that that decision suffers from a *process cost deficiency*. Controlling process costs is clearly the impetus for much of the decision aid development work we witness today. Take the case of expert systems for, say, evaluating bank loan applications. There is at best minimal evidence that such systems lead to markedly better decisions than traditional approaches. Nevertheless, they are often considered highly useful because they are less costly to banks than human loan officers (see Article 44, HEURISTICS AND SATISFICING).

Future directions

The course of scholarly inquiry is strongly affected by the problems which investigators set out to solve and by the yardsticks they use to gauge their success. Suppose that, as we propose, many of the goals and success metrics of future decision behavior research are embodied in the conception of decision quality described above. Then such research is likely to be quite different from that of the past. In many respects, we can expect it to be broader and to adopt or at least experiment with perspectives and methods used in other domains, including other areas of cognitive science.

Over the years, we have analyzed countless decision episodes described by our informants. Those accounts indicate that the typical decision problem entails most if not all of several *cardinal decision issues*. These issues are *cardinal* in the sense of having great importance. That is, if they are not resolved successfully, there is a good chance that the resulting decision will be seriously deficient, exhibiting one or more of the shortcomings we described above. We expect that, in their efforts to help decision-makers avoid the broad panoply of common deficiencies, researchers will seek an understanding of how, and how well, people typically address the cardinal issues and try to devise means of enhancing those activities. Here, in the context of descriptions of the cardinal issues, we offer modest forecasts for how the future course of decision behavior scholarship can be anticipated to develop.

Cardinal issue 1: instigating need We earlier defined the instigating need for a decision episode as the source of the discomfort that tells the decision-maker that he or she in fact has a decision problem to solve. We use the term *need* rather than, say, *wish*, because if the impetus for the discomfort is not addressed, it will not simply go away – for instance, through forgetting. People often assume that the instigating need is obvious, that it is equivalent to the decision-maker's stated aims. This can be a costly mistake. Sometimes decisions are deficient because the instigating need is misidentified; the decision-maker's attempt to resolve the first of the cardinal issues, the *instigating need issue*, fails. The eventual decision is deficient because the decision-maker for all practical purposes tries to solve the wrong problem. Some of the more interesting examples of such cases involve marriage. Marriage experts often cite instances in which unions fail because one or both partners enter the marriage out of a mistaken belief that the unhappiness they were experiencing in their single lives was an indication that they ought to marry someone. Thus, from their perspective, the decision problem was misread as one of determining whom to marry. Current mainstream decision research seldom acknowledges the instigating need issue. But this will change as investigators are more often faced with practical decision problems in ill-defined personal and organizational arenas.

Cardinal issue 2: options If the decision-maker never recognizes (or creates) some alternative, then that alternative obviously cannot be selected. The second cardinal issue, the *options issue*, is resolved successfully if the decision-maker identifies an option that is not inferior on the whole to any other that exists or *could* exist. In practice, it is virtually impossible to tell when absolute success has been achieved. Yet, after the fact, people often realize when they have done especially poorly in addressing the options issue. Affairs of the heart again, sadly, provide ready examples. Each of us can recall a friend who one day is positive he has found *the one* but shortly thereafter discovers another *true love* who seems even better. There have been a number of recent attempts to improve decision-makers' option identification. Some of the more exciting and promising ones, such as electronic brainstorming (Dennis and Valacich, 1993), seek to exploit computing technology and social interaction. The success of efforts like these, modest as they have been, will almost certainly inspire more and better ones.

Cardinal issue 3: possibilities Once a decision alternative has been selected, any number and variety of outcomes can occur as a result. The decision-maker is indifferent to most of them, but some actually matter to the decision-maker, positively or negatively. The decision-maker successfully resolves the third cardinal issue, the *possibilities issue,* to the degree that he or she acknowledges the personally significant consequences that really are potential outcomes of the options under consideration. Of all the mishandled cardinal issues that appear to have contributed to the hundreds of deficient decisions people have described to us, none has manifested itself more often than the possibilities issue. In fact, we have yet to encounter anyone who cannot readily cite instances in their personal lives where they were *blind-sided* by unrecognized possibilities. That is, they chose some course of action and then experienced a nasty – perhaps even devastating – consequence of that action. When they had deliberated about the decision, the possibility of that consequence had never even crossed their minds. Yet, with the benefit of hindsight, they recognized that the possibility was *obviously* there all the time, and was perhaps even a certainty.

Consider a case in point from the public arena. Richard Nixon was driven from the presidency of the United States by the Watergate scandal. Many incidents played roles in the drama. Among them was the playing of incriminating tape recordings of conversations between Nixon and his aides. It is inconceivable that, when he made the decision to have all his office conversations recorded, it ever occurred to Nixon that the tapes might contribute to his loss of the presidency. Because of its significance, we expect decision research implicating the possibilities issue to be especially prominent in the future. Indeed, current artificial intelligence work in medicine already pointedly addresses possibilities concerns (Warner, 1989).

Cardinal issue 4: realization Not every possible consequence of a given alternative actually occurs; only some of them are realized in any particular episode. Our fourth cardinal issue, the *realization issue,* entails judging whether the acknowledged potential consequences will indeed come about. Clearly, an investor or horse-racing bettor who repeatedly predicts that losers will be winners and acts accordingly soon goes broke. More generally, no decision can turn out well if it is predicated on misjudgments of the chances of actual occurrences. From the outset, decision scientists, like everyone else, have recognized the critical importance of the realization issue. And that is why realization issues remain among the hottest in the field, although approaches to them are different, given the limited success of older strategies focusing on principles like Bayes's theorem. A representative contemporary issue is illustrated by a physician whose average probability judgment that her diagnoses are correct is 85 percent, but whose actual percentage of correct diagnoses is only 75 percent. Such excessively extreme judgments are often characterized as *overconfident.* Overconfidence is not uncommon, and its implications seem obvious. Close scrutiny of overconfidence and other realization phenomena is thus quite understandable.

The realization issue is a context in which the differences between the perspectives of behavioral decision scholars and other cognitive scientists stand in especially sharp relief. Cognitive scientists generally have concluded that acknowledged experts (e.g., in physics and chess) perform their work to an extraordinary level of excellence, in absolute as well as relative terms – that is, in comparison to nonexperts (cf. Chi et al., 1988). But this is not the case for the kinds of judgments required in common decision problems. For years, behavioral decision researchers have been aware of, and intrigued by, the fact that the judgments of recognized experts (such as experienced physicians making diagnoses and educators predicting student success) are often outperformed by simple algebraic models. Indeed, there have been several replicable instances in which the judgments of presumed experts are actually inferior to those of novices. Contrasts like these dramatize the significance of the differences in the problems which decision researchers, as compared to other cognitive scientists, have chosen to study. There are also implicit challenges. Why, for instance, do judgment experts look so *bad* by comparison with other experts? Are judgment tasks inherently *harder* than those required of other experts, and if so, exactly how and why are they harder? Or is the performance difference a reflection of different metrics used by decision researchers and other cognitive scientists, implying that the experts studied by the latter might not really be as good as they have seemed?

Cardinal issue 5: value We observed that subjective value is arguably the most distinctive and challenging feature of decision problems. Although one person might consider some outcome highly pleasing, the next could regard it as revolting. (Tastes in music and food are prosaic examples.) This fact of life implicates the fifth cardinal issue that arises routinely in decision episodes, the *value issue*: To what degree would the decision-maker find the actual experience of potential consequences satisfying or repugnant? People's preferences are highly diverse, and all of us have great difficulty predicting what others will like and dislike. (There is considerable truth in the common lament, "There's no accounting for taste.") But it is even worse. Sometimes we cannot even forecast our *own* values. You can surely recall numerous instances in which, say, you anticipated enjoying some activity immensely (e.g., the subject matter in a certain course) but found yourself bored silly with the actual experience. For a long time, it was impossible to find anything in the decision literature even acknowledging those aspects of the value issue highlighted here. (For instance, various utility functions used in expected utility analysis can *describe* individual differences in values, yet they can neither explain nor predict them.) But the practical problems which decision researchers must address these days (e.g., in medicine, marketing, and counseling) have fueled a recent upsurge of scholarship on such value questions.

Cardinal issue 6: conflict In virtually every decision episode, there comes a time when the decision-maker is confronted with the final cardinal issue, the *conflict issue*: Although any serious alternative excels in some respects, it is inferior in others. As a concrete example, consider the problem of deciding whether to accept admission to a graduate program that is slightly worse than its competitors with respect to faculty ratings yet offers a more generous financial aid package. In practice, how are such agonizing conflicts resolved, thereby essentially dictating ultimate choices? Moreover, how *should* they be resolved?

Expected utility theory is largely irrelevant to most of the other cardinal issues we have discussed. From that perspective, it is now more apparent why decision aiding efforts built on the EU concept have been so disappointing. But EU theory pointedly prescribes a resolution of the conflict issue as it arises in a restricted class of decision situations. Those circumstances are ones that can be represented in a manner analogous to the gamble G example we described earlier. In that illustration, note the nature of the conflict. Gamble G entails a $9 outcome that is highly attractive compared to the $3 which the decision-maker could have for sure. But the chances of actually acquiring that $9 (i.e., 1/3) are markedly worse than the 100 percent guarantee of the $3. Expected utility theory offers a recommendation for how to resolve the conflict: namely, pick the option with the better expectation.

Rightly so, many decision-makers find such recommendations less than fully satisfying. Whether the recommendation ultimately is for gamble G or for the $3 depends not only on the expectation operator, but also on the form of the utility function applied to the amounts of money – for example, how strongly a person's desire for more money is sensitive to how much he or she has already. And, as we saw in the discussion of the value issue, EU theory per se provides no guidance on that score. The decision-maker is left hanging, wondering, "Where does the balance really lie?" Interestingly, every other scheme for conflict resolution that has been discussed in the decision literature, such as *multi-attribute utility analysis*, is likely to leave practical decision-makers with the same sense of disappointment as EU theory. Those *consumers* of decision research can be expected to demand that future scholars come up with something better.

Final remarks

We have suggested that behavioral decision research is so different from other areas of cognitive science at least in part because the problems themselves are qualitatively different (e.g., regarding emphases on value and uncertainty). But the methods are different, too. To an extent, this is the outcome of historical accident. The intellectual roots of mainstream decision scholarship are mainly in fields such as economics and statistics. By contrast, traditional cognitive science traces much of its heritage to computer science and process-oriented experimental psychology. The modes of inquiry in these disciplines are quite distinct.

Related to methodological differences, what constitute satisfying or even acceptable *explanations* are different in traditional cognitive science and behavioral decision research. In mainstream cognitive science (particularly when it has addressed *high-level* cognition, such as problem solving, rather than *low-level* processes like visual attention), the goal historically has been to arrive at a detailed, step-by-step account of how some cognitive task is literally achieved. In some sense, the ideal has been a computer model that actually runs, performing the task in real time, just as a person would. By contrast, traditional decision researchers have had little concern with – or use for – procedural details. Instead, their purposes have been served adequately by things like mathematical models, and sometimes even verbally expressed ones. Expected utility theory and subjective counterparts of Bayes's theorem are simple illustrations. Few researchers take seriously the idea that decision-makers literally do things like compute and compare expectations in their heads. Rather, they have been content with indications that people decide *as if* they were doing that.

In view of the differences we have described, it is unrealistic to expect decision behavior scholarship to ever become indistinguishable in character from other work in cognitive science. But significant convergence should be anticipated, with behavioral decision research moving more toward traditional cognitive science than the other way around (e.g., Busemeyer and Townsend, 1993). This is partly because – rightly or wrongly – people find typical cognitive science explanations (e.g., real-time running models) inherently compelling. (We exclaim, "How marvelous! This machine acts just like a human being, with a personality and everything!") More importantly, the convergence will be driven by the broader demands being placed on decision scholarship – for instance, to help people resolve all the cardinal decision issues, not just those addressed by traditional, limited paradigms. The prior success of other cognitive science approaches with similar problems will no doubt encourage decision scholars to try them as well.

References and recommended reading

Busemeyer, J. R. and Townsend, J. T. 1993: Decision field theory: a dynamic-cognitive approach to decision making in an uncertain environment. *Psychological Review*, 100, 432–59.

Chi, M. T. H., Glaser, R. and Farr, M. J. (eds) 1988: *The Nature of Expertise*. Hillsdale, NJ: Erlbaum.

Dennis, A. R. and Valacich, J. S. 1993: Computer brainstorms: more heads are better than one. *Journal of Applied Psychology*, 78, 531–7.

Kahneman, D. and Tversky, A. 1979: Prospect theory: an analysis of choice under risk. *Econometrica*, 47, 263–91.

Klein, G. A., Orasanu, J., Calderwood, R. and Zsambok, C. E. (eds) 1993: *Decision Making in Action: Models and Methods*. Norwood, NJ: Ablex.

Simon, H. A. and Kaplan, C. A. 1989: Foundations of cognitive science. In M. I. Posner (ed.), *Foundations of Cognitive Science*, Cambridge, Mass.: MIT Press, 1–47.

von Neumann, J. and Morgenstern, O. 1944: *The Theory of Games and Economic Behavior*. Princeton: Princeton University Press.

Warner, H. R., Jr. 1989: ILIAD: moving medical decision making into new frontiers. *Methods of Information in Medicine*, 28, 370–2.

*Yates, J. F. 1990: *Judgment and Decision Making*. Englewood Cliffs, NJ: Prentice-Hall.

—— *Decision Tools* (forthcoming).

11

Emotions

PAUL E. GRIFFITHS

Emotions are an extremely salient and important aspect of human mental life. However, until recently they have not attracted much attention in cognitive science. Despite this neglect by cognitive scientists, other investigators have been actively studying emotions and developing theoretical perspectives on them. These theoretical perspectives raise a number of important questions that cognitive scientists will have to address as they bring emotions into their purview: (1) Is it the physiological or the cognitive aspects of an emotional experience that primarily determine *which* emotion is being experienced? (2) Are emotions culturally specific or widely shared across cultures? (3) Are either emotions themselves or the causes that elicit them *innate* in one or more of that word's several senses? I will address these issues in the course of surveying the different theoretical perspectives that have directed research on emotions.

The physiology of emotion

The scientific study of emotion began with Charles Darwin's *The Expression of the Emotions in Man and Animals* (1872/1965). Darwin used posed photographs to show that observers can reliably identify emotions from facial expression. He analyzed the muscle movements in each expression and argued that human expressions are sometimes homologous (descended from a common ancestor) with those of primates, despite differing superficial appearances, because the underlying muscle contractions are the same. Darwin identified several expressions still recognized today as pan-cultural human behaviors with affinities to the behaviors of other primates.

Darwin argued that expressions of emotion typically evolve from behaviors with some direct value to the organism in the situation that elicits the emotion. In surprise the eyes are widely opened and the head oriented to the stimulus. This serves to obtain as much information as possible. Chimpanzees expose their teeth in subordinate threat displays, signaling the intention, and perhaps the ability, for biting attack. Darwin argued that the reliable link between these behaviors and emotional states gave the behaviors a secondary adaptive value as signals of emotional state. The behaviors might even be modified to make them clearer signals (later ethologists called this *ritualization*). This secondary communicative function allowed the behaviors to be retained when their original role declined. A human confronted in a bar brawl may display an expression homologous to that of the chimpanzee. The behavior signals the emotion of anger, rather than the intention or ability to bite.

Like most nineteenth-century writers, Darwin thought of the physiology of emotion as a mere manifestation of private emotion feelings. His modern followers have been more inclined to identify emotions with their associated physiology. Both views imply that an emotion can be reidentified across cultures as long as the physiology is present. Until quite recently, as we shall see, most philosophers and psychologists would have rejected this conclusion.

Another early theory of emotion also linked emotions very strongly to their attendant physiology. In the 1890s William James proposed that a conscious emotion feeling is the perception of autonomic nervous system changes caused directly by an external stimulus via a reflex arc. According to the famous James–Lange theory of emotion, the perception of a fearful object directly precipitates the autonomic nervous system (ANS) changes of the flight response. The later perception of these changes constitutes the feeling of fear. At the present time the James–Lange theory is undergoing a revival. Antonio Damasio's research into the neural basis of emotion embraces James as an intellectual ancestor. Damasio (1994) argues that emotion feeling is the perception in the neocortex of bodily responses to stimuli mediated through lower brain centers.

Cognitive and "noncognitive" theories of emotion

The pioneering neuroscientist Walter D. Cannon campaigned strongly against the James–Lange theory in the 1920s and 1930s. He tried to show that emotional responses involving the ANS were just another example of the control of the body by limbic areas of the brain, particularly the hypothalamus, that had been revealed by his research into bodily homeostasis. Among Cannon's many powerful empirical criticisms, his claim that any stressful stimulus produces roughly the same set of ANS responses gave rise to a continuing controversy in emotion theory. If this finding is correct, then differences in the feelings associated with various emotions cannot be the result of different ANS feedback.

The idea that ANS arousal does not differentiate between emotions has been used to support the wider conclusion that emotions are not individuated by their attendant physiology at all. In perhaps the most widely cited single study on emotion, Stanley Schachter and Jerome Singer (1962) suggested the alternative *cognitive labeling* theory of emotion. Physiological arousal is a necessary condition of emotion, but the very same arousal can be *labeled* as many different emotions. Emotions are individuated by the cognitions that accompany them. Schachter and Singer put forward three specific hypotheses for experimental test: (1) a subject will label a state of ANS arousal for which they have no other explanation in terms of the cognitions available to them at the time; (2) if subjects are offered an immediate physiological explanation of their arousal, they will not label the arousal as an emotion; (3) an individual will report emotion only if physiologically aroused.

Schachter and Singer divided their subjects into four groups. One group was injected with a placebo. The remaining three groups were injected with adrenalin. One of these three groups was told the genuine physiological effects that they would experience, another was told nothing, and the third group was misinformed about what they would experience. Half the participants in each group were subjected to conditions designed to produce happiness or euphoria, and the other half to conditions designed to produce anger. These emotions were to be induced by the behavior of stooges placed with the subject and, in the latter case, by the use of impertinent questionnaires. Schachter and Singer gathered results by making secret observations of their subjects during the anger and euphoria conditions and by asking them to fill in questionnaires after the event. The effects found in the experiment were weak, but broadly supportive of the three hypotheses: (1) subjects in the euphoria condition reported euphoria, and subjects in the anger condition reported anger; (2) the group fully informed about the effects of the injection of adrenalin showed and reported the least signs of emotional arousal, the group misinformed about the effects showed and reported the greatest degree of arousal, and the group told nothing fell in between; (3) the placebo group showed and reported relatively little emotion.

Not only were the effects in Schachter and Singer's experiment weak, but there have been problems with replication. More importantly, it is unclear that they succeeded in simulating the normal experience of emotion. People unable to account for their own behavioral or physiological responses (e.g., after brain damage) often invent demonstrably incorrect explanations of their symptoms. This phenomenon is known as *confabulation*. One would *expect* Schachter and Singer's uninformed subjects to confabulate in order to explain the abnormal arousal caused by adrenalin injections. The results obtained do not discriminate between this hypothesis and the hypothesis that the experiment simulated normal emotion.

The question as to whether emotions are individuated by the cognitions that accompany them was the focus of a pointed dispute in the 1980s between R. B. Zajonc, who denied that emotions need involve cognitions at all, and Richard Lazarus, who vigorously defended the cognitivist view. (See Ekman and Scherer, 1984, for a review.) Lazarus started from the uncontroversial premise that emotion requires processing of information concerning the stimulus. The cognitivist claims that this processing is sufficiently sophisticated to be called *cognition*. Zajonc opposed this claim, citing a large number of empirical findings which suggest that there are direct pathways from the perceptual system to limbic areas implicated in emotional responses. He argued that the processes linking perception and emotion should not be regarded as *cognition*.

Despite appearances, this is *not* a trivial semantic dispute. Although the term *cognition* is used very loosely in contemporary psychology, there are certain traditional paradigms of *cognitive* processes, such as problem solving, and certain traditional paradigms of *noncognitive* processes, such as reflexes. Lazarus claimed that the triggering of emotions resembles paradigm cognitive processes, whereas Zajonc claimed that emotions are *modular*. They are reflex-like responses controlled by a dedicated system largely independent of the processes underlying long-term, planned action. His arguments in favor of this view are threefold. First, experiments by Zajonc and others show that emotions can be produced by subliminal stimuli. No information about these stimuli seems to be available to paradigm higher cognitive processes such as conscious recall and verbal report. Second, the affect program emotions are homologous with responses in far simpler organisms and are localized in brain areas shared with those simpler organisms. Finally, the modularity hypothesis explains the anecdotal data about the *passivity* of emotion. Like reflexes or perceptual inputs, emotions *happen* to people, rather than being planned and performed.

Social constructionism about emotion

Cognitivists have frequently assumed that emotions are reidentifiable across cultures because the cognitions that define them can occur in different cultures. However, in recent years the view that emotions are culturally specific has gained popularity as part of a broader interest in the social construction of mind. Social constructionists have characterized emotions as *transitory social roles* (Averill, 1980). People adopt an emotion as one might a theatrical role, in situations in which that role is culturally prescribed. These roles have been compared to culturally specific categories of mental or physical illness. Medieval people expressed psychological distress through the myth of spirit possession. Eighteenth-century gentlewomen negotiated their demanding social role by being subject to fits of the vapours. In a parallel fashion, romantic love is a pattern of thought and action produced by a person who wants to receive the treatment appropriate to a *lover* from their society. This pattern is interpreted by the lover and by society as a natural, involuntary response. Like illness roles, emotion roles differ across time and culture and are acquired by example and by exposure to stories and other cultural products. Constructionism suggests that emotions must be investigated by looking at the cultural context of thought and behavior. A conventional cognitivist approach would overlook the wider social context that makes sense of individual cognitions. A physiological investigation of love or *feelings of disempowerment* would be misguided in the same way as a search for the physiological basis of a medieval man's *honor*. Like having honor, having an emotion is not an individualistic property.

The ethological tradition

Cognitivist and constructionist theories of emotion stand in stark contrast to Darwin's interest in pan-cultural physiology. Darwin's work had little influence on psychology in the first half of this century. It emphasized the inheritance of complex behavior patterns, in contradiction to the main thrust of behaviorist research. It was also rejected in anthropology, where the consensus was that emotions are culturally specific. Darwin's work was finally revived by classical ethologists like Konrad Lorenz (preface to Darwin, 1872/1965). Like Darwin, ethologists looked at behavior *comparatively*, using resemblances across species to diagnose the function and evolutionary causes of behaviors. They also believed, perhaps mistakenly, that evolved behaviors should be seen in all human cultures. Ethological work caused a revival of interest in Darwin's ideas in the 1960s. In one of the best-designed studies, Paul Ekman and Wallace Friesen (1971) studied members of the Fore language group in New Guinea. These people understood neither English nor pidgin English, had seen no movies or magazines, and had not lived or worked with Westerners. Subjects were shown three photographs of faces and told a story designed to involve only one emotion. They were asked to pick the person in the story. Forty photographs were used in experiments with 189 adult and 130 child subjects. Subjects reliably chose the pictures representing Westerners' expressions of the emotion in the story. In one experiment the photographs represented sadness, anger, and surprise. The New Guineans were asked to select the face of a man whose child has died. Some 79 percent of adults and 81 percent of children selected the sadness photograph. These results suggest that some facial expressions of emotion are pan-cultural.

Darwin's other experimental technique, analyzing expressions into component movements, was revived by Ekman and a large group of collaborators (Ekman, 1971). Twenty-five subjects from Berkeley and the same number from Waseda University in Tokyo were shown a stress-inducing film known to elicit similar self-reports of emotion from Japanese and Americans. Subjects were alone in a room, aware that skin conductance and heart rate measures were being made, but unaware that their facial expressions were being videotaped. The facial behavior of the two sets of subjects was classified using a standard atlas of facial expressions. Correlations between the facial behavior shown by Japanese and American subjects in relation to the stress film ranged from 0.72 to 0.96, depending upon whether a particular facial area was compared or the entire face. This result also supports the view that some facial expressions of emotion are pan-cultural.

The ethological tradition crystallizes in the *affect program* theory of emotions. This is very similar to the modular theory of emotions suggested by Zajonc. Certain short-term human emotional responses, often labeled surprise, anger, fear, disgust, sadness, and joy, are stereotypic, pan-cultural responses with an evolutionary history. They involve coordinated facial expression, skeletal/muscular responses (such as flinching or orienting), expressive vocal changes, endocrine system changes, and autonomic nervous system changes. Emotion feelings and cognitive phenomena such as direction of attention are obvious candidates for addition to the list. Affect programs are sometimes conceived as literal, neural programs. There is considerable evidence that control of these behaviors is localized in the limbic system. However, the term *affect program* can be used to refer simply to the coordinated set of changes observed.

The current *evolutionary psychology* movement has suggested that there may be many more specific emotional adaptations, such as a specific cognitive-behavioral response of sexual jealousy. The methodology of these recent authors is very different from that of the ethological tradition. Rather than seeking evolutionary explanations for pan-cultural behaviors observed in the field, they use *adaptive thinking* as a heuristic whereby to search for such behaviors. Robert Frank derives a theory of emotions from a game-theoretic model of the *commitment problem*: the problem of convincing another organism that you will follow through a signaled intention (Frank, 1988). Amongst other emotions, Frank predicts a *sense of fairness* that would motivate agents to forgo profit in order to punish trading partners for exploiting their competitive position. In contrast to the classical ethologists, he looks for evidence that this behavior exists *after* he has adaptively explained why it *should* exist.

Are emotions innate?

The ethological tradition has stressed the pan-cultural and inherited nature of emotion, something that has been hotly denied by other researchers. This dispute has been caused in part by the fact that different theorists discuss different parts of the overall domain of emotion. However, much of the nature–nurture dispute in emotion theory is due to a failure to distinguish between the *output side* and the *input side* of emotional responses. The thesis that people are everywhere afraid in the same way and the thesis that they are everywhere afraid of the same things are almost always conflated. Evidence for the first thesis is produced to show that fear is *innate*, and evidence against the second thesis to show that fear is not *innate*.

The ethologist Irenäus Eibl-Eibesfeldt (1973) applied one of the fundamental experimental paradigms of classical ethology – the deprivation experiment – to facial expressions of emotion. He showed that the pan-cultural expressions of emotion develop in infants born deaf or blind. He concluded that these expressions are *inborn*, and that they *mature*, as opposed to being learned. It is not necessary to accept these particular theoretical constructs to recognize that the six affect programs develop in a way more akin to classic anatomical structures like organ systems than to classic psychological structures like beliefs. However, both this deprivation experiment and Ekman's cross-cultural studies reviewed above concern the *output side* – the behavior displayed in emotion. It is a separate question whether the *input side* – the stimuli that give rise to emotions – have the same developmental pattern and/or are pan-cultural.

The behaviorist John Broadus Watson found support for his extreme environmentalist view of mental development in the fact that newborns are sensitive to very few emotion stimuli. They respond to loud sounds and to loss of balance with fear, to prolonged restraint with rage, and to gentle forms of skin stimulation with pleasure. In addition, neonates are extremely responsive to the facial expressions of care-givers (Trevarthen, 1984). Sensitivity to a broader range of emotional stimuli does not *mature* in any very rigid fashion. At best, there is some evidence of biased learning (e.g., fewer trials may be needed to form negative associations with classic phobic stimuli than with arbitrary stimuli). In general, however, the emotions are produced in response to stimuli that, in the light of the individual's experience, have a certain general significance for the organism. On the input side, cultural and individual diversity are the norm.

Conclusion

A great deal is known about the psychological phenomena grouped under the heading of *emotion*, but no single *theory of emotions* commands widespread allegiance. Theories of emotion are typically not applicable to emotional phenomena other than those which initially inspired them. The affect program account of emotion, for example, seems applicable to only a tiny fragment of human emotional experience. Disputes between emotion theorists often seem to concern whose favored phenomena constitute the most *important* part of the domain of emotion! Both social constructionists and cognitivists try to push ethological and neurological findings about emotion into the background for fear that their favored research programs will be abandoned in favor of more *biological* programs. Overall, the state of the field strongly suggests that the emotions are a collection of very different psychological phenomena, and that they cannot all be brought under a single theory. Surprise may have no more in common with love than it does with many nonemotional psychological states. The same may apply to individual emotions, such as contempt or anger. These *single emotion* categories may contain everything from phylogenetically ancient reactions realized in the limbic brain to complex social roles requiring a very specific cultural upbringing. On one occasion *anger* may be a rigid, involuntary affect program, and on another a *strategic behavior* adopted to manipulate other people. A successful theory of one of these phenomena should not be rejected because it cannot deal with the others and hence fails as a general theory of *emotion*.

References and recommended reading

Averill, J. R. 1980: A constructivist view of emotion. In R. Plutchik and H. Kellerman (eds), *Emotion: Theory, Research and Experience*. vol. 1: *Theories of Emotion*, New York: Academic Press, 305–40.

*Damasio, A. R. 1994: *Descartes' Error: Emotion, Reason and the Human Brain.* New York: Grosset/Putnam.

Darwin, C. 1872/1965: *The Expression of the Emotions in Man and Animals.* Chicago: University of Chicago Press.

Eibl-Eibesfeldt, I. 1973: Expressive behavior of the deaf and blind born. In M. von Cranach and I. Vine (eds), *Social Communication and Movement*, New York: Academic Press, 163–94.

Ekman, P. 1971: Universals and cultural differences in facial expressions of emotion. In J. K. Cole (ed.), *Nebraska Symposium on Motivation 4*, Lincoln, Nebr.: University of Nebraska Press, 207–83.

Ekman, P. and Friesen, W. V. 1971: Constants across cultures in the face and emotion. *Journal of Personality and Social Psychology*, 17, 124–9.

Ekman, P. and Scherer, K. (eds) 1984: *Approaches to Emotion.* Hillsdale, NJ: Erlbaum.

Frank, R. H. 1988: *Passions within Reason: The Strategic Role of the Emotions.* New York: Norton.

*Griffiths, P. E. 1997: *What Emotions Really Are: The Problem of Psychological Categories.* Chicago: University of Chicago Press.

Lyons, W. 1980: *Emotion.* Cambridge: Cambridge University Press.

*Parkinson, B. 1995: *Ideas and Realities of Emotion.* London and New York: Routledge.

Schachter, S. and Singer, J. E. 1962: Cognitive, social and physiological determinants of emotional state. *Psychological Review*, 69, 379–99.

Trevarthen, C. 1984: Emotions in infancy: regulators of contact and relationship with persons. In Ekman and Scherer (eds), *Approaches to Emotion*, 129–61.

12

Imagery and spatial representation

RITA E. ANDERSON

Take a moment to use mental imagery to perform the following tasks: (1) decide whether an apple is more similar in shape to a banana or an orange, (2) determine how to rearrange the furniture in your bedroom to make room for a new dresser, and (3) drive home during rush hour. Although we take our ability to perform tasks such as these for granted, they raise a host of interesting questions about imagery. For instance, what is the relationship between imagery and perception? What types of processes are needed to account for our ability to generate, maintain, transform, and inspect images? How do we characterize individual differences in imagery ability? What is the relation between imagery and spatial representation? What can we do with mental imagery? This essay will focus on visual imagery, because we know more about visual imagery than auditory, haptic, gustatory, or olfactory imagery (Reisberg, 1992, focuses on auditory imagery). The expectation is that many principles, although not all details, of emerging theory will be applicable to imagery in other modes.

Imagery in perspective

Mental images are the ultimate in the subjective. An image can be directly experienced only by the imager. For instance, the images you generated to perform the above tasks cannot be experienced by other people or compared directly with images generated by different people. Consequently, developing a cognitive understanding of imagery has been fraught with difficulties. Because mental images are neither directly shareable nor directly measurable, researchers have been forced to develop experimental procedures that allow them to make inferences from behavioral and neuropsychological data about the nature of both the mental representations and the processes involved in the generation, maintenance, transformation, and inspection of images.

Imagery has been a central cognitive concept, for good or for bad, since antiquity. Greek orators used imagery-based mnemonic devices to help them remember the sequence of events when reciting long oral traditions and other pieces, just as modern-day speakers rely on notes and multimedia props to guide their presentations. For instance, to use the method of loci to encode a sequence of events, an orator would create a mental image of each event in each location in a series. To remember the events, the orator would then mentally revisit each location in sequence. Over the centuries, philosophers from Aristotle and Plato to the British empiricists have examined the role of the mental image in thought. In the 1880s, Sir Francis Galton provided the first psychological documentation that individuals varied greatly in the vividness of their visual imagery. To his surprise, some people could not (or would not) even attempt to form an image of their breakfast table when asked to do so, much less introspect upon the image to answer questions about it, such as the color of the tablecloth.

Psychology was originally defined as the study of the mind, and many early experimental psychologists between 1889 and 1913 focused on the study of imagery. However, that research stumbled badly, largely because the techniques for studying imagery were seriously flawed. Data obtained from the highly fallible technique of introspection led to increasingly vituperative arguments between those who believed that thinking was based on imagery and those who believed in imageless thought. Ultimately, the lack of satisfactory techniques for studying imagery led to the downfall of the experimental study of the mind and the rise of radical behaviorism by the 1920s.

By the early 1970s, cognitive psychologists had developed a variety of reliable behavioral techniques to investigate the role of imagery in cognition. The results of several studies suggested that under certain circumstances imagery was critically involved in various aspects of memory and on-line cognitive processing. Although the results of any single study could be challenged, taken together, the evidence from a wide variety of experimental procedures converged to reinstate imagery as a valid cognitive concept.

The modern-day study of imagery originated in studies of MEMORY. In the 1960s, Alan Paivio initiated a programmatic study of the facilitating effects of imagery-related variables on memory performance (Paivio, 1986). To account for observations that memory is often enhanced when imagery is involved, Paivio formulated his dual coding theory. This theory, which is fundamentally about the representation of knowledge in (semantic) memory, posits two coding systems. The verbal system is specialized for the processing of linguistic materials, while the nonverbal system is specialized for the processing of nonverbal objects and events. The representational units for the verbal system (*logogens*) and the nonverbal system (*imagens*) are richly interconnected. Hence, seeing a dog activates the dog-imagen, which can activate the relevant dog-logogen, thereby ensuring that objects can be named. Likewise, hearing the word *cat* activates the cat-logogen, which can activate the cat-imagen. Dual coding theory explains the facilitating effect of imagery on retention with reference to the beneficial effects of coding in two systems. That is, if verbal materials are concrete, or the conditions are conducive to the use of imaginal strategies, they activate processing in the nonverbal system as well as the verbal system. Because nonverbal codes are assumed to be more memorable than verbal codes, conditions that increase nonverbal coding will enhance memory performance. Although the explanatory adequacy of dual coding theory has been challenged on some fronts (de Vega et al., 1996), its legacy endures, as some version of dual coding theory is evoked, implicitly or explicitly, whenever the nature of the representation of knowledge is discussed.

Imagery and perception

Mental imagery appears to be closely linked to the relevant perceptual system. Sydney Segal and Vincent Fusella found that people have more difficulty perceiving something while simultaneously imaging something else in the same sensory system than they do when a different sensory system is involved. For instance, in one of their studies, detecting the presence of a faint visual stimulus (e.g., a small blue arrow) was impaired by concurrent visual imagery (e.g., a tree) but not by concurrent auditory imagery (e.g., a telephone ringing), and vice versa.

Within the visual system, Martha Farah demonstrated that visual imagery can facilitate processing of particular content. In her study, detecting a letter (e.g., H) was enhanced when people imaged that letter rather than an alternative letter (e.g., T) during the test interval. In yet other studies, perceptual and imaginal versions of the same task yield the same pattern of results. For instance, Roger Shepard and Peter Podgorny found that the pattern of response times for determining if a presented dot fell on a target block letter was the same whether the letter was presented visually or was imaged.

Neuropsychological evidence is also consistent with the presumed links between visual perception and visual imagery. For instance, Edoardo Bisiach and his colleagues have shown that patients with unilateral visual neglect in visual perception also suffer from comparable neglect in imagery. The advent of neuroimaging techniques, such as positron emission tomography (PET) to measure regional cerebral blood flow, has allowed researchers to demonstrate that visual imagery activates those portions of the brain used in visual perception. For instance, using PET scans, Stephen Kosslyn and his colleagues have shown that the primary visual cortex is activated when people image objects with their eyes closed.

Making the link between visual imagery and visual perception allows us to use our knowledge of visual perception to guide our thinking about the why and the how of visual imagery. That is, we can use our knowledge of the evolutionary history and adaptive significance of visual perception to think about the functions of visual imagery in cognition. In addition, we can use our knowledge of how visual perception operates to formulate effective models of visual imagery.

An evolutionary perspective suggests that visual imagery may be more prevalent than other forms of imagery due to the centrality of vision in primate cognition. Furthermore, visual imagery may reflect the way in which ecological conditions have shaped the abilities of our visual system through evolutionary time. Among other things, visual perception allows us to inspect, reach for, and manipulate objects, as well as to navigate in space. Visual imagery presumably allows us to simulate these perceptual-motor activities in the service of solving problems, whether in ongoing behavior (e.g., anticipating accurately who will be in what lane while driving home during rush hour on a freeway) or in advance of the behavior (e.g., planning which route to take to avoid traffic while still at home). En route to discussing some functions of visual imagery, I will first describe our current conception of the nature of imaginal representations and of the processes that operate on them.

Visual and spatial information in percepts and images

Visual perception provides information about objects and their spatial relations. Neuropsychological research on visual perception has determined that different brain systems process information about object properties (i.e., *what*) and spatial relations (i.e., *where*). The ventral system, involving pathways from the primary visual cortex to the parietal lobe, is dedicated to the processing of object properties, such as shape, color, and texture, independently of the location of the object. The dorsal system, involving pathways from the primary visual cortex to the temporal lobe, is dedicated to processing spatial properties of objects needed for manipulation or navigation, such as the size, location, and orientation of objects.

Although historically there has been controversy as to whether the information in visual images should be characterized as visual or spatial, the evidence today is consistent with the conclusion that visual images represent both types of information. Much of this evidence stemmed from the ingenious efforts of Roger Shepard and his colleagues to develop behavioral tasks that could reveal the nature of both mental images and the transformations that can be performed on these representations (Shepard and Cooper, 1982). He and Susan Chipman demonstrated that people can use visual images to make decisions, using a task similar to the shape comparison task you performed. They found that people could use remembered information about visual appearances (e.g., shapes of states in the USA) to make similarity judgments comparable to those judgments made when they actually viewed the shapes. Hence, visual images preserve visual information about shape such that the image is analogous to the percept of that shape. Success in object comparison tasks depends upon how well the object properties are represented in the image, independent of the location of the image of the object.

A vast number of studies have used variations on the mental rotation task, originally developed by Roger Shepard and Jacqueline Metzler in the early 1970s, to examine the mental transformation of spatial information in imagery. They asked people to determine whether a test stimulus was the same as, or a mirror image of, a comparison stimulus. The comparison and test stimuli were presented side by side, and the absolute angular disparity between the two stimuli varied from 0 to 180 degrees. The time taken to make the decision increased with increasing angular disparity, suggesting that people mentally rotate visually presented shapes just as they would rotate a stimulus physically. These and many other studies have demonstrated that a visual image is analogous to a visual percept, in that it can be used to represent and process information about object properties and spatial relations. In addition, the processes used to operate on mental images appear to be functionally analogous to those used to operate on actual objects in space.

Neuropsychological evidence also supports the claim that visual imagery can represent both objects and spatial relations. For instance, Martha Farah and her colleagues have described two patients showing a double dissociation between the ability to recognize objects and spatial relations in vision and in imagery. One person had difficulty identifying objects in vision and in imagery but could process spatial information in both modes, while the other could recognize objects in vision and in imagery but could not process spatial relations in either.

The imagery debates and theoretical development

In the 1970s, Zenon Pylyshyn challenged the *picture-in-the-head* metaphor of imagery, arguing that images, although real, are epiphenomenal, that they do not serve any cognitive function, and that all mental work is done using a single type of mental representation (e.g., amodal language-like propositions). His arguments focused attention on two substantive questions. First, do images depict object properties and spatial relations such that they can be accessed for further processing, or are images the epiphenomenal by-products of propositional processing? Second, if images are depictive, how are they generated, maintained, transformed, and inspected? Stephen Kosslyn and his colleagues rose to Pylyshyn's challenges on numerous occasions, armed with empirical data and progressively more sophisticated computational models, to show that mental images can depict visual and spatial information that can be further interpreted by the cognitive system.

Yet another challenge was posed by John Anderson in the late 1970s. He pointed out that because neither theoretical structures nor processes can be firmly anchored by behavioral data alone, any depictive theory of imagery can be reformulated as a propositional theory. This argument led Kosslyn to look to neuropsychological evidence to guide theory development. Consequently, his most recent model, which convincingly addresses both substantive issues, is not only grounded in behavioral data from many studies of visual imagery and visual perception but also reflects our current understanding of the neural mechanisms of visual perception.

Kosslyn's (1994) theory of imagery is based on his theory of high-level visual PERCEPTION, which incorporates the use of knowledge about objects and events and top-down hypothesis-testing mechanisms to recognize objects under varying (less than optimal) viewing conditions. To account for the perception of object properties and spatial relations, Kosslyn's model includes a visual buffer where images are formed, a variety of subsystems devoted to deploying attention and processing information about object properties and spatial relations, and an associative memory which coordinates information about objects and locations. Kosslyn argues that imagery is an integral part of high-level perception. That is, the top-down processes that support perceptual priming, thereby providing for the facilitating effects of expectations on perception, are the same processes used for visual imagery. The subsystems postulated to account for high-level perception are used to model our ability to generate and maintain images; motor programming subsystems are additionally implicated to model our ability to inspect and transform images.

In Kosslyn's model, visual images are depictive representations that can be examined, manipulated, and (re)interpreted so as to allow people to extract novel information about object properties and spatial relations. Both percepts and images are patterns of activity in a visual buffer, a topographically mapped area in the primary visual cortex. The spatial extent of the buffer is limited, with greatest resolution in the center. As a part of the visual system, the pattern of activity in the buffer is temporally transient. Gestalt-like perceptual processes operate on the patterns of activity in the visual buffer to fill in missing details, complete contours, and so on.

There are many ways to generate visual images in the visual buffer. Visual memory codes stored in associative memory can be activated in whole or in part. Pattern codes of previously stored parts can be activated and combined in novel ways using either categorical or coordinate spatial information. In addition, the attentional subsystems can be used to draw an image mentally. To maintain an image in the visual buffer, pattern codes from the stored representations must be reactivated continuously, or, in the case of mental drawing, attention must be constantly re-engaged. Images can be inspected using the processes used to encode and interpret objects in perception, while images can be transformed (e.g., rotated, scanned) by replaying a stored transformation or by altering the representation of the spatial properties of the imaged object.

An exciting aspect of Kosslyn's model is its potential for unraveling the mystery of individual differences in imagery ability. Traditionally, imagery has been viewed as a global, undifferentiated ability that individuals have to a greater or a lesser degree.

Perhaps because there have been many ways to conceive of imagery ability (e.g., vividness, control, spatial ability, visual memory) and because there are many ways to assess it (e.g., self-report, performance on various behavioral tests of spatial ability or memory), most investigations have failed to provide convincing evidence that individual differences in imagery relate to performance on any task of consequence. By contrast, Kosslyn's model allows for a quite different conception of individual differences. If people differ in the operational efficiency of the various subsystems involved in imagery, then individuals will differ in their profile of imagery strengths and weaknesses. Because Kosslyn's model can be used to identify the subsystems implicated in performing different imagery tasks, it is possible to estimate the degree of similarity between them. In one study, people performed 13 different imagery tasks ranging from mental rotation, mental scanning, and memory tasks to filling out questionnaires designed to assess the vividness of images and the ability to control images. The correlations between performance on various imagery tasks varied widely, and estimated task similarity accounted for almost 40 percent of the variance in the obtained correlations. These results suggest that visual imagery is not one global ability but is composed of several subsystems.

Quite a different approach to understanding imagery has been taken by Robert Logie. Whereas Kosslyn's model reflects its perceptual origins, Logie's (1995) model stems from research on how visual and spatial information is processed in MEMORY and REASONING tasks. Logie points out that even the most mundane task, such as finding an object on one's desk, reaching for it, and then taking it to another room, requires a system that provides on-line processing capabilities and temporary representations of the current state of our environment and interactions with it. Such a system is called "working memory." Working memory is composed of a limited-capacity central executive responsible for reasoning, decision making, and coordinating various slave systems which hold and rehearse material relevant to the task at hand. Research has implicated working memory in the processing of language, mental arithmetic, and mental imagery.

In Logie's model of visuo-spatial working memory, visual images are generated and manipulated by the central executive, while the relevant visual and spatial information accessed from long-term memory representations is held in separate temporary stores. The central executive is a flexible resource that hosts the conscious visual image and controls the operations performed on the image. Additional visual information about the properties of objects and scenes is held temporarily in the visual store, ready to be incorporated into the visual image as needed. The contents of the visual store are maintained by an active spatially based rehearsal mechanism, the inner scribe.

Taking account of evidence from behavioral and neuropsychological studies that memory for spatial information, memory for movements, and the planning of movements rely on overlapping cognitive resources, Logie explicitly incorporated movement into his concept of a spatial representation. He points out that scanning an object or an environment, moving from one location to another, and moving an object all involve space, whether the action is performed physically or mentally. Spatial information can be gathered by scanning from object to object; the relative location of objects can also be established by touching them or by moving between them physically or mentally. In the latter cases, visual perceptual input need not even be involved. Hence, in Logie's model, the spatial component is separate from the visual store. In addition to serving as an inner scribe, this spatial component can extract information from the visual store to plan targeted movements.

The models developed by Logie and Kosslyn complement one another. Although they were developed in response to different problems, there appears to be a great deal of overlap between them. As yet, it is not clear how visual imagery in Logie's visuo-spatial working memory relates to visual imagery in Kosslyn's system. Clarifying the relation between the two models will greatly expand our understanding of visual imagery.

And what of spatial representation? Although Edward Tolman introduced the concept of a cognitive map as early as 1948, research on how we mentally represent past, current, and future locations of the self, other selves, and objects in large-scale environments has only recently begun to flourish. Many techniques have been developed to examine how we acquire and mentally represent spatial information gleaned from actual experience, maps, or verbal descriptions (de Vega et al., 1996; McNamara, 1991; Tversky, 1991). To determine how people represent space, we need to determine how spatial relations are processed, from what perspective, by which processing system. Despite the fact that imagery research faces the same types of problems, research on spatial representation rarely connects with research on imagery. Hence, although we know that understanding spatial representation is central to understanding visual imagery, it is difficult to assess the role of imagery in spatial representation. Making the connection between the two areas of research should benefit theory development in both areas.

Imagery and discovery

Although visual imagery and visual perception share some neural subsystems, visual imagery must differ from visual perception in significant ways. After all, we rarely mistake images for percepts. A current, theoretically interesting problem concerns the extent to which mental images can be used to make discoveries (Finke, 1990; Cornoldi et al., 1996; Roskos-Ewoldsen et al., 1993). Despite a rich anecdotal record suggesting the centrality of mental imagery to the creative discoveries of many celebrated individuals (e.g., Kekulé, Tesla, Einstein), research shows that it is generally easier to reconstrue percepts or drawings of images than images themselves. For instance, Stephen Reed showed that people were more likely to find a hidden part of a complex figure (e.g., a triangle in the Star of David) in perception than in imagery. Likewise, Deborah Chambers and Daniel Reisberg demonstrated that people were much more likely to find the alternative interpretation of classical ambiguous figures (e.g., the Jastrow duck–rabbit) in a percept of the figure than in an image.

Despite the apparent limitation on the ability to reinterpret images, most people are able to use imagery to make discoveries in the open-ended mental synthesis task devised by Finke and Slayton. This task requires people to use imagery to synthesize a novel pattern (not one specified by the experimenter) from three randomly selected simple geometric shapes and alphanumeric characters. For instance, given the parts, *circle, triangle, and capital letter X*, a person might report and then draw *a ferris wheel*. Curiously, Anderson and Helstrup found that providing perceptual support does not seem to facilitate discovery performance in the mental synthesis task as much as it does in reconstrual tasks.

In conclusion

Mental imagery today is truly on the verge of fulfilling its promise as a central cognitive concept. Given the subjective nature of mental imagery, we must remember to balance theoretical enthusiasm with experimental rigor. Nonetheless, current conceptions suggest that the processes underlying visual imagery play a major role in everyday perception and support the on-line cognitive processing of visual and spatial information. Furthermore, it appears that visual imagery can be used to simulate future plans and actions, and even to make discoveries.

References and recommended reading

Cornoldi, C., Logie, R. H., Brandimonte, M. A., Kaufmann, G. and Reisberg, D. 1996: *Stretching the Imagination: Representation and Transformation in Mental Imagery*. New York: Oxford University Press.

de Vega, M., Intons-Peterson, M. J., Johnson-Laird, P. N., Denis, M. and Marschark, M. 1996: *Models of Visuospatial Cognition*. New York: Oxford University Press.

*Finke, R. A. 1990: *Creative Imagery*. Hillsdale, NJ: Erlbaum.

*Kosslyn, S. M. 1994: *Image and Brain: The Resolution of the Imagery Debate*. Cambridge, Mass.: MIT Press.

Logie, R. H. 1995: *Visuo-Spatial Working Memory*. Hove, UK: Erlbaum.

McNamara, T. P. 1991: Memory's view of space. In G. H. Bower (ed.), *The Psychology of Learning and Motivation*, vol. 27, New York: Academic Press, 147–86.

*Paivio, A. 1986: *Mental Representations: A Dual-Coding Perspective*. New York: Oxford University Press.

Reisberg, D. (ed.) 1992: *Auditory Imagery*. Hillsdale, NJ: Erlbaum.

Roskos-Ewoldsen, B., Intons-Peterson, M. J. and Anderson, R. E. (eds) 1993: *Imagery, Creativity, and Discovery: A Cognitive Approach*. Amsterdam: Elsevier Science.

Shepard, R. N. and Cooper, L. A. (eds) 1982: *Mental Images and their Transformations*. Cambridge, Mass.: MIT Press.

Tversky, B. 1991: Spatial mental models. In G. H. Bower (ed.), *The Psychology of Learning and Motivation*, vol. 27, New York: Academic Press, 109–45.

13

Language evolution and neuromechanisms

TERRENCE W. DEACON

Neuromechanisms of language processing

The first major advances in the understanding of the neurological bases for language abilities were the results of the study of the brains and behaviors of patients with language impairments due to focal brain damage (for historical review see Lecours et al., 1983). The two most prominent pioneers in this field are remembered because their names have become associated with distinctive aphasia (language loss) syndromes and the brain regions associated with them. In 1861 Paul Broca described the damage site in the brain of a patient who had lost the ability to produce articulate speech as a result of a stroke. Broca's aphasia, as it is now commonly called, is produced by damage that involves the inferior left frontal lobe of the brain. It tends to produce both a loss of articulatory skill (sometimes but not always with paralysis) and (as discovered much later) problems with grammar and syntax, including grammatical comprehension under certain conditions. Despite speech production impairments, Broca's aphasics do not have basic word- and sentence-comprehension difficulties, when comprehension is not dependent on syntactic analysis. In 1874 Carl Wernicke demonstrated that damage to the left superior temporal lobe produced a different syndrome of language impairments. Wernicke's aphasia generally spares fluent, syntactic speech but impairs the use and comprehension of content words. Wernicke's aphasics' speech includes inappropriate word substitutions, neologisms, and semantic confusions, and their comprehension is similarly disturbed. These early studies also demonstrated that these aphasic disturbances were almost exclusively associated with damage to the structures of the left cerebral hemisphere, even though anatomically corresponding structures are present on both sides of the brain. However, this is not true of all language functions. Some left-handed individuals can suffer aphasia from right hemisphere damage, and recent studies of patients with right hemisphere damage without aphasia have demonstrated impairment of their comprehension above the sentence level (e.g., interpretation of jokes and stories) and in the production and interpretation of prosodic features of speech (rhythmic and tonal patterning that conveys emphasis and emotion).

These classic theories of language processing were predicated on the idea that neural function could be understood in terms of processing centers and their interconnections. The proximity of Broca's area to the oral motor cortex and of Wernicke's area to the auditory cortex suggested that language processing might be organized in terms of input–output centers and the links between them, and *higher* cortical centers (see figure 13.1). Analyzing word meaning might thus be understood as a form of auditory association, and word production and syntax might be understood as programming of articulatory sequences. Damage to intermediate links between these centers could be expected to interrupt the translation from sound to speech, and so undermine the ability to repeat speech. Damage linking these two major language centers to other cortical areas could be expected to disturb the conceptual links between language and other cognitive or sensory functions. For example, the links between Wernicke's area and visual areas of the cerebral cortex should cause loss of reading abilities (alexia), and disconnection of Wernicke's area from other *higher association areas* should render patients capable of speaking and repeating speech but leave them unable to use it otherwise.

Much of the basic logic of this approach has been retained in contemporary approaches; however, during the more than a century since these initial discoveries, there have also been many fundamental criticisms of its logic and an accumulation of data that are difficult to assimilate to these sorts of models. The chief alternative accounts have treated language processing as far more distributed within the brain, and have suggested that the classic language *centers* are rather final common pathways, where signals converge, not loci for language knowledge. Four new tools have begun to provide a more detailed picture, in which language-associated regions are subdivided according to cell structure, connection patterns, and function into clusters of areas that handle component computations underlying language. These tools are electrical cortical stimulation (ECS), radiolabeled regional cerebral blood flow (rCBF, now little used), positron emission tomography (PET), and functional magnetic resonance imaging (fMRI).

Analysis of brain function based on electrical stimulation began in the nineteenth century, though its systematic use for the study of human brains was brought from curiosity to clinical research tool by neurosurgeon Wilder Penfield in the 1950s, and has been refined with respect to language studies in recent decades by George Ojemann and his colleagues. The procedure involves low-level electrical stimulation of the exposed cortical surface of locally anesthetized neurosurgical patients while they are performing certain tasks. The electrical stimulation essentially introduces noise at this point and interrupts local functions. Contrary to expectations, stimulation of either Broca's or Wernicke's area disrupts the same language functions, and these disruptions do not correspond to aphasic symptoms (see figure 13.1). The likely reason for this difference is that stimulation indirectly sends noise to other regions connected to the stimulation site as well. For this reason, stimulation may activate highly distributed circuits that utilize long intercortical connections (see figure 13.2A). Electrical stimulation studies demonstrated that many widespread cortical regions, including prefrontal cortex, parietal cortex, and medial frontal cortex (supplementary motor cortex) are involved in language processes.

NEUROIMAGING techniques have recently added another dimension: the dimension of metabolic *costs* of cognition. All three of the major functional imaging techniques (rCBF, PET, and fMRI) provide signals that reflect relative metabolic demand, either above or below some baseline or control level. Significantly elevated metabolic activity strongly suggests that a structure is specially recruited for a correlated cognitive processing task, while reduction of activity below baseline in a structure may indicate an equally necessary complementary role to decrease a competing or inhibitory function. The shifting and complicated patterns of relatively elevated metabolic activity correlated with different language tasks (see examples summarized in figure 13.1) demonstrate that many different sectors of the left temporal, parietal, and frontal cortex contain

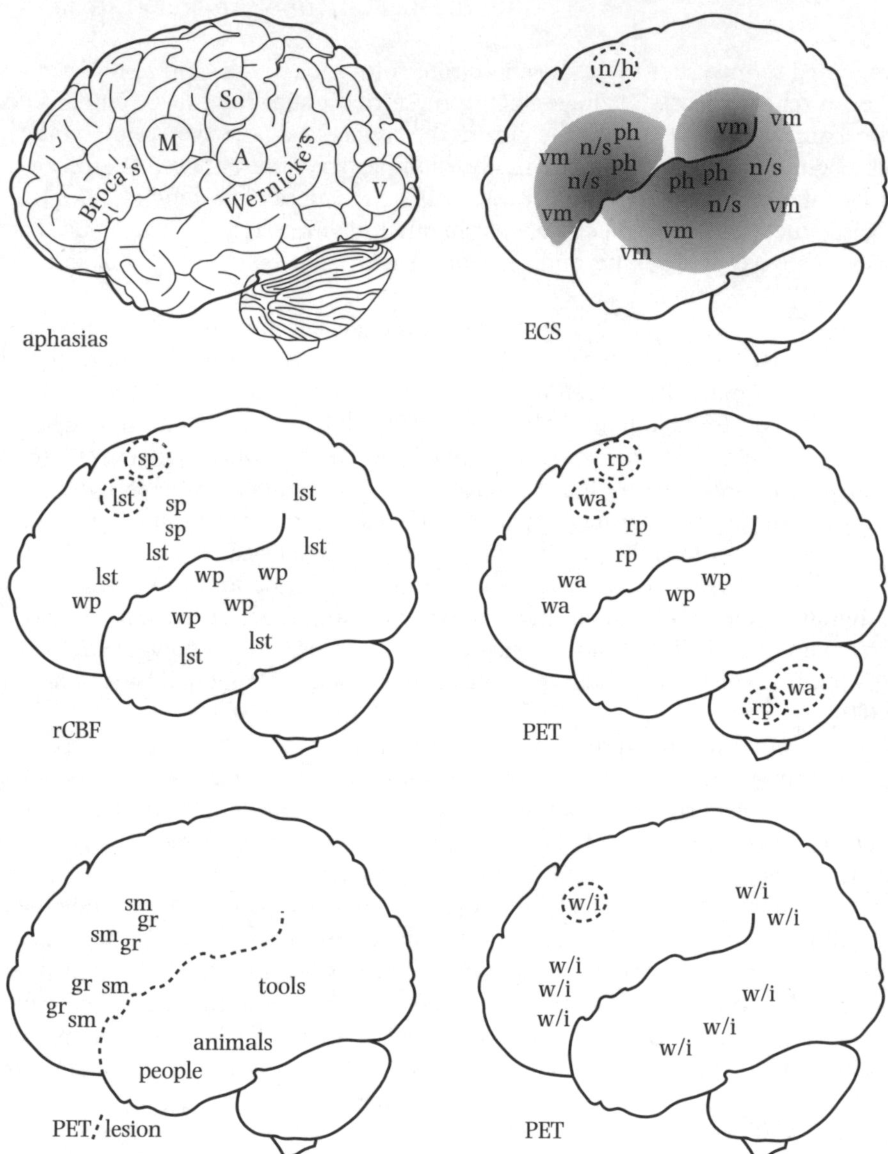

Figure 13.1 Patterns of language representation in cerebral cortical areas as demonstrated by different neuropsychological approaches. *Top left*: regions associated with Broca's and Wernicke's aphasias and locations of primary auditory (A), visual (V), and somatosensory (So) and motor (M) areas for the facial-oral-vocal muscles. *Top right*: site-specific functions interrupted by electrical cortical stimulation (ECS): ph = phoneme production/identification; n = naming errors; s = syntactic errors; vm = verbal short-term memory errors; h = hesitation to speak (see Ojemann, 1991). Shading indicates decreasing probability of effect on language functions further from core perisylvian areas. *Middle left*: regional cerebral blood flow (rCBF) increases in different tasks: wp = word perception; lst = word list generation; sp = repetitive speech (see Lassen et al., 1978). *Middle right*: results from a positron emission tomography (PET) study, showing metabolically activated sites: wp = word perception; rp = repeating words just heard (wp signal subtracted); wa = word association task (generate verb for noun, with wp and rp subtracted) (see Posner and Raichle, 1994; and Article 32, NEUROIMAGING). *Bottom left*: PET study of the process of monitoring

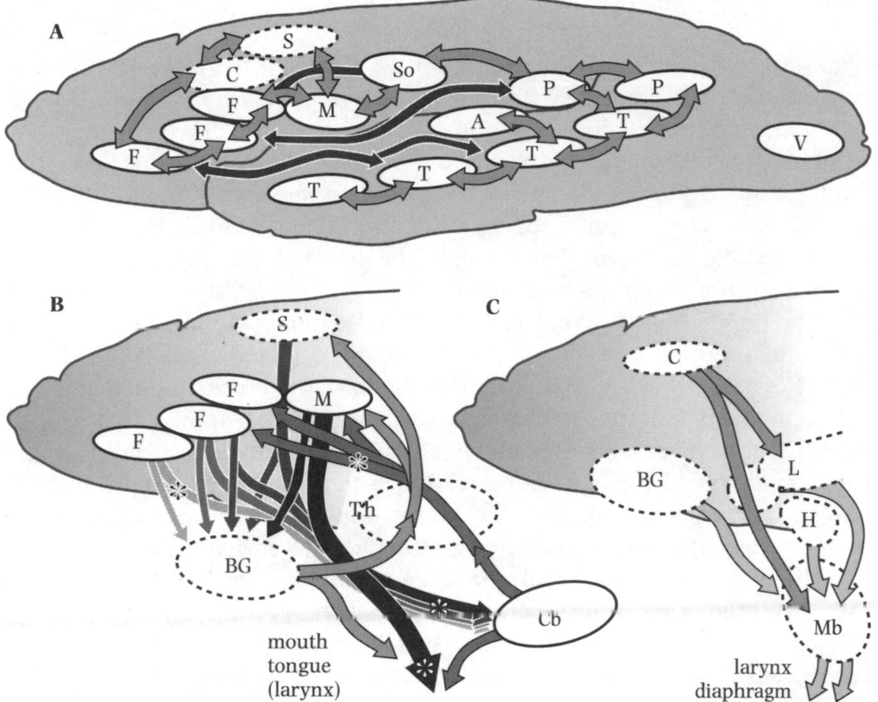

Figure 13.2 Schematic diagram of connections linking cortical language sites in A, speech output systems in B, and the general mammalian nonspeech vocalization systems in C, all as inferred from studies in other primate brains. Left hemisphere neocortical areas associated with language functions include ventral prefrontal (F), middle temporal (T), and inferior parietal (P) sites; midline sites, including supplementary motor (S), and cingulate cortex (C). Hidden structures are shown by dashed lines. Cortical regions involved in language processing are interconnected in tiers radiating outward from primary auditory (A), motor (M), and somatosensory areas (So). Prefrontal regions are also linked to parietal and temporal association areas, though only the anterior ventral prefrontal sites are linked to the auditory association cortex. Intercortical connections are almost always reciprocated (though to different cortical layers). All prefrontal, premotor, and motor areas send outputs to the basal ganglia (BG) and cerebellum (Cb) which integrate these many signals, feed them back to motor cortex via the thalamus (Th), and play critical roles controlling highly learned and rapidly produced language skills. Asterisks indicate connections that do not exist in primates but may be uniquely expanded to these territories in human brains (see below). The organization of innate call systems (in both humans and other species) is almost entirely subcortical, with forebrain cingulate cortex (C), limbic (L), and hypothalamic (H) outputs activating motor program networks in the midbrain (Mb) which have indirect outputs to brain stem (BS) motor nuclei for the larynx and respiration. In humans, motor outputs from cortex may also send direct outputs to these visceromotor motor nuclei (see next section and figure 13.4D).

a story for either semantic categories (sm) or grammatical errors (gr), showing closely linked but slightly different activation in ventral prefrontal sites (Nichelli et al., 1995); also data from a lesion study (separated by horizontal lines) investigating naming shows distinct temporal activation for different classes of objects named: people, animals, and tools (Damasio et al., 1996). *Bottom right*: PET study showing sites activated by both image identification and identification of words for the same images (w/i) (Vandenberghe et al., 1996).

regions that can be recruited, depending on the details of the task. The coordinated involvement of multiple brain systems is required to accomplish each distinct language task, as opposed to the activity of just one or two centers. This does not undermine the notion of local specialization of function, but suggests that what is localized may not be so much a linguistic operation per se but rather a component computation supporting a more distributed linguistic operation. Also, involvement of many systems not specific to language is evident. For example, the anterior cingulate cortex is probably associated with the sustained attentional/intentional effort involved in most on-line language processes, and the contralateral cerebellum may contribute to some of the rapid shifts of attention and actions that many language processes demand. In addition, many *language* areas may simultaneously serve the processing streams of other linked functions, such as image identification (figure 13.1, bottom right).

There appear to be considerable individual differences and sex differences, and remarkable plasticity of language representation in the brain. For example, men appear more prone than women to permanent aphasia due to left hemisphere damage alone, and women often show more language-related effects with right hemisphere damage. Localized brain damage to bilingual speakers often impairs the use of one language far more than the other, suggesting separate localization. Interestingly, disturbance of one's first language with spared second language abilities may be correlated with subcortical damage sites such as the basal ganglia (Salvatore and Fabbro, 1993). Hemispheric plasticity for language is demonstrated by significant recovery (not always complete) of young children's language abilities after nearly complete left hemispherectomy, and also by hemispheric shift in second language representation to the right during training to perform simultaneous translations (Fabbro, 1992). Finally, there are curious differences in the correlation between aphasic syndromes and brain damage sites in speakers of different languages. For example, speakers of highly inflected languages with relatively free word order syntax (e.g., Italian) show more loss of grammatical abilities after left posterior lesions (Wernicke's aphasia), whereas speakers of uninflected languages with more strict word order constraints (e.g., English) show more grammatical disturbance after left frontal damage (Broca's aphasia) (Bates, 1991).

In summary, though language processing recruits specialized localized structures in the left cerebral hemisphere, there may be no one-to-one correlation of individual brain regions with functions defined in linguistic terms. Instead of neural modules being specialized for specific classes of linguistic functions, it appears that many language processes are distributed as component neural computations, performed in concert in many different brain structures. Specific classes of linguistic operations might better be identified with specific signature patterns of distributed activity.

Explaining language uniqueness

Human language is an unprecedented adaptation. Though a large number of other species use sounds to communicate, the vocal complexity, rule-governed structure, and referential functions of language set it apart. Other species' gestural, olfactory, and call repertoires can nevertheless be highly sophisticated and capable of conveying diverse contents in a very precise manner. Humans also have a repertoire of calls and gestures that are the homologues of those produced by our closest primate relatives. Among these are laughter and sobs, smiles and grimaces, and postures universally associated with bluster and courtship. Our innately predisposed communicative calls and gestures, as well as their counterparts in other species, comprise a relatively closed repertoire; they require minimal experience to be produced or interpreted; they are often produced with little or no awareness or intention; there is a tendency to join in and echo another's call or gesture; and they are largely dependent on subcortical brain regions, particularly the limbic system, midbrain, and brain stem, for production. By contrast, language communication is dependent on the learning of an open-ended set of non-innate associations between certain sounds, meanings, and referents; is an acquired skill in both the perceptual-motor and cognitive domains; often requires considerable mental effort to produce and interpret; and is critically dependent on cerebral cortical structures as opposed to limbic and midbrain structures (see below). These structural, functional, and neuroanatomical differences pose some perplexing cognitive and biological questions about its origins.

Hypotheses proposed to explain the origins and physiological basis for our unique language abilities (reviewed in more detail in Deacon,1997) are diverse and difficult to test. There are three classic approaches to the biological basis of human linguistic abilities. These can be designated in terms of what each proposes as the critical difference between humans and nonhuman species that makes humans capable of language. They are (1) a greater general intelligence, (2) the presence of a language instinct or innate knowledge of language, and (3) a capacity for skilled vocal behaviors. Though not mutually exclusive, there is considerable disagreement over which, if any, of these represent the critical threshold that was crossed to produce human linguistic abilities. To these, I would add a fourth hypothesis: (4) the cognitive difficulty of symbolic reference.

Intelligence and encephalization The most commonly cited theories argue that the human language capacity is the indirect consequence of a major increase in general intelligence in our species with respect to other great apes and mammals in general. This view is supported by long-recognized evidence for unusually enlarged human brains. From approximately 2 million years ago to the near present (200,000 years ago) brain size in hominid fossil species increased from a little over 500 cc to nearly 1,500 cc in archaic members of *Homo sapiens*, with only modest increase in average body size. The modern mean is approximately 1330 cc. This increase in brain size with respect to body size is most commonly interpreted to represent the effects of selection for increased intelligence (e.g., learning abilities, memory, etc.) during this period, though independent evidence for brain size correlates of general intelligence either within humans or between species is still hotly debated. In this view, evolution of a more complex, more flexible communicative repertoire – language – is presumed to be a correlate of increased intelligence. Specific language specializations of the brain are unnecessary, because language learning is assumed to be secondary to more general cognitive mechanisms.

Innate knowledge of grammar The major competition for general intelligence theories comes from theories that argue that language is too complex and too unusual in structural organization to be acquired by any general learning procedure. These theories argue, furthermore, that a general learning theory cannot explain the many commonalities of grammatical logic that underlie all the world's languages, particularly since language behavior is highly idiosyncratic compared with other behaviors. Both, however, can be explained if one assumes the existence of a sort of innate language template that serves as a guide for children's *guesses* about language structure during early acquisition and a constraint on the range of structures that can produce humanly usable languages. This foreknowledge might be compared to the predispositions we refer to as *instinct* in other species (see Pinker, 1994). Instincts can be remarkably complex and specific. To choose just one example, weaver birds build large flexible nests that are suspended below tree branches. These hanging nests derive their remarkable structural integrity from an elaborate pattern of intertwining carefully selected grasses and twigs. Weaver birds spontaneously produce this complex pattern of behaviors without training (though observation may play a role, and with time they may improve their nest-building ability). Like this and other such "species-typical" behaviors, language too has a predictable timing of appearance, develops sophisticated capabilities despite relatively limited experience, and produces a product with an underlying architecture shared throughout the species. To support the contention that such uniquely human predispositions exist independent of general intelligence, most linguists cite the paradoxical sophistication of human children's language acquisition at a time in their lives when many other learning capacities appear still poorly developed, and the fact that the information provided to young children does not appear to contain sufficient learning clues (e.g., explicit correction of errors in use of grammar or syntax). The neurological implication of language instinct theories is that human brains must include uniquely human language-specific processing modules. However, although localized brain regions are involved in language use (as discussed above), they do not exhibit a one-to-one correlation with linguistic functions, and there are compelling reasons to doubt that these regions are uniquely human brain structures (discussed below).

Vocal skill and the descent of the larynx Though vocal skill is a critical feature of language function, it has not occupied as central a position in discussions of language uniqueness as have these cognitive issues. Nevertheless, skilled mimicking and rapid articulation of speech sounds is an unprecedented ability among mammals. Other mammals, including great apes, are surprisingly poor at vocal learning, showing almost no ability to learn to produce novel non-innate vocal sounds, though many bird species exhibit quite sophisticated vocal mimicry. This unusual human capacity almost certainly is directly relevant to the origin of human language abilities.

One notable correlate of our vocal abilities is the fact that the human larynx occupies an unusually low position within the throat. This provides an increased range and flexibility of sound production, especially with respect to vowel sounds of speech, and decreased nasality of vocalization. This anatomical oddity evolved despite a correlated increase in the probability of choking due to airway obstruction. Though the position of the larynx cannot alone account for human vocal skills, it is difficult to imagine how such a costly anatomical change could have evolved without selection for its usefulness in vocalization. A significant neurological difference in control of this system must also be involved, but the position of the larynx may serve as an index of the fossil age and importance of vocal communication. Unfortunately the hyoid bone – the only bony structure directly associated with the larynx – tends not to be preserved in fossil remains (only one has been recovered from a Neanderthal skeleton, and none from earlier hominid species), and so most assessments of the position of the larynx in human evolution are based on extrapolations from other features of the base of the skull. These suggest that this anatomical change was relatively late to appear and may not have reached current proportions until the appearance of *Homo sapiens*, approximately 200,000 years ago. Some have taken this to indicate a relatively recent origin of speech. However, because laryngeal position is probably a consequence and not a precondition of speech, the origins of speech would have to significantly precede this date.

Recognition of poor vocal abilities in great apes, despite their apparent high intelligence, prompted numerous studies of apes' abilities to learn nonvocal language-like systems over the last three decades, including the hand gestures of American Sign Language and a number of token-based systems (for an insightful example see Savage-Rumbaugh and Lewin, 1994). If lack of a sufficiently facile and flexible medium were the major constraint limiting language acquisition in these species, then these supports should have helped animals approach human linguistic capacities far more closely. These studies have not produced unequivocal successes, and claims and criticisms of these claims have produced intense debates (see Article 2, ANIMAL COGNITION). Apes perform considerably better on visual-manual communication problems than on auditory-vocal language tasks; however, the extent to which they approach human levels of performance with these same systems is difficult to assess due to many ambiguities in methods and interpretations. One thing is clear, however, from the only modest results of even the most successful of these studies: vocal abilities are not the only difference.

Symbolic abilities There is a fourth major factor that has generally been ignored, yet which is at least as problematic and fundamental. Language enables us to communicate in abstract, nonconcrete terms by virtue of its unusual mode of reference: symbolic reference. The problem of explaining reference in language has long challenged philosophers, so it should come as no surprise that it plays a critical role in the language origins mystery. All three of the hypotheses described above address the issue by means of language complexity, but if the complexity of language were the major problem blocking other species, we might expect there to be dozens of simpler animal languages in the natural world. As far as we know, there are none. More generally, the symbolic features of language have no obvious counterparts in the rest of nature. As far as we can tell, symbolic reference evolved once, in only one species, and persists in only one highly specific and elaborate form. Moreover, other species experience nearly insurmountable difficulties with even understanding word meanings (as opposed to producing *rote* responses). None of the other hypotheses explain why such an apparently simple and basic feature of human communication and cognition should pose such a formidable impediment.

The distinction between language and nonlanguage is not one of referential versus nonreferential signs. There are many examples of very specific referential forms in animal communication, including the directional information in the foraging "dance" of the common honeybee and the predator-specific alarm calls of birds and vervet monkeys (see Hauser, 1996, for extensive review of animal communication and neuromechanisms; see also Article 28, COGNITIVE ETHOLOGY). Our own laughter and sobbing calls not only communicate information about the state of the caller; they also provide information about the type of events external to the caller and immediately prior or concurrent with the call. These calls refer by virtue of a direct causal correlation (physiological link) between the call, the state of arousal (in both producer and interpreter), and the event. This mode of reference is generally called *indexical* reference. This difference between language and nonlanguage is also not simply a matter of whether the response is arbitrary. Consider a pigeon conditioned to peck at a spot in response to a specific sound (for example, a spoken word) in order to get a food reward. The pigeon takes advantage of arbitrary reference when it learns that the sound *indicates* the availability of food in the apparatus, but it does not symbolize food.

By contrast, words refer to things indirectly, by virtue of an implicit system of relationships between them. This requires that they work in combination (even if only implicitly), referring to one another and modifying one another's reference, to produce a kind of *virtual reference* in which each is associated not so much with some specific concrete object or event, but with kinds, abstract classes, or predicates that can be applied to things. This systemic or combinatorial form of reference confers considerable flexibility and enables almost unlimited referential variety (often referred to as generativity). But this open-ended scope of possible outputs also raises the requirement for cognitive effort and skill (as well as perceptual and motor skills) and complicates the learning process because of the need to acquire a whole system of symbols before they acquire referential power.

Language training studies in other species have demonstrated that it is possible under certain circumstances for them to acquire limited symbolic abilities, though these are far more limited than at first suspected. In these studies, the difficulties the subjects exhibit and the special circumstances of their successes are highly informative. It appears that concrete learning predispositions in these nonhuman species interfere with symbol-learning tasks and may be difficult to inhibit. This suggests that the abstract and indirect nature of symbolic associations may pose very difficult, though not insurmountable learning problems for other species, and that human children may approach symbol learning with predisposed learning biases evolved to overcome this (the nature of these difficulties is discussed further in Deacon, 1997).

Human neuroanatomical uniqueness

Many classic theories of language origins suggested that such unprecedented cognitive capacities could be explained only by the evolution of some newly added language-specific brain structure (which might incidentally also account for the larger human brain size). The classic language areas have been prime candidates for this role. Many have proposed that Broca's area might be a novel human brain structure. But evidence for the existence of unprecedented (nonhomologous) human brain structures is dubious on many grounds. In general, accretive conceptions of brain evolution are nineteenth-century anachronisms – evolution proceeds not by addition of new parts or developmental stages, but by modification of preexisting structure and function. With respect to classic language areas, cellular and connectional homologies have also been identified recently in other primate brains, and the distribution of language operations across multiple cortical areas (reviewed above) argues against newly evolved language centers. The alternative is that subtle modifications of ancestral brain structure relationships support the human language adaptation.

One might better compare the evolution of language to the evolution of flight. Flight was an unprecedented change of function. It arose independently in dinosaurs, birds, and mammals as a result of modifications of existing limb structures, primarily changes in bone length, bone shape, muscle attachments, and skin or feather growth. The adaptations underlying linguistic abilities might similarly reflect a more ancient neural logic that only recently (in evolutionary terms) was recruited for this highly atypical task, supported by tweaks and adjustments of preexisting relationships to manage these otherwise unprecedented functions. The human neurological language adaptation is not a single difference, but a suite of adaptations, including auditory and articulatory specializations, supports for rapid syntactic analysis and production, predispositions for social interactions that facilitate vocal mimicry of speech, and learning biases that aid the analysis of symbolic relationships. These almost certainly have diverse representation in the brain, and many of these should be reflected in divergent features distinguishing human brains from nonhuman brains.

Part-by-part volumetric analysis of primate brains suggests that the comparatively large size of the human brain is the result of neither uniform expansion nor expansion along lines predicted from trends shared by other primate species. Instead, there are global deviations in the sizes of human brain structures with respect to size trends that are otherwise shared by all other monkeys and apes. The greatest divergence from typical primate brain structure is exhibited in the sizes of the cerebral and cerebellar cortices relative to other brain structures. The global pattern of these deviations suggests that they can be traced back to cell production differences that correspond to early gene expression domains in the embryonic neural tube (see Deacon, 1997).

An embryonic shift in cell production in cerebral and cerebellar cortices may be the initial step in a cascade of developmental events that ends in a substantially altered human brain. Research into mammal brain development has demonstrated that the patterning of the connections that link brain regions is determined by a dynamic competitive interaction between growing axons. After neurons have become differentiated, their axons begin to grow toward distant brain targets that are only generically predetermined (e.g., they grow toward particular classes of brain structures or with spatial biases in growth tendency), branch into a number of potential targets, then compete for a limited number of permanent synaptic sites against other axons which have also grown to that target (schematized in figure 13.3). This Darwinian-like process allows a sort of *post hoc* fine tuning of connection patterns and cell numbers (significant neuronal elimination also results from this competition) as the developing nervous system adapts to itself and the rest of the body. As a result, evolutionary modifications of one part of the nervous system can be accommodated by the rest of the brain without any additional genetic changes. One particularly instructive example (depicted in figure 13.4A and B) is the blind mole rat (*Spalax*), a species of rodent in which the brain has been adapted to a radical alteration of peripheral organs (vestigialized eyes and hypertrophied head and neck muscles) (Doron and Wollberg, 1994). This natural experiment parallels laboratory experiments in which input from the eyes is interrupted at an early stage of brain development. Under both conditions the normal targets for visual inputs are instead invaded by axons from other senses, and the resulting displacement of function ramifies throughout the brain (see Deacon, 1997).

A parallel modification may be of special importance for understanding the organization of human brains, but whereas brain changes follow peripheral changes in *Spalax*, the locus of reorganization is within the brain itself in humans. In the human brain relative overproduction of cells in the cerebral cortex, cerebellar cortex, and other dorsal neural tube structures has probably altered innumerable connection patterns

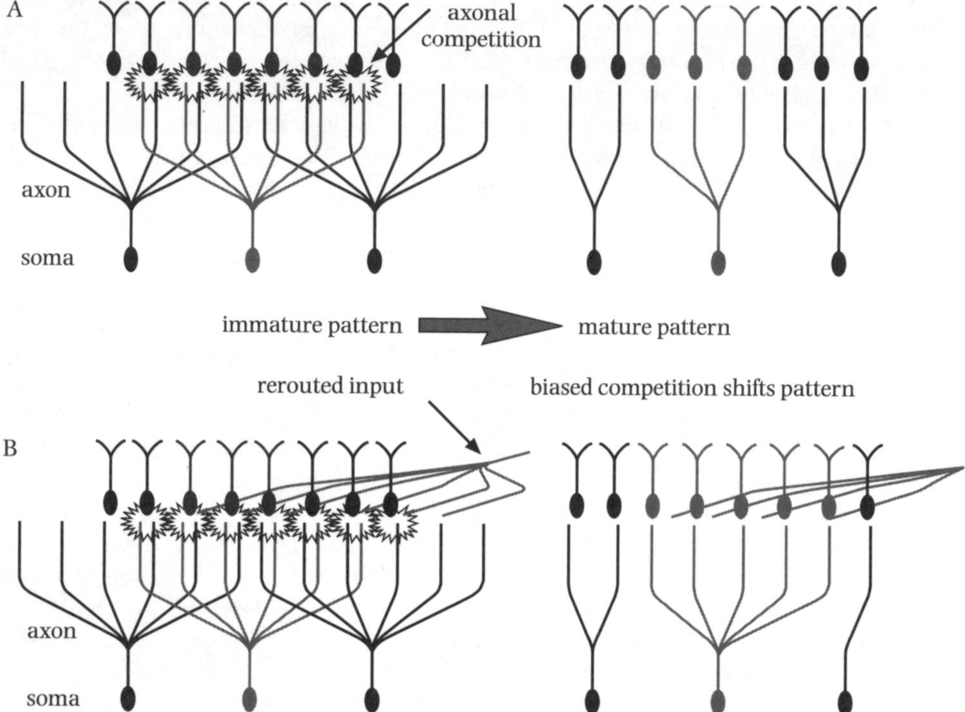

Figure 13.3 A: development of neuronal network architecture begins with initial relatively unspecific axon pathfinding followed by competition for synapses based on correlation patterns between converging signals. Mature networks are produced by pruning of axonal connections as well as by neuronal cell death. B: Because axonal competition is mediated by signal processing, changes in the distribution, numbers, or information carried by competing axons can produce systematic biases in mature network organization.

analogous to the way this occurs in *Spalax*. In general, we can predict with some confidence that human brain structures that are significantly enlarged will be disproportionately favored in the developmental axonal competition, and so will end up recruiting a wider range of targets than would typically be the case in other primate brains. Two likely examples of this that might be relevant to human language abilities are depicted in figure 13.4C and D: increases in the extent and distributions from the enlarged cerebral cortex to motor output structures and to other cortical areas.

The evolutionary novelty of language communication is clearly reflected by the dependence of language on very different neural substrates than are nonlinguistic vocal communications, including such human calls as laughter and sobbing. Pioneering studies using electrical stimulation and lesions of different brain regions in other species have demonstrated that core regions in the central midbrain and brain stem are critical loci for most if not all of the stereotypic motor programs that generate the varieties of mammal *calls* (see figure 13.2C). Stimulation of a variety of basal forebrain and limbic structures, but essentially no neocortical areas, can also initiate call production secondary to arousal and emotion, whereas midbrain stimulation produces normal vocalizations, but out of context from other emotional correlates. In general, the neural

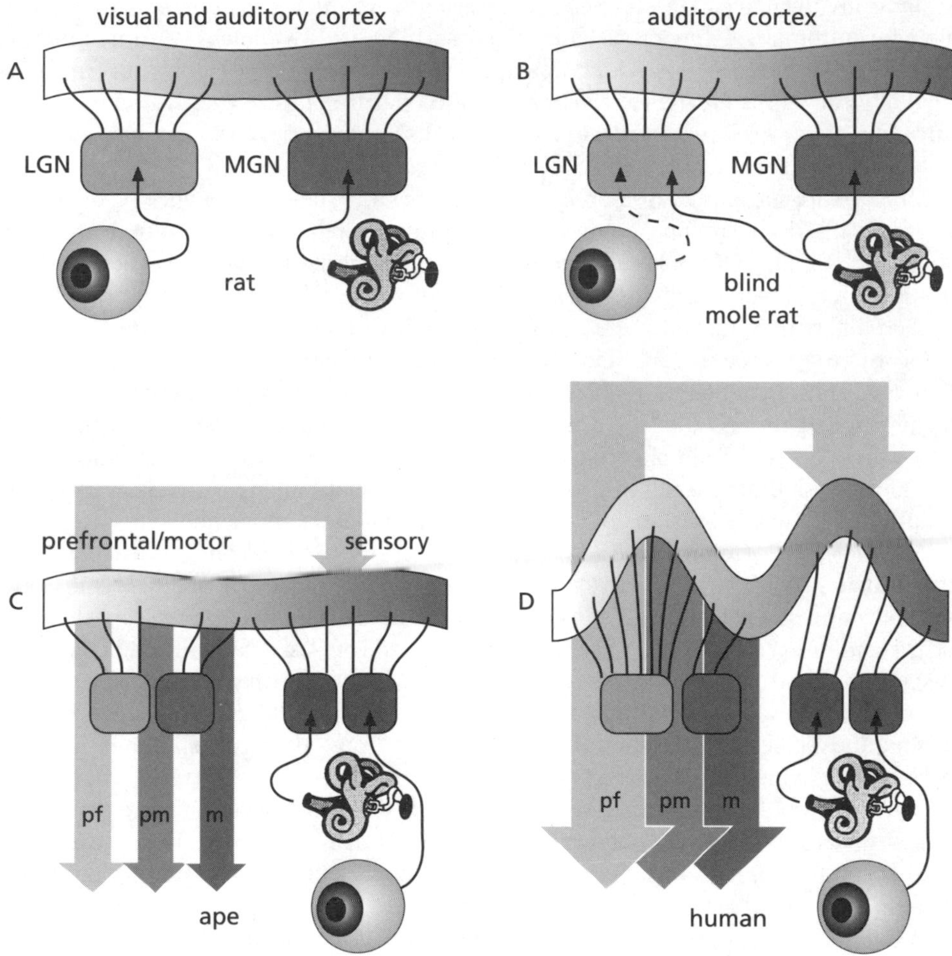

Figure 13.4 Evolutionary displacement of neural connections due to competitive biases in the brain of the blind mole rat (B), *Spalax*, and as predicted in the human brain (D), both as a result of quantitative changes. A: vestigialization of the eyes in the blind mole rat allows its CNS target (lateral geniculate nucleus, LGN) to be competitively invaded by other inputs (auditory) that normally would recruit only the neighboring medial geniculate nucleus (MGN) (Doron and Wollberg, 1994). B: In the human brain there is embryonic overproduction of precursor cells in the cerebral cortex, cerebellar cortex, and a few other dorsal brain structures with respect to others that have not enlarged. Two predicted biases of axonal growth and competition are shown diagrammatically: descending arrows: increased prefrontal, premotor, and motor cortical outputs to midbrain, brain stem, and spinal cord motor nuclei (including to visceromotor nuclei of the larynx and respiratory system); arrow above: increased proportion and extent of prefrontal inputs to all other cortical areas.

systems involved in call production are termed visceral motor systems and are responsible for automatic behavior systems, like breathing and swallowing – functions that must operate autonomously and are capable of overriding consciously driven activities. This accounts for the relatively automated nature of these vocalizations as well, and why they are largely untrainable (for a discussion of exceptions such as birds, dolphins, etc., see Deacon, 1997).

Shifts of proportions in human brain structures may help to explain this difference in vocal skill. The disproportionate size of the human cerebral cortex compared to brain stem and spinal cord nuclei for oral-vocal muscles would promote invasion of these output nuclei by descending cortical projections during development (figure 13.4D). Speech differs from most other sorts of vocalizations in both its skilled voluntary character and the degree to which mouth and tongue movements are used to modify the flow of sounds. Laughter and sobbing exemplify the relatively repetitive respiratory and laryngeal activity and the lack of tongue articulation characteristic of most mammalian calls. The expansion of cortical outputs to include laryngeal and respiratory control along with oral musculature may be the source of our ability to precisely integrate breathing, vocal fold tension, and oral-lingual articulatory movements in speech.

Another size-correlated shift in connectivity within the forebrain may account for the unusual cognitive differences associated with symbol learning abilities. One of the most striking size deviations is demonstrated by the unusually large prefrontal region of the human cerebral cortex (about twice the predicted size for human brain size). Because prefrontal regions have only very indirect links to peripheral sensory or motor systems, they are least constrained by the comparative size difference between human brains and bodies. In the competition for connections, the proportionately smaller peripheral organs recruit proportionately less of the cerebral cortex (and other linked systems) for input–output functions and thus cede cortical space to this more centrally linked structure. As a consequence, human brains have come to be unusually dominated by prefrontal connections and the specific classes of neural computational processes performed there.

What sort of cognitive bias might this produce? Answers to this question can be proposed only in the most general and speculative terms, since there are essentially no data with which to directly address the relationship between such neural proportions and cognitive abilities. However, we can extract a few clues by comparing neuropsychological accounts of damage effects. The prefrontal cortex has been called the center for executive functions and for working memory. Damage to this structure causes people to have difficulty regulating behavior that must be sensitive to changes in context and in shifting between patterns of responding. For example, card-sorting tasks that require prefrontal patients to shift from one sorting strategy to another show their susceptibility to perseveration (inability to give up a previous strategy despite influences which suggest shifting). This demonstrates a feature common to prefrontal functions in all species. It is critical in tasks where previously acquired tendencies are themselves the basis for specifically inhibiting their activation. This kind of processing is not relevant to most learning contexts, where immediate stimuli can activate appropriate responses, but it is essential in cases where information acquired in one context would be usefully applied to entirely different contexts. This aspect of learning is probably nowhere more critical than in symbol acquisition tasks, since symbolic reference is entirely dependent upon recoding learned associations into an entirely different realm of relationships between symbols. A significant bias that favors this class of cognitive operations may allow humans to be more natural symbol-learners, whereas a bias against this otherwise rarely needed learning tendency may be responsible for other species' symbol-learning difficulties (Deacon, 1997).

Hominid fossil evidence may provide indirect clues to the timing of this shift in vocal skill and learning strategy. Since brain and body proportions during embryogenesis are major factors influencing the relative size of these brain systems with respect to other sensory and motor areas (via biases in axonal competition), fossil brain and body size estimates can serve as an index of the evolution of these cortical proportions and cognitive biases. The evolution of larger relative brain size that began with the fossil species *Homo habilis* just over 2 million years ago and persisted until the dawn of our own species suggests that selection favoring this cognitive bias persisted for this entire period. Thus, some form of symbolic communication may be traced 2 million years into the past, roughly contemporaneous with the first appearance of stone tools. This poses the possibility that brain structure and languages have long been bound together in a coevolutionary feedback and have each significantly affected the form of the other. This may help to explain why human cognitive abilities have become so divergent from those of other species.

References and recommended reading

Bates, E. (ed.) 1991: Special issue: crosslinguistic studies of aphasia. *Brain and Language* 41(2).

Damasio, H., Grabowski, T. J., Tranel, D., Hichwa, R. D. and Damasio, A. R. 1996: A neural basis for lexical retrieval. *Nature*, 380, 499–505.

*Deacon, T. W. 1997: *The Symbolic Species: The Coevolution of Language and the Brain*. New York: W. W. Norton, Inc.

Doron, N. and Wollberg, Z. 1994: Cross-modal neuroplasticity in the blind mole rat *Spalax Ehrenbergi*: a WGA–HRP tracing study. *NeuroReport*, 5, 2697–2701.

Fabbro, F. 1992: Cerebral lateralization of human languages: clinical and experimental data. In J. Wind, B. Chiarelli, B. Bichakjian, and A. Nocentini (eds), *Language Origin: A Multidisciplinary Approach*, Proceedings of NATO Advanced Institute, Cortona, Italy, 1988, Amsterdam: Kluwer, 195–224.

*Hauser, M. D. 1996: *The Evolution of Communication*. Cambridge, Mass.: MIT Press.

Lassen, N. A., Ingvar, D. H. and Skinhöj, E. 1978: Brain function and blood flow. *Scientific American*, 239, 62–71.

*Lecours, A. -R., Lhermitte, F. and Bryans, B. 1983: *Aphasiology*. London: Bailliè re Tindall.

Nichelli, P., Grafman, J., Pietrini, P., Clark, K., Lee, K. Y. and Miletich, R. 1995: Where the brain appreciates the moral of a story. *NeuroReport*, 6, 2309–13.

Ojemann, G. A. 1991: Cortical organization of language. *Journal of Neuroscience*, 11, 2281–7.

*Pinker, S. 1994: *The Language Instinct: How the Mind Creates Language*. New York: William Morrow and Co.

Posner, M. I. and Raichle, M. E. 1994: *Images of Mind*. New York: Scientific American Library.

Salvatore, A. and Fabbro, F. 1993: Paradoxical selective recovery in a bilingual aphasic following subcortical lesions. *NeuroReport*, 4, 1359–62.

*Savage-Rumbaugh, E. S. and Lewin, R. 1994: *Kanzi: The Ape at the Brink of the Human Mind*. New York: John Wiley and Sons, Inc.

Vandenberghe, R., Price, C., Wise, R., Josephs, O. and Frackowiak, R. S. J. 1996: Functional anatomy of a common semantic system for words and pictures. *Nature*, 382, 242–4.

14

Language processing

KATHRYN BOCK AND SUSAN M. GARNSEY

Imagine a telephone conversation between a presidential aide and a wealthy supporter, shortly after news breaks that the president plans to veto a bill that the supporter strongly favors. The nervous aide opens with "I'm calling to let you know that the president regrets his, uh, his decision. . . ." The supporter's hopes rise at the intimation that the president changed his mind. But when the aide continues: ". . . did not meet with your apparel, I mean, your approval," the crestfallen (and, perhaps, former) supporter realizes that the president is sorry merely about the impact of his decision, not the decision itself. The aide's utterance seems to mean one thing at first, but something entirely different in the end. Unintentional verbal about-faces of this kind are common in everyday language use – the outspoken owner of an American baseball team recently said that she regretted her remarks offended many people – yet listeners rarely notice the miscues, the wealthy supporter notwithstanding. This is one of the chief motivations behind the psycholinguistic study of language use.

The presidential aide's utterance also illustrates what speakers do and a little about what can go wrong. Her entire utterance lasts less than 10 seconds, during which she retrieves from memory, arranges, and produces more than 20 words. This rate of speech is well within normal bounds, and its maintenance requires that words be recalled, sequenced, and converted into muscle movements so rapidly that talking would be an Olympic event were it not mastered so universally. Problems arise when words do not come directly to mind (leading to the "ers" and "uhs" and other disfluencies that dot everyday conversation), when the wrong word comes, when words or sounds emerge in the wrong order, and so on. From the standpoint of human communication, it is all the more remarkable that the athletic event of talking serves to convey a thought, turning incidents from a speaker's mental life into noises in the air.

When communication succeeds, the details of a speaker's thought are re-created in some analogous form in the mind of the listener. Looked at in terms like these, the processes of understanding and speaking rival extrasensory perception in mystique. Yet language is so mundane that its natural wonders are easily overlooked. These wonders are the province of theories which aim to explain language production and comprehension in terms of the interplay between cognitive processes and the language knowledge possessed by virtually everyone over the age of three.

Like cognitive science itself, this enterprise is relatively young. The interest in language as a cognitive ability emerged at mid-century from Noam Chomsky's emphasis on language as a biologically rooted form of individual knowledge, not a merely conventional cultural code. Together with contemporaneous developments in cognitive psychology, this inspired research on how individuals put their knowledge of language to use. The research focused on the nature of the interpretations, the ideas, that result from reading or hearing sentences. A major finding was that comprehenders remember little about the wording of a message. Rather, they remember what the message meant.

From these early revelations about the products of language understanding, two new themes have emerged during the last 25 years. In the late 1970s, interest shifted to the moment-by-moment unfolding of meaning during comprehension. Leading this shift were Lyn Frazier, William Marslen-Wilson, David Swinney, and Michael Tanenhaus. At the same time, research emerged on how speakers formulate what they say, pioneered by Merrill Garrett, W. J. M. Levelt, Gary Dell, and Herbert Clark. This paved the way toward a balanced view of language ability, which comprises both talking and understanding.

Speakers and listeners juggle three kinds of demands in order to produce and understand comprehensible utterances. One set of demands is cognitive, involving rudimentary mental, neuromotor, and perceptual processes: Speakers must retrieve language elements from memory and produce them in sequence, articulating sounds that listeners perceive and recognize as the elements intended. A second set of demands is linguistic: To convey a specific meaning, sounds and words must be strung together by speakers and parsed by listeners in accordance with the vocabulary and grammar of the language being spoken. A third set of demands is communicative: For the meaning of an utterance to register appropriately, the utterance must make contact with what the listener knows, including what the listener knows about the speaker. So, speakers take their listeners' frames of reference into account, and vice versa.

What makes a juggling act of these requirements is the intricacy of their coordination, as reflected in some brute numbers associated with each demand. Words can be comfortably articulated at a rate of four per second, through the combined efforts of more muscle fibers than may be required by any other mechanical performance of the human body, and audited at the same rate, by the 16,000 hair cells of the cochlea. Adult speakers know somewhere between 30,000 and 80,000 words; the average for a high-school graduate has been estimated at about 45,000 words. These words can be arranged in any of an infinite number of ways that conform to the grammar of the language. The ramifications of this can be at best dimly appreciated in the approximate number of English sentences with 20 or fewer words: 10^{30}. Using these resources in conversation in the course of a typical summer day, the authors and three of their colleagues at the University of Illinois each spoke with 19 other individuals on average, including family, friends, co-workers, and total strangers. For each interlocutor, the speakers tacitly evaluated the suitability of English for the exchange, the amplitude of their voices, the appropriateness of the vocabulary, topics of mutual interest, the circumstances of the encounter, motives for speaking, and more.

In the sections that follow, the complexities of language production and comprehension are surveyed further in terms of their cognitive, linguistic, and communicative demands. We begin as a conversation might, with some of the problems of producing a sentence.

Language production

An oversimplified but still commonplace standard for knowing a language is the ability to speak it fluently. Although this standard underestimates the challenges of language comprehension, it reflects a widespread intuition that it is hard to produce language. What makes talking seem so difficult, perhaps, are its evident perils.

Speaking is vulnerable to a variety of mishaps. One consequence is that an average speaker spends about half of her speaking time in *not* speaking, hemming and hawing, lapsing into disfluency between three and twelve times per minute. These mishaps are sometimes interesting in themselves, but they can also reveal fundamental features of the activity of speaking.

The study of errors in speech has shown that the effort to say what one means does not always lead to meaning what one says. Speakers can detect mismatches between what they said and what they meant, implicating cognitive mechanisms that are not under the direct control of the communicative intention. The goal of the psycholinguistic study of language production is to explain these mechanisms and how they mediate the transition from thought to language. For each of the three demands of processing, we can describe one phenomenon that has illuminated what goes on.

Cognitive processes in production Foremost among the cognitive demands of speaking is the need to rapidly retrieve words from the mental lexicon. All too often, speakers find themselves unable to recall from the 45,000 or more words they know the one that is needed on a particular occasion. This precipitates the mental agony described as having a word on the tip of one's tongue. The word is often rare, since the more frequently a word is used, the faster and more accurately it can be retrieved.

The phenomenology of the tip-of-the-tongue state was evoked long ago by William James:

> Suppose we try to recall a forgotten name. The state of our consciousness is peculiar. There is a gap therein: but no mere gap. It is a gap that is intensely active. A sort of wraith of the name is in it, beckoning us in a given direction, making us at moments tingle with the sense of our closeness, and then letting us sink back without the longed-for term. If wrong names are proposed to us, this singularly definite gap acts immediately so as to negate them. They do not fit into its mould. And the gap of one word does not feel like the gap of another, all empty of content as both might seem necessarily to be when described as gaps. (James, 1890, vol. 1, pp. 251–2)

Most notable about the condition is its specificity. The speaker knows that she knows a word that perfectly expresses an intended meaning, yet the word fails to materialize. Occasionally some of the sounds can be recovered, the number of syllables, or the stress pattern. But often there is only the gap. Recent research has shown that speakers who are unable to recall anything about the sounds of an elusive word can nonetheless report specific grammatical properties of the word that cannot be derived from its meaning (Vigliocco et al., 1997).

The implication is that our knowledge about words comes in pieces, and these pieces need not be recovered all at once. In fact, converging evidence from experimental and observational studies of language production indicates that in normal speaking, the semantic, syntactic, and phonological properties of words are called upon in quick succession, not simultaneously. One illustrative finding is that electrophysiological indicators of semantic retrieval processes precede the indicators of phonological retrieval processes by about 100 milliseconds (van Turennout et al., 1997). Thus, what may feel like a simple, unitary act of finding-and-saying-a-word is actually a complicated but fast assembly of separate, interlocking features.

Although there is fairly good agreement among psycholinguists about the kinds of information that must be retrieved and assembled, theorists differ over the dynamics of retrieval. One point of contention is the discreteness of retrieval operations. Discrete-stage views argue that retrieval proceeds in punctuated steps from meaning to grammatical features to sound, whereas cascade views posit a continuous flow of information. A second question is whether information that is retrieved and used early in processing (e.g., the grammatical features of a word) can be influenced by the transient memory state of information that is used later (e.g., the word's phonology). For instance, a speaker might find it easier to retrieve the grammatical features of the word *cat* (noun, singular, count) if its component sounds are unusually active in memory, perhaps because of recently hearing similar sounds. This possibility is endorsed on the *interactive* view, but denied in the *modular* view (see Article 49, MODULARITY).

Linguistic processes in production The grammar of a language can be likened to a code for the conversion of meanings into sounds, and vice versa. The code is known to native speakers, who likewise know how to use the code to convey and receive messages. The cognitive operations that implement the coding scheme constitute the linguistic processes of production.

As a coding scheme, a grammar involves symbols and rules for arranging symbols. The symbols of spoken language, words and sounds, are tacitly assembled by the linguistic processes of production in accordance with grammatical rules. The rules are not so much obeyed as they are simply realized in the assembly: Linguistic processes develop during language learning to embody all but the most effete rules of grammar in automatic operations.

The normally hidden machinations of linguistic processes are revealed in many kinds of speech errors. Consider errors in which two symbols exchange. In word exchanges like "room in your phone" (when "phone in your room" was intended), the speaker inadvertently reverses two words, putting each in the other's place. In sound exchanges like "accipital octivity" (when "occipital activity" was intended), the initial vowels reverse. Such errors vividly illustrate that speaking is more than reciting lines. It involves active, ongoing construction of utterances from rudimentary linguistic parts.

Exchange errors also disclose the embodiment of rules in the arrangement process. When words exchange, they exchange almost exclusively with other words from the same syntactic class (noun, verb, adjective, and so on). When sounds exchange, they exchange almost exclusively with other sounds from the same phonological class (consonant or vowel). The net result is that erroneous utterances are almost always grammatical, albeit often nonsensical.

The preservation of grammaticality at the cost of sense in speech errors is a touchstone in the argument that there is a cognitive architecture to language production. Evidently, a speaker's conscious intention to communicate such-and-such a message falls victim to automatic and normally efficient operations which proceed fluidly, despite losing track of the symbols they are manipulating. What suffers is what the speaker meant to convey, not the basic schemes of grammatical arrangement.

Communicative processes in production All the processes of language production serve communication, but we will single out the activities that tailor utterances to the needs of particular listeners at particular places and times. The tailoring requirements are extraordinary in their diversity. They range from such patent demands as choosing language (English? Dutch?) and gauging loudness (whisper? shout?) to the subtle need to infer what the listener is likely to be thinking and what he is likely to be able to readily recollect.

The problems a speaker routinely confronts in assessing a listener's knowledge can be seen in the simple act of opening a telephone conversation. For some listeners (e.g., an anonymous clerk in a municipal services department), the speaker must provide introductory information to establish a relevant frame of reference: "Hello. My name is June Irwin, and I live on the north side of the 1000 block of West University Avenue in Champaign." For casual acquaintances, a first and last name will suffice ("Hello, this is June Irwin"), while friends need only a first name. Intimates and relatives can manage with one's voice alone: "Hi, it's me." The risk for the speaker is in how much to presume. Virtually everyone has experienced the momentary panic prompted by a caller who presumes too much ("Hi, it's John") and the uneasiness engendered by a caller who presumes too little ("Hello, this is your mother").

Speakers rely heavily on overt and covert feedback from listeners in making communicative adjustments, and use this information in efforts to bring their speech into line with the listener's frame of reference. Even so, speakers' efforts often fall short. Perhaps the most common shortcoming is presuming too much, failing to anticipate the myriad ways in which a listener can construe any given utterance or expression. Even the simplest of pronouns can create confusion:

Kay: It's too long!
Susan: Which *it* do you mean?

The source of the speaker's presumptuousness is transparent: Speakers know what they intend. For them, there is no ambiguity in the meaning of their utterances. And in this one crucial respect they have little in common with listeners, as we shall see.

Language comprehension

The listener in our hypothetical conversation has a task at least as complex as the speaker's. Just as rapidly as speakers talk, listeners must understand what they are saying. And listening is only one of a literate comprehender's problems. Reading presents different challenges, along with some advantages. Unlike listeners, readers can control their rate of intake, slowing down when prose is hard and speeding up when it is easy. This helps skilled readers to read a little faster (averaging 4–5 words per second) than the average speech rate (averaging 3–4 words per second). However, readers lack information from intonation in speech, which is conveyed only crudely with punctuation.

Despite such differences, the goal in both modalities is to understand the meaning behind the language. The problem that stands in the way is the ambiguity of the information arriving at every instant in any episode of listening or reading. Explaining how these inexorable ambiguities are successfully (and mostly unconsciously) resolved is the goal of a theory of comprehension.

Perceptual and cognitive processes in comprehension The first of many ambiguities presents itself to listeners immediately. The acoustic signal seldom contains enough information to uniquely identify spoken words. A simple demonstration consists of asking listeners to identify single words excised from recorded speech. Words that sound perfectly clear in the full recording can be unrecognizable outside the original context. The contrast is dramatic in highly predictive compared to nonpredictive environments: Identical recordings of the word *nine* after "A stitch in time saves . . ." or "The next number you hear will be . . ." give very different impressions about what number was heard.

Compounding the ambiguity of speech sounds in the acoustic signal is the difficulty of locating word boundaries. There are few actual gaps in the stream of speech. In a familiar language the words may sound separate, but anyone listening to an unknown language quickly realizes the impossibility of hearing where one word ends and the next begins. In an unfamiliar environment even a known language can trip us up: There are famous instances of mishearings like "holy imbecile tender and mild" for *holy infant so tender and mild.* Readers of English (and of English hymnals) are spared these mistakes by the spaces that flank words.

Misunderstandings of speech are made rarer than they might otherwise be by listeners' efficient use of knowledge to disambiguate the acoustic information. The relevant knowledge spans the context specifically and the world and language generally. The rapidity with which knowledge can be used is illustrated in speech shadowing. Shadowing requires the immediate repetition of natural continuous speech as one hears it. Some people can do this at lags of fewer than two syllables (about a quarter of a second), initiating a word even before it is completely heard. Yet the effects of context are discernible. When small errors are included in the speech being shadowed, the shadower typically corrects the errors in ways consistent with the preceding context, without noticing or delaying.

Vivid testimony to the importance of context for understanding speech comes from what happens when the ability to use knowledge about the context is absent or rudimentary. The weakness of knowledge-based support has plagued the development of artificial speech recognition systems, which remain notoriously limited and unreliable despite Herculean efforts to reap the economic rewards that await viable utilities. In the meantime, crude remedies substitute for what listeners do naturally. For example, in order to get around the problem of segmenting the speech stream, most artificial speech recognition systems require speakers to pause between words.

The root of the difficulty, whether in explaining human speech perception or implementing artificial speech recognition, is that the mechanisms that bring knowledge to bear remain incompletely understood. There is controversy about whether context affects the perception of speech sounds directly, or affects later processes that select a word from a set of candidates that are consistent with the perceptual input. A system of the former kind requires interaction between sensory or perceptual and higher-level processes, while the latter does not. This is again a question of interactivity versus modularity.

Linguistic processes in comprehension The successful identification and segmentation of wordlike stretches in speech makes way for another set of ambiguities. These ambiguities develop when comprehenders retrieve knowledge about words and construct interpretations of word sequences. They occur irrespective of the modality, demanding tacit choices by listeners and readers alike.

The most familiar ambiguities involve alternative meanings (*plant* as factory or *plant* as flora) and grammatical classes (*plant* as noun or *plant* as verb). Understanding the word demands a decision about which meaning and grammatical class are relevant. Another, less obvious type of ambiguity is illustrated in the utterance of our fictitious aide, ". . . the president regrets his decision. . . ." The words may be unambiguous, but a choice must be made about how to combine the words: that is, about the sentence structure. Is the president sorry about his decision? As it happened, no: He regretted his supporter's unhappiness. Examination of eye movement patterns and electrophysiological activity in the brain indicates that readers and listeners deal with many of these linguistic choices as the choices present themselves, without awaiting further information. What determines the options that comprehenders take?

There are several ways in which they might proceed. They might entertain all possibilities at once and choose among them, or evaluate one option at a time. If the latter, which options are considered first? How quickly does higher-level knowledge influence the choice? Which kinds of choices are affected by this knowledge? Do different sorts of knowledge contribute equally, or are some weighted more heavily than others?

These questions have been addressed in research on the time course of disambiguation. In a typical experiment, listeners hear ambiguous words in sentences that provide different types or amounts of disambiguating context (e.g., "The leaves of the plant . . ." or "The smoke from the plant . . ."). Before, during, or shortly after the ambiguous word, another word that is related to one or the other of the ambiguous word's meanings may be presented visually (e.g., *bush* or *factory*). Relative to control conditions, the speed of response to the visual word (e.g., saying it aloud) provides a measure of the mental activation of the alternative meanings.

This research has shown that on some occasions there is a brief instant when multiple meanings are subconsciously active prior to interpretation. On other occasions, when one meaning is more common in the comprehender's experience and the context supports it, it alone may influence ongoing comprehension – for better or worse. Less certain is whether the processes for understanding words are dynamically similar to the processes for interpreting phrases (i.e., sequences of words). Of special interest is whether words and phrases are equally vulnerable to the simultaneous emergence of conflicting interpretations, and whether the mechanisms for resolving upcoming, ongoing, and emerging conflicts are the same for words and phrases (MacDonald et al., 1994). Complicating an already complex picture is the likelihood that people differ in their ability to consider multiple interpretations simultaneously and to make use of relevant knowledge quickly.

Apart from these uncertainties, there is little disagreement about something that was uncertain until recently: Understanding unfolds fluidly in time, as incoming sensory information triggers stored knowledge of what words and their arrangements mean. The speed with which this occurs, largely outside awareness, has challenged the ingenuity of investigators to invent increasingly sensitive probes of mental processing.

Communicative processes in comprehension Ambiguity also abounds in the relationships between linguistic expressions and the things they denote. To know that the flora sense of *plant* is intended is not enough when someone shrieks "That plant has tarantulas!" One surely wishes to know which plant. Successful communication demands referential grounding, the identification of the entities and events a speaker intends.

The simpler the referring expression, the greater the potential for ambiguity. Consider pronouns again. Their successful interpretation involves complicated inferences about what the speaker or writer intends, guided in part by conventions of use. Many of these conventions involve shared attention to something in the world or to a recent or prominent entity in the discourse. However, the conventions give way to specific linguistic rules in some environments, to causal reasoning (compare the interpretation of *she* in "Kay annoys Susan because she snores" and "Kay hit Susan because she snores"), or to wracking of memory and pointed query ("Which 'it' do you mean?").

Throughout our discussion of comprehension, we have emphasized the ambiguity that threatens comprehension. In fact, the threat is mostly unrealized, since comprehenders rarely notice how treacherous the language is. It is the semblance of effortlessness that has to be explained. Because the potential for ambiguity in language is unremitting, the explanation must lie in the cognitive processes of comprehension.

Conclusion

Finally, we must consider how the processes of production and comprehension are related to one another. Logically, they could be entirely separate, entirely the same except in their sensorimotor apparatus, or something intermediate. Research on language disorders suggests a degree of independence in the processing systems, because people with disorders of production can display seemingly normal comprehension abilities, and vice versa. Still, the architecture of the systems may be comparable except in direction of flow of the information: From meaning to sound in production, and from sound to meaning in comprehension.

Beyond differences in architecture or processing dynamics, at an abstract level production and comprehension have to draw on the same linguistic knowledge. We speak as well as understand our native languages. Communication occurs because speakers and listeners know the same code, a grammar, that governs how arrangements of sounds and words convey meaning.

For both modalities, the issues of interactivity and modularity offer springboards for debate. These distinctions may become blurred with advances in the understanding of language processing, since the measure of such advances will be the scientific adequacy of explanations for the detailed events of production and comprehension. Still, the stakes in the debate are high, because of their bearing on fundamental questions about human nature and human knowledge. These questions have to do with whether language and its component parts are in essence the same as other forms of cognition and, more broadly, whether all types of knowledge are in essence the same in their acquisition and use. Different answers to these questions lead to different conceptions of what we know, how we learn, and who we are, as a consequence of their implications for the generality of intelligent thought and intelligent behavior.

References and recommended reading

*Garnham, A. 1985: *Psycholinguistics: Central Topics*. London: Methuen.

*Gleitman, L. R. and Liberman, M. (eds) 1995: *An Invitation to Cognitive Science*. Vol. 1: *Language*. Cambridge, Mass.: MIT Press.

James, W. 1890: *The Principles of Psychology*, vol. 1. New York: Dover.
*Levelt, W. J. M. 1989: *Speaking: From Intention to Articulation*. Cambridge, Mass.: MIT Press.
MacDonald, M. C., Pearlmutter, N. J. and Seidenberg, M. S. 1994: The lexical nature of syntactic ambiguity resolution. *Psychological Review*, 101, 676–703.
*Miller, J. L. and Eimas, P. D. (eds) 1995: *Handbook of Perception and Cognition*. Vol. 11: *Speech, Language, and Communication*. Orlando, Fl.: Academic Press.
*Pinker, S. 1994: *The Language Instinct*. New York: Morrow.
van Turennout, M., Hagoort, P. and Brown, C. M. 1997: Electrophysiological evidence on the time course of semantic and phonological processes in speech production. *Journal of Experimental Psychology: Learning, Memory, and Cognition*, 23, 787–806.
Vigliocco, G., Antonini, T. and Garrett, M. F. 1997: Grammatical gender is on the tip of Italian tongues. *Psychological Science*, 8, 314–17.

15

Linguistic theory

D. TERENCE LANGENDOEN

Goals of linguistic theory

The goals of linguistic theory are to answer such questions as "What is language?" and "What properties must something (an organism or a machine) have in order for it to learn and use language?" Different theories provide different answers to these questions, and there is at present no general consensus as to what theory gives the best answers. Moreover, most linguists, when pressed, would say that these questions have not yet been answered satisfactorily by any theory.

In order to try to answer these questions, one strategy, originally employed by Joseph Greenberg (1963), is to undertake a comprehensive study of the languages of the world, to determine what properties they have in common and what distinguishes them from things that everyone agrees are not languages. Another, advocated by Noam Chomsky (1980), is to examine a few particular languages in depth to determine which of the intricate details that are found in one language turn up in all the others. As each of these approaches is extended, they merge into one another, and can be expected, ultimately, to converge on the same answer.

Expression and meaning

Although we do not yet know enough to provide a definitive answer to the question "What is language?", what we do know enables us to say with certainty that every language is a system with sufficient resources for communicating its speakers' intentions, desires, and beliefs, no matter how complex and unusual they may be. Let us call the spoken, signed, or written vehicle of communication *expression*, and what is communicated *meaning*. For example, an American English speaker can communicate the desire to find out what the people he or she is talking to talked about on a particular occasion in the past by saying [ˌwʌ̄ˌʤʲɔ̄l'tʰɔ́k.əˌbɑ̀ʷt˺], which is a transcription using the International Phonetic Alphabet (IPA) of what would ordinarily be written *What did you all talk about?* In this case, [ˌwʌ̄ˌʤʲɔ̄l'tʰɔ́k.əˌbɑ̀ʷt˺] is the expression, and the desire to find out what the people one is talking to talked about on a particular occasion in the past is the meaning.

An expression, when spoken, can be analyzed as a sequence of *syllables*, each said with a particular degree of loudness (or stress) and pitch (or intonation). For example, the expression [ˌwʌ̄ˌʤʲɔ̄l'tʰɔ́k.əˌbɑ̀ʷt˺] consists of five syllables, the third of which has strongest stress and the fourth the weakest. This stress pattern is indicated by the sequence of marks [ˌˌ' . ˌ], in which ['] indicates the strongest stress, [.] the weakest stress, and [ˌ] an intermediate stress. The highest pitch also falls on the third syllable, the lowest pitch on the second, fourth, and fifth syllables, and the first syllable has an intermediate pitch. This intonation pattern is indicated by the marks [¯ ˋ ˊ ˋ ˋ], in which [ˊ] indicates the highest pitch, [ˋ] the lowest pitch, and [¯] an intermediate pitch. The stress and intonation pattern of an expression helps convey part of the meaning of that expression, as discussed further below.

A syllable, in turn, consists of a nucleus, usually a vowel, possibly flanked by consonants fore and aft. For example, the syllable [ˌwʌ̄] contains the vowel [ʌ] preceded by the consonant [w], but with no consonant following, whereas the syllable [.ə̀] contains the vowel [ə] alone, and the syllable [ˈtʰɔ́k] contains the vowel [ɔ] which is both preceded and followed by a consonant. Consonants and vowels can be further analyzed as bundles of phonetic features, specifying the movement, position or activity of articulators, such as the lips, tongue, and vocal cords, or their acoustic effects. See Ladefoged and Maddieson, 1996, for discussion of phonetic analysis and notation.

A meaning can also be analyzed into component parts, though there is much less agreement about semantic structure (the structure of meaning) than there is about phonological structure (the structure of spoken expressions). Whatever representation is chosen for meaning, it must meet certain criteria of adequacy, including the following. First, it must provide a way to determine what things it may be used to talk about: for example, situations in a world of the speaker's and hearer's experience or imagination. Second, it must provide a way to determine the logical properties of meaning, including what it implies and what it is implied by. Third, it must indicate what act the speaker is performing when expressing it (e.g., making a statement, asking a question, issuing a command). Fortunately, for our purposes, it is not necessary to formulate meanings precisely, using a notation that meets these (and other) criteria of adequacy. Except as needed, we represent the meaning of an expression by enclosing its ordinary spelling in single quotes; for example, we represent the meaning of the expression [ˌwʌ̄ˌʤʲɔ̀lˈtʰɔ́k.ə̀ˌbɑ̀ʷt˥] as "What did you all talk about?" See Saeed, 1997, and Larson and Segal, 1995, for discussion of semantic analysis and notation.

Given that a language provides a means of expressing meaning, we may add to the goals of linguistic theory that it answer two additional questions, which we call the question of language *perception* and the question of language *production*. The question of language perception is: "How does one determine a meaning for an expression one has heard?" The question of language production is: "How does one determine an expression for a meaning one intends to convey?"

Morphs and morphemes

The simplest expressions of a language are those that cannot be divided into meaningful parts, other than the entire sequence of syllables that makes up each such expression and its stress and intonation contour. For example, the expressions [ˈnő:] "No?" and [ˈnȍ:] "No!" are among the simplest expressions of English, since their only meaningful parts are the syllable [no:] itself and the stress and intonation contours [ˈ˝] (rising) and [ˈ˥] (falling). The instances of the consonant [n] and vowel [o:] that combine to form the syllable [no:] are not meaningful, even though in other expressions, that consonant by itself, and that vowel by itself, are meaningful (e.g., in

[ˌō:'sīs.t^{h}ɚz.ǹ'bɾʌ́ð.ɚz] “Oh sisters and brothers”). The fact that [no:] is meaningful only in its entirety is the basis of the challenge “What part of ‘no’ don’t you understand?” emblazoned on T-shirts and bumper stickers throughout the English-speaking world.

The syllable [no:] with the associated meaning “no” in English is a theoretical construct known as a *morph*. A morph is a specific pronunciation associated with a specific meaning such that the pronunciation cannot be broken down into meaningful parts whose meanings combine to form the meaning of the whole. Behind the morph is an even more abstract theoretical construct known as a *morpheme*. A morpheme is an association of pronunciation and meaning such that the pronunciations and meanings of an entire class of morphs can be determined from it. In the case of [no:] “no”, there is no distinction between the morph and the morpheme, since there is no other determinable pronunciation with the meaning “no”, and no other determinable meaning associated with the pronunciation [no:]. However other morphemes differ phonologically or semantically from one or more of the morphs which are derivable from them. For example, the morpheme /ju:/ “you” underlies a dozen or more morphs including [ju:] “you”, [ju:] “you (sg.)”, [ju:] “you (pl.)”, [jə] “you”, [jə] “you (sg.)”, [jə] “you (pl.)”, [ju:] “you”, [ju:] “you (sg.)”, [ju:] “you (pl.)”, []“you”, [] “you (sg.)”, and [] “you (pl.)”. In the expression [jù:'k^{h}ɔ̃ld] “Did you call?”, the morph [ju:] “you” represents /ju:/ “you”, just as [no:] represents /no:/. However, in [jə̀'θɪ̃́ŋk'só:] “Do you think so?”, the morph [jə] “you” represents /ju:/ “you”; in [ˌwʌ̄ˌʤ^{j}ɔ̀l't^{h}ɔ́k.əˌbɑ̀^{w}t˺] “What did you all talk about?”, the morph [^{j}] “you (pl.)” represents the morpheme /ju:/ “you”. Finally, in [ˌhɛ̄lp.jɚ'sɛ̃lf] “You (sg.) should take what you (sg.) want”, the morph [] “you (sg.)” (occurring at the very beginning of the expression) represents /ju:/ “you”.

In describing the relation between morphs and morphemes, we have used the terms *underlie* and *derivable from* to emphasize the fact that the relation is systematic both phonologically and semantically. Given the phonological form of the morpheme /ju:/ “you”, the various phonological forms of the morphs derivable from it can be determined by well-understood processes of vocalic weakening (from [u:] to [ə]) and deletion, and consequent consonantal weakening (from [j] to [^{j}]) and deletion. Given the semantic form of the morpheme /ju:/ “you”, the various semantic forms of the morphs derivable from it can be determined by similarly well-understood processes of contextual delimitation (in the expression [ˌwʌ̄ˌʤ^{j}ɔl't^{h}ɔ́k.əˌbɑ̀^{w}t˺], the interpretation “you (pl.)” is required for consistency with the co-occurring morpheme /ɔl/ “all”, whereas in the expression [ˌhɛ̄lp.jɚ'sɛ̃lf], the interpretation “you (sg.)” is required for consistency with the co-occurring word *yourself*).

On the other hand, even though the morph [ju:] “ewe” has the same phonological form as the morpheme /ju:/ “you”, it is not derivable from that morpheme, because the relation between “ewe” and “you” is not systematic semantically. Similarly, even though the morph [ðɑ^{w}] “you (sg.)” has the same semantic form (in those dialects in which that morph occurs) as the morph [ju:] “you (sg.)”, it is not derivable from the morpheme underlying the latter morph, because the relation between [ðɑ^{w}] and /ju:/ is not systematic phonologically. We conclude that /ju:/ “you”, /ju:/ “ewe”, and /ðɑ^{w}/ “you (sg.)” are distinct morphemes in English. Henceforth, whenever neither the exact phonological nor the exact semantic form of a morpheme is at issue, we represent it by its ordinary spelling in italics, for example *no* for [no:] “no”, *you* for /ju:/ “you”, *ewe* for /ju:/ “ewe”, and *thou* for “you (sg.)”.

The lexicon

The particular pairings of expression and meaning represented by individual morphemes have to be learned individually. The totality of all the pairings in a language that must be learned individually is known as its *lexicon*, and its individual members are called *lexemes* (or *lexical items*). The lexemes of a language consists of all of its morphemes together with those combinations of lexemes whose meaning or expression cannot be systematically determined from the lexemes of which it is comprised. For example, the meanings of the word *unusual* and of the phrase *touch base* cannot be systematically determined from the meanings of their constituent lexemes; hence those combinations are also English lexemes. Similarly, the word *went* is an English lexeme, even if it is analyzed as containing the two lexemes /go:/ "go" and /d/ "in the past"; although its semantic form can be predicted from that combination, its phonological form cannot be. (Compare the word *planned*, consisting of the lexemes /plæn/ "plan" and /d/ "in the past"; it is not an English lexeme, since both its phonological and semantic forms are derivable from those of its component lexemes.)

Morphology and syntax

The vast majority of the pairings of expression and meaning represented by combinations of lexemes, whether these pairings are words, phrases, or sentences, do not have to be learned, because they can be systematically determined from the lexemes of which they are composed and the way they are combined and arranged. For example, the expression and meaning of the sentence *What did you all talk about?* can be determined from the expression and meaning of the individual lexemes *about, all, did, talk, what*, and *you*, and the way they are combined and arranged. The importance of how the component lexemes are combined and arranged can be seen from the fact that when they enter into other combinations or are rearranged, the result is always a different expression, which may have an entirely different meaning, or no meaning at all. For example, the sentences *What all did you talk about?* and *What did you talk all about?* consist of the same lexemes, but differently combined and arranged, and have different meanings from the original. The sentence *About what did you all talk?* also has a different arrangement, but the same meaning as the original. Finally, the arrangement in *You what talk did about all* is meaningless. Linguists generally call the expressions which are meaningless in a language *ungrammatical*, and indicate that status by prefixing an asterisk to the expression, as in **You what talk did about all.*

The study of how lexemes combine to form words is called *morphology* (see Spencer, 1991), and of how lexemes (typically, but not necessarily words) combine to form phrases and sentences is called *syntax* (see Culicover, 1997). The boundary between morphology and syntax varies widely from language to language; what can be expressed in a word in one language requires a phrase in another. For example, the meaning "from our hands" can only be expressed by a phrase made up of at least three words and four lexemes in English (*from, hand-s*, and *our*; we use a hyphen to separate lexemes within a word), but can be expressed by a single word made up of five lexemes in Turkish (*el-ler-im-iz-den*).

Recursion

The combination of lexicon, morphology, and syntax gives every language its expressive power, its ability to express any desired meaning. That power is, as far as linguists have been able to determine, the same for all languages. If a fluent speaker of a language lacks a word or simple phrase to express a particular meaning, he or she can do so by means of a more elaborate phrase. Moreover, if a particular meaning which can only be expressed in a complicated way by the members of a community becomes important to them, they will come up with simpler expressions for that meaning.

Since lexemes must be learned individually, no language can have more than a finite number of them, and a relatively small number at that. (No language has been observed to have more than 10^6 lexemes, though the problem of deciding exactly how to individuate lexemes makes estimating a more exact upper bound extremely difficult.) Hence in every language there must be meanings which can be expressed only by words, phrases, and sentences which are not single lexemes. Moreover, it also appears to be the case for every language that there are meanings which can be expressed only by phrases or sentences which are not single words. That is, the full expressive power of every language appears ultimately to depend on its syntax.

How this task is accomplished is most easily explained by means of examples. First, the syntax of English permits us to form two-word phrases such as *these cats, these dogs*, and *these hamsters*, as well as five-word phrases in which these two-word phrases are joined by the lexeme *and*, such as *these cats and these dogs, these dogs and these hamsters*, and (allowing for repetition of the two-word phrases) *these cats and these cats*. Let us call the two-word phrases *simple* and the five-word ones *coordinate*. In addition, the syntax of English permits us to form longer coordinate phrases by joining a simple phrase to a coordinate one with a short intonation break in between, such as *these hamsters, these cats and these dogs*. Then, since the latter phrase is itself coordinate, it can be joined with another simple phrase, as in *these cats, these hamsters, these cats and these dogs*, and so on without limit. By this means, we can form an infinite number of coordinate phrases in English, each of which is distinct both phonologically and semantically from the others. This syntactic device of coordinate phrase formation does not by itself provide a means for expressing all possible meanings in English, but it's a start.

Second, the syntax of English permits us to form simple sentences such as *I could fly*, as well as *complex* sentences in which a sentence is subordinated to a larger one, by the addition of the structure *I thought that* or *You thought that* at the beginning, as in *I thought that I could fly* and *You thought that I could fly*. Then since any sentence, not just simple ones, can be subordinated to a larger one, the syntax of English also permits us to form longer complex sentences in which complex sentences are subordinated, such as *You thought that I thought that I could fly* and *I thought that you thought that I thought that I could fly*, and so on without limit. Again, this syntactic device of complex sentence formation does not by itself provide a means for expressing all possible meanings in English, but now we're on the road.

Our two examples illustrate *recursion*, the formation of phrases of a certain type (we assume from this point on that a sentence is simply a certain kind of phrase) out of phrases of exactly the same type. Each illustration starts with *base cases*: phrases of a given type which do not contain any phrase of that type within it. In the coordination illustration, *these cats and these dogs* is a coordinate phrase which does not contain any coordinate phrase within it. In the subordination illustration, *I could fly* is a sentence which does not contain any sentence within it (i.e., a simple sentence). It then proceeds to a *recursive step*, according to which a larger phrase of a given type can be constructed out of parts which include a smaller phrase of the same type. In the coordination illustration, the recursive step consists of adjoining a simple phrase to a coordinate phrase to form a larger coordinate phrase. In the subordination illustration, the recursive step consists of subordinating a sentence to a larger sentence by adjoining the structure *I thought that* or *You thought that* to the original sentence.

The expressive power of language depends on the existence of many different recursive devices in its syntax and their ability to freely combine. For example, the coordination and subordination processes just illustrated can be combined to permit the formation of coordinate complex sentences such as *I thought that I could fly, you thought that I thought that I could fly and I thought that you thought that I thought that I could fly* (coordination of complex sentences) and *I thought that you could navigate, she could steer and he could make dinner* (subordination of coordinate sentences). Each of these sentences expresses a meaning not expressible by coordination or subordination alone.

Movement and deletion

Languages employ various syntactic devices to prevent expressions for complex meanings from becoming inordinately long. Two devices which appear in the grammars of all languages are known as *movement* and *deletion*. We have already given an example which illustrates movement: namely, the question *What did you all talk about?*, in which the lexeme *what* is displaced from its normal position following *about*, as in *You all talked about what?* In this case, the movement of *what* serves the purpose of focusing or highlighting what kind of answer the speaker is expecting (the name of a thing rather than of a person, for example). In more complex sentences such as *What did she tell him that you all talked about?* and *Did she tell him what you all talked about?*, it makes a difference semantically where the lexeme is moved to. By the same token, it also makes a difference semantically where the lexeme originates; compare *What did you all put the baskets in?* (which is equivalent to *You all put the baskets in what?*) and *What did you all put in the baskets?* (which is equivalent to *You all put what in the baskets?*).

Deletion may be employed in many situations in which the repetition of phrases can be *reconstructed*, as in the coordinate sentence *You thought that I could fly, but I didn't*, which has two meanings depending on what is reconstructed, either *think that I could fly*, resulting in the meaning "You thought that I could fly, but I didn't think that I could fly," or simply *fly*, resulting in the meaning "You thought that I could fly, but I didn't fly." If we exchange the second occurrence of *I* with *You* in the original sentence, yielding *I thought that I could fly, but you didn't*, the result has three meanings: "I thought that I could fly, but you didn't think that I could fly"; "I thought that I could fly, but you didn't think that you could fly"; and "I thought that I could fly, but you didn't fly." The additional meaning results from the fact that the second occurrence of the lexeme *I* in *I thought that I could fly* can *depend* on the first. If this dependency is reconstructed in determining the meaning of *but you didn't*, the result is "but you didn't think that you could fly." See Fiengo and May, 1995, for analysis of dependency and its interaction with deletion.

Universal Grammar

A complete analysis of the lexicon, morphology, and syntax of a language is called its *grammar*. The theory of grammar, which is a significant part of linguistic theory as a whole, has as one of its goals the answer to the question "What is the class of all possible grammars?" The properties of that class are what is called *Universal Grammar*. Related to the goal of understanding universal grammar is the solution to the problem of language acquisition: "What enables a child upon limited exposure to a language to learn the grammar of that language?" One widely held assumption is that people are innately endowed with knowledge of universal grammar, so that the task of language acquisition is really one of selecting the best grammars for the languages of one's experience.

Generative grammar

The first complete statement of the theory of grammar in the form presented here was the *standard theory of generative grammar* presented in Chomsky, 1965, henceforth ST. (Chomsky's 1957 presentation of the theory of grammar, which is widely recognized as having launched the modern enterprise of linguistic theory, was less well developed, inasmuch as it neither attempted to account for the meaning of expressions nor dealt with language acquisition.) According to ST, a grammar consists of a number of interacting components including a lexicon, two syntactic components, a semantics, and a phonology. For each meaningful expression, the grammar provides a *derivation*, which starts with a member of a set of *axioms* such as #*S*# (for "Sentence") and terminates with a pair <Mng, Exp>, where "Mng" is its semantic structure (representation of meaning) and "Exp" is its phonological structure (representation of expression).

The derivation of a meaningful expression in ST begins in one of its syntactic components called *phrase structure*, which constructs a hierarchical representation of the underlying syntactic structure of the expression. The phrase structure component is responsible for, among other things, recursion. The output of phrase structure is the input to the lexicon, which inserts lexical items into the underlying syntactic structure. Chomsky called the result of that operation *deep structure*. The deep structure of an expression is the input both to the semantics and to the second of the syntactic components, known as the *transformational component*. The semantics is responsible for constructing the meaning of the expression, whereas transformational structure is responsible for, among other things, movement and deletion. Finally, the output of the transformational structure, called *surface structure*, is the input to the phonology, which determines the form of the expression proper, including its stress and intonational contours. The overall structure of ST is outlined in figure 15.1.

ST does not attempt to answer the questions of language perception and production posed at the end of the second section. Rather than relating meaning and expression directly, it relates them indirectly, through deep structure. Thus, given an expression, the only way to determine its meaning according to ST is first to recover its deep structure, then to compute its meaning, using semantics. However, the recovery of deep structure from expression is not an operation which is well defined within ST. The operations of the transformational and phonological components of ST map deep structure onto expression but are not capable of mapping expression onto deep structure.

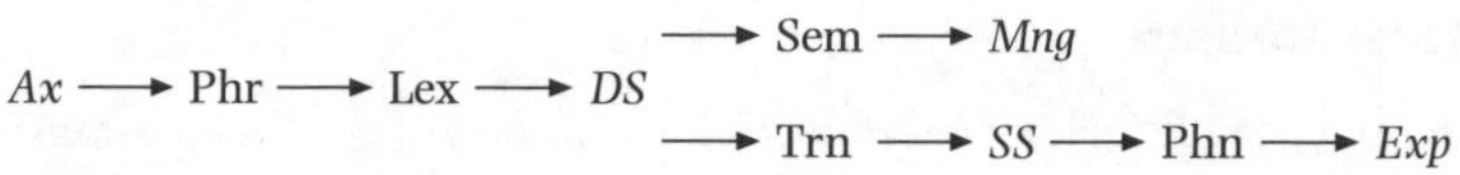

$$Ax \longrightarrow \text{Phr} \longrightarrow \text{Lex} \longrightarrow DS \begin{cases} \longrightarrow \text{Sem} \longrightarrow Mng \\ \longrightarrow \text{Trn} \longrightarrow SS \longrightarrow \text{Phn} \longrightarrow Exp \end{cases}$$

Figure 15.1 Outline of the standard theory of generative grammar (Chomsky, 1965). Legend: *Ax* = axiom, *Phr* = phrase structure, *Lex* = lexicon, *DS* = deep structure, *Sem* = semantics, *Trn* = transformational structure, *Mng* = semantic structure (meaning), *SS* = surface structure, *Phn* = phonology, *Exp* = phonetic structure (expression). Inputs and outputs are italicized.

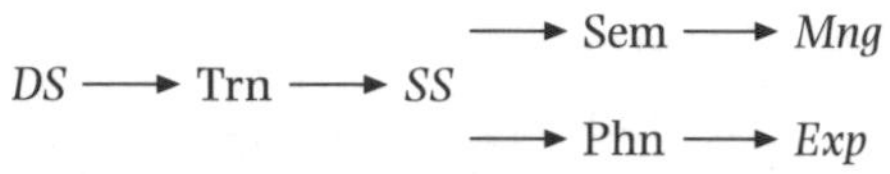

$$DS \longrightarrow \text{Trn} \longrightarrow SS \begin{cases} \longrightarrow \text{Sem} \longrightarrow Mng \\ \longrightarrow \text{Phn} \longrightarrow Exp \end{cases}$$

Figure 15.2 Outline of the government and binding theory of generative grammar (Chomsky, 1981).

Similarly, given a meaning, the only way to determine its expression according to ST is again to recover its deep structure. However, just as the operations which map deep structure onto expression are not reversible, neither are the operations that map deep structure onto meaning. Chomsky justifies his decision to limit linguistic theory to merely providing a means of relating meaning and expression, rather than to providing a means of determining one from the other, by maintaining that the first is a matter of linguistic *competence* (what people know about language) and hence is properly within the scope of linguistic theory, whereas the second is a matter of linguistic *performance* (how people use language) and hence lies outside its scope. Not everyone agrees with Chomsky on this point, and in any event a total theory of language (whatever it is called) would have to encompass both linguistic competence and linguistic performance.

The theory of generative grammar has evolved considerably over the past thirty years, but continues to retain its *forked* structure, in which meaning and expression are related indirectly through an intermediate construct. For example, the version of generative grammar known as *government and binding* theory (Chomsky, 1981) treats deep structure (renamed "D-structure") as the starting point of a derivation and surface structure (renamed "S-structure") as the intermediate construct from which both meaning and expression are determined, as in figure 15.2.

Finally, the *minimalist program* of generative grammar (Chomsky, 1995) treats the lexical items which contribute to both meaning and expression as the starting point of the derivation, eliminating both D-structure and S-structure entirely. Nevertheless, following application of transformational rules of "merge," "move," and "check," derivations are still branched, with strictly phonological rules (now called "spellout") applying on the branch leading to phonological representation (now called "phonetic form" or "PF"), and strictly semantic rules applying on the branch leading to semantic representation (now called "logical form" or "LF"), as in figure 15.3.

It is of course not a necessary feature of the theory of generative grammar that it fail to provide answers to the questions of linguistic perception and production. One theory that maps meaning to expression in the grammar, and hence attempts to answer the question of linguistic production, is *generative semantics*, which originated in the 1960s as an alternative to ST (see Huck and Goldsmith, 1995). As diagrammed in figure 15.4, the mapping from meaning to surface structure is carried out by transformational

$$\begin{array}{l} \longrightarrow \text{Sem} \longrightarrow \textit{LF} \,(= \textit{Mng}) \\ \textit{Lex} \longrightarrow \text{Trn} \\ \longrightarrow \text{Phn} \longrightarrow \textit{PF} \,(= \textit{Exp}) \end{array}$$

Figure 15.3 Outline of the minimalist theory of generative grammar (Chomsky, 1995).

$$\textit{Mng} \longrightarrow \text{Trn} \longrightarrow \textit{SS} \longrightarrow \text{Phn} \longrightarrow \textit{Exp}$$

Figure 15.4 Outline of the generative semantics theory of generative grammar.

a. $\textit{Mng} \longrightarrow \text{Con} \longrightarrow \textit{Exp}$
b. $\textit{Exp} \longrightarrow \text{Con} \longrightarrow \textit{Mng}$

Figure 15.5 Outline of the optimality theory of generative grammar.

rules, including rules which replace semantic substructures by lexical items represented in terms of their underlying phonological properties, and the mapping from surface structure to expression is carried out by phonological rules.

Optimality theory

However, only very recently has a theory of generative grammar been proposed that attempts to answer the questions of both linguistic perception and linguistic production: namely, *optimality theory* (Archangeli and Langendoen, 1997; Prince and Smolensky, 1997, in press). Optimality theory assumes that the universal classes of expressions and meanings (i.e., the class of all expressions of all possible languages and the class of all meanings) can each be defined by a *generator*, which works rather like the phrase structure component of ST, or the rules for constructing well-formed formulas in a system of logic. Then, given a particular meaning, one determines its expression in a particular language by evaluating the members of the universal class of expressions against a universal set of *constraints* (henceforth Con). Similarly, given a particular expression, one determines its meaning (if any) in a particular language by evaluating the members of the universal class of meanings against Con. That is, the problems of language production and language perception are considered problems of selection, just like the problem of language acquisition as described above. Figure 15.5 provides an outline of optimality theory.

If, as optimality theory proposes, the classes of meanings, expressions, and constraints are universal, how do languages differ? The answer is in the *ranking* of the constraints, with each language corresponding to a possible ranking of the members of Con; for numerous illustrations of how this works, see Archangeli and Langendoen, 1997. Presumably included among the members of Con is the universal class of lexical items, so that the determination of which lexical items belong to which languages is also a matter of constraint ranking.

Optimality theory has had its greatest success so far in accounting for the relation between lexical items and their expression, and is only now beginning to deal with the major phenomena of syntax, including recursion, movement, deletion, and dependency; whether it will succeed in extending the scope of linguistic theory from the narrow concerns of linguistic competence to the broad concerns of linguistic performance remains to be seen.

References and recommended reading

*Archangeli, D. and Langendoen, D. T. (eds) 1997: *Optimality Theory: An Overview*. Oxford: Blackwell.

Chomsky, N. 1957: *Syntactic Structures*. The Hague: Mouton.

—— 1965: *Aspects of the Theory of Syntax*. Cambridge, Mass.: MIT Press.

—— 1980: *Rules and Representations*. New York: Columbia University Press.

—— 1981: *Lectures on Government and Binding*. Dordrecht: Foris.

—— 1995: *The Minimalist Program*. Cambridge, Mass.: MIT Press.

*Culicover, P. 1997: *Principles and Parameters: An Introduction to Syntactic Theory*. Oxford: Oxford University Press.

Fiengo, R. and May, R. 1995: *Indices and Identity*. Cambridge, Mass.: MIT Press.

Greenberg, J. (ed.) 1963: *Universals of Language*. Cambridge, Mass.: MIT Press.

Huck, G. and Goldsmith, J. 1995: *Ideology and Linguistic Theory: The Deep Structure Debates*. London: Routledge.

Ladefoged, P. and Maddieson, I. 1996: *The Sounds of the World's Languages*. Oxford: Blackwell.

*Larson, R. and Segal, G. 1995: *Knowledge of Meaning*. Cambridge, Mass.: MIT Press.

Prince, A. and Smolensky, P. 1997: Optimality: from neural networks to universal grammar. *Science*, 275, 1604–10.

—— In press: *Optimality Theory: Constraint Interaction in Generative Grammar*. Cambridge, Mass.: MIT Press.

Saeed, J. I. 1997: *Semantics*. Oxford: Blackwell.

*Spencer, A. 1991: *Morphological Theory*. Oxford: Blackwell.

16

Machine learning

PAUL THAGARD

Machine learning is the study of algorithms that enable computers to improve their performance and increase their knowledge base. Research in machine learning has taken place since the beginning of ARTIFICIAL INTELLIGENCE in the mid-1950s. The first notable success was Arthur Samuel's program that learned to play checkers well enough to beat skilled humans. The program estimated the best move in a situation by using a mathematical function whose sixteen parameters describe board positions, and it improved its performance by adjusting weights on the parameters in order to generalize about what moves were successful.

Machine learning emerged as a special area of research around 1980, and since then has been the subject of workshops, conferences, and the journal *Machine Learning*, which started in 1986. Numerous techniques have been developed for enabling computers to learn in different ways; although there is no unified theory of learning, various approaches have proved computationally powerful and cognitively interesting. This article briefly summarizes six major approaches: learning from examples, artificial neural networks, genetic algorithms, explanation-based learning, evaluating hypotheses, and case-based reasoning (analogy). Although together these approaches do not offer a full theory of how humans and machines learn, they have provided rigorous descriptions of learning that have been tested for computational power and psychological relevance.

Each of these approaches can be characterized in terms of inputs, outputs, algorithms, and applications. The inputs to the learning process are the representations that learning operates on; the outputs are the representations that result from the learning. These new representations are produced by means of algorithms – mechanical procedures that modify the input representations. Different approaches to machine learning have had different applications to practical problems and cognitive modeling. The first three approaches I will describe can all function in systems that have very little information and need to learn from experience without guidance from previously acquired knowledge. By contrast, the final three approaches involve learning in systems that have well-developed knowledge bases.

The most active area of research in machine learning has been the learning of concepts and other representations from examples. The input to such learning programs consists of descriptions of examples, and the output consists of different kinds of representations that encode generalizations about those examples. These output representations can be new concepts, new rules, or decision trees that provide a convenient means of classifying new examples.

One of the first machine-learning projects on concept learning was Patrick Winston's program that learned the concept of an arch. Given both examples of arches and examples of things that were not arches, the program's task was to produce a general description of arches that would correctly classify additional examples of arches while excluding non-arches. Various algorithms have been developed for producing such concept descriptions. Some of these are incremental, handling training instances one at a time, while others process many instances at once. Some concept-learning algorithms start from specific descriptions that are then expanded to be more general, while other algorithms first produce general descriptions and are then corrected to handle more specific information. For example, a program might learn the concept of an arch by supposing that every arch is exactly like the first arch described to it, and then gradually generalize as it receives more examples that are different from previous ones. Alternatively, a program might be designed to make an immediate broad generalization, such as that an arch consists of two vertical blocks and an object on top of them, and then correct the generalization to fit new examples as they are provided. While early concept-learning programs attempted to produce concepts that provided strict definitions, more recent ones have produced more flexible concepts that approximately match the examples they classify.

Another kind of learning from examples produces general RULES instead of concepts. Language acquisition is an important domain for investigating rule learning: given a set of grammatical sentences from some language, the machine-learning problem is to generate a set of rules for recognizing as grammatical other sentences in that language. In English, for example, the sentence "The computer is expensive" is grammatical, but the sentence "The expensive is computer" is not. Programs can also be given the task of generating mathematical rules from data, as when the BACON program of Pat Langley and Herbert Simon generated scientific laws about the motions of the planet from numerical descriptions of their locations.

A third method of learning from examples produces decision trees rather than rules or concepts. The best-known such program is J. R. Quinlan's ID3, whose input is a list of positive and negative instances of some concept, and whose output is a decision tree for sorting new instances. For example, the program can be given the task of predicting the weather given attributes such as outlook (with values sunny, overcast, and rain), temperature (with values cool, mild, and hot), humidity (with values high and normal), and windy (with values true and false). A particular morning might be described as sunny, cool, high humidity, and windy. ID3 can take many such examples and produce a decision tree that classifies new mornings and yields predictions such as that if it is sunny, cool, and humid, then it will be windy. It does this by choosing a subset of the examples at random and constructing a tree that classifies all of them using the attributes and values. Subsequent examples are then classified, and if errors are made (e.g., predicting that a morning that is sunny, cool, and not humid will not be windy), then the decision tree is modified to produce improved classifications. ID3 uses an information-theoretic evaluation function to determine how to organize the attributes and values in the decision tree in a way that best classifies examples. ID3 and similar programs have been applied to many domains, such as determining the value of chess positions.

Learning concepts, rules, and decision trees from examples all produce symbolic descriptions; but a very different kind of output is produced by another approach to machine learning using artificial neural networks (see Article 38, CONNECTIONISM,

ARTIFICIAL LIFE, AND DYNAMICAL SYSTEMS). In the 1960s, some of the earliest work on machine learning used simple artificial neural networks, but the kinds of network used were too simple to perform very complex kinds of learning. In the 1980s, new techniques were developed for learning in artificial neural networks that provided increased power for cognitive modeling and engineering applications. The inputs to a neural network learning program are a network that consists of a set of nodes connected by excitatory and inhibitory links, along with a set of training examples. The links represent the strengths of the connections, positive and negative, between the nodes. Learning algorithms modify these strengths to improve the performance of the network at some task – for example, predicting the weather or classifying animals. The output of the learning algorithm is not a new representation such as a concept or a rule, but rather a modification of the strengths of the connections in the network.

The most commonly used learning algorithm in artificial neural networks is called "backpropagation," which trains a layered network by adjusting the weights that connect the different nodes. The layers in backpropagation network consist of input nodes representing features of examples (e.g., for hot, humid, windy days) and output nodes representing a conclusion (e.g., sunny or rainy). These input and output nodes are connected by a small number of hidden nodes that have no initial interpretation. After weights are randomly assigned to the links between nodes, input nodes are activated for features that the network is supposed to learn about. Then activation is spread forward through the network to the hidden nodes and then to the output nodes. Errors are determined by calculating the difference between the computed activation of the output nodes and the desired activation of the output nodes. For example, an input of hot and dry may erroneously predict that the day will be rainy. To improve future performance, errors are propagated backwards down the links, changing the weights in such a way that the errors are reduced. Eventually, after many examples have been presented to the network, it will correctly classify different kinds of features.

Backpropagation learning requires a supervisor to determine whether the output is correct, but non-supervised learning can also be performed by neural networks. One simple kind is Hebbian learning, in which if two nodes are both active at the same time, then the weight on the link between them should increase, establishing an association between the nodes. This kind of learning is unsupervised, in that it does not require any teacher to tell the network when it has right or wrong answers.

In the 1970s, John Holland developed a novel approach to machine learning called "genetic algorithms," using an analogy between learning and biological adaptation. Species become better adapted to their environments by genetic variation, which introduces new genetic combinations, and by natural selection, which favors members of the species who have genes that enable them to survive and reproduce in their environments. Similarly, the inputs to a genetic algorithm program are structures for performing tasks; they may be simple structures such as strings of 0's and 1's that represent the presence or absence of features, or more complex structures such as computer programs. The outputs from a genetic algorithm are modified structures that are intended to perform the desired task more effectively. To get this improvement, genetic algorithms modify the input structures by randomly modifying them (mutation) and by combining them (crossover). The resulting structures are then evaluated for their effectiveness, and variation and selection are repeated many times. Although it is not known whether genetic algorithms function in human learning, they have had many useful technological applications, such as generating expert systems for designing computer circuits.

Examples, neural networks, and genetic algorithms are all mechanisms that can generate new knowledge using very little background knowledge. By contrast, explanation-based learning (sometimes called analytic learning) employs rich knowledge of a domain to add additional knowledge about the domain. Explanation-based learning can start from a single example instead of the large training sets that are typically used in learning from examples. The input to an explanation-based learning algorithm consists of an example plus a data base of general rules or schemas. The output is a new concept that is formed from the example, first by constructing an explanation for why the example is an instance of the goal concept, and second by generalizing the explanation to obtain the goal concept. The goal concept *kidnap*, for example, can be formed by trying to understand a piece of text describing the abduction of someone whose family is then asked for money. The new concept arises, not from generalization from numerous examples, but from the knowledge-intensive attempt to understand what is going on in a particular example.

Hypothesis evaluation and causal reasoning also usually take place in information-rich contexts such as medical diagnosis. A doctor whose patient has multiple symptoms needs to generate and evaluate hypotheses that can explain why the patient has those symptoms. The inputs to a learning problem involving hypotheses and causal reasoning are a set of facts to be explained and a knowledge base that can be used to generate hypotheses to explain those facts. The outputs are judgments about the acceptability of various hypotheses. The algorithms required need first to generate various hypotheses by searching the knowledge base for various explanations, and second to choose among competing hypotheses to pick the best ones. Some systems for causal reasoning use Bayesian networks, in which facts and hypotheses are represented by nodes connected by links that represent conditional probabilities. Other systems for evaluating hypotheses proceed more qualitatively: for example, by construing hypothesis choice as a process of parallel constraint satisfaction that can be computed by artificial neural networks.

In the machine-learning literature, analogical inference is often called *case-based reasoning*. The inputs to analogical learning are (1) a problem to be solved (the target) and (2) a memory base of previously solved problems. Problems to be solved can include plans, such as how to get to the airport; designs, such as how to build an airport; and explanations, such as why a plane crashed. A new plan, design, or explanation can be formed by adapting a previous one (the source) that exists in memory. To carry this out, algorithms are needed for retrieving a potentially relevant case from memory, mapping it to the target problem to establish correspondences between them, and adapting the case to provide a solution to the target problem. Once the target problem has been solved, another kind of learning can take place by abstracting the common properties of the target and source to provide a schema that can be used for subsequent problem solving.

Analogical inference is useful when general information about a problem is not available but there is a stock of similar, previously solved problems. (See Article 1, ANALOGY, for additional discussion of the psychology of analogical inference, and Article 36, CASE-BASED REASONING, for further computational discussion.) Analogy programs have been used to model the results of many psychological experiments, and numerous expert systems have been developed that employ case-based reasoning.

These six approaches (learning from examples, neural networks, genetic algorithms, explanation-based learning, hypothesis evaluation, and analogy) do not exhaust the kinds of learning studied in artificial intelligence. Other topics of current investigation include mathematical analyses developed as part of computational learning theory, studies of learning in PRODUCTION SYSTEMS, and multi-strategy learning which combines two or more methods of learning. One of the most useful kinds of investigation in recent machine-learning research is comparison of different methods on standard problems. It has been found, for example, that many different algorithms for learning from examples have similar performance. Notably, the cognitive performance of artificial neural network models using backpropagation can often be duplicated by programs such as ID3 that use very different kinds of representation and algorithms. Such comparative computational experiments are important for evaluating claims about the psychological plausibility of connectionist and other approaches to understanding human learning.

The study of machine learning has provided ideas concerning how adults learn concepts and rules and also concerning how children's abilities develop, thus contributing to developmental and cognitive psychology. Some of these ideas concern language acquisition, making machine learning relevant to linguistics. In addition, some research in machine learning addresses problems concerning the nature of inductive inference and scientific discovery that have traditionally been the concern of philosophy. Not all work in machine learning is directly relevant to psychology, however, since artificial intelligence researchers sometimes develop effective algorithms that bear little relation to human learning. Such work is aimed at the engineering goal of artificial intelligence to make powerful computers rather than to mimic human learning. The most successful machine-learning programs produce expert systems that perform even better than humans. Recently, machine learning has become a useful technology for mining large data bases for information of commercial and scientific importance.

References and recommended reading

*Carbonell, J. (ed.) 1990: *Machine Learning: Paradigms and Methods*. Cambridge, Mass.: MIT Press.

Holland, J. H., Holyoak, K. J., Nisbett, R. E. and Thagard, P. R. 1986: *Induction: Processes of Inference, Learning, and Discovery*. Cambridge, Mass.: MIT Press/Bradford Books.

Langley, P. 1996: *Elements of Machine Learning*. San Francisco: Morgan Kaufmann.

Ling, C. and Marinov, M. 1993: Answering the connectionist challenge: a symbolic model of learning past tenses of English verbs. *Cognition*, 49, 235–90.

*Michalski, R. and Tecuci, G. (ed.) 1994: *Machine Learning: A Multistrategy Approach*, Vol. 4. San Francisco: Morgan Kaufman.

*Mitchell, T. 1997: *Machine Learning*. New York: McGraw-Hill.

Quinlan, J. R. 1986: Induction of decision trees. *Machine Learning*, 1, 81–106.

Samuel, A. L. 1959: Some studies in machine learning using the game of checkers. *IBM Journal of Research and Development*, 3, 210–29.

*Shavlik, J. W. and Dietterich, T. G. (eds) 1990: *Readings in Machine Learning*. San Mateo, Calif.: Morgan Kaufmann.

Winston, P. H. 1975: Learning structural descriptions from examples. In P. H. Winston (ed.), *The Psychology of Computer Vision*, New York: McGraw-Hill, 157–209.

17

Memory

HENRY L. ROEDIGER III AND LYN M. GOFF

Memory is a single word that refers to a complex and fascinating set of abilities which people and other animals possess that enables them to learn from experience and retain what they learn. In memory, an experience affects the nervous system, leaves a residue or trace, and changes later behavior. Types of memory are tremendously varied; so, too, are the techniques used in cognitive science to investigate them. The aim of the present chapter is to give an overall sense of types of memory as well as of techniques used in the experimental study of memory. The process of remembering will be broken up into two main components: encoding and retrieval. We shall then discuss the factors which determine the effectiveness of each component and some of the ways in which memory both succeeds and fails.

Varieties of learning and memory

Biologists, philosophers, and psychologists have described and discussed dozens of types of memory. We consider here some of the most frequently used categories.

Procedural memory refers to the knowledge of how to do things such as walking, talking, riding a bicycle, tying shoelaces. Often the knowledge represented is difficult to verbalize, and the procedures are often acquired slowly and only after much practice. (Imagine someone trying to learn how to swim from reading about swimming, but not practicing the skill.) The types of conditioning to which most species of animals are subject – classical (or Pavlovian) conditioning and instrumental (or operant) conditioning – are other examples of procedural memory.

Procedural memory is often contrasted with *declarative memory*, or knowing facts about the world and about one's past (Squire, 1987). A major distinction within declarative memory is that between episodic and semantic memory. *Episodic memory* refers to the remembering of episodes of our lives and is contextually bound; that is, the time and place of occurrence are inextricable parts of memory for episodes. This type of memory enables the mental time travel in which we engage when we think back to an earlier occasion; because it constitutes every individual's personal history, it is sometimes called autobiographical memory. *Semantic memory* (or generic memory) refers to our general knowledge of the world (that NaCl is the symbol for salt, what the word *platypus* means, and so on). This knowledge is not tied to one episode, and we need not refer to the time or place in which we learned these facts to know that they are true.

This is not the only way to distinguish types of memory. Another, important difference is between short-term and long-term memory. *Short-term memory* (or *primary memory*) refers to our ability to hold in mind a relatively small amount of information that is rapidly forgotten if we stop attending to it. A good example is remembering a telephone number for a brief period after looking it up. This ability is also referred to as *working memory*, because it permits us to perform the mental work of manipulating symbols and thinking. *Long-term memory* (or *secondary memory*) is a rather imprecise term that is used to refer to retention of various kinds over long time periods; depending on content, *long* may mean anywhere from 30 seconds to many years (hence the fuzzy nature of the term).

We will leave aside many other distinctions that psychologists have made and confine our remarks to what most people have in mind when they refer to *remembering* or *memory*, which is long-term episodic memory. How do we remember what we read in the paper, where we parked our car this morning, the earliest event from our childhood, and the myriad other events of our lives? We often need to recall events from the past as accurately as possible, and this process can be effortful. The process of recognition (when we are asked to judge whether something has been presented to us previously) appears easier than recall. We have often had the experience of being able to recognize some event when we failed to recall it. Below we consider (a) the study of memory, (b) some of the major factors that have been shown experimentally to affect memory, (c) two critical principles of remembering, (d) the problem of forgetting, and (e) memory illusions and false memories.

The study of memory

Philosophical speculation about how people learn and remember dates from Aristotle and Plato, whose prescient comments on the topic are still worth reading, and continues to the present day. Empirical research on memory is much more recent, usually dated from the great experiments of Hermann Ebbinghaus (1850–1909), published in book form in 1885. Ebbinghaus made up lists of nonsense syllables composed of two consonants and a vowel, such as TEIF, and placed them in long lists which he painstakingly memorized and on which he later tested himself. Despite the fact that he was the only subject in all his experiments, he succeeded in obtaining very regular results by testing himself many times in each experimental condition. All of his findings have stood the test of time. His work was quite important, not only for the original findings about memory, but also because he showed that the higher mental processes could be subjected to careful experimental study, at a time when other psychologists were arguing that experimental methods could be applied only to the study of sensory processes.

Ebbinghaus advocated careful laboratory research as a sure path to knowledge, and the laboratory research tradition begun by Ebbinghaus still exists, albeit in radically different form. The development of alternative approaches has enriched today's cognitive science, however. Some researchers advocate more naturalistic methods (the *everyday memory* tradition). Others seek the biological underpinnings of memory in studies of animals or in the tradition of cognitive neuroscience (measuring neural activity through modern NEUROIMAGING techniques while people are engaged in memory tasks, or studying the DEFICITS AND PATHOLOGIES of memory in brain-damaged patients). Yet another approach takes inspiration from ARTIFICIAL INTELLIGENCE and asks how much human memory resembles computer memories. Some researchers seek to simulate and to understand memory processes by creating neural network models. Each of these approaches makes a contribution, but we will focus on the perspective on learning and memory that began with Ebbinghaus and continues in today's experimental psychology, employing behavioral methodologies as the primary tool of study. We will introduce findings from other traditions when they are appropriate.

The learning/memory process can be divided into three hypothetical stages: encoding (original acquisition of information), storage (retaining information over time), and retrieval (gaining access to information when it is desired) (Melton, 1963). Any time someone accurately remembers an event, all three stages are successfully completed. If someone forgets or misremembers an event, we can ask at what stage or stages the process went awry. However, answering this question is not as straightforward as it seems, because the three stages are interlocked, and psychology experiments cannot give clean answers to the question of what stage in the process suffered a breakdown.

A standard psychology experiment on learning and memory has two stages. In a first stage people are exposed to information to be learned, be it sets of words, numbers, pictures, sentences, a story or prose passage, or a videotape of a complex event. In the second stage, a test is given some time later in which people may be asked to recall or to recognize the material. The first stage of memory experiments corresponds to the encoding of material, but of course there is no way to tell if material was actually encoded unless it is tested. The second stage corresponds to the retrieval stage, but of course it does not measure retrieval per se – information can only be retrieved if it was encoded and stored.

Since the work of Tulving and Pearlstone (1966), psychologists have distinguished between *availability* and *accessibility* of information in memory, where availability refers to the information about events that a person has encoded and stored and accessibility refers to the information that can be retrieved on any particular test occasion. The holy grail for psychologists interested in memory would be a test or procedure that accurately measured the contents of a person's knowledge – what the person had encoded and stored. At one time it was argued that procedures for measuring recognition represented this perfect indicator of knowledge, but this hope was ill-founded, and recognition procedures are subject to the same vagaries as are recall procedures. Every test of memory is an imperfect indicator of knowledge, whether in the classroom, in standardized tests, or in the psychology laboratory. We can never measure what information is encoded and stored; we can only measure what information is accessible or retrievable under a particular set of test conditions.

Despite these problems, the division of the learning/memory process into three stages can still be useful. We can still sometimes ascribe forgetting to failures (say) of retrieval. Imagine people studying a list of 100 words on which *umbrella* is the fifty-first word. If people were tested by being asked to recall in any order on a blank sheet of paper (a procedure called *free recall*), the probability of recalling *umbrella* would be vanishingly small. Was the word not encoded? not retained? or just not retrieved? There is no way to know from this one condition. However, if the people were then given retrieval cues to prompt memories for the words and the cue *parasol* elicited recollection of *umbrella*, then clearly the word had been encoded and stored, and the failure on the free recall test was one of retrieval. (It would be necessary to safeguard against the possibility that people are merely guessing the words from the cues, but in practice insuring this is relatively easy.)

Most experiments on memory can be classified as encoding experiments or retrieval experiments. *Encoding experiments* involve manipulation of some factor during the encoding stage (e.g., the type of material, the way the material is processed), with other

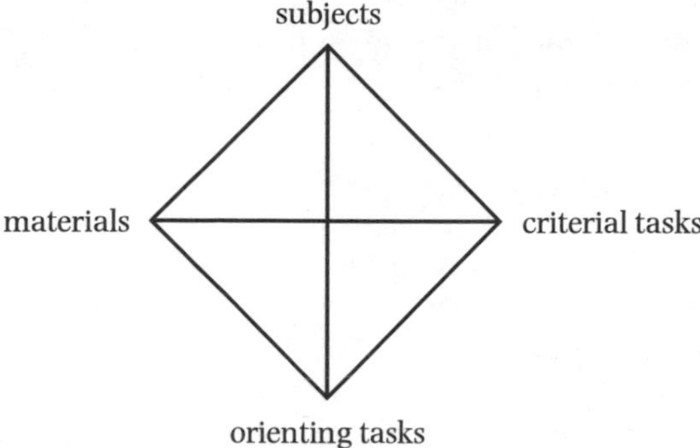

Figure 17.1 Modified version of Jenkins's (1979) tetrahedral model of memory experiments. The four points represent the four factors that affect the outcome of all experiments.

factors (e.g., the type of test that is used to assess knowledge) held constant. *Retrieval experiments* hold constant the encoding factors but manipulate the retrieval factors, such as the type of test given or the particular instructions given before the test. One particularly useful research strategy in investigating memory combines these two types of experiments and has been called the *encoding/retrieval paradigm* (Tulving, 1983). In this type of analysis, of which we will provide examples below, one manipulates both encoding and retrieval conditions within the same experiment. For example, two different strategies for studying material might represent the encoding manipulation, and two different forms of test might be used to assess knowledge. The encoding/retrieval paradigm is efficient, because it permits several questions to be asked at once. For example, will the outcome of the encoding manipulation generalize across more than one kind of test? Similarly, will different types of test show different patterns of knowledge acquired and stored during the earlier phases of the experiment?

Besides specific encoding and retrieval factors, memory is affected by many other conditions. The outcome of any experiment is, in a sense, affected by the factors that were *not* manipulated but which the experimenter chose to hold constant. Jenkins (1979) developed a tetrahedral model of memory experiments, shown here in figure 17.1, to illustrate the fact that remembering is always contextually conditioned. The four general factors emphasized are orienting tasks (or factors that are used to orient the learner to the situation), materials (the stuff that the learner will be exposed to and how this will occur), the type of subjects or participants in the research (e.g., adults, children, brain-damaged patients), and the criterial tasks (or the type of test used to assess knowledge). These four factors can interact in complex ways, but Jenkins's main point was that any experiment on memory involves making choices and holding some factors constant while varying others. A question that always needs to be considered is how the results of an investigation might change if other variables were manipulated. Would the findings generalize across these other factors? It is useful to keep the analysis represented in figure 17.1 in mind while reading (below) about factors that have been shown to critically affect remembering.

In the sections below we organize some of the main factors that have been shown to affect remembering by reference to two critical principles: effectiveness of encoding and the effectiveness of retrieval cues. These factors are studied through encoding and retrieval experiments, respectively. We will see that these two principles are not independent of one another and will illustrate their interaction with experiments employing the encoding/retrieval paradigm.

Effectiveness of encoding

One critical aspect of the learning and memory process is the original acquisition or learning of information. Many experiments have documented the importance of a general principle: namely, that the more effectively information is encoded, the better later recall is. Of course, such a statement runs the risk of being tautologous, unless we can specify a way (independently of level of recall or recognition) of defining *effectiveness of encoding*. Frequently, that is impossible to do. However, we will persevere despite that difficulty and show how this general principle can at least order many findings from the experimental study of remembering. In general, all the research in this section conforms to an encoding paradigm, as discussed above: experiments are conducted in which a variable is manipulated during the study phase of an experiment, and the interest is in seeing how it affects performance on a later test.

Meaning

All else being equal, more meaningful information is better remembered than less meaningful information. For example, coherent passages are remembered better than chaotic ones (created, for example, by keeping the words from the coherent passage the same but rearranging them). Similarly, new information about bridge, chess, or baseball will be better remembered by experts in those domains than by novices. The new information can be better assimilated (encoded) in terms of the expert's knowledge base.

Even very simple materials – such as words studied in a long list – can reveal this effect. Craik and Tulving (1975) reported an experiment showing a *levels-of-processing effect* in remembering. Before describing the experiment itself, we identify their guiding assumptions. The basic idea that Craik and Tulving were exploring is that the cognitive system processes information to different levels, or depths, and that the depth of processing determines later retention. For example, in reading the German word *Gedächtnis*, a reader of English (with a knowledge of the orthography of Western alphabets) could apply at least an orthographic or graphemic analysis and identify the graphemes of the word. A person with some knowledge of phonology in German could sound the word out, even if he or she did not know its meaning. Finally, a person fluent in reading German could know the meaning of the word too. (And a German–English bilingual could translate it as *memory*.) To comprehend the word, the reader must progress through graphemic (visual), phonemic (sound), and semantic (meaning) codes. The levels-of-processing approach predicts that remembering depends on the level to which it has been processed, with deeper (meaningful) processing leading to better retention.

Craik and Tulving (1975) manipulated experimentally the depth to which subjects, college students, had to process words on a list of 60 common words, such as *bear*, by requiring them to answer different questions about the words. Some questions directed attention to the word's appearance ("Is it in upper-case letters?"); others directed

Table 17.1 Results of a Craik and Tulving (1975) levels-of-processing experiment presented as proportions of correct recognition. Subjects responded *Yes* or *No* to orienting questions requiring graphemic, phonemic, or semantic analysis.

	Encoding Condition		
Response	*Graphemic*	*Phonemic*	*Semantic*
Yes	0.42	0.65	0.90
No	0.37	0.50	0.65

analysis to the word's sound ("Does it rhyme with *chair?*"); whereas others required consideration of the word's meaning ("Is it an animal?"). For half the questions the answer was *yes* and for the other half it was *no*. Subjects saw each word for five seconds while answering the question. The questions were designed to encourage graphemic, phonemic, or semantic levels of processing.

To measure the effect of this manipulation, subjects were given a recognition test in which the 60 items studied were randomly intermixed with 120 nonstudied words; subjects were told to go through the words and pick exactly 60 that they believed were previously studied. Chance performance on the test was 0.33 (60 out of 180 could be obtained by someone who had not studied the list at all). The recognition results are shown in table 17.1, which gives the proportion of words correct in each of the six conditions (three levels of processing x two answers). Clearly, the levels-of-processing manipulation had a dramatic effect on recognition. Following a graphemic analysis, recognition was barely above chance, whereas a semantic analysis produced extremely accurate retention (especially when the answer to the question was *Yes*). Keep in mind that the subjects viewed the words for five seconds in all conditions and that they could answer the questions in each case in under a second. What the results show is that, with all else held constant, retention could be dramatically affected by the split-second cognitive processes engendered by the questions that were asked. How well people remember events depends partly on what the events are, but also on how they are encoded: deep (meaningful) processing of information surpasses other, phonemic or graphemic analyses in its effect on later retention. We shall see this same principle in operation in the next two sections.

Imagery and vividness

Since the time of the ancient Greeks, scholars have known that imagery can aid remembering. Instructors of rhetoric taught speakers mnemonic devices, which was critical for people who could not use written reminders. Modern experimental psychologists have confirmed the wisdom of using imagery in several types of controlled experiments. In most types of test, pictures are better retained than words; this is true even in tests that would seem to favor verbal encoding. For example, if a long series of pictures and concrete words (words that refer to concrete, hence picturable objects) are presented, and people are asked to recall them by writing either the words presented or the names of the pictured objects, pictures are better remembered than words. This occurs despite the fact that the verbal mode of response would seem to favor verbal over pictorial encoding. In addition, concrete words (*hyena, trampoline*) are remembered better than words that refer to abstract concepts (*freedom, beauty*) when other factors such as word frequency are controlled. Both these outcomes (superior recall of pictures versus concrete words and better recall of concrete than abstract words) can be explained by Allan Paivio's (1986) *dual coding theory* (see Article 12, IMAGERY AND SPATIAL REPRESENTATION). The basic idea is that people can represent information in both verbal and imaginal form, and that if information is represented in both verbal and imaginal codes, its retention will be better than if only one code is used. Therefore, when people study pictures in preparation for a verbal recall test, they code them both imaginally and verbally, whereas they encode concrete words in a verbal code (although some imagery may be evoked by the vivid words). Because pictures are strongly encoded in both modes, they are better recalled than are concrete words. However, concrete words do weakly activate a nonverbal code relative to abstract words, so they are better recalled than are abstract words. Other experiments show that if people are given verbal material and instructed to form images representing the material, they retain more than control subjects who are not so instructed. The role of imagery in memory is unquestioned. Many mnemonic devices, such as those discussed below, use imagery to promote retention.

Distinctiveness

Another factor strongly affecting remembering is the distinctiveness of the event to be remembered. In general, distinctive, surprising, emotional events are well remembered. Suppose a person saw a series of 100 pictures and the fifty-first picture was of a standard desk chair. If asked to recall the names of the items later, few would remember *chair* having been in the series. Now imagine that the same picture of a chair was placed in the same position in a series of 99 words. Now the same object, being distinctive, would be recalled almost perfectly. The general point is that recall of any event depends not only on its nature, but on the context in which it occurs.

This same principle extends to remembering events from our lives. Most of us can recall with more accuracy what we did on some salient occasion (New Year's Eve, our birthday) than a day occurring a week earlier or later. A special name, *flashbulb memories*, is employed for memories of occasions that are emotionally very powerful, such as witnessing the birth of a child or participating in some great national tragedy (an assassination). The analogy is that our memories are so clear as regards details surrounding the place of occurrence, our feelings, and even fine details of the event (or our reaction to it), that they seem to have been caught as in a photographic flash and indelibly imprinted in memory. People have great certitude about such memories, even though studies show that some of the retained information is false. There is debate about whether flashbulb memories must be explained by some special mechanism, or if they are simply strong variants of particularly distinctive events working through the same general mechanism that makes a picture well remembered when placed in the context of many words (see Conway, 1995).

The three factors listed above – endowing events with meaning, using imagery, and making events distinctive – are all examples of how factors manipulated at encoding can powerfully affect memory. However, as we shall see below, just because the manipulation occurs during encoding does not mean that retrieval processes are not important. In most cases, the interaction between encoding and retrieval factors critically determines retention.

Effectiveness of retrieval cues

If you reflect on experiences you have had in trying to remember events from the distant past, the importance of retrieval conditions for remembering will become obvious. You see someone familiar but cannot remember her name; a bit later the name comes to you. Or someone asks you who starred in a particular movie and you draw a blank; when several possibilities are mentioned, you immediately know which one is correct. In another case, you return to a place where you used to live, and the sights and sounds bring back memories of events that you had not thought of for years. All of these common experiences show that having information encoded and stored in memory is no guarantee that it will be remembered; in addition to good encoding, appropriate retrieval conditions must exist for the events to be remembered.

Psychologists have studied the critical role of retrieval processes by manipulating the conditions and the types of cues provided to people during retrieval. In one common technique, people are given long lists of words belonging to common categories (e.g., birds – *pigeon, sparrow*; furniture – *dresser, hat rack*, etc.) with instructions to remember the objects in the category. Afterwards, some people are given a free recall test in which they receive a blank sheet of paper with instructions to recall as many words as possible from the list. In one experiment people remembered 19 of 48 studied words under these conditions (Tulving and Pearlstone, 1966). What happened to the missing 29 words? Were they not well encoded and stored? Another group of people received a cued recall test with the category names given as retrieval cues. In this condition, subjects recalled about 36 words, or almost twice as many as in free recall. This shows that the failure to recall words under free recall conditions was due not solely to problems in encoding or storage, but also to retrieval factors. When supplied with strong retrieval cues, people can remember events that seemed forgotten under other conditions.

Many studies, using many different types of materials, have revealed the same general point: It is impossible to make absolute statements about how much or what kind of information has been encoded and stored in memory. We might like to know what information is *available* (or stored) in memory, but all we can ever know is what information is *accessible* (retrievable) under a particular set of test conditions. Change the retrieval conditions (or the nature of the test), and a different estimate of accessible information will be produced.

What determines the effectiveness of retrieval cues? The general rule that is supported by considerable research is the *encoding specificity principle*, which states that retrieval cues are effective to the extent that they match the way the original events were encoded (Tulving, 1983). In the experiment just described, the category names served as effective cues because they helped to re-create the encoding of the presented words, at least relative to free recall conditions. Similarly, the context in which events occurred can serve as an effective cue, which is why returning to a place from which one has long been absent can bring back memories of old experiences. The encoding specificity principle indicates that it is a mistake to consider either encoding factors or retrieval factors in isolation when discussing memory. Rather, the interaction between encoding and retrieval is critical.

Interaction of encoding and retrieval

Remembering is best conceived as the successful interaction of encoding and retrieval. Consider, for example, the effects of distinctiveness on recall discussed in the section above. If a person sees a picture in a list of 99 words, it will be well recalled, but the same picture would be poorly recalled after being embedded in a list of 99 other pictures. Although the manipulation of distinctiveness occurs during the encoding stage of the memory experiment, the reason for its effectiveness probably depends critically on retrieval. The retrieval cue "picture in the list" identifies only one item in the list, helping to remind the person as to that one distinctive item, but the same cue is essentially useless when a huge number of pictures has been studied. The same argument can be made with the other *encoding* factors described in the earlier section; understanding how each affects retention would necessitate consideration of retrieval factors too.

As another illustration of the interaction between encoding and retrieval factors, consider the effects of drugs on memory. Most drugs that depress activity in the central nervous system harm memory. Drinking alcohol or inhaling marijuana, for example, create poor recall of events that occur while the person is under the influence of the drug. The traditional explanation has been that these drugs harm the brain's ability to encode and store events, so retention is poor. Although this explanation in terms of encoding factors is probably partly correct, it is not the whole story, because retrieval factors (in interaction with encoding) come into play in an interesting way. This is observed in the phenomenon of *state-dependent retrieval*: how well an event is remembered depends on the person's pharmacological state both during encoding and during retrieval. Matching states during both phases aids retention relative to mismatching states.

In the most common type of experiments on state-dependent retrieval, four groups of people are tested in various conditions, as in an experiment by Eich, Weingartner, Stillman, and Gillin (1975). Two groups studied words in a categorized list like the one described above under conditions when they were sober at study, whereas two other groups were given a drug prior to study. A day after studying the material, the people returned and were then tested either sober or intoxicated, with all four possible combinations of conditions between study and test being used (sober at study, sober at test, etc.). People were given a free recall test followed by a cued recall test. The conditions and results from the Eich et al. experiment are shown in table 17.2; because these researchers used categorized word lists, the retrieval cues were category names. First examine the free recall results. The first two rows show the standard effects of marijuana on memory: People who were intoxicated during encoding remembered less of the information when tested sober than did people who were sober on both occasions. The results in the third row show that intoxication during only the retrieval phase also hurts recall, although not as badly. The interesting case is the last row: People who were intoxicated during study actually recalled the information better if they were intoxicated again during the test! The advantage of the drug–drug condition (10.5 words recalled) relative to the drug–sober condition (6.7 words recalled) defines the phenomenon of state-dependent retrieval: Matching the pharmacological state during

Table 17.2 Results of the Eich et al. (1975) experiment on state-dependent retrieval. Subjects were given a drug (marijuana) or were sober at the time of learning and testing of materials under free recall and cued recall conditions.

Condition		*Mean number of words recalled*	
Study	*Test*	*Free recall*	*Cued recall*
sober	sober	11.5	24.0
drug	sober	6.7	22.6
sober	drug	9.9	23.7
drug	drug	10.5	22.3

study and test improves recall. (The same point can be made between the sober–sober condition and the sober–drug condition, but that outcome is not as surprising.) Of course, the findings in table 17.2, which have been replicated many times, do not argue that depressive drugs aid memory. The sober–sober condition always produces the best retention.

Another interesting point to emerge from table 17.2 is that the phenomenon of state-dependent retrieval occurs only under free recall conditions. With the powerful category name retrieval cues, recall improved in every condition relative to free recall, and the state-dependent retrieval effect vanished. The general pharmacological *state* of the person seems to serve as a weak retrieval cue. Under conditions of free recall, where there are no specific overt cues that come into play, *state* cues have an effect.

These same general principles also seem true of mood and memory research. People who learn information while depressed, for example, remember it better when they are depressed rather than happy (and conversely). Again this outcome occurs in free recall but not cued recall.

The phenomena just discussed show the powerful interaction of encoding and retrieval conditions: Proper understanding of all memory phenomena depends on considering encoding factors, retrieval factors, and their interaction. This is true even of mnemonic devices, discussed next.

Mnemonics

Mnemonic devices are memory-improvement techniques, which have been of great interest to scholars throughout recorded history. The most common techniques have been repeatedly discovered and employed. All mnemonic techniques employ the general principles described above: They supply strategies for both effective encoding and effective retrieval, as we illustrate here through two examples.

The simplest mnemonic is the link method. If you have several items you wish to remember (say, in a grocery list), each item is converted into an image, and then the images are linked one to another (*milk* splashing on *bananas*, *bananas* placed on a loaf of *bread*, etc.). To recall the items, the rememberer must retrieve the first item, which enables retrieval of the second through the interactive image, and so on. This chain of images aids recall of the series, but of course the chain is only as strong as its weakest link. Forget one of the items, and it may be hard to break back into the chain. But the link method tries to provide both effective encoding (through images) and good retrieval cues (by linking the images). Still, the next method is more effective.

The method of loci is the oldest mnemonic, described clearly by the ancient Greeks and not really changed since then. A person establishes a path of landmarks which he or she knows well, such as a path leading from the front yard of one's house through each room in the house. The number of salient places or locations (*loci* in Latin) can be as many as desired. These places serve as markers for encoding and (later) as retrieval cues. The set of items one needs to remember are then imaged and mentally placed at each location throughout the house and yard. For example, *milk* on the front sidewalk, *bananas* on the front porch, *bread* inside the front door, etc. When a person needs to retrieve the information, he or she mentally walks through the permanently memorized locations, which serve as retrieval cues, and looks up the item stored there. The use of interactive imagery again provides a good encoding strategy, and the locations serve as the effective retrieval cues.

There are many other mnemonic devices, but all are based on similar principles to those described above: They encourage effective encoding strategies and employ powerful retrieval cues that help recreate that original encoding.

Errors of memory

Our memories are remarkable for being as accurate as they are. People who are rendered amnesic as a result of brain damage must be institutionalized or receive complete care at home, because our ability to remember affects everything that we do and every aspect of our being. (Imagine not being able to remember names, faces, where you put things, who told you facts, and so on.) Yet, as good as our memories are under normal circumstances, we are acutely aware that they are not perfect. We forget where we parked our car, our friend's telephone number, and important appointments. More surprisingly, we can systematically misremember events. That is, we do not forget that some event occurred, but we get the details, or even the gist of what happened, wrong. We consider these two issues – forgetting and false memories – in the next two sections.

Forgetting

The oldest problem in the psychology of memory is the nature of forgetting. Ebbinghaus (1885/1964) first described how information is lost over the course of time. He found a logarithmic relation, with greatest forgetting occurring soon after learning, with decreasing losses over time, shown in figure 17.2. Later research has confirmed this outcome. The key feature in standard research on forgetting is that different groups of people learn the same material and then are tested (using some standard test) at various times since original learning, and the forgetting curve is plotted from the various groups' performance. So, forgetting here means loss of information over time. However, as we have already discussed, forgetting in this sense does not necessarily imply that the forgotten information has vanished from the brain; testing at any interval with more powerful retrieval cues would show recovery of the *forgotten* information. Nonetheless, it remains useful to speak of forgetting as loss of information over time when tested in a particular constant way.

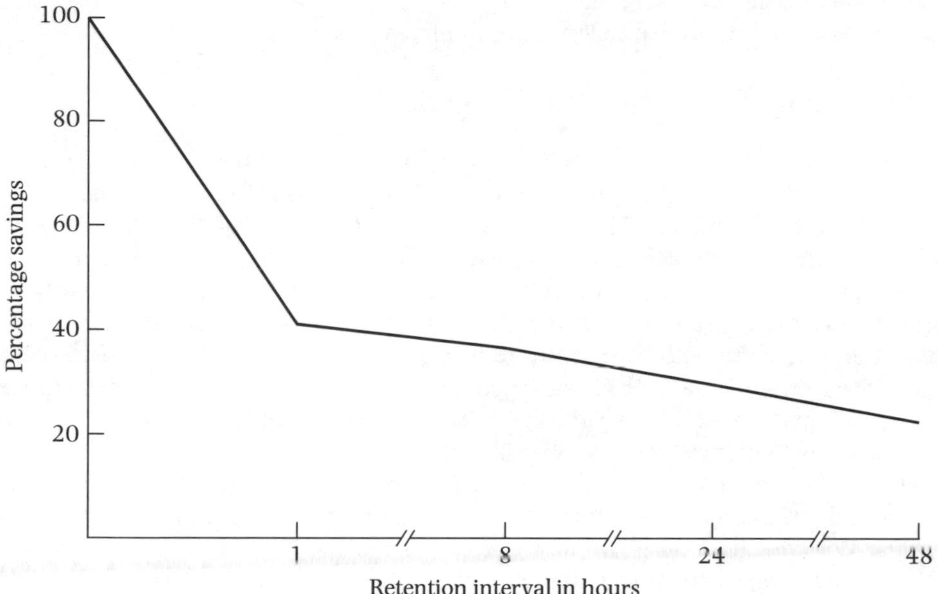

Figure 17.2 Ebbinghaus (1885/1964) examined the rate of forgetting by learning a list of nonsense syllables and measuring the savings in relearning the list after various periods of time.

The nature of the forgetting function is relatively clear, but the explanations for forgetting are more unsettled. The earliest idea was simply that memories decay over time. Just as muscles atrophy without use and become weaker, memory traces were thought to have a certain strength that decayed over time if they were not used. However, this notion has been discredited as a general explanation of forgetting (McGeoch, 1932). No mechanism is postulated; further, decay is occasioned by time, but time is not an explanatory construct. (Suppose a child asked why her bicycle rusted when left outside in the rain for a long time; telling her that *time* caused the rust would not do, whereas an explanation in terms of oxidation – the process operating over time – would be more accurate.) In addition, empirical evidence showed that forgetting could be greater or lesser over time depending on the intervening conditions. In particular, if the time between learning some event and being tested on it is filled with similar events, greater forgetting occurs. This fact turned psychologists away from decay as an explanation of forgetting and toward interference.

Interference is undeniably critical to forgetting, but there is still no complete explanation of interference effects. Two classes of interference exist: proactive and retroactive interference. Suppose you try to remember the exact spot where you parked your car when you arrived at work on Monday, two weeks ago. This represents a difficult task for most of us because of interference. We park our car in different locations every day. All the times you parked your car before the day in question produce *proactive interference* for the target memory; all the places you parked your car after the day in question exert *retroactive interference*. The names indicate that interference can either have effects on retention of events coming later, a proactive effect; or later events can interfere with earlier ones, a retroactive effect. These two classes of interference have been systematically examined for almost a hundred years, and both can be quite potent in causing forgetting under appropriate circumstances.

False memories

Forgetting usually refers to the omission of information: We try to remember something, and either nothing comes to mind, or what does come to mind can be rejected as the wrong information. The issue raised under the rubric of "false memories" is whether we can vividly remember an event and its surrounding details, but either the event never actually occurred, or it happened in a way very different from the way it is remembered. This issue of erroneous memories has been investigated, sporadically, since the turn of the century, and this research has occasionally played a large role in the wider world, such as in legal cases where the accuracy of eyewitnesses' memories of crimes is at stake. Psychologists have now identified several factors that reliably lead to creation of false memories. We consider several here.

One of the most potent factors creating false memories is retroactive interference. We considered the role of interference in forgetting in the preceding section, but interference does not lead simply to omissions of memories, but also to false memories. People can become confused about the source of material and can incorporate information that they read or heard about after an event's occurrence into their recollection of an event. E. F. Loftus (1991) has reported many experiments documenting this phenomenon. In the basic paradigm, people witness a simulated accident or crime (say, a robbery) presented on videotape or in a series of slides. At some later point, they read a passage or answer a series of questions. In an experimental condition, the passage or questions contain some erroneous information about the original scene, such as the statement that the robber had a mustache (when in fact he did not). Subjects in a control condition read the passage without the misleading information. Later, subjects in both conditions receive a recognition or recall test in which they are asked about the crime or accident. Interest centers on memory for the misleading information that was planted later. The outcome in dozens of experiments is that people will frequently remember the erroneous information as having actually happened in the original event, although the magnitude of the misinformation effect (as it is called) depends on many factors. The misleading information not only causes forgetting of what really happened, but seems to replace the correct information with erroneous information. One practical implication is that suggestive questioning of witnesses to a crime by police or lawyers can undermine the witnesses' accurate retention of what really transpired.

A second method of creating false memories is through presentation of related information. If people read a list of related words, or hear a prose passage, they will often mistake another related word or sentence as actually having occurred when in fact it did not. In one straightforward paradigm for creating such a memory illusion, people hear lists of words that are all associatively related to a word that is not presented. For example, they hear "hill, valley, climb, summit, top, peak, . . ." all of which are associates of the nonpresented word *mountain*. Subjects frequently recall the word *mountain* as having occurred in the list and recognize it as often as they do words that actually were presented (Roediger and McDermott, 1995). These illusory memories may be due to failures of reality monitoring, as Johnson and Raye (1981) call them: Did I hear something? Or did I only imagine it?

As the previous question indicates, a third potent source of false memories is imagination. Just as imagery can boost retention of events that actually did occur, as described above, so can imagination create false memories. If people imagine events, they are more likely to think they really happened when they are tested later. In addition, imagining events can inflate one's estimate of the frequency that the events actually occurred.

The three factors listed here – interference, relatedness, and imagery – represent only some of the factors known to produce illusions of memory. The issue is a critical one to understanding memory and will be the focus of continued research in years to come.

Conclusion

The present essay has focused on the experimental study of memory: study which involves manipulation of the environment of the rememberer. Such study can be complemented by constructing computational and neurobiological models (see Article 41, NEUROBIOLOGICAL MODELING) of memory as well as by consideration of the role of an agent's personal context or situation in determining what and how well she remembers (see Article 39, EMBODIED, SITUATED, AND DISTRIBUTED COGNITION). One of the advantages of locating the study of memory within multidisciplinary cognitive science is that this encourages viewing a cognitive phenomenon like memory from a variety of different points of view.

References and recommended reading

*Bjork, E. L. and Bjork, R. A. (eds) 1996: *Memory*. San Diego, Calif.: Academic Press.

Conway, M. 1995: *Flashbulb Memories*. Hillsdale, NJ: Erlbaum.

Craik, F. I. M. and Tulving, E. 1975: Depth of processing and retention of words in episodic memory. *Journal of Experimental Psychology: General*, 104, 268–94.

Ebbinghaus, H. 1885/1964: *Memory: A Contribution to Experimental Psychology*. New York: Dover.

Eich, E., Weingartner, H., Stillman, R. C. and Gillin, J. C. 1975: State-dependent accessibility of retrieval cues in the retention of a categorized list. *Journal of Verbal Learning and Verbal Behavior*, 14, 408–17.

*Higbee, K. L. 1988: *Your Memory: How it works and How to Improve it*, 2nd edn. Englewood Cliffs, NJ: Prentice-Hall.

Jenkins, J. J. 1979: Four points to remember: a tetrahedral model of memory experiments. In L. S. Cermak and F. I. M. Craik (eds), *Levels of Processing in Human Memory*, Hillsdale, NJ: Erlbaum, 429–46.

Johnson, M. K. and Raye, C. L. 1981: Reality monitoring. *Psychological Review*, 88, 67–85.

Loftus, E. F. 1991: Made in memory: distortions in recollection after misleading information. In G. H. Bower (ed.), *The Psychology of Learning and Motivation*, New York: Academic Press, 187–215.

McGeoch, J. A. 1932: Forgetting and the law of disuse. *Psychological Review*, 39, 352–70.

Melton, A. W. 1963: Implications of short-term memory for a general theory of memory. *Journal of Verbal Learning and Verbal Behavior*, 2, 1–21.

Paivio, A. 1986: *Mental Representations: A Dual Coding Approach*. New York: Oxford University Press.

*Roediger, H. L. 1996: Memory illusions. *Journal of Memory and Language*, 35, 76–100.

Roediger, H. L. and McDermott, K. B. 1995: Creating false memories: remembering words not presented in lists. *Journal of Experimental Psychology: Learning, Memory, and Cognition*, 21, 803–14.

*Schacter, D. L. 1996: *Searching for Memory: The Brain, the Mind, and the Past*. New York: Basic Books.

Squire, L. R. 1987: *Memory and Brain*. New York: Oxford University Press.

Tulving, E. 1983: *Elements of Episodic Memory*. New York: Oxford University Press.

Tulving, E. and Pearlstone, Z. 1966: Availability versus accessibility of information in memory for words. *Journal of Verbal Learning and Verbal Behavior*, 5, 381–91.

18

Perception

CEES VAN LEEUWEN

A systems view versus constructivism

Our actions are continuously shaped by what we perceive. When a hammer is picked up, this is done in one smooth movement, the hand adjusting in anticipation to the scale and position of the target. But the opposite is also true: what we perceive is determined by our actions. We tend to see those properties of the world which are meaningful for our everyday actions. Interpretation figures in normal seeing: we see an object as a hammer, a person as a friend.

The immediacy of these experiences makes it easy to take perception for granted. Yet, perception requires the flexible cooperation of complex neuro-anatomical resources. The eye, the optic nerve, and also a significant portion of the brain are involved in vision. We may further consider the eye muscles that are used for focusing and targeting of the gaze to be part of the visual system, as well as the muscles of the neck and shoulders with which postural adjustments are made.

Thus conceived, the major problem in explaining perception is: how do all these resources get coordinated to let the system as a whole perform its function in the relevant circumstances? Flexibility requires that resources can be used in a variety of ways. Just as a hand can be used as a shovel or a hammer depending on the situation, so neural tissues can operate as line detectors and as components of a mental image as well.

What is remarkable is that it doesn't take a mysterious demon or homunculus to set up a perceptual system. There are self-organizing processes that allow the appropriate response to a situation. These processes permit rapid switches in response due to minimal changes in circumstances, as long as these are important enough. Thus perception may appear immediate, but it is achieved through a variety of adaptational, learning, developmental, and evolutionary processes. These should form an essential part of the description of the system. The quest for such a description constitutes the *systems* approach to perception.

The systems approach, which is grounded in the experience of everyday perception, is not the standard approach within psychology. For several decades, the predominant approach had a different experiential basis. This is the constructivist approach, which models perception after reasoning about what is believed to be the case in the world. There are important differences between the systems and the constructivist approach of perception. These differences will be the main topic in this chapter.

The constructivist approach assumes that evolution, learning, and development have no direct significance for how the perceptual system works; if the system were created

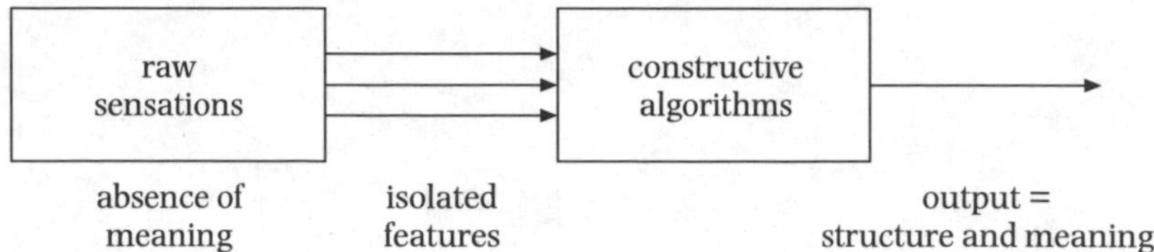

Figure 18.1 According to constructivism, a sensory stage preceding perception provides isolable features as a basis for constructing perceptual structure and meaning.

just a split second ago, it would behave in exactly the same way. If changes in the system beyond the time scale of perception can be ignored, components of the process can be regarded as fixed, and their function can be studied in isolation. According to this view, a first broad outline of what the components of the perceptual system are can be given from general design considerations. These start from the observation that perceptual processes mediate between the physical world and what we believe to be the case. Perception starts from a pattern of external, physical stimulation (e.g., the photons that reach the eye) and is completed when this pattern is matched to an internally kept set of beliefs or representations of the world. A conceptual distinction is therefore needed between sensory processing and an inferential reasoning stage, which could be called perceptual in a more narrow sense of the word (figure 18.1).

Sensory processes are involved in measuring the physical stimulation. Employing linear, semi-linear, or threshold functions, they faithfully represent certain relevant aspects of physical signals such as light intensity and hue or sound intensity and pitch. Physical stimulation will arrive in a particular spatiotemporal pattern. The sensory process, however, is indifferent to this pattern. For instance, suppose a detector measures the light intensity in a certain area on the retina. This patch of light will be registered as the same sensory feature regardless of whether it is part of a triangle, a square, or just a random configuration. Further sensory processing will combine the output of earlier detectors into higher-order ones in order to identify features of increasing complexity. Thus, there will be detectors for features such as contours, line elements, and curvature. Nevertheless, in sensory processing the identification of each of these features will still not be influenced by the overall pattern of which it is a component.

In the constructivist account, sensory processes provide only the lines and angles of intersection; perception tells you what object you are looking at. Perceptual processes operate on the sensory features to construe a perceptual representation. Unlike sensory features, perceptual representations do not depend faithfully on stimulation: ambiguous patterns (such as the one shown in figure 18.2) illustrate this. The existence of alternative responses to the same pattern of sensory stimulation requires two alternative perceptual representations for the same pattern of sensory stimulation.

Different patterns of sensory stimulation may also elicit the same perceptual response. In particular, it is important that the perceiver recognizes an object as the same under different orientations. An elephant is an elephant whether one is looking at the front, back, or side. For this reason, perceptual representations are often assumed to have a viewpoint-independent frame of reference. Even in nonstable circumstances, such representations will provide a stable basis for further evaluation against the background of what we know about the world.

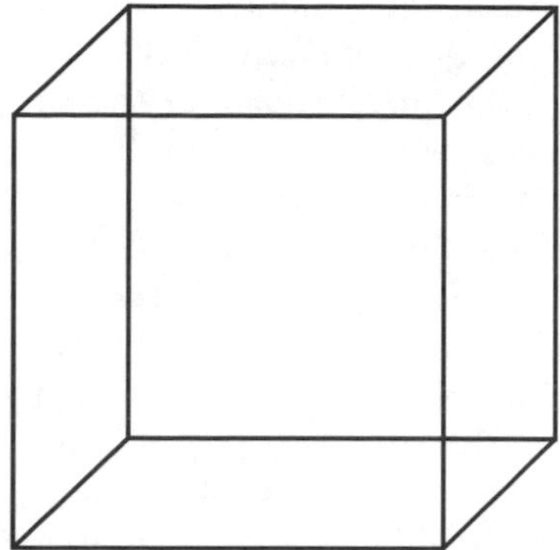

Figure 18.2 The Necker cube has two, rival interpretations, referring to alternative views from below and from above, which are mutually exclusive.

The major problem from the constructivist point of view is how to get from objects and events in the world to perception of them. The fact that sensory processes, being indifferent to object structure and meaning, mediate between the world and experience imposes severe restrictions on perceptual models. By contrast, the need for mediation is denied by a systems account. On this view, perceptual systems operate and have evolved in close interaction with the world. So the perceptual system fits like lock and key with the patterns of the environment. A crucial distinction between systems and constructivist approaches to perception concerns the construal of sensory processes.

Sensory processing: an evaluation

The notion of sensory processes has its historical root in the concept of sensation. A sensation is the phenomenal awareness of a primary quality (the brightness and hue of a color, the loudness and pitch of a tone). *Phenomenal awareness* means that the perceiver experiences what it is like to sense the color or the tone; *primary* refers to the fact that these are the operands, presupposed in the notion of constructive operation. The concept of sensation has found its justification in classical a priori conceptions about the perceptual process, which may well be based on false assumptions. The first question that should be answered, therefore, is: Do sensations exist?

The study of sensation has evolved as a separate domain with its own research methods. Classical psychophysics, which started in nineteenth-century Leipzig with Gustav Theodor Fechner, tries to establish lawful connections between how perceivers judge their experience, on the one hand, and physical quantities, on the other (brightness as a function of intensity). Exponential (Fechner/Weber) or power functions (S. S. Stevens) of the signal have been proposed to describe sensory quantity. Fechner's proposal results from his assumption that just noticeable differences, proportional to physical intensity, are the units of sensation. This involves subjects detecting a weak signal (a light flash, a sound) or discriminating between two signals. But what if a perceiver is just cautious, for instance, in reporting the observed signal? The next question, therefore, is: How much are sensations a by-product of judgmental factors? Signal Detection Theory (Green and Swets, 1966) has provided a technique for distinguishing sensory sensitivity from judgmental bias.

The application of all these techniques, however, seems restricted to extremely simple items. With meaningful objects, interpretation will influence how we judge elementary sensory qualities such as the hue of a color or the pitch of a tone; the same color patch is judged to be redder if a person is told that it belongs to a tomato than if told that it belongs to an apple. When the pitch (high versus low) of a word has to be distinguished, speed and accuracy of classification are influenced by the semantic meaning of the word. For instance, the word *high* can be said in a low pitch or the word *low* in a high pitch. In these incongruent conditions it is more difficult to classify the pitch than when *high* is said in a high pitch or *low* in a low pitch (this is an instance of what is known as incongruence, or the *Stroop effect*). Sensations appear to be influenced by semantic meaning. For this reason, the status of sensations as independent components of perceptual experience is doubtful. Interpretation is an inextricable part of sensation.

If there is no genuine basis for the notion of elementary sensation in experience, we must define sensory processes independently. I propose that true sensory processes must fulfill two conditions. First, the outcome of a sensory process must depend faithfully on the signal. Faithful representation should not be misunderstood as involving accuracy about the state of affairs in the world. A potential sensory feature like retinal extension can be considered a cue to the size of the object, but retinal extension is not related in a lawful way to object size. So if anything like a retinal extension sensor were to exist somewhere in the nervous system, it might faithfully represent optical extension but not true object size.

Second, the outcome of a sensory process must have a fixed functional significance for the perceptual process. Suppose a perceptual system uses retinal extension as a cue to infer object size. Let us assume that the inference process ascribes a specific degree of validity to this cue, proportional to the correlation between extension and true object size (we will not worry about how this correlation was ever established). It is crucial for the inference that the degree of cue validity doesn't change during the inference. This is not to deny that the sensory detector could have different functions – for a different perceptual process! Even the next time that extension is used as a size cue, its validity may have changed as a result of experience. But within the context of the perceptual inference, the function of the cue will have to be fixed. An account of sensory processing that fulfills the above two conditions reduces sensory processes to faithful feed-forward propagation of a set of independent signals.

The neurosciences have provided a classical description of the visual system, which is in good agreement with this notion of sensory processing and is therefore frequently discussed as *the* view of the neurosciences. On this view the visual system is a feed-forward processing hierarchy which exhibits convergence. Rods and cones, the receptor cells involved in the registration of light intensities at neighboring positions on the retina, combine their signals to generate on–off patterns in ganglion cells. These are projected through relay stations in the thalamus called *lateral geniculate nuclei* onto cells of the visual cortex. These cells respond most actively to contours or line segments in a specific orientation, resulting from the combined projection of several partially overlapping lateral geniculate cells. Some of these cells are more specific to the retinal position of the stimulus than others. Sensory processing, therefore, seems to combine physical signals into features of increasing complexity (Hubel and Wiesel, 1962) but still without global information. The global pattern is not represented in the individual cells of the cortex but is available for further processing, because each retina projects to the visual cortex in a systematic manner that respects the topographical organization of the retina.

The classical view is an oversimplification from the perspective of more recent developments in the neurosciences (Zeki and Shipp, 1988). Besides convergence, divergence occurs in the visual pathways. From the earliest stage, the retina, a division into specialized pathways can be observed. For instance, two routes to the lateral geniculate nuclei with different cortical projections can be distinguished. One operates in a slow and sustained manner, has a high spatial resolution and restricted detection sensitivity, and is specific to wavelength (color); the other operates fast and in a transient manner and has restricted spatial resolution, but greater detection sensitivity.

Modern neuroscience in general suggests a division of labor in the brain according to different *sensory modules*, each specialized for a certain modality (color, contrast, odor, temperature, and pitch). Many important attributes of perception, however, are amodal (duration, rhythm, shape, intensity, and spatial extent) or multi-modal (such as being a brushfire, which involves the heat, the smell, and the glow). So the notion of sensory modularity increases the need for perceptual integration.

This apparently is still in agreement with the principles of constructivism, which maintains that integration is achieved by processes of a post-sensory, inferential nature. Unimodal perception will, therefore, precede integration across the modalities in development. According to a systems point of view, it is the other way around. Amodal and multi-modal aspects of perception are primary properties, precisely because of the importance of these structures in the environment. The child will therefore start responding to the multi-modal structures, and development is aimed at differentiation.

David Lewkowicz and his colleagues have, over several years, collected ample evidence that young infants (4 months old) perceive inputs in different modalities as equivalent if the overall amount of stimulation is the same. These infants, due to the immaturity of their nervous system, appear to react to the lowest common denominator of stimulation, which is quantity. Quantity is, therefore, modality-unspecific – that is, not associated with a specific sensory quality or process. Lewkowicz proposes that these early equivalences may form the basis for later, more sophisticated equivalency judgment processes. For the attributes of time, for instance, infants differentiate according to synchrony first, and this differentiation forms the basis for the subsequent differentiation of responsiveness to duration, rate, and rhythm.

Research in sensory development suggests that perceptual integration is not achieved according to the constructivist picture of sensory processing as feed-forward signal propagation. Rather, significance of amodal and cross-modal information early in development suggests that integration between the sensory modules occurs early in processing. Such a notion of intersensory processing is in accordance with a systems account of perception, which emphasizes the role of coordination between the components of the system, rather than their isolated contributions to perception.

The neurosciences support the notion of intersensory perception at all possible levels of description. At the smallest scale, this is realized through interneurons, which provide individual cells within the visual pathway with lateral, mostly inhibitory connections. Lateral inhibitions are useful, for instance, to selectively enhance boundaries in the pattern of sensory stimulation, because identically stimulated neighbors will cancel out each other's activity. This example illustrates that integration of sensory stimulation into a coherent pattern does not wait until sensory processing is completed but begins in the earliest stage. Lateral connections also occur between different sensory modules and may serve to flexibly enhance or reduce the contribution of a sensory module to the process.

In addition to feed-forward and lateral connections, there are also backward connections, which are likely to play an important role in perception – for instance, from the higher visual areas back to the primary visual cortex and from there back to the thalamus. This is in accordance with the downstream operation of semantic information. Pattern code could be mapped downward in the sensory detection system to correct its output. This would make sensation dependent on background knowledge and meaning. Constructivism would deny that such context has great significance for early sensory processes. But the effects mentioned above of categorization (tomato versus apple) on the shade of the red color patch perceived and of word meaning on perceived pitch suggest otherwise. The interactive, intersensory character of early processing is in accordance with the notions of self-organization favored by the systems approach.

Flexibility of perception and the inference of meaning

It no longer appears possible to maintain that sensory features exist in isolation from their context. This severely complicates the constructivist problem of fitting a perceptual representation to a sensory pattern. In order to find a way around this problem, constructivists propose that processing takes place in specialized *perceptual modules* (see Article 49, MODULARITY). These use mandatory processing heuristics, which embody certain assumptions about the structure of the world. Because processing is mandatory, perceivers have no direct access to their sensations, but report the output of a module. Context dependence is restricted to the information encapsulated within the module. This precludes the influence of large-scale semantic knowledge on perception. Even veridical beliefs such as the knowledge that the horizontal segments in the Müller–Lyer illusion are of equal length (figure 18.3), are disregarded in the inference of a perceptual representation.

David Marr (1982) exemplifies this approach. He proposes that there are perceptual modules which have built-in assumptions about the structure of sensory stimulation in real-world conditions. For instance, he proposes that the knowledge that most natural objects are continuous except at their boundaries is used in finding the structure of the object. According to Marr, processing starts from a retinal image. This is a spatial distribution of intensity values across the retina corresponding to a mosaic of sensory impressions. From this image, the primary sketch is computed, which is simply a spatial distribution of intensity changes, or zero-crossings, in the retinal image. These are combined to detect lines, edges, and bars, leading to the $2\frac{1}{2}$-D sketch: a description of surfaces and their orientations with respect to the viewer. Then a 3-D model, a canonical object representation in a viewer-independent frame of reference, is matched to the sketch. 3-D models are adjusted in scale and orientation to yield an optimal fit.

Marr's approach bears all the characteristics of a constructivist approach. It involves matching of abstract representations, the 3-D models, against the sensory input of the retinal image. The sensory input is a snapshot of the world, rather than something that evolves over time. It contains no intrinsic structure and meaning. Canonical representations are abstract, though, strictly speaking, not viewpoint-independent; they possess a preferred scale and frame of reference. However, they are used to code an

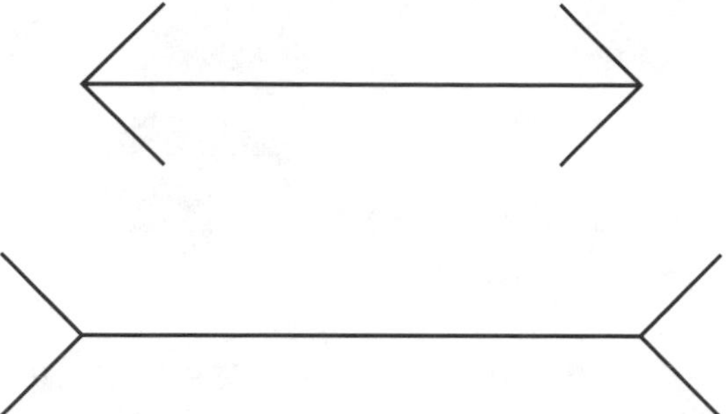

Figure 18.3 Müller–Lyer illusion. The horizontal middle segment in the upper figure appears smaller than that in the lower figure, due to the flanking context.

object independently of scale and orientation. Finally, the model is constructed from general, design principles, rather than situated in a context of learning and evolution.

Although issues of learning and evolution are not usually addressed by constructivists, it may certainly be useful to consider Marr's model from this perspective. It is extremely unlikely that the abstract 3-D models have been acquired through learning. The snapshot retinal images simply do not provide the information from which the 3-D models could have been obtained; this could only be provided by an extremely powerful generalization process, which would be even more difficult to explain. So this knowledge must be innate, shaped by evolution. Evolution may explain the emergence of a function such as recognizing patterns. But reference to evolutionary processes is an empty gesture if applied to individual components of a function which may not be necessary for the whole function to be performed. Some theorists have thought that constructing abstract, viewpoint-independent representations is necessary in order to behave appropriately with respect to objects, but recent work outside the constructivist position disputes this (Ballard, 1996).

The need for viewpoint-independent representations has been disputed within the constructivist tradition by the late Irvin Rock and his colleagues; a square and a rhombus, for instance, are treated as distinct objects, despite the fact that they differ only in orientation. More recent approaches have thus turned away from abstract, viewpoint-independent representations. Neural network theory suggests that it may be sufficient to represent objects in terms of between-view associations. A related notion is that of an aspect graph, which describes the structure of an object as a graph connecting a set of its representative 2-D aspects (van Effelterre, 1994). In both cases a viewer-independent 3-D model of an object is replaced by a connected set of 2-D views. It can be concluded that constructivism has been forced to compromise its original distinction; elementary sensory features turn out to be not so elementary, and abstract representations turn out to be not so abstract after all.

New concepts to overcome old dichotomies

The systems point of view offers a radically alternative approach. The objects of perception are taken to be invariant, higher-order properties of the environment of an organism. The patterns of intensity of the reflected light in the ambient optic array (for

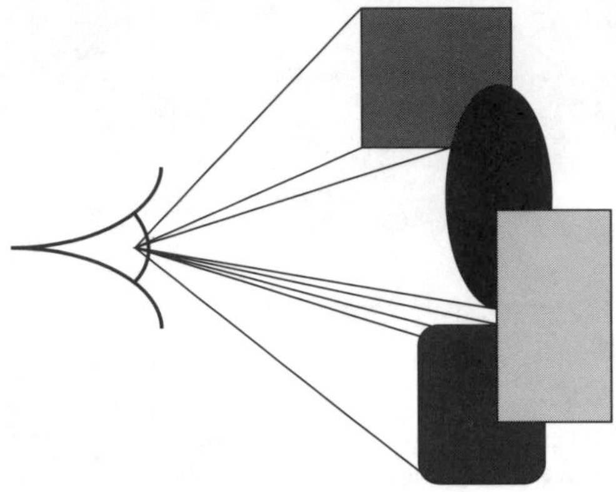

Figure 18.4 The optic array contains structures of reflected light.

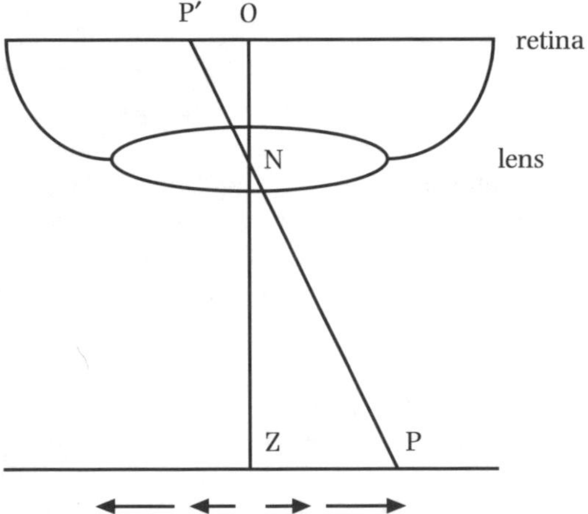

Figure 18.5 Time to contact as calculated by Lee and Reddish (1981). At time t, the distance from the animal to the surface equals NZ_t. The animal is moving with a velocity $V(t)$ toward the water. Texture elements P of the water surface project an optic flow pattern onto the retina. The image P′ of P spreads outward from O with velocity v_t. From similar triangles, $NZ_t/ZP = 1/OP'_t$. Taking the first derivative with respect to time yields $V_t/ZP = v_t/(OP'_t)^2$. Dividing the first equality by the second yields $NZ_t/V_t = OP'_t/v_t = \tau_t$.

an example, see figure 18.4) contain several higher-order invariants which are relevant for action. A key example is the quantity τ (see figure 18.5). τ is a parameter of flow in the optic array. The value of this parameter specifies time to contact for a perceiver approaching a target. The quantity τ is not calculated by the perceiver; a perceptual system has direct sensitivity to it, much like a velocity meter is sensitive to velocity without measuring meters and seconds. The perceptual system sensitive to τ is calibrated to the needs of the perceiver. It reaches a critical state at the moment τ reaches a value at which an action can be performed just in time.

Anyone who has ever seen a gannet (a seabird in the same family as sea gulls) circle above the water and then suddenly dive down to catch a fish under the surface should be familiar with the idea. For these birds, it is important to close their wings just before splashing into the water. Lee and Reddish (1981) showed that plummeting gannets use τ to do so. Use of τ has also been observed in athletes to determine the moment and position of their jump.

James Gibson (1979) coined the term *affordance* for properties like τ. Affordances are higher-order properties of the environment that enable the perceiver to perform a certain action (e.g., a flower affords landing to a bee, a mailbox affords mailing letters, and an aperture of a certain size affords walking through). Perceptual systems pick up the invariants in the ambient light that specify features of the environment that are significant for the perceiver.

But in Gibsonian perceptual theory, there is a *zombie* problem that severely restricts its applicability: conscious perceptual experience is completely irrelevant in the Gibsonian approach; the perceptual system functions just as well without it. Gibson treated it as irrelevant for the perception–action cycle, because he identified awareness with having sensations. Invariants, like the critical value of the famous time-to-contact parameter τ, may be complex, in terms of the physical arrays in which they are exhibited; but in their causal role for perception they act as simple quantities. To require that whatever can be perceived directly must be perceptually simple may be an unnecessary restriction on perceptual directness. It could be motivated only by the tradition to which Gibson was opposed but seemingly unable to relinquish, which held that only elementary qualities are perceived directly.

There is no reason why complex perceptual structures could not be perceived directly, given that both the detection of simple quantities like τ and that of more complex perceptual structures are the result of self-organization processes in the perceiver's nervous system. It may even be necessary within the Gibsonian framework to allow direct perception of complex structures. The zombie problem in Gibsonian theory may be a result of the fact that action is identified with immediate action. The bread says "Eat me" to the zombie, and the zombie eats. But conscious beings are more inclined to suspend the action – I could make the conscious decision not to eat the bread before I finished typing this section.

The actions of conscious, intentional beings transcend the here and now. For instance, we may explore an object without an immediate purpose in mind. In our exploration, we may discover properties which may or may not be relevant for action at some later point in time. For instance, hominids once broke a stone and discovered that this resulted in sharp edges, which could be useful for cutting. They probably did not have the intention of using a stone for cutting before they made their discovery. Their intention would not have gone beyond exploration for its own sake. The same holds true for us today, when we stand still to admire a work of abstract art in a gallery. What would Gibsonian theory have to say about the affordances of such suspended intentions? Probably that we should investigate the structure of the actions, in which one is suspended in favor of the other. Structures of interrelated affordances may correspond to these structures of actions.

The paradigm case of mediated action from a Gibsonian point of view is tool use. Which perceptual processes are involved in the selection of a tool? To begin, a tool is selected not just for being a good extension of the hand. At least as important, the hand-cum-extension must be a good fit to a third, target object. This implies that some immediate action must have been suspended; perceivers just looking for an object that nicely fitted the hand probably wouldn't find the screwdriver. As Eleanor Gibson has stressed, a tool can only be selected on purpose (i.e., not by trial and error) if the affordance that fits the hand-cum-extension is already anticipated in the selection of a tool.

A great deal of MEDIATED ACTION is culturally conditioned. The theory of affordances has to deal with this. A door handle will usually afford opening of the door to satisfy the intention to leave a room. But the handle of a prison door will not. The prisoner will suspend his inclination to leave his prison cell by using the door handle in favor of considering more indirect actions to the same effect. This may or may not involve the construction of tools; what matters is that prisoners' awareness of the culturally determined constraints of the situation allows them to see affordances they otherwise would not see; thus a sheet will come to be seen as affording what a door handle normally would.

Not all perception involved in mediated action, including tool and cultural affordances, can be understood if one construes affordances as perceptually simple entities. Instead, perception must be evaluated in a context of interrelated affordances. The fact that these complex structures determine what is perceived in a situation does not entail that perception involves construction. The direct perception of complex structures may be an important theoretical option. This notion was the starting point for Gestalt psychology. The Gestalt principles, such as the laws of good continuation, common fate, and proximity, are the experiential correlates of the spontaneous structuring of the visual field (figure 18.6). These laws are all subsumed under the principle of Prägnanz, or global figural goodness.

Such radically holistic concepts are notoriously hard to define. It is therefore worthwhile to consider whether the holism of Gestalt psychology – in particular, of its most renowned protagonists, the Berlin school of Koffka, Köhler, and Wertheimer – was perhaps too radical. Beyond the Berlin school, Gestalt psychology took more moderate positions. Gaetano Kanizsa (1979), for instance, provided numerous demonstrations that perceptual organization is actually more local than Gestalt holism would expect. In figure 18.7 the holistic structure is a checkerboard, of which one square is colored black. But this regularity does not enter our awareness. Instead, a black cross is seen lying on a checkerboard.

Recent investigations suggest that the perceiver can, to some extent, influence strategically whether or not whole object structure is predominant. This may lead to a state of perception in which isolated components within the same figure are attended to selectively and the rest of the object is ignored (Peterson and Hochberg, 1983). This state of perception, which is in fact the end product of a quite sophisticated and effortful intentional process, is phenomenally similar to the state of perceiving elementary features – what constructivism wanted us to believe was the primitive stage of perception. The sophisticated strategy to intentionally shift from holistic to piecewise perception may be useful to artists who have to draw details of an object in a piecewise manner. Strategic intention could also lead to a change in another direction. For instance, in figure 18.7 one may be able to strengthen the holistic organization by concentrating on the black diagonals so as to come to see a checkerboard with one black square extra.

The notion that components of a local object structure can be attended to separately runs counter to the radical holism of the Berlin school of Gestalt psychology. At least one branch of Gestalt psychology, called the Graz school (Von Ehrenfels, Meinong,

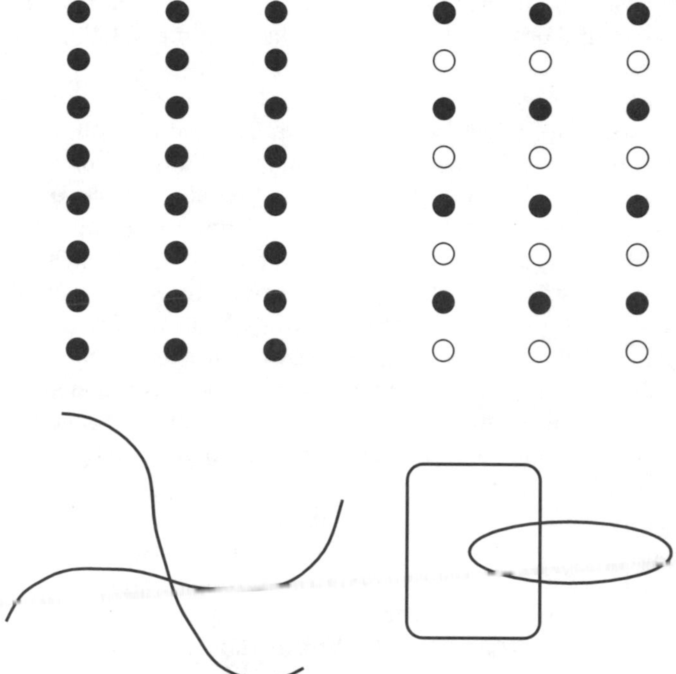

Figure 18.6 Illustrations of Gestalt principles of organization: proximity, similarity, good continuation, and closure.

Figure 18.7 Kanizsa figure: a cross instead of a square occludes the checkerboard. Adapted from Kanizsa, 1979.

Benussi, Metelli, Kanizsa), never committed itself to the radical holism of the Berlin school. In the Graz view, whole object structures are the primary entities of perception. Nevertheless the components can, in principle, be attended to separately. For this reason, whole object structure is not enough to specify a percept. A full structural description will contain multiple levels of components. Following this line of thought, the notion of structural description was reintroduced into psychology in the 1980s by Stephen Palmer.

The view of object structure pioneered by the Graz school still provides a challenge today: to deal with part–whole relationships in perception without reducing them either to constructions out of elements or to emanations from a whole. The nonreductive attitude of the Graz school accords with modern attempts to describe part–whole relationships in algebraic terms, using topological descriptions for object structure (e.g., Smith, 1988). The same spirit is found in the theory of hierarchical organization in nonlinear dynamic systems proposed by the German physicist Hermann Haken. This approach identifies a restricted number of collective variables which are responsible for the macroscopic behavior of a complex system characterized by many microscopic variables. In this approach, the microscopic variables (the components) and the macroscopic order (the whole) mutually determine each other. Application of Haken's approach to perception and motor behavior has been pioneered by Scott Kelso and his group at Florida Atlantic University.

As far as the dynamics of perception is concerned, it may be illustrative to take a historical perspective again. Whereas the Graz school of Gestalt psychology was most concerned with description of object structure according to experience, another school of Gestalt psychology studied how it evolved in time. This was the Leipzig school initiated by Friedrich Sander, which has also become known as the microgenesis movement. "Microgenesis" is a translation of the German word *Aktualgenese.* Sander used this term to indicate the dynamic processes giving rise to the formation of a percept. The developmental psychologist Heinz Werner, who introduced Sander's work in the 1950s in the USA, was responsible for its English translation. Sander believed that a percept originates as an undifferentiated whole (a pre-Gestalt) in the deep, primordial layers of the brain. A series of rapid transformations leads to the full articulation of the object structure in the newer structures of the brain, in particular the cortex. Thus, the most important hierarchy of processing is actually inside-out.

Apparently, an empirical basis for this notion is provided by the fact that large shape is recognized before detail (e.g., Navon, 1977). This effect, however, has nothing to do with a hierarchy in processing but should be understood in terms of the earlier-noted distinction between fast but coarse and slow but fine-grained sensory processing channels operating in parallel. Effects of parallel processing, which relate to size, are easily confused with the hierarchical aspects of processing. When we wish to focus exclusively on these aspects, we must study how the whole and its local details are integrated to form a fully articulated object structure. For these processes, the contemporary experimental literature suggests a great uniformity of results, which, however, is contrary to the intuitions of the microgenesis movement. Let us first consider the assembly of components at the smallest possible level: that of individual line segments. Irving Biederman and his colleagues showed how line segments of a figure are joined into figural components. They presented two kinds of degraded objects. As in figure 18.8, some objects had whole components deleted (object A), while others had all the components present, but segments of the components were deleted (object B). Identification of these degraded figures was compared as a function of presentation time before a mask. On brief visual presentations (65/100 msec), subjects recognized figures with component deletions more accurately. At longer presentations (200 msec)

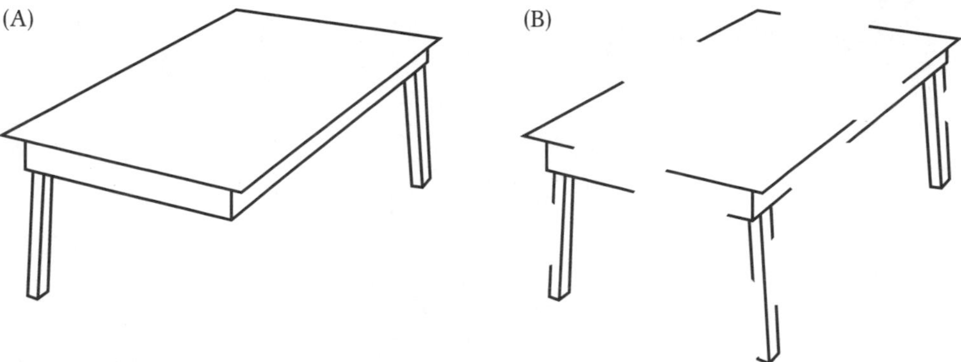

Figure 18.8 The leftmost figure (A) has component deletion, the rightmost (B) has segment deletion. Visual processing below 100 msec is impaired more by segment deletion, processing after 200 msec by component deletion, suggesting that segment processing starts before component processing. Adapted from unpublished work by Biederman et al., as cited by Anderson, 1995.

they were more accurate on the segment deletions. Thus, loose segments appear to be processed before they are joined to form integral components of the object structure.

On a slightly higher level of object structure, Sekuler and Palmer (1992) presented occluding figures (figure 18.9A) for short durations (50–400 msec) before interrupting viewing by presenting the next figure. The occluding figures were perceived as a mosaic (figure 18.9B) for the shorter durations (below 200 msec) and as an occlusion (figure 18.9C) for the longer ones. Components of a mosaic could still be thought of as independent of each other, but for an occlusion a relation between the two is necessary. In this sense, this experiment illustrates the further increment in whole object structure.

Beyond the time scale of this experiment we reach a level where information is sampled across fixations. Eye fixations last approximately 200 msec and are interrupted by saccades, quick jumps to another location in the visual field. These can be both automatic or strategically planned. Eye fixations are needed for reading or for scanning an image, because the spatial resolution of the eye is optimal around the fovea. The interesting question is, therefore, how information yields the impression of whole object structure across saccades. Objects could be perceived as integral wholes across saccades in terms of interconnected partial aspects (as in the earlier-mentioned aspect graph representations), although the possibility of more abstract representation (e.g., Marr's 3-D models) cannot be excluded from what is currently known about this process.

From these illustrations it could be argued that the micro development of the percept is a *hologenetic* process; it starts off from components and joins these to form integral wholes of increasing complexity. It might seem, therefore, that we have come back to observations which support a constructivist account of perception. After all, this approach argued that a percept is synthesized from independent components. But only on a very superficial level do these results resemble their constructivist description. The essential difference between hologenesis and feature integration in the constructivist sense is that in hologenesis the parts interact nonlinearly to form the whole object structure. There are no fixed feature representations in the hologenetic process, so the

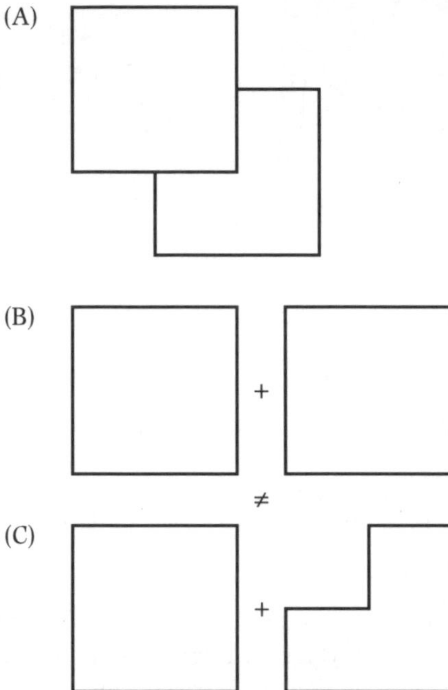

Figure 18.9 Mosaic versus occlusion interpretation. In Sekuler and Palmer, 1992, mosaic interpretations preceded occlusion interpretations. Mosaic interpretations were seen below 200 msec, completed after 300 msec. Although a strict sequential ordering between mosaic and occlusion interpretations cannot be concluded from these experiments, the results suggest the predominance of increasingly complex perceptual structures as a function of processing time, in accordance with the principle of hologenesis.

components of an organization that are obtained later do not resemble the ones that are predominant earlier in processing. So the structural composition of the percept does not reflect its history. For instance, neither the line segments in Biederman's experiment nor the mosaic components in Sekuler and Palmer's experiment are present any longer in the later percepts. Components are not isolable features, but dynamically evolving substructures which can be corrected as object structure emerges.

What brain mechanisms might underlie the dynamics of perceptual processing? The way piecemeal object structures get coordinated resembles a process of mutual constraint satisfaction. The process will be nonlinear, to allow for correction of components once they are given a role within the configuration. This would imply a role for the primary visual cortex as a sketch pad of perception.

Hologenetic development, however, appears not to be limited to the perceptual time scale. Not only is hologenesis found in micro development as in the previous studies; it is also observed in the formation of perceptual categories and in the case of perceptual pattern learning. Similar phenomena occur at the scale of perceptual development and in the learning of syntactic structure (as in the work on language acquisition by Elissa Newport and Jeff Elman). The growth of object structure through a process of self-organization among the components could therefore be proposed as a principle for perceptual dynamics across a variety of time scales.

The brain principle we are looking for, therefore, must encompass both short-term and long-term processing loops. Christoph von der Malsburg, Wolf Singer, and other theorists have proposed the synchronization of oscillatory activity as a mechanism for selective component binding within cortical areas (see Article 43, THE BINDING PROBLEM). Synchronization of brain activity has been observed, using a technique which involves inserting microelectrodes in the visual cortices of primates and cats. Synchronization is more pronounced with coherence in the input. Using noninvasive techniques, synchronization of EEG activity has been studied in humans. The synchronization observed could originate from within the visual cortex or in interaction with other areas, in particular the thalamic nuclei. The synchronization is observed mainly in the 40 Hz frequency band but appears distributed across a wider spectrum. It is therefore possible that the synchronization mechanism uses coupling of chaotic activity. Chaotic activity has important advantages, such as intrinsic instability, by means of which it is possible to model the intrinsic instabilities of perception, such as switching between alternative organizations when presented with an ambiguous figure (like figure 18.2). Neural network research has merely begun to explore the potentials of chaotic activity. Because it comprises a number of desiderata, such as nonlinear flexibility, cooperation, hierarchical self-organization, and control, this line of research carries an important promise for the future.

Epilogue

The central problem of constructivism – how to get from isolated sensory features to the representation of integral object structure – appears to be a misconceptualization. Isolated sensory features do not seem to exist. The close interactions observed, both within and between the sensory modules, appear more in accordance with the view that the sensory system communicates with the world on the level of patterns than that communication is on the level of isolated signals. On the other hand, perceptual object structure doesn't appear to have the abstract characteristics that constructivism attributed to it. It may therefore seem that a systems approach to perception could provide a better explanation for perceptual phenomena. But the systems approach is not without problems of its own. From a systems point of view, it may appear a miracle that perception functions so well in situations where the conditions require us to go beyond the information given, like limited vision conditions or conditions where the goal of the action is way beyond the horizon of visual stimulation. The constructivist approach explains this from the overall tendency of perception to make sense of a situation. Pictures and films exploit this tendency of perception, including that of being misled by expectation. We *see* a bank robbery where, in fact, there is only a filmset of a bank robbery.

A systems account cannot simply treat this as a separate class of phenomena beyond the domain of perception. So, as James Cutting stresses, perceptual processes operate in culturally mediated contexts, of which the significance obviously goes beyond the here and now. By denying a sensory–perceptual distinction, perception is guided by structure and meaning. This implies that there is no strict separation between the phenomena of perception and cognition. Major figures in Gestalt psychology like Wertheimer and Kanizsa have accordingly applied the generic organizational principles of perception to issues of problem solving. For that purpose, insight into the structure of a problem was treated as an analogy to the perception of the Gestalt structure of an object. Gibsonian ecological realism insists that cognitive and proto-cognitive forms of behavior are approached with the same notions and methodology; the Gibsonian principle of affordance, for instance, is applied to tool use and to symbolic social interaction. Some progress has been made in attempts to characterize rigorously affordances of tools and of social situations. Although noncognitive intuitions constituted their point of departure, these approaches seem to reject a methodological distinction between perception and other information processes. The domain of perception is the bulwark from which other domains of mental functioning are to be conquered.

In the process of setting up a systems approach to perception, brain processes cannot be neglected. The problem is to find a broad, general characterization of these processes. In accordance with the systems approach, the dynamics of perceptual organization in the brain could be approached from the perspective of self-organization. I proposed hologenesis as a uniform principle of self-organization in the perception of object structure and suggested that this principle is embodied in the chaotic activity of the brain (van Leeuwen et al., 1997). Hologenesis is the nonlinear counterpart of a notion we are all familiar with: the idea that the brain is an instrument for stepwise creative synthesis. This notion forms the basis for the constructivist approach, which requires that inference processes be posited to explain how the perceiver makes sense of a situation. Alternatively, the principle of hologenesis illustrates that a systems account of these phenomena is possible.

References and recommended reading

Anderson, J. R. 1995: *Cognitive Psychology and its Implications*. San Francisco: Freeman.

Ballard, D. H. 1996: On the function of visual representation. In K. A. Akins (ed.), *Perception*, Vancouver Studies in Cognitive Science, 5, New York: Oxford University Press, 111–31.

Gibson, J. J. 1979: *The Ecological Approach to Visual Perception*. Boston: Houghton Mifflin.

*Gordon, I. E. 1989: *Theories of Visual Perception*. Chichester: Wiley and Sons.

Green, D. M. and Swets, J. A. 1966: *Signal Detection Theory and Psychophysics*. New York: Wiley.

Hubel, D. H. and Wiesel, T. N. 1962: Receptive fields, binocular interaction and functional architecture in the cat's visual cortex. *Journal of Physiology*, 166, 106–54.

Kanizsa, G. 1979: *Organization in Vision: Essays on Gestalt Perception*. New York: Praeger.

Lee, D. N. and Reddish, P. E. 1981: Plummeting gannets: a paradigm of ecological optics. *Nature*, 293, 293–4.

*Lockhead, G. R. and Pomerantz, J. R. (ed.) 1991: *The Perception of Structure*. Washington, DC: American Psychological Association.

Marr, D. 1982: *Vision*. San Francisco: W. H. Freeman and Co.

*Masin, S. C. (ed.) 1993: *Foundations of Perceptual Theory*. Amsterdam: North-Holland.

Navon, D. 1977: Forest before trees: the precedence of global features in visual perception. *Cognitive Psychology*, 9, 353–83.

Peterson, M. A. and Hochberg, J. 1983: Opposed-set measurement procedure: a quantitative analysis of the role of local cues and intention in form perception. *Journal of Experimental Psychology: Human Perception and Performance*, 9, 183–93.

Rock, I. and DiVita, J. 1987: A case of viewer-centered object perception. *Cognitive Psychology*, 19, 280–93.

Sekuler, A. B. and Palmer, S. E. 1992: Perception of partly occluded objects: a microgenetic analysis. *Journal of Experimental Psychology: General*, 121, 95–111.

Smith, B. (ed.) 1988: *Foundations of Gestalt Theory*. Munich: Philosophia Verlag.

van Effelterre, T. 1994: Aspect graphs for visual recognition of three-dimensional objects. *Perception*, 23, 563–82.

van Leeuwen, C., Steijvers, M. and Nooter, M. 1997: Stability and intermittency in large-scale coupled oscillator models for perceptual segmentation. *Journal of Mathematical Psychology*, 41, 319–44.

Zeki, S. and Shipp, S. 1988: The functional logic of cortical connections. *Nature*, 335, 311–17.

19

Perception: color

AUSTEN CLARK

A neighbor who strikes it rich evokes both admiration and envy, and a similar mix of emotions must be aroused in many neighborhoods of cognitive science when the residents look at the results of research on color perception. It provides what is probably the most widely acknowledged success story of any domain of scientific psychology: the success, against all expectation, of the opponent process theory of color perception. Initially proposed by a Ewald Hering, a nineteenth-century physiologist, it drew its inspiration from the existence of opposing muscle groups. Hering thought that analogous opposing processes could explain some aspects of color perception, but the resulting theory was more complicated and less intuitive than that proposed by the great Hermann von Helmholtz. Helmholtz carried his day, but in the long run Hering turned out to be right.

The opponent process model

How opposing muscles might cast light on color perception takes some explaining. It helps first to allocate descriptive vocabulary to three distinct levels: the physics of stimuli, the physiology of receptors, and the psychology of post-receptoral processes. The visible spectrum is linearly ordered by wavelength, ranging (for humans) from approximately 400 to 700 nanometers (nm). Newton's well-known experiments with prisms yielded the spectral hues, or hues each produced by a particular wavelength of electromagnetic radiation within the spectrum. Sunlight is a mixture of light of all those different hues. Hues which, when mixed, yield white are called *complements*. But the ordering of colors is complicated immediately by the existence of extra-spectral hues – hues not found in the rainbow, such as the purple needed to connect the end points, or colors such as brown. Even the so-called unique red – a red which is not at all yellowish and not at all bluish – is nowhere to be found in the spectrum. Furthermore, we find that a given spectral hue can be matched by light composed of many different combinations of wavelengths, and that there is no simple rule of physics that yields all and only the combinations that match in hue. Those matching combinations are called *metamers*. The existence and constitution of metamers is an entry-level puzzle that any theory of color perception must explain.

Details of the physiology of receptors can help. The optics writers after Newton confirmed that, if one chose carefully, any spectral hue could be matched using just three different lights – three different primaries – in different intensities. One had to take care that none of the primaries was a complement of the others, and that none could be matched by a combination of the others. Thomas Young took cognizance of this trichromatic character of human color vision, noted that the retina of the eye was limited in surface area, and proposed in 1801 that the retina contained exactly three different types of color-sensitive elements. The many different hues manifest in visual experience could be produced by suitable combinations of outputs of those three. Young's deduction was basically correct; there are three classes of cones in the normal human retina, which differ in the parts of the spectrum to which each is optimally sensitive. Short wavelength (S) cones are optimally sensitive to radiation of about 430 nm, middle wavelength (M) cones to 530 nm, and long wavelength (L) cones to 560 nm. Each photo pigment will absorb photons of other wavelengths, but with less reliability.

The properties of retinal receptors can explain many of the facts of color mixing and matching. Knowing the absorption spectrum for each of the three classes of cones and the energy spectrum of light entering the eye, one can calculate the likely number of absorptions in each of the three systems S, M, and L. This yields a point in a three-dimensional wavelength-mixture space, whose axes are numbers of absorptions in the three cone systems. Since the visual system has no inputs other than the absorptions in its receptors, stimuli that yield the same point in wavelength-mixture space will match. Complex combinations of wavelengths can be treated as vector sums; as long as the combination, no matter how complex, eventually arrives at the same point, you have a metamer. The various laws of color mixing – Grassman's laws – have an algebraic flavor, with "+" standing for *mix together* and "=" for *match*. For example, if A = B and C = D, then A + B = C + D. With retinal physiology better understood, all these laws can be interpreted literally, with "=" now meaning equal numbers of absorptions in S, M, and L cone systems. Sums become vector sums. Grassman, who was a mathematician, would be pleased. Physics fails us here, but the physiology of the retina allows us to write simple rules for the constitution of metamers.

Retinal details do not constitute a theory of perception, but they do suggest a simple, intuitive model. From the retina there proceed three channels of chromatic information (three "fibres"), one corresponding to each class of cone. These three channels are combined centrally to yield sensations of color. Such was the proposal of the Young–Helmholtz theory. References in the older literature to the S cones as "blue cones" are probably holdovers from this long-dominant theory.

By contrast, Hering's opponent process theory is complex and counterintuitive. Hering thought that there were four fundamental colors, organized in two pairs: red versus green and blue versus yellow. Hue information could be carried in just two channels, one for each such pair. Each channel takes inputs from at least two of the classes of cones, and has an opponent organization, being excited by inputs from some classes of cones and inhibited by inputs from the others. In this model, no cone is a blue cone, since blue only arises in a more central process, requiring inputs from at least two classes of cones. In addition, the model proposes a third, achromatic channel, which sums inputs from all three cones.

It is important to recognize that Helmholtz and Hering could agree that the retina contains three distinct classes of cones, and could agree on all the facts about color mixing and matching. Any similarities between colors that could be explained by similarities of retinal processes would also fail to distinguish between the theories. They agree on what matches what. Their dispute concerns only processes that commence beyond the retina; they propose differing organizations for post-receptoral processes. How might one distinguish between such theories?

The answer lies in other aspects of the qualitative similarities among colors. If, for convenience, one shifts to colored paint chips and tries to arrange them so that their relative distances correspond to their relative similarities, one finds that hues are not ordered linearly, as along the spectrum, but rather form a circle (a hue circle), with extra-spectral purples connecting the spectral reddish blues to the long-wavelength reds. The center of the circle will be achromatic – some point on the gray scale, which matches the lightness of all the chips in the circle, but is neither red nor green nor yellow nor blue. The distance of a chip from the center reflects the saturation of the hue – roughly, the extent to which the hue of the chip is mixed with white. Colors at the end points of a diameter are complements; their hues cancel to yield the achromatic center. Each hue circle is two-dimensional with hue as the angular coordinate, saturation as the radius. To capture the entire gamut of colors that humans can perceive, one must construct hue circles of different lightness levels, from white to black, and then stack them one on top of the other. The entire order is hence three-dimensional, with dimensions of hue, saturation, and lightness.

Hering hypothesized that hue cancelation was due to opposing physiological processes. Processes set in motion by some stimuli could be inhibited by others. This requires that each opponent process receive inputs from more than one class of cone, and that some are excitatory, others inhibitory. Instead of using angular coordinates, the organization of the hue circle could be captured by two orthogonal opponent processes: one running from red through the achromatic center to green, the other from blue to yellow. The neutral point – baseline activation – of the red-green process yields a hue neither red nor green, found at the achromatic center point; and similarly for the yellow-blue process. If one of the opponent processes is neutral, excitation of the other yields one unique hue, and inhibition yields its complement. So if yellow-blue is quiescent, we get a color sensation of either unique red or unique green – the hues at the end points of that opponent process axis. Yellow and blue are the other unique hues; the remaining hues are binary, or produced by combinations of activation and inhibition of the two opponent processes.

None of these facts about the qualitative similarities among the colors follows from the facts of color mixing and matching. From retinal-based explanations we get at best receptoral similarity, or proximity within wavelength-mixture space. But the perceptual similarities of colors do not map in any simple way onto such receptor-based similarities. The orientations of the opponent axes in color space and the consequent identities of the unique hues are not determined by color mixing and matching, or even by the structure of perceptual similarities among colors. Many pairs of colors are complements, and so far as mixing and matching data go, could serve equally well as end points of the opponent axes. Even though the model is a model of color perception, it proposes principles of organization that lie rather deep within the physiology of the organism, remote from direct empirical test.

Contemporary successes

Largely for this reason, the simpler Young–Helmholtz theory continued to dominate until the 1950s, when the team of Leo Hurvich and Dorothea Jameson began formulating quantitative versions of the opponent process model (see Hurvich, 1981). They demonstrated the robustness of hue cancelation, devised a technique to derive *chromatic response functions* from such experiments, and used them to predict the appearances of broadband stimuli. Quantitative links were proposed between opponent processes and the absorption spectra of the three cone systems. The model gives a simple explanation for the various patterns of color vision deficiency: why, for example, dichromats (who can match any hue they can see with mixtures of just two primaries) either confuse reds and greens or (more rarely) confuse yellows and blues. They lose one or another opponent process. If you carve nature at the joints, you have also picked the places where things are most likely to break, and so the theory does.

Besides hue cancelation, the best evidence for the identity of unique hues came from research in color naming and from cross-cultural linguistic evidence (the Berlin Kay hypothesis; see Article 25, WORD MEANING). Individuals can readily describe all the hues in the color circle using just the four terms for the four unique hues: red, green, yellow, and blue. If prevented from using those terms, description requires a larger, more complex vocabulary. Orange is a reddish yellow, but it is hard to see red as an orangish purple.

Opponent process theory thus had some compelling success in explaining psychological data, but it was physiology that finally gave investigators confidence that the theory described real processes in the nervous system. Russell De Valois and collaborators found cells in the lateral geniculate nucleus (LGN) of the macaque monkey whose spiking frequencies were spectrally opponent – excited by some wavelengths, inhibited by others. With this the race was on, and physiological details burgeoned. Now it is known that spectrally opponent cells are rife through the parvocellular pathways of the LGN and through the termination points of those pathways in the primary visual cortex (area V1). These spectrally opponent cells, as predicted, fall into discrete types, with differing inhibition/excitation curves and neutral points.

With all this progress we have yet to identify cells anywhere in the brain that behave in precisely the fashion proposed by opponent process theory. One difficulty lies in reconciling spectral opponency with the spatial organization of receptive fields of the cells that have been found. For example, in V1 the typical spectrally opponent cell has a center/surround receptive field, with the center excited by just one type of cone and the surround inhibited by outputs from the other two, or even from all three. They might be L+ in the center, M– and S– in the surround. In fact, one finds a bewildering variety of different arrangements. Furthermore, all the opponent cells found so far respond to achromatic stimuli. (Interestingly, in the early flush of enthusiasm, L+M– cells in the LGN were often labeled "R+G–," as if the locus of opponent processes had been found. This label will probably go the way of the earlier Helmholtz-inspired *blue cone, green cone* terminology.) Opponent cells of various sorts have been identified in secondary visual areas as well, through at least V4, but even these fail one or another of these tests. In the near future, watch for the identification of the cortical locus of pure opponent processes. Failing such identification, watch for revisions in the model. The notion of opponency might itself be modified, so as to take into account the noted spatial organizations. Loci proposed so far have all been too peripheral; as they are marched inwards, the complexity of the processes identified can only increase.

The sudden receipt of confirming evidence from a different discipline is one way to endow what were merely theoretical entities with reality. Another is to start using them as tools in other experiments. We find color researchers doing the latter as well. Stimuli constructed using the assumptions of opponent process theory are used to test other perceptual models. One example is provided by the use of *equiluminant* stimuli – stimuli that match in luminance and differ only in wavelength composition. The borders between such stimuli are visually indistinct, or at least much less distinct than those across which there is also a change in luminance. Perhaps some of the modules involved in edge detection, perception of shape, or spatial perception are *color-blind*, or insensitive to pure differences in wavelength composition. Equiluminant stimuli can be used in this way to test for modularity – although results to date are subject to ongoing debate (see Article 18, PERCEPTION). Another example is provided by the construction of special stimuli to test the identities of opponent process channels. Such channels are made of neurons, and neurons generally *adapt*, or slow their response to constant input. Perhaps one could construct a stimulus that would arouse selective adaptation of the neurons in a post-receptoral channel without adaptation of the receptors. To do this, the stimulus would have to change in wavelength over time, so as not to affect any one class of cones in a constant fashion yet constantly affect opponent cells that take input from all those classes of cone. The strategy is similar to that of isolating a specific muscle group when working out with weights; here the muscles are remote from the periphery. Such *second site* adaptation effects have been found and have been used to test for channels.

As a final testimony to the reality of formerly theoretical entities, recent developments in genetics have provided a totally unexpected route to the confirmation of some opponent process claims. In the past few years it has become possible to sequence the genes for the various photo pigments in the cones of many different species. Color sensations leave no fossils, but the similarities of such sequences among existing species can reveal evolutionary kinships and divergences. Genes for the various photo pigments diverged at different times. The photo pigments in M and L cones are relatively similar to one another and diverged relatively recently, perhaps 30 million years ago (mya). But both the S cone photo pigment and the common ancestor of M and L are much older, diverging some 530–670 mya. The earliest mammals probably had two cone photo pigments. Our ancestors in Paleozoic and Mesozoic times were at best dichromats. They could have had only one opponent process system, corresponding to yellow-blue; adding the red-green system required a third cone photo pigment. It appears among mammals as a relatively recent acquisition, found only in some primates (see Jacobs, 1993).

That gene sequencing can contribute to a model of color perception testifies to the robustness of the model. It has led to an intriguing recent revision in Young's trichromatic retina: several variants of human M and L photo pigments have been found. Their peak sensitivities vary in discrete steps. DNA sequencing of an individual can help to identify that individual's particular variants of M and L photo pigments and are in turn highly correlated with that individual's color matching in red/green (see Neitz et al., 1993). Gene sequencing and better techniques for measuring photoreceptor sensitivity have also led to an explosion of results in the study of comparative color vision. Many species have much better color vision than any of the rather impoverished Old World primates. Some have color *hyperspaces* of more than three dimensions and are tetra- or even penta-chromatic (see Thompson, 1995).

Future directions

One fly in the ointment has already been mentioned: the failure so far to identify the precise neural locus for opponent processes. It should be emphasized that opponent process theory continues to change and grow, partly to explain some of the puzzling aspects of the standard model. The spatial organization of spectrally opponent receptive fields has already been mentioned. Why spectrally opponent cells should have this spatial organization is a puzzle; their responses would seem to confound luminance and chromatic information. Furthermore, the sheer variety of different spectral and spatial organizations is bewildering, and it is difficult to see how to fit them all into a single stage model.

Russell and Karen De Valois (1993) show how sophisticated recent variants of opponent process theory have become. They make a virtue out of the variety of opponent organizations, demonstrating that if there is a third stage of processing, at which outputs of various types of cone opponent cells are combined, one can completely separate chromatic and luminance information. This also solves quite neatly the puzzle about the spatial organization of receptive fields. The model is quantitative and tied directly to neuro-anatomy. Research prompted by such models is the wave of the future.

Scanning those waves, one spies a final, philosophical puzzle. It may or may not be flotsam. Some contemporary philosophers urge that models which purport to explain color appearance in fact do no such thing. They all fail to explain the qualitative character of chromatic experience. The failure is allegedly simple to demonstrate: all those models would be true, it is said, of a functionally equivalent zombie – an entity that makes the same discriminations as a person and has internal machinery that functionally mimics the human nervous system, but whose internal states lack qualitative character – so-called *qualia* – altogether. Or perhaps those internal states have qualitative character, but of a different sort than ours, so that those models would be true of someone who suffers from inverted qualia. But if the models were true of someone whose internal states lack qualitative character or have a qualitative character that differs from ours, then such models would not explain the qualitative character that our states have. They might explain how humans discriminate this stimulus from that stimulus, but not what it is like to see red. This problem, like any philosophical problem, is easier to get into than to get out of (see Article 9, CONSCIOUSNESS). Part of the difficulty lies in understanding what it would mean to explain what it is like to see red, and this conceptual issue is one that current empirical models are unlikely to touch. Various answers to it have been proposed, but they all suffer from the rhetorical disadvantage of being much more complicated than the question itself. Perhaps a simple answer to this seemingly simple question can be devised. Or perhaps intuitions can be altered, so that the question no longer seems simple. Either would be progress.

References and recommended reading

De Valois, R. L. and De Valois, K. K. 1993: A multi-stage color model. *Vision Research*, 33, 1053–65.

Hardin, C. L. 1988: *Color for Philosophers*. Indianapolis: Hackett Publishing Company.

*Hurvich, L. M. 1981: *Color Vision*. Sunderland, Mass.: Sinauer Associates Inc.
Jacobs, G. H. 1993: The distribution and nature of colour vision among the mammals. *Biological Reviews*, 68, 413–71.
Neitz, J., Neitz, M. and Jacobs, G. H. 1993: More than three different cone pigments among people with normal color vision. *Vision Research*, 33, 117–22.
Thompson, E. 1995: *Colour Vision*. London: Routledge.

20

Problem solving

KEVIN DUNBAR

In the movie *The Gold Rush* Charlie Chaplin and his friend are stranded in a log cabin in the middle of winter while a blizzard rages. The cabin is isolated, and they have a *very* big problem – there is nothing to eat. They pace around wondering what to do. Charlie's friend starts to see Charlie as a chicken, and he tries to kill him. He chases Charlie around the cabin many times. Eventually they hit upon the idea of boiling an old boot and eating it for dinner. With great delicacy they sit at the table and eat the boot as if it were a gourmet meal. They solved the problem of having nothing to eat. While their solution to the problem did not result in a culinary feast, this example reveals two crucial features of problem solving. First, a problem exists when a goal must be achieved and the solution is not immediately obvious. Second, problem solving often involves attempting different ways to solve the problem. Put more formally, a problem has four components. First, there is an initial state. This is the person's state of knowledge at the start of a problem. Second, there is the goal state: the goal that the person wishes to achieve. Third are the actions or operations that the problem-solver can use to get to the goal state. Fourth is the task environment that the solver is working in. The task environment consists of the features of the physical environment that can either directly or indirectly constrain or suggest different ways of solving a problem. I will sketch out the main currents of thinking in research in this area, beginning by reviewing the history of research on problem solving and then focusing on a number of important issues in problem-solving research. Finally, I will give an overview of some recent developments.

A miniature history of problem solving

Research on problem solving has a long and varied history. Many of the early psychologists at Würzburg, such as Oswald Külpe, Karl Bühler, and Otto Selz, investigated the mental processes that are engaged in during complex reasoning and problem solving. While they made a number of interesting discoveries on the nature of thinking and problem solving that had a large effect on the later Gestalt school of psychology, their research has been almost forgotten by contemporary researchers. The Gestalt psychologists overlapped with and continued research on complex thinking and problem solving. In the 1940s and 1950s Gestalt psychologists investigated how people solve difficult problems. This research resulted in a number of classic problems that have been used extensively in problem-solving research. For example, in Karl Duncker's radiation problem subjects were asked to find a way to destroy a stomach tumor without harming the surrounding tissue. Gestalt psychologists often used *insight* problems, in which subjects must discover a crucial element, and once this element is discovered, all the other elements fall into place, and the problem is solved. These psychologists argued that rather than problems being solved by trial and error (as the behaviorists argued), problem-solvers need to gain *insight* into the problem. They often referred to the four different stages that a person might go through in solving a problem (preparation, incubation, insight, and verification). Answers to the questions of what insight is and what mechanisms underlie it had to wait until more adequate language for characterizing problem solving and detailed accounts of it were invented. One further approach to problem solving is the one taken by Sir Frederick Bartlett in his 1958 book *Thinking*. In this book, Bartlett characterizes problem solving as a form of exploration.

It was not until the 1960s, when Herbert Simon and his colleagues began investigating how human subjects solve difficult problems, that problem-solving research took its current form. There were several distinctive, influential features that characterized Simon's approach. First, he used complex problems in which there was no one key element that led to the solution of a problem; thus, the focus was not on insight, but rather on characterizing the processes underlying all problem solving. Second, Simon used concurrent verbalizations (rather than introspections) obtained from subjects to identify the mental operations, representations, and strategies that people use when they solve problems. Third, Simon and his colleagues built a series of computer programs that simulate human problem solving. Using both protocols and computational modeling, Newell and Simon (1972) were able to propose a comprehensive theory of problem solving that continues to be at the heart of contemporary theorizing about problem solving.

Since the 1970s researchers in problem solving have tended to use the approaches of Simon or have used a more descriptive approach of problem solving in the Gestalt tradition. One key aspect of research on problem solving has been the use of verbal protocols (see Article 33, PROTOCOL ANALYSIS). Using this approach, subjects are asked to state out loud what they are thinking while they are solving a problem. These *think-aloud protocols* become the data for formulating models of problem solving. The researchers use the protocols, and whatever actions the subjects took, to build a model of the problem-solving strategies used. The early work on problem solving was concerned with problems that were puzzles or games, such as the Tower of Hanoi task (see below for a description of the task). Later research has tended to focus on more complex *real-world* tasks taken from domains such as science and writing.

Understanding problem solving: searching problem spaces

Newell and Simon (1972) proposed that problem solving consists of a search in a problem space. A problem space has an initial state, a goal state, and a set of operators that can be applied that will move the solver from one state to another. Thus, the problem space adds the notion of an operator to the definition of problem solving presented earlier. The complete set of states that can occur when the operators are applied is known as the problem space. A classic task that has been used to investigate problem solving is the Tower of Hanoi task. The initial states and goal states for this task are shown in figure 20.1. In this task, a subject is given a board with three rods on it. Three disks of decreasing size are placed on the leftmost rod. The goal of the subject is to place all the disks on the rightmost rod. There are two rules for moving the disks: only one disk can be moved at a time, and a larger disk can never be placed on a smaller disk.

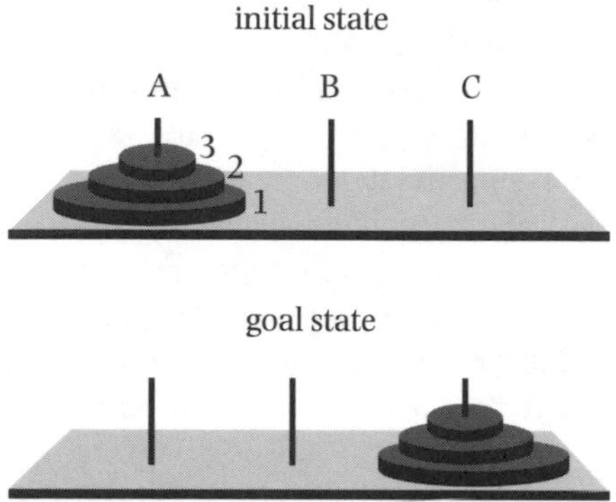

Figure 20.1 The Tower of Hanoi task.

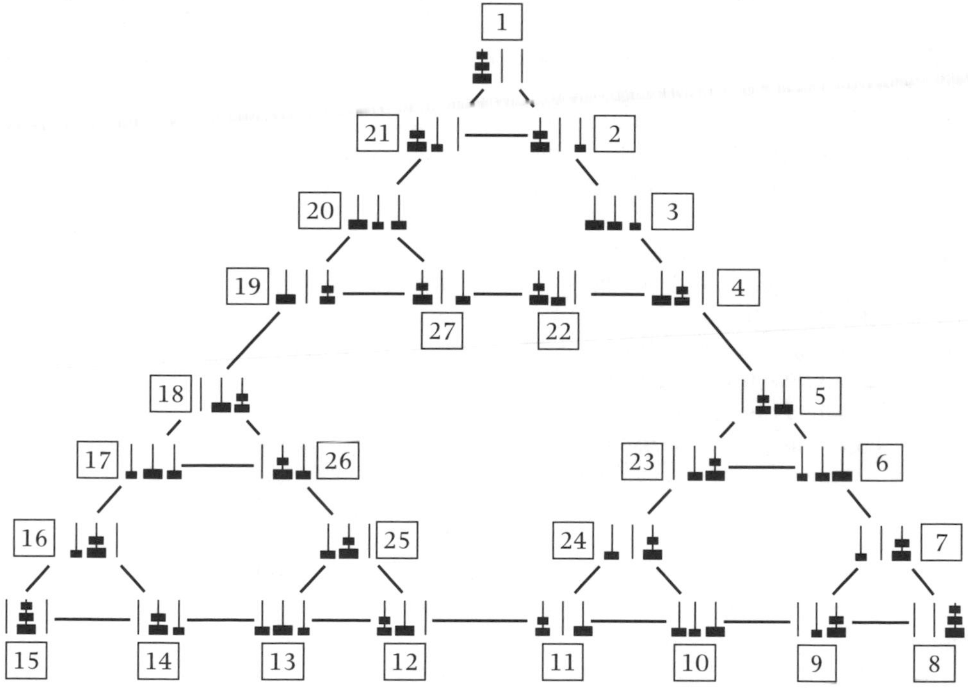

Figure 20.2 The problem space for the three-disk version of the Tower of Hanoi task.

The problem-solver will have to perform many actions (either physical actions, or mental operations) to get from the initial state to the goal state. In the Tower of Hanoi example, the only operation that a solver can perform is to move disks.

In the Tower of Hanoi problem, the problem space consists of the initial state, when all the disks are on the peg A; the goal state, when all the disks are on peg C; and all other possible states that can be achieved when the operation of moving a disk is applied. The complete problem space is given in figure 20.2. The complete problem space consists of 27 states. Note that each state is linked to each other state by the movement of one disk. In the standard version of this task, the initial state is state 1, and the goal state is state 8. This problem can be most efficiently solved by going from states 1 to 8 in increasing order. The problem space shown in figure 20.2 also reveals that there are many different ways of getting from state 1 to state 8. The problem space thus provides a way of understanding the different ways that a problem can be solved.

One important comment about problem spaces is that the problem-solvers are not presumed to have the entire problem space represented in their mind when they are solving a problem. Often, problem-solvers will only have a small set of states of the problem space represented at any one time. Furthermore, some problem spaces, such as that for chess, are so large that it is impossible to keep the entire space in mind. In many problems, the problem-solver will not be able to consider all possible problem states and will have to search the problem space to find the solution. Thus, one of the most important aspects of problem solving becomes one of searching for a path through the problem space that will lead to the goal state. Problem-solvers will use strategies or heuristics that allow them to move through a problem space (see Article 44, HEURISTICS AND SATISFICING).

In problem-solving research, a heuristic is a rule of thumb that will generally get one to the correct solution but does not guarantee the correct solution. An example of a heuristic might be: "If I start playing tic-tac-toe by putting an X in the middle square, I will win." This heuristic does not always work; sometimes I lose even with this strategy! Heuristics can be contrasted with algorithms, where application of the algorithm always guarantees the correct answer (e.g., the rules of addition). I will discuss some of the different heuristics for searching a problem space in order of increasing complexity. The most simple search technique is to randomly pick a next step. People often use this strategy when they have no idea what will lead them to the goal state. A slightly more complex strategy is to move to the state that looks most like the goal state. In this situation, the solver looks just one move ahead and chooses the state that most closely approximates the goal state. This is known as a hill-climbing technique. This strategy can be useful if it is impossible to look more than one move ahead, but it can lead subjects astray. That is, a move that, locally, may look as if it is bringing the solver closer to the goal state may in fact be taking the solver further away from the goal state. For example, in the Tower of Hanoi task, moving from state 5 to state 23 may look like the best move, as state 23 more closely resembles the goal state than state 6. However, state 6 would lead the problem-solver to the goal state faster than state 23.

For problems such as this, a more effective strategy is means–ends analysis. Using this strategy, a solver looks at what the goal state is and sees what the difference is between the current state and the goal state. If the solver cannot apply an operator that will get to the goal state, because the operation is blocked or cannot be executed, then the solver sets a subgoal of removing the block. Using means–ends analysis, the problem-solver then decomposes the difference between the current state and the goal state into another subproblem and sets a goal of solving that problem. In this situation, removing the blocked state becomes the new goal. If that subproblem cannot be solved using the current operators, the problem is further decomposed until an operator can be applied. This strategy can be applied recursively until the problem is solved. When the solver can solve one of the subproblems, the solver can then solve the higher-level problem and ultimately reach the goal state. Means–ends analysis is a particularly useful strategy for solving the Tower of Hanoi problem. An initial goal might be to get the largest disk onto the rightmost rod, but this goal is blocked by the medium-sized disk. A subgoal is set to get the medium-sized disk out of the way, but this is also blocked, so a new subgoal is set of removing the smallest disk. This goal can be achieved, and the solver can then achieve the goal of removing the medium-sized disk.

Strategies such as hill climbing and means–ends analysis involve an incremental search of a problem space. For example, the problem-solver moves from state 1 to state 8 via the other six states in the Tower of Hanoi problem. Not all search heuristics are like this. For certain problems it is possible to jump from one part of a problem space to another part, bypassing many of the intermediate states. One heuristic for jumping from one part of a problem space to another is to reason analogically. If the problem-solver has solved a similar problem in the past, she or he can go directly to the solution by mapping the solution to the old problem onto the current problem. A good example of this is the way that subjects solve Duncker's radiation problem (mentioned above) when they have been preexposed to a problem with a similar solution. Mary Gick and Keith Holyoak (1983) told subjects about a problem in which an army is attacking a fortress. All the roads leading to the fortress have landmines that will explode when the weight of an entire army walks on it. The general breaks up the army into smaller groups to avoid setting off the landmines. Each small group takes a different road, and they converge at the fortress. Subjects were then given Duncker's radiation problem to solve. Subjects who had been exposed to the fortress problem mapped the army onto the radiation and solved the radiation problem by proposing to break up the radiation into a converging set of rays. We can interpret the Gick and Holyoak results from a problem-solving perspective. In their task, subjects are shown the correct path through the problem space by being given the solution to the fortress problem. The subjects can then use a previous problem (the fortress problem) to solve a new problem (Duncker's radiation problem) and bypass conducting an incremental search of the problem space. Thus, analogical reasoning can be seen as a powerful strategy for making a more efficient search of large problem spaces. Using analogy to search large problem spaces is very efficient and is frequently used in science (Dunbar, 1997).

Types of problems

Researchers have distinguished between two main types of problems: well-defined and ill-defined. Well-defined problems have a definite initial state and goals and operators that are known. Examples of well-defined problems might be solving an equation or adding numbers. Ill-defined problems are ones in which the solver does not know the operators, the goal, or even the current state. Examples of ill-defined problems might be finding a cure for cancer or writing the first great twenty-first-century novel. For ill-defined problems the solver must discover operators, define more specific goal states, or perhaps even the initial state. Most research has been conducted using well-defined problems. Researchers such as Simon have argued that many of the heuristics that are used to solve well-defined problems will also be used for ill-defined problems, and that the distinctions between the two types of problems may not be as important as they appear at first glance.

Is problem solving restricted to solving puzzles? While much of the early work on problem solving was concerned with solving puzzles, much of human thinking and reasoning can be regarded as a form of problem solving. Problem-solving strategies can be used in many different domains, regardless of what the domains are. The goal of much research in problem solving has been to identify these domain-general heuristics, as they are clearly of great importance and can be used to understand problem solving generally. Domain-general strategies are contrasted with domain-specific strategies, which are applicable only in a very specific domain such as chess or designing an experiment using a particular hormone in endocrinology. Herbert Simon and his colleagues have characterized scientific thinking and concept acquisition as forms of problem solving and have identified a number of domain-general strategies, such as *finding a constant*, that can be applied in a large number of different domains (Kaplan and Simon, 1990). The problem-solving framework has been applied to many domains, ranging from architecture to medical reasoning to scientific reasoning (see Article 60, SCIENCE). The reason why it has been possible to regard these different activities as forms of problem solving is due to the concept of searching a problem space. Thus, Simon and Lea (1974) and Klahr and Dunbar (1988) have argued that much of concept acquisition and scientific thinking can be thought of as a search in two problem spaces: a space of hypotheses and a space of experiments. What researchers have done is to identify the structure of these spaces and the heuristics that are used to search them.

Problem solving and representation

One of the key elements in solving a problem is finding a good way of representing the problem. When a problem is represented, the solver brings to the forefront certain features of the problem and uses these features to choose what to do when searching a problem space. This is often the case in politics. Frequently, rival political parties form different representations of what is the cause of an economic problem and propose very different operators to apply in solving the problem. Thus, tax cuts will be a solution offered by one party, and tax increases will be the solution offered by the rival party. What is at the root of these different solutions is an underlying difference in the way in which the problem is represented. What this means is that the same problem will have a number of different representations, and that some representations may be more beneficial to solving a problem than others. This often happens in science. For example, molecular biologists recently discovered the breast cancer gene (BRCA1). When different laboratories were in search of this gene, each laboratory represented the problem in different ways, and it turned out that one way of representing it was the most efficient in discovering the gene. Thus rival laboratories having different representations of the problem space that they were searching resulted in the use of different strategies for finding the breast cancer gene.

How do problem-solvers find a representation? In experiments using fairly simple problems such as the Tower of Hanoi, subjects construct their representation based upon the problem statement and features of the task environment. The solver isolates what she or he thinks are the relevant features of the problem and then constructs a representation using those features. By varying the types of instructions given and monitoring their effect on problem-solving behavior, it is possible to discover the effects of different types of representation on solving a problem. For example, in the Tower of Hanoi problem, researchers have used different isomorphs of the problem. One version might be the disk version discussed above. Another version that has been used is the tea ceremony problem. In this problem, three people must perform an oriental tea ritual. There are a set of rules that specify the order in which steps can be carried out, and the problem-solver must discover the sequence of steps that can be used to complete the ritual. The underlying structure of the tea ceremony problem is identical to that of the Tower of Hanoi problem; however, the cover story and task environments are different. What researchers have found is that the cover story can have a very large effect on the way in which a person represents a problem; even minor differences in the wording of a problem can lead problem-solvers to very different types of representations of that problem.

General models of problem solving

Researchers have proposed many different models of how particular problems are solved. However, there have been few general models of the problem-solving process. The general models that have been proposed have been based upon the problem space hypothesis and have been instantiated as PRODUCTION SYSTEMS. Knowledge is represented in these models as symbols, and the production systems operate on the symbols to produce new knowledge and to solve problems. The main place that problem solving occurs is in *working memory*. Working memory is the part of memory in which computations on the currently active symbols take place. What happens is that a production (i.e., a rule) will replace one symbol with another and will incrementally search through the problem space. When solving a problem, production systems construct many temporary representations in working memory as a problem space is searched.

One of the most influential early models was the GPS model of Newell, Shaw, and Simon. They proposed a comprehensive model of problem solving that incorporates many different search strategies and can solve problems from many different domains. One central component of GPS is means–ends analysis. When the GPS program is presented with a Tower of Hanoi problem, it uses means–ends analysis to solve it. In 1990 Newell proposed a Unified Theory of Cognition using the Soar architecture, in which all human behavior can be thought of as a search in a problem space. Newell and his colleagues have applied the Soar architecture to a wide variety of domains and have shown how search in a problem space can be used to understand many different aspects of cognition. John Anderson (1993) has proposed the ACT family of models that he has used to account for cognition in general and problem solving in particular.

Other approaches to problem solving

Two other computational approaches to modeling problem solving have been the connectionist approach and a hybrid approach that is a combination of a symbolic system and a connectionist system (see Article 38, CONNECTIONISM, ARTIFICIAL LIFE, AND DYNAMICAL SYSTEMS). There are few connectionist models of problem-solving processes, more than likely due to the fact that problem solving frequently involves the use of many temporary representations. These types of process have proved somewhat difficult to model in connectionist systems, though they constitute an important topic of contemporary research. When connectionism has been used to model problem solving, it has been used in hybrid models in which there is a symbolic level that is connected to a connectionist layer. The Barnden and Holyoak (1991) book has an interesting collection of hybrid models. This is an area of problem-solving research that will rapidly change over the next few years.

Recently, there has been much debate in the problem-solving literature regarding the role of the task environment in problem solving. Some researchers, such as James Greeno and Lucy Suchman, have argued that one of the most important factors in problem solving is the task environment, and that this is the major determinant of how a person will solve a problem. This *situated* viewpoint stresses the role of the objects and physical features of the environment and how the environment constrains what a problem-solver can and will do. A summary of different views on this topic can be found in a special edition of the journal *Cognitive Science* edited by Norman (1993). (See Article 40, MEDIATED ACTION.)

Recent developments in problem-solving research

One important issue in problem solving is how previous experience with a problem, or related problems, influences current performance on a problem. Gestalt psychologists proposed the concept of functional fixedness to account for the negative effects of previous experience with a problem. However, it is only recently that the effects of experience have been integrated into a problem-space view of problem solving. Marsha Lovett and John Anderson (1996) have looked at the way in which previous experience with a particular class of problems might influence performance on the current problem. They have shown that problem-solvers use both the current state of the problem and their previous history of success at using specific operators when deciding what to do next while solving a problem. For example, if applying an operator, such as moving a piece in chess, theoretically leads one closer to the goal state, but past experience with that particular operator leads to failure, a problem-solver must incorporate both these items of knowledge when attempting to solve the problem. They have developed a model in ACT-R that combines experience with the particular operators and the current state of the problem in an additive manner to predict performance on a problem-solving task. This combination of previous experience and state of the problem is a consequence of the cognitive architecture and is not necessarily accessible to consciousness. Indeed, Schunn and Dunbar (1996) have investigated how solutions to one problem can be primed by solving a similar problem, and have shown that subjects are often unaware that earlier experience on a problem is having a predictable effect on their current problem-solving efforts.

Much cognitive research on problem solving has focused on the ways in which an individual problem-solver sets about solving a problem. Although the research just reviewed, and indeed much of the work on problem solving has traditionally involved the use of puzzles in very artificial domains, recently a number of researchers have started to investigate more complex real-world problems and problem solving in groups (e.g., software design, engineering, and science). These researchers have found that much real-world problem solving is done by groups rather than individuals. Work on problem solving in groups indicates that groups encourage the generation of alternative representations of a problem. When solving a problem, a group can potentially examine a number of possible representations and decide which appears to be the best. Furthermore, in group problem solving, steps such as inductions, deductions, and causal reasoning can be distributed among individuals. Thus, an important component of group problem solving is distributed reasoning (Dama and Dunbar, 1997). (See Article 39, EMBODIED, SITUATED, AND DISTRIBUTED COGNITION.) Group problem solving is not always successful. If all the members of a group are from the same background, they tend to represent the problem in the same way. If their representation is incorrect, they fail to solve the problem. If, however, the members of the group are from different backgrounds but also share similar goals and have overlapping knowledge bases, they will generate many representations of a problem (Dunbar, 1997).

Research on problem solving has tended to take place independently of research on other higher-level cognitive activities, such as concept validation, decision making, induction, deduction, and causal reasoning. However, each of these areas could be regarded as a form of problem solving (Simon and Lea, 1974). While a number of the models, such as John Anderson's and Allan Newell's, have incorporated problem solving into a general account of cognition, problem solving and search in problem spaces have been regarded as applicable only to well-defined problems such as puzzles. Consequently, a challenge for researchers in problem solving is to integrate problem solving with other cognitive activities such as remembering, reasoning, and decision making.

Overall, research on problem solving has centered on the notion of the representation of knowledge and the concept of working in a problem space. Early research focused on puzzles, and more recent research has focused on more complex domains. The shift to more complex domains has necessitated the postulation that problem-solvers search in multiple problem spaces rather than one problem space (Klahr and Dunbar, 1988) and has forced researchers to give much more explicit accounts of the role of the task environment in problem solving. Thus, one of the goals of current research is to determine how people generate new representations and problem spaces as they work on a problem (Kaplan and Simon, 1990). The next decade of research should see models of problem solving that will incorporate theoretical constructs from other aspects of cognition. Finally, researchers in problem solving are now beginning to tackle the question of the role of the brain in problem solving and which different parts of the brain mediate which aspects of problem solving. Many new discoveries await researchers in this field.

References and recommended reading

*Anderson, J. A. 1993: *Rules of the Mind.* Hillsdale, NJ: Erlbaum.

Barnden, J. A. and Holyoak, K. J. (eds) 1991: *High-Level Connectionist Models: Advances in Connectionist and Neural Computation Theory*, vol. 3. Norwood, NJ: Ablex Publishing Corp.

Dama, M. and Dunbar, K. 1997: Distributed reasoning: when social and cognitive worlds fuse. In *Proceedings of the Eighteenth Annual Meeting of the Cognitive Science Society*, Hillsdale, NJ: Erlbaum, 166–70.

Dunbar, K. 1997: How scientists think: online creativity and conceptual change in science. In T. B. Ward, S. M. Smith and S. Vaid (eds), *Creative Thought: An Investigation of Conceptual Structures and Processes*, Washington, DC: APA Press.

Gick, M. and Holyoak, K. 1983: Schema induction and analogical transfer. *Cognitive Psychology*, 15, 1–38.

Kaplan, C. A. and Simon, H. A. 1990: In search of insight. *Cognitive Psychology*, 22, 374–419.

*Klahr, D. and Dunbar, K. 1988: The psychology of scientific discovery: search in two problem spaces. *Cognitive Science*, 12, 1–48.

Lovett, M. and Anderson, J. A. 1996: History of success. *Cognitive Psychology*, 31, 168–217.

Newell, A. 1990: *Unified Theories of Cognition*. Cambridge, Mass.: Harvard University Press.
*Newell, A. and Simon, H. A. 1972: *Human Problem Solving*. Englewood Cliffs, NJ: Prentice-Hall.
*Norman, D. A. (ed.) 1993: Special issue on situated action. *Cognitive Science*, 17, 1–147.
Schunn, K. and Dunbar, K. 1996: Priming, analogy, and awareness in complex reasoning. *Memory and Cognition*, 24, 271–84.
Simon, H. A. and Lea, G. 1974: Problem solving and rule induction: a unified view. In L. W. Gregg (ed.), *Knowledge and Cognition*, Hillsdale, NJ: Erlbaum, 105–28.

21

Reasoning

LANCE J. RIPS

To a first approximation, cognitive science agrees with everyday notions about reasoning: According to both views, reasoning is a special sort of relation between beliefs – a relation that holds when accepting (or rejecting) one or more beliefs causes others to be accepted (rejected). If you learn, for example, that everyone dislikes iguana pudding, that should increase the likelihood of your believing that Calvin, in particular, dislikes iguana pudding. Reasoning could produce an entirely new belief about Calvin's attitude toward the pudding, or it could modify an old one. In either case, accepting the second idea on the basis of the first exemplifies reasoning of the simplest sort. More complex reasoning results from chains of such changes. (Café Maudit serves everything Calvin dislikes. So, since everybody dislikes iguana pudding, Calvin does; since Calvin does, Café Maudit serves it.)

Reasoning is a central component of cognition, in that many other cognitive processes depend on it. Theories of comprehension invoke reasoning to explain how people predict upcoming information in a text or a conversation and how they link unexpected information to what they have gathered so far. Theories of memory invoke reasoning to explain how people reconstruct past events from the fragmentary clues they are able to recall. Theories of learning invoke reasoning to explain how people decide which properties of a new experience are relevant generalities rather than mere accidents. Theories of visual PERCEPTION invoke reasoning to explain how people reconstruct three-dimensional scenes from the two-dimensional information that falls on their retinas. Theories of planning, PROBLEM SOLVING, and DECISION MAKING all invoke reasoning to explain how people construct strategies, recognize when they are applicable, and carry them out. In all these applications, it is controversial how much reasoning occurs – sometimes people fail to make relevant inferences – but it is hardly controversial that reasoning plays important roles in these domains. In each case, the motives for reasoning are to enlarge the scope of information available to an individual, to reconcile information already in hand, and to make this information available to other mental processes.

These interconnections make it difficult to provide a simple summary of cognitive research in reasoning. Since we cannot pursue all these applications, let us focus on issues that may cut across them. As a start, we can look first at the range of internal reasoning mechanisms and at reasoning as it occurs in external, social contexts. We will then turn to two key problems that face reasoning theories: the problem of accounting for modal inferences (e.g., inferences involving the concepts of necessity and possibility) and the problem of arriving at lawlike generalizations (e.g., all water molecules contain hydrogen).

The mechanics of reasoning

According to the everyday conception, reasoning occurs when accepting one belief causes acceptance of another. But how can believing something *cause* you to believe something else? To explain this mysterious causal link, most cognitive scientists have tried to cash out the reasoning process as a causal connection between mental REPRESENTATIONS. Mental representations have a physical embodiment in you, according to this view; hence, a mental representation can stand in a causal relation to another and thus implement reasoning. (Similarly, lines of code in a computer program have a physical embodiment in the computer; hence lines of code cause calculations and other events.) The exact form of the mental representations that participate in reasoning is subject to debate, however, with diagram-like representations and sentence-like representations (and their extensions, such as schemas, MOPs (memory organization packets), and frames), being the most popular choices.

Explaining reasoning in terms of representations rather than beliefs has an advantage in its explicitness, but it also has problematic features. For one thing, it is possible that reasoning is a product of internal events that do not involve explicit (symbolic) mental representations. This is especially likely when the external environment contains cues that guide a reasoner (see Barwise, 1989). For another, not all mental representations are beliefs, since you can presumably represent the possibility that Calvin likes iguana pudding without believing it. Likewise, not all effects that one mental representation exerts on another count as reasoning. According to many theories, for example, activating one representation can activate a related one, but this change can constitute merely an increase in the ease with which you can retrieve the second representation (see Article 17, MEMORY), rather than a process of reasoning. The notions of belief and of reasoning belong to a different level of description of a cognitive system than do the notions of representation and inter-representation change. Moreover, belief and reasoning seem to belong to the *same* level of description, since they appear to be tightly interdependent concepts: Whether a process counts as reasoning seems to depend on whether it produces a change in belief, and whether something is a belief seems to depend on what inferences it leads to and what inferences lead to it. Cognitive science puts its money on the possibility that researchers will be able to work out the relations between the level of reasoning and the level of causal change to representations.

One way to view theories of reasoning, then, is in terms of the ways in which changes in representations implement reasoning. What relations have to hold among mental representations in order for one of them to be a reason for another? According to some current theories, reasoning takes place when one set of representations triggers special mental rules that produce or modify other representations. Here, the rules may be either innate, learned as part of knowledge of the lexicon, or learned on the basis of past predictive success, either for the individual reasoner (Holland et al., 1986) or for the enterprise of knowledge to which the belief belongs (e.g., law or auto mechanics). According to other theories, reasoning is a representational change caused by recall and mapping of relevantly similar cases (see Article 36, CASE-BASED REASONING; Article 1, ANALOGY; and Kolodner, 1993). A current puzzle or issue reminds the reasoner of (representations of) similar situations, and these earlier representations yield new ones about the current situation. These theories usually explain the similarity responsible for recall in terms of the properties and relations shared by the two situations. According to still other theories, reasoning takes place when representations pass to each other some continuous quantity associated with acceptance, such as a probability or a degree of activation. These options are, of course, not mutually exclusive, in that several types of reasoning mechanisms may exist in the same individual, perhaps even for the same inference.

Different choices of reasoning mechanisms will give rise to different empirical predictions. In theories based on mental rules, for example, the difficulty of solving a reasoning problem will depend in part on the total number of rules the problem requires, on the difficulty of retrieving and applying each rule, on the number of false starts that occur before a sequence of correct rules is found, and on the number of alternative correct sequences. By estimating the parameters associated with rule application and search, investigators can generate predictions about the likelihood that people will solve the problem correctly, the amount of time they will take to do so, the solution path that successful reasoners will follow, and the dead ends that unsuccessful reasoners will reach. Investigators typically generate such predictions for sets of problems that embody the rules in varying combinations. They then test their theory by comparing its predictions to data from experiments in which people either produce their own solutions or decide whether stated solutions are correct. (See Osherson, 1976, and Rips, 1994, for examples of this method applied to problems in logic.) Of course, the empirical difficulty of reasoning problems may also depend on factors other than reasoning mechanisms per se. If people fail to understand the statement of the problem in the intended way, fail to understand the task they are supposed to perform, fail in formulating a correct natural-language response, or fail in remembering the key parts of the problem, then their answers will tend to be incorrect. These extraneous sources of error present both theoretical and methodological difficulties for the study of reasoning. The theoretical difficulties concern the proper way to partition reasoning from other mental processes (e.g., sentence interpretation and production, memory, and ATTENTION); the methodological difficulties lie in isolating these disparate types of error in the data.

Social reasoning

The theories just mentioned take reasoning to be a process that an individual engages in. And so it is. As a process of accepting and rejecting beliefs, however, reasoning also occurs in more interactive ways. At the most obvious level, if I claim that Calvin dislikes iguana pudding, and you say "That's right," then each of us is publicly committed to this claim. But we can also carry out more complex negotiations over its status. For example, you can challenge the original claim ("What makes you say he dislikes it?"), placing the burden on me to justify it. Then, if you come to accept my justification, you should also accept the initial claim (assuming you choose not to attack it on other grounds). You can also attempt to defeat the claim ("No, I saw him enjoying a bowl of it"). If I subsequently accept your rebuttal, then I am obliged to retract my initial claim.

These dialogues present some intricate puzzles about reasoning and rationality. As the dialogue unfolds, claims that participants contest can later become mutually accepted or mutually rejected. Since these claims may support others, the status of the latter claims may have to be amended in ways that are not always obvious. Suppose that you attempt to rebut my first claim, and I attempt to rebut your rebuttal. If you then accept my rebuttal, do you thereby accept my first claim? On the one hand, one could argue that I have merely averted an attack on the claim rather than bolstered it. On the other hand, my removing a reasonable doubt might be taken as a kind of indirect support for my point of view. (The latter approach is tempting from a connectionist perspective, since inhibiting an inhibitory connection generally produces excitation in these models – see Article 38, CONNECTIONISM, ARTIFICIAL LIFE, AND DYNAMICAL SYSTEMS.)

Understanding the nature of these reasoning dialogues is of interest for a couple of reasons. First, a number of concepts associated with reasoning seem easiest to explicate in these contexts (as Hamblin, 1970, argued). *Burden of proof*, for example, hardly makes sense without the notion of opposite sides or points of view in an argument. Similarly, *question begging* seems easiest to understand in terms of whether a reasoner's argument presupposes beliefs that his or her partner rejects – that are not part of the reasoners' common ground. Second, as many authors have noted, even solo reasoning is often a matter of justifying or defeating claims from multiple points of view, and dialogue-based reasoning provides a natural model for this process. Third, rules governing dialogues can interact with other changes in belief to produce novel inferences (Grice's (1989) *conversational implicatures*), including inferences about the capabilities of other people (e.g., Hilton and Slugoski, 1986). This does not mean that social reasoning is a substitute for individual reasoning or that individual reasoning is nothing but internalized social reasoning. On the contrary, it is an intriguing problem to provide a cognitive theory for the type of mental processes that allow us to comprehend and engage in these debates and arguments.

Modal involvement

A crucial issue for reasoning theories is explaining how people recognize certain relations as holding across all states of affairs. People's reasoning sometimes reflects not only how things currently stand, but also how things might be or might have been. In a deductively valid argument – for example, the argument from the premise "Everyone dislikes iguana pudding" to the conclusion "Calvin dislikes iguana pudding" – the second statement holds in all states of affairs in which the first statement does. There is no doubt that people can draw simple inferences that match these arguments (see Johnson-Laird and Byrne, 1991, and Rips, 1994, for reviews of research on deduction). It is less clear whether people appreciate deductive validity as such (see Harman, 1986) – whether they can discriminate arguments that are deductively valid from those in which the premise merely makes the conclusion more probable (e.g., the argument from "Most people dislike iguana pudding" to "Calvin dislikes iguana pudding"). But if they do, they must be able to recognize, at least indirectly, the stability of the premise–conclusion relationship across situations.

It is worth noticing that validity can't be approximated by high probability: A deductively valid argument isn't simply one whose premises lend a probability of 1 to the conclusion. The premise "Amber has a backbone" gives the conclusion "Amber has a heart" a probability of 1, but the corresponding argument is not deductively valid. (In some logically possible world or state of affairs, there may be creatures with backbones but no hearts.) It is possible to get from probability to validity, but doing so requires generalizing over different probability assignments, which presents much the same problem as generalizing over different states of affairs. So we cannot account for validity simply by regarding it as an extreme case of probabilistic reasoning.

Reasoning seems to involve alternative states of affairs, not only indirectly via validity, but more directly in inferences that use modal notions such as necessity or possibility. For example, even children appreciate that "It's possible that Calvin doesn't like iguana pudding" entails (and is entailed by) "It's not necessarily true that Calvin likes iguana pudding" (Osherson, 1976). These inferences pose an obvious problem for theories that extract reasoning from the past experiences of the reasoner. It is difficult to see, for example, how theories based on remembered cases could have sufficient power to account for our confidence in them. Similarly, theories based on revisions of probabilities also have difficulty here, for the reasons mentioned in the preceding paragraph. Modal thinking of this sort also occurs in reasoning about time, knowledge, and obligation, so the problems it raises are important and general ones. Although there are numerous formal models of modal reasoning in philosophy and artificial intelligence, there are few proposals about how to adapt them to human styles of reasoning.

Lawlike generalizations

A related problem for both scientific and everyday reasoning is that inferences often involve generalizations that specify how things would be under conditions that do not currently exist. For example, a generalization like "All humans learn a native language" allows you to predict that if Amber were human, then she would learn her native language. Reasoning to these generalizations is problematic, as Nelson Goodman (1955) pointed out, because past and present instances are equally consistent with many rival generalizations that you would not rely on – for example, that all humans but Amber learn to speak, that all and only humans born before next month learn to speak, or that all and only humans born within 50,000 miles of Fargo learn to speak. (Non-lawlike generalizations – such as "All current residents of Fargo have learned their native language" – do not have these problems, since present instances completely determine their truth or falsity.)

The difficulty that lawlike generalizations pose is clear in the case of theories based on memories of past experiences. By hypothesis, past experiences are equally consistent with the unreasonable generalizations just mentioned as with the reasonable ones. The theories have to posit further constraints to decide among them. Probabilistic or activation-based theories are in the same boat, because the degree of probability or activation provided for a generalization is typically a function of past cases. Rule-based theories also stumble here, because it is difficult to see how the rules could be other than generalizations of the same sort, begging the question of why we use these reasonable rules rather than unreasonable ones. To make matters worse, related problems arise for rule theories in understanding what constitutes following one rule rather than another. Is the rule you are following "Predict speech if human" or "Predict speech if human and born less than 50,000 miles from Fargo"? You may consciously adopt the first policy rather than the second, but all current and past evidence is consistent with the possibility that your rule-applier interprets *human* as "human-and-born-less-than-50,000-miles-from-Fargo" (See Kripke, 1982, for an elaboration of this predicament, and Osherson et al., 1986, for an examination of rival theories of inductive reasoning.) It is difficult to see how it is possible to have a theory of inductive reasoning based solely on representations. Representation is supposed to account for inferences, because the form of the representation permits some inferences (the intuitively reasonable ones) and not others. Any proposed form, however, can be made to countenance wild inferences that no one would find reasonable, as Goodman (1955) also demonstrated.

Commonalities

The issue of modality tends to appear in discussions of deduction, and the issue of lawlikeness in discussions of induction; nevertheless, the problems they pose for cognition are similar. In both cases, the inferences we perform often accord well with normative standards, at least within the limits imposed by time, memory, and other constraints on resources. Yet attempts to explain these inferences on cognitive grounds become stymied, since the inferences themselves seem to demand access to information that is not available either to an individual or to a culture – information about all (or all normal) situations or possible worlds or future states. This threatens the project of accounting for reasoning via prior cases, empirical probabilities, learning, evolution, and other methods that stick close to past experience; in the case of lawlike generalizations, it may even threaten the entire project of accounting for reasoning via mental representations. One reaction could be "So much the worse for mental representations and cognitive science"; but we still need to explain how people are able to deal with these inferences, and no good substitutes for representations are in sight that comport with other data in cognitive psychology and AI.

To get around these difficulties with purely representational theories, however, we may need to appeal to other factors that could shape human reasoning. Perhaps what is missing in current theories of reasoning is a proper account of the role played by cognitive architecture, cultural practices, natural language, neural substrates, or world causal structure. Merely naming these factors, however, is obviously not enough, since it is perfectly possible to ask what it is about them that permits us to reason as we do. A satisfactory understanding of reasoning entails working out the connections between these factors and specific inference skills.

References and recommended reading

Barwise, J. 1989: Unburdening the language of thought. In *The Situation in Logic*, Menlo Park, Calif.: CSLI, 155–71.

*Goodman, N. 1955: *Fact, Fiction, and Forecast*. Cambridge, Mass.: Harvard University Press.

*Grice, H. P. 1989: *Studies in the Way of Words*. Cambridge, Mass.: Harvard University Press.

Hamblin, C. L. 1970: *Fallacies*. London: Methuen.

*Harman, G. 1986: *Change in View: Principles of Reasoning*. Cambridge, Mass.: MIT Press.

Hilton, D. J. and Slugoski, B. R. 1986: Knowledge-based causal attributions. *Psychological Review*, 93, 75–88.

Holland, J. H., Holyoak, K. J., Nisbett, R. E. and Thagard, P. R. 1986: *Induction: Processes of Inference, Learning, and Discovery*. Cambridge, Mass.: MIT Press.

Johnson-Laird, P. N. and Byrne, R. M. J. 1991: *Deduction*. Hillsdale, NJ: Erlbaum.

Kolodner, J. 1993: *Case-based Reasoning*. San Mateo, Calif.: Morgan Kaufmann.

Kripke, S. A. 1982: *Wittgenstein on Rules and Private Language.* Cambridge, Mass.: Harvard University Press.

Osherson, D. N. 1976: *Logical Abilities in Children,* vol. 4: *Reasoning and Concepts.* Hillsdale, NJ: Erlbaum.

Osherson, D. N., Smith, E. E. and Shafir, E. B. 1986: Some origins of belief. *Cognition,* 24, 197–224.

Rips, L. J. 1994: *The Psychology of Proof.* Cambridge, Mass.: MIT Press.

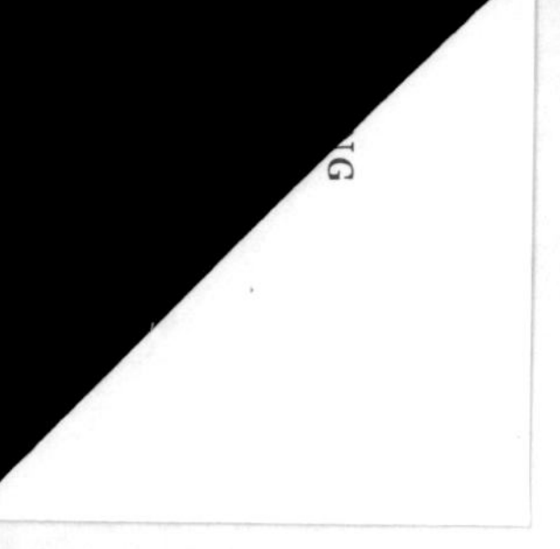

22

Social cognition

ALAN J. LAMBERT AND ALISON L. CHASTEEN

Introduction

Social cognition refers to a discipline in which researchers seek to understand social phenomena in terms of models which emphasize the role of cognitive processes (e.g., attention, encoding, cognitive organization, storage, MEMORY) in mediating social thought and action. Although social cognition is a relatively new field, it is important to note that social psychologists have long been concerned with many of the same issues that are central to cognitive science, such as how people store and retrieve information about their environment in memory. Thus, it would be inaccurate to portray the emergence of social cognition as the first time that social psychologists have focused on the role of cognition in mediating social behavior. Nevertheless, it is true that in the mid-to-late 1970s, a branch of social psychology emerged which was unique in the extent to which it overlapped with the theoretical orientation of, and the methodological tools employed by, cognitive psychologists. It is this time period that most scholars peg as the era in which social cognition grew and matured as a discipline in its own right.

Our review of social cognition is organized around four main topics: (a) person memory, (b) categorization, (c) concept accessibility, and (d) automatic versus controlled processes. We chose this organization to illustrate how researchers in social cognition have explored traditionally *social* phenomena (such as impression formation) in terms of these basic processes. It is important to note, however, that the discussion to follow is highly selective and in no way represents an overview of the entire field of social cognition. Rather, it is our hope to convey how social cognition differs from previous theoretical/methodological approaches in social psychology and to give the reader a brief taste of the sorts of theoretical advances that have been, and are currently being made, in social cognition.

Person memory

Although social psychologists have maintained an interest in memory processes for many years, this area represented, until relatively recently, a rather minor aspect of research and theory in social psychology. No longer. With the emergence of social cognition, researchers have devoted an enormous amount of attention to the process by which people cognitively organize, store, and retrieve their impressions of persons and events. Interest in social memory can be traced, in large part, to an important shift in the way that researchers study and conceptualize the process of impression formation. For many years, social psychologists (including Solomon Asch and Norman

Anderson) studied the process of impression formation by asking participants to form an impression of another person on the basis of a long list of trait adjectives (*intelligent, industrious, impulsive, critical, stubborn, envious*) which ostensibly described this person. Researchers in this area were concerned with such issues as the effects of varying the order of these traits on judgments of the target and whether the underlying meaning of any given trait adjective (e.g., *critical*) varied with the context in which it appeared.

Research in this area did not make any assumptions, however, about the underlying mental representations that people form of others. Nor did these early approaches specify the underlying processes that mediate the judgments and decisions formed about persons. This state of affairs changed rather dramatically in the early 1980s, with the development of comprehensive models of person memory and judgment that devoted far greater attention to the content and structure of people's cognitive organization of social information with a strong emphasis on specifying the underlying *processes* involved in impression formation, such as encoding, organization, representation, retrieval, and output (Hastie et al., 1980). In many cases, these theoretical models were developed and tested using many of the experimental procedures and tasks long familiar to memory researchers in the cognitive area, including free recall, clustering analyses, and reaction time.

The theoretical models that were developed in this line of research have been remarkably successful in accounting for a broad range of judgment and memory phenomena, including the way that people process information about social categories, social events, and the self. These models have provided theoretical accounts of a number of counterintuitive effects, such as the fact that people often have better recall of information about another person if it is evaluatively inconsistent, rather than consistent, with their initial expectations about this person. Moreover, the models originally developed in the person memory area served as a basis for the development of more general theoretical frameworks that have been applied to a number of different substantive domains (e.g., stereotyping, impression formation). Indeed, most of the research described in the remaining three areas covered in this chapter build either directly or indirectly on these previous models.

Social categorization

The antecedents of social prejudice

In an early effort to understand the roots of social prejudice, work by T. W. Adorno and his colleagues proposed that prejudice stemmed from a fundamental flaw in moral development characterized by an overly rigid thinking style and a tendency to punish others who violate cherished societal norms. It was in this historical context that a much different account of racial prejudice appeared which owed far more to then-recent developments in cognitive psychology. In his classic book *The Nature of Prejudice* (1954), Gordon Allport drew liberally on work by a number of cognitive psychologists (most notably, Jerome Bruner) in proposing what has since been referred to as the *cognitive approach to stereotyping*. This account was remarkably different from that proposed by Adorno, in that Allport saw stereotyping as part of a larger, intrinsic need to reduce the overwhelming complexity of the social environment, rather than as evidence for some sort of breakdown of moral development.

The writings of Allport have had a profound impact on current research and theory in social cognition. For one thing, although Allport's writings were well known throughout the 1950s and 1960s, it was not until the late 1970s that stereotyping researchers began to empirically test and extend many of the basic principles that he first proposed. For the most part, these efforts have been quite successful. For example, a large amount of research has shown that the extent to which people rely on stereotypes about others can be manipulated experimentally by placing participants in conditions of relatively high versus low cognitive loads, with the typical result that (as the cognitive economy perspective would suggest) greater levels of prejudice are exhibited in the former than the latter condition.

This is not to say, however, that all stereotyping processes are perfectly captured by cognitive models, and that stereotyping theorists have abandoned altogether the role of affect and motivation in social perception. For example, theorists have focused on the kinds of *integral* affect that may be associated with and triggered by social stereotypes and on the role which this category-based affect plays in the way that people encode and retrieve information about single group members. Theorists have also focused on how the use versus disuse of stereotypes may be moderated by *incidental* affect: that is, temporary moods or emotions (happiness, sadness) elicited by fortuitous events completely unrelated to the stereotype or to the intergroup context (Mackie and Hamilton, 1993).

How do people store information about social groups in memory?

Until very recently, social cognition researchers were in near-universal agreement that perceivers store social information in the form of abstract representations (e.g., schemata, prototypes, scripts), with little, if any, attention to the possibility that social information may be represented in the form of exemplars. Exemplar-based representations differ from abstract representations in that the former type of knowledge representation retains far more *specific* information about individual members of the category. Over the past five years, researchers have argued that exemplar-based models (many of which were developed originally in the cognitive domain) provide a better framework for understanding how people store and retrieve information about social categories (Smith and Zarate, 1992). More recently, researchers have also shown interest in integrating recent connectionist models with current theories of social behavior and thought (Smith, 1996).

Finally, several recent efforts have focused on the possibility that the kinds of categorical representations that people form can be quite different, depending on the nature of the perceiver's underlying goals and motivation. For example, a provocative line of work by Lawrence Barsalou has recently argued that people often form ad hoc categories that are oriented towards achieving a particular goal or outcome (e.g., things to eat on a diet). As Barsalou notes, the nature of such categories is often quite different from that of non-goal-directed categories (e.g., politicians). For example, whereas the best fitting or most representative member of the category *politicians* is likely to be based on the features that occur most frequently in this group, the best-fitting member of the category *things to eat on a diet* may be based on the *ideal* example of the category (e.g., foods that contain no calories), which may not occur very frequently at all.

Ingroup versus outgroup membership

Although social and cognitive psychologists share a general interest in categorization processes, they differ in at least one important respect. For cognitive psychologists, the category membership of the *perceiver* is rarely of interest. This seems largely attributable to the fact that in most cognitive experiments, the stimuli presented to participants are symbols or inanimate objects whose features have little, if any, personal relevance to the perceiver. A much different picture emerges from the social domain. In this case, the perceiver's own category membership, and its relation to that of the person being judged, is critical. Several lines of research have shown, in fact, that people process information about social categories in memory very differently, depending on the degree of overlap between their own category membership and that of the persons being judged. The best-documented effect of this sort (shown in a classic series of studies by Henri Tajfel) is that people show greater favoritism to the members of ingroups as opposed to outgroups. In addition, more recent work by a number of researchers (e.g., Charles Judd, Patricia Linville, Bernadette Park) has shown that people typically perceive greater variability among ingroups than outgroups, an effect usually referred to as the *outgroup homogeneity effect*. For example, whereas liberal perceivers are likely to view conservatives in largely homogeneous terms (e.g., "They're all alike"), the reverse is likely to be true among conservative perceivers. For the most part, previous research has focused on the *antecedents* of perceived group variability, such as the extent to which perceived variability is mediated by differences in familiarity with the target group or whether more motivational factors come into play as well. In recent years, however, there has been greater interest in the *consequences* of group variability, such as how this factor plays a role in category-based inferences.

Category accessibility

It is self-evident that before one can react to a stimulus, one must make an initial determination of its meaning. According to most cognitive theorists, people construe the meaning of a stimulus by first accessing some kind of internal representation that matches the properties of the presented stimulus. However, people typically do not engage in an exhaustive search in memory for all possible representations that match the features of the target. Rather, they select the representation that is most accessible in memory that sufficiently matches its features. Thus, when people are asked to interpret the ambiguous phrase "Mary sat up all night waiting for her husband to return," people tend to interpret the phrase in terms of one concept (*devoted wife*) or the other (*jealous wife*), depending on the relative accessibility of these concepts in memory. The observant reader may have already noted that this is another manifestation of the *cognitive economy* notion discussed earlier. That is, just as people tend not to focus on every individuating feature of others when forming impressions of single category members, so too they do not consider every possible knowledge representation when making decisions about the world around them. This notion pervades much of the highly influential work by Daniel Kahneman and Amos Tversky on the various heuristic strategies (e.g., the availability heuristic) that people rely on when making potentially complex decisions about self and others.

What are the factors that determine whether social constructs become relatively accessible?

The above considerations naturally raise the more general issue of precisely what sorts of factors might make a concept more accessible. Although work in the cognitive area has focused mainly on temporary differences in cognitive accessibility (e.g., as driven by recent exposure to *priming* stimuli), researchers in social cognition have also explored more *long-term* differences in cognitive accessibility. For example, work by John Bargh and Tory Higgins has explored the notion that certain trait concepts may, by virtue of their relevance to life goals, vocational choice, or personal disposition, become *chronically* accessible due to their frequent activation over time and situations. Such models of chronic accessibility (which have their roots in the early writings of George Kelly) provide possible explanations for why people might tend to *see* the same traits in others that they see in themselves.

When do activated constructs lead to assimilation versus contrast effects?

To the extent that impression formation involves a series of interrelated categorization processes that occur at both the behavioral ("Tom acted in a kind manner") and the person level ("Tom is a kind person"), it is not such a large leap to consider how issues of category accessibility could play an important role in the way that people form impressions of others (Wyer and Lambert, 1994). Beginning with a seminal article by Higgins, Rholes, and Jones (1977), researchers in the social area showed interest in how differences in the accessibility of trait concepts can drive the sorts of impressions that people form of others. In the Higgins et al. study, participants who were asked to construe the meaning of an ambiguous passage (e.g., "Donald was well aware of his abilities to do things well") were significantly more likely to interpret this description in terms of *confidence* (as opposed to an equally applicable concept, such as *conceit*) if this concept had been primed in an earlier task. The findings of Higgins et al. represented a kind of assimilation effect, in that increasing the accessibility of a particular concept made it more likely that subsequently presented information would be interpreted in terms of that primed concept. However, subsequent studies have sometimes reported *contrast*, in which making a trait concept more accessible leads people to judge stimuli in ways *opposite* to the implications of the activated concept. According to one interpretation, people are likely to use the primed concept only if they are unaware that it has been activated as a result of the previous priming episode. When they *are* aware of this fact, however, they actively take steps to avoid using it. However, there is currently some debate over the precise mechanisms underlying such effects and the viability of this explanation in contrast to competing models.

Automatic versus controlled processes

Interest in automaticity dates back at least to William James, who noted that certain kinds of mental and/or behavior processes (e.g., ones that we have repeated over time) appear to occur relatively effortlessly and run to completion in the absence of conscious control. Although it is generally accepted that these sorts of *automatic* processes play a large role in guiding human thought and behavior, psychologists have disagreed as to the precise definition of automaticity. The most influential model proposed to date has been a conceptualization developed by Schneider and Shiffrin (1977), although more recently a number of alternative conceptualizations have been proposed in the literature.

Do people "spontaneously" form trait inferences about others?

One of the core assumptions of classic work in attribution theory was that dispositional attributions are more informative and useful than situational attributions. For example, global, dispositional attributions (e.g., "John is an angry person") allow one relatively greater confidence in predicting the sorts of behaviors that people will perform in future compared to situational attributions (e.g., "John was made angry by the rude patron"), which are, by definition, more circumscribed. Relatedly, evidence from the social domain suggests that people are quite willing to form dispositional attributions, even when the available information about the target is scanty, and even when participants know that the information available is not particularly credible. In light of this and other evidence showing the ubiquitous nature of dispositional attributions, a line of work by James Uleman and his colleagues has recently explored the notion that people *spontaneously* form trait inferences about other individuals, in the sense that such processes are unintentional and occur even when people are not trying to form impressions of these persons. There is, however, some disagreement among different researchers as to the boundary conditions of these effects.

Do personal attitudes moderate the automatic activation of stereotypes?

Another issue engendering considerable debate concerns the extent to which mere exposure to and/or thoughts about members of racial minorities (e.g., Blacks) is sufficient to automatically activate traits that are part of the negative group stereotype (e.g., hostile), and whether such effects are contingent upon whether the perceiver personally endorses the stereotype or not. Intuitively, one might infer that such automatic activation effects should arise only among prejudiced, not nonprejudiced, people. In a widely cited article, Devine (1989) recently proposed that these intuitions may, in fact, be incorrect. In particular, she argues that early patterns of socialization lead to relatively strong, well-formed associations between certain minority groups and stereotypically negative traits. These linkages are sufficiently strong that, years later (and even among people who do not personally endorse the stereotype), mere exposure to stimuli related to the category of Blacks may be sufficient to automatically activate the negative stereotype. According to Devine, personal attitudes play a role in controlled processes *after* initial activation of the stereotype, in which people may correct for its influence, provided that they are motivated and have the resources to do so. Although Devine presented data in support of this conceptualization, more recent work by Russell Fazio and his colleagues has challenged certain aspects of her model and its generalizability.

Suppression processes

The fact that certain types of processes can arise without conscious intent is not always adaptive. In particular, our efforts to complete successfully many different types of tasks are often hampered by the intrusion of unwanted thoughts that spring to mind precisely when they are most unwelcome. Cognitive psychologists have long been interested in the processes underlying thought suppression and the conditions under which these efforts might be relatively successful or unsuccessful. The familiar Stroop effect, for example, illustrates how people are unable to ignore certain features of a word (i.e., its meaning) that can interfere with their ability to attend to other features of the stimulus (i.e., the color in which the word is printed).

As social psychologists have increased their attention to the role of automatic processes in social contexts, this interest has naturally extended to the processes by which people attempt to suppress these automatic processes and why these efforts sometimes fail. For example, in many social contexts, we find it necessary to disregard or ignore information, either because we have been explicitly requested to do so and/or as a result of the kinds of personal goals or objectives we have at the moment. Such situations are likely to arise, for example, when judges direct the members of the jury to disregard some piece of inadmissible evidence. We have already noted some recent research that bears on these issues, such as people's attempts to avoid being biased by the information with which they have recently been primed, or the efforts that some people exert to avoid being biased by their own stereotypes.

What are the consequences of thought suppression?

Social/personality theorists have long been aware that efforts to suppress unwanted thoughts and emotions sometimes cause a *rebound effect*, in that the thoughts and emotions soon flood back into consciousness even more than if no attempts at suppression had been made in the first place. Although a number of different explanations (some rooted in psychodynamic theory) have been offered for such effects, a provocative line of research by Daniel Wegner and his colleagues has recently developed a model of thought suppression which provides an alternative account. According to Wegner, thought suppression is mediated by two complementary mechanisms: namely, a *monitoring* and an *operating* process. The monitoring process is relatively automatic and searches memory for the unwanted thought or concept. When an unwanted thought is detected, the operating process (which is relatively controlled) is activated and then seeks to replace the unwanted thought with some other, more desirable thought. An ironic outcome of this dual process system is that under conditions of limited cognitive capacity, the unwanted thought becomes hyper-accessible. This is because the operating process is rendered ineffective due to limited cognitive resources, but the monitoring process automatically continues to search for signs of the unwanted thought and brings the thought to consciousness when there is no operating process to prevent it surfacing.

Conclusion

Social cognition represents an enormous field of research and, apart from its obvious links to cognitive psychology, currently reflects an active integration with a number of intellectual disciplines, including political science, clinical psychology, and health psychology, to name but a few. The diversity of these areas further illustrates the point that social cognition is not so much a substantive area but a particular *approach* to studying human behavior and cognition. In presenting this brief overview, it has been our goal to show how this approach has been, and continues to be, a valuable way of gaining insight into the processes mediating social thought and behavior.

References and recommended reading

Allport, G. 1954: *The Nature of Prejudice*. Reading, Mass.: Addison-Wesley.

Devine, P. G. 1989: Stereotypes and prejudice: their automatic and controlled components. *Journal of Personality and Social Psychology*, 56, 5–18.

*Fiske, S. T. and Taylor, S. E. 1991: *Social Cognition*, 2nd edn. New York: McGraw-Hill.

Hastie, R., Ostrom, T. M., Ebbesen, E. G., Wyer, R. S., Hamilton, D. L. and Carlston, D. E. 1980: *Person Memory: The Cognitive Basis of Social Perception*. Hillsdale, NJ: Erlbaum.

Higgins, E. T., Rholes, W. S. and Jones, C. R. 1977: Category accessibility and impression formation. *Journal of Experimental Social Psychology*, 13, 141–54.

*Kahneman, D., Slovic, P. and Tversky, A. 1979: *Judgment under Uncertainty: Heuristics and Biases*. New York: Cambridge University Press.

Mackie, D. M. and Hamilton, D. L. (eds) 1993: *Affect, Cognition, and Stereotyping: Interactive Processes in Group Perception*. San Diego, Calif.: Academic Press.

Schneider, W. and Shiffrin, R. M. 1977: Controlled and automatic human information processing: I. Detection, search, and attention. *Psychological Review*, 84, 1–66.

Smith, E. R. 1996: What do connectionism and social psychology offer each other? *Journal of Personality and Social Psychology*, 5, 893–912.

Smith, E. R. and Zarate, M. A. 1992: Exemplar-based model of social judgment. *Psychological Review*, 99, 3–21.

Wyer, R. S. and Lambert, A. J. 1994: The role of trait constructs in person perception: a historical perspective. In P. Devine, T. Ostrom and D. Hamilton (eds), *Social Cognition: Impact on Social Psychology*, San Diego, Calif.: Academic Press.

*Wyer, R. S. and Srull, T. K. 1994: *Handbook of Social Cognition*, 2nd edn. Hillsdale, NJ: Erlbaum.

23

Unconscious intelligence

RHIANON ALLEN AND ARTHUR S. REBER

There is no dispute over the existence of functions and processes that operate outside consciousness. No one knows what his or her liver is doing, and we all shed a tear when Lassie comes home, even though we know we are watching a movie with a dog who responds to off-camera signals. Where matters become interesting (and contentious) is over such issues as whether unconscious processes are routinized and inflexible – in a word, stupid – or whether they can be seen as sophisticated, flexible, and adaptive – in a word, intelligent. In the cognitive sciences, until recently, the former perspective dominated for reasons which are rooted not in any objective data base, but in philosophical considerations. The long-standing presumption of a not-very-bright unconscious is, we suspect, the end product of a continuing adherence to Lockean and Cartesian epistemic traditions. While proponents of these two orientations didn't agree on much, they were of one mind with regard to the proposition that that which is mental is conscious. Contemporary psychologists and philosophers who feel comfortable with this position tend to view unconscious cognitive processes as low-level, crude, primitively connectionistic, and rather stupid, since they equate sophisticated control operations with the functions of consciousness.

However, there exists ample evidence that human phenomena we would like to label "smart" are not always accessible to, or have their origins in, conscious contents or procedures. These phenomena range from the most basic perceptual processes to the learning of complex systems and the production of creative ideas. We will begin with an exploration of the evidence for unconscious components of basic perceptual processes, then move on to examine like evidence in the operations of memory, emotion, motivation, learning, and creativity. Finally, we will suggest that basic principles of evolutionary biology entail the ontology of a sophisticated unconscious. Throughout, we will admittedly skirt the issue of how exactly one delineates unconscious from conscious processes. While this is a critical issue that we and others have addressed directly elsewhere, here we will raise appropriate definitions within the context of particular experiments rather than holding all studies to a single criterion (see Article 9, CONSCIOUSNESS).

Unconscious perception

This issue is best known as *subliminal perception*. The topic exploded on the scene a few decades ago when it was reported that messages like "Buy popcorn, drink Coke" could influence behavior despite exposure times so brief that the messages could never be consciously decoded. Actually, the initial report was a total fraud, having been constructed wholesale by a self-promoting advertising agent. It did, however, attract a great deal of attention. To complicate matters, early experimental tests were so poorly designed that one could not conclude that subliminal messages influence behavior or emotional state.

Nonetheless, more recent, carefully designed experiments have revealed that subtle, subliminal perception effects are probably real, although small. For example, Anthony Marcel used *two-alternative forced choice* (2AFC) and priming procedures to demonstrate that words can indeed enter long-term memory without conscious awareness; in priming procedures, participants are exposed to stimuli which predispose them to respond in particular ways later on (see Article 27, BEHAVIORAL EXPERIMENTATION). Marcel first flashed words like *nurse* subliminally, Then he showed participants two words, one of which had been presented subliminally and one of which was new. As part of this 2AFC procedure, participants were asked to identify which word was seen previously. Participants were unable to pick out the presented words at rates greater than chance. Thus, according to the 2AFC procedure, there appeared to be no significant conscious awareness of the words.

Marcel then showed the participants words like *doctor* and *chair* under partly occluded conditions and asked them to read them. Participants were reliably better at reading words like *doctor* than unrelated words, because the former had been primed subliminally by the previous presentation of *nurse*. There are actually two important points here: first, that unconscious perception does occur, and second, that subliminally perceived words can be processed semantically and hence *intelligently*.

Since Marcel's studies were first published in the 1970s, scores of experiments have supported the basic finding that people can indeed pick up information without conscious awareness that they are doing so (as evidenced by priming). For example, Eyal Reingold, using a variation of Larry Jacoby's dissociation procedure with the word-stem-completion priming technique, showed participants lists of words either subliminally (i.e., for only 5 msec) or supraliminally (500 msec). Participants were then presented with a three-letter stem such as MOT ____ and asked to complete the stem with the first word that comes to mind.

It is well known that under supraliminal exposure learning, people tend to use words that they have just seen. For example, if the original list contained *motor*, then this is what participants tend to give. In order to ensure that participants did not use words actually perceived consciously, Reingold asked them to make sure that they never used any of the words that they had seen. The results were quite startling. Participants avoided words that had been presented supraliminally, but consistently completed stems with words from the subliminal list.

Thus, while it is now clear that subliminally presented messages do not influence macro behavior, there is evidence that they might influence the contents of memory and consciousness in subtle ways. They appear to shape states of perceptual and productive readiness, and make one more, or possibly less, likely to respond to subsequent events in a particular way. The effect of subliminal presentation on perceptual fluency, or the enhanced ease and accuracy of conscious perceptual identification, seem especially adaptive, since rapidity and fluency of conscious identification of environmental objects is a prerequisite for survival.

A related phenomenon of unconscious perception in which stimuli are registered without awareness can be seen in the neurological condition of blindsight. Many human and nonhuman primates with extensive damage to the visual striate cortex nonetheless can and do respond to objects and movement in the blind field, although human subjects will report that nothing is seen. Indeed, some human patients can achieve 90–100 percent accuracy in a 2AFC procedure when encouraged to guess or rely on intuition. Lawrence Weiskrantz (1995) has demonstrated that in one patient correct discriminations of object trajectories could in fact be made in complete absence of even vague awareness of the stimulus event. The phenomenon of blindsight suggests that conscious awareness of visual events depends on later-evolving cortical structures, but that some ability to adapt to environmental events is mediated by subcortical connections independent of consciousness. The existence of subliminal perception and blindsight indicate that consciousness of visual events is not necessary for some basic (albeit subtle) adaptations to those events.

Unconscious memory

The question here is whether the vaguely oxymoronic notion of an implicit memory is possible. That is, do people establish implicit memory representations of events that they are not aware of having formed and are unable to retrieve on command? While the evidence of subliminal perception just discussed certainly hints at an implicit or unconscious memory system, there are even earlier indications. The first reports go back to the work of the Russian neurologist Sergei Korsakoff, who studied alcoholic patients whose many years of nutritional neglect had left them with the amnesic syndrome that now bears his name. Korsakoff noted that several patients seemed to react to situations in ways that clearly showed knowledge about events despite avowal of no memory for these events. For example, one patient participated in a study that used electrical shock. Although he claimed never to have been in the study and indeed never to have met the neurologist, he was suspicious of Korsakoff and refused to cooperate with him.

Similar clinical anecdotes from Edouard Claparède and other pioneers of neurocognition were unfortunately not systematically followed up on until Daniel Schacter began his experimental work on unconscious or implicit memory in the 1980s. Although there are dozens of procedures in use for studying implicit memory, the most commonly used methods are similar to those used in studies of subliminal perception; the difference is that all stimuli are presented supraliminally and hence must enter awareness at least temporarily. In a typical implicit memory word-stem study, participants are shown a list of words and asked to memorize them. After a short break, they are then asked to recall as many of the words as possible. Later they are given a stem-completion task. On this stem-completion task, they typically provide many words from the original list that they were unable to recall consciously.

Schacter and others working with amnesic patients have demonstrated this priming phenomenon to an exaggerated degree. If one asks amnesic patients to memorize words, then after an hour requests that they recall those words, no words will be consciously recalled. In fact, the patients are unlikely to remember learning the list or even the learning session itself. However, they complete the stem-completion task with words from the learning list at rates equal to those of normal, nonamnesic subjects.

In an interesting variation dubbed the dissociation procedure, Jacoby and his colleagues (1992) instructed participants not to use any words from the memorized list on the stem-completion task. As in Reingold's work, participants tended to use words from the original list, suggesting the existence of unconscious memorial contents.

In fact, the evidence is rather clear that many more events and facts exist in memory than we can recall consciously at any one time. Evidence accumulates almost every day, ranging from the relatively straightforward experience of hypermnesia described by Matthew Erdelyi, whereby information inaccessible to consciousness one minute can surface with minimal elicitation in the next, to the more esoteric phenomena of unconscious memories acquired in hypnotic and anesthetized states. Given the costs of consciousness, there should be an evolutionary adaptive pressure behind this ability to register vast amounts of information without permanently clogging consciousness. There is no need for Korsakoff's patient to remember the shock apparatus as long as he refuses to reenter the room.

Accessibility of acquired skills

Unlike subliminally perceived information, implicit memories generally originate in explicit or conscious awareness. They share this feature with most types of routine skills, such as bicycling or typing. Motor skills, or at least those acquired after infancy, are generally under at least partial conscious control during the early acquisition phase. In fact, acquisition of a complex skill can make enormous demands on consciousness. As the skill becomes practiced, however, attention and conscious control gradually absent themselves. Eventually, most highly practiced and automatized skills – even those which are highly complex and involve continual adaptation to environmental cues and feedback, such as driving a car – become resistant to consciousness. Learning curves for skills such as word processing show that the time taken to execute an action comes to equal the time for motor execution alone, thus demonstrating that *thinking time* disappears. Such automatization of skill represents a successful adaptation that frees consciousness from the task of constant monitoring of action. In fact, we see this in everyday occurrences of what Ellen Langer has called "*mindlessness*" such as engaging in conversations and responding to requests without effortful control or attentive monitoring.

There are many examples of dissociation between skilled action and conscious awareness of those actions analogous to the dissociation between explicit and implicit knowledge seen in amnesics. Jean Piaget (1976) described many such problem spaces (i.e., using a catapult to hit targets) in which forced awareness of physical principles can actually disrupt success, thereby causing a failure of adaptation. Annette Karmiloff-Smith (1992) has also argued that apparent regressions in such skills as the use of articles and possessives in speech are a result of dawning explicit representation. A similar dissociation between cognizance and action was illustrated by Dennis Carmody and colleagues when they showed that experienced radiologists do not follow their own explicit advice for scanning X-rays. Finally, Elizabeth Glisky and her colleagues have documented individual amnesic patients who have been able to learn new skills such as how to operate a computer without the assistance of explicit memory. These examples are particularly intriguing because they suggest that not only do skills become freed from conscious control, but that they can evolve independently of, and be largely nonisomorphic with, conscious beliefs about how the skill is performed. This problem of unconscious representation is a familiar one to those who teach diagnostics and to those who construct expert artificial intelligence systems, since traditionally they have relied upon the introspections of expert practitioners to guide them. Alas, when the experts lack awareness of their own operations and processes, problems emerge.

Unconscious emotion and motivation

Before we abandon consideration of low-level complexity systems and progress to more complex unconscious cognitive phenomena, it is apt to briefly discuss one of the original approaches to unconscious phenomena – that associated with the psychoanalytic perspective on unconscious emotions and motivations (see Article 11, EMOTIONS). Curiously, this facet of the unconscious is simultaneously the most widely believed by both laypersons and professionals, yet the most weakly supported by objective evidence.

The study of the unconscious began in earnest with the work of Sigmund Freud and the group of psychoanalysts who gathered around him in Vienna. A staple of their approach was that our deep, inaccessible needs, wishes, and fears have an impact on our behavior through the workings of a complex and sophisticated unconscious system that functions to protect our vulnerable conscious selves. Fresh debates in the 1994 and 1995 issues of the *American Psychologist* centered on the adaptive nature of, and the cognitive substrates of, these unconscious motivational forces. Even in these modern debates, however, what we like and dislike and much of what we say and do are seen as dictated by forces that lie outside our awareness.

Despite the near universality of the acceptance of this hypothesis in Western nations, there are virtually no reliable data collected under controlled laboratory conditions to support it. One problem is that it is extraordinarily difficult to study the influence of unconscious processes on emotional and motivational facets of behavior. One primary complication is that it is difficult to define laboratory situations and procedures that can rule out conscious emotional determinants with the same certainty that the 2AFC and learning curve techniques rule out conscious content and control.

However, there are carefully controlled studies that suggest that emotions or motivations can be influenced by unconscious factors. Unfortunately, there are not many – and they do not approach the deep emotional levels that concern psychoanalysts. First, there exists a variety of studies in which there appears to be at least a partial dissociation between verbal/questionnaire reports of emotional-attitudinal states and physiological, behavioral, or projective assessments of states. However, it is dangerous to equate active verbal reports with all of consciousness – a problem that also plagued early experiments on subliminal perception, and one still seen in the arguments of many philosophers and linguists.

More objective evidence comes from the work of William Kunst-Wilson and Robert Zajonc (1980), using a subliminal perception technique. Participants were briefly exposed to geometric shapes, then tested for awareness using 2AFC, in which they were shown pairs of one novel and one old geometric shape. Participants were at chance levels in indicating which of the pair they had seen earlier, suggesting no conscious awareness of the original stimuli. They were then shown the pairs again, but this time were asked which of the two they *preferred*. Now participants reliably selected the ones that had been shown earlier. This *mere exposure effect* suggests that we come to prefer the familiar, even while we are unaware of what the familiar is.

While this might be an analog of the effects of subliminal exposure on priming and stem completion, one should be rather cautious in ruling out simple demand characteristics. Unlike the subliminal priming and stem-completion studies, in which a response must be actively generated, the Kunst-Wilson and Zajonc request for a preference can be seen by the experimental participant as an opportunity to relax the criterion for responding on the basis of a hazy but explicit knowledge base, rather than as a technique that taps totally unconscious knowledge.

However, work in the laboratories of Keith Holyoak, Marcia Johnson, Louis Manza, and Arthur Reber suggests that implicit preferences for stimuli can be influenced by supraliminal exposure, and that the mere exposure effect cannot be reduced to a slight modification of the response criterion. For example, Johnson and her colleagues demonstrated that amnesics developed preferences for melodies based on Korean musical patterns that they were unaware of having heard in a previous exposure session. Thus, people come to prefer stimuli that conform to structures they have learnt, even though they may not be able to explain or even recall exposure to those structured stimulus displays.

Unconscious learning

Here the primary research issue is whether people can learn without knowing *what* it is they have learned or *how* they learn, even though they may be aware that they are learning. For example, we all speak and understand English and know that we do. In order to use English, we must know the underlying phonological, morphemic, semantic, syntactic, and pragmatic rules of the language. Yet few of us can articulate those rules clearly or know how we learned them. Similar arguments can be applied to systems of social conduct. We have all learned to follow an exceedingly complex set of rules and principles for socially appropriate behavior – again without being fully aware of the process or products of learning.

Although we are surrounded by examples of complex adaptive learning in the apparent absence of consciousness, we still need to ask whether there is hard scientific evidence to support this notion of implicit learning. Under controlled laboratory conditions, can one demonstrate that humans learn rule systems without total cognizance?

The signature experiment to demonstrate implicit or unconscious learning is known as the *artificial grammar (AG) learning* experiment. An artificial grammar is a highly simplified but nonetheless complex analog of a natural language, wherein elements (typically letters or other symbols) are arranged in displays according to a rule system. In the classical AG learning experiment, participants are told that they are in a memory study, are given what appear to be nonsense strings of letters, and are asked to memorize them. They are subsequently informed that the strings were generated by (unspecified) rules and are asked to judge whether or not new stimuli can be *grammatical*. To paint with a wide brush, learners are able to make grammaticality judgments, show only limited evidence of accurate conscious problem solving, retain acquired implicit knowledge for long periods of time, and generally outperform participants encouraged to engage in conscious problem solving. The mass of evidence is in favor of concluding that unconscious processes are involved in the learning of such structured contingencies.

Evidence of unconscious learning has also been obtained through serial reaction time (SRT) studies. In the typical SRT experiment, the learner sits at a computer screen, across the bottom of which four places are illuminated. An asterisk suddenly appears above one of the places. The learner's task is to press the computer key that corresponds to that position and to do so as quickly as possible. When the button is pressed, the asterisk blinks out and reappears at a different location, and so on for many trials. Unbeknownst to the learner, the order of locations is either determined by a complex set of rules or is a repeating pattern. Careful analysis of the speed with which learners respond shows that they become faster, suggesting that they are learning to exploit the underlying rules to anticipate location of subsequent stimuli. This interpretation is supported by evidence that learners slow down dramatically if the computer shifts to a random sequence or one that follows a different pattern. That some of this rule knowledge is unconscious is suggested by learners' failure to generate sequences that conform to the rules during subsequent testing.

Further evidence that complex learning can occur without awareness of the underlying rules comes from work with persons with amnesic syndrome. In a series of studies by Barbara Knowlton and Larry Squire, it has been shown that amnesic patients in the classic AG learning experiment make grammaticality judgments in a manner indistinguishable from normals, despite having virtually no explicit memory of having even been in the learning phase of the experiment. Only if asked to try to carry out the task by referring back to the previously memorized strings do amnesics drop below the normals in performance.

Debates in implicit learning have centered on the degree to which such learning is unconscious, the conditions that elicit or inhibit unconscious processes, the nature of the representations generated by unconscious processing, and specific criteria for identifying unconscious learning. Despite these controversies, most cognitive scientists would agree that apprehension of structure in the environment can take place without development of direct conscious knowledge of the structure. Even many strong advocates of explicit expectancies in learning have acknowledged that the acquired contingencies that undergird action can be implicit and accurate under some conditions.

Exploitation of the environment

Most of the unconscious processes described above are limited to rather passive reception of information. In that sense, we have failed to provide evidence that humans are uniquely adaptive in their unconscious processing, except perhaps in degree. After all, all animal life forms learn contingencies and covariations that guide their reactions to the world, even though these may be somewhat limited in comparison to humans. Surely what makes a species more intelligent is an ability to exploit and manipulate the environment. Here we will briefly consider three facets of adaptive exploitation of the environment – anticipation, cognitive problem solving, and insight – and argue that unconscious functions play a role in each.

An intelligent being thrives by anticipating and exploiting the actions and structure of the environment, as a predator occasionally heads off prey or lies in wait rather than merely chases. Anticipation of location, in the probable absence of conscious apprehension, has been shown in human probability learning studies that require participants to predict the location of a light.

Humans have also been shown to utilize implicitly gained knowledge of artificial grammars to generate anagram solutions. This problem solving is observed in conjunction with a partial inability to explain the basis of the solution. In addition, Herman Spitz has also argued, primarily on the basis of evidence from savants, that people often create spontaneous organizations and solutions to problems without full cognizance of the techniques they develop.

Finally, anecdotal evidence of unconscious problem solving exists in the form of insight and incubation processes that take place on the subliminal *fringe* of consciousness. For example, Henri Poincaré insisted that he gained a fundamental mathematical insight without conscious processing, and August Kekulé is reputed to have solved the problem of the structure of benzene upon waking from a dream. If one accepts Morton Prince's notion that the processes underlying dreams are subconscious but can provide the fuel for the conscious creative process, then one can view the font of some creative insights as residing in the periphery of consciousness. While these insights do not always lead to creative endeavors, they can provide the constituents and epistemic foundations for further conscious elaboration and exploitation.

While the evidence for the role of the unconscious here is somewhat weaker and more anecdotal than in the areas of learning and perception, it is worth taking a brief look at the perspective put forward by Michael Polanyi. Polanyi argued that at its core, the process of scientific creativity is the elevation of tacit or personal knowledge from its initial unconscious representational form to an articulable, conscious form in which it can be communicated to others. In Polanyi's epistemology, the implicit system is always *ahead*, as it were, of the explicit system. The unconscious acquisition of knowledge is the natural and normal mode – it is the articulation and expression of it that is cognitively demanding and unnatural.

The necessity of unconscious processing

Animal species evolved consciousness over millennia, and verbalizable cognizance with top-down conscious control must be a late arrival on the evolutionary scene. That explicit functions are relatively recent developments is suggested by their fragility in the face of injury and disease in adult humans, as reflected in the studies with amnesics. Other species must be aware of their environments in some way, but it is unlikely that this awareness takes the form that we so closely ally with the varieties of consciousness typically displayed by undergraduate students in cognitive psychology experiments. Nor is it likely that their awareness takes the form of cognizance or consciousness of one's internal processes. And yet, as Donald Griffin (1981) and others have emphasized, those species are exquisitely adapted to their environments by virtue of an implicit knowledge acquisition system that yields accurate representations of experience.

While the brain structures of humans and closely related species might afford us unique adaptations, species as different (indeed, brainless) as the sea slug *Aplysia californica* are capable of learning simple and differential associations which allow them to adapt to new conditions. While *Aplysia* might be *aware* of the association between the pressure and shock applied in a laboratory experiment, there is no reason to assume that it is conscious of this association in the same sense that a student might be conscious that a particular word was presented in a learning list. Indeed, there is some debate on the issue of whether humans are conscious of the simple contingency inflicted in eye-blink conditioning. Work on hippocampal amnesics also suggests that some facets of cognitive awareness are mediated by specific brain structures such as the hippocampus, yet destruction of those centers leaves implicit learning capacity intact.

Further, adult consciousness is unlikely to be present in human infants. Human infants lack some neurological and cognitive functions that exist in adults. If consciousness exists in the human neonate, it is radically different from that which we hold as the standard in adult studies. In fact, Jean Piaget and Jerome Bruner have argued that infant adaptations are qualitatively different from those found in older children and in adults and are based on sensorimotor adaptations fundamentally different from symbolically mediated thought (see Article 6, COGNITIVE AND LINGUISTIC DEVELOPMENT). Further, Piaget and several modern researchers of metacognition and representation have argued that consciousness cannot be directed towards one's own cognitive processes until relatively late in development and mastery.

Finally, we raise the point that consciousness probably has intrinsic limits. As William James argued over a century ago, we are aware of but a small slice of our existence, memories, and activity at any one time. Consciousness is a limited-channel processor, apprehending but a fragment at a time. Recent formal models which are less tied to the constraints of linear processing (e.g., connectionist) map a number of phenomena more accurately than older processing models which were more closely allied to a model of linear or quasi-linear consciousness. In fact, it is difficult to see how a connectionist model could operate if the organism must be aware of all processes at all times.

In summary, there is ample evidence that unconscious processes are involved in adaptation and intelligent functioning. Indeed, it might well be the case that humans could not adapt to the demands of life quite as well as they do without relegating some tasks and knowledge to the unconscious.

References and recommended reading

Berry, D. C. and Dienes, Z. 1993: *Implicit Learning: Theoretical and Empirical Issues*. Hillsdale, NJ: Erlbaum.

Griffin, D. R. 1981: *The Question of Animal Awareness: Evolutionary Continuity of Mental Experience*. New York: Rockefeller University Press.

Jacoby, L. L., Toth, J. P., Lindsay, D. S. and Debner, J. 1992: Lectures for a lay person: methods for revealing unconscious influences. In R. F. Bornstein and T. Pittman (eds), *Perception without Awareness*, New York: Guilford Press, 81–120.

Karmiloff-Smith, A. 1992: *Beyond Modularity: A Developmental Perspective on Cognitive Science*. Cambridge, Mass.: MIT Press.

*Kihlstrom, J. F. 1990: The psychological unconscious. In L. A. Pervin (ed.), *Handbook of Personality: Theory and Research*, New York: Guilford Press, 445–64.

Kunst-Wilson, W. R. and Zajonc, R. B. 1980: Affective discrimination of stimuli that cannot be recognized. *Science*, 207, 557–8.

Marcel, A. J. 1983: Conscious and unconscious perception: experiments on visual masking and word recognition. *Cognitive Psychology*, 15, 197–237.

Piaget, J. 1976: *The Grasp of Consciousness: Action and Concept in the Young Child*. Cambridge, Mass.: Harvard University Press (1st pub. 1974).

*Reber, A. S. 1993: *Implicit Learning and Tacit Knowledge: An Essay on the Cognitive Unconscious*. Oxford: Oxford University Press.

Reber, A. S., Allen, R. and Reber, P. J. forthcoming: Implicit learning. In R. J. Sternberg (ed.), *The Concept of Cognition*, Cambridge, Mass.: MIT Press.

Reingold, E. M. and Merikle, P. M. 1990: On the inter-relatedness of theory and measurement in the study of unconscious processes. *Mind and Language*, 5, 9–28.

*Schacter, D. 1987: Implicit memory: history and current status. *Journal of Experimental Psychology: Learning, Memory, and Cognition*, 13, 501–18.

*Shanks, D. R. and St John, M. F. 1994: Characteristics of dissociable human learning systems. *Behavioral and Brain Sciences*, 17, 367–447.

*Squire, L. R. 1994: Declarative and nondeclarative memory: multiple brain systems supporting learning and memory. In D. Schacter and E. Tulving (eds), *Memory Systems 1994*, Cambridge, Mass.: MIT Press, 203–31.

Weiskrantz, L. 1995: Blindsight: not an island unto itself. *Current Directions in Psychological Science*, 5, 146–51.

24

Understanding texts

ART GRAESSER AND PAM TIPPING

Introduction

Adults spend most of their conscious life speaking, comprehending, writing, and reading discourse. It is entirely appropriate for cognitive science to investigate discourse especially as transmitted texts or printed media, such as books, newspapers, magazines, and computers. However, there is another reason why text understanding has been one of the prototypical areas of study in cognitive science: Interdisciplinary work is absolutely essential. As cognitive scientists have unraveled the puzzles of text comprehension, they have embraced the insights and methodologies from several disciplines.

One of the central challenges has been to understand how meaning is constructed during text understanding. When a short story is comprehended, the reader constructs a fictitious micro world of characters, objects, spatial layouts, actions, events, emotions, conflicts, and so on. What sort of information is normally put in a micro world? How is the micro world represented? What cognitive mechanisms participate in the process of constructing the micro world? Cognitive scientists have identified the mental representations and processes that appear to participate in these constructive mechanisms. The computational feasibility of these mechanisms has sometimes been tested by building a computer model that constructs the multilayer mental representations. Models of comprehension have been tested by collecting data from humans, such as the tracking of eye movements during reading and the measurement of text recall after comprehension is completed.

Multiple levels of text representation

During the 1970s and 1980s, Teun van Dijk (a text linguist from the University of Amsterdam) and Walter Kintsch (a cognitive psychologist from the University of Colorado) identified most of the levels of representation that are constructed during text comprehension (van Dijk and Kintsch, 1983). Suppose, for illustration, that a college student read the following excerpt from a novel: "When June woke up on Sunday, she discovered she was pregnant. He didn't say a word." What levels of representation would be constructed when this excerpt is comprehended? According to van Dijk and Kintsch, there would be at least five levels: the surface code, the textbase, the situation model, the pragmatic communicative context, and the text genre.

Surface code

The surface code preserves the exact wording and syntax of sentences. The first sentence of the example could have been worded somewhat differently, yet preserved approximately the same meaning: for example, "On Sunday June woke up and discovered that she was pregnant." The surface code is different, but the meaning is approximately the same. The surface code does not hang around the cognitive system very long. Memory experiments have shown that readers normally remember the surface code of only the most recent clause, unless the wording and syntax have some important repercussions for meaning (van Dijk and Kintsch, 1983). Readers would have a great deal of difficulty remembering whether the text stated "June woke up on Sunday" or "On Sunday June woke up" when tested 20 seconds after comprehending the text.

Textbase

The textbase contains explicit propositions in the text in a stripped-down form that preserves the meaning, but not the surface code. For example, the first sentence would have the following explicit propositions:

PROP 1: wake-up (June, TIME = Sunday)
PROP 2: discover (June, PROP 3)
PROP 3: pregnant (June)
PROP 4: when (PROP 1, PROP 2)

A proposition refers to a state, event, or action that may have a truth value with respect to the micro world. Each proposition contains a *predicate* (e.g., main verb, adjective, connective) and one or more *arguments* (e.g., nouns, embedded propositions). The arguments are placed within the parentheses, whereas the predicates are outside the parentheses. In most propositional representations, each argument has a functional role, such as agent, patient, object, time, or location. The textbase also includes a small number of inferences that are needed to tie together the explicit propositions and thereby establish text coherence. For example, there would be a bridging inference that designates that the *she* in the second clause refers to *June* in the first clause. It takes additional processing time to construct these bridging inferences. When researchers have measured the amount of time that it takes to read words and sentences in a text, they have found that an additional 100–300 milliseconds is needed to construct a bridging inference (van Dijk and Kintsch, 1983).

Situation model

The situation model is the content or micro world that the text is about. In stories, the situation model includes the characters, objects, spatial setting, actions, events, knowledge states of characters, emotional reactions of characters, and so on. In an expository text about the human heart the situation model would probably specify the causal mechanisms that explain the functioning of the heart and the spatial layout of the components of the heart. The situation model is constructed inferentially on the basis of the explicit textbase, background world knowledge, and the other levels of representation. An understanding of the second clause in the example requires the construction of a situation model. The textbase for "He didn't say a word" is not directly related to the four propositions associated with the first sentence. In order to connect sentence 2 with sentence 1, the reader must construct a situation model that fills in plausible information that connects the two sentences. For example, a reader might activate a marriage script. The fact that the husband does not talk suggests that he is either unhappy or extremely thrilled about the pregnancy. The reader would need to infer that the wife told the husband about the pregnancy, or that the husband perceived it without being told. Readers go well beyond the explicit information when they construct situation models. The claim that readers construct these situation models is supported by memory data, think-aloud protocols, reading times, and other types of behavioral data. When readers later recall a text, they sometimes have *memory intrusions* (i.e., content that was not explicitly mentioned) which match the content of the situation model (van Dijk and Kintsch, 1983). When readers are asked to *think aloud* while comprehending a text, the content of the situation model is exposed (Chi et al., 1994; Trabasso and Magliano, 1996). Reading times for sentences increase when readers need to construct situation inferences in an effort to fill coherence gaps (Graesser et al., 1994; Zwaan et al., 1995).

Pragmatic communicative context

There is an implicit dialogue between the author and the reader when text is comprehended. Pragmatic principles facilitate the communicative exchange between them. Many of these pragmatic principles were discovered by those who analyzed oral conversation. Philosopher Paul Grice (1975) pointed out that speakers should make every attempt to be truthful, relevant, clear, and succinct when composing speech acts in conversation. Herbert Clark (1993), a cognitive psychologist at Stanford University, has analyzed how speech participants monitor the *mutual knowledge* that they assume the other shares. When a listener fails to understand the meaning of a term of the referent of a noun that is expressed by a speaker, the listener gives negative *backchannel feedback* (i.e., shakes head, frowns, has puzzled facial expression) or asks a clarification question (e.g., "What does that mean?" or "Who are you talking about?").

Many of the pragmatic principles from oral discourse also apply to written text. Thus, a writer should keep track of content in the textbase and the situation model that the reader already knows. The text should have surface cues that distinguish between *given* (known) information and *new* information. A sentence should be relevant to the previous discourse context, unless there are explicit cues that flag the initiation of a new topic (e.g., "On a separate matter, . . ."). This principle of relevance is critical for understanding the second sentence in the example. The reader assumes that "He didn't say a word" is relevant to the previous sentence, because there was the pronoun *he*, and there were no markers flagging a new episode. Given that the second sentence was relevant to the first, the reader had to figure out a plausible situation that would explain the connection. Without the principle of relevance, the reader might not bother constructing a situation model that builds a conceptual bridge between the two sentences.

Text genre

Texts are frequently classified into four major categories: description, narration, exposition, and persuasion. Each category can of course be subdivided into subcategories that capture more fine-grained distinctions. The structural components, features, and pragmatic ground rules are quite different among the various text genres. For example, newspaper articles are different from literary short stories. The facts and events expressed in a newspaper article are supposed to be true propositions about the world, whereas this constraint does not apply to fiction. A newspaper article conveys the main point of the text near the beginning, whereas the main theme of a short story may not emerge until the end. It should be noted that some texts are amalgamations of two or more text genres, as in the case of historical fiction.

Most discourse researchers agree that the above five levels of representation are constructed during comprehension. However, there are disagreements about the time course of constructing these levels of code and the interactions among the various levels. For example, some researchers believe that the parsing of sentence syntax is completed immediately by a separate syntax module that is very quick and detached from other discourse levels; the other discourse levels may subsequently override the initial product of the syntax module. However, other researchers adopt an *interactive* model rather than a *modular* model (see Article 49, MODULARITY). According to an interactive model, discourse context can exert its influence very early in the parsing process. The separation and interaction among discourse levels is likely to be a pervasive concern of cognitive scientists for many decades.

The construction of knowledge-based inferences during comprehension

What inferences do readers normally construct during the process of comprehending text? This question has stimulated lively debates in cognitive science. There is an extreme position, which we will call "the promiscuous inference generation hypothesis." According to this hypothesis, the reader would encode the following classes of inferences while reading a novel: the goals and plans that motivate characters' actions, character traits, characters' knowledge, character emotions, causes of events, consequences of events, properties of objects, spatial settings and layouts among entities, the global theme or point of the text, the referents of nouns and pronouns, the attitudes of the writer, and the appropriate emotional reactions of the reader. In essence, the mind constructs a high-resolution mental videotape of the situation model, along with details about the mental states of characters and the communicative exchange between the writer and reader. The reader thus ends up generating thousands of inferences for short text segments.

Cognitive scientists do not believe that the promiscuous inference generation hypothesis is plausible. There would be a serious computational explosion problem if a researcher attempted to develop a computer model that simulated this approach to inference generation. In essence, the volume of inferences would be far too large to manage the computation in real time. In the 1970s and early 1980s, Roger Schank (a researcher in computational linguistics and artificial intelligence at Yale University) and his research associates built a series of computer models that simulated inference generation during the comprehension of natural language (Schank and Reisbeck, 1981). Some of the texts were organized by highly conventional *scripts*, such as narrative about eating at a restaurant, a story about infidelity, or a newspaper article about a car wreck. Schank and his associates proposed that inference generation is accomplished by accessing and using information that is stored in natural packages of world knowledge, such as scripts, stereotypes, roles, spatial structures, and previous autobiographical experiences. Analyses of memory intrusions, reading times, and other forms of behavioral data have indeed confirmed that many inferences in a situation model are inherited from the natural packages of generic knowledge (Graesser et al., 1994). In order to minimize the inference explosion problem, Schank proposed some specific cognitive heuristics that reduced the space of inferences to a manageable set. For example, one cognitive heuristic is to focus on inferences that are interesting, whereas another heuristic is to focus on inferences that explain why anomalous events occur. However, most of these heuristics were not systematically tested on human comprehenders.

A *minimal hypothesis* was proposed by Gail McKoon and Roger Ratcliff (1992) as a radical alternative to the promiscuous inference generation hypothesis and the computational models proposed by Schank. McKoon and Ratcliff conducted psychological experiments that tested the minimalist hypothesis at Northwestern University in the late 1980s and early 1990s. According to the minimalist hypothesis, comparatively few inferences are generated on-line when text is read at a normal pace, about 150–400 words per minute. The only inferences that are encoded are those that are needed to establish local text coherence (such as linking a pronoun to an argument in the previous clause) and those that are readily available in working memory. Other classes of inference are merely elaborations of the text and are not consistently generated, although they may be generated if the reader has special goals that are tuned to particular inference classes. Unfortunately, the minimalist hypothesis was tested only on experimenter-generated *textoids* that have minimal global coherence. The hypothesis was never tested under conditions in which readers read naturalistic, coherent text at a comfortable pace.

The most plausible model of inference generation lies somewhere between the extremes of the promiscuous inference generation hypothesis and the minimalist hypothesis. In the 1990s, a constructionist theory was proposed by Art Graesser at the University of Memphis, Murray Singer at the University of Manitoba, and Tom Trabasso at the University of Chicago (Graesser et al., 1994). The theory developed by these three discourse psychologists appears to be compatible with most of the psychological research that has been collected on textoids and on coherent naturalistic texts. According to the constructionist theory, four assumptions determine the classes of inference that are generated on-line. The first two assumptions are not particularly interesting because they would be adopted by virtually any theory. The last two are not uniformly shared by other theoretical positions, so they are more interesting.

(1) *Reader goals* The reader generates those inferences that are relevant to his or her comprehension goals. For example, if the reader has the goal of tracking the personality of a particular character, then he or she generates inferences about the traits of the character.

(2) *Convergence and constraint satisfaction* Readers encode inferences that are activated by several information sources (e.g., the explicit text, activated scripts, text genre, stereotypes) and that satisfy the constraints imposed by the various information sources.

(3) *Explanations* Readers generate inferences that explain why actions and events occur in the text and why the author explicitly mentions information in the text. Explanation-based inferences include the causal antecedents of actions and events (e.g., the husband didn't say a word in the example because he was upset about the

pregnancy) and superordinate goals that motivate character actions (e.g., the husband wanted to communicate his anger or wanted to prevent his wife from being hurt by harmful words). Explanation-based inferences are the most prevalent verbalizations that emerge in *think aloud* protocols collected while readers read stories (Trabasso and Magliano, 1996) and while good comprehenders study expository texts (Chi et al., 1994). The content of think aloud protocols reflects the content of consciousness during normal comprehension (see Article 33, PROTOCOL ANALYSIS). Explanation-based inferences are also highly activated in tasks that tap unconscious processes during on-line comprehension. For example, in word-naming experiments, test words are periodically presented to readers while they read texts, and the readers are instructed to name the test words as quickly as possible whenever they receive them. The test words should be named very quickly when they match an inference that is generated consciously or unconsciously. Studies have shown that a test word is named very quickly if it matches an explanation-based inference that is associated with the previous sentence in the text, whereas there is no facilitation in naming time for many other classes of potential inferences (Graesser et al., 1994).

(4) *Local and global coherence* Readers encode inferences that establish text coherence at both local and global levels. Local coherence is achieved if the incoming sentence can be related to information in the previous sentence or in working memory. Global coherence is achieved if the incoming sentence can be related to a global theme of the text or to knowledge structures that do not reside in working memory (e.g., information three pages earlier in a novel).

It should be noted that most classes of inference are not generated on-line according to the constructionist theory. For example, readers do not automatically encode properties of objects, spatial relationships among entities, and sub-plans of character actions (Graesser et al., 1994). These details are not encoded unless they help to explain why actions and events occur in the text. Similarly, readers do not normally forecast multiple, hypothetical situation models in working memory. Most causal consequences that are forecasted end up being wrong, so readers would be spinning their wheels to no avail. The content of think aloud protocols during comprehension includes many more explanations of explicit statements than forecasts about the future and associative elaborations (Trabasso and Magliano, 1996). Word-naming latencies (and other measures that tap on-line activation of information) show facilitation when they match explanation-based inferences, but not when they match inferences that refer to predictions about the future and to associative elaborations (Graesser et al., 1994). An important objective for text researchers is to identify the particular classes of inference that are generated on-line and the conditions under which they are generated.

Computational models of text comprehension

Cognitive scientists have developed some sophisticated quantitative and computational models of text comprehension during the last decade. These models specify the mental representations, processes, and interactive mechanisms in sufficient detail to simulate complex patterns of data in psychological experiments. The most fine-grained models simulate the creation, activation, and inhibition of particular information units as text is comprehended, word by word or clause by clause. Two cognitive models have dominated most of the efforts in simulating text-comprehension data. One of these models is the CAPS/READER model, which was developed by Marcel Just and Patricia Carpenter (1992), two cognitive psychologists at Carnegie–Mellon University. The CAPS/READER model adopts a production system computational architecture for creating, updating, and removing information units from working memory and long-term memory (see Article 42, PRODUCTION SYSTEMS). The other model is Walter Kintsch's construction–integration model (1988), which adopts a connectionist architecture (see Article 38, CONNECTIONISM, ARTIFICIAL LIFE, AND DYNAMICAL SYSTEMS). Actually, when these two models are dissected more carefully, both are hybrids of the production system and connectionist architectures. These complex models have had some impressive success in simulating complex patterns of results in experiments that collect eye-tracking data, text recall, reading times for text segments, and many other tasks.

References and recommended reading

*Britton, B. K. and Graesser, A. C. (eds) 1996: *Models of Understanding Text.* Mahwah, NJ: Erlbaum.

Chi, M. T. H., de Leeuw, N., Chiu, M. and La Vancher, C. 1994: Eliciting self-explanations improves understanding. *Cognitive Science,* 18, 439–77.

Clark, H. H. 1993: *Arenas of Language Use.* Chicago: University of Chicago Press.

*Gernsbacher, M. A. (ed.) 1994: *Handbook of Psycholinguistics.* New York: Academic Press.

Graesser, A. C., Singer, M. and Trabasso, T. 1994: Constructing inferences during narrative text comprehension. *Psychological Review,* 101, 371–95.

Grice, H. P. 1975: Logic and conversation. In P. Cole and J. L. Morgan (eds), *Syntax and Semantics: Speech Acts,* San Diego, Calif.: Academic Press, 41–58.

Just, M. A. and Carpenter, P. A. 1992: A capacity theory of comprehension: individual differences in working memory. *Psychological Review,* 99, 122–49.

Kintsch, W. 1988: The role of knowledge in discourse comprehension: a constructive-integration model. *Psychological Review,* 95, 163–82.

McKoon, G. and Ratcliff, R. 1992: Inference during reading. *Psychological Review,* 99, 440–66.

Schank, R. C. and Reisbeck, C. K. 1981: *Inside Computer Understanding.* Hillsdale, NJ: Erlbaum.

Trabasso, T. and Magliano, J. P. 1996: Conscious understanding during comprehension. *Discourse Process,* 21, 255–87.

van Dijk, T. A. and Kintsch, W. 1983: *Strategies of Discourse Comprehension.* New York: Academic Press.

*Weaver, C. A., Mannes, S. and Fletcher, C. R. (eds) 1995: *Discourse Comprehension: Essays in Honor of Walter Kintsch.* Hillsdale, NJ: Erlbaum.

Zwaan, R. A., Magliano, J. P. and Graesser, A. C. 1995: Dimensions of situation model construction in narrative comprehension. *Journal of Experimental Psychology: Learning, Memory and Cognition,* 21, 386–97.

25

Word meaning

BARBARA C. MALT

Questions about the nature of word meaning have drawn attention across the cognitive science disciplines. Because words are one of the basic units of language, linguists working to describe the design of human language have naturally been concerned with word meaning. Perhaps less obvious, though, is the importance of word meaning to other disciplines. Philosophers seeking to identify the nature of knowledge and its relation to the world, psychologists trying to understand the mental representations and processes that underlie language use, and computer scientists wanting to develop machines that can talk to people in a natural language have all worked to describe what individual words mean, and, more generally, what kind of thing a word meaning is.

At first glance, one may wonder why there is enough mystery to this topic to have held the attention of scholars in all these fields over the years. After all, dictionaries are filled with definitions of words. Aren't these definitions word meanings, and couldn't the nature of word meaning be determined just by examining these definitions? If the answer to these questions were *yes*, the job of cognitive scientists in this domain would be much simpler. As we consider a number of issues that have been raised in the study of word meaning, it will become clearer why dictionaries don't tell cognitive scientists all they need to know about word meaning.

Approaches to the study of meaning

The two major questions for theories of meaning – How can the meaning of individual words be described? and What kind of thing in general is a meaning? — are difficult to discuss independently. Although ideas about how to describe individual meanings overlap across different views of the nature of meaning, the relative pros and cons of these ideas depend in part on the larger view in which they are embedded. The discussion that follows is therefore organized around views of the general nature of meaning, with ideas about how to describe specific meanings addressed under them.

Meaning belongs to individuals

Many people intuitively think of word meanings as something that they have in their heads. Not surprisingly, since psychologists are interested in how knowledge is represented and used by humans, this view of meaning is consistent with how most psychologists treat word meaning. That is, they consider a word meaning to be a mental representation, part of each individual's knowledge of the language he or she speaks. In fact, psychologists typically have not distinguished between the meaning of a word and a concept; for instance, they treat the meaning of *bachelor* as equivalent to a person's concept of bachelorhood. This approach is also shared by linguists in the *cognitive linguistics* camp, who view knowledge of language as embedded in social and general conceptual knowledge.

Given this view of word meanings, the central question becomes: What is the nature of the meaning representation? What kinds of information do word meanings (or concepts) consist of? An answer adopted by many psychologists in the 1970s (see Smith and Medin, 1981), dating back to Plato's quest to define concepts like piety, justice, and courage, came into psychology by way of a linguistic theory that we will touch on later. This answer is that what a person knows when he or she knows the meaning of a word is a set of defining (or necessary and sufficient) features: that is, features that are true of all things the person would call by that name and that together separate those things from all things called by other names. For instance, defining features for the word *bachelor* might be *adult*, *male*, and *unmarried*. If someone's representation of the meaning of *bachelor* consisted of this set of features, then he or she would consider all and only people with those features to be bachelors. Although this sort of analysis was most often applied to nouns, psychologists George Miller and Philip Johnson-Laird, in their 1976 book, applied a similar kind of analysis to a large number of verbs.

A problem for this possibility, though, is raised by an earlier analysis by the philosopher Ludwig Wittgenstein in 1953. He argued that for many words, there is no single set of features shared by all and only the things that the word refers to. His famous example is the word *game*. Some games involve boards and movable markers, others involve balls and hoops or bats, still others involve singing; furthermore, some involve a winner and some don't, some are purely for fun and others are for monetary reward, and so on. The psychologists Eleanor Rosch and Carolyn Mervis, drawing on Wittgenstein's analysis, suggested in 1975 that what people know about many common nouns is a set of features having varying strengths of association to the category named by the word. For instance, most fruits are juicy, but a few (like bananas) are not; many fruits are sweet, but some (like lemons and limes) are not; some fruits have a single large pit, while others have many small seeds. The most common features, like *sweet* and *juicy* for fruit, are true of prototypical examples but do not constitute necessary and sufficient conditions for using the word. In support of their suggestion, they found that a sample of college students could not list features shared by all the members of several categories, but the students' judgments of how typical the objects were as members of a category were strongly correlated with how many of the more common category features each had. Linguists Linda Coleman and Paul Kay argued in 1981 that verbs such as *lie* may work in a similar way: They found that the lies considered most typical by their subject sample involved deliberate falsehoods with the intent to deceive, but some acts that subjects verified as lies lacked one or more of these features.

This prototype view, although capturing more of the apparent complexity associated with many common words, shares with the defining features view an assumption that the meaning of a word is a relatively constant thing, unvarying from situation to situation. Yet it has long been noted that the same word can have more than one meaning. For instance, *foot* can refer to a human body part, the end of a bed, or the base of a mountain, which are uses distinct enough to warrant thinking of them as involving different, albeit related, meanings. Further, it is clear that the context in which a word occurs may help to determine how it is interpreted. In the 1980s, Herbert

Clark argued that context does more than just select among a fixed set of senses for a word: It contributes to the meaning of a word on a particular occasion of use in a deeper way.

Specifically, Clark argued that many words can take on an infinite number of different senses. For instance, most people have the knowledge associated with the word *porch* that it refers to a structure used for enjoying fresh air without being completely outdoors. But in the context of the sentence "Joey porched the newspaper," a new meaning is constructed: namely, "threw onto the porch." And in "After the main living area was complete, the builders porched the house," the meaning "built the porch onto" is constructed. Because there is no limit to the number of contexts that can be generated for a word, there can be no predetermined list of meanings for a word. Other authors have made related points for less unusual cases of context, arguing, for instance, that the meaning of the word *line* is subtly different in each of many different contexts (e.g., "standing in line," "crossing the line," "typing a line of text" (Caramazza and Grober, 1976)), and that the variants are constructed at the time of hearing/reading the word from some core meaning of the word in combination with the context in which it occurs.

Although this last view differs from the defining features and prototype views in that it doesn't treat word meanings as things that are stored in their entirety in someone's head, all three approaches share the basic assumption that *some* critical knowledge of meaning is held by individuals. Several issues arise from this assumption. One is how people understand each other, since meanings must somehow be shared among people in order for communication to take place. The defining features view can easily account for how meanings are shared by assuming that everyone will have the same set of defining features for a word. The prototype approach, though, in proposing that meaning is a much broader set of features with varying strengths of association to the word, opens the possibility that individuals will differ from one another in the features that they represent and the strength of the associations to the word. Each person's experience with bachelors will be slightly different. One person may think of them as driving fast cars and partying; another may think of them as more like the Norwegian bachelor-farmers of Lake Wobegon. Similarly, this version of meaning opens the possibility that each person's meanings will change over time as his or her experiences change. The third view of meaning, by taking meaning to be context-dependent, likewise implies that a word meaning may differ from person to person and, notably, from situation to situation. And if meaning is person- and situation-dependent, then it is difficult to know if anything should be called *the* meaning of a word and what the mental representation of a word consists of. The idea that there is some *core* part of meaning that is invariant across all contexts or instances of a category offers a useful solution to this problem in principle, but in practice, cores for many words may be difficult or impossible to identify, just as were defining features. Thus the assumption that meaning is something that belongs to individuals, while having intuitive appeal, at the same time raises a number of difficult issues which must be resolved.

Meaning as publicly held

Most linguists and many philosophers view word meanings not as something inside individual people's heads, but as part of a language in a more abstract sense. Many computer scientists likewise seem to take this view of meaning, though they are typically less explicit about such assumptions. Meanings, on this view, are treated as attached to words regardless of the individuals who use them or what they know about them. The most extreme way of formulating this position is to consider meanings to be part of a system that can be characterized in terms of its properties without reference to language-users at all, just as the properties of the solar system might be described without reference to its relation to humans (a view expressed, for instance, in the title of linguist Jerrold Katz's 1981 book, *Language and other Abstract Objects*). A more moderate formulation is to think of meanings as things fixed by convention within a language community. A word can then be characterized as having some particular meaning within the linguistic community even if some, or even many, members of the community do not know that meaning or have incomplete knowledge of that meaning. For example, the word *turbid* might be characterized as meaning muddy, cloudy, or dense in English, even if not all people who speak English know its meaning.

In the 1960s and 1970s, substantial effort was made by linguists (and also anthropologists) to describe meanings in terms of features that define the conditions under which something would be labeled by the word. This effort, by investigators such as Jerrold Katz, Jerry Fodor, and others, is in fact the source of the example of defining *bachelor* as *male, adult,* and *unmarried* used by psychologists (adapted there to a more psychological perspective). Although primarily applied to nouns, this sort of defining features analysis was also applied to verbs by a number of linguists such as James McCawley and Ray Jackendoff.

A major benefit of this approach is its usefulness in attempting to specify how words are related to other words. Within linguistics, doing so has often been taken to be a major goal for a theory of meaning. Thus linguists have wanted to capture meanings in a way that would allow them to identify what words are synonymous with other words, what words are antonyms (opposites), what words name things with part–whole relations (as, for example, *arm* and *body*), what ones name things with inclusion relations (as, for example, *dog* and *animal*), and so on. Characterizing meanings in terms of defining features provides a way of doing this: Two words are synonymous if they have the same defining features; two words have an inclusion relation if the defining features of one are included in the defining features of the other; and so on. The defining features approach has also provided a convenient way of representing meanings and their relation to each other for use in computer programs that attempt to deal with natural language input, and featural approaches along these lines have been widely used within artificial intelligence.

Another benefit of this approach is that we can then treat some of the individual differences in knowledge about word meanings by saying that a person might not fully grasp whatever the meaning of the word actually is. So, someone who doesn't understand *bachelor* to mean *adult, male,* and *unmarried* but only *adult* and *unmarried*, doesn't fully grasp the meaning of *bachelor*. To the extent that successful communication and consistency in individual representations of meaning occur, they are presumably achieved because people aim to acquire the meaning given to the word by linguistic convention.

Nevertheless, several potentially serious problems arise for the defining features version of meanings as public entities. A major one is that, as discussed earlier, it seems impossible to provide an analysis of many words (such as *game*) in terms of defining features. Another is that, also along lines discussed earlier, we might want to include other features such as *likes to party* and *drives a sporty car* as part of the meaning of *bachelor*. One solution to these problems is to expand the notion of meaning to encompass a broader range of features, as discussed for the more psychological perspective. Such solutions have sometimes been proposed and have been incorporated in some artificial intelligence systems for representing meaning. However, these solutions create the problem of trying to decide where word meanings end and general knowledge begins; they also undermine the attempt to provide an account of relations like synonymy and antonymy between words. Another solution, adopted in the 1980s by the linguist George Lakoff and others, is to view a word as having a set of distinct but specifiable meanings that may have a variety of relations, including metaphorical relations, to one another. This solution likewise makes it more difficult to see how relations like synonymy can be specified, and it requires enumerating a potentially very large number of meanings for each word.

Meaning as a relation between a word and the world

So far, we have been talking about meaning in the way in which it is used in everyday language: as something that can be described in conceptual terms. So, whether we want to say that the meaning of *bachelor* resides in individual heads or belongs to a language in some more abstract sense, we can describe its meaning in terms of concepts like *adult* and *male*. However, scholars of meaning since the philosopher Gottlob Frege in the late 1800s have distinguished between two components or aspects of meaning. One, the *sense* or *intension* of a word, is the conceptual aspect of meaning that we have discussed so far. The other is the *reference* or *extension* of a word, the set of things in the world that the word refers to. For the word *bachelor*, for instance, the reference of the word is the set of all (real or possible) bachelors in the world. In other words, the reference aspect of meaning is a relation between a word and the world.

Psychologists, linguists, and computer scientists holding any of the views of meaning discussed so far would generally consider the sense of a word to be the primary concern for a theory of meaning, although they would also agree that the theory should account for what entities the word is used to refer to. A view of meaning quite distinct from this perspective, though, has recently been influential, and that is a view that says, essentially, that the meaning of a word *is* its relation to things in the world; that is, meaning *is* reference.

An important argument for this view, derived primarily from analyses of meaning by philosophers Hilary Putnam and Saul Kripke, is based on the observation that the features that one thinks of as constituting the meaning of word could turn out not to be true. For example, a person (or a language) might specify features like *sour* and *yellow* as the meaning of the word *lemon*, but it could turn out that these features don't accurately reflect the truth about lemons. Research could reveal that pollution makes lemons yellow and sour, but normally they would be green and sweet. The word *lemon* would still refer to the set of things in the world that it did before everyone revised their knowledge of the properties of lemons. Similarly, new scientific discoveries could add to or alter beliefs about the properties of many objects, but those changes in the properties associated with the words would not change the set of things correctly named by the word. Putnam suggested, on the basis of these and other arguments, that words function simply to pick out sets of things in the world. On this referential view, the properties constitute a stereotype of what the object is like (or seems to be like), but they do not constitute the meaning of the word. As Putnam wrote in advocating this view in 1973: "Cut the pie any way you like, 'meanings' just ain't in the *head*!" (And likewise, according to this view, they "just ain't" definitions held by a linguistic community.)

A benefit of this referential view of meaning is that it provides an account of stability in meaning and communication: A word refers to the same set of things in the world regardless of variations in knowledge among people, and use of a word to refer to a particular set of things can be passed from generation to generation regardless of changes in beliefs about properties of the objects. However, it also has weaknesses, and one prominent one is that the analysis does not seem to apply to many common words. For instance, the word *bachelor* seems intrinsically to involve the property of being unmarried. Although we can imagine researchers discovering that lemons really are green, it just isn't possible for researchers to discover that bachelors really are married people. Even if all men previously thought to be unmarried turned out to be married, we wouldn't change the properties associated with *bachelor*; we would say that these men weren't bachelors after all. Likewise, *island* seems to intrinsically refer to things with the property of being surrounded by water; research can't change that property. And although discussion of this view is usually restricted to nouns, the same point would apply to verbs: *Run*, for instance, seems to intrinsically refer to a certain kind of motion, and any activity not involving that motion just wouldn't be running. In such cases, having the associated properties *does* seem to be critical to whether or not the word can be applied to the object. If the referential view is correct for some words, this observation raises the interesting possibility that the nature of the meanings may differ for different words, and one analysis of meaning may not be appropriate for all words.

Conclusion

This discussion has illustrated some of the fundamental issues facing cognitive scientists who want to understand the nature of word meaning. It should now be clear why dictionary definitions don't tell cognitive scientists all that they need to know. At the broadest level, dictionaries don't address what kind of thing a meaning is: a mental representation in someone's head, something that is part of a language in some more abstract sense, or a relation between words and the world. At a more detailed level, dictionaries don't reveal the status of the pieces of information they offer about a word. Is a given property truly defining? Is it associated with only some of the things labeled by the word? Or is it part of a stereotype that is not the actual word meaning? Nor do they address the role of context in meaning and the extent to which words may take on new meanings in new situations, or the full extent to which words may have many subtly different uses. Finally, they don't reveal whether some words, like *lemon*, may differ fundamentally in the nature of their meaning from other words, like *bachelor* or *island*.

One historical stumbling block to a full account of meaning has been that scholars in each discipline often were not aware of issues raised by the other disciplines and so were satisfied with proposals that were relatively narrow in scope. However, the emergence of the multidisciplinary cognitive science effort has already increased shared awareness of some of the complexities to be dealt with. In doing so, it has provided a push toward broader perspectives on how to tackle these issues: Psychologists have begun to incorporate aspects of philosophical theories into their views of mental representation and processing; linguists have begun to make use of the information provided by laboratory experiments on meaning; and so on. Although it is not yet clear what form a more integrative theory of meaning will take, progress may be on the horizon.

References and recommended reading

Caramazza, A. and Grober, E. 1976: Polysemy and the subjective structure of the lexicon. In C. Rameh (ed.), *Georgetown University Roundtable on Language and Linguistics* (Semantics: Theory and Application), Washington, DC: Georgetown University Press, 181–206.

Clark, H. H. 1983: Making sense of nonce sense. In G. B. Flores d'Arcais and R. J. Jarvella (eds), *The Process of Language Understanding*, London: Wiley, 297–333.

Coleman, L. and Kay, P. 1981: Prototype semantics: the English word "lie." *Language*, 57, 26–44.

Frege, G. 1892: Über Sinn und Bedeutung. *Zeitschrift für Philosophie und philosophische Kritik*, 100, 25–50. (Translated as "On sense and reference" in P. T. Geach and M. Black (eds), *Translations from the Philosophical Writings of Gottlob Frege*, Oxford: Blackwell, 1952, 56–78.)

Jackendoff, R. S. 1972: *Semantic Interpretation in Generative Grammar*. Cambridge, Mass.: MIT Press.

Katz, J. J. 1981: *Language and other Abstract Objects*. Totowa, NJ: Rowman and Littlefield.

Katz, J. J. and Fodor, J. A. 1963: The structure of a semantic theory. *Language*, 39, 190–210.

Kripke, S. 1972: Naming and necessity. In D. Davidson and G. Harman (eds), *Semantics of Natural Language*, Dordrecht: Reidel, 253–355.

Lakoff, G. 1987: *Women, Fire, and Dangerous Things: What Categories Reveal about the Mind*. Chicago: University of Chicago Press.

McCawley, J. D. 1968: Lexical insertion in a transformational grammar without deep structure. In B. J. Darden, C-J. N. Bailey, and A. Davison (eds), *Papers from the Fourth Regional Meeting of the Chicago Linguistics Society*, Chicago: Department of Linguistics, University of Chicago, 71–80.

Miller, G. A. and Johnson-Laird, P. N. 1976: *Language and Perception*. Cambridge, Mass.: Harvard University Press.

Putnam, H. 1970: Is semantics possible? In H. E. Kiefer and M. K. Munitz (eds), *Language, Belief, and Metaphysics*, Albany: State University of New York Press, 50–63.

—— 1973: Meaning and reference. *Journal of Philosophy*, 70, 699–711.

Rosch, E. and Mervis, C. B. 1975: Family resemblances: studies in the internal structure of categories. *Cognitive Psychology*, 7, 573–605.

*Smith, E. E. and Medin, D. L. 1981: *Categories and Concepts*. Cambridge, Mass.: Harvard University Press.

Wittgenstein, L. 1953: *Philosophical Investigations*, trans. G. E. M. Anscombe. New York: Macmillan.

PART III

METHODOLOGIES OF COGNITIVE SCIENCE

26

Artificial intelligence

RON SUN

An old theme

The field of artificial intelligence (AI) can be characterized as the investigation of computational systems that exhibit intelligent behavior (including algorithms and models used in these systems). The emphasis is not so much on understanding (human) cognitive processes as on producing models, algorithms, and systems that are capable of apparently intelligent behavior by whatever means available. The idea of AI has had a long history that can be traced all the way back to, for example, Leibniz. The idea was furthered through the development, early in this century, of mathematical logic, cybernetics, and information theory, all of which contributed to AI theorizing. Alan Turing's bold claim that machines can think, advanced in his famous 1950 paper, and the "Turing test" he proposed therein as a means to test the intelligence of machines predated and prompted the formation of AI. Six years later, the Dartmouth conference in 1956, the participants of which included Allen Newell, Herbert Simon, Marvin Minsky, and John McCarthy, inaugurated AI as an academic discipline. This contemporary version of AI was facilitated by the invention and rapidly widening use of digital computers. AI embodies an extremely diverse set of ideas that are often mutually conflicting (as in the case of symbolism versus connectionism), and its relationship with other fields studying cognition (such as psychology and linguistics) has been a close but uneasy one.

The approach that AI takes to produce models of intelligent behavior is mainly computational analysis: examine the requirements of a task domain in which an AI model is to be produced; analyze the behavioral requirements (i.e., the input–output relations) and also intermediate steps when such steps are discernible through analysis of necessary steps and prior theories/meta-theories (which, especially in the early days, drew ideas from other fields in cognitive science, such as psychology and philosophy); perform computational analyses with regard to what kind of computation is likely to be underlying the task performance, in accordance with the above-identified requirements and the steps to be captured; synthesize the above information into a relatively complete working algorithm, and through repeated revision, debugging, and testing, crystallize the algorithm into a well-defined coherent model that has a well-delineated scope, a clearly defined input–output format, and tangible behavioral regularities. In addition, introspective *phenomenological* analyses are often used to identify processes and representational constructs/entities that can be used in models.

For example, in Sun, 1994, I performed such an analysis of everyday commonsense reasoning. From the analysis, a model named CONSYDERR was produced, based on ideas drawn from an understanding of the reasoning requirement (based on analyzing human reasoning data), the study of underlying computational processes (based on a number of different paradigms), meta-theoretical insight, and *phenomenological* introspection (see Sun, 1994).

Such an analysis is likely to involve search and representation, two fundamental ideas that originated in the earliest days of AI and have played central roles throughout its history. Let us discuss the idea of search space first. In any problem to be tackled by AI, there is supposed to be a large space of *states*, each of which describes a step in problem solving (i.e., inference), including the original state where problem solving (inference) starts and the goal state that is to be reached in the end; at each moment in problem solving, a number of *operators* are applied to reach a new state from the current state; an AI system is supposed to search through this state space of possible states resulting from successive application of operators, starting from the original state, so as to get to the goal state efficiently. Search techniques adopted in the early days of AI included depth-first search and breadth-first search. In depth-first search, from the current state, the system examines one alternative ("path") at a time by applying one of the operators to the current state, which leads to a new state. Then from the new state, the same process is repeated, until it hits a dead end where there is no operator that can be applied. In that case, the system backs up to a previous state and tries a different alternative. In breadth-first search, by contrast, the system examines all alternatives ("paths") at once, by applying each of the applicable operators to the current state in turn, resulting in a number of new states. Then the system selects one new state at a time and examines all alternatives from it. This process continues until all reachable states have been reached. Such exhaustive search techniques are frequently applied, but they are inefficient. To speed up the search, many heuristic search algorithms have been proposed, some of which are generic, others domain-specific. One generic idea is to use an evaluation function that estimates the promise of each path and then select the most promising path to explore further (i.e., the hill-climbing search). The idea of search space has been applied in all areas of AI, including PROBLEM SOLVING (see Article 20), natural language processing, robotics and vision, knowledge representation/reasoning, and MACHINE LEARNING (see Article 16; and Russell and Norvig, 1995, for more details regarding search).

Another important idea in AI is representation (see Article 51, REPRESENTATIONS). This idea embodies the belief that knowledge should be expressed in an internal form that facilitates its use, in relation to the requirement of the task at hand and mirroring the external world (in some way). The AI idea of representation was influenced by the philosophical notion of representation, which has been an essential ingredient of epistemology in some segments of modern philosophy. A variety of representation forms have been developed in AI, most of which are used in conjunction with search algorithms for inference. One of the earliest representation forms is rule-based reasoning, in which discrete rules are used to direct search (inference). Rules are composed of both conditions, which specify the applicability of rules, and conclusions, which specify actions or outcomes. Rules are modular: ideally, each rule can be added to or deleted from a system, without affecting the other parts of the system (modularity may, however, inadvertently hamper computational flexibility and dynamic interaction in reasoning).

A popular form of rule-based reasoning is the production system, evolved from some psychological theories which emerged in the 1960s and 1970s. A production system

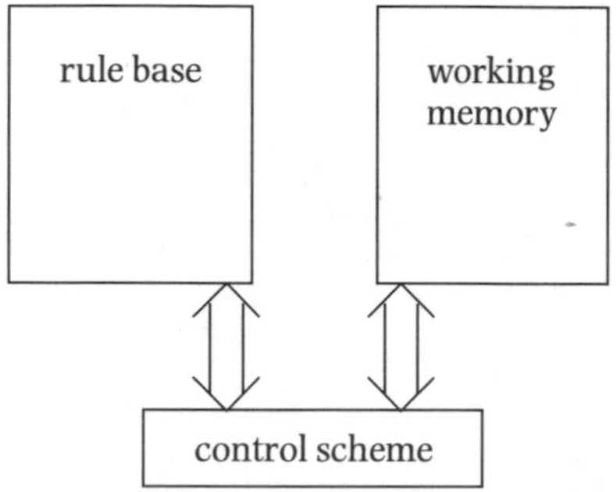

Figure 26.1 The structure of a production system.

consists of (1) a production rule base (for storing all the rules), (2) a working memory (for storing initial, intermediate, and final results), and (3) a control structure for coordinating the running of the system (see figure 26.1). The inference process in a production system can be either forward chaining or backward chaining (two different search directions). In forward chaining, if the condition of a rule matches some facts in the working memory, we derive new facts from the conclusion of the rule. In backward chaining, we have a list of hypotheses to prove. If a hypothesis matches facts in the working memory, the hypothesis is proved. Otherwise, if a match between a hypothesis and the conclusion of a rule is found, we replace the original hypothesis with the conditions of the matching rule. The process keeps going until all the hypotheses on the list are proved. (See Article 42, PRODUCTION SYSTEMS.)

Formal logics constitute an alternative approach in rule-based reasoning, as advocated chiefly by John McCarthy and associates (see McCarthy, 1968). They are relatively simple, formally defined languages capable of expressing rules in a rigorous way. Logic inference is performed in formally defined ways that guarantee the completeness and soundness of the conclusions, by using a variety of algorithms more complicated than simple forward and backward chaining. So the advantage of logics is their clarity and correctness. The logic approach is not overly concerned with detailed internal organization, as are production systems, but rather with the formal properties of systems. There has been a tradition of psychological study of human REASONING (see Article 21) using logic approaches. Overall, rule-based reasoning has only limited success vis-a-vis the goal of accounting for all types of human reasoning. Formal logics and most production systems are restrictive: one needs to have all the conditions specified precisely in order to perform one step of inference. Thus rule-based systems are unable to deal with partial, incomplete, and approximate information. There is also no built-in way for similarity-based processes, which are evident in human reasoning (as indicated by the work of Allan Collins and Ryszard Michalski, 1989, and of Ron Sun, 1994).

Another type of representation aims to capture the aggregate *structures* of knowledge (instead of dispersing such structures throughout, for example, a production system). The general idea is that knowledge resides in structured chunks, each of which is centered around a particular situation or object, and each can contain, or be contained in, other chunks, as well as call up other chunks during processing. Different from rules, each chunk contains all the pieces of information regarding a certain situation or object, as well as their interrelations. Such representations have roots in psychology, as

Frame: kitchen	
Slot	Value
refrigerator	(3, 5)
dishwasher	(5, 5)
stove	(4, 6)
pantry	(4, 6)

Figure 26.2 An example of a frame.

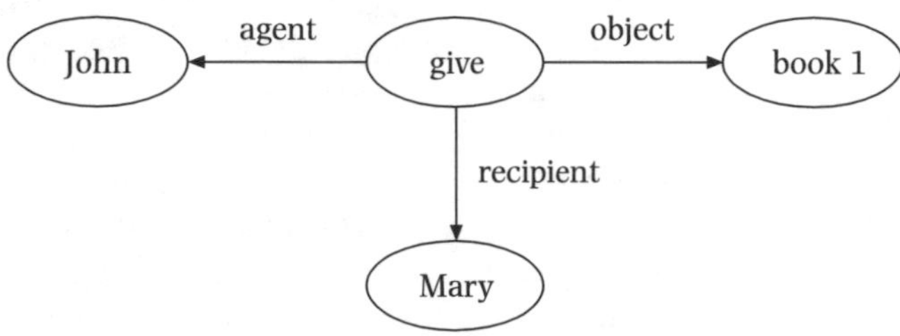

Figure 26.3 An example of a semantic network.

advocated by, among others, Ross Quillian (1968), who started this line of investigation. They include frames, scripts, and semantic networks, and are often grouped together under the umbrella term *schemas*.

For instance, a *frame* (which was proposed by Marvin Minsky in the mid-1970s) represents a concept in terms of its various attributes (slots), each of which has a name (label) and a value (or a set of values). By organizing knowledge in frames, access is made easier, because all the relevant pieces of information can be obtained in a structured way. With frames, it is also easy to add or remove information relevant to a situation or object (see figure 26.2). A *semantic network* (proposed by Ross Quillian in the late 1960s) consists a set of nodes, each of which represents a particular (primitive) concept, and labeled links, each of which represents a particular relation between nodes. As a simplified example, to represent "John gave Mary a book," we can build a semantic network as in figure 26.3. Semantic networks also allow efficient and effective access to knowledge, albeit in a different way, by following links that go from one concept to all the others that have a particular relation to the original one. A useful method for accessing information in semantic networks is "spreading activation." A node is initially activated when we want to retrieve all the information related to the concept it represents, and the activation can then be spread out to other nodes representing related concepts via the labeled links. We can gather all the useful information by identifying activated nodes and their relations to the original concept through tracing the link labels. *Scripts* (proposed by Roger Schank and associates in the late 1970s) are used for representing prototypical sequences of events in stereotypical situations

– for example, eating in a restaurant. In such a situation, there is an almost invariable sequence that happens every time. The purpose of scripts is to help with the efficient recognition and handling of these situations. A script may be divided into multiple tracks, each of which handles a subtype of situation (e.g., fancy restaurants versus fast food restaurants). A script-based system can recognize different situations (using an appropriate script and an appropriate track) and handle them accordingly. Despite their popularity in AI systems, these aggregate representations have shortcomings. (1) Structures often need to be determined a priori and hand-coded; (2) structures are usually fixed and cannot be changed dynamically; and (3) they can be too costly and unwieldy in capturing the full extent of complex real-world situations.

Some new variations

There are many new variations on the old AI theme discussed above that attempt to address some major problems with the early approaches. We will look into a few of them. One significant feature of current AI is the existence of a variety of different and often conflicting paradigms. The two main strands are:

- The symbolic paradigm. The field of AI, since its inception, has been conceived mainly as the development of models using symbol manipulation. The computation in such models, as discussed earlier, is based on explicit representations that contains symbols organized in some specific ways, and aggregate information is explicitly represented with aggregate structures that are constructed from constituent symbols with syntactic combinations.
- The connectionist paradigm. Emergence of the connectionist paradigm results from dissatisfaction with symbol-manipulation models, especially their inability to handle flexible, robust processing in an efficient manner. This paradigm aims at massively parallel models that consist of simple and uniform processing elements interconnected by extensive links. The functionality of such models emerges from the interaction of the constituent simple processing elements. (See Article 38, CONNECTIONISM, ARTIFICIAL LIFE, AND DYNAMICAL SYSTEMS.) In many connectionist models, representation is distributed throughout a large number of processing elements (corresponding to the structure of such models). Sometimes the constituent symbolic structures of aggregate information are embedded in a network and are difficult to identify. Due to their massively parallel nature, such models are good at flexible, robust processing, and are showing promise at dealing with some tasks that have proved difficult for the symbolic paradigm.

Note that, up to now, no single existing paradigm can fully handle all the major AI problems. Each paradigm has its strengths and weaknesses and excels at certain tasks while falling short with some others. This situation indicates the need to integrate these existing paradigms (see, e.g., Sun and Bookman, 1994).

Symbolic knowledge representation

There are developments in the symbolic paradigm that aim to remedy some of the problems with traditional symbolic representation, especially with logic-based approaches, as discussed earlier. These extensions include default logic, circumscription, autoepistemic logic, and causal theory, among many others.

For example, *default logic* is proposed to model default reasoning: it deals with beliefs based on incomplete information, which might be modified or rejected later on the basis of subsequent observations. When no exact conclusions may be reached, people customarily make assumptions that are consistent with their current knowledge of the world. These assumptions are *defaults*, which lead to nonmonotonic reasoning – that is, reasoning in which conclusions drawn earlier may be revised later on. *Circumscription* is an attempt at capturing a range of issues in modeling default commonsense reasoning, especially the conjectural reasoning regarding properties of objects, by assuming that the objects that one can determine to have a property are the only objects that do. Like default logic, it is nonmonotonic, in that when new information becomes available, previous conclusions may be revised. *Autoepistemic logic* aims to model a rational agent's belief about its own knowledge. An autoepistemic theory that describes the beliefs of a rational agent builds in certain conditions that such beliefs must satisfy. A modal operator L is introduced to denote *belief*. This logic is nonmonotonic, allowing an agent to draw tentative conclusions based on the lack of information until information becomes available. *Causal theory* can be viewed more or less as a subset of the above logic, but with efficient inference algorithms. However, this logic is intended to model causal reasoning, so the modal operator L is interpreted as *necessity*. There are two types of conditions: in reasoning, if *necessary conditions* (denoted by L) are true and there is no information that the *possible conditions* are false, one can draw the corresponding conclusions. This makes the logic nonmonotonic, because if later on, one of these possible conditions turns out to be false, the inference has to be retracted. These logics are normative models and deal with partial/incomplete information. However, they have some shortcomings, including the lack of capabilities for dealing with (1) approximate information, (2) inconsistency, and (3) most of all, reasoning as a complex, interacting process (Sun, 1994). Thus they are somewhat deficient from the standpoint of capturing human reasoning. For an overview of these models, see Davis, 1990.

In an effort to amend some deficiencies, Lotfi Zadeh proposed *fuzzy logic* (which caught on in the 1980s; see Zadeh, 1988), primarily to capture vagueness or approximate information in linguistic expressions, such as "tall" or "warm," which have no clear-cut boundaries. The basic idea is as follows: for each concept, there is a set of objects satisfying that concept to a certain degree, which form a subset: namely, a fuzzy subset. This fuzzy (sub)set contains as its elements pairs consisting of an object and its grade of membership, which represents the degree to which it satisfies the concept associated with the set. Given this idea, it is easy to construct a logic with which one can reason about the fuzzy truth values of concepts. Fuzzy logic has been linked to advanced logics such as Shoham's logic (Sun, 1994).

The *probabilistic approach* is another mixed symbolic/numeric approach, which treats beliefs as probabilistic events and utilizes probabilistic laws for belief combinations. This approach is viable because beliefs are formed as a *distillation* of our experiences, and for reasons of storage economy and generality, we forget the actual experiences and retain the mental impressions in the forms of averages. However, human reasoning may not always conform to the assumptions and the laws of probability theory, as discovered by psychologists (see Article 10, DECISION MAKING and Article 21, REASONING), probably in part because of the complexity of the formal models. In addition, it is not always possible to obtain precise probability measures in practice.

Connectionist knowledge representation

Recently, there have been significant developments of knowledge representation within the connectionist paradigm. Many high-level connectionist models were proposed that employ representation methods that are comparable to or surpass symbolic representations, and thus they remedy many problems associated with traditional representation methods.

First of all, logics and rules can be implemented in connectionist models in a variety of ways. For example, in one type of connectionist system, reasoning is carried out by constraint satisfaction through minimizing an error function. By constructing the error function in a way that captures the underlying relations between logic formulas, this process can produce a near-optimal interpretation of the inputs. But the process is extremely slow. Another type of system, proposed by Lokendri Shastri and many others in the early 1990s, uses more direct means, representing rules by links that directly connect nodes representing conditions and conclusions respectively; reasoning in these models amounts to forward activation propagation. They are thus more efficient. They also deal with the so-called variable binding problem in connectionist networks. Those advanced logics mentioned earlier that go beyond classical logic can also be incorporated in connectionist models (see, e.g., Sun, 1994).

Aggregate information (schema structures) can also be incorporated into connectionist models. A system developed by Risto Miikkulainen and Michael Dyer in 1991 encodes scripts through dividing input units of a backpropagation network (see Article 16, MACHINE LEARNING) into segments, each of which encodes an aspect of a script in a distributed fashion. There are also localist alternatives (such as those proposed by Trent Lange and Michael Dyer in 1989 and by Ron Sun in 1992), in which a separate unit is allocated to encode an aspect of a schema.

A system developed by John Barnden and associates around 1992 utilizes CASE-BASED REASONING (see Article 36) in connectionist networks. The system has multiple configuration matrices, each of which contains a case for short-term processing; a relatively small set of gateways provides the interface between short-term processing and long-term memory; current cases compete to have their contents copied into gateways, where they can cause similar previous cases to be retrieved from long-term memory, and some symbol substitutions take place after the retrieval, to adapt the cases to the current situation, thus accomplishing case-based reasoning.

There are also multi-module connectionist systems that utilize multiple representation methods and explore their interaction to achieve a synergistic outcome. For example, the system I developed (Sun, 1994), CONSYDERR, consists of two levels: the top level is a connectionist network with localist representation, and the bottom level is a connectionist network with distributed representation: concepts and rules are diffusely represented in the bottom level by sets of feature units overlapping each other. This is a similarity-based representation, in which concepts are "defined" in terms of their similarity to other concepts in these representations. The localist network is related to the distributed network by linking each node in the top level representing one concept to all the feature nodes in the bottom level representing the same concept.

Through a three-phase interaction between the two levels, the model is capable of both rule-based and similarity-based reasoning with incomplete, inconsistent, and approximate information, and accounts for a large variety of seemingly disparate patterns in human reasoning data.

Search, the main means of utilizing knowledge in representation, is employed or embedded in connectionist models. Either an explicit search can be conducted through a settling or energy-minimization process (as discussed earlier), or an implicit search can be conducted in a massively parallel and local (in-place) fashion. The method of global data storage, retrieval, and reorganization during search, as performed in many symbolic AI models, requires such an overhead that it has been considered implausible as a model for cognition. Global energy minimization, as in some connectionist models, may also be too time-consuming. Local (in-place) computation in connectionist models is a viable alternative: computation is done right where the data are; pieces of knowledge are stored in a network connected by links that capture search steps (inference) directly. Search in such a network amounts to activation propagation (by following links similar in some ways to those in semantic networks, without global control, monitoring, or storage).

The advantage of connectionist knowledge representation is that such representation can not only handle symbolic structures but goes beyond purely symbolic models in dealing with approximate information, uncertainty, inconsistency, incompleteness, and partial match (based on similarity), and in treating reasoning as a complex dynamic process. On the other hand, developing representation in restricted media such as connectionist networks has been a difficult task. (For an overview of connectionist knowledge representation, see Sun and Bookman, 1994.)

Learning

Learning is another thrust of AI research. It is not only of practical importance, but it is also of importance to the theoretical understanding of cognition, in relation to, for example, the genesis of representation. Although some incremental algorithms exist, most work in symbolic machine learning focuses on "batch learning." Typically, the learner is given all the exemplars/instances, positive and/or negative, before learning starts. Most of these algorithms handle the learning of concepts with simple rules or decision trees.

There are some recent developments also worth mentioning. Some learning algorithms have been extended to deal with noisy and inconsistent data. In addition, in inductive logic programming, a learner tries to induce more powerful first-order *Prolog* rules (logic rules that involve predicates, functions, and variables, as in the logic programming language Prolog) from data. In explanation-based learning, a learner tries to extract specific rules regarding a certain situation by examining the inference that leads to a conclusion and identifying pertinent conditions in the inference.

Recently, there have been some learning approaches that handle dynamic sequences, which necessarily involve temporal credit assignment – i.e., attributing properly success/failure to some preceding steps. They include reinforcement learning methods, such as Q-learning and real-time dynamic programming, and evolutionary algorithms. Many of these algorithms are incremental and are more compatible with human learning processes than some traditional algorithms. For dealing with dynamic sequential environments, some investigators seek to combine rule learning with reinforcement learning, to perform on-line incremental learning that acquires a variety of different types of knowledge simultaneously.

Within the connectionist paradigm, there are also models that perform the aforementioned learning tasks. These connectionist learning algorithms combine the advantages of their symbolic counterparts with the connectionist characteristics of being noise/fault-tolerant and being capable of generalization (through function approximation). (For an overview of symbolic and connectionist learning, see Shavlik and Dietterich, 1990.)

AI models and cognitive processes

AI models can be made to correspond to cognitive processes in a variety of ways and thus shed light on cognition. In modeling cognition, AI models can be (1) broad or narrow (covering a large set of data or being very specialized), (2) precise or imprecise, and (3) descriptive or normative. There are at least the following types of correspondence between AI models and cognitive processes, in increasing order of precision:

- Behavioral outcome modeling: in this case, an AI model produces roughly the same types of behavior as humans do, under roughly the same conditions. For example, given a set of scenarios for decision making, an AI model makes roughly the same kind of decisions as a human decision-maker (e.g., in commonsense reasoning models). Or, given the same piece of text, an AI model extracts roughly the same kind of information that a human reader would extract (e.g., in AI natural language processing systems).
- Qualitative modeling: an AI model produces the same qualitative behaviors that characterize human cognitive processes under a variety of circumstances. For example, the performance of human subjects improves/deteriorates when one or more control variables are changed; if a model shows the same changes, given the same manipulations, we may say that the model captures the data qualitatively.
- Quantitative modeling: an AI model produces exactly the same quantitative behaviors as exhibited by humans, as indicated by quantitative performance measures. For example, we can match point by point the learning curve of an AI model with that of humans; or we can match the step-by-step output of a model with the corresponding performance of humans, in a variety of situations.

AI models have had some successes in modeling a wide variety of cognitive phenomena in one of the above three senses (especially the first two). These phenomena include learning (such as learning concept description), cognitive skill acquisition, and decision tree or connectionist models of development. They also include everyday commonsense reasoning, such as logic-based models (see McCarthy, 1968; Collins and Michalski, 1989), case-based models, and connectionist models (see Sun, 1994) that combine some of the important features of the first two types. Other phenomena being tackled include word-sense disambiguation, analogical reasoning, learning the past-tense forms of verbs, playing games (such as Go or Tic-tac-toe), expertise, and so on.

It is essential to consider the source of power in accounting for the success of an AI model in modeling cognitive phenomena. The source of power can include (1) algorithms; (2) parameters of algorithms; (3) task representation (including the scope, the covered aspects, and the way the human data to be modeled are collected and interpreted); (4) learning/training regimes, especially their underlying assumptions and biases, including data sampling and selection, data presentation (the order and the frequency), and data representation; and (5) most of all, paradigms used for modeling (including the underlying theories), which can have a profound effect on the perception of success and failure in practice. Some of these aspects may not, or should not, be legitimate sources of power. Understanding the source of power can be useful in zeroing in on central aspects of a successful AI model and in identifying commonalities among models tackling the same problems, as well as in comparing different models. Up to now, it has been a matter of case-by-case considerations in the context of specific models and tasks, without any generic criterion that can be applied universally.

It is evident by now that any sufficiently powerful AI model, symbolic or connectionist, can be made to capture data in any narrow domain (especially those sanitized, isolated data within a fixed representation). Therefore, we need to look into deeper issues beyond simple measures of fit. Such issues might include (1) degrees to which data are accounted for by a model – especially important are real-world situations, not just sanitized, isolated data, because real-world situations may be vastly different from laboratory situations; (2) the source of power, and whether it is appropriate in the particular context of the model and the task tackled; (3) connections with theoretical and meta-theoretical principles and insights; (4) compatibility with models (or principles) in other, related domains; (5) compatibility with evidence from other disciplines (especially psychological and neurobiological evidence); (6) computational power and the correspondence of the model with the human data in this regard; (7) computational complexity; and (8) aesthetic quality of the model. There is always a many-to-many mapping between computational models and cognitive phenomena to be modeled, so development and application of abstract criteria in analyzing and comparing models are essential.

Beyond comparing individual models, we also need to compare and contrast various AI paradigms, mainly the symbolic paradigm and the connectionist paradigm, to reveal their respective strengths and limitations, as well as underlying assumptions, which can lead to better understanding and more rapid advance of the field of artificial intelligence.

References and recommended reading

Collins, A. and Michalski, R. 1989: The logic of plausible reasoning: a core theory. *Cognitive Science*, 13(1), 1–49.

*Davis, E. 1990: *Representations of Commonsense Knowledge*. San Mateo, Calif.: Morgan Kaufmann.

McCarthy, J. 1968: Programs with common sense. In M. Minsky (ed.), *Semantic Information Processing*, Cambridge, Mass.: MIT Press, 403–18.

Quillian, M. R. 1968: Semantic memory. In M. Minsky (ed.), *Semantic Information Processing*, Cambridge, Mass.: MIT Press, 216–20.

*Russell, S. and Norvig, P. 1995: *Artificial Intelligence: A Modern Approach*. Englewood Cliffs, NJ: Prentice-Hall.

Shavlik, J. and Dietterich, T. 1990: *Readings in Machine Learning.* San Mateo, Calif.: Morgan Kaufmann.
*Sun, R. 1994: *Integrating Rules and Connectionism for Robust Reasoning.* New York: John Wiley and Sons.
Sun, R. and Bookman, L. (eds) 1994: *Computational Architectures Integrating Neural and Symbolic Processes.* Boston: Kluwer Academic Publishers.
Turing, A. 1950: Computing machinery and intelligence. *Mind,* 59, 433–60.
Zadeh, L. 1988: Fuzzy logic. *Computer,* 21, 83–93.

27

Behavioral experimentation

ALEXANDER POLLATSEK AND KEITH RAYNER

How might one study the complex processes of the mind? The method favored by early philosophers and psychologists was introspection. While introspection is still used today, perhaps the major source of evidence used by cognitive scientists to understand cognition is data collected from experiments in which subjects are engaged in some type of relevant task. While these data all come from some type of *experiment*, the methods differ widely, and, as we shall see, the type of method is strongly influenced by the area of inquiry.

The establishment of the first major psychological laboratory in the late 1800s by Wilhelm Wundt led to numerous experimental investigations of basic processes associated with, among other things, cognition, memory, reading, speech perception, and attention. However, the onset of the behaviorist revolution in American psychology in the early 1900s resulted in psychologists not studying cognition for a long period of time. During the height of the behaviorist movement (about 1920 to 1960), rigorous experimental methods were devised to study various topics of interest to behaviorists. However, studies of mental processes were pretty much taboo during this period, as evidenced by the fact that most experimental psychology used animals such as white rats, rather than humans, as subjects.

The cognitive revolution, which occurred during the mid-1960s, led to a resurrection of interest in basic cognitive processes. While most of the theoretical tenets of the behaviorists were rejected by cognitive psychology, the rigorous experimental methods that they employed were adopted by the field. Cognitive scientists from disciplines other than cognitive psychology (such as computer science, philosophy, and neuroscience) now have a tendency to refer to research in cognitive psychology as *behavioral experimentation* (and the resulting methods as *behavioral methods*). Our own preference is to refer to the work as *experimental cognitive* research (and the methods as *experimental methods*).

In this chapter, we will first describe some basic experimental methodologies and then point out how they are used to address specific research questions. It is impossible for us to review thoroughly all the experimental methods in existence. Thus, we will selectively discuss methods which we believe are used most frequently. For the most part, we will describe methodologies used by cognitive psychologists. However, many of the methodologies have also been adopted by other cognitive scientists, particularly in interdisciplinary research. We will start with methodologies that were developed for studying *low-level* cognitive processes such as sensation and perception and then move to those used for studying *higher-level* cognitive processes such as memory and problem solving. One reason for this is that many of the tools that were developed for studying sensation and perception have been adapted for studying higher-level processes as well.

Basic experimental methodologies

Psychophysical methods

Psychophysical methods are among the oldest experimental techniques in psychology. Originally, these methods were developed to measure the psychological correlates of physical quantities, such as the subjective perception of the loudness of a tone as a function of the physical intensity of the stimulus. The goal was to understand the *psychophysical function*: the functional relationship between the physical and psychological quantities (Fechner, 1860). Basically, psychophysical methods fall into two categories: *subjective judgment* methods and *discrimination* methods. In the former, the subject reports his or her perception either by verbal report or by adjusting a comparison stimulus so that it bears some desired relation to the experimental stimulus. In the latter, subjects are asked to discriminate between two or more stimuli, and the percentage or pattern of errors is used as the basic data.

Subjective judgment methods As indicated above, these methods often rely on verbal reports: either a numerical judgment for a quantitative variable such as loudness or a categorical judgment, such as the color of a stimulus or which way a Necker (reversible) cube appears to be facing (see figure 27.1a). In the early research, it was assumed that (a) people's introspections were synonymous with perception, (b) people could reliably report their introspections, and (c) the numerical values which subjects reported were linearly related to the actual internal states that were giving rise to these perceptions. It is still generally agreed that there are a reasonable number of circumstances in which these subjective judgments are reliable and may be the only way to get at the question of interest (e.g., what color does the stimulus appear to be?); but there is general agreement that assuming that a numerical judgment literally measures an internal perceptual intensity is problematic. Among other things, these numerical judgments have been shown to be sensitive to details of the instructions and to the composition of the actual set of stimuli presented in the experiment.

As a result, these methods have been refined by getting people to judge the relationship between the experimental or *test* stimulus and a standard stimulus. For example, in studying the Müller–Lyer illusion, experimenters might measure the strength of the illusion by giving subjects one of the two *arrowhead* stimuli in figure 27.1b and then have them adjust a standard (a horizontal line segment) so that it is the same subjective length as the test stimulus. This kind of procedure has two different variants. In one, the subject has control of the length of the standard and adjusts it until subjective equality is reached. In the other, the experimenter presents a sequence of standards, and the subject has to make a forced choice between the standard and the test stimulus (in this example, saying whether the standard is longer or shorter than the test stimulus). The typical method is to present a sequence of standard stimuli either increasing or decreasing in length by a fixed amount, and when the subject's judgment changes, the experimenter reverses the direction of change and also the size of the increment. In the example above, the experimenter might start out with a standard appreciably bigger than the test stimulus and then present standard stimuli that are successively 1 cm smaller than the prior standard until the subject judges the standard to be smaller than the test stimulus. Then the experimenter increases the standard

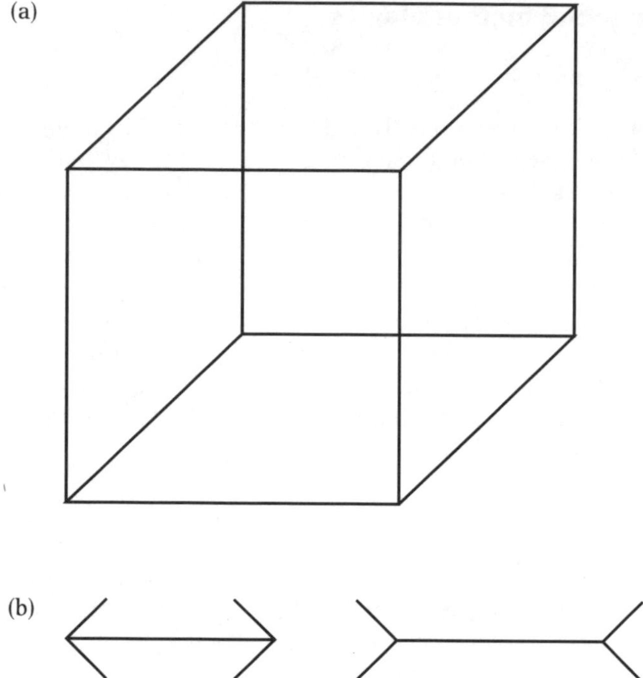

Figure 27.1 Two classical perceptual figures: (a): a reversible (Necker) cube; (b): the Müller–Lyer illusion.

stimuli by (say) 0.5 cm increments until the subject judges the standard to be bigger, at which time the standards would start to decrease in size by 0.25 cm, and so on, until the subject appears to be judging *longer* or *shorter* essentially at random. This is commonly called *the staircase method.* (Much the same procedure is used in the typical eye examination.)

In these comparison methods, as in the example above, the subject is usually asked to make a judgment of equality on a particular dimension. Other examples would be to judge whether two sources of light had the same amount of red in them or whether two tones were subjectively the same pitch. However, these methods were expanded in such a way as to try to reproduce the psychophysical functions without requiring absolute judgments (i.e., requiring numbers). In these methods, subjects would be asked to adjust a standard so that a certain relationship existed between two stimuli (e.g., adjust the standard tone so that it sounds half as loud as the test tone). Psychophysical functions can also be constructed from these data.

Discrimination data Almost all the above methods were widely criticized as being too subjective when the behavioristic revolution came to psychology. Hence, the emphasis changed to more objective methods, which were usually variations of the same–different judgment. Some of the time, these methods weren't too different from those mentioned in the previous two paragraphs, although the usual question was the more *objective* one of whether a subject could discriminate between two stimuli, rather than the more subjective one of having the subject judge whether two stimuli were equal on a particular dimension. The purpose of these methods was originally to determine *thresholds*, or stimulus values below which subjects perceived nothing. Thus, on some trials, subjects were given a stimulus, such as a faint tone, and on other, *catch* trials, they were given no stimulus but were asked if they detected the tone. The basic data would be the *hit rate*, or the percent of times a subject would say that he or she detected the tone when it was presented, and the *false alarm rate*, or the percent of times a subject *detected* a tone on the catch trials. If the hit rate is no higher than the false alarm rate for a given intensity stimulation, the stimulus intensity is deemed as being below the threshold. Above the threshold, the difference between the hit rate and the false alarm rate served as another index of the subjective intensity of a stimulus. There is considerable controversy, however, about how to compare the hit and false alarm rates. The most commonly used value is called *d′*, which is the difference between the normal transforms of the two probabilities and comes from *signal detection theory* (Swets et al., 1961).

A second, and perhaps more general, paradigm entails measuring *discrimination thresholds*, or assessing how well people can discriminate stimuli that are similar to each other. Originally, the method most widely used was a simple same–different discrimination, in which two stimuli are presented (e.g., tones that differ by a small amount in either pitch or volume), and the subject simply decides if they are the same or different. Because subjects differ in how often they choose to respond *same*, most modern experiments use more sophisticated variants. Perhaps the most common is called the "ABX" paradigm. For example, the subject gets tones A and B successively (let's say A = 1,000 Hz, B = 1,005 Hz), followed by tone X, which is the same as either tone A or tone B. The subject's task is to judge whether the third tone is the same as the first tone or the same as the second. (X is randomly varied to be the same as A or B from trial to trial.)

In the limit, if the subject is at chance on the task, one would say that his or her discrimination was below threshold. However, these methods are used more widely to get at discrimination functions. That is, the probability of discriminating A from B is used as a measurement of *how far apart* the two stimuli are subjectively. As a result, these methods have also been used to create psychophysical functions, using the discrimination measure to index the psychological distance on a continuum. One commonly used term here is the *just noticeable distance*, or *JND*, which (by convention) is the point at which the subject can correctly discriminate between two stimuli 75 percent of the time. To create a psychophysical function from these data, people often make the assumption that all JNDs are subjectively equal. That is, if someone can discriminate A = 1,000 Hz and B = 1,010 Hz tones correctly 75 percent of the time and can correctly discriminate C = 2,000 Hz and D = 2,015 Hz tones correctly 75 percent of the time, then the subjective difference between A and B is assumed to be the same as the subjective difference between C and D. This assumption, of course, need not be true and has been the source of considerable controversy.

Tachistoscopic methods As indicated above, the *classic* uses of psychophysical methods typically involved presenting stimuli for reasonably extended periods of time, with no time pressure on the subject's response. Thus they presumably measured the subject's perception of a stimulus after a reasonably extended period (often several seconds). However, a major issue that arose in perception was how percepts developed over time (sometimes called the "microgenesis" of perception). This question most naturally arises in the area of visual form perception. For example, consider the perception of a visual word such as *work*. Researchers were interested (for example) in whether individual letters were perceived before the word as a whole. (Another question of interest is whether a word's sound is perceived before or after its meaning.) A method that is widely used to study this issue employs brief presentation of a stimulus. Machines were developed about 100 years ago called *tachistoscopes* or *T-scopes* (for short), which present a stimulus for a very brief, precisely controlled amount of time. The older machines show stimuli, such as film transparencies or cards, which are illuminated by lights that can be turned on or off within a millisecond (msec). An easier, more modern version controls the presentation of the stimulus by opening or closing a shutter.

The method involves presenting a stimulus such as a word quite briefly (perhaps for 20 msec) and asking a subject to report what he or she saw. Most crudely, one could vary the presentation duration and determine how much time subjects needed in order to identify the word without error. This *temporal threshold* would indicate how quickly the visual information was being processed. Of perhaps greater interest, however, would be to compare the *temporal threshold* of two different stimuli such as a word and a letter in isolation. (Surprisingly, the threshold for a word is lower than for a letter in isolation.) One important detail, however, needs to be made clear: that is, if a visual stimulus is presented briefly (say for 20 msec), there is visible persistence after it is turned off. This visible persistence is commonly called *iconic memory* (Sperling, 1960). The duration of the persistence depends on the lighting conditions, but usually lasts at least 200 msec. Hence, if experimenters literally ran experiments as described in the previous paragraph, they would not identify the time course of early perception, because the stimulus would be effectively present for substantially longer than its actual duration. As a result, in the usual experiment, a *masking stimulus*, a stimulus which immediately follows the test stimulus in the same location, is used to terminate this visible persistence. When words are the target stimulus, the most effective masks, called *pattern masks*, are stimuli that have letter-like features (e.g., bits and pieces of letters). Given that masks are usually used, a computer is often involved as a tachistoscopic presentation device. Using a computer display is somewhat less satisfactory than using a real tachistoscope because of *phosphor persistence*: that is, the phosphor on the computer screen decays rapidly initially, but then has a relatively slow subsequent decay, which means that the stimulus is actually physically present an indeterminate amount of time after it has nominally been turned off. The assumption in the masking literature, however, is that the presentation of a mask directly on top of the stimulus makes extraction of further information from the relatively faint phosphor persistence virtually impossible.

One more psychophysical method, which has its analogs in subsequent sections, is employed in an attempt to get at the nature of the percept. This method involves presenting a *degraded* stimulus – a briefly presented stimulus or a stimulus with *noise* added, such as speech with auditory noise superimposed, or a visual stimulus with random dots superimposed – and having the subjects attempt to identify the stimulus presented. Of major interest is the pattern of errors or *confusions* which presumably index the similarity of stimuli to each other (e.g., in speech, *b* would be falsely recognized as *p* more often than as *t*). This method can be considered to be an extension of the discrimination threshold method introduced above. The former method suffers from the problem that, in many cases, subjects rarely make errors in identification unless stimuli are not degraded in some way; hence discrimination thresholds can only be used to scale the psychological distance of *normal* stimuli that are highly confusable (such as tones differing only slightly in pitch or loudness). On the other hand, the problem with degrading stimuli to uncover perceptual similarities is that masks are not neutral (e.g., the pattern of confusions for speech stimuli could easily depend on the particular form of *noise* used).

Reaction time methods

Perhaps the most commonly used method in current human experimental psychology is measuring *reaction time* (RT), or the time which elapses between the onset of the stimulus presentation and the initiation of the subject's response. Reaction time is measured precisely, usually to the nearest millisecond. Moreover, it involves averaging over many trials; usually somewhere between 10 and 50 trials per condition need to be run on a single subject. One use of RT is as a measure of the similarity of two stimuli: the longer it takes to judge that two stimuli are different, the more similar they are. This is sometimes used as an alternative to threshold methods, because RT is often sensitive to differences in stimuli even when the subject can discriminate them without error.

The major use of reaction time methods, however, is to measure the time course of mental events in order to understand the components of mental processing. Let us illustrate with examples from what is one of the most popular uses of RT methods: visual word identification. One obvious question is how long the process of word identification takes. One could have subjects either *name* the word (i.e., read it aloud), perform a *lexical decision* on the word (i.e., judge whether it is a word or not), or perform a *semantic categorization task* (e.g., decide whether it represents a living thing). While the RTs in the three tasks can all be used to make inferences about the time of word identification, it is clear that the total time between presentation of the stimulus and the response in any of the tasks could not be taken as the time to identify a word. This is because the total RT includes not only the time to identify the word but also subsequent processes relevant to formulating a decision and response relevant to the specific task the experimenter sets after the word has been identified. Moreover, there are concerns in the naming and lexical decision tasks about whether subjects truly have to identify the stimulus in order to either name it or judge whether it is a word or not. However, the absolute RTs in some tasks can be used as upper bounds on the total time of the operation of interest.

More commonly, RTs are used to assess the relative time course of a process. Thus, for example, one could contrast the RTs to high-frequency (common) words with those to low-frequency (less common) words in one of the tasks mentioned in the previous paragraph. If the frequency of a word affects the word identification process, it should have an effect on the time to name the word (it does). Moreover, one can interpret the difference in RTs as reflecting the difference in identification times for high- and low-frequency words if one assumes that word frequency doesn't affect *stages of processing* other than word identification (see Article 53, STAGE THEORIES REFUTED). Often experimenters seek converging evidence by assessing whether the pattern of RTs is the same across different tasks.

Four major paradigms involving RTs need to be mentioned. The first is the use of RT to a stimulus as a *probe* of ongoing mental activity. For example, subjects could be cued to attend to a region of space (while holding their eyes fixed). If that region of space is in fact attended to, subjects should process stimuli in that region of space better than stimuli that appear in other regions. Hence one could present a probe and ask subjects to respond to it; if RTs are faster when the stimulus is in that region of space than when it is outside it, one would have evidence that subjects can in fact attend to regions of space.

The second paradigm is often known as *priming* (Meyer and Schvaneveldt, 1971). A typical priming experiment involves two words (the *prime* and the *target*, respectively) which are presented in succession, and subjects are asked to respond to the target, making a lexical decision about it. The key manipulation is the relation between the prime and the target; in the original experiments, the prime and the target were either semantically related (e.g., *CAT–DOG*) or unrelated (e.g., *HAT–DOG*). The major result was that the RT to the target *DOG* was faster when preceded by the related prime *CAT*, indicating that (a) a memory trace of the prime lingers beyond its presentation, and (b) the relationship between target and prime is influencing processing of the target. One can thus view the above priming paradigm as a probe paradigm, where the target is the probe indicating (a) whether the prime is *active* and (b) which codes of the prime are active. Priming methods can also be used in more complex situations. For example, one can have a passage of text in which the word *it* appears, referring to a previously mentioned dog. One can then probe a subject immediately after reading *it* with *CAT* (e.g., requiring the subject to name it) to determine whether the concept *dog* has been reactivated. (The naming time would be compared to that of an unrelated control word.) One could also investigate the time course of this phenomenon by probing at different points in the text.

A third general paradigm using RTs is *search* (Neisser, 1964). For example, in visual search, an array of *display elements* (e.g., letters) is presented, and subjects press one key when a target (e.g., a *B*) is found and a second key when the target is not present in the display. One of the key manipulations is the number of *distractor* (i.e., nontarget) elements. If subjects need to search serially through the elements of the array to find the target, one would expect RT to be a linear function of the number of distractor elements. On the other hand, if subjects could process all the elements at the same time (*in parallel*), then one might even expect RTs to be unaffected by the number of distractor elements. In this research, it is often found that when discrimination of the target element depends on a single feature (e.g., a slanted line target with vertical line distractors), the search appears to be parallel; whereas when the discrimination of the target relies on integrating several features (e.g., a B target with various letter distractors), the search appears to be serial. (The pattern of data, however, has turned out to be far from simple.)

A fourth basic paradigm using RTs is the dual task paradigm (Posner and Boies, 1971). This paradigm is used to study attentional processes, but the basic logic is somewhat different from the probe paradigms above. The basic assumption of the method is that people have a finite processing capacity or pool of cognitive resources. If so, then one would expect that doing two tasks at once would be more difficult than doing either one alone if both took up processing resources, because the two tasks would have to share these resources. Thus, for example, one could have a subject name a visually presented word while simultaneously have them make a manual response to a tone (e.g., respond with the right hand if it is high in pitch and with the left hand if it is low in pitch). The key question is whether the response times for these tasks are slower when the subject must do the tasks simultaneously than when they are done in isolation. If so, one has evidence that both the tasks are using some shared processing resources. Moreover, the extent of the slowdown can be taken as a measure of the amount of resources that a task takes.

Another example of the dual task logic is the *phoneme monitoring* task, used to study speech perception. Here, the subject has two tasks to perform, both on the same stimulus. The first is to comprehend the meaning of the orally presented sentences. The second is to detect the presence of the phoneme /b/ and to press a key as soon as a /b/ is detected. The logic of the paradigm is that when lexical, semantic, or syntactic processing is easy, the target phoneme will be easy to detect. However, if the processing is difficult, it should take longer to detect the target phoneme.

We should close this section with a brief discussion of two general methods for using RTs to uncover basic components of mental processing. The earlier, dating from the nineteenth century, is the *subtractive method.* The original research by Donders (1868), for example, contrasted RTs in a *simple RT* task (e.g., the subject presses a single key when a stimulus appears, regardless of its form) with those in a *choice RT* task (e.g., the subject presses one of two keys, depending on whether it is a square or a circle). Assuming that the same stimuli were used in the two tasks, Donders assumed that the difference in RT between the two – about 250 msec – reflected the time course of mental components involved in the choice RT task that were not entailed by the simple RT task (e.g., discrimination of the two stimuli, deciding which of two keys to press).

The subtractive method has been criticized because it assumes *pure insertion* of processing stages (i.e., that the mental processes in the choice RT task are identical to those in the simple RT task except that certain additional processing stages have been added). This assumption is hard to justify, leading to the development of the more widely used *additive factors* method (Sternberg, 1969). The key assumption that this method rests on is that there are discrete mental stages of processing between stimulus and response (i.e., that one stage must be finished before another begins). If there are discrete stages, and if variable A affects one stage of processing and variable B affects a different stage of processing, then one should expect the effects of A and B to be additive; by contrast, if two variables affect the same stage of processing, then it is likely that the variables will interact.

An example of the use of the additive factors logic would be to answer the following question: In processing a printed word in visual *noise* (i.e., random dots added to the word), is there a prior *cleanup* stage of perceptual processing before the word is identified? Variable A, which presumably influences the *cleanup* stage could be whether the stimulus is *intact* (with no visual noise) or *degraded* (with visual noise), and variable B, which presumably influences word recognition, could be the frequency of the word in the language. If the effects are additive, this means that the addition of noise has the same effect on high- and low-frequency words, which would be evidence that the cleanup process precedes word identification. In fact, the noise affects recognition of lower-frequency words more, indicating that the cleanup process is not a separate stage, but intimately connected with the process of word recognition.

Processing time methods

Processing time methods are closely related to reaction time methods and tachistoscopic methods, but the distinction that we make in this chapter is that the unit of analysis in processing time methods is much larger than is typically the case in either reaction time or tachistoscopic experiments. As indicated above, most reaction time or tachistoscopic experiments require subjects to respond to a single stimulus or event, such as a single word which they must name. By contrast, with processing time methods, subjects either (a) may be asked to read a sentence or short paragraph of text, and the amount of time that it takes them to do so is measured; or (b) they may be given a fixed amount of time to read something, and their accuracy is measured.

Processing time methods have been used quite frequently in studies of language processing. Numerous studies have been conducted in which subjects are asked to read a sentence, and the amount of time that it takes to do so is recorded. By varying characteristics of target sentences, it is possible to infer something about the mental processes occurring during language processing as a function of differences between target sentences. For example, holding the number of words constant but varying the number of *semantic propositions* that are contained in the sentence has revealed that reading time is longer when there are more propositions.

While the *reading time* method just discussed provides information about how long it takes to read a sentence or a passage, many experimenters desire more accuracy in terms of how long it takes to process a given segment of text. Thus, the *self-paced reading* paradigm was developed to obtain greater precision about the time course of processing. In this paradigm, segments of text are presented on a screen, and subjects push a button to obtain the next segment of text. The size of the segment typically varies as a function of the issue being investigated: sometimes only a single word is presented as a segment, while at other times three words or an entire phrase or clause may be presented. In some versions of the self-paced reading paradigm, the new segment of text is always presented in the center of the video monitor from which the subject is reading. In another version (called the *self-paced moving window* paradigm) the text is presented sequentially across the screen, and usually dashes are presented as place-markers for the yet-to-be-presented words.

A major problem with the self-paced reading paradigm is that reading is slowed significantly when subjects must push a button to obtain more text. This is undoubtedly because the reaction time of the hand is slower than the reaction time of the eyes (in normal reading, new text is obtained by moving the eyes; see below). The question then arises as to whether subjects engage in different strategies than usual when reading is slowed down.

Another attempt to study language processing involves holding the amount of time that a segment of text is presented constant. In the *RSVP* (rapid serial visual presentation) paradigm, successive segments of text are presented for a set period of time (such as 200 msec) in the same spatial location on the video monitor (Potter et al., 1980). The presentation rate may be varied, so that single words are presented for 50, 100,

150, 200, or 250 msec, for example, and the rate of comprehension (measured as the number of words from a sentence that can be recalled) is then examined. This method is a generalization of the tachistoscopic method.

An analogous method for studying listening is the *time-compressed speech* paradigm, in which the rate of speech is increased by eliminating silent periods in the speech record to determine if listeners can comprehend speech when it is presented at fast rates. Finally, in the *auditory moving window* paradigm (which is analogous to the self-paced moving window paradigm discussed above), listeners push a button to get each new segment of speech.

These methods, of course, are not limited to studying reading or language processing. The time taken in other extended tasks, such as problem solving, are also used to make inferences about problem solving. Like the reading experiments, such experiments also attempt to investigate substages of processing by presenting only part of the problem and then having subjects press a key when they are ready for the next piece of information relevant to it.

Eye-movement methods

As we noted above, most of the processing time methods have an inherent problem wherein normal processing is slowed by the requirement, for example, to push a button to advance the next segment of text. The recording of eye movements allows researchers information about fine-grained processing without slowing down the process. For example, while the amount of time that it takes to read a sentence can be ascertained via processing time methods, recording of eye movements enables researchers to know how long it took to process individual words in the sentence. Furthermore, subjects can read at their own rate, so there is much less chance that specialized strategies will be adopted in experiments in which eye movements are recorded. Like the self-paced reading paradigm, the unit of analysis can be either a word or a series of words (usually referred to as regions of analysis). However, exactly where subjects look when eye movements are recorded is under their own control (unlike the self-paced method, where subjects are not free to look back in the text).

Eye movements have been utilized very effectively to study questions about reading, language processing (such as how readers process syntactically ambiguous sentences or sentences containing lexically or phonologically ambiguous words), visual search, scene perception, face perception, and so on. In many experiments dealing with reading and language processing, for example, readers are asked to read text, and their eye movements are recorded. Where readers look and how long they look at individual words have been documented to be related to the ease or difficulty associated with processing those words.

One particularly informative eye-movement method has been the *eye-contingent display change* paradigm (McConkie and Rayner, 1975). In this paradigm, changes in the text or scene that a subject is viewing are made contingent upon the position of the eye. For example, in the *eye-contingent moving window* paradigm, a window of text around a reader's fixation point contains normal text, but outside the window the text is perturbed in some way (all of the letters outside the window could be replaced with *x*'s). In this paradigm, an eye-movement recording system is interfaced with a computer, which in turn is interfaced with a video monitor from which the subject reads.

By sampling the position of the eye every millisecond and then quickly updating the contents of the text on the monitor, changes in the text can be made very quickly (within 7 msec). In this paradigm, then, wherever the reader looks, normal text is displayed, but outside the window the text is perturbed. The size of the window can be varied to determine how much useful information the reader is able to process on each eye fixation. Another variation of the eye-contingent paradigm is the *boundary* paradigm. In this paradigm, the contents of a critical word location are changed during an eye movement that brings the reader's eye onto the target word. This is accomplished by identifying a boundary location (e.g., the last letter of the word preceding the target word) in the computer, and when the readers' eye movement crosses the boundary, the initially displayed stimulus is changed to the target word. So, if the sentence initially read is "The man picked up an old map from the chart in the bedroom," and the boundary location is the letter *e* in *the*, when the readers' eye movement crosses the boundary location, the word *chart* will change to *chest*. By varying the similarity of the initial stimulus to the target stimulus, it is possible to make inferences about what type of information was obtained about the target location prior to fixating on it. In the example presented here, the stimuli are orthographically similar, but semantically different; orthographic, lexical, semantic, and phonological differences can be exploited.

Eye-movement and analogous paradigms have also been used extensively in studying children's processing of stimuli. In these experiments, however, the main focus is on using what the child looks at as a measure of preference. This has been employed with quite young children, to uncover processing capabilities. For example, very young children (a month or two old) are shown two stimuli side by side (one a human face and one containing features of a human face rearranged), and the amount of time that the child looks at each in a fixed period of time (such as a minute or two) is recorded. In these experiments, it is found that infants look longer at the faces than the comparison stimuli, indicating that they apparently can discriminate between faces and nonfaces. It is worth adding that such experiments need to be run carefully to avoid experimenter bias. For example, the infants usually have to be placed in their mother's lap, so the situation has to be arranged so that the infant, but not the mother, can see the stimuli; otherwise the mother might subtly guide the infant. Similarly, the videotapes of the infant have to be scored by someone who does not know whether the face is the stimulus on the left or the right.

These preferential viewing experiments have been used to examine a whole range of intellectual activities in children. The logic is that if two stimuli differ on a key attribute, then if viewing time is not random (i.e., significantly different from equal viewing time for the two alternative stimuli), the child must be able to process the key attribute. This has been extended to more cognitive attributes, such as whether an action is physically possible. There are two chief problems with the method. The first is that in many experiments it is not clear which way the effect should go. For example, in the above example, if children had looked at the face less often than the comparison stimulus, the researchers would have drawn the same conclusion. This makes the method fairly weak, as there is little good theory about why one stimulus is preferred. A second problem is that there is often uncertainty about what the basis for the discrimination is. For example, in the face experiment, it is hard to construct the comparison stimulus so that one is really certain that the preference is for *faceness* rather than some lower-level feature of the face stimulus.

An example of the preferential looking paradigm that is more extended in time is studying children watching television. In these experiments, the percentage of time that children look at the television (rather than something else in the room) is used as an index of their processing and their interest in the television message. For example, it has been shown that even relatively young children (two–three years old) will look at the television less often when the program "doesn't make sense" (e.g., when segments of a cartoon are randomly sequenced) than when it does. It should be pointed out that these methods with children are measuring attention in a grosser manner than those with adults; the children are not only moving their eyes – they are usually moving their heads and bodies as well.

"Physiological" methods

In the last 20 years, there has been an increasing emphasis on studying mental processes by obtaining various *physiological* measures of processing. In all these methods, a recording device placed on the subject's skin records some electric or magnetic activity that is thought to be a measure of some bodily process indicative of mental activity. Sometimes such records are used as indirect measures of mental activity. For example, measures of heart rate are thought to index effort (among other things) and thus are used in studying attention (especially with infants); measures of lip muscle activity may be used to index *covert speech*; palm sweating may also be used to index attention. However, the current focus is largely on measures thought to index actual brain activity. These methods will be discussed more fully elsewhere in this volume, so our discussion will be brief.

The most common method used is *event-related potentials* (*ERPs*), which are recorded from electrodes placed on the scalp. Although the electrical signals that are observed are small, by averaging the signal over many trials per condition (time-locked to the presentation of the stimulus), one can obtain a fairly reliable pattern of voltages that indexes brain activity. The voltages vary continuously both in polarity and in amplitude, resulting in a series of peaks and valleys that are reasonably interpretable. The following illustrates the method of interpretation. There is a peak (P200) that varies in amplitude with whether a stimulus is attended to or not, but not with whether or not it is the target in a detection experiment. Thus, it is claimed that this peak indexes the process of identifying a stimulus, but precedes the process of deciding what response to make to the stimulus.

While the logic is uncomfortably close to being circular, the method does offer some advantages over RT methods. First, if the ERP peak or valley does reflect the actual brain activity corresponding to identification of the stimulus, then this method might allow one to record the time course of mental events. In this case, the latency of the brain component (less than 200 msec) is appreciably less than the RTs measured in the previous section for identification of stimuli, so they give a tighter upper bound on when stimuli are identified. However, there is no guarantee that the ERP wave is reflecting the mental processing of interest (rather than being a later reflection of mental processing). Indeed, some ERP effects can occur after the response is made.

Another advantage of ERPs is that they have the potential to locate the activity in the brain. Currently, when spatial localization is attempted, arrays of 64 or more electrodes are placed on the scalp. With these methods, one can get quite precise spatial patterns of electrical activity on the scalp. However, there is considerable controversy as to how well the pattern of brain activity can be inferred from the pattern of scalp activity. Fortunately, the inferences made from ERP recordings appear to be reasonably consistent with those obtained from methods that are better spatial indices of brain activity. (A method related to ERPs uses magnetic signals and gives better spatial localization measures, but is far more expensive.)

The two most widely used measures for localizing brain activity during mental operations are *functional magnetic resonance imaging* (*fMRI*) and *positron-emission tomography* (*PET*). Both may be somewhat risky to the subject (unlike ERPs); the former involves subjecting the head to strong magnetic fields, and the latter involves ingestion of small amounts of radioactivity. They also require access to extremely expensive equipment. They both index physiological processes in the brain relating to ongoing neural activity (in one case, blood flow), and one can obtain a fairly precise record of which areas in the brain are the most active.

There are two major problems with these latter two methods. The first is that they record the total activity over quite extended temporal intervals (often minutes). Thus, at present, they can index only the location of brain activity, not the time course. The second is that the brain undergoes all sorts of activity unrelated to the task at hand. Thus, a variant of the subtractive method (discussed earlier) is often used to interpret the patterns of activities. For example, in one condition a subject passively looks at words and doesn't respond to them, whereas in another condition the subject decides whether each word represents a living thing. The difference in activity between the two patterns is then taken as indicating the brain regions used for processing the meaning of words (assuming that such processing is not going on automatically during the passive condition.) A similar subtractive logic is often used in ERP experiments as well.

The evolving methodology is to use both ERPs and either PET or fMRI on the same subjects, combining the temporal precision of ERPs with the spatial precision of PET or fMRI. Together, these methods hold out the promise of providing patterns of brain activity that are indices of mental processes. However, we should emphasize that, even in the best cases, these are merely indices of mental activity. For example, if one finds that region X of the brain is reliably more active at time *t* when subjects perform a semantic classification of a word than when they look passively at the same word, one cannot assume that that region of the brain is performing the classification at that time. The increased activity might merely reflect (for example) the fact that some other region of the brain has performed the computation of classification, and that region X is maintaining the result some time after the original computation for the purposes of responding or storing the information in memory. Hence, it is unlikely that these methods will replace the methods described elsewhere in the chapter. Instead, they can provide a useful supplement, enabling us to make some educated guesses about the relation between mental processing and brain activity.

Memory methods

Most experiments designed to assess what the memory representation is after someone has learned something are variations of a few basic paradigms. In these paradigms, subjects are typically presented with some stimuli in a learning phase of an experiment and then tested to see what they remember. For example, subjects may be presented with a list of 25 unrelated words and then asked to write down as many of the words as they can remember. Such a task is typically referred to as a *recall* task – subjects are asked to recall as many of the words as they can. The number of words recalled is used as a measure of memory. Another measure of interest is the pattern of errors. (This is related to the stimulus confusion methods described earlier.) For example, if errors tend to be semantically related to the test words, one might infer that the memory that they are being retrieved from is semantically organized.

A second basic memory paradigm involves a *recognition* judgment regarding the words. In such an experiment, subjects may see a study list of words (say 50 words) and then be presented with a test list containing 100 words, 50 of which were in the original list and 50 of which were not. They are then asked to judge which of the words in the test list are *old* (i.e., were in the study list) and which are *new* (i.e., not in the study list). The recognition memory paradigm is conceptually similar to several of the psychophysical methods we discussed earlier; in the former, the question is whether the sensory representation is above threshold, and if so, how strong it is; whereas in the latter, one is asking analogous questions about a memory representation.

A third method for testing memory is *savings*. In such a paradigm, subjects may be shown a list of words to remember and be given a large number of trials until they can remember them all correctly. They would then be brought back several days later, and their memory tested by examining the number of trials needed to learn the same list correctly. If the number of trials needed to relearn the list is less than the number of trials in the original learning, then one can infer that the subjects must have some memory of the original list. (Of course, controls must be run to ensure that the improvement isn't due to some improvement in the ability to memorize.) The savings method or variants thereof have often been shown to be the most sensitive measures of memory.

Modern variants of the savings method are variations of the probe methods described earlier. One example would be to present subjects with a study list of 50 words and then later present a series of 100 words tachistoscopically and have subjects identify them. Note that this is not a recognition task; the test phase merely asks subjects to identify the test words. What is usually found is that subjects identify test words that were in the study list more accurately and/or more rapidly than test words that were not in the study list. That is, the memory of prior learning is tested indirectly, or implicitly.

Interestingly, there is evidence of amnesic patients with damage to the limbic system in the brain (the hippocampus and related regions) who show relatively intact memory when tested on such implicit memory tasks, but extremely impaired memory on both recognition and recall tasks. This has led to theorizing that there are two fundamentally different kinds of memory systems, usually called *procedural* and *declarative*, in which these implicit memory tasks tap procedural memory, and recall and recognition tasks tap declarative memory. However, there is considerable controversy on this issue, and it is by no means clear that any method of testing memory is a *pure* index of one type of memory.

Question-answering methods

These methods are similar to memory methods, but the focus is somewhat different, in that these methods are directed more to nonliteral use of material. Two common examples are text comprehension and problem solving. In the former, subjects are given text to read and then questions to answer based on their understanding of the text. In the latter, subjects are given a problem (such as a mathematics *story problem*) to solve. As indicated above, the focus in text comprehension is not on literal memory of the text, but rather on what the subject *has done to* the text (e.g., what inferences he or she has made, how the text has been *organized* in memory). Subsequent sections will discuss more introspective, *process-oriented* types of data obtained from these activities (such as having subjects talk out loud while trying to solve problems), but at present, we will discuss more objective, *product-oriented* types of data.

As with the simpler, perceptual tasks discussed earlier, one can use simple accuracy or latency measures to get at mental activity. For example, one could index comprehension of a passage of text by some combination of speed in answering questions about the passage and number of errors. Or one could use speed of solution of a problem or the probability that subjects solve it as an index of the difficulty of the problem. These could be used to assess the effects of certain variables (e.g., comparing comprehension of the same ideas expressed in prose in two different ways).

In addition to measuring the number of errors, these methods often rely on examining the pattern of errors. For example, in text comprehension, a common question is whether the subject has made inferences of certain types. One method of probing this is to ask subjects whether they have seen a certain phrase in the text literally. If they falsely *recognize* inferences with a high probability, one may infer that the inference was drawn and stored as part of the process of reading the text. Conversely, one can examine the pattern of responses to the more usual pattern of comprehension questions (which ask the subject to answer on the basis of both the actual text and everything that can reasonably be inferred from it). For example, it might be hypothesized that children do not draw certain types of causal inferences from text (especially those relating to motivations of people in a story). Accordingly, one can draw up a battery of questions that probe different kinds of inferences and examine the pattern of results on this battery of questions.

We should point out that recent work has used probe methods related to priming to get at similar issues. One simple example is the following. If a text has a sentence in it, such as "The assailant stabbed the woman with his weapon," but *knife* is never explicitly mentioned in the paragraph, one might want to know whether the reader inferred that the weapon was a knife. In such an experiment, the subject would read the passage, and immediately afterwards, they would name a target word aloud (*knife* in this case). The RT to name the target word would be a measure of the prior activation of *knife*. (This RT would have to be compared with other conditions, one of which didn't involve stabbing, and one which explicitly mentioned *knife*.)

A problem with all the above methods which attempt to assess processing of a complex stimulus such as a text is that they involve responding to a question (or some other probe). Thus, one often cannot be sure whether the inference drawn from the material occurred during the original processing or only at the time at which the question or probe was presented. As a result, there is considerable controversy about how to interpret such results. However, the use of probe tasks such as naming a single word, minimizes the chances that the inferential processes occur merely at the time of test.

The above controversy, however, does not extend to the use of these methods for studying problem solving and inference, because in these cases, one is examining thought processes explicitly provoked by the question. We will give two brief examples of using the pattern of responses to make inferences about mental operations. First, a common paradigm used to study inferential reasoning involves the presentation of an assertion like "Robins can get edomosis" (a fictitious disease); the subject is then to judge whether some other animal (e.g., a bat or an ostrich) is likely to get the disease. Commonly, subjects are asked to estimate the likelihood that the inference is true (scaling it from 0 to 100). Presumably, this method can be used to infer which aspects of similarity are used to support inference (e.g., superficial visual similarity versus deeper biological similarity). Some research has used this method cross-culturally to determine whether different cultures use similarity differently in making such judgments.

The second example uses a question-answering technique to examine the mental processes underlying probabilistic judgments. For example, when given the question, "The mean IQ of the population of eighth-graders in a city is known to be 100. You have selected a random sample of 10 children for a study of educational achievements. The first child tested has an IQ of 150. What do you expect the mean IQ to be for the whole sample?," a majority of subjects answer 100 (even though the correct answer is 105). This indicates that subjects are not using correct statistical logic in answering the question. Moreover, the answer is consistent with a theory called *representativeness*, which asserts that many people believe that all samples should be representative of the population (and hence have the same mean). Thus, these data are taken as evidence for the theory. Needless to say, there could be other explanations for this single piece of information. The logic used by good research in this area is to develop a set of substantive questions in which representativeness (or whatever theory is being examined) predicts a specific pattern of wrong and right answers, and then to examine whether the observed pattern of answers is consistent with the predicted pattern.

A problem with this research is to know what *consistent* means. Clearly, it would be unrealistic to expect that all subjects (or even a sizable subset of subjects) exhibit a pattern exactly like the predicted one. A subject might misread a question, make a slip of the pen in answering a question, or otherwise fail to attend; moreover, subjects may not consistently use the same logic or *heuristic* on every problem. Yet the theory might be of value in uncovering a mental process that a reasonable number of subjects use reasonably often in thinking about a particular domain. The question of how consistency should be measured, however, is still largely unresolved.

Observational methods

All of the methods described above typically involve subjects being tested in a laboratory setting. Thus, stimuli can be presented for precise periods of time; exact measurements can be taken of how long it takes subjects to respond or process the material; the number of errors made can be counted; and so on. Observational methods, on the other hand, typically involve examining behavior in a natural setting, which could be outside a laboratory setting (though it need not be).

Observational methods often utilize videotaping of subjects, and data analysis is based on trained raters analyzing the behavior of interest. Videotaping is not always done, and sometimes the ratings take place directly on the behavior (though obviously videotaping is preferred because when the raters disagree, the behavior can be reanalyzed). Interviews are also frequently used in observational studies.

These methods can be considered generalizations of the preferential looking studies described earlier to examine face perception and television viewing. The distinction here is that these observational methods often use significantly less objective coding schemes for characterizing what the subject has done. For example, the record of someone engaged in problem solving or in a social situation might be coded in terms of the types of activity they are engaged in and how much time they are involved in each activity. In many cases, these behavioral categories are not nearly as clear cut as deciding whether a child is looking at a TV set or not. Hence, it is usually important to have different people independently code the record, to determine whether such a coding scheme is reasonably reliable.

Often, the analysis rests heavily on a content analysis of the subject's verbalizations during the activity. Two methods for provoking verbalization – both developed to study problem solving and reasoning – are worth special mention. In one, the subject is encouraged to *think out loud* almost continuously as he or she is attempting to solve a problem. A key aspect of this method is that the experimenter does not interact with the subject while he or she is verbalizing.

A second way of provoking verbalization is an *interview*. This could be characterized as an idealized Socratic tutoring session. In the interview, a subject is again encouraged to *think out loud*, but not necessarily continuously. However, the experimenter/interviewer periodically gives *probes* (either explicit questions or subtle hints to the subject that further elaboration of a thought is needed) that presumably help to draw out the mental processes that subjects bring to bear on questions. Obviously, the fact that the experimenter is probing the subject's response makes it hard to characterize what the effective stimulus is for the subject.

A key controversy that runs through this research is how *objective* or *scientific* it is. Usually, there is some variable being manipulated in the situation, such as the type of problem presented, some detail of the problem, or some aspect of a social situation. One then examines whether this manipulation produces any differences in behavior, as examined by either an objective coding scheme or a more subjective or intuitive evaluation scheme. An experiment in which the subject behaves naturally in the situation and in which the coding is objective is to be preferred as long as one has confidence that the behavior itself and the coding scheme are really indicative of mental processes that are going on. However, in examining activities like problem solving, objective indices, such as how often the subject looks up at the ceiling, draws pictures to represent the problem, etc., are informative, even though they are unlikely to reveal a great deal about what mental structures the subjects are bringing to bear on the problem. This is why many researchers have used verbalization as a key part of the data base. But there is considerable controversy at present about the best way to collect and interpret such a record.

As indicated above, a widely used method involves subjects *thinking out loud* (with no intervention by the experimenter). The verbalizations obtained in this manner are usually assumed to be a relatively direct record of the subject's actual thought processes. There are two principal problems with this method. The first, obviously, is that the verbal *protocol* may in fact be only tangentially related to the subject's actual thought processes. The second is that *thinking out loud* is unnatural for many subjects. Spontaneously, subjects often stop talking out loud quite soon after starting to solve a problem. As a result, they often need to have preliminary training in *thinking out loud* (i.e., to keep talking throughout the problem-solving activity). This raises the concern that the act of continuous verbalization is causing subjects to solve the problem in a substantially different way than they would if left to their own devices. That is, this procedure may be influencing how subjects think by identifying thinking with verbal report.

The interview technique also asks subjects to verbalize, but not in a continuous stream. Instead, the emphasis is on having subjects give justifications for their answers or explanations of their reasoning. Unlike the *thinking out loud* paradigm, where verbalizations are assumed to reveal steps in thinking, the interview technique generally treats verbal reports as products of the thought process (i.e., a thought may be verbalizable only after the subject has solved a problem or a part of a problem). Probing by the experimenter is used, because it is felt that subjects' spontaneous verbalizations may be quite ambiguous when it comes to revealing actual mental processes. Of course, this results in a loss of objectivity and reproducibility in method. As a result, these methods are often used in conjunction with the more objective question-answering methods described above, where the objectively formulated question evolves from the interview situation.

Summary

The methods we have discussed have progressed from more objective methods, in which the experimenter has complete control of the stimulus, and the measurement of the subject's response is quite objective, to methods in which both control of the stimulus and characterization of the response are quite loose. This progression is largely a result of the problem area being studied.

The initial methods evolved for studying the perception of simple stimuli that are easy to characterize objectively (simple tones and visual stimuli) and are not extended in time. We then moved to methods that are used to study more complex ecological stimuli, such as reading text or processing extended speech or television, but in which the task was still largely one of processing the stimulus. However, such tasks soon involve memory and higher-level processes, as processing text or television involves having not only a literal memory, but also imperfect memory and interpretation and condensation. We then moved to methods used for explicitly studying memory, and then to methods for studying problem solving and thinking.

The questions then move from ones that are easy to characterize, such as "Did the subject perceive the stimulus correctly?" and "What did the subject remember about the study list or text?," to much looser questions, such as "What mental processes did the subjects use when solving this problem?" We have indicated that somewhat less objective methods have often been used to study these more complex mental processes and that these less objective methods may be justified. This latter characterization is somewhat controversial, however, and there are many psychologists who object to the use of such methods. For instance, in our example of people's views of random sampling, many researchers believe that only *objective* (i.e., reproducible) question-answering methods are valid. These involve giving people several variants of a problem and looking at the pattern of responses across these variants.

There are several problems with these more objective methods, however. The major one is that the responses are often difficult to interpret. One symptom of this is that it is often the case that minor differences in wording may cause large differences in patterns of response. Thus, for example, it is often unclear whether the subject's difficulties with a particular problem are a superficial function of the wording or due to the *logic* of the problem (the latter is presumably what is of interest to the experimenter). One possible solution here is to present many different wordings and to look at the pattern of responses over all these wordings. On the other hand, this is likely to lead to a combinatorial explosion of different problems and/or different versions of a single problem needing to be presented. This may require either an impractical number of subjects or asking a subject to solve an impractically large number of problems. As a result, there appears to be a real need for *less objective* methods (i.e., ones which may not be so easily reproducible) as well for studying complex mental processes. In our opinion, research in these areas should attempt to combine both more and less objective methods.

References and recommended reading

Donders, F. C. 1868: Over de snelheid van psychische processen. Onderzoekingen gedaan in het Psychologisch Laboratorium der Utrechtsche Hoogeschool. *Tweede reeks*, 2, 92–120. Translated by W. G. Koster for *Attention and Performance*, vol. 2, which was published in *Acta Psychologia*, 30, 412–30.

Fechner, G. T. 1860: *Elemente der psychophysik*, vols. 1 and 2. Leipzig: Breitkopf and Härtel.

McConkie, G. W. and Rayner, K. 1975: The span of the effective stimulus during a fixation in reading. *Perception and Psychophysics*, 17, 578–86.

Meyer, D. E. and Schvaneveldt, R. W. 1971: Facilitation in recognizing pairs of words: evidence of a dependence between retrieval operations. *Journal of Experimental Psychology*, 90, 227–34.

Neisser, U. 1964: Visual search. *Scientific American*, 210 (June), 94–102.

Posner, M. I. and Boies, S. W. 1971: Components of attention. *Psychological Review*, 78, 391–408.

Potter, M. C., Kroll, J. F. and Harris, C. 1980: Comprehension and memory in rapid, sequential reading. In R. S. Nickerson (ed.), *Attention and Performance*, vol. 8, Hillsdale, NJ: Erlbaum, 395–418.

Sperling, G. 1960: The information available in brief visual presentations. *Psychological Monographs*, 74 (no. 498).

Sternberg, S. 1969: The discovery of processing stages: extensions of Donders' method. In W. G. Koster (ed.), *Attention and Performance*, vol. 2, also published in *Acta Psychologica*, 30, 276–315.

Swets, J. A., Tanner, W. P. Jr. and Birdsall, T. G. 1961: Decision processes in perception. *Psychological Review*, 68, 301–40.

28

Cognitive ethology

MARC BEKOFF

Naturalizing animal minds

Cognitive ethology is the comparative, evolutionary, and ecological study of nonhuman animal (hereafter animal) minds, including thought processes, beliefs, rationality, information processing, and consciousness. It is a rapidly growing interdisciplinary field of science that is attracting much attention from researchers in numerous, diverse disciplines, including those interested in animal welfare (Cheney and Seyfarth, 1990; Ristau, 1991; Griffin, 1992; Allen and Bekoff, 1995, 1997; Bekoff and Allen, 1997; Bekoff and Jamieson, 1996). Cognitive ethology can trace its beginnings to the writings of Charles Darwin, an anecdotal cognitivist (Jamieson and Bekoff, 1993), and some of his contemporaries and disciples. Their approach incorporated appeals to evolutionary theory, interests in mental continuity, concerns with individual and intraspecific variation, interests in the worlds of the animals themselves, close associations with natural history, and attempts to learn more about the behavior of animals in conditions that are as close as possible to the natural environment where selection has occurred. They also relied on anecdote and anthropomorphism to inform and motivate more rigorous study. In addition, cognitive ethologists are frequently concerned with the diversity of solutions that living organisms have found for common problems. They also emphasize broad taxonomic comparisons and do not focus on a few select representatives of limited taxa. Many people inform their views of cognitive ethology by appealing to the same studies over and over again (usually those done on nonhuman primates) and ignore the fact that there are many other animals who also show interesting patterns of behavior that lend themselves to cognitive studies.

Comparative cognitive ethology is an important extension of classical ethology, because it explicitly licenses hypotheses about the internal states of animals in the tradition of classical ethologists such as Nobel laureates Niko Tinbergen and Konrad Lorenz. However, although ethologists such as Lorenz and Tinbergen used terms such as *intention movements*, they used them quite differently from how they are used in the philosophical literature. *Intention movements* refers to preparatory movements that might communicate what action individuals are likely to undertake next, and not necessarily to their beliefs and desires, although one might suppose that the individuals did indeed want to fly and believed that if they moved their wings in a certain way they would fly. This distinction is important, because the use of such terms does not necessarily add a cognitive dimension to classical ethological notions, although it could.

In his early work Tinbergen identified four overlapping areas with which ethological investigations should be concerned: namely, evolution (phylogeny), adaptation (function), causation, and development (ontogeny), and his framework also is useful for those interested in animal cognition (Jamieson and Bekoff, 1993). The methods for answering questions in each of these areas vary, but all begin with careful observation and description of the behavior patterns that are exhibited by the animals under study. The information provided by these initial observations allows a researcher to exploit the animal's normal behavioral repertoire to answer questions about the evolution, function, causation, and development of the behavior patterns that are exhibited in various contexts.

Donald R. Griffin and modern cognitive ethology

The modern era of cognitive ethology, with its concentration on the evolution and evolutionary continuity of animal cognition, is usually thought to have begun with the appearance of Donald R. Griffin's (1976/1981) book *The Question of Animal Awareness: Evolutionary Continuity of Mental Experience*. Griffin's major concern was to learn more about animal consciousness; he wanted to come to terms with the difficult question of what it is like to be a particular animal (for critical discussion of Griffin's agenda see Jamieson and Bekoff, 1993). While Griffin was concerned mainly with the phenomenology of animal consciousness, it is only one of many important and interesting aspects of animal cognition (Allen and Bekoff, 1997). Indeed, because of its broad agenda and wide-ranging goals, many view cognitive ethology as being a genuine contributor to cognitive science in general. For those who are anthropocentrically minded, it should be noted that studies of animal cognition can also inform, for example, inquiries into human autism.

Methods of study

Ethologists interested in animal minds favor research in conditions that are as close as possible to the natural environments in which natural selection occurred or is occurring. When needed, research on captive animals can also inform the comparative study of animal cognition; but cognitive ethologists are resistant to suggestions that (1) field studies of animal cognition are impossible (difficult, yes, but certainly not impossible); (2) they should give up their attempts to study animal minds under natural conditions; and (3) studies of learning and memory are all that are needed to learn about animal cognition. Naturalizing the study of animal cognition and animal minds in the laboratory and in the field should lead to a greater appreciation for the cognitive skills of animals living under natural conditions. Animal minds can be studied rigorously using methods of natural science and will not ultimately have to be reduced or eliminated.

The tractability of cognitive questions involves application of a diverse set of comparative methods in order to draw inferences about cognitive states and capacities. Cognitive research may include staged social encounters, playback of recorded vocalizations, the presentation of stimuli in different modalities, observation of predator–prey interactions, observation of foraging behavior, application of neurobiological techniques, and studies of social and other sorts of learning. Computer analyses are also useful for those who want to learn what kind of information must be represented in an adequate computational model.

There are no large differences between methods used to study animal cognition and those used to study other aspects of animal behavior. Differences lie not so much in what is done and how it is done, as in how data are explained. Thus Allen and Bekoff (1997) argue that the main distinction between cognitive ethology and classical ethology lies not in the types of data collected, but in the understanding of the conceptual resources that are appropriate for explaining those data.

Perhaps one area that will contribute more to the study of animal minds than to other areas of comparative ethology is neurobiology and behavior. Those interested in the cellular or neural bases of behavior and animal cognition and consciousness may use techniques such as positron emission tomography (PET) that are also employed in other endeavors. In general, studies using neuroimaging have provided extremely valuable data for humans engaged in various sorts of activities, whereas the use of these and other imaging techniques on animals has not been pursued rigorously for individuals engaged in activities other than learning or memory in captivity. Furthermore, while neurobiological studies are extremely important to those interested in animal cognition, there remains an explanatory gap between neurophysiological processes and behavior.

Behavioral studies usually start with the observation, description, and categorization of behavior patterns of animals. The result of this process is the development of an ethogram, or behavioral catalog, of these actions. Ethograms present information about an action's form or morphology and its code name. Descriptions can be based on visual information (what an action looks like), auditory characteristics (sonograms, which are pictures of sounds), or chemical constituents (output of chromatographic analyses of glandular deposits or urine or feces, for example). It is essential that great care be given to the development of an ethogram, for it is an inventory that others should be able to replicate without error. Permanent records of observations allow others to cross-check their observations and descriptions against original records. The number of actions and the breadth of the categories that are identified in a behavioral study depend on the questions at hand; but generally it is better to split, rather than lump together, actions in initial stages, and then lump them together when questions of interest have been carefully laid out.

In studies of behavior it is important to know as much as possible about the sensory world of the animals being studied. Experiments should not ask animals to do things that they cannot do because they are insensitive to the experimental stimuli or unmotivated by them. The relationships among normal ecological conditions and differences in the capabilities of animals to acquire, process, and respond to information constitute the domain of a growing field called *sensory ecology*. A good ethologist asks what it is like to be the animal under study and develops an awareness of the senses that the animals use singly or in combination with one another. It is highly unlikely that individuals of any other species sense the world in the same way we do, and it is unlikely that even members of the same species sense the world identically all of the time, so it is important to remain alert to the possibility of individual variation.

Stimulus control and impoverished environments

While carefully conducted experiments in the laboratory and in the field are often able to control for the influence of variables that might affect the expression of behavioral responses, it is usually the case that there is a possibility that the influence of some variable(s) cannot be accounted for. Field studies may be more prone to a lack of control, because the conditions under which they are conducted are inherently more complex and less controllable.

An illustration of the concern for control is found in the excellent cognitive ethological field research of Cheney and Seyfarth (1990) on the behavior (e.g., communication and deception) and minds of vervet monkeys. In their studies of the attribution of knowledge by vervets to each other, Cheney and Seyfarth played back vocalizations of familiar individuals to other group members. These researchers were concerned, however, about their inability to eliminate "all visual or auditory evidence of the [familiar] animal's physical presence" (p. 230). Actually, this inability may not be problematic if the goal is to understand "how monkeys see the world." Typically, in most social situations the physical presence of individuals and access to stimuli from different modalities may be important to consider. Vervets, other nonhumans, and humans may attribute mental states using a combination of variables that are difficult to separate experimentally. Negative or inconclusive experimental results concerning vervets' or other animals' attribution of mind to other individuals may stem from impoverishing their normal environment by removing information that they normally use in attribution. Researchers may also be looking for complex mechanisms involved in the attribution of minds to others and thus overlook relatively simple means for doing so. Just because an animal does not do something does not mean that it cannot do it (assuming that what we are asking the animal to do is reasonable: i.e., within their sensory and motor capacities). Thus, insistence on absolute experimental control that involves placing and maintaining individuals in captivity and getting them accustomed to test situations that may be unnatural may greatly influence results. And the resulting claims, if incorrect, can wreak havoc on discussions of the evolutionary continuity of animal cognitive skills. Cheney and Seyfarth recognize some of these problems in their discussion of the difficulties of distinguishing between alternative explanations, maintaining either that a monkey recognizes another's knowledge or that a monkey monitors another's behavior and adjusts his own behavior to the other.

Although control may be more of a problem in field research than in laboratory work, it certainly is not the case that cognitive ethologists should abandon field work. Cognitive ethologists and comparative or cognitive psychologists can learn important lessons from one another. On the one hand, cognitive psychologists who specialize in highly controlled experimental procedures can teach something about the importance of control to those cognitive ethologists who do not perform such research. On the other hand, those who study humans and other animals under highly controlled and often contrived and impoverished laboratory conditions can broaden their horizons and learn about the importance of more naturalistic methods: they can be challenged to develop procedures that take into account possible interactions among stimuli within and between modalities in more naturalistic settings. For example, among those who are interested in important, *hot* questions about animal minds that are typically studied in controlled captive conditions (e.g., inquiries into the possibility of self-recognition), there is a growing awareness that more naturalistic approaches are needed. The use of single tests relying primarily on one modality – for example, vision – for comparative studies represents too narrow an approach. Ultimately, all types of studies should be used to exploit the behavioral flexibility or versatility of the animals under study.

Some criticisms of cognitive ethology

A balanced view of cognitive ethology requires consideration of critics' points of view. Criticisms of cognitive ethology come in many flavors but usually center on: (1) the notion that animals do not have minds; (2) the idea that (many, most, all) animals are not conscious, or that so little of their behavior is conscious (no matter how broadly defined) that it is a waste of time to study animal consciousness (cognitive ethology is really a much broader discipline than this suggests, see below); (3) the inaccessibility to rigorous study of animal mental states (they are private) and whatever (if anything) might be contained in them; (4) the assumption that animals do not have any beliefs, because the contents of their beliefs are not similar to the contents of human beliefs; (5) the lack of rigor in collecting data; (6) the lack of large empirical data bases; (7) the nature of the (merely instrumental) soft, nonparsimonious, yet complex explanations that rely heavily on theoretical constructs (e.g., minds, mental states) that are offered for the behavioral phenotype under study (they are too anthropomorphic, too folk-psychological, or too *as if-fy* – animals act *as if* they have beliefs or desires or other thoughts about something; and (8) the heavy reliance on behavior for motivating cognitive explanations (but this is not specific to cognitive inquiries).

While most criticism comes from those who ignore the successes of cognitive ethology, to those who dismiss it in principle because of strong, radical behavioristic leanings, or those who do not understand the basic philosophical principles that inform cognitive ethology, it should be pointed out that more mechanistic approaches to the study of animal cognition are not without their own problems. For example, comparative psychologists often disregard the question of how relevant a study is to the natural existence of the animals under consideration and pay too much attention to the logical structure of the experiments being performed, without much regard for more naturalistic approaches. Noncognitive, mechanistic *rules of thumb* can also be very cumbersome and nonparsimonious and often appeal to hard-to-imagine past coincidences. Furthermore, it is not clear whether the differences between noncognitive rules of thumb and cognitive explanations are differences in degree or differences in kind. Both noncognitive and cognitive explanations can be *just so* stories (just like many evolutionary explanations) that rely on hypothetical constructs, and neither type applies in all situations.

Three case studies

Three case studies that support the application of a broadly comparative cognitive ethological approach can be found in recent field research of anti-predator behavior in birds (Ristau, 1991; Bekoff, 1995b) and field and laboratory research on social play behavior in various canids (domestic dogs, wolves, and coyotes; Bekoff, 1995a). Many other examples can be found in Cheney and Seyfarth, 1990; Ristau, 1991; Griffin, 1992; Allen and Bekoff, 1995, 1997; Bekoff and Jamieson, 1996; and references therein. Although Griffin has included the results of many excellent studies of the possibility of language in nonhuman primates, cetaceans, and birds in his broad discussions of animal minds, they do not fall squarely within the primary domain of cognitive ethology as I envision it: the study of natural behaviors in natural settings from an evolutionary and ecological perspective (see Article 2, ANIMAL COGNITION). (Of course, this is not to discount the importance to cognitive ethology of research on captive animals.) Only future research will tell if the behavior of the few captive individuals who have been intensively studied in *language studies* (and those captive individuals observed in other endeavors) is related to the behavior of wild members of the same species, or if the data from captive animals are more a demonstration, admittedly important, of behavioral plasticity and behavioral potential.

Ristau (1991) studied injury feigning in piping plovers (the broken-wing display), wanting to know if she could learn more about deceptive injury feigning if she viewed the broken-wing display as intentional or purposeful behavior ("The plover wants to lead the intruder away from her nest or young") rather than as a hard-wired reflexive response to the presence of a particular stimulus, a potentially intruding predator. She studied the direction in which birds moved during the broken-wing display, how they monitored the location of the predator, and the flexibility of the response. Ristau found that birds usually enacted the display in the direction that would lead an intruder who was following them further away from the threatened nest or young, and also that birds monitored the intruder's approach and modified their behavior in response to variations in the intruder's movements. These and other data led Ristau to conclude that the plover's broken-wing display lent itself to an intentional explanation: that plovers purposely lead intruders away from their nests or young and modify their behavior in order to do so.

In another study of anti-predator behavior in birds, Bekoff (1995b) found that western evening grosbeaks modified their vigilance or scanning behavior in accordance with the way in which individuals were positioned with respect to one another. Grosbeaks and other birds often trade off scanning for potential predators and feeding: essentially (and oversimplified), some birds scan, while others feed, and some birds feed, when others scan. Thus, it is hypothesized that individuals want to know what others are doing and learn about others' behavior by trying to watch them. Bekoff's study of grosbeaks showed that when a flock contained four or more birds, there were large changes in scanning and other patterns of behavior that seemed to be related to ways in which grosbeaks attempted to gather information about other flock members. When birds were arranged in a circular array, so that they could see one another easily, compared to when they were arranged in a line, which made visual monitoring of flock members more difficult, birds who had difficulty seeing one another were (1) more vigilant, (2) changed their head and body positions more often, (3) reacted to changes in group size more slowly, (4) showed less coordination in head movements, and (5) showed more variability in all measures. The differences in behavior between birds organized in circular arrays and birds organized in linear arrays were best explained in terms of individuals' attempts to learn, via visual monitoring, about what other flock members were doing. This may say something about if and how birds attempt to represent their flock, or at least certain other individuals, to themselves. It may be that individuals form beliefs about what others are probably doing and predicate their own behavior on these beliefs. Bekoff argued that cognitive explanations were simpler and less cumbersome than noncognitive rule-of-thumb explanations (e.g., "Scan this way if there are this number of birds in this geometric array" or "Scan that way if there are that number of birds in that geometric array"). Noncognitive rule-of-thumb explanations did not seem to account for the flexibility in animals' behavior as well or as simply as explanations that appealed to cognitive capacities of the animals under study.

Social play behavior is another area that lends itself nicely to cognitive inquiries. The study of social play involves issues of communication, intention, role playing, and cooperation, and the results of this type of research may yield clues about the ability of animals to understand each others' intentions. Play is also a phenomenon that occurs in a wide range of species and affords the opportunity for a comparative investigation of cognitive abilities, extending the all-too-common narrow focus on primates that dominates discussions of nonhuman cognition. A recent study of the structure of play sequences in canids (Bekoff 1995a) showed that an action called the *bow* (an animal crouches on its forepaws, elevates its hind end, and may wag its tail) is often used immediately before and immediately after an action that can be misinterpreted and disrupt ongoing social play.

The social play of canids (and of other mammals) includes actions, primarily bites, that are used in other contexts (e.g., agonistic or predatory encounters) that do not involve bows. It is important for individuals to tell others that they want to play with them and not fight with them or eat them, and this message seems to be sent by play-soliciting signals, such as the bow, which occur almost only in the context of social play. In canids and other mammals, actions such as biting, accompanied by rapid side-to-side shaking of the head, are used in aggressive interactions and also during predation and could be misinterpreted when used in play. Bekoff hypothesized that if bites accompanied by rapid side-to-side shaking of the head or other behavior patterns could be misread by the recipient and could result in a fight, for example, then the animal who performed the actions that could be misinterpreted might have to communicate to its partner that this action was performed in the context of play and was not meant to be taken as an aggressive or predatory move. On this view, bows would not occur randomly in play sequences; the play atmosphere would be reinforced and maintained by performing bows immediately before or after actions that could be misinterpreted. The results of Bekoff's study of different canids supported the inference that bows served to provide information about other actions that followed or preceded them. In addition to sending the message "I want to play" when performed at the beginning of play, bows performed during social play seemed to carry the message "I want to play despite what I am going to do or just did – I still want to play" when there might be a problem in the sharing of this information between the interacting animals. The noncognitive rules-of-thumb, "Play this way if this happens" and "Play that way if that happens" seem to be too rigid an explanation for the flexible behavior that the animals showed.

Where to from here?

There are other examples that could have been chosen, but these three make the case that chauvinism on either side(s) of the debate as to how to study animal behavior and how to explain animal behavior is unwarranted; a pluralistic approach should result in the best understanding of the nonhumans with whom we share the planet. Sometimes some nonhumans (and some humans) behave as stimulus–response machines, and at other times some nonhumans (and some humans) behave in ways that are best explained using a rich cognitive vocabulary. Methodological pluralism is needed:

species-fair methods need to be tailored to the questions and the animals under consideration, and competing hypotheses and explanations must always be considered.

Those interested in animal cognition should resist the temptation to make sweeping claims about the cognitive abilities (or lack thereof) of all members of a given species. A concentration on individuals and not on species should form an important part of the agenda for future research in cognitive ethology. There is a lot of individual variation in behavior within species, and sweeping generalizations about what an individual ought to do because she is classified as a member of a given species must be taken with great caution. Furthermore, people often fail to recognize that in many instances sweeping generalizations about the cognitive skills (or lack thereof) of species, not individuals, are based on small data sets from a limited number of individuals representing few taxa, individuals who may have been exposed to a narrow array of behavioral challenges. The importance of studying animals under field conditions cannot be emphasized too strongly. Field research that includes careful, well-thought-out observation, description, and experimentation that does not result in mistreatment of animals is extremely difficult to duplicate in captivity. While it may be easier to study animals in captivity, they must be provided with the complexity of social and other stimuli to which they are exposed in the field; in some cases this may not be possible.

Cognitive ethologists should also strive to make the study of animal cognition tractable by carefully operationalizing the processes under study. Cognitive ethology can raise new questions that may be approached from various levels of analysis. For example, detailed descriptive information about subtle behavior patterns and neuroethological data may be important for informing further studies of animal cognition and may be useful for explaining data that are already available. Such analyses will not make cognitive ethological investigations superfluous, because behavioral evidence takes precedence over anatomical or physiological data in assessments of cognitive abilities.

To summarize, those positions that should figure largely in cognitive ethological studies include: (1) remaining open to the possibility of surprises concerning animal cognitive abilities; (2) concentrating on comparative, evolutionary, and ecological questions and sampling many different species, including domesticated animals – going beyond primates and avoiding talk of *lower* and *higher* animals, or at least laying out explicit criteria for using these slippery, value-laden terms; (3) naturalizing methods of study by taking the animals' points of view (talking to them in their own languages) and studying them in conditions that are as close as possible to the conditions in which they typically live; often animals do not do what we expect them to (sometimes prey will approach predators), and knowledge of their natural behavior is important in the development of testable, realistic models of behavior; (4) trying to understand how cognitive skills used by captive animals may function in more natural settings; (5) studying individual differences; (6) using all sorts of data, ranging from anecdotes to large data sets; and (7) appealing to different types of explanations as best explanations of the data under scrutiny. Cognitive ethology need not model itself on other fields of science such as physics or neurobiology in order to gain credibility. Hard-science envy is what led to the loss of animal and human minds in the early part of the twentieth century.

We are a long way from having an adequate data base from which claims about the taxonomic distribution of various cognitive skills or about having a theory of mind can be put forth. Consider studies that show that some monkeys cannot perform imitation tasks that some mice can. If the point is to answer the question whether monkeys are smarter than mice, it is misleading, for there is no reason to expect a single linear scale of intelligence. In the world of mice it may be more important to be able to do some things than it is in the world of monkeys, but in other respects a monkey may have a capacity that a mouse lacks. There is also much variation within species, and this too must be documented more fully.

It is unlikely that science will make complete contact with the nature of animal minds at any single point. Both *soft* (anecdotal) information and *hard* (empirical) data from long-term field research are needed to inform and motivate further empirical experimental research. So, questions such as "Do mice ape?" or "Do apes mice?" are premature. Does this mean that many, some, or no animals have a mind or a theory of mind? It would be premature to attempt to answer these questions definitively at this time, given our current state of knowledge.

References and recommended reading

*Allen, C. and Bekoff, M. 1995: Cognitive ethology and the intentionality of animal behaviour. *Mind and Language*, 10, 313–28.

—— —— 1997: *Species of Mind: The Philosophy and Biology of Cognitive Ethology*. Cambridge, Mass.: MIT Press.

Bekoff, M. 1995a: Play signals as punctuation: the structure of social play in canids. *Behaviour*, 132, 419–29.

—— 1995b: Vigilance, flock size, and flock geometry: information gathering by western evening grosbeaks (Aves, fringillidae). *Ethology*, 99, 150–61.

Bekoff, M. and Allen, C. 1997: Cognitive ethology: slayers, skeptics, and proponents. In R. W. Mitchell, N. Thompson, and L. Miles (eds), *Anthropomorphism, Anecdote, and Animals: The Emperor's New Clothes?*, Albany, NY: SUNY Press, 313–34.

*Bekoff, M. and Jamieson, D. (ed.) 1996: *Readings in Animal Cognition*. Cambridge, Mass.: MIT Press.

Cheney, D. L. and Seyfarth, R. M. 1990: *How Monkeys See the World: Inside the Mind of Another Species*. Chicago: University of Chicago Press.

*Griffin, D. R. 1976/1981: *The Question of Animal Awareness: Evolutionary Continuity of Mental Experience*. New York: Rockefeller University Press.

—— 1992: *Animal Minds*. Chicago: University of Chicago Press.

Jamieson, D. and Bekoff, M. 1993: On aims and methods of cognitive ethology. *Philosophy of Science Association*, 2, 110–24.

*Ristau, C. (ed.) 1991: *Cognitive Ethology: The Minds of Other Animals. Essays in Honor of Donald R. Griffin*. Hillsdale, NJ: Erlbaum.

29

Deficits and pathologies

CHRISTOPHER D. FRITH

Introduction: a brief history of neuropsychology

The systematic examination of the relationship between brain and behavior is generally considered to have begun in 1861 with Broca's description of a patient with a specific language deficit associated with a circumscribed lesion of the left frontal cortex. This observation was taken to show that there was a specific region in the brain concerned with language which was relatively independent of other regions concerned with other abilities. In the next decade a number of other patients were described with various circumscribed deficits. Wernicke described aphasic patients who, in contrast to Broca's patients, were better at speech production than they were in understanding speech. Harlow described the case of Phineas Gage, who showed little sign of loss in his intellectual abilities but suffered a disastrous change in personality after bilateral damage to his frontal cortex. The aim of many of these early neuropsychologists was to associate particular psychological functions with specific brain regions by detailed study of neurological cases.

At the beginning of the twentieth century, however, this whole enterprise fell into disrepute (see Shallice, 1988, ch. 1). This was largely due to lack of appropriate technology. On the one hand, the lesions in the patients studied could not be localized accurately. On the other hand, there was a dearth of experimental procedures and well-formed theories to describe the behavior of the patients. In consequence, much of the early work seemed very speculative, and the notion of localization of function was widely rejected. Alternative concepts, such as *mass action*, were proposed. According to this, there is a general loss of intellectual ability proportional to the volume of tissue damaged, irrespective of its exact location (Lashley, 1929). If brain regions are massively interconnected and the brain functions *as a whole*, then it is impossible to work out what the functioning of a particular part of the brain is on the basis of what happens when that part is damaged.

In the second half of the twentieth century the situation changed dramatically. First, a number of brain-imaging techniques were developed, which permitted the precise localization of lesions while the patient was still alive. Second, experimental techniques and standardized tests were developed, which allowed the patient's intellectual functioning to be characterized in detail. Third, behaviorism was replaced by cognitive psychology as the dominant theoretical approach. Cognitive psychologists attempt to understand behavior in terms of the hidden processes involved in the representation and manipulation of information and knowledge. This level of discourse provides an ideal link between brain and behavior, since the function of the brain can also be described in terms of the representation and manipulation of knowledge. Cognitive

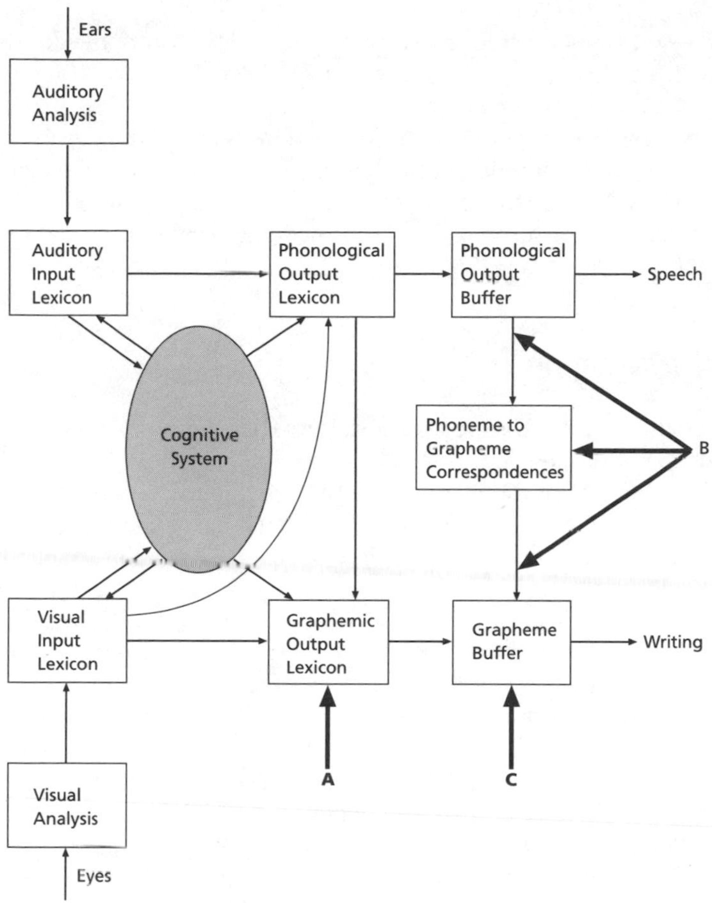

Figure 29.1 John Morton's logogen model of a system for recognizing and producing single words, showing the many different stages involved. A, B, and C represent the regions in the model where damage would produce specific problems in writing (agraphia).

neuropsychology, which is the application of the cognitive approach to the study of patients with damaged brains, now flourishes, especially in Europe. Many practitioners, like their forebears at the end of the nineteenth century, consider that the study of single cases is particularly informative.

Cognitive psychologists are noted for their use of *box and arrow* diagrams. The boxes in these diagrams indicate the type of information or process required by the system, and the arrows indicate how this information flows through the system. A major purpose of these models is to indicate *functional segregation* within the system that the model is illustrating. Processes that occur in different boxes can function independently of other processes, whereas processes that occur within a box cannot be separated. Given this framework, patients with brain damage provide an important source of data for testing and developing these models. If there are relatively independent components within some cognitive system, then it should be possible to find patients in whom one component is damaged while others remain intact. Cognitive neuropsychologists have devoted much time to the examination and discussion of a few such special patients.

In these discussions, at least until recently, the exact location of the damage has not been important. The behavior of the patient is used solely to demonstrate the existence of independent cognitive components.

Of course, it is possible to reveal the existence of independent cognitive modules and identify their function in normal people. However, such studies are difficult, because the human brain takes advantage of its parallel construction to solve problems by many processes simultaneously. It is only in the damaged brain that these processes can be seen functioning in isolation.

Functional segregation

My concern in this chapter is to convince readers that the study of people with brain abnormalities can inform our understanding of normal cognitive function. Those who have argued against this position have used the metaphor of the television set that has been attacked with an axe. By studying the behavior of your set after it has been randomly damaged in this way, they ask, how would it be possible to infer the normal function of its various components? Gregory (1961) has specified the problem very precisely: "In a serial system the various identifiable elements of the output are not separately represented by the discrete parts of the system. . . . The removal, or activation, of a single stage in a series might have almost any effect on the output of a machine, and so presumably also for the brain" (p. 321). The key word here is *serial*. If, as seems most likely, the brain is not a serial system, but a collection of many modules which function in parallel and with a great deal of independence, then it *is* possible to discover the function of its components on the basis of circumscribed damage.

The principle that the brain contains many independent modules is known as *functional segregation*. Evidence for this principle is both empirical and theoretical. Of particular importance is the evidence from anatomy. At the turn of the century Korbinian Brodmann undertook the prodigious task of showing that the cortex of the human brain could be mapped into 52 discrete areas on the basis of cytoarchitectonics: the structure of cells and the characteristic arrangement of these cells in layers. The assumption is that, if areas of the brain have different anatomical structure, then they must also have a different function. However, this does not mean that a single *Brodmann area* has the same function throughout, even if there is no obvious difference in cytoarchitecture. Recently Brodmann's work has been confirmed and refined using modern techniques (see Article 4, BRAIN MAPPING).

During the last few decades direct evidence about brain function has come from techniques such as recording electrical activity in single cortical cells. For example, the work of Semir Zeki and his colleagues has demonstrated the functional segregation of the extra-striate visual cortex into areas specializing in color, form, and motion (Zeki, 1993).

As is so often the case, after the empirical demonstration, a number of powerful intellectual arguments have been put forward to show that we should expect functional segregation on purely theoretical grounds. First, there is the argument from evolution (Simon, 1969). The brain is an enormously complex system which has been *designed* by evolution. In other words, this system has developed through a series of very small, random changes. If the brain functioned as a single, heavily interconnecting entity, then a small change in one part would require balancing changes throughout the system. In these circumstances it is very unlikely that any small change is going to produce an improvement in the functioning of the system as a whole. However, if the brain contains a number of relatively independent modules, then a small change in one module will hardly affect the functioning of the other modules, and improvement of local functioning is more likely to occur. On the basis of this argument, a modular brain will evolve more rapidly and be more successful than one that works by mass action.

Second, there is the argument from economy. Representation of environmental features like contours depends upon mechanisms such as lateral inhibition which require interactions between neurons concerned with particular sub-features (e.g., orientation). Such interactions depend upon connections between neurons. These connections will be achieved more economically if the relevant neurons are grouped together. Thus neurons are likely to be grouped in terms of what they represent – that is, their function.

Third, there is the argument from complexity. It is possible to derive a measure of the complexity of systems like the brain, which consist of a large number of heavily interconnected units. Simulations show that this measure of neural complexity is low for systems that are composed either of completely independent parts (maximum segregation) or of parts that show completely homogeneous behavior (maximum integration, mass action). Neural complexity is highest for systems that conjoin local specialization with global integration. This result emphasizes that although functional segregation is an important principle of brain design, we must also consider *functional integration*. Brain function also depends on the interactions between the modules.

In the rest of this chapter I shall present some examples of studies of deficits and pathologies which have major implications for our understanding of functional segregation in the normal brain.

Memory

Memory can be defined as a mechanism whereby events in the past influence our behavior in the future. However, the study of patients with memory problems has shown that there are so many different kinds of memory that this term may cease to have much value.

The most obvious features for categorizing the different kinds of memory are content and time. We can remember many different kinds of things: telephone numbers or patterns, the meanings of words or what we had for breakfast yesterday, a poem by Blake or how to ride a bicycle. The length of time we can remember something also varies markedly. We may remember a telephone number for a few minutes, remember what we had for breakfast for a few days, and how to ride a bicycle for the rest of our lives.

Short-term memory and long-term memory

Early theoretical accounts of memory suggested that there was an important distinction between short-term memory (minutes) and long-term memory (hours, days, years), but they assumed that material was passed through the short-term store into long-term memory. However, in the 1970s a series of patients who had a severe, specific impairment of short-term memory and no impairment of long-term memory were described by Elizabeth Warrington and her colleagues. One of these cases is J.B. J.B. had a meningioma removed from the temporo-parietal region of her left hemisphere when she was in her twenties. Some 20 years later she still has a severe deficit of short-term memory for spoken words. She cannot remember a string of random letters or numbers longer than about two items. In spite of this handicap she has no problem producing or understanding speech, and her long-term memory is perfectly normal. Furthermore, her span for strings of letters or numbers presented visually is normal. Because her short-term memory deficit is so circumscribed, she has no difficulty with her responsible job as a medical secretary. Practical problems arise for her only when she is given a telephone number or a long, unusual name over the telephone.

What this and similar cases suggest is that there are different short-term stores for different kinds of material, and that these short-term stores function independently of long-term memory. J.B. has a specific problem with short-term storage of verbal material (words, numbers, letters) presented through the auditory modality. This phonological store seems to be located in the left inferior parietal region of the brain.

Evidence that short-term memory involves a number of modality-specific stores which can function independently of one another can also be found in normal people, but this demonstration depends upon subtle and ingenious experimentation. Alan Baddeley and his colleagues have developed a model of working memory in normal people and have demonstrated the existence of independent components by using the dual task paradigm. For example, concurrent articulation (saying *blah blah blah blah*) interferes with short-term memory for words but not with short-term memory for visual patterns. We presume that this interference occurs because the brain system involved in articulation overlaps with the brain area concerned with short-term memory for words but not with the system concerned with short-term memory for patterns.

Learning skills and remembering episodes

Someone with amnesia has a severe impairment of memory, while other intellectual functions remain largely intact. However, this memory impairment is quite different from that observed in patients like J.B. An amnesic patient usually has no problem with short-term memory tasks such as repeating back a telephone number. As long as the material stays in mind, as in a working memory or an active memory task, he or she has no problem. But once the material is no longer in mind, due to distraction or the passage of time, amnesic patients find it very difficult to recall that material. Thus they cannot remember what they had for breakfast or who they saw yesterday. However, this impairment of long-term memory is also somewhat circumscribed. In spite of their obvious long-term memory impairment, amnesic patients can acquire new information and retain and use it over long periods (Parkin, 1987). I investigated such a patient and had him practice a computer game every day for a week. His skill at this game increased from day to day at the same rate as with normal people, and he showed complete retention of his skill after a gap of a week. In marked contrast to his normal acquisition of the skill was his recollection of previous episodes. Every time I saw him, he claimed that he had never played the game before and could not remember who I was. Many such cases are described in the literature, the most famous being

H.M. who was first studied by Brenda Milner in 1957. H.M.'s amnesia dated from the time when he had parts of his temporal lobe surgically removed from both hemispheres for relief from intractable epilepsy.

The learning of a skill such as playing a computer game or riding a bicycle is often called "procedural learning." Performance of the skill is elicited when we are in the right context (e.g., sitting on a bicycle), but we have very little insight into how we do it. Procedural memory is often contrasted with declarative memory, thanks to which we can talk about what we remember. However, a more important categorization in relation to amnesia is episodic memory. There is an important distinction between remembering facts about the world (e.g., the names of breakfast cereals) and remembering what it was like to have breakfast this morning. These are both examples of declarative memory, but only the latter is an example of episodic memory: that is, remembering a particular episode in one's life rather than a fact. It is episodic memory, in particular, that is impaired in amnesic patients.

Familiarity and recollection

An important principle, derived in part from the study of brain-damaged patients, is that normal performance depends upon the parallel operation of more than one process (dual process models). Pathological cases can be found in which one process has been damaged, so the operation of the remaining process can be observed in isolation. In the case of amnesia this phenomenon occurs in recognition memory tasks. Such patients may not recollect that they were shown a list of words a few minutes ago, but they are often quite good at distinguishing between words that were presented in this list (old words) and new words. An influential theory (Mandler, 1980) suggests that there are two processes engaged in this word recognition task. First, the presentation of an *old* word acts as a cue eliciting recollection of the original experience of hearing the list. If this happens, then the subject *remembers* that the word was presented. Second, the presentation of an *old* word gives rise to a feeling of familiarity which is not associated with presentation of an old word. In this case the subject will *know* that the word was presented, without actually remembering the experience. For an amnesic patient there is no recollection of the original experience, but the sense of familiarity remains. This is why such patients can perform recognition memory tasks quite well. Various sophisticated experimental techniques have now been developed for separating out these two processes in the performance of normal subjects also. This spared ability in amnesic patients is an example of information being available to guide behavior (so that they can recognize the *old* word), while this information is not sufficient to provide a conscious recollection of when the word was first presented.

Describing and reaching

A particularly striking example in which information is available, but not to consciousness, is provided by certain neurological cases with damage to the posterior parts of the temporal lobe. These patients can still see, in the sense of distinguishing light from shade, but can no longer recognize objects from their shape. This loss of shape awareness applies even to low-level aspects like the orientation of a line. One such patient is unable to adjust a stick so that it lines up with a narrow slot. Her performance on this task is hardly better than chance. However, if she is asked to push her hand through the slot, she can perform this task perfectly well and adjusts her hand to the appropriate orientation before it reaches the slot (Goodale and Milner, 1992). Clearly, information about the orientation of the slot is available to control reaching movements of the hand, but this information is not available to consciousness. In addition, information about the shapes of objects is not available to permit recognition of the identity of the objects. Patients have also been reported who have precisely the opposite problem. These patients have lesions in the parietal lobe. They can easily tell you what the orientation of a slot is but cannot orient their hand appropriately when pushing it through the slot. They can readily distinguish between a square and a rectangle from the lengths of the sides, but they cannot use this information to choose the appropriate distance between thumb and finger when picking up such objects.

The existence of these two groups of patients is an example of a *double dissociation*. Within each group we have a single dissociation; the patients are selectively impaired on one type of task – for example, they can reach for objects but not describe them. A double dissociation occurs when two patients have complementary deficits; one patient can reach but not describe, while the other can describe but not reach. Double dissociations are very important to cognitive neuropsychologists, because they are considered to be strong evidence of underlying functional segregation. However, this conclusion is not without its critics (see Shallice, 1988, ch. 10). Although the existence of segregated areas would lead to double dissociation, it does not follow that the existence of a double dissociation proves that certain functions are segregated.

In the double dissociation we are considering here, it appears that the same information (about orientation) is available to describe or identify an object but is not available to control reaching (or vice versa). However, it is possible that this information is in different forms. In order to reach for something, information about orientation or distance or size must be in egocentric coordinates that relate directly to one's own current position relative to the object or, more particularly, to the position of the limb that is going to do the reaching. By contrast, in order to describe or recognize an object, we need information concerning its shape, such as size, orientation, and distance, in terms which are independent of our current position or the current position of the object (i.e., object-centered coordinates). If this account is correct, it is not that the brain damage has prevented access to consciousness of information that still exists; rather, it is that the intact region that controls reaching contains information that never reaches consciousness even in the undamaged brain. In this case it should be possible to demonstrate the existence of this *unconscious* control of action in normal volunteers (see Article 23, UNCONSCIOUS INTELLIGENCE).

In one such experiment subjects look at a screen on which there is a target surrounded by a frame. In certain circumstances it is possible to make the subject report that the target has moved when, in fact, it was the frame that moved. However, if the subject is asked to point to the target rather than report its position, then no error occurs (Bridgeman, 1992). This example shows that there is a separation between the information that the subject reports (this is conscious information that is in error) and the information that he uses to control his pointing response. This latter information is not conscious.

Neuropsychiatry

Capgras's syndrome

A patient with Capgras's syndrome believes that a member of his immediate family (e.g., his wife) has been replaced by a double. This symptom is rare but can be found in both psychiatric cases (e.g., schizophrenics) and in patients with an obviously organic disorder (e.g., Alzheimer's disease). Similar disorders have also been reported in which the patient believes that, for example, his house has been replaced by a duplicate. When asked to explain the reason for these strange beliefs, patients often report that the person (or the house) looks slightly different from usual, but they have some difficulty specifying how so. In the past, such symptoms were explained in psychodynamic terms. However, in recent years, explanations have been proposed in terms of loss of access to certain types of information about faces (or houses). As we shall see, such explanations have many parallels with our discussion in the previous section about the distinction between reaching for and identifying objects.

Lesions in posterior regions of the brain can sometimes lead to a disorder of face perception known as *prosopagnosia*. This disorder has been studied in some detail, and its cognitive basis is well described, although there is still argument about the precise location of the lesion or lesions that are necessary and/or sufficient to produce it. A patient with prosopagnosia can no longer recognize familiar faces. He can usually distinguish one face from another, but he no longer knows whose face it is. However, he is still able to recognize who the person is from other cues. He does not recognize his wife's face, but, as soon as she speaks or moves, he knows her. It appears that the perceptual appearance of a face no longer provides the necessary information for identifying that face. However, in a number of cases it has been shown that information about the identity of the face is available, even though the patient remains unaware of this. For example, a patient can identify which faces are familiar even though he cannot say who they belong to. In one study it was shown that patients could be primed by the identity of a face, even though they were unaware of this identity. Thus, a photograph of Hillary Clinton would be unrecognized, but the subsequent recognition of the name "Bill Clinton" (presented in written form) would be faster. It has also been observed that patients show autonomic responses to familiar faces, even though they do not recognize them. These results suggest that different kinds of information about identity are available when looking at faces. One kind of information leads to conscious identification of the face, while other kinds do not. This distinction is analogous to the one we have just drawn between reaching for and identifying objects. In prosopagnosia the information that permits conscious identification of faces is no longer available. What kind of information is available in the intact channel? The answer is still unclear, but it seems plausible that aspects such as familiarity and emotional responses might be included.

How can this account help us to understand Capgras's syndrome? The interesting suggestion has been made that Capgras's syndrome is the mirror of prosopagnosia (Ellis and de Pauw, 1994). The conscious channel through which faces are identified is intact, but the unconscious channel which brings a sense of familiarity and an emotional response is no longer transmitting information. The patient is confronted

Box 29.1 Examples of passivity experiences (from Mellor, 1970)

Passivity of thought I look out of the window and I think the garden looks nice and the grass looks cool, but the thoughts of Eamonn Andrews come into my mind. . . . He treats my mind like a screen and flashes his thoughts onto it.

Passivity of emotion I cry, tears roll down my cheeks and I look unhappy, but inside I have a cold anger because they are using me in this way, and it is not me who is unhappy, but they are projecting unhappiness into my brain.

Passivity of will When I reach my hand for the comb it is my hand and arm which move, and my fingers pick up the pen, but I don't control them. . . . I sit there wanting them to move, and they are quite independent, what they do is nothing to do with me.

with a person who clearly looks like his wife, but who does not elicit in him either a sense of familiarity or the usual emotional response. There is clearly something different about this person. If the patient has also developed a psychotic tendency to adopt bizarre beliefs and detect conspiracies, then he may conclude that his wife has been replaced by another.

Just as in the case of reaching for or identifying objects, in the case of face perception also, results from the study of pathological groups are highly revealing. In both cases they strongly suggest that parallel, independent pathways are involved in the recognition of the identity of faces. It is difficult to see how this conclusion could have been reached from the study of normal volunteers alone.

Psychosis

By far the commonest brain disorders are those seen by psychiatrists. Of these, schizophrenia is the most devastating. In the last two decades ample evidence has emerged that there is a biological basis for this disorder, but it has also become clear that it is not associated with localized brain lesions. One of the problems in studying schizophrenia is the variability of its presentation and course. A popular current approach is to study particular symptoms, such as auditory hallucinations, just as neuropsychologists have studied amnesia in patients with various lesions. Experimental studies of certain symptoms associated with schizophrenia provide my final example of how studies of psychopathology can throw light on normal mind–brain relationships.

Passivity experiences There is a class of symptoms associated with schizophrenia called *passivity phenomena*. The patient believes or, more accurately, experiences that his actions are no longer fully controlled by himself but are controlled by some outside force. Alien thoughts are *inserted* into his mind. Simple movements are *made* for him by outside forces. Even emotions may be imposed upon him (see box 29.1).

I believe that auditory hallucinations also belong to this category. There is evidence, at least in some cases, that the voices the patient hears are his own thoughts or subvocal speech. In all these cases the patient's thoughts, actions, or emotions are attributed to some outside force. These abnormal experiences are telling us something about our normal sense of self: that is, the sense that gives us the feeling that we have direct control over our own actions.

I believe that these symptoms too can be explained in similar terms to our previous examples: information is available from one source but not another. It is the mismatch between the sources which causes the abnormalities of experience (Frith, 1995).

Every action has sensory consequences. When we speak, we hear the sound of our own voice. When we move our limbs, we feel the new angles in our joints. It is important to distinguish this self-generated sensation from sensations caused by external events. Some of the symptoms I have described may occur because this distinction is no longer being properly made. As a result, the sensory consequences of our own actions are perceived as being caused by external influences.

Basic studies of motor control suggest a relatively straightforward mechanism by which the critical distinction can be made. In principle, the sensory consequences of an action can be predicted from our prior knowledge of the precise form the action is going to take. This prediction could be achieved by taking a copy of the commands that were sent to the motor system to cause the action (efference copy or corollary discharge). Using this copy of the motor commands, the sensory consequences could be computed, producing a *forward model* of the consequences of the action. If in the model, the expected consequences matched the actual, observed consequences, then we would know that there were no external influences. Given this mechanism, we would have to conclude that there were external influences if the predicted and the observed sensation did not match. Clearly, such mismatches could occur if parts of the system were damaged.

How do we test this conjecture? One way is to study motor control in patients with passivity experiences. The model I have outlined was not developed to explain our sense of self with regard to our own actions. Rather, it was constructed to explain phenomena directly concerned with motor control, such as error correction and learning. The problems that lead to misperceptions of action should also lead to more basic problems of motor control. Experimental studies have generated some evidence that patients reporting passivity experiences have various specific problems with more basic aspects of motor control.

What kind of neuropathology can lead to this problem? The answer to this question still remains very speculative. However, if the normal mechanism depends upon a comparison of sensations predicted on the basis of motor commands with actual sensations, then the underlying neurophysiology must involve interactions between frontal areas concerned with generating motor commands and posterior areas concerned with sensation. Recent brain-imaging studies suggest that there may be functional disconnections between such areas in certain patients with schizophrenia.

Conclusions

In this essay I have presented a number of examples of pathological behavior and experience which have strong implications for normal mental function and for the relations between mind and brain. These examples suggest that normal functioning depends upon many parallel processing streams, each involving subtly different kinds of information. When some of these processing streams are damaged, the nature of those that remain can be seen with much greater clarity. Such studies are beginning to throw light even on the brain systems that underlie *deep* mental properties such as the sense of self. I am confident that in a relatively short time other aspects of the physiological basis of consciousness will also be understood, and that this will come largely from studies of psychopathology.

References and recommended reading

Bridgeman, B. 1992: Conscious vs unconscious processes: the case of vision. *Theory and Psychology*, 2, 73–88.

Ellis, H. D. and de Pauw, K. W. 1994: The cognitive neuropsychiatric origins of the Capgras delusion. In A. S. David and J. C. Cutting (eds), *The Neuropsychology of Schizophrenia*, Hove: Erlbaum, 317–36.

*Frith, C. D. 1992: *The Cognitive Neuropsychology of Schizophrenia*. Hove: Erlbaum.

—— 1995: Schizophrenia: functional imaging and cognitive abnormalities. *Lancet*, 346, 615–20.

Goodale, M. A. and Milner, A. D. 1992: Separate visual pathways for perception and action. *Trends in the Neurosciences*, 15, 20–5.

Gregory, R. L. 1961: The brain as an engineering problem. In W. H. Thorpe and O. L. Zangwill (eds), *Current Problems in Animal Behaviour*, Cambridge: Cambridge University Press, 310–30.

Lashley, K. S. 1929: *Brain Mechanisms and Intelligence*. Chicago: University of Chicago Press.

Mandler, G. 1980: Recognising: the judgement of previous occurrence. *Psychological Review*, 87, 252–71.

*McCarthy, R. A. and Warrington, E. K. 1990: *Cognitive Neuropsychology*. London: Academic Press.

Mellor, C. S. 1970: First-rank symptoms of schizophrenia. *British Journal of Psychiatry*, 117, 15–23.

Parkin, A. 1987: *Memory and Amnesia: An Introduction*. Oxford: Blackwell.

*Shallice, T. 1988: *From Neuropsychology to Mental Structure*. Cambridge: Cambridge University Press.

Simon, H. A. 1969: *The Sciences of the Artificial*. Cambridge, Mass.: MIT Press.

Zeki, S. 1993: *A Vision of the Brain*. Oxford: Blackwell.

30

Ethnomethodology

BARRY SAFERSTEIN

Ethnomethodology examines how people develop and apply their understanding of the world in routine daily activities. Ethnomethodological studies point out that during every social interaction, participants collectively work to understand not only the explicit topics of their interaction, but also power relations, appropriate behavior, and options or constraints that affect subsequent actions in a particular setting. Participants generally do not identify the latter topics as the subjects of their interaction. However, how their routine tacit or *transparent* social practices relate to those topics affects the social organization in which they interact (Cicourel, 1968; Garfinkel, 1967; Mehan and Wood, 1975; Sharrock and Anderson, 1986).

Researchers applying ethnomethodological approaches generally accept the importance of studying everyday interaction, often in the form of language use. Researchers also generally share an emphasis on understanding and examining the perspectives of the regular participants in a particular social setting (the members' perspectives). They do this through more or less rigorous comparison of the members' perspectives with their own analytical perspectives toward the subject of study. Of significance to cognitive science is the consistent finding that social interaction does not just reflect internal mental processes but involves social cognitive processes that affect the organization and recall of information (cf. Cicourel, 1974; Latour, 1988; Mehan, 1979; Saferstein, 1994; Suchman, 1987).

Ethnomethodological studies do not share a unified approach to research. Some ethnomethodologists, for example, reject cognitive science as being primarily a reductivist enterprise imposing mechanistic models of mind derived from computer science and neurobiology. So construed, cognitive science runs counter to the ethnomethodological project (Coulter, 1983).

By contrast, other researchers find that studies in cognitive sociology, cognitive psychology, COGNITIVE ANTHROPOLOGY, and ethnomethodology contribute to cognitive science by developing an understanding of the relation between interaction, sense-making processes, and social organization. From this perspective, issues of memory and recall are central to explaining the basis of socially constructed understandings of the world (Mehan, 1979; Saferstein, 1992). For example, in a series of studies in medical settings undertaken during the 1970s, 1980s, and 1990s, Aaron V. Cicourel has explicitly used cognitive science research (e.g., schema theory, neural network models) as a comparative referent for observable social cognitive activities in everyday settings (Cicourel, 1974).

Between these contrasting perspectives are a number of researchers who study the social aspects of interpretation and memory in order to understand decision processes and power relations in institutional settings. Among researchers who have applied ethnomethodological approaches to major studies of the social cognitive activities of work in organizational settings are Robert Anderson, Charles Goodwin, Marjorie H. Goodwin, Christian C. Heath, Paul Luff, David Middleton, Jonathan Potter, and Wes Sharrock. Karin Knorr-Cetina and Michael Lynch have studied the decision processes of scientists. Lucy A. Suchman has developed an ethnomethodological approach to distributed cognition in her studies of decision processes and sense-making activities related to the use of technology in her studies at Xeroxparc in California.

All this ethnomethodologically informed research presents theory and method for understanding cognition through the study of its social aspects in nonlaboratory settings.

Interaction and interpretation

Researchers applying ethnomethodological and cognitive sociological approaches to the study of organizational activities have analyzed the cognitive and communicative processes of work, emphasizing that:

- cognition is a social process rather than an individual mental process;
- the understandings that participants apply to their work and positions in organizations are developed collectively through interaction in the work place.

In the following example, I present an analysis of routine work interaction that is based on the type of research on language and social interaction inspired by ethnomethodology. A sound effects spotting session during production of the television program "Hill Street Blues" exemplifies how participants with differing expertise develop shared perspectives about materials and objectives (see also Saferstein, 1992, pp. 70–80). During sound effects spotting sessions, the participants reviewed the edited film and dialogue of an episode, in order to determine the quality and placement of background sounds and voices. The participants were the associate producer, who supervised the production of each episode's sound track subsequent to the filming; a production assistant (a liaison between the associate producer and participants at other phases of production); a sound editor, who developed and edited background sounds other than dialogue; and an ADR editor, who worked on the recording and editing of background voices and dialogue added after the filming (ADR stands for automated dialogue replacement). After the spotting session, different combinations of these people worked with others to complete the sound track.

This is part of the background knowledge that I brought to the textual and audio data that I compiled. During the "Hill Street Blues" sound effects spotting session, the participants construct a shared interpretation of the materials on which they work by presenting, discussing, and arguing over their individual interpretations of materials and immediate ideas for subsequent work to clarify a scene. The production participants' interaction reveals some of the bases of their interpretations. Of significance for studies of cognition is how participants' individual and social cognitive activities are interrelated.

Below, as I examine these social cognitive activities, I also discuss my own understandings of the activities and materials I encountered and the role they played in forming models of collaboration and collective cognition. In the process of developing models, I relied on my own situated interaction with the participants as well as on the background knowledge that I brought to the setting. The background knowledge included my own experiences as scholar, researcher, former film-production participant, and television-viewer. My analysis of the production activities that I observed, recorded, and studied involved continually comparing this background knowledge with the information and understandings expressed by the production participants as they worked. My recall was prompted by the production activities I observed or by records of that activity (including audiotape recordings and transcripts of the work interaction). The analytical process was not a matter of simply applying categories developed prior to the research activity, although both everyday and academic categories were part of my sense-making process. Regular comparison and evaluation of the researcher's perspective with the perspectives that participants express about their activities helps to expand the analytical framework which scientists apply to data.

Analyzing work discourse

Analysis of the audiotaped data involved repeatedly examining the recorded work discussions, in order to identify common patterns of interaction. In this case, the environment, a dark screening room, precluded videotape recording, so I audiotaped the work discourse. Audiotape data provide a record of topics of conversation as well as participants' linguistic patterns. Such data contain information about social interaction, including participants' preferences, occupational concerns, expertise, construction of meaning, and negotiation of authority. Audio data do not provide a record of movements, gestures, facial expressions, or participants' use of technology. A videotaped record of such activities is often useful for understanding activities in a particular setting. However, technical capabilities and choices of what to record may also limit the scope of video data. Thus, whether audiotaping or videotaping, the researcher must supplement the recorded data with notes that point out significant happenings.

Again, in this case, after identifying relevant patterns of interaction from the audiotape, I transcribed those segments of the recorded conversations. In other cases, depending on the amount of recorded data, transcription resources, and the scope of the research project, linguistic data may first be transcribed and then examined for relevant patterns of behavior.

It is difficult to represent in text everything that participants express linguistically. For example, often two or more people speak at once, participants speak in muffled voices, or microphones do not pick up voices. In order for analysts to distinguish significant patterns of discourse from technical limitations, transcriptions must indicate such aspects of recordings. For example, I have used the following transcription devices for the work discourse discussed below:

/two words/	Words between slashes overlap words between slashes of the following speaker.
.	Each period represents a pause of one second or less.
?	Each question mark not used as punctuation represents one unclear word.
("word"?)	Parentheses surrounding words in quotation marks followed by a question mark represent approximations of unclear utterances, based on sound.

(laughs) Words inside parentheses without quotation marks represent the analyst's descriptions of sounds or actions.

These aspects of the recorded discourse helped me to study the issues of decision making and collaboration in which I was interested. I decided to emphasize these aspects of the recordings by comparing my ethnographic knowledge of the setting (developed through observations and discussions with the participants) with my field notes and with audits of the recorded work interaction. In other settings, with other types of recording equipment or different research concerns, one may find it useful to note in transcriptions other aspects of recordings (e.g., vocal inflections, aspirated sounds, hand gestures, head movements). After transcribing recorded discourse, one can then analyze the form and content of the transcribed discussions to identify how the participants interact to organize their work activities and to deal with organizational constraints.

For example, as the participants in the sound effects spotting session watched the edited version of the "Hill Street Blues" episode, they expressed confusion about what was happening in a scene. They indicated that the positions of various characters in relation both to other characters and to the setting were not always clear. They discussed adding screams and dialogue in order to clarify the scene. However, the dramatic purpose and placement of screams and dialogue were not immediately clear to all the participants in the spotting session:

32 ADR editor: where are we here? We're in a different place now. We're in, uh
33 sound editor: But you wouldn't know it's her
34 ADR editor: until you know – until /you see her/
35 sound editor: /I mean the/ scream would be ahead of /that ???/
36 associate producer: /But you know/ it would be her inside the room
37 sound editor (challenging): Where? . . . Where would the scream go? . . .
38 ADR editor: /???/
39 sound editor: /The scream has to go/ before the shot I think, too.

By communicating their responses to the edited scene and to each other's comments, the participants organized their interpretations of the scene's confusing complexities in order to draw upon their various types of expertise and to develop plans for applying that expertise. For instance, the discussion between the sound and ADR editors in lines 32–9 and lines 46–9 led to the notion of a "reaction scream" and clarified the potential use of a scream:

46 ADR editor: /she/ didn't react to what he's done to kill himself, /right?/
47 sound editor: /yeah, yeah. That's what I'm talking about/
48 production assistant: /the other one ??/
49 ADR editor: /it would be a reaction/ scream for that, but she ??/

Through their discussion, the editors developed the notion that a scream would serve the purpose of letting the audience know that the character of the secretary, who is not always seen, is near the antagonist as he pulls his gun. Thus, viewers would not be confused when the secretary suddenly appears as a hostage.

During the spotting session, I viewed the "Hill Street Blues" scene and the work interaction as activities to interpret. Thus, my concerns in regard to making sense of both the production activities and the unfinished scene were similar to the participants' concerns with making sense of the scene. In both cases, the products we developed would encounter the institutionally sanctioned interpretations of professional colleagues. However, while I was in a position to delay the consequences of my interpretations, the spotting session participants had to sort through, clarify, and act on their interpretations of materials at the session. The multiplicity of interpretations and the distributed work of constructing interpretations is a resource rather than a complication in completing that task.

For instance, whether or not the *reaction scream* would become part of the final sound track, both the discussion leading to that notion and the resulting models of the scene contributed to the participants' abilities to discern, formulate, and express other issues, such as whether to add dialogue to clarify the scene. As they formed models for immediate production objectives, the participants expressed interpretive frameworks that reflected the occupational expertise and organizational activities of those involved in the production process: for example, types of story grammars they would apply, the use of short-term memory to work through the materials, and the models they expected others to develop as a consequence of encountering the work products. In both cases, work conventions are a type of collective memory of day-to-day processes and representations that participants developed and encountered.

Products of collective cognition: models of creative and administrative activities

After the participants finished their sound effects spotting work on the entire edited version of the episode, they reviewed the notes that they had made about the ensuing work to be done elsewhere. At this point, they were able to refer to a model for an improved version of the scene. However, they did not produce a detailed verbal or written record of their understandings of the scene and the projected changes. Rather, each participant set a course of administrative actions to accomplish work toward his or her interpretation of the shared objectives. Their notes index specific changes that the participants developed during the session to bring the scene closer to that new model.

The following discussion took place during their review of the spotting session, 12 minutes, 30 seconds, after they had discussed adding screams and additional dialogue to clarify the scene involving the criminal and the secretary. At this point, the review returns to the segment of the scene that the ADR editor had indexed earlier as a reaction scream:

501 production assistant: Ok. /Now reel five/
502 ADR editor: /Now reel five/ we have this, uh, we have
503 production assistant: /secretary/
504 ADR editor: /ugh, the secretary/ doing something /? of ?/
505 sound editor: Screaming and /probable added line./
506 production assistant: /an additional line./
507 ADR editor: Alright, then you wanted /to – you've got one cut of the whimper-scream at the
508 beginning of it/

509 sound editor: /(this may be part of separate conversation with the production assistant) ??? I think
510 . . . hates (?), hates doing it/
511 production assistant: /(starting at first "hates" in line 510) We'll find out./ We're bringing in
512 associate producer or sound editor: (barely audible, overlaps "We're bringing in," line 511)
513 ADR editor: Let me know if they're going to do that for sure. You want me to check before the other
514 screams are in ??
515 production assistant: Screams and assorted /work/
516 sound editor: /I'd/ like a scream
517 ADR editor: ?? her
518 sound editor: right after the shot or before/ – after – yeah. It's like he's – he's got a gun/
519 ADR editor: /No, before it . . . Well he's gotta go ?/ and grabs her
520 production assistant: Well, she has to scream to set /the fact/
521 associate producer: /To set/
522 production assistant: that she's kneeling over there
523 associate producer: Give her some /presence . . . We have to give her some presence/
524 production assistant: /that she's coming . . . ?????/
525 sound editor: /(with "We have," in line 523) You need a scream/ /before the gun, but after ???/
526 associate producer: /and – and she's kneeling down there/ – There's a one-shot looking down at
527 Buntz, where you see the back of her head
528 ADR editor: ?
529 associate producer: have her whimpering a bit there.

Earlier in the spotting session, the participants exhibited difficulty in making sense of the actual scene as they viewed it. Yet, by the end of the session – without observing the scene again – they were able to refer to a sequence of actions that make dramatic and narrative sense. The various participants contributed utterances that converged to form compatible models of objectives and work procedures. They located the immediate topic in the course of the day's sound effects spotting work as well as its position in the product, an episode of "Hill Street Blues." They typified a kind of sound effect in terms of its link to a dramatic character (the secretary), its position in the scene (the beginning), and its dramatic quality (whimpering). They also acknowledged that the production and subsequent use of the scream were contingent on activities that the associate producer would pursue separately.

In this discussion, participants often spoke simultaneously or continued each other's utterances, displaying their respective models of objectives and procedures (503–4, 504–6, 507–10, 515–16, 518–19, 520–1, 523–5, 525–6). The overlapping utterances of the production assistant, sound editor, ADR editor, and associate producer were voiced not so much as responses or interruptions but as completions and continuations of statements about the scene and the work on it. Models of objectives are explicitly stated in terms of shared responsibility or the contributions of others:

- "we have" (ADR editor, line 502);
- "the secretary doing something" (ADR editor, line 504);
- "then you wanted" (ADR editor, line 507);
- "We'll find out. We're bringing in" (production assistant, line 511);
- "Let me know if they're going to do that for sure. You want me to check before the other screams are in ??" (ADR editor, lines 513–14).

Recognizing the activities indexed by these utterances requires background knowledge of the production process. Applying that knowledge is a matter of both practical reasoning and occupational expertise, for both the production participants and the researcher. Once researchers recognize sets of particular resources and constraints that affect people who act in a specific organizational and occupational context, they can also code the recordings and field data in regard to those sets of resources and constraints. They can then gather subsequent data in the form of additional recorded work interaction or recorded discussions with the original participants, in which they comment on the initial recordings or transcripts. This *indefinite triangulation of data* (Cicourel, 1974) helps researchers to further refine judgments about the differences between their own perspectives and the members' perspectives, while providing additional detail about the social processes that occur in a particular setting and their consequences for the social organization in that setting.

For example, in lines 507 and 511 above, the ADR editor and production assistant recall an earlier discussion regarding the source of the new scream (lines 25–31, discussed in detail in Saferstein, 1992, pp. 73–6), and they correlate their resulting models of administrative activities. The ADR editor projects his review of the immediate work into the setting of the upcoming ADR session: "Alright, then you wanted to – you've got one cut of the whimper-scream at the beginning of it." By using the term *one cut*, he indexes an activity related to his subsequent work, the editing of the ADR track. He also suggests that he has a model of dramatic objectives and work activities in mind when he mentions the "whimper-scream" (the term *whimper* was used by the production assistant prior to verbally acting out a potential line for the secretary (line 78)). The ADR editor also locates a scream to be added at the beginning of the scene. The ADR editor's classification of the scream as a "whimper-scream" and mention of its placement not only have relevance to the immediate work context, separating his model of the scream and its use from different models suggested and acted out earlier, but also relate to activities at the subsequent ADR session, where he and the associate producer work together with the actress to produce a scream compatible with the objectives just developed during the sound effects spotting session (though the collaboration at the ADR session may further modify such objectives).

Overlapping the latter part of the ADR editor's utterance ("Alright, then you wanted /to – you've got one cut of the whimper-scream at the beginning of it," lines 507–8), the production assistant states: "We'll find out." Thus, the ADR editor and the production assistant reiterate the model of contingency developed in an earlier exchange between the associate producer and ADR editor about how and where to obtain the scream (Saferstein, 1992, pp. 73–6), and they solidify that model by linking it to work procedures (editing, ADR) and dramatic conventions (the situated portrayal of affect to signify the consequences of actions – i.e., whimpering). In lines 513–14, the ADR editor explicitly displays the administrative aspect of his emergent model of objectives for the scene: "Let me know if they're going to do that for sure. You want me to check before the other screams are in ??"

Analysis of the interaction through which the spotting session participants developed models for changing the scene reveals a number of activities central to studies in cognitive science. The participants' work interaction involves a form of distributed cognition (see Article 39, EMBODIED, SITUATED, AND DISTRIBUTED COGNITION) as they display their responses and interpretations of materials and work constraints. It also involves the formation of cognitive schemata: that is, links between categories and perceptions. Key researchers who have studied distributed cognition include Edwin Hutchins and Donald A. Norman. Important studies of cognitive schemata have been done by Danny Bobrow, Roy D'Andrade, Donald A. Norman, and David Rumelhart.

Reflexivity and research

Reflexivity, in the form of attention to the ways in which one makes sense of a setting or activity, is central to ethnomethodologically informed theory and method. In early ethnomethodological studies, Aaron V. Cicourel and Harold Garfinkel separately emphasized the importance of reflexivity to social science research. More recently, Melvin Pollner in the United States and Steven Woolgar in Great Britain have reemphasized the methodological importance of reflexivity. Two aspects of reflexivity are:

- attention to the mutual influence between subjects' interpretation activities and social organization;
- attention to the ways in which researchers' interpretations of the settings and activities they study reflect practical reasoning related to the research enterprise.

Reflexive research approaches examine how people organize perception through interaction. They also examine the social constraints on expressing and interpreting information.

One foundation of ethnomethodology is the finding that analysts of social activity develop their understandings of the people they study through the same processes of practical reasoning used by those people. Early ethnomethodological studies pointed out that the activities and interpretations of researchers are but a context-specific case of sense-making activity. During the 1960s and early 1970s, Aaron V. Cicourel, developing Harold Garfinkel's descriptions of situated sense-making processes, pointed out that all people use virtually the same few basic social interpretive procedures in making sense of communicative interaction. Thus, by examining the expressed differences between what participants pay attention to in a setting and what researchers pay attention to (while recognizing that both researchers and participants apply the same interpretive procedures), we learn about significant social constraints of the setting as well as methods of sense making which people apply to that setting. The check for validity lies in the ongoing comparison of the researcher's perspective, actions, and products with the participants' perspectives, actions, and products. This differs from the traditional hypothesis-testing model of research in which scholars try to conform data about social activities to preconceived analytical constructs. The hypothesis-testing model treats the reasoning leading to hypotheses as universally applicable. However, ethnomethodological studies have found that such reasoning is itself embedded in researchers' practices and constraints, which are often unexamined. These practices and constraints often impede researchers from considering many practical activities and categorical distinctions that are significant for understanding behavior and social organization in a particular setting.

I noticed links between the review discussion and other phases of production by comparing the work discourse with my knowledge regarding the particulars of the participants' activities at other work sites. My general model of the subsequent work on the "Hill Street Blues" episode did not call for the level of specificity expressed in the review discussion (e.g., lines 513–29). Having noticed that specificity, I then compared it with the information about the television production process which I had gained through fieldwork.

By comparing discourse at several points in the spotting session, I was able to recognize differences in the ways in which participants described the filmed scene. My recognition of the different narrative and semantic meanings of the participants' utterances (meanings indexed in academic and critical discourse by terms such as *dramatic character, position,* and *dramatic quality*) was contingent on the background knowledge I brought to the data. My concern about why and how the participants reviewed the scene pointed me to the importance of the participants' notes and short-term memories in their process of forming models of objectives and procedures.

Examining social cognitive activities

Reflexive analysis of this case elaborates and presents a practical example of cognitive science models of distributed cognition. The ability of the various participants to continue each other's utterances, while maintaining congruence with both the local agenda and a general model of objectives, is the result of their previous discourse work – the collective cognition. They had agreed on the use of a scream in a way that the ADR editor indexed as a "reaction scream" (lines 46–9). Although they did not use the term "reaction scream" during the review at the end of the spotting session, their discussion indicates that their models were consistent with the notion of a reaction scream that they had developed earlier. The discussion shows that, through their collaboration, participants developed models of the scene as a dramatic, narrative representation. Moreover, they developed models of specific administrative activities and work to be done in order to realize such a representation.

However, lines 516–19 of the review show that the earlier agreement between the ADR editor and the sound editor about the added scream being a reaction scream did not result in congruent or complete models of how to use the scream. Should the secretary scream when the antagonist displays the gun, when he grabs her, or after he shoots? The general topic of a reaction scream became an element of the local agenda for developing the background sound of the scene, because participants associated it with their respective administrative activities. This helped them to continue working on the scene during the spotting session, even though each participant may have associated different activities and outcomes with the term "*reaction scream.*"

The participants' review of the planned changes to the scene also indicates that the discussion, memory of the work just done, and actions taken during the meeting involved differing though related models for each participant. The discourse demonstrates that participants' differing models of objectives and procedures were linked to their different work activities, both at the session and subsequent to it. During the review at the end of the meeting, the collectively expressed and constructed discursive models of materials, objectives, and procedures continued to impinge on the participants' individual mental models, indexed by the notes and schedules they made. However, when participants compared and tested their perspectives at the review, they did not work through their differences as they did in the initial discussion of the scene. The unresolved, unaddressed differences at the review may have had the cognitive effect of reinforcing participants' differing occupational perspectives and their respective schemata of their subsequent work on the scene.

Participants in the spotting session simultaneously processed information relevant to both group and individual activities. In this instance, we find that participants constructed at least two types of related schemata. One type involved administrative activities. Another involved creative models of the scene. By means of their collaboration, participants organized information (both recalled and expressed) into these types of schemata by verbally constructing the difference between creative models and administrative activities (Saferstein, 1992).

Participants' respective, occupationally specific administrative activities reflexively helped them to interpret materials and create models of creative objectives. The social cognitive processes of developing models of the scene and constructing models of professional roles overlapped when participants developed objectives and plans for their subsequent work away from the spotting session. Yet, participants also separated each of these aspects of their knowledge in order to apply them to their work activities at other settings and times. Their work interaction exemplifies how people develop and reinforce models of objectives through the discourse practices constituting collective cognition.

The situated social interaction through which the production participants constructed work-related understandings did not merely reveal internal mental processes. The discussion and the resulting artifacts (notes, plans, schedules) were significant components of the cognition at the spotting session. They played a key role in the organization and recall of information, in the formation of cognitive schemata, and in establishing mental links between domains of information. An ethnomethodologically informed model of the participants' practical reasoning and distributed cognition derives from reflexively examining and comparing the social processes of understanding of the researcher and of the participants.

References and recommended reading

Cicourel, A. V. 1968: *The Social Organization of Juvenile Justice*. New York: Wiley.

—— 1974: *Cognitive Sociology: Language and Meaning in Social Interaction*. New York: Free Press.

—— 1990: The integration of distributed knowledge in collaborative medical diagnosis. In J. Galegher, R. E. Kraut, and C. Egido (eds), *Intellectual Teamwork: The Social and Technological Foundation of Cooperative Teamwork*, Hillsdale, NJ: Erlbaum.

Coulter, J. 1983: *Rethinking Cognitive Theory*. New York: St Martin's Press.

Garfinkel, H. 1967: *Studies in Ethnomethodology*. Englewood Cliffs, NJ: Prentice-Hall.

Latour, B. 1988: Visualization and cognition: thinking with eyes and hands. *Knowledge and Society*, 6, 1–40.

Mehan, H. 1979: *Learning Lessons*. Cambridge, Mass.: Harvard University Press.

Mehan, H. and Wood, H. 1975: *The Reality of Ethnomethodology*. New York: Wiley.

*Pollner, M. 1991: Left of ethnomethodology: the rise and decline of radical reflexivity. *American Sociological Review*, 56, 370–80.

Saferstein, B. 1992: Collective cognition and collaborative work: the effects of cognitive and communicative processes on the organization of television production. *Discourse and Society*, 3, 61–86.

—— 1994: Interaction and ideology at work: a case of constructing and constraining television violence. *Social Problems*, 41(2), 316–44.

*—— 1995: Cognitive sociology. In J. Verschueren, J-O. Östman, and J. Blommaert, (eds), *The Handbook of Pragmatics*, Amsterdam: John Benjamins, 140–7.

Sharrock, W. and Anderson, B. 1986: *The Ethnomethodologists*. London: Tavistock.

Suchman, L. A. 1987: *Plans and Situated Actions*. Cambridge: Cambridge University Press.

*—— 1988: Representing practice in cognitive science. *Human Studies*, 11, 305–25.

—— 1995: Making work visible. *Communications of the ACM*, 38, 56–64.

*Woolgar, S. (ed.) 1988: *Knowledge and Reflexivity: New Frontiers in the Sociology of Knowledge*. London: Sage.

31

Functional analysis

BRIAN MACWHINNEY

The functional approach to language holds that the forms of natural languages are created, governed, constrained, acquired, and used in the service of communicative functions. To evaluate this claim, we need to examine both the strengths and the weaknesses of the functional approach.

No one would deny the importance of functions in human language. We constantly use language to communicate intentions between one person and the next. For example, we can use language to tell another person how to drive a car, where to look for edible mushrooms, and how to avoid falling into crevasses when walking over glaciers. We can also use language to foster social solidarity by greeting and acknowledging other people with salutations and standardized phrases. Yet another use of language is to represent our thoughts and goals internally. Both inner speech and external written expression allow us to talk to ourselves in ways that help foster creativity, invention, and memory. Additional artistic functions of language include drama, poetry, and song.

Given the importance of these various functions of human language, it may be surprising to learn that there is a major debate in linguistic and psycholinguistic circles regarding the extent to which functions determine the shape of language. To the outsider, it would seem almost obvious that the shapes and forms of human language are determined by the functions being served. We use nouns to refer to things and verbs to refer to actions. By choosing one word order over another, we distinguish who did what to whom. In this way, the most basic forms of human language are functionally determined. But exactly how does function have its impact on form? Is the impact direct and immediate, or only indirect and delayed? Is there only one basic way in which functions determine forms, or are there various types of form–function relations? Is it even possible that the system of forms could become freed from linkage to function and take on some type of autonomous existence?

The antithesis to functionalism is formalism. The formalist position holds that, although language may serve a variety of useful functions, the actual shape of linguistic form is determined by abstract categories that have nothing to do with particular functions or meanings. On this view, language is a special gift to the human species, whose formal contours reflect the abstract, reflective, and impractical nature of the human mind. Categories such as "verb" or "subject" are abstract objects that are processed and represented in a separate mental module devoted to grammar. The objects of this module are universal and derive not from functional pressures or ongoing conceptualizations of the world but from the innate language-making capacity. The language module is informationally encapsulated. This means that it relies only on its own abstract category and rule information to process and represent language; it does not depend upon information from other aspects of cognition. According to this view, the liberation of linguistic form from any tight linkage to function has led to the modular architecture that produces the power inherent in the human mind. Because language is being used inside a separate module in the mind, it is not subject to the functional pressures of communication (see Article 49, MODULARITY).

Functionalists recognize that language plays an important role in supporting "inner speech." However, following the lead of Vygotsky, they view this inner speech as social speech that has been captured inside the mind. Although its form is abbreviated and modified in various ways, inner speech still obeys the functional communicative pressures that operate to shape the social uses of speech. In effect, we serve as our own conversational partner when we use language as a medium for thought (see Article 40, MEDIATED ACTION).

The core issue on which functionalism and formalism disagree is that of autonomy versus modularity. Formalists claim that the shape of language is minimally constrained by functional pressures, since language basically follows its own rules, in a separate, informationally encapsulated autonomous cognitive module. Functionalists claim that language is continually subject to the need to express conceptual and social messages, and that these pressures govern the processes of language change, language learning, and language processing.

Naive functionalism

A major stumbling block in understanding the extent to which we want to emphasize the functional determination of language has been the existence of a variety of naive functionalist analyses. Formalists find it easy to dismiss these naive analyses as prescientific and empirically flawed. Unfortunately, formalist critiques of functionalism tend to focus exclusively on these naive formulations, while ignoring more complex and powerful versions of functionalism. Perhaps the oldest naive approach to the relation between form and function is the notion of sound symbolism that we find first expressed by Plato in the *Cratylus*. Asking why the word for table has the sound it does in the Greek language, Socrates replies that this sound is inherent in the nature of the thing itself. The problem with Plato's approach to the relation between sound and meaning is that different languages use radically different sounds to name the same object. If the English word *table* had some privileged relation to the object being named, we would have to conclude that the Spanish word *mesa* and the German word *Tisch* are simply impoverished or degenerate attempts to capture a relation that is best expressed by the English word *table*.

In fact, the relation between a word and its meaning is an excellent example of the limits of functional determination. As the Swiss linguist Ferdinand de Saussure argued at the beginning of this century, the relation between a word and its referent is entirely arbitrary. Saussure elevated this arbitrariness of the linguistic sign to a fundamental principle of psycholinguistic dogma, viewing the word as an association between a phonetic sign and a semantic signification or function. To be sure, some words reflect a bit of phonetic symbolism. Words such as *bump, dump, thump, slump,* and *sump* all express a certain lowering of material. And words such as *bright, tight, bite,* and *light* all express a high intensity or brightness. Sound symbolism of this type allows us to guess at the meanings of words in a new language with something better than chance success in a simple yes–no judgment task. But these occasional correspondences hardly form the backbone of our understanding of the vocabulary of our language.

Although Saussure was correct in viewing the relation of form to function as arbitrary, his analysis should not be used to dismiss functional determination. On the contrary, Saussure thought of words as mappings between cognitive functions and phonological form. The word for *peach* is used to express ideas about a particular type of fruit, just as the word *however* can be used to express the contrary juxtaposition of two ideas. Each sign stands in a strong indexical relation to the function being signified. Although the word *peach* is not peachlike in itself, its use in sentences is completely determined by the underlying function that it expresses. In this sense, although language has limited iconicity, the mapping relation between the sign and the signified makes language rich in functional expressiveness.

Functionalism in syntax

But does functional expressiveness extend beyond the lexicon? Is there also functional determination for systems such as syntax and morphology? When we look at the forces that govern the order of words in a sentence, we find some obvious candidates for functional determination. One of the most pervasive functional relations was captured in the last century by the linguist Behaghel. According to Behaghel's first law, words that belong together mentally are placed close together syntactically; conversely, words that appear next to each other in sentences are usually related conceptually. Virtually any sentence can be used to illustrate this effect. Consider a simple sentence such as "All my friends like to eat goat cheese." Here, the word *goat* is not closely related to *friends* but is mentally highly related to *cheese*, which is why it appears next to *cheese* and not next to *friends*. In a sense, we can think of sentence structure as arising from the compression of a three-dimensional graph structure onto a one-dimensional linear chain. This compression results in a great deal of ambiguity, but the basic impact of conceptual determination is still clearly evident. Languages like Classical Latin that maintain a rich set of inflectional markers manage to transcend Behaghel's law for stylistic effects by separating related words. However, this can be done only when the markings are clear enough to allow the reader to recover the original relations. Less fully inflected languages like English or even Vulgar Latin are more strictly governed by Behaghel's first law.

Going slightly beyond Behaghel's first law, we can look at the serial order in sentences such as "Travel over the bridge and through the forest" as evidence of the way in which sentences tend to map the order of real-life procedures onto the left-to-right order of words in a sentence. This sentence provides us with instructions to first go over the bridge and then through the forest, rather than the reverse. In general, language tends to provide instructions for action by putting first things first. These principles of natural ordering and iconicity represent a certain level of basic functionalism in language that no one would deny. But we cannot push syntactic iconicity too far. Some languages use basic subject–verb–object (SVO) word order, whereas others use subject–object–verb (SOV) or verb–subject–object (VSO) order. It would be a mistake to think that one of these orders represents the true flow of human thought, since no one of the three dominates in the languages of the world. Although word order has an important iconic expressive function, this function interacts with many other factors in complex and flexible ways.

Another example of a functional grammatical universal is the tendency to mention the topic, or thing we are talking about, before making a comment about that topic. In English, we can topicalize a newly mentioned referent in causal utterances such as "You know Betty's friend, she came all dressed in pink and green." Here "Betty's friend" is being introduced as a new topic. In such forms, we use the initial position of the sentence to introduce the new topic about which we then make an explicit comment. Other languages, like Chinese, Hungarian, or Czech, elevate this ordering of topic before comment into a fundamental grammatical principle. In these languages, the basic word order of sentences is a direct reflection of the functional value of marking topics and comments. Another linguistic device for marking topics is post-topicalization. Often we begin an utterance without making the topic completely clear and find that we need to tack on a topic statement as an afterthought at the end. An example of this would be a sentence such as "She likes those diamonds, Mary does."

Functional linguists have explored a wide variety of interesting correlations between form and function. Some examples of functional syntactic relations that have been studied include the grounding of relative clauses on deictic elements such as *that* and *there*, the development of aspectual systems from generalized auxiliary verbs such as *have*, *go*, or *be*, and the evolution of temporal conjunctions from analogous spatial prepositions. Among the most intriguing patterns studied by functionalist grammarians are the patterns that give rise to ergative syntactic and inflectional marking. This fairly exotic alternative to the nominative–accusative form found in Indo-European languages occurs in languages such as Samoan and Mayan. Ergative syntax arises in a fairly straightforward functional fashion from the fact that people tend to delete subjects when they are well known and topical. The more a given participant has been mentioned in a narrative sequence or a conversational exchange, the more likely we are to delete or pronominalize that participant. If we were to take an English sentence like "The boy chased the girl" and delete the subject, we would end up with a phrase like "chased the girl," in which the patient is elevated to the primary unmarked case role. In this way, functional conversational pressures can force a fundamental reorganization of the shape of the grammatical system. It is also interesting to find that many languages that have developed some form of ergativity have confined use of ergative marking to cases in which the patient is in the third person. These split ergative systems retain nominal marking for first- and second-person subjects, but ergative marking in the third person. Other split ergative systems mark ergativity differentially across tenses and aspects. These complex interactions between ergativity, tense, evidentiality, and person are excellent grist for the mill of functionalist analysis.

The presence of ergative syntax in some languages, not others, raises still other important questions that must be addressed. If the functionalist pressures arising from conversation and narration are similar in different cultures, why do languages have such widely varying grammatical systems? Perhaps the formalists are correct in saying that grammar takes on an autonomous life of its own inside the syntactic module, without any direct linkage to functional pressures. The functionalist answer to this is much like the answer to similar questions in biology. One can argue that all species of birds instantiate particular adaptations to the functional pressures of food source, territorial competition, predation, and reproduction. The fact that all species do not look alike does not mean that these functional pressures are not operative in all cases. It simply means that the exact form of the functional pressures varies from one ecological niche to the next. The same must be true of human languages and human cultures. Although all languages are functionally determined, the exact form of the complex interacting pressures varies in detail from culture to culture.

The Competition Model

Although topicalization and post-topicalization are clearly important functional determiners of syntactic structure, particularly in languages like Chinese or Hungarian, it is not the case that the first noun in an English sentence is always the topic. Often the first noun appears in the position before the verb not because it is a topic, but because it expresses the role of the agent of the verb. But how can we know in a given case whether the noun is in first position in English because it is expressing the function of agency or because it is expressing the function of topic? This reflects a basic problem in functionalist analysis. If we try to link each form to a single function, we quickly find that we have constructed a type of naive functionalism that fails to reflect the multifunctionality of grammatical forms. As soon as we try to model the interaction of functions and forms, we soon find that we need to consider radically more complicated types of models.

One model that attempts to deal with multiple functional determination is the Competition Model (MacWhinney and Bates, 1989). Let us look at how this model analyzes the functional forces that motivate preverbal positioning in English. In active sentences, the agent is the noun before the verb. However, in passive sentences, the agent is expressed by a prepositional phrase with the word *by*. Thus, in an active sentence such as "The man bit the cat," agency is expressed by preverbal positioning of *man* before *bit*, whereas in a passive sentence such as "The cat was bitten by the man," agency is expressed by placement of *man* after *by*. Why does English provide this alternation? The reason is that sometimes the agent is not the topic. When the agent is identical with the topic, we can use the active form. However, when the patient is the topic, we must use the passive. If we ask "What happened to the cat?," we must reply that "The cat was bitten by the man," rather than "The man bit the cat."

One can wire up a network grammar to control the activation of these competing form–function relations. Figure 31.1 attempts to show how agency and topicality can activate preverbal positioning. However, when the agent is nontopical, a nonlinear combination unit must be activated, which then activates the use of the *by* clause.

Figure 31.1 is a rather clumsy way of expressing the competition between preverbal positioning and the *by* clause as alternative ways of mapping agency. The formalism of neural networks (see Article 38, CONNECTIONISM, ARTIFICIAL LIFE, AND DYNAMICAL SYSTEMS) provides a better way of understanding this type of nonlinear combination of forms and functions. Figure 31.2 shows how agency and topicality can combine in a nonlinear way in a neural network through what are called *hidden units* to activate either preverbal positioning or the *by* clause or both.

Neural network models of the type shown in figure 31.2 differ from hand-wired networks of the type shown in figure 31.1 in that one can formulate a general learning rule for the neural network that can allow a child to learn the system of form–function relations without any hand-wiring of the network. However, a simple neural

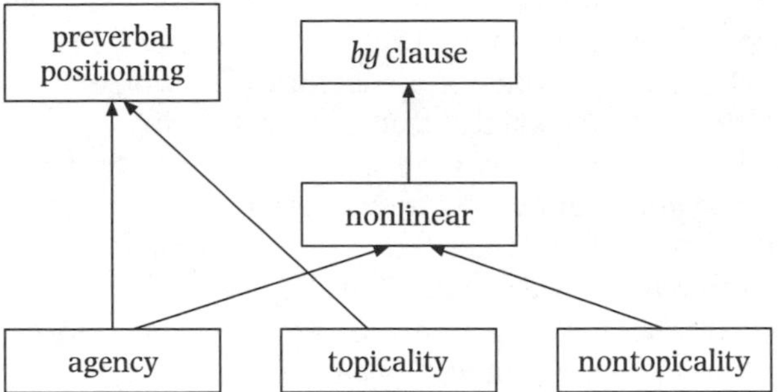

Figure 31.1 Hand-wired form–function relations for English subject marking.

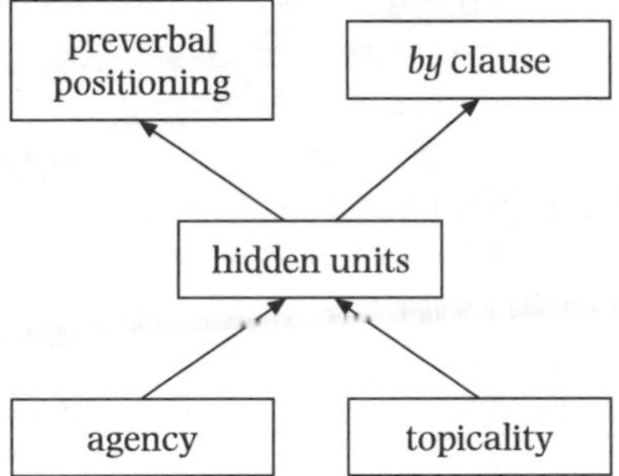

Figure 31.2 A neural network for form–function relations for English subject marking.

network of the type shown in figure 31.2 has some other limitations. First, this figure only conveys the ways in which functions activate forms during language production. Second, this particular figure only looks at two forms and two functions. A fuller network model of language processing would include many more forms and functions and would provide separate models for comprehension and production. Third, a fuller model would also represent the levels of competition and cooperation between and among forms and functions. In the actual use of language, certain functions tend to co-occur to a large extent. For example, it is typically the case that the topic is definite, and it is also nearly always true that a human is animate. This type of natural coalition between forms has its reflection on the level of forms. For example, in English, the preverbal noun almost always agrees with the verb in person and number. A fuller network model would express all of these additional complexities of the language processing and representation system.

Mapping and the limited channel

The Competition Model view of sentence processing tends to emphasize the extent to which our limited information processing capacities in both production and comprehension force us to channel a rich set of cognitive functions through a very narrow temporal funnel. As we strive to shape our thoughts into meaningful utterances, we are constantly asked to choose between a variety of options, each of which points in a slightly different direction. Should we say "The refrigerator has a pie on top of it" or "There's a pie on top of the refrigerator"? Our choice of one starting point over another leads us into a further set of grammatical commitments, which may turn out, in the end, to be incomplete or unacceptable. Ideally, we would always like to be able to plan our utterances well before we produce them. But, in the real world of real time processing, we must produce and comprehend utterances in a highly incremental word-by-word fashion. This incremental approach to processing limits the load on our system, but it can also lead to errors and even dead ends.

Grammatical systems have evolved under the functional pressure of actual language use. An important pressure on language is our need to map a large set of cognitive functions onto a small set of competing forms. A general functional principle that all grammars must follow is the principle of *peaceful coexistence*. Because certain functions tend to co-occur, the forms that express these functions also tend to co-occur. Processing can take advantage of these co-occurrences. A good example of a system of peaceful coexistence is the system of subject marking in English. The opposite side of the coin of peaceful coexistence is the principle of *dividing the spoils*. Whenever a coalition governed by peaceful coexistence falls apart, it is important to still be able to express the opposing functions. The English passive is an example of a system designed to properly divide the spoils. After the function of topicality has won out in the competition for the form of preverbal positioning, the function of agency has to be satisfied with the *booby prize* of expression through the *by* clause. Typically, booby prizes occur later in utterances, after the basic functions have already been expressed.

Cue reliability and form reliability

During sentence comprehension, the processor must learn to rely more on some cues or forms than others. The exact identity of the cues or forms that are most reliable varies markedly from language to language. For example, in Hungarian, the direct object of the verb is almost always marked with a final /-t/. Hungarian children learn to depend on the presence of this marker and the absence of any marking on the subject noun as sure cues to sentence interpretation. Because of this, we can say that Hungarian is a strong case-marking language. The actual order of the main constituents of a Hungarian sentence can vary greatly, as long as the presence or absence of the case-marking suffixes is clear. In English, on the other hand, there are no suffixes that reliably mark the subject or the object. Children learning English soon come to realize that the placement of a noun before the verb is a reliable cue to subjecthood. Although Hungarian and English differ radically in the forms they use to mark the subject function, both languages provide the listener with highly reliable cues to subject and object identification.

The counterpart to cue reliability in sentence production is form reliability. When a speaker needs to express a particular function, it is important to be able to rely consistently on a standard form to express that function. For example, we may want to take the point of view of a subject noun that is modified by an indefinite article. However, we tend to prefer sentences with initial nouns that are definite, rather than indefinite. One way around this particular problem is to use the *presentational* construction, such as "There is a pie on the refrigerator." By using the presentational *there*, we can reliably express both indefiniteness and subjecthood at the same time. By increasing fine-tuning of this system of systemic options for production, we can develop increasingly strong control over the maximally expressive use of our language.

Lexical influences on syntax

So far, we have been treating major grammatical categories such as subject, verb, and object as monolithic, unanalyzed units that float free in cognitive space. In fact, the exact ways in which functions and forms compete is heavily linked to the actual identity of the words in the sentence. For example, consider this pair of sentences:

1 (a) Although John frequently jogs, a mile is a long distance for him.
  (b) Although John frequently smokes, a mile is a long distance for him.

On-line processing studies have shown that listeners tend to entertain the possibility in 1(a) that *a mile* is the direct object of *jogs*. This occurs because a verb like *jog* can occur in either a transitive or an intransitive frame. A verb like *smoke*, on the other hand, would not take a noun like *a mile* as a complement. In fact, the details of the various lexical expectations between verbs and nouns are quite complex. A strongly transitive verb like *hit* has a preference for animate subjects but can take either animate or inanimate objects. On the other hand, an action verb like *chase* tends to expect an animate object.

The attachment of prepositional phrases is also governed by fairly specific lexical expectations. Consider this pair of sentences:

2 (a) The ladies discussed the dogs on the beach.
  (b) The ladies discussed the tennis match on the beach.

In 2(a) we can imagine either that the dogs are actually on the beach or, alternatively, that the ladies who are walking on the beach are discussing their dogs who are back home. However, in 2(b) the difficulties involved in conducting a tennis match on a sandy beach tend to preclude the interpretation in which the tennis match is actually occurring on the beach. In these examples, there is competition between attachment of the prepositional phrase "on the beach" to the preceding noun or the verb.

The interaction of verbs, nouns, and prepositions in these systems is governed on line by the general features of the words involved. As comprehension of the sentence deepens with time, the specific aspects of the words involved start to come more and more into play. The resolution of syntactic attachments is not different in principle from many other types of grammatical ambiguity resolution. For example, in a phrase like "The container held the apples," we tend to think about a large basket or box. However, in a phrase such as "The container held the beer," we tend to think about a bottle or an aluminum can.

These examples, and many others like them, emphasize the extent to which words are forced to adapt their meanings when they combine with other words. We can think of these patterns of between-word adaptation as involving "pushy polysemy." The notion of pushy polysemy is much like the notion of ambiguity or homonymy. If we look up common words like *run* or *take* in the dictionary, we will find that they have dozens of alternative meanings, or *readings*. What makes the selection of one of these readings over another a matter of pushy polysemy is the fact that some words tend to push other words into particular polysemic pathways. Consider a phrase such as "another apple." Here, the word *another* tends to expect a count noun, and the word *apple* nicely fulfills that expectation. If, however, we encounter the phrase "another sand," we have to make significant extra effort to develop an interpretation of the word *sand* that conforms with the mass noun expectation from the *pushy* operator word *another*. One way of making *sand* fit this expectation is to conceptually convert the mass noun into a count noun by packaging the quantity into bags. For example, when we are at the lumber yard ordering sacks of concrete for a construction job, we might well ask for someone to toss *another sand* onto our pickup, meaning by this that they should toss on another bag of sand. Yet another way of fulfilling the expectations deriving from the word *another* is to treat *sand* like a noun derived from a verb. Just as we could say to the masseur at the massage parlor that we would like *another rub*, we could ask the cabinet maker for *another sand* on our newly purchased kitchen cabinets.

Pushy polysemy is only one consequence of the dynamic functioning of groups of lexical items. When we look more generally at the ways in which lexical forms influence syntactic patterns, we see that nearly all syntactic constructions emerge from interactions of lexical items. A good example of this type of syntactic emergence is the double object construction in English. Most verbs of transfer can take either prepositional dative forms or double object forms. We can say either "John threw the ball to Tim" or "John threw Tim the ball." However, some verbs that seem to involve transfer cannot take the double object form. For example, we cannot say "Sue recommended the library the book." By conducting a thorough lexical analysis, we can see that verbs share additional semantic features that help us to understand why some permit the double object construction and others do not. We can think of the emergent properties of these groups as representing *extensional pathways*. Some of these pathways are quite general. For example, we can extend the name of any written work of art to refer to a particular book that contains that work. This allows us to say, "I think I left my *Hamlet* on top of my *Iliad*." Extensional uses of this type are motivated by general principles of lexical function.

Functionalism and abstract paradigms

Although many linguists would agree in assigning a major role to communicative function in determining forms such as lexical extensions, word-order patterns, syntactic constructions, or case role marking, they would assign a much more peripheral role to functional determination of complex grammatical paradigms. It would be difficult to find an area of language that involves more nonfunctional arbitrariness than the marking of declensional paradigms in languages like Latin, Russian, or German. As Mark Twain complained in his essay on "The Aweful German Language," it seems unfair for the German language to decide that the sun (*die Sonne*) should be feminine and the moon (*der Mond*) masculine, while relegating a beautiful young girl (*das Mädchen*) to the neuter gender. However, even in this hotbed of anti-functionalism, we find a rich set of cues or determinants at work to assign nouns to one of the three genders of German. Some of these cues are semantic in nature. For example, alcoholic beverages are masculine, as are rocks and minerals. But the major determinants of assignment of gender are not semantic but phonological cues. Words ending in *-e* are typically feminine, whereas words ending in *-er* or containing umlauts are typically masculine. Using neural network models based on these cues, we can show that the system is a complex, but predictable lattice of interlocking cues.

But why should such complexities exist at all, if the goal of language is to express communicative functions? Although it is true that the gender contrast in German often provides useful cues for grammatical role and sentence interpretation, the same effect could easily be achieved through a simpler gender system. For example, Spanish marks many masculine nouns with *-o* and many feminine nouns with *-a*. Spanish achieves the same functional effect, using a smaller set of cues than does German. Perhaps we should view the German system as an example of formal determination run amok. However, we need to bear in mind the fact that the linkage of nouns to gender class is bought at a minimal processing cost. Although these systems are difficult for foreigners to learn, they cause very little trouble to German children. What this means is that acquisition of meaningless form classes is a basic part of our language-making capacity, as long as the assignment of words to form classes can be achieved on the basis of superficial features such as phonological structure or minor semantic features. Thus, although grammatical gender is predictable, we would certainly not want to say that it is fully functionally motivated.

How far can we push functionalism?

Much of the evidence supporting the functionalist position derives from studies of language typology or surveys of lexical patterning. This work has repeatedly demonstrated correlations between linguistic form and linguistic function. Given the pervasiveness of these patterns, it is fairly easy to accept the notion that languages evolve in ways that maximize the ability of forms to express communicative functions. But is it possible that the appearance of these form–function correlations in languages is essentially epiphenomenal? Perhaps, as some formalists would argue, language is fundamentally a liberated, autonomous structural engine whose operation occasionally produces form–function correlations as an accidental by-product. Maybe the functional grounding of forms in conversation and narration is something that only a few speakers realize at occasional rare intervals. Just as schoolchildren seldom stop to think about the deeper nature of the pledge of allegiance to the flag, we as speakers may only rarely appreciate the functional determination of linguistic forms. Rather, in our daily language usage, we tend to rely on abstract, functionless, modular syntactic rules whose functional determination is seldom really called into play.

In order to refute this type of formalist claim, a functionalist account needs to look at the processing of functional cues during on-line sentence processing. This means that the eventual analysis of the claims of functional linguistics rests on the shoulders of psycholinguists. So far, the evidence collected by psycholinguists regarding the use of functions during on-line processing has been supportive of the functionalist position. The major formalist position in this area has been the modular processing approach developed by workers such as Frazier, Fodor, Clifton, Perfetti, and others (see Article 49, MODULARITY). This modular approach assumes that sentence processing depends on computations that occur in a separate syntactic module that does not rely initially on input from nonsyntactic factors. This module implements a highly deterministic process that builds abstract syntactic parse trees without initially paying regard to the meaningful relations between words or the listener's general understanding of the situation. Recent work has shown that the formalist model cannot account for the details of reaction time patterns for the processing of sentences involving alternative syntactic attachments. For example, when we hear a sentence such as "The spy saw the cop with the revolver," we immediately realize that the revolver is being held by the cop and not by the spy. In other words, we immediately attach the prepositional phrase to the preceding noun rather than to the verb. We do this because of our understanding of the meaningful relations between these words, even though this attachment violates the claimed modularity and autonomy of the syntactic processor proposed by formalists.

Although there is good evidence that on-line sentence processing is driven by functional factors, one might still argue that the core of the grammar remains modularly separated from the impact of these functional considerations. One way of maintaining this formalist position would be to claim that the learning of language by the child relies not on functional cues but on more abstract grammatical principles. For example, one could argue that nature provides an underlying set of neurobiological tools that determine a set of abstract modules that function during language development. This view of language development would tend to emphasize the predetermination of linguistic form through the operation of individual neural structures.

Here, again, the weight of evidence seems to favor the functionalist position. The process of language learning seems to be heavily determined by the exact cue validities of the language being learned. Children do not come to the language-learning task with some abstract set of formal categories to be matched. Instead, they use the input they receive from words embedded in rich situational contexts to guess at the ways in which syntactic constructions map linguistic functions.

The debate between functionalism and formalism is perhaps the single most important issue in linguistics and psycholinguistics. Moreover, this debate has further important consequences for cognitive neuroscience, developmental psychology, philosophy, and artificial intelligence. Given this, a clarification of positions on these issues must be viewed as an important item for cognitive science. Although the two camps have stuck closely to their respective positions, researchers are now becoming increasingly aware of the need for an ongoing dialogue. Once this dialogue has begun in earnest, we will be able to better understand how we can formulate a theoretical perspective that reconciles these two sharply contrasting positions.

References and recommended reading

Chafe, W. 1994: *Discourse, Consciousness, and Time*. Chicago: University of Chicago Press.
Gernsbacher, M. 1990: *Language Comprehension as Structure Building*. Hillsdale, NJ: Erlbaum.
Givon, T. 1995: *Functionalism and Grammar*. New York: John Benjamins.
MacWhinney, B. and Bates, E. 1989: *The Crosslinguistic Study of Language Processing*. New York: Cambridge University Press.

32

Neuroimaging

RANDY L. BUCKNER AND STEVEN E. PETERSEN

Introduction

A growing number of scientists have become interested in the relation between cognitive processes and their biological basis. This growth in interest has led to the creation of a subfield within psychology called *cognitive neuroscience*, which has now spawned its own scientific journal, a conference, and several graduate programs around the United States. One reason for recent enthusiasm is the development of several methods that allow researchers to observe brain activity in healthy, awake subjects while they perform cognitive tasks. These methods, which are referred to as *neuroimaging techniques*, provide a unique window into the function of the human brain (Posner and Raichle, 1994).

The goal of this chapter is to illustrate the kind of information which neuroimaging can provide and how that information can be used to help unravel the biological basis of cognition. Two areas of research, one dealing with the study of speech production and one dealing with memory, will be discussed. Advances in both these areas have been primarily driven by findings from neuroimaging studies. Not surprisingly, little attention was given to the findings, and they were presented with meager interpretation. This is because they were novel and did not fit well within existing frameworks.

The ability of neuroimaging data to stimulate original ideas makes them an important tool in cognitive neuroscience. It also means that most of these findings are just beginning to be understood. In this respect, both the areas discussed represent bodies of work still very much in progress. Ultimately, it will probably require the combined contributions of many techniques (BEHAVIORAL EXPERIMENTATION, analysis of patients with lesions, DEFICITS AND PATHOLOGIES, studies of animal models, and electrical scalp recordings) to fully understand the implications of the neuroimaging findings.

Before discussing the results themselves, a brief description of current neuroimaging techniques will be given. Where possible, examples are provided, to illustrate how the techniques are implemented.

How neuroimaging works

Basis of neuroimaging

Neuroimaging techniques indirectly measure local changes in neuronal activity. These measurements can be made in healthy, awake human subjects, with little risk and only minimal discomfort. The reason for using neuroimaging techniques is the belief that all cognitive processes are carried out by the combined activities of multiple brain

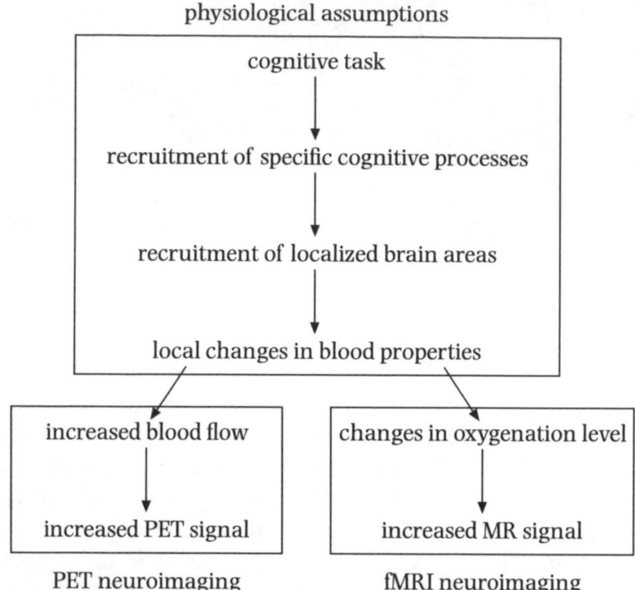

Figure 32.1 A schematic diagram showing the physiological assumptions that PET and fMRI neuroimaging techniques rely on. As can be seen, the two techniques have many properties in common but differ with regard to the specific signal they measure.

areas. Each brain area provides a processing function, which, when combined with those of other brain areas, performs the operations necessary to complete a given cognitive task (Posner et al., 1988). Thus, neuroimaging can be used to determine which brain areas are active during the performance of that task. Combined neuroimaging studies can then be used to identify a class of tasks which activate a given brain area. This latter approach is particularly powerful, because it allows researchers to clarify the processing role of certain brain areas. In practice, this can be accomplished only by conducting thorough task analyses and identifying task demands that are common across the class of tasks that activate a given brain area.

Two neuroimaging techniques are currently in widespread use: positron emission tomography (PET) and functional magnetic resonance imaging (fMRI). Both are based on the same general principle: *when areas of the brain increase their activity, a series of local physiological changes take place that correlate with that activity.* PET and fMRI can both measure these changes, but they use slightly different physiological responses as their basis (figure 32.1). Most PET studies rely on the observation that increases in brain activity lead to increases in blood flow (Raichle, 1987), just as blood flow increases in our muscles when we exercise. Increases in brain activity also lead to changes in the concentration of oxygen in the blood, and the most commonly used fMRI methods are capable of measuring this change (DeYoe et al., 1994).

Limitations

There are a number of challenges which all current neuroimaging techniques must meet. First, the physiological measures that correlate with brain activity have baseline values that are always present in the brain. For example, there is always a baseline level of blood flow in the brain. This means that for PET to detect changes in brain activity, it must rely on detecting changes in blood flow *relative* to that baseline. In practice, this is usually accomplished by directly comparing blood flow from one task state to another. fMRI uses similar methods to compare task-induced changes in blood oxygen concentration.

At a task design level, this means that multiple behavioral tasks must be imaged and compared to one another to determine which brain areas are differentially activated across the tasks. For example, in order to isolate brain areas related to vision, a PET neuroimaging experiment might acquire blood flow images while subjects view a simple visual display, perhaps a flickering checkerboard or a set of random moving dots. These blood flow images could be compared to images while subjects are presented with no visual stimulation. As the main difference between the two tasks is demand on visual processing, local changes in blood flow could be used to determine the brain areas activated during certain kinds of visual processing (figure 32.2).

In using such task comparisons, two assumptions are made that are worth discussing. First, it is assumed that the processing demands of the tasks being compared are well understood; and second, it is assumed that most of the processing demands do not change with the specific task context. A number of debates have arisen because some researchers have correctly argued that it is difficult to know if these two assumptions are valid. The practical solution to this problem is straightforward. Cognitive task analysis must be used on all tasks being examined, to understand them as completely as possible. This helps to create interpretational power. Then, knowing that task analyses are imperfect, assume that any single task analysis (or pair of task analyses) is only partially correct and seek convergent results across multiple task comparisons and studies. In this manner, the use of neuroimaging can progress as efficiently as possible in an experimental arena that contains only an approximate understanding of the underlying cognitive phenomena being studied.

A final important point about neuroimaging methodology is that, although neuroimaging techniques are good at identifying the location of active brain areas (to within 3–5 mm), current techniques have poor temporal resolution. Under ideal circumstances, fMRI can acquire data in a matter of a few seconds, while PET requires at least 30 seconds. This means that if a task is being studied that involves brain areas increasing and decreasing activity within shorter time periods, the resultant image will be an average of this activity. For example, during a word-reading task it is likely that activity in brain areas being used to perceive a visual word precedes activity in brain areas being used to guide the spoken output. However, because it only takes about 500 msec to read a word, both visual and motor areas would appear in the same PET or fMRI image. Methods for combining PET and fMRI neuroimaging techniques with high temporal resolution techniques are being developed (Heinze et al., 1994). This combination might ultimately provide a set of methods that enable both high spatial resolution and high temporal resolution.

Examples of neuroimaging research

What follows is a description of two areas of research that are based on the neuroimaging techniques just discussed. Each represents an area where progress has been

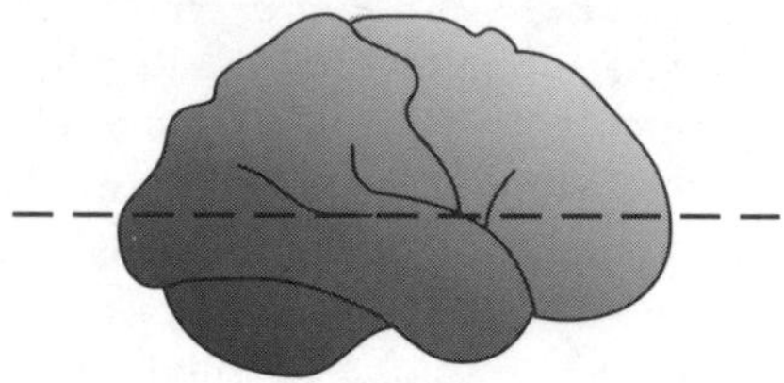

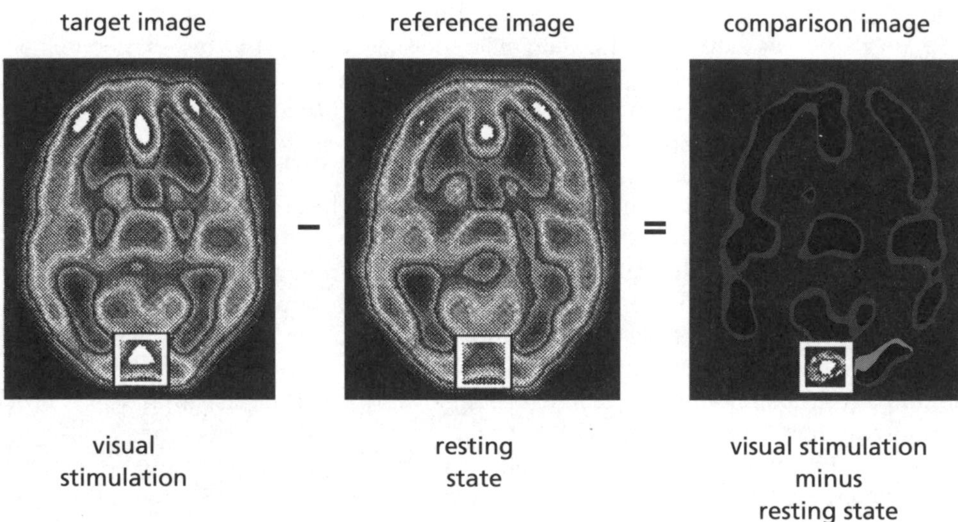

Figure 32.2 The basis of neuroimaging techniques is the detection of changes in brain activation from one task state to another. One example of such a change is depicted for a set of PET neuroimaging data. The lateral view of the brain at the top shows the approximate level of the brain that is being imaged. The three images below show actual PET blood flow data. The images are presented in the horizontal plane, with the bottom of the images being the back of the brain and the top being the front. The box in each of the images highlights the visual cortex. As can be seen in the leftmost section, the image appears bright in the visual cortex when the subjects are performing the target task – visual stimulation. However, the relative activation of that area can only be appreciated when it is compared to a second task state, such as the one depicted in the middle section. In this state, the subjects are performing a reference task in which they receive no visual stimulation. By directly comparing data from the two task states, as shown in the rightmost section, differences in brain activation are isolated. In this image, it is easy to see that the visual stimulation task activates the visual cortex.

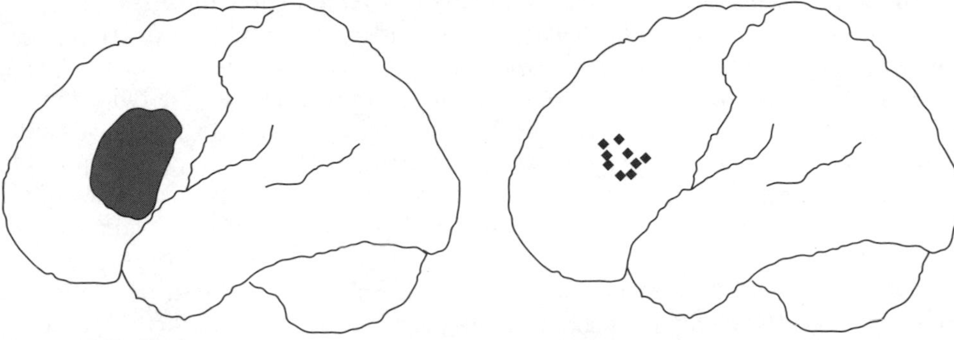

Figure 32.3 Lateral views of the left hemisphere of the brain, illustrating the approximate location of Broca's area – a region that causes speech production impairments when damaged (*left*) – and the locations of neuroimaging activations observed during a task related to speech production (*right*).

made through the use of neuroimaging techniques. The first area is concerned with the study of speech, and we will see how neuroimaging has helped to reveal the processing roles of left prefrontal brain areas that are important for speech production. The second area is concerned with new ideas about the functional anatomy of long-term memory, which has been extensively studied using neuroimaging.

Left inferior prefrontal cortex: beyond Broca's area

One of the best-established neuropsychological findings is that damage to certain parts of the brain can cause speech production impairments, commonly referred to as "production aphasia." The part of the brain believed to be damaged in most forms of production aphasia is called "Broca's area," after Paul Broca, a famous French surgeon and anthropologist. Broca's primary observation was that speech impairments could be caused by damage to the prefrontal cortex – and later, coincident with the insights of Marc Dax, that the left prefrontal cortex is particularly important for speech production, as compared to the right. Since that time, numerous detailed studies of aphasia have been documented in the literature and, in general, Broca's initial observations have held up quite well. However, in spite of acceptance of the aphasic syndrome and its general localization to Broca's area, it has been difficult to determine the exact processes that have been lost in aphasia. And it has been even more difficult to determine which *specific* prefrontal areas might underlie those processes. Neuroimaging techniques provide a set of tools to aid in the characterization of left prefrontal cortex's role in speech production.

Some of the earliest neuroimaging work on the functional anatomy of speech was done by observing brain areas activated in subjects doing various tasks involving the processing of single words (Petersen et al., 1988). In the simplest task that was studied, subjects were asked to passively view words (e.g., see *house*). This task encouraged subjects to process the visual form of words and perhaps to read the words subvocally. In a second task, subjects were asked to read words aloud (e.g., see *house*, say *house*) which additionally required them to produce overt speech. Finally, in the most demanding task, subjects were asked to generate verbs that related to presented nouns (e.g., see *house*, generate and say *build*). This last task required subjects to access word meaning and generate appropriate words from memory, in addition to producing speech. Relevant to this chapter is whether left prefrontal areas were activated in the two overt speech tasks. If both the speech tasks were found to activate Broca's area, then previous ideas about Broca's area being generally related to speech output would have been supported. However, the results were not so simple and led to new ideas about speech production and the role of the left prefrontal cortex at (and around) Broca's area.

The surprising result was that, although both speech tasks activated areas involved in motor production, only the task involving the generation of verbs activated extensive areas in the left prefrontal cortex (near Broca's area). The other, more automated speech task involving noun reading showed only minimal activation of the left prefrontal cortex but activations in the bilateral Sylvian-insular cortex – a distinct set of brain areas. These findings suggested the possibility that all speech-production tasks do not rely on the same areas of the left prefrontal cortex. It seemed possible that there might be at least two pathways that could be utilized during speech production. One pathway, relying on more prefrontal cortex, became activated during the verb-generation task, in which considerable effort is demanded. A second pathway, relying on the bilateral Sylvian-insular cortex, became activated during the more automatic word-reading task. At a conceptual level, such an idea is appealing. The prefrontal pathway might provide a more flexible, but relatively inefficient, pathway for generating verbal responses. The Sylvian-insular pathway might constitute an efficient, highly automatic pathway able to take over for those verbal responses that are programmed repetitively.

Motivated by this possibility, Marc Raichle and colleagues (Raichle et al., 1994) set out to test the idea that elaborate speech-production tasks use a pathway involving the left prefrontal cortex, while more automated speech tasks utilize a separate pathway involving the Sylvian-insular cortex. They did this in two ways. First, they replicated the findings of Petersen et al. (1988) that simple reading, a highly over-learned task, activated the Sylvian-insular cortex but not the left prefrontal cortex. Second, and more importantly, they rehearsed subjects on the verb-generation task discussed above and imaged them during this well-practiced state as well as during the naive state used by Petersen et al. (1988). At a behavioral level, the naive state was characterized by long reaction times in producing the verbs, and the verbs produced for a given noun varied across trials. In the practiced state, verbs were produced much more quickly, and the same verb was usually given for each noun. In essence, the practiced verb-generation task was automatic, much as the simple word-reading task was.

And what did they find? As before, unpracticed verb generation activated the left prefrontal cortex, while practiced verb generation, a task automated through repetition, activated the Sylvian-insular cortex but not the left prefrontal cortex – confirming the hypothesis that two separate pathways can be used for speech production.

The study of patients with a subtype of aphasia called *transcortical motor aphasia* shows a corollary to these findings, even though the syndrome has not traditionally

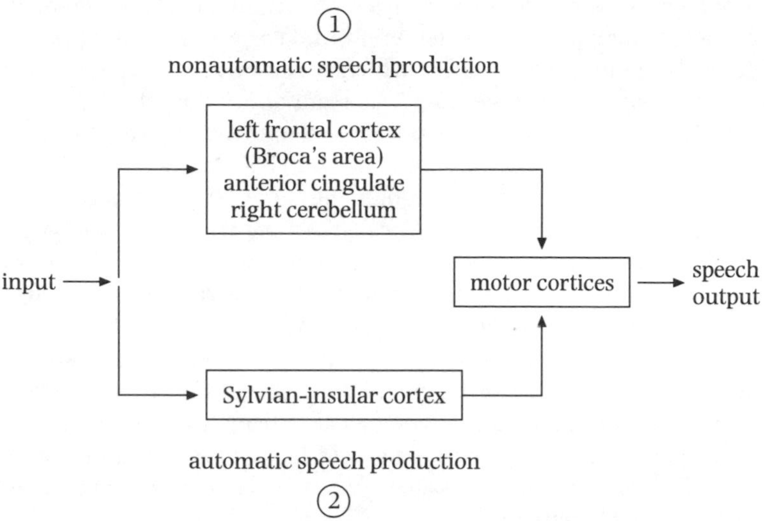

Figure 32.4 A plausible model based on neuroimaging data that depict two separate pathways that might be used for speech production (Posner and Raichle, 1994). The first pathway, which fits well with traditional ideas of speech production, robustly activates the prefrontal cortex and appears to be used during tasks that require elaborate, nonautomatic processing. The second pathway, which has only been appreciated recently, based on neuroimaging data, involves the bilateral Sylvian-insular cortex and appears to be used for speech tasks that have been well-automated through practice.

been interpreted as reflecting the existence of two separate speech pathways. Transcortical motor aphasia is characterized by difficulty in generating spontaneous speech with the preservation of repetition and simple reading. Perhaps patients with this kind of aphasia have suffered damage to the nonautomatic prefrontal pathway but not the Sylvian-insular pathway. Interestingly, patients with the most severe speech-production deficits typically have larger lesions, which may (and sometimes are known to) include damage to the Sylvian-insular cortex as well as the premotor and motor areas (Mohr et al., 1978). Thus the neuroimaging data combined with the lesion data suggest a new idea about the role of the left prefrontal cortex in speech production. *The idea is that the left prefrontal cortex plays a role in nonautomatic speech production and is part of just one of potentially several pathways that can be used to guide different kinds of speech production.*

The data discussed so far leave open the question of what role the left prefrontal region, including Broca's area, might be playing in the nonautomatic left prefrontal pathway. To gain insight into possible answers to this question, it is necessary to look at a large number of studies and see which, if any, found this left prefrontal region activated. Figure 32.3 displays such data. A large number of speech-production and related tasks show activation in similar left prefrontal areas. In all these tasks, subjects were instructed to generate words or make decisions about what words mean. Thus, a common demand of the tasks was that they required subjects to access information about words. Surprisingly, a demand that was not common to the tasks was overt speech. Several of the tasks required subjects to simply think about the words they generated or to make a key-press response rather than say them aloud. The finding that tasks showing activation in the left prefrontal cortex do not always require overt speech, while automated speech tasks, which require overt speech, often show minimal activation there tells us something important about the role of the left prefrontal cortex in speech production: namely, that it plays a processing role that is at a higher level than simply programming the speech per se. This illustrates a second idea about left prefrontal involvement in speech production: *The left prefrontal cortex appears to underlie processes related to accessing and/or representing verbal information, regardless of whether that information is used to guide overt speech output.*

Not surprisingly, on the basis of this kind of data, several researchers using neuroimaging have suggested a variety of possible roles for the left prefrontal cortex in speech production, including accessing meaning (Petersen et al., 1988), retrieving words from memory (Buckner and Tulving, 1995), generating internal representations (Frith et al., 1991), and accessing phonological information (Fiez et al., 1995). This may make it seem as if the neuroimaging field has generated more confusion than answers. However, very recently, possible answers to even this debate have been proposed, based on neuroimaging data.

The emerging idea is that the diversity of functional roles proposed for the left prefrontal cortex may reflect multiple areas within the left prefrontal cortex underlying distinct processing functions. Although this idea has been explored in work with nonhuman primates, characterization of the human prefrontal cortex has been held back because of the absence of techniques for observing functional subdivisions in humans. Pioneering this work, Petrides and colleagues (1995) have shown that there are multiple functionally distinct subdivisions of the human prefrontal cortex. Recently (Buckner et al., 1995a), we were able to show that at least two separable prefrontal regions were differentially activated across speech-production tasks, with one region being recruited across multiple tasks, while a second region was recruited when the task demanded access to word meaning. It seems quite possible that multiple prefrontal areas are activated in differing numbers and combinations, depending on the specific demands of the tasks. Again, these very recent findings suggest another idea about left prefrontal involvement in speech production: *Multiple areas within the left prefrontal cortex may play different roles in speech production.* Most importantly, neuroimaging techniques offer a means to differentiate the functional roles of these areas.

But what about areas outside the left prefrontal cortex? It seems equally possible that, just as multiple prefrontal areas participate in certain forms of speech production, areas outside the prefrontal cortex might also be involved. One important advantage of neuroimaging techniques is that they often survey the entire brain for activity. This enables researchers to see the multiple areas of the brain that might be concurrently activated during a task. When one looks beyond the prefrontal cortex in the tasks just discussed, one is confronted with two observations. First, there are several areas that are often activated concurrently with the left prefrontal cortex. These include the anterior cingulate gyrus, an area well known to have connections to lateral prefrontal areas, and the right lateral cerebellum, an area that has only recently been explored as being responsible for cognitive functions (e.g., Fiez et al., 1992; Leiner et al., 1989). And second, depending on the specific task demands, quite distinct and often multiple additional brain areas are recruited. The manner in which these distributed brain areas collectively contribute to speech production and the separate roles of the areas constitute a hot topic of study but, unfortunately, are well beyond the scope of this chapter. What is remarkable, and what should be taken as the main point of this explication, is that neuroimaging offers a powerful method for gaining insight into the multiple areas involved in these pathways and what the separable roles of these constituent areas might be.

Frontal lobe involvement in memory retrieval

Cognitive psychology has yielded a rich literature describing MEMORY processes. One major success in the cognitive domain is the understanding that memory relies on temporally distinct stages: encoding, storage, and retrieval. Encoding refers to the initial acquisition of information, storage to the transfer of this information into a more enduring form, and retrieval to the process of accessing the stored information. In behavioral settings, these distinct stages can be separated and studied (see Article 27, BEHAVIORAL EXPERIMENTATION).

At the neurobiological level, however, methods for separating different stages of memory have been more problematic. Studies of patients who have brain lesions prove difficult, because lesions are permanent and thus are necessarily present during the encoding, storage, and retrieval stages of any new memories. This produces an interpretation problem: if a patient with a brain lesion cannot retrieve information in an experimental setting, it is difficult to know whether the brain lesion disrupted the encoding, storage, or retrieval of the information, or any combination of the three (see Article 29, DEFICITS AND PATHOLOGIES).

Neuroimaging provides a method that can be used to study the different stages of memory. By having subjects study information before a scan period, encoding and storage can be made to take place prior to the scan. Then, brain areas selectively active during memory retrieval can be monitored while subjects retrieve information within the neuroimaging device. A number of laboratories have used such an approach to study intentional memory retrieval.

What has emerged from these studies is that areas within the prefrontal cortex are activated across a wide range of memory retrieval tasks. Prior to these findings, the prefrontal cortex was not thought to play a prominent role in long-term memory processes. Rather, considerable attention was given to brain areas in the medial temporal lobe believed to be involved in long-term memory storage. The neuroimaging findings, by focusing attention on another set of brain areas, has helped to expand the field's area of inquiry. Currently, researchers are asking the question of what role(s) prefrontal areas might be playing in memory retrieval and how this role(s) might interact with other systems that facilitate memory encoding and storage.

Even more provocative is the finding that *specific* areas of the prefrontal cortex are activated during memory retrieval. For example, an area in the right anterior prefrontal cortex has been activated across a wide range of tasks that require memory retrieval. This was initially noted in a series of experiments using verbal materials (see Buckner et al., 1995b). In these experiments, subjects were presented with study words before a PET scan and then, during the PET scan, were asked to intentionally recall them, using word stems as cues. A consistent right anterior focus of activation was found across all three experiments. Additional brain areas outside the right prefrontal cortex, including the left inferior prefrontal cortex, were also activated; but these areas were

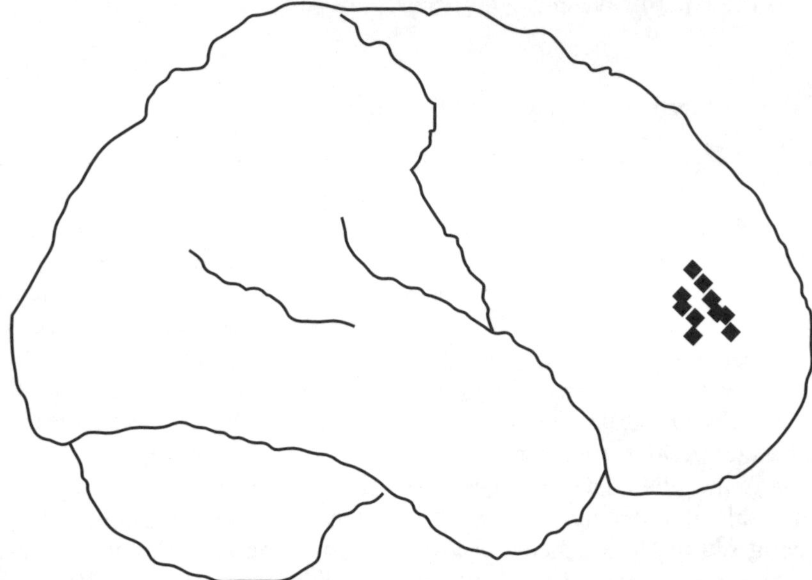

location of activations across PET
studies involving episodic memory retrieval

Figure 32.5 A lateral view of the right hemisphere of the brain, illustrating the locations of a number of activations that have been detected in subjects performing episodic memory retrieval tasks.

also activated by other tasks that did not require retrieval of recently studied words. Several other similar tasks involving retrieval of verbal information have also activated the right anterior prefrontal cortex, convincingly demonstrating that the finding is highly reliable (see Buckner and Tulving, 1995).

Another set of retrieval tasks relying on nonverbal information has also activated this area, indicating that it is generally used in retrieval tasks in which different kinds of information are being accessed. Most notably, Haxby and colleagues at the National Institutes of Health have reported activation of the right anterior prefrontal cortex during a task requiring the recognition of recently learned faces (learned 15 minutes before the PET scan). This finding has subsequently been generalized to several other studies by the same group. In one case, faces were learned one day prior to the scan, and in another, older subjects were used (Grady et al., 1995).

This remarkably consistent pattern of right anterior prefrontal cortex activation was surprising, considering little was known about this brain area, and, as already discussed, only a minority of researchers suggested that the prefrontal cortex played a primary role in long-term memory retrieval. The neuroimaging data thus provided a novel insight: the right anterior prefrontal cortex is activated during memory retrieval.

This finding is robust and reliable, as the anterior right prefrontal cortex has been found across many forms of retrieval for recently learned information. The tasks varied the modality of the studied information and test cues (auditory and visual), the length of time between study and test (3 minutes to one day), and the kind of information being recalled (words, pictures, faces, and sentences).

Moreover, although many surface variables were changed across the several studies discussed, all the studies relied on a similar form of memory. Tulving (1983) has called this form of memory "episodic" memory, because information is being retrieved from a specific study episode. Other neuroimaging studies of memory retrieval relying on forms of memory considered to be distinct from episodic memory have shown many other brain areas to be active, but usually not this right anterior prefrontal area. In this respect, the neuroimaging data demonstrate a biological dissociation between episodic memory retrieval and other forms of memory retrieval.

The exact processing role that this right anterior prefrontal brain area is playing in episodic memory is still unclear. Some researchers have proposed that some specific component of the episodic memory search is being facilitated by the right prefrontal cortex, while others have suggested that the role of the area will be found to be more general than purely episodic memory retrieval. It is also unclear how this finding will relate to information gathered about this brain area from other methods, such as the study of patients with lesions.

Nonetheless, the finding that a large number of laboratories and testing procedures have produced converging results illustrates the power of neuroimaging. This set of studies suggests that the right anterior prefrontal cortex plays a role in episodic memory retrieval. We would not have even begun to explore in more detail what its role might be if it had not been for the collective body of neuroimaging studies discussed.

Conclusions

Two areas of neuroimaging research have been discussed, with the goal of demonstrating that neuroimaging can be used to gain insights into the biological basis of cognition. As such, this explication has focused almost exclusively on data from neuroimaging and has not attempted to tie the results to data from other methodologies. This is not meant to imply that other methodologies are not equally important. On the contrary, many techniques are actively being used in the field of cognitive neuroscience, and each has its limitations and its strengths (see Article 41, NEUROBIOLOGICAL MODELING). Neuroimaging methods (both PET and fMRI) provide tools for detecting brain areas active in human subjects while they perform cognitive tasks and thus provide an important set of methods for examining the functional anatomy of human cognition. It is hoped that such neuroimaging techniques, used in conjunction with all the other techniques available to neuroscientists, will provide an array of methods to confront the infinitely complex study of human cognition.

References and recommended reading

*Buckner, R. L. and Tulving, E. 1995: Neuroimaging studies of memory: theory and recent PET results. In F. Boller and J. Grafman (eds), *The Handbook of Neuropsychology*, Amsterdam: Elsevier, 439–66.

Buckner, R. L., Petersen, S. E. and Raichle, M. E. 1995a: Dissociation of human prefrontal cortical areas across different speech production tasks and gender groups. *Journal of Neurophysiology*, 74, 2163–73.

Buckner, R. L., Petersen, S. E., Ojanann, J. G., Mitzin, F. M., Squire, R. L. and Raichle, M. E. 1995b: Functional anatomical studies of explicit and implicit memory retrieval tasks. *Journal of Neuroscience*, 15, 12–29.

*DeYoe, E. A., Bandettini, P., Neitz, J., Miller, D. and Winans, P. 1994: Functional magnetic resonance imaging (fMRI) of the human brain. *Journal of Neuroscience Methods*, 54, 171–87.

Fiez, J. A., Petersen, S. E., Cheney, M. K. and Raichle, M. E. 1992: Impaired non-motor learning and error detection associated with cerebellar damage. *Brain*, 115, 155–78.

Fiez, J. A., Tallal, P., Raichle, M. E., Miezin, F. M., Katz, W. F. and Petersen, S. E. 1995: PET studies of auditory and phonological processing: effects of stimulus characteristics and task demands. *Journal of Cognitive Neuroscience*, 7, 357–75.

Frith, C. D., Friston, K., Liddle, P. R. F. and Frackowiak, R. S. J. 1991: Willed action and the prefrontal cortex in man: a study with PET. *Proceedings of the Royal Society of London*, B244, 241–6.

Grady, C. L., McIntosh, A. R., Horwitz, B., Maisog J. M., Ungerleider, L. G., Mentis, M. J., Pietrini, P., Schapiro, M. B. and Haxby, J. V. 1995: Age-related reductions in human recognition memory due to impaired encoding. *Science*, 269, 218–21.

Heinze, H. J., Mangun, G. R., Burchert, W., Hinrichs, H., Scholz, M., Münte, T. F., Gös, A., Scherg, M., Johannes, S., Hundeshagen, H., Gazzaniga, M. S. and Hillyard, S. A. 1994: Combined spatial and temporal imaging of brain activity during visual selective attention in humans. *Nature*, 372, 543–6.

Leiner, H. C., Leiner, A. L. and Dow, R. S. 1989: Reappraising the cerebellum: what does the hindbrain contribute to the forebrain? *Behavioral Neuroscience*, 100, 998–1008.

Mohr, J. P., Pessin, M. S., Finkelstein, S., Funkenstein, H. H., Duncan, G. W. and Davis, K. R. 1978: Broca aphasia: pathologic and clinical. *Neurology*, 28, 311–24.

*Petersen, S. E. and Fiez, J. A. 1993: The processing of single words studied with positron emission tomography. *Annual Review of Neuroscience*, 16, 509–30.

Petersen, S. E., Fox, P. T., Posner, M. I., Mintun, M. and Raichle, M. E. 1988: Positron emission tomographic studies of the cortical anatomy of single-word processing. *Nature*, 331, 585–9.

Petrides, M., Alivisatos, B. and Evans, A. C. 1995: Functional activation of the human ventrolateral frontal cortex during mnemonic retrieval of verbal information. *Proceedings of the National Academy of Sciences*, USA, 92, 5803–7.

*Posner, M. I. and Raichle, M. E. 1994: *Images of Mind*. New York: Scientific American Library.

*Posner, M. I., Petersen, S. E., Fox, P. T. and Raichle, M. E. 1988: Localization of cognitive operations in the human brain. *Science*, 240, 1627–31.

Raichle, M. E. 1987: Circulatory and metabolic correlates of brain function in normal humans. In F. Plum and V. Mountcastle (eds), *The Handbook of Physiology* vol. 5: *Higher Functions of the Brain*, Bethesda, Md: American Physiological Association, 643–74.

Raichle, M. E, Fiez, J. A., Videen, T. O., MacLeod, A. M. K., Pardo, J. V., Fox, P. T. and Petersen, S. E. 1994: Practice-related changes in human brain functional anatomy during nonmotor learning. *Cerebral Cortex*, 4, 8–26.

Tulving, E. 1983: *Elements of Episodic Memory*. New York: Oxford University Press.

33

Protocol analysis

K. ANDERS ERICSSON

The central problem which cognitive scientists face in studying thinking is that *thinking cannot be observed directly by other people.* The traditional solution has been to rely on introspective methods, where individuals observe their own thinking and reflect on its characteristics. In everyday life, the most common technique involves asking people questions about their thinking, knowledge, and strategies. Psychologists have refined the methods for questioning individuals by designing questionnaires and structured interviews. However, these two ways of obtaining information about thinking share two fundamental methodological problems. First, the accuracy of the reports cannot be assessed, because in naturally occurring situations in everyday life the investigator doesn't have any other empirical evidence against which to evaluate the validity of the subjects' reported information. Hence, psychologists are forced to trust the subjects to provide valid information. The second issue concerns whether even those individuals who strive to give accurate reports are able to access and supply valid information on the cognitive processes that mediate their behavior (Nisbett and Wilson, 1977).

To avoid these seemingly unresolvable problems, most experimental psychologists in the first part of this century turned away from the study of thinking and towards basic processes of learning and memory. However, with the emergence of the human information processing theory (Newell and Simon, 1972), investigators began to propose detailed computational models of thinking for well-defined tasks. These models made empirically testable predictions for observable indicators of thinking, such as latencies, eye movements, physiological indicators (GSR, EEG), and verbal reports. These developments made it possible to address the earlier issues about the validity of verbal reports (Ericsson and Simon, 1993).

Since the turn of the century introspection has been dismissed as scientifically unacceptable, and many scientists have extended the repudiation of introspection and have rejected all types of verbal reports of thinking. This unfortunate confusion will be addressed in the first part of this chapter, where the history of the use of verbal reports on thinking will be briefly reviewed to show that earlier controversies surrounding introspection did not concern the validity of verbal reports of thinking as a sequence of thoughts but reflected introspective efforts to go beyond this type of readily accessible information. In the second, main part, models for the generation of verbal reports will be presented to identify under what conditions subjects' verbal reports have been found to yield valid evidence on their thought processes.

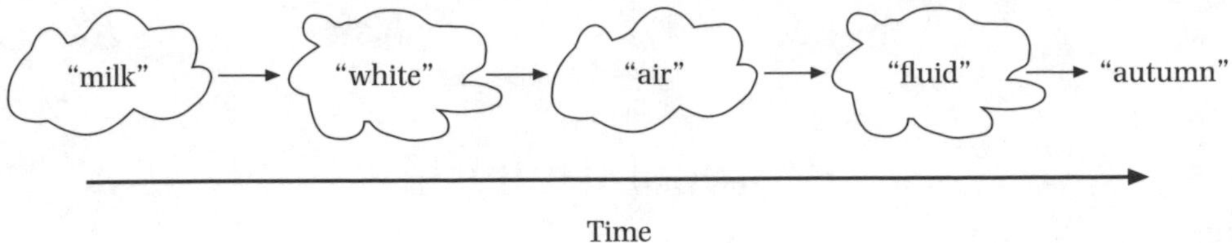

Figure 33.1 An illustration of Aristotle's view of thinking as a sequence of states.

Historic background: from self-observation of naturally occurring thought to laboratory studies of reproducible cognitive performance

The observation of one's own spontaneous thinking has a long history, which can be traced back at least as far as the Greeks. Aristotle is generally given credit for the first systematic attempt to record and analyze the structure of thinking, in particular the process of recalling a specific piece of information from memory. He even gave the following description of a specific episode corresponding to the recall of "autumn": "from milk, to white, from white to air, and this to fluid, from which one remembers autumn, the season one is seeking." Based on an analysis of many observed episodes of thinking, Aristotle argued that thinking corresponds to a sequence of thoughts (see the boxes in figure 33.1), where the brief transition periods between consecutive thoughts (see the arrows in figure 33.1) do not contain any reportable information. Hence, the processes determining how one thought triggers the next could not be observed directly but had to be inferred by a retrospective analysis of relations between consecutive thoughts. By examining his memory of which thoughts tended to follow each other, Aristotle concluded that previously experienced associations were the primary determining factor.

Aristotle's account of thinking as a sequence of thoughts has never been seriously challenged. However, such a simple description of thinking was not sufficiently detailed to answer the questions about the nature of thought raised by philosophers in the seventeenth, eighteenth, and nineteenth centuries (Ericsson and Crutcher, 1991). For example, could all thoughts be described as mixtures of sensory images derived from past experiences? In order to evaluate these claims, philosophers would typically relax and daydream, allowing their thoughts to wander. Once a thought emerged, it would be inspected carefully and studied to assess its sensory modality and components. The British philosopher David Hume even argued that these thoughts were as detailed as original perceptions (e.g., we cannot image a printed page of text without imaging every letter on the page at the same time).

However, toward the end of the nineteenth century, philosophers generally agreed that thoughts and mental images differed fundamentally from external objects. External objects can be inspected, and details about them can be noticed without changing their content and structure. By contrast, when thoughts and images are inspected – for example, when focusing on the petals of the image of a rose – the mental image also changes. Hence, extended introspective analysis changes the content and structure of originally experienced thoughts and cannot be used to assess the structure of thoughts as they emerge during spontaneous thinking.

In order to address these criticisms, around the turn of this century investigators at the University of Würzburg invited trained introspective observers to their laboratory and asked them questions. The observers were asked to give their answers to the questions as fast as possible, and then after each answered question they gave very detailed retrospective reports about their thoughts which mediated the question-answering process. The retrospective reports were extensive and detailed. Most reported thoughts consisted of visual and auditory images, but some subjects claimed to have experienced thoughts without any corresponding imagery (imageless thoughts). The existence of imageless thoughts had far-reaching theoretical implications and led to a heated exchange between the psychologists who claimed to have observed them and others who argued that these reports were artifacts of the reporting methods and the poor training of the observers. The most devastating aspect of this controversy was that it showed that the issue of imageless thoughts could not be resolved empirically, which in turn raised doubts about analytic introspection as a scientific method.

In the behaviorist reaction to the methodological and theoretical problems of introspection, psychologists redirected their research away from thinking and toward processes that were unaffected by prior experience and knowledge. These processes were presumed to reflect more basic general processes of learning and memory. Subjects were given well-defined tasks, such as memorizing lists of nonsense syllables – e.g., XOK, ZUT – where memory performance could be measured objectively by the accuracy of recall. Given the assumed noncognitive nature of the formation of basic associations in memory, subjects were not expected to report any mediating thoughts, and the issue of trusting subjects' verbal reports was no longer relevant.

The cognitive revolution in the 1950s and 1960s brought renewed interest in higher-level cognitive processes, and with it came renewed concerns about how to describe the structure of thinking. Information processing theories (Newell and Simon, 1972) sought computational models that could duplicate human performance on well-defined tasks by the application of explicit procedures. One of the principal methods of that approach is *task analysis*, where the investigator identifies many different procedures that people could possibly use, in light of their prior knowledge and limited information processing capacities, to generate correct answers to tasks in a domain. Hence, the task analysis provides a set of possible thought sequences for the performance of a task, where the application of each alternative procedure is associated with a different sequence of thoughts (intermediate steps). Let me illustrate how this general method can be applied to the task of mental multiplication. Most adults have only a limited mathematical knowledge: they know their multiplication tables and only the standard "pencil and paper" procedure taught in school for solving multiplication problems. Accordingly, one can predict that they will solve a problem such as 36×24 by first calculating 4×36, then adding 20×36. However, this specific problem can be solved using alternative methods, which are more efficient for subjects who know some of the squares of 2-digit numbers. More sophisticated subjects may recognize that 24×36 is equivalent to $(30 - 6) \times (30 + 6)$ and use the formula $(a + b) \times (a - b) = a^2 - b^2$, thus calculating 36×24 as $30^2 - 6^2 = 900 - 36 = 864$. Other subjects may recognize other shortcuts, such as $36 \times 24 = (3 \times 12) \times (2 \times 12) = 6 \times 12^2 = 6 \times 144 = 864$.

A task analysis allows the investigators to identify a limited number of different sequences of thoughts and intermediate products that any subject could have relied

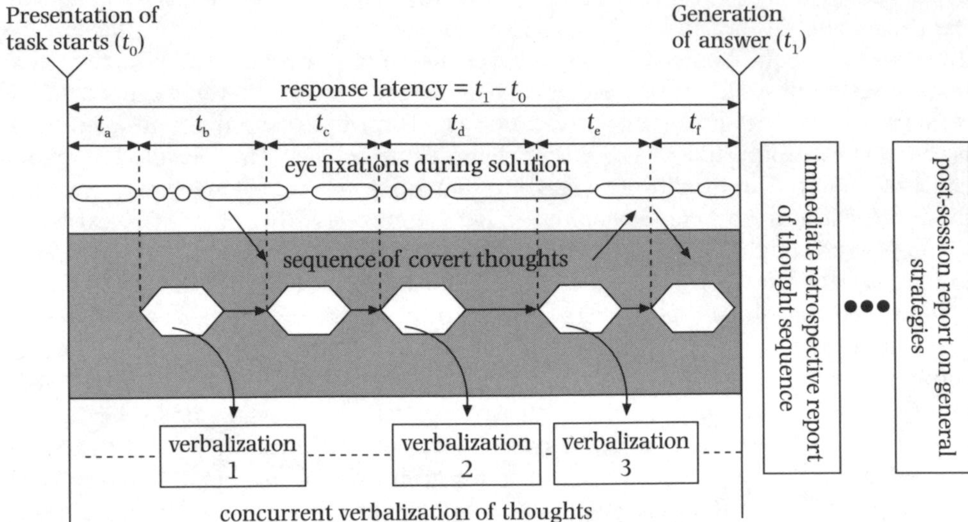

Figure 33.2 An illustration of a covert sequence of thoughts (center) generated in response to a presented task and its associated observable nonverbal indicators at the top and verbal reports at the bottom.

on to generate the correct answer to a specific problem. By collecting and analyzing empirical evidence on how a particular subject reached the solution, it is often possible to discard most of the alternative thought sequences as inconsistent, thus leaving often only a single acceptable description of that subject's sequence of thoughts.

Although a covert sequence of thoughts generated during the performance of a task (illustrated in the center of figure 33.2) is never directly observable, it is associated with several types of observable indicators. On the top, the total time required by subjects to generate their answer (response latency) is shown to correspond to the sum of the component durations for the generation of each thought (intermediate product). Given that procedures differ in the number of intermediate steps that they require for different problems, each hypothesized procedure will lead to a distinguishable pattern of expected latencies across a set of different problems that will differentially agree with the subject's observed latency pattern. In addition, to produce the correct answer to most problems, the subjects must sequentially direct their visual gaze to the essential presented information, and an eye-tracking device can register their sequences of eye fixations. Given that different hypothesized procedures typically predict different sequences of eye fixations, it is possible to assess the procedure with the highest agreement with the observed eye fixations. Finally, there are three common types of method to elicit verbal reports from subjects. First, the subjects can be instructed to give a concurrent report of their thinking while they perform the task. Alternatively, the subjects can be asked to give a retrospective report of their thought sequence immediately after the completion of the task. Finally, the subject can give a general description of their strategies (post-session reports) once they have completed numerous tasks during the testing session.

Models of generation of verbal reports

Based on their theoretical analysis, Ericsson and Simon (1993) argued that the closest connection between thinking and verbal reports is found for concurrent verbal reports where subjects verbalize most of the thoughts and end products of processing as these enter their attention – illustrated in the lower part of figure 33.2. For example, when one subject was asked to think aloud while mentally multiplying 36 by 24 on two test occasions one week apart, the following protocols were obtained:

> OK, 36 times 24, um, 4 times 6 is 24, 4, carry the 2, 4 times 3 is 12, 14, 144, 0, 2 times 6 is 12, 2, carry the 1, 2 times 3 is 6, 7, 720, 720, 144 plus 720, so it would be 4, 6, 864.

> 36 times 24, 4, carry the – no wait, 4, carry the 2, 14, 144, 0, 36 times 2 is, 12, 6, 72, 720 plus 144, 4, uh, uh, 6, 8, uh, 864.

In these two examples, the reported thoughts are not analyzed introspectively into their perceptual or imagery components but merely verbally expressed and referenced, such as "carry the 1," "36," and "144 plus 720." Hence, when subjects are instructed to simply verbalize thoughts after these have spontaneously emerged in attention (think aloud), the sequence of subsequent thoughts should not change as a result of such verbalization.

Similarly, Ericsson and Simon (1993) argued for a tight connection between thinking and its verbal report when the subject's recall of the corresponding sequence of thoughts (immediate retrospective verbal report) is given immediately after completion of the task. The instructions for thinking aloud and retrospective recall focus on the report of the spontaneous sequences of thoughts induced by the task. By contrast, when subjects are instructed to explain or carefully describe their thoughts, they are not able to merely verbalize each thought as it emerges but need to engage in additional cognitive processes to generate the required explanations and descriptions. This additional cognitive activity is likely to influence subsequent thoughts and thus affect the performance of subjects. Similarly, if subjects are asked to recall their thoughts after a long delay or after many intervening solutions to other problems, the completeness and accuracy of recall will be impaired, and subjects may try to infer their thoughts as opposed to recall them from memory.

The empirical evidence concerning the major factors influencing the validity of verbal reports will be discussed in the next three sections.

Does the generation of verbal reports change the cognitive processes?

The act of verbalizing one's thoughts was believed by many investigators to cause a change in the associated thoughts and thought processes. To demonstrate the effects of verbalization, the investigators compared the performance of subjects who verbalized with that of silent control subjects performing the same tasks. A comprehensive review of over 40 such studies (Ericsson and Simon, 1993) shows that only some types of verbalization instructions influence performance. For think-aloud and retrospective reports, they found no evidence for changes in thought processes and the corresponding accuracy of performance for well-defined tasks. The only systematic effect was that subjects took longer to complete some of the tasks, due to the additional time required to verbalize the thoughts. By contrast, when subjects were asked to explain or to provide detailed descriptions, the accuracy of their performance often improved; however, in some cases, the performance was systematically worse than the silent control condition. In sum, after a few minutes of instruction, subjects have been found to be able to give think-aloud and retrospective verbal reports without any systematic changes to their thought process (see Ericsson and Simon, 1993, for detailed instructions and associated warm-up tasks recommended for laboratory research). The fact that subjects must already possess the necessary skills for verbalization of thoughts is consistent with the extensive evidence on the acquisition of self-regulatory private speech during childhood (Diaz and Berk, 1992) and on the spontaneous vocalization of inner speech by adults (Ericsson and Simon, 1993).

Under what circumstances does verbally reported information reflect cognitive processes?

The validity of the reported information depends on the type of verbal report requested of the subjects. Even in the case of thinking aloud, where the connection between thoughts and reports is the closest, there is no perfect mapping. This lack of one-to-one correspondence is primarily due to the fact that thoughts that pass through attention are not always verbalized. However, the evidence for validity is consistently strong for those thoughts that are verbalized (Ericsson and Simon, 1993). First, the verbalized information is consistent with one of the sequences of thoughts derived from the task analysis, which was completed prior to the observation of the verbal reports. For example, the sample protocols on mental multiplication (reported above) are consistent with only one of the methods uncovered in the task analysis – namely, the "paper-and-pencil" method. Even if the observed protocol had been more sparse for a highly skilled subject and had only contained "144" and "720," the reported information would still have been sufficient to reject each of the alternative multiplication methods, because neither of those methods involved generating both the reported intermediate products. The general finding that a task analysis can identify, a priori, the specific intermediate products that are later verbalized by subjects during their problem solutions provides the strongest evidence that concurrent verbalization reflects the processes that mediate the actual generation of the correct answer.

Second, the verbal reports are only one of several different kinds of indicators of the same thought process (cf. figure 33.2). Given that each kind of empirical indicator can be separately recorded and analyzed, it is possible to compare the results of such independent data analyses. Ericsson and Simon (1993) found that the solution methods derived from an analysis of the verbal reports were consistent with those derived from analyses of response latencies and sequences of eye fixations. Furthermore, the verbal reports often contain more information about the thought processes than the other indicators and can be used, for example, to identify changes in subjects' solution methods on different problems within an experimental session. Although an analysis of associated latencies can confirm a reliable difference (a change in the subjects' methods), the latencies are rarely sufficiently informative to allow an independent verification of the specific solution strategies involved in the change.

Finally, the validity of verbal reports on thinking depends on the time interval between the occurrence of a thought and its verbal report, where the highest validity is observed for concurrent verbalization (think-aloud). For tasks with relatively short response latencies (less than 5–10 seconds), validity of retrospective reports remains very high. However, for cognitive processes of longer duration, the problems of accurate recall of prior thoughts increase with a corresponding decrease in validity of the associated verbal reports.

In sum, think-aloud and immediate retrospective reports provide the most informative data available on thinking during cognitive tasks. Those aspects of verbal reports that can be validated against other sources suggest a close correspondence between the reports and the cognitive processes. On the other hand, the reports are not infallible; subjects may occasionally make errors of recall or speech errors in their verbalizations of thoughts. Similarly, experimenters may incorrectly encode the transcripts of the subjects' reports. Such problems, however, are inherent in any analysis of observed behavior. Most important, it would be very difficult for a devious subject to give intentionally invalid verbal reports that met the reviewed checks on validity, especially in the case of think-aloud reports. Hence, trust is no more an issue for these types of verbal reports than for any other type of observation.

Types of reporting instructions that compromise the validity of verbal reports

There appear to be two principal factors that contribute to the decreased validity of verbally reported information. The first arises when the investigators have tried to obtain more information than the subjects' recall of their thought sequences provides. For example, some investigators ask subjects *why* they responded in a certain manner. Sometimes the subjects' recall of their thoughts may provide a sufficient answer, but typically the subjects need to go beyond any retrievable memory of their processes to give an answer. As subjects can only access the end products of their cognitive processes during perception and memory retrieval, they cannot report why only one of several logically possible thoughts occurred to them and must thus resort to confabulation to answer such questions. In support of this argument, Nisbett and Wilson (1977) found that subjects' responses to why questions in many circumstances were as inaccurate as those given by other subjects who merely observed and explained another subject's behavior. There are also many types of factors that influence retrieval and speed of access to information from memory, such as priming, where the corresponding processes do not leave intermediate and thus reportable products in consciousness (Ericsson and Kintsch, 1995; Wilson, 1994).

Second, investigators have asked subjects to describe their methods after solving a long series of different problems. If subjects generated and consistently applied a general strategy for solving the problems, they should be able to answer such requests easily with a single memory retrieval. However, subjects typically employ many methods and shortcuts and even change their strategies during the experiment through learning. Under such circumstances, subjects have great difficulties describing a *single* strategy used consistently throughout the experiment, even in the unlikely event that they were motivated and able to recall most of the relevant thought sequences. It is therefore not surprising that subjects' descriptions are imperfectly related to their average performance during the entire experiment. There are some related issues encountered during knowledge-extraction interviews with experts (Hoffman, 1992), where the experts have difficulties in accurately describing the general methods they employed in their professional activities. However, their think-aloud protocols given during their performance of specific representative problems from their domain of expertise reveal a much more valid and informative account of their complex thinking and planning (Ericsson and Lehmann, 1996).

Protocol analysis of concurrent and retrospective verbal reports has emerged as one of the core methods for studying theoretical issues in thinking (Crutcher, 1994). As a further recognition of its validity, protocol analysis now plays a central role in applied settings, such as in the design of surveys and interviews (Sudman et al., 1996) and user testing of computer tools and applications (see Article 56, EVERYDAY LIFE ENVIRONMENTS). Finally, several interesting adaptations of verbal report methodology are emerging in the study of text comprehension (Pressley and Afflerbach, 1995) and education. Of particular note are recent efforts to promote a deeper integration of instructional material by requiring subjects to report self-explanations during text comprehension (Chi et al., 1994), thus capitalizing on the beneficial changes in cognitive processes induced by directed verbal reporting.

References and recommended reading

Chi, M. T. H., de Leeuw, N., Chiu, M. H. and LaVancher, C. 1994: Eliciting self-explanations improves understanding. *Cognitive Science*, 18, 439–77.

*Crutcher, R. J. 1994: Telling what we know: the use of verbal report methodologies in psychological research. *Psychological Science*, 5, 241–4.

Diaz, R. M. and Berk L. E. 1992: *Private Speech: From Social Interaction to Self-regulation*. Hillsdale, NJ: Erlbaum.

Ericsson, K. A. and Crutcher, R. J. 1991: Introspection and verbal reports on cognitive processes – two approaches to the study of thought processes: a response to Howe. *New Ideas in Psychology*, 9, 57–71.

Ericsson, K. A. and Kintsch, W. 1995: Long-term working memory. *Psychological Review*, 102, 211–45.

Ericsson, K. A. and Lehmann, A. C. 1996: Expert and exceptional performance: evidence on maximal adaptations on task constraints. *Annual Review of Psychology*, 47, 273–305.

*Ericsson, K. A. and Simon, H. A. 1993: *Protocol Analysis: Verbal Reports as Data*, rev. edn. Cambridge, Mass.; MIT Press.

Hoffman, R. R. (ed.) 1992: *The Psychology of Expertise: Cognitive Research and Empirical AI*. New York: Springer-Verlag.

Newell, A. and Simon, H. A. 1972: *Human Problem Solving*. Englewood Cliffs, NJ: Prentice-Hall.

Nisbett, R. E. and Wilson, T. D. 1977: Telling more than we can know: verbal reports on mental processes. *Psychological Review*, 84, 231–59.

Pressley, M. and Afflerbach, P. 1995: *Verbal Protocols of Reading: The Nature of Constructively Responsive Reading*. Hillsdale, NJ: Erlbaum.

Sudman, S., Bradburn, N. M. and Schwarz, N. (eds) 1996: *Thinking about Answers: The Application of Cognitive Processes to Survey Methodology*. San Francisco, Calif.: Jossey-Bass.

Wilson, T. D. 1994: The proper protocol: validity and completeness of verbal reports. *Psychological Science*, 5, 249–52.

34

Single neuron electrophysiology

B. E. STEIN, M. T. WALLACE, AND T. R. STANFORD

Sensory input as a cognitive initiator

All of our information about the world is derived from the function of our senses, and thus they are the principal source of all our knowledge. This was recognized explicitly by early Greek philosophers, remained an important point of discussion for nineteenth-century philosophers, and continues to be a key issue for present-day philosophers, psychologists, and neuroscientists. It is a key issue in cognitive science because, by initiating the processes that store and evaluate information, sensory information transmission can be considered a cognitive initiator. In a very real sense, who we are and how we see the world are the result of the experiences that are mediated by our sensory systems. These are some of the reasons why the study of sensory systems has occupied such a prominent place in modern cognitive neuroscience (see Gazzaniga, 1995).

A fundamental task of cognitive systems is to interpret incoming sensory information so that behaviors appropriate to the situation can be produced. While this is certainly not the only function of cognitive systems, it is one that is essential for survival. Because the consequences of inefficient information transmission are no less severe than those of improper use of this information, the task of accurately representing external *reality* has been a powerful driving force in evolution. The result is that extant organisms possess an impressive array of highly specialized senses for which unique peripheral organs have evolved. Each of these sensory organs is a marvel of bioengineering and contains receptors that transduce a specific kind of energy into a code which the brain can use to detect and interpret external events. More impressive still is that while each of our sensory systems has its own specific neural machinery, which provides us with unique sensory impressions that have no counterparts in other senses (e.g., the perception of pitch is specific to the auditory system, tickle to the somatosensory system, hue to the visual system, etc.), the brain regularly integrates these inputs to provide a coherent perceptual experience.

The fact that all multicellular organisms possess multiple sensory channels with which to monitor their external environments is not serendipitous. Rather, the coexistence of several sensory channels, which can function in concert or substitute for one another when necessary, significantly enhances an organism's potential for survival. The combination of inputs from different sensory systems can often provide information that is not available from any one of them, and this will be discussed further below (also see Stein and Meredith, 1993).

Electrophysiological recording techniques

One of the best means for assessing how the brain deals with information about external events is to record directly from neurons responsible for transmitting and evaluating this information at each way station in the brain. One can examine how signals from the external transducer (i.e., the receptor organ that transforms environmental energy into the neural code the brain reads) are segregated and distributed to different areas of the brain. This segregation is essential, because different regions of the brain are responsible for different aspects of information processing. Armed with a detailed picture of how different structures segregate and process this information and an understanding of how these structures are interconnected, one can then make some reasonable suppositions about how different brain structures accomplish their tasks and how the information they deal with is finally reassembled to produce a comprehensive percept that identifies the nature of the external event. This approach is equally adaptable to examining how the brain initiates and controls overt responses. These are Herculean undertakings, not only because of the number of areas of the brain that are closely interconnected and share some responsibility for the same general tasks, but because many neurons within each brain area are involved in multiple circuits that may play different roles in sensory and/or sensorimotor behaviors in different circumstances. Given such formidable problems, it is impressive to note how much information electrophysiological techniques have already revealed about the functional organization of the brain.

There are many electrophysiological recording techniques that are used to study information processing in the brain, but, by definition, all depend on the ability to record the electrical activity of neurons. Electroencephalographic and event-related (or *evoked*) potential approaches deal with brain responses in terms of the synchronous behavior of large populations of neurons. Because many electrical potentials can be recorded through the surface of the skull and the skin, these techniques need not be invasive, so are very well adapted to studies of human subjects. However, in order to obtain some insight about what is happening at the level of the basic unit of information processing in the nervous system, the individual neuron, an electrode must enter the brain and come within a neuron's electric field (extracellular techniques) or actually penetrate (intracellular techniques) its membrane. These invasive single neuron recording techniques are usually conducted on nonhuman subjects, though single neuron recordings and electrical stimulation techniques are sometimes used during human brain and spinal cord surgery to help guide the surgeon in determining the focus of interest and/or dysfunction.

In the space available here we will attempt to illustrate how electrophysiological techniques are used to understand how the brain codes information about the external world. We will deal primarily with examples of sensory neurons obtained from extracellular single neuron recordings. Before doing so, however, it is important to note that modern neuroscientists are not limited in their investigations by a single approach; they are problem-oriented rather than technique-oriented. Thus, despite the power of the single neuron recording technique, it is only one method for evaluating brain function. The information obtained from this technique is generally combined with that from other physiological, behavioral, and anatomical observations, in order to obtain a more comprehensive view of any functional issue (see Article 32, NEUROIMAGING).

The brain is an active contributor to its own perceptual processes

It is also important to note that seemingly logical assumptions about how the brain must deal with information in order to identify external events can sometimes be quite misleading. For example, there is little doubt that the external information which the brain can receive is limited by the sensitivity of its receptors. But to make the assumption that the information derived from the receptors is the only information that goes into identifying the initiating event would be incorrect. The brain does not passively reflect incoming signals but actively interprets and alters this information. Thus, what one *knows* may, in some circumstances, bear less similarity to the external event than one might imagine. This is particularly evident in the fascinating array of illusions, which demonstrate quite dramatically that incoming sensory information may be only the starting point for a perceptual experience. For example, consider the following: a subject is seated in a dark room facing two small spots of light that are illuminated in sequence. Generally, if there is a long interval between the first light going off and the second light coming on, the subject sees two independent lights. If the interval is short, the lights no longer appear as two independent events but rather as a single light that moves from one position to another. Clearly, the brain *sees* the event very differently if the timing of the inputs is altered. Similarly, we have no difficulty identifying the direction of movement (e.g., clockwise) of a wheel moved at low velocities. However, at substantially higher velocities the wheel appears to move counterclockwise. Many factors, such as luminance, color, timing, etc., affect this illusion of apparent motion. Nor is it restricted to the visual system. Although these are classified as *illusions* because they do not accurately reflect the physical properties of the stimulus in the external world, they do accurately reflect what is happening in the brain. Indeed, electrophysiological correlates of some illusions have already been observed.

Coding stimulus features at the level of the single neuron

In the hands of the scientist, single neuron recording techniques are most often directed toward an in-depth evaluation of the capabilities of the neuron under study. Whether the objective is to understand the codes which the brain uses to represent the features of a stimulus or a complex percept, the procedures are much the same. They involve recording a neuron's responses, in the form of action potentials, to a wide variety of different stimuli (or during different decisions, in different behavioral states, etc.). For example, in the somatosensory system, stimuli such as cooling, warming, indenting, stretching, or moving across the skin at different velocities are often presented to determine the *trigger* features of a given neuron. This gives the researcher some idea of the selectivity of that neuron. By also determining whether the neuron can systematically vary its responses to parametric manipulation of some feature of the stimulus, such as intensity, one can begin to construct its physiological profile. The same test

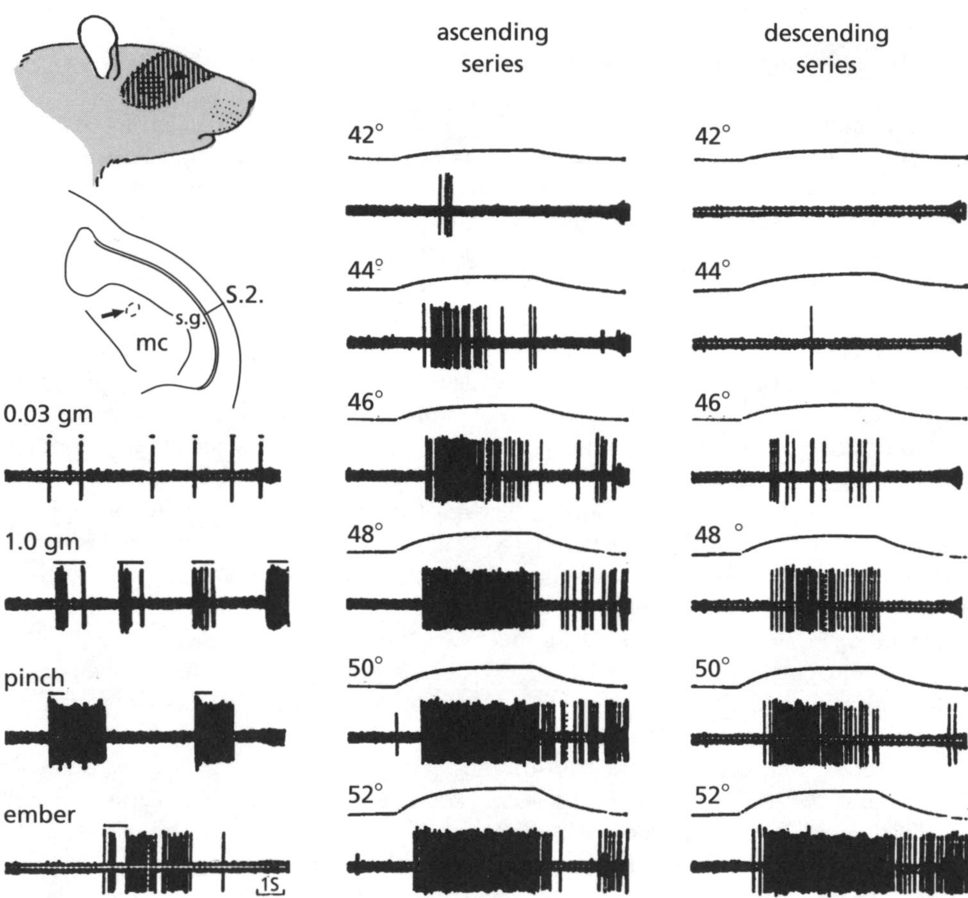

Figure 34.1 Responses of a wide dynamic range neuron in the spinal cord of a rat to ascending and descending shifts in skin temperature. This neuron was located (arrow) in the subnucleus magnocellularis (mc) of the spinal trigeminal nucleus. It had a receptive field (*vertical hatching*) with a region of greatest sensitivity near its center (*cross hatching*). Responses to mechanical stimuli (0.03 gm and 1.0 gm von Frey filaments and noxious pinch) and noxious heat (ember) are shown in the oscillograph traces of neuronal discharges (*far left*) and demonstrate the graded increase in the number and frequency of discharges evoked with increasing stimulus intensities. In the initial ascending series of stimulus temperatures (*left center*), a threshold response was evoked at 42°C. As stimulus intensity was increased, the number of impulses evoked also increased, with an after-discharge that became evident at 46°C and was more pronounced at higher temperatures. A subsequent descending series (*right center, bottom to top*), initiated 5 minutes following completion of the ascending series, evoked responses that consisted of fewer impulses at equivalent skin temperatures. Adapted from McHaffie et al., 1994, p. 416.

can be presented to a neuron in the spinal cord as to one in the cerebral cortex; thus the profiles of neurons at different levels of a system can be compared.

An example of a neuron in the spinal cord of a rat that codes the intensity of mechanical and thermal stimuli on the skin is shown in figure 34.1. The neuron responds only to stimuli within its area of sensitivity on the skin, or its *receptive field*. It is classified as a *wide dynamic range* neuron because it is capable of signaling the intensity of a stimulus in both the innocuous and noxious ranges. Hence, it responds with increasing numbers of action potentials (i.e., impulses) to different stimulus intensities over a wide response range.

Wide dynamic range neurons represent only one of several classes of neurons responsive to painful stimuli, and these *nociceptive* neurons have been studied in many species. They are much the same in primates and rodents and are found broadly distributed in those areas of the brain concerned with pain (or, more properly, *nociception*), and though all wide dynamic range neurons process information in similar ways, their impact on perception and behavior varies, depending on the particular circuit they are involved in. The same information is as critical to signaling a noxious sensation as it is to triggering defensive behaviors, but initiating a sensory-discriminative process versus a motivational-affective experience versus a nocifensive behavior depends on activating very different patterns of brain connections. Because the information from wide dynamic range neurons (and other nociceptive neurons) in the spinal cord is distributed broadly to higher brain centers involved in nociception, electrically activating them should evoke the appropriate sensations. This was tested by Mayer and colleagues (1975) in patients undergoing surgery of the spinal cord. The electrical stimulation parameters used in this study were thought to selectively activate a large population of wide dynamic range neurons, and a variety of different painful sensations were elicited. The patients reported that the qualities of these sensations were similar or identical to those evoked by naturally occurring stimuli.

Encoding the spatial features of touch

In a series of elegant experiments, Kenneth Johnson and colleagues (Phillips et al., 1988) examined the encoding of stimulus features by single neurons at various stages of the somatosensory system. These experiments, conducted in awake behaving monkeys, used raised letters on a rotating drum to study neural activity. In one set of experiments the researchers recorded from nerve fibers arising from receptors embedded in the finger pads. In this paradigm, as the drum was rotated, a portion of the embossed letter passed through the region of the skin from which responses could be elicited in the nerve fiber (i.e., its receptive field). After each rotation, the drum was moved slightly, and the trial was repeated. The activity of the individual nerve fiber was recorded for each pass of the drum, and after the letter had passed through the entire receptive field, the total activity was transformed into a plot that mapped the neuron's activity onto the spatial features of the stimulus (figure 34.2a). Such a plot, known as a *spatial event plot*, although derived from a single neuron, can be assumed to represent a population of neighboring neurons with similar response properties. In these experiments, spatial event plots from nerve fibers arising from one class of receptors (i.e., slowly adapting) faithfully reproduced the spatial characteristics of the letters (i.e., had high spatial resolution), whereas those from a different class of receptors (i.e., rapidly

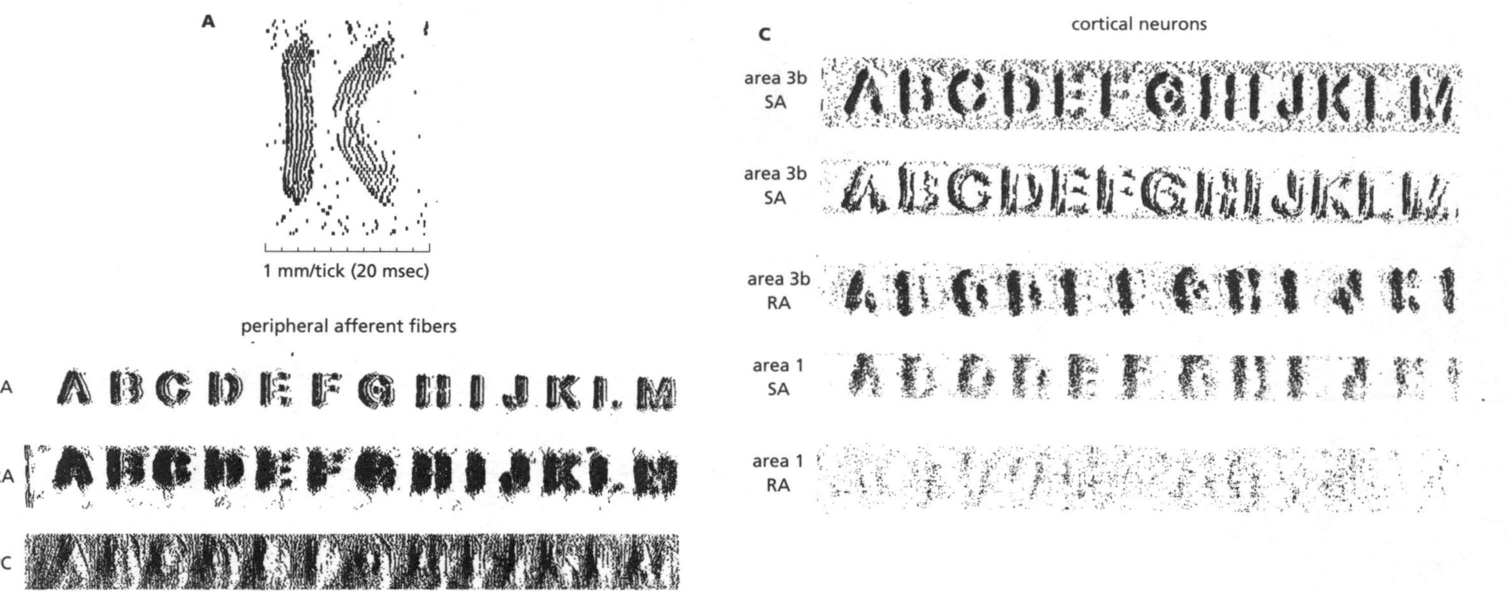

Figure 34.2 The spatial characteristics of the tactual world are encoded in the discharges of individual neurons at various stages of the somatosensory system. A: Spatial event plots are produced by sweeping embossed letters (in this case the letter *K*) across the receptive field of a single neuron innervating the finger pad, while recording the discharges of individual neurons at several stages of the somatosensory pathway. Each tick in the plot represents a single action potential. For each trial (horizontal line of action potentials), the drum is rotated, and the stimulus is swept across the receptive field. After each sweep, the drum is lowered 200 μm and swept through the field again. The time of occurrence of each action potential relative to adjacent stimulus position-markers is recorded and ordered from top to bottom in order to assign a spatial location relative to the stimulus. B: Spatial event plots constructed for afferents innervating three different receptor types: slowly adapting (SA), rapidly adapting (RA), and Pacinian corpuscles (PC). Note the high fidelity of spatial information encoded by slowly adapting fibers, and the poor fidelity for Pacinian fibers. C: Spatial event plots for individual neurons in the primary somatosensory cortex. Note that whereas neurons innervated by slowly adapting fibers in the primary thalamo-recipient zone of somatosensory cortex, Brodmann's area 3b, continue to code spatial characteristics with good fidelity, those at other stages (area 1) reproduce spatial characteristics with much less fidelity. Adapted from Phillips et al., 1988, pp. 1318, 1319.

adapting Pacinian corpuscles) produced a very low-resolution spatial representation (figure 34.2b).

Though the nerve fibers innervating the skin might be expected to contain such information, how faithful is this spatial representation at various stages of the somatosensory pathway? To examine this, Johnson and his colleagues used the same paradigm in another series of experiments, recording from two areas of the cerebral cortex in which neurons responsive to somatosensory stimuli are found. They found that the information was faithfully represented in some classes of neurons, whereas other areas dealt with other features of the stimulus and did not maintain high-resolution spatial images of these letters. This sort of segregation of function is a hallmark of the central nervous system, and as information *ascends* to higher-order regions of the brain, more abstract stimulus features and even stimulus *meaning* seem to be encoded. These issues will be dealt with somewhat later.

Because the signals of a single neuron may have little impact among the host of neural signals occurring in the brain at any moment, it is important to understand how its actions are pooled with those of other neurons so that, together, their activity identifies a stimulus attribute, initiates a percept, or organizes a motor response. This requires that different neurons code information in similar ways, as noted in the example of coding the spatial properties of tactile stimuli presented above. The concept that neurons fall into certain categories with respect to how they process information has been confirmed by recording serially from many neurons using the same electrode or by recording many neurons through many electrodes simultaneously. Although neurons tend to fall into rather discrete categories, as we become more sophisticated in the questions we ask, we find larger numbers of categories and note that the same neuron may change its behavior in different circumstances. This variability of a neuron's responses is evident in the example shown in figure 34.1, where a second series of stimuli (i.e., the descending thermal series) evokes the same functional profile as does the ascending series but is composed of far fewer impulses per stimulus. Examples of neurons whose fundamental response profiles are altered by circumstances are dealt with below in the section entitled "Mechanisms of attention."

How the brain obtains more information than is contained in the individual sensory inputs

The brain has many ways of using incomplete information to categorize a stimulus or event. One technique is to fit the information into previously established categories based on experience; a sort of *best fit* decision is then made. Another technique is to combine the minimal information obtained from multiple sensory inputs. Because the information carried along each sensory channel (e.g., visual, auditory, somatosensory) will reflect different features of the stimulus, the synthesis of their inputs can provide far more information than might be available from them individually. A synthesis of this sort requires that inputs from different sensory modalities converge somewhere in the brain. In fact, there are many areas of the brain that contain *multisensory* neurons (see Article 43, THE BINDING PROBLEM).

Through synthesis of multisensory information, events are more readily perceived and have less ambiguity, and responses to them are produced far more rapidly. The process is evident electrophysiologically as an enhancement in the discharges evoked from multisensory neurons, an enhancement that not only exceeds the response levels produced by either of the unimodal stimuli alone but often far exceeds their sum (see below). Although ultimately we are concerned with how this process affects the human CNS and our own behaviors and perceptions, it is interesting to note that multisensory convergence and integration per se are evolutionary ancient schemes that antedate mammals and even the evolution of the nervous system. They are present in unicellular organisms and have been retained throughout multicellular speciation and the evolution of humans. In fact, we know of no animal in which there exists a complete segregation of sensory processing. It seems logical that during the process of developing differentiated sensory systems, mechanisms were preserved and/or elaborated for using their combined action to provide information that would be unavailable from their individual operation.

One of the best-studied areas of the brain containing multisensory neurons is a midbrain structure, the superior colliculus. This structure receives visual, auditory, and somatosensory inputs and plays an important role in attentive and orientation behaviors (see Stein and Meredith, 1993). Each of these sensory modalities is represented in the superior colliculus in a maplike, or topographic, manner. Thus, we say that the retina (and thus, the visual world) is *mapped* onto the superior colliculus. Whereas neurons in the front part of the superior colliculus have receptive fields representing the front (i.e., nasal) part of visual space, those in the back part of the structure represent more peripheral (i.e., temporal) regions of visual space. This organizational feature is shared by each of the three modalities, and thus the *maps* are said to be in correspondence with one another. Not only do the maps correspond, but the individual receptive fields of a single multisensory neuron are in good correspondence. Thus, a multisensory visual-somatosensory neuron whose visual receptive field is in front of the animal will have a somatosensory receptive field near the center of the face. If the visual receptive field is in upper space, the somatosensory receptive field will be in upper space; if it is in lower space, the somatosensory receptive field will be in lower space. Similarly, a visual receptive field in the far periphery (i.e., temporal in visual space) will be coupled with a somatosensory receptive field far back on the body. This relationship can be seen in figure 34.3, and the same sort of correspondence is present among the receptive fields of visual-auditory (see figure 34.4), auditory-somatosensory, and trimodal neurons.

The ability of a superior colliculus neuron to integrate two different sensory inputs is illustrated for a visual-auditory neuron in figure 34.4. In this example the visual stimulus and the auditory stimulus are not very effective when presented individually within their respective receptive fields. However, when presented at the same time, their combined response is enhanced far beyond the sum of the two unimodal responses. This sort of neural amplification at the level of the single neuron in the superior colliculus is reflected as an amplification of superior colliculus-mediated behaviors. Multisensory stimuli evoke substantially more rapid and more reliable orientation movements, as well as an enhancement in the detectability and apparent intensity of weak stimuli. Multisensory neurons with similar properties are found in many areas of the brain, and, just as was noted for nociceptive neurons earlier, their impact on behavior depends on the specific circuits in which they are involved.

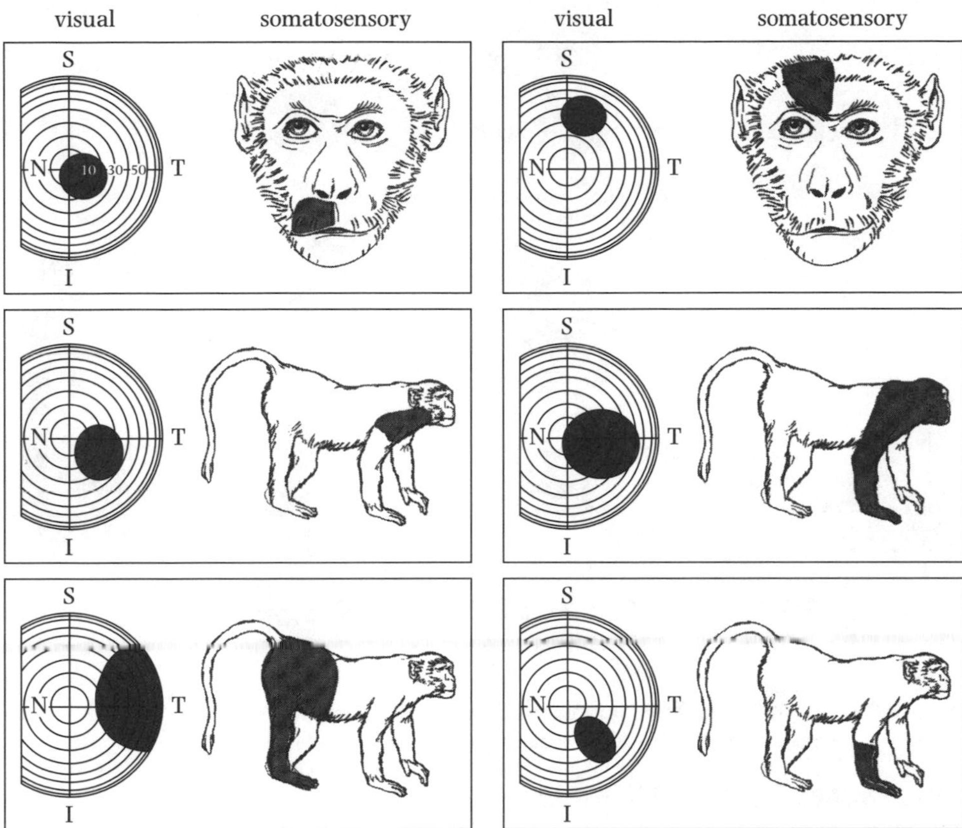

Figure 34.3 The receptive fields of six bimodal visual-somatosensory neurons. The visual receptive fields were plotted on a hemisphere during experimentation and were transferred to this two-dimensional map of visual space as viewed by the monkey. The fovea is at the intersection of the vertical (S = superior, I = inferior) and horizontal (N = nasal, T = temporal) meridians, and the circles represent each 10° of eccentricity in visual space. In the first example (*upper left*) an inferior and nasal visual receptive field (*dark circle*) in the right visual hemi-field is coupled with a somatosensory receptive field (*dark region*) on the right upper lip. As the visual receptive fields move more peripheral (*rightward*) in the first column, the somatosensory receptive fields move further back on the right side of the body. As the visual receptive fields in the right column move from superior to inferior, the somatosensory receptive fields move from upper to lower portions of the body. From Wallace et al., 1996, p. 1259.

Combined approaches: single neuron recording and behavioral psychophysics

The preceding discussion emphasized the way in which the brain encodes the sensory information that shapes perceptions of the physical world (e.g., visual, thermal, tactile, auditory) and provided demonstrations of how the activities of single neurons within the CNS encode specific attributes of physical stimuli (e.g., temperature). To understand how the physical stimulus, the neural activity it evokes, and the behaviors that

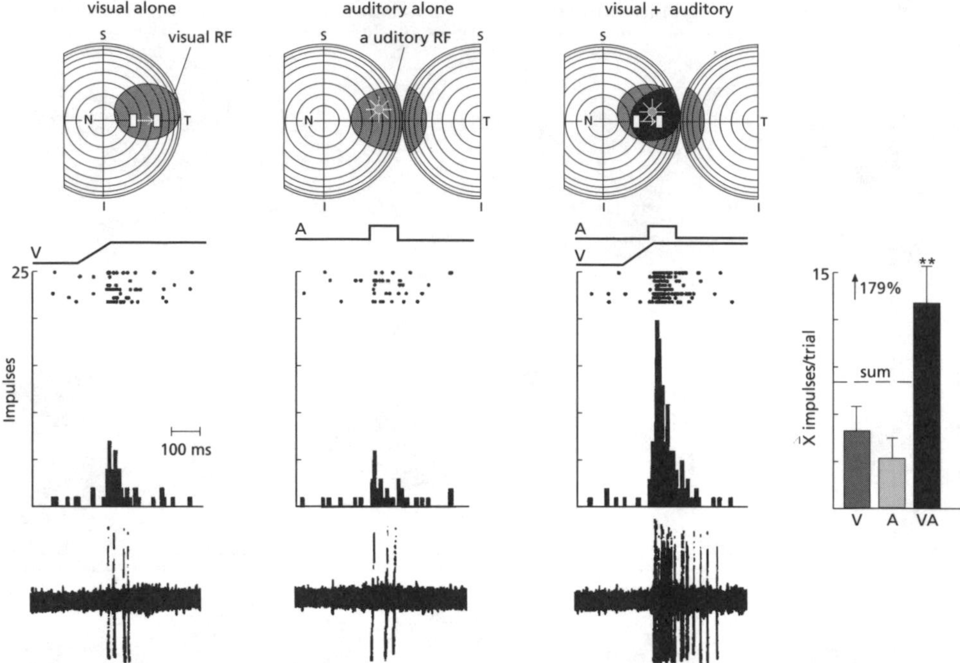

Figure 34.4 Multisensory enhancement. *Upper row*: the receptive fields (RF) of this visual-auditory neuron are shown, as are the locations of the test stimuli (bar of light, icon of speaker). The visual stimulus is a bar of light moved from left to right. The diagram of auditory space (*middle*) contains a half circle that represents space behind the right ear. It is folded forward so that the entire auditory receptive field can be viewed here. The auditory stimulus was a broadband noise burst delivered through a stationary speaker. The right-hand panel shows the area of overlap of the visual and auditory receptive fields (*black*). *Central row*: illustrated here are neuronal responses in the form of rasters, peristimulus time histograms, and bar graphs for each of the three stimulus conditions (visual alone, auditory alone, visual + auditory). The ramp labeled "V" represents the electronic trace of the visual stimulus, and the line labeled "A" represents the trace of the auditory stimulus. Each dot of the raster represents a single neuronal impulse, with each horizontal series of dots representing the response to a single stimulus repetition (a total of eight trials for each stimulus condition). Peristimulus time histograms depict the summed activity for each of the stimulus conditions, and bar graphs on the right illustrate the average response to each condition. A statistically significant ($p < 0.01$) response enhancement was evoked by the spatially coincident stimulus combination that far exceeded the response predicted by a simple summation of impulses (sum). *Bottom row*: shown here are representative oscillographs (neuronal discharges) from each of the stimulus conditions. From Wallace et al., 1996, p. 1261.

result are interrelated, a melding of techniques is required, so that electrophysiological experiments can be conducted in alert, behaving animals. The combination of single-cell electrophysiology and behavioral psychophysics is a comparatively recent innovation in the history of electrophysiological methodology, but one that has proved to be a powerful tool for cognitive neuroscientists.

Experiments that combine observation of behavior and electrophysiology are often conducted in the rhesus monkey, a primate species with a highly developed cognitive capacity that readily learns complex behavioral tasks. The studies described below, which investigate the neural mechanisms underlying visual perception and attention, illustrate the approach of combining perceptual psychophysics and single neuron electrophysiology to examine cognitive function.

Neural activity and the perception of visual motion

Movement within the visual scene is an extremely powerful cue. Thus, relative motion and differences in the speed and/or direction of motion are important cues for distinguishing objects of interest from one another, as well as for distinguishing features in an otherwise motionless background. Because motion perception is so important (consider trying to avoid oncoming traffic without it), it is no surprise that many brain regions are specialized for it. In these areas, visually responsive neurons are more sensitive to moving than to static stimuli and are differentially sensitive, or *tuned,* to both the speed and the direction of a target's motion.

There is perhaps no more dramatic description of the consequences of disrupting motion perception than that described in a case study provided by Zihl and colleagues (1983). The patient, referred to as L.M., suffered an unfortunate cerebrovascular accident that produced a lesion in an area of the cerebral cortex known as visual area 5 (V5). She became *motion blind* and, for example, could not effectively pour coffee into a cup, because she could not perceive the movement of the coffee out of the pot or the cup becoming progressively fuller. Instead, the coffee would appear frozen, like a glacier. Similarly, L.M. had the odd sensation that people or objects would appear first in one place, then, abruptly, in another.

In monkeys, area V5, also known as the mesio-temporal area (MT), is composed of neurons that are sensitive to the direction of visual motion. To determine how the activity of direction-selective neurons in this area might underlie the perception of visual motion, Newsome and colleagues (1989) recorded from MT neurons in rhesus monkeys trained to perform a motion-discrimination task. The key issue was to determine whether the perceptual and neuronal responses in this task were related.

In these experiments, monkeys were trained to discriminate between directions of motion that differed by 180 degrees (i.e., up/down, left/right, etc.). A monkey watched a display of moving dots and indicated its judgment of motion direction by simply looking at one of two light-emitting diodes that it had learned to associate with the two possible directions of motion. Eye movements were monitored electronically, so that the experimenters would know where the animal was looking. The animal was rewarded for each correct response. In order to modulate the monkey's behavioral performance, the experimenters had to vary the detectability of the direction of motion. This was accomplished by systematically varying the proportion of dots that moved in the same direction, with the remainder of the dots moving randomly. In this way, the

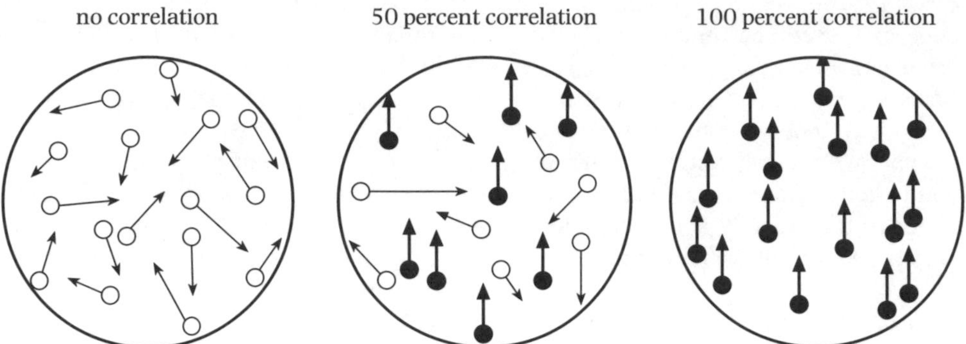

Figure 34.5 Schematic examples of visual motion stimuli varying from 0 to 100 percent coherence. Actual stimuli were dots presented on an oscilloscope. Each dot persisted for 45 msec before being replaced by another at a new location. For uncorrelated stimuli (*left*), each new dot was placed randomly and could have any direction of motion. For fully correlated stimuli (*right*), each new dot moved uniformly in a fixed direction. Partially correlated stimuli (e.g., 50 percent – *middle*) were composed by varying the percentage of random and uniformly moving dots in the display. From Newsome and Paré, 1988, p. 2202.

stimulus could be varied from 100 percent coherence, in which all the dots moved in the same direction, to 0 percent coherence, in which all the dots moved randomly (figure 34.5). As expected, monkeys performed at near 100 percent for highly coherent stimuli, at chance (near 50 percent) for completely random stimulus motion, and somewhere in between for intermediate levels of correlated stimulus motion.

How can the behavioral performance be related to the activity of direction-selective MT neurons? Figure 34.6 compares the responsiveness of an MT neuron to stimulus motion in opposing directions for each of three degrees of stimulus correlation. In figure 34.6a, for example, the neuron almost always responded more vigorously to one of the two directions of stimulus motion (i.e., its *preferred* direction – hatched bars). So, on the basis of the number of action potentials, this neuron reliably *discriminated* between motion in the two directions. Note, however, that the tails of the two distributions overlap, so that there is a small probability that using the number of neural impulses to make a judgment of stimulus direction will produce errors. Indeed, as the degree of motion correlation declines, the distributions overlap to a greater extent, and the neuron's activity becomes much less reliable in correctly identifying the direction of stimulus motion. Is there an analogous decline in behavioral performance? In part b of the same figure, the behavioral performance is directly compared with neural performance: that is, the probability that the animal makes a correct response is compared with the probability that the neuronal response will differentiate between motion in opposite directions. As argued by Newsome et al., the similarity of the functions is consistent with the hypothesis that the behavioral judgment is based on the direction tuning of motion-sensitive neurons in MT.

Mechanisms of attention and single cell recording

Our cognitive resources have a limited capacity. Thus, the ability to select a small fraction of the information that is available to our senses is critical if we are to effectively

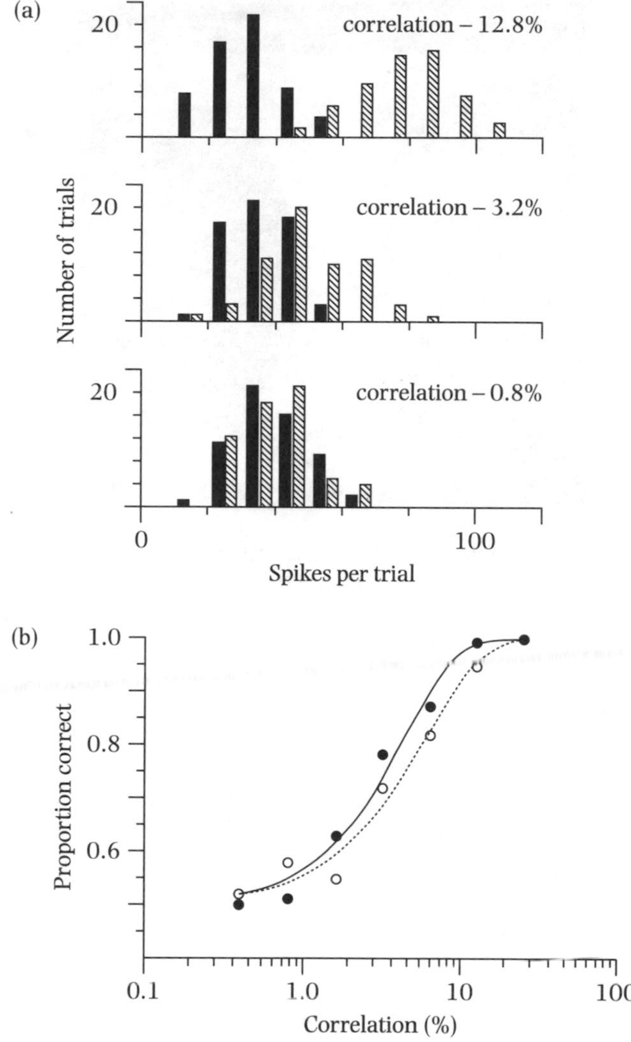

Figure 34.6 Comparison of neural and behavioral performance for an MT neuron. a: Histograms comparing the responsiveness of an MT neuron to stimulus motion along its preferred direction (hatched bars) to that for motion 180 degrees opposite to the preferred direction (null direction – filled bars). From top to bottom, the three plots show that decreasing motion coherence diminishes differences between the responses to stimulus motion along the preferred and null directions. b: Comparison of the neurometric (filled circles) and psychometric (open circles) functions for the same set of stimulus trials. See text for details. From Newsomc et al., 1989, p. 53.

process information that is relevant within a specific context. A familiar example is the *cocktail party effect*, in which we selectively attend to a speaker in a noisy room, and other conversations seem to recede into the background despite the fact that the physical environment has not changed. Figure–ground alternation is another example of selective attention. This phenomenon was recognized by the Danish psychologist Edgar Ruben, who described how a black-and-white image may be seen either as a pair of

Figure 34.7 Opposing silhouettes or a white vase? This example of figure–ground alternation is attributed to Danish psychologist Edgar Ruben (1915), who noted that one perceives either a white vase or a pair of black silhouettes when viewing this graphic. Which of these is perceived as the object and which becomes background seems to depend on the object of one's attention.

opposing faces in black silhouette or as a single white vase on a black background (figure 34.7). Attending to the white area allows one to *see* the vase, while the remainder is perceived as background. Alternatively, a redirection of visual attention allows one to *see* the faces to the exclusion of the vase. Many are familiar with the potency of this illusory effect through the works of the well-known artist Maurits Escher (figure 34.8) (see Article 3, ATTENTION).

Studies of single neurons have sought to examine the neural mechanisms of attention by examining the influence of behavioral context on neural activity. In perhaps the simplest demonstrations of the attentional modulation of a sensory response, neurons in some of the brain's visual areas have been shown to respond more vigorously when a visual stimulus has some behavioral significance. A good example is the enhancement of a visual neuron's response when an animal is required to make an eye movement (i.e., a saccade) toward a visual stimulus in order to receive a reward. The possibility of the reward gives the response significance and is directly linked to the neuron's enhanced responses. The response enhancement precedes the actual eye movement, because attention shifts to the location of the stimulus (see Posner, 1995). As pointed out by Posner (1980), visual response enhancement can also be found independently of orienting movements if the tasks assigned require the monkey to identify significant features of a peripheral stimulus. Under these circumstances, the neural activity is presumed to reflect a covert redirection of attention.

One brain region in which visual responsiveness has been examined in different behavioral contexts is an area of the cerebral cortex known as visual area 4 (V4). One of the visual stimulus features to which V4 neurons are sensitive is orientation. Thus, for a linear or rectangular stimulus, a given cell responds best over a restricted range of stimulus orientations. Given this orientation *tuning*, these cells may be important

Figure 34.8 An illusory work by M. Escher takes advantage of the figure–ground alternation principle. This painting is alternately perceived as a pattern of birds or a pattern of fish.

for encoding the orientation of stimuli within the visual scene. Robert Desimone and colleagues have recorded from orientation-selective V4 cells, to find out if responsiveness can be modulated by the amount of attention required to perform an orientation-discrimination task (Spitzer et al., 1988). In their experiment, monkeys were trained to discriminate differences between the orientation of rectangular visual stimuli. Each monkey was first presented with the sample stimulus, and later with the test stimulus. The task was simply to judge whether the test stimulus was the same or different from the previously viewed sample and to indicate this decision by releasing (if they are the same) or continuing to hold (if they are different) a lever. Each correct response was rewarded. To perform the task correctly, the monkey had to pay attention to the orientation of the sample, so that it could be compared with the subsequent test stimulus. To modulate the attentional demands of the task, the investigators devised both an easy and a difficult variation on the task.

Did the increased attentional demand of the difficult task produce correlated changes in the activities of orientation-selective V4 cells? Yes. In figure 34.9, for example, it is apparent that the response of a V4 neuron to the sample stimulus is both more vigorous and more selective for orientation (i.e., more sharply affected by changes in stimulus orientation) when presented in the context of the difficult task as compared to the easy task. Thus, these results show that the responses to physically identical sensory stimuli can be modulated by attentional effort. Presumably, the effect of attentional demand on the responses of V4 cells contributes to the monkey's ability to perform the finer discrimination.

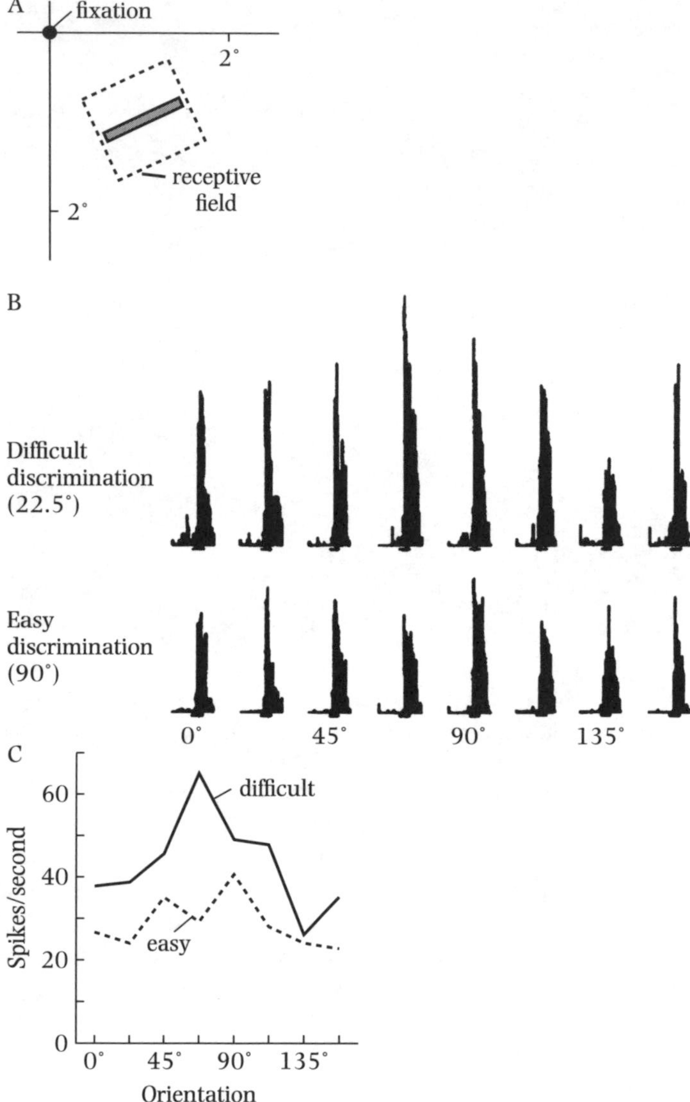

Figure 34.9 Modulation of sensory responses by attentional effort. A: An example of an oriented bar stimulus (*solid rectangle*) placed within the receptive field (*dotted square*) of a V4 cell. B: Histograms of the response of a V4 neuron to the sample stimulus as a function of its orientation for both the difficult (*above*) and easy (*below*) tasks. In the difficult task, the sample had to be discriminated from a test stimulus that, when nonmatching, had an orientation that differed from the sample by 22.5 degrees. In the easy task, nonmatching test stimuli differed from the sample by 90 degrees. C: Plot of the discharge rates associated with the histograms shown in B. Note that responses obtained in the context of the difficult task are shown to be more vigorous and more sensitive to orientation than those recorded in the context of the easy task. From Spitzer et al., 1988, p. 339.

Summary

Throughout this chapter we have provided examples showing how neuronal recording can be used to investigate the neural bases of cognitive function. In doing so, we have been able to touch on only a small number of areas in which these techniques have been useful. The immense contributions of these techniques in furthering our understanding of learning, memory, and sensorimotor integration are but a few of the issues that we have omitted for the sake of brevity. Also omitted are the substantial contributions made by the various other electrophysiological methods. Yet, despite these necessary limitations of scope, it is hoped that the examples provided here give the reader some insight into the role played by electrophysiology in cognitive science.

References and recommended reading

Gazzaniga, M. S. 1995: *The Cognitive Neurosciences*. Cambridge, Mass.: MIT Press.

Mayer, D. J., Price, D. D. and Becker, D. P. 1975: Neurophysiological characterization of the anterolateral spinal cord neurons contributing to pain perception in man. *Pain*, 1, 51–8.

McHaffie, J. G., Larson, M. A. and Stein, B. E. 1994: Response properties of nociceptive and low-threshold neurons in rat trigeminal pars caudalis. *Journal of Comparative Neurology*, 347, 409–25.

Newsome, W. T. and Paré, E. B. 1988: A selective impairment of motion perception following lesions of the middle temporal visual area (MT). *Journal of Neuroscience*, 8, 2201–11.

Newsome, W. T., Britten, K. H. and Movshon, A. J. 1989: Neuronal correlates of a perceptual decision. *Nature*, 341, 52–4.

Phillips, J. R., Johnson, K. O. and Hsiao, S. S. 1988: Spatial pattern representation and transformation in monkey somatosensory cortex. *Proceedings of the National Academy of Sciences, USA*, 85, 1317–21.

Posner, M. I. 1980: Orienting of attention. *Quarterly Journal of Experimental Psychology*, 32, 3–25.

—— 1995: Attention in cognitive neuroscience: an overview. In M. Gazzaniga (ed.), *The Cognitive Neurosciences*, Cambridge, Mass.: MIT Press, 615–24.

Spitzer, H., Desimone, R. and Moran, J. 1988: Increased attention enhances both behavioral and neuronal performance. *Science*, 240, 338–40.

*Stein, B. E. and Meredith, M. A. 1993: *The Merging of the Senses*. Cambridge, Mass.: MIT Press.

Wallace, M. T., Wilkinson, L. K. and Stein, B. E. 1996: Representation and integration of multiple sensory inputs in primate superior colliculus. *Journal of Neurophysiology*, 76, 1246–66.

Zihl, J., von Cramon, D. and Mai, N. 1983: Selective disturbance of movement vision after bilateral brain damage. *Brain*, 106, 313–40.

35

Structural analysis

ROBERT FRANK

A major objective of cognitive science is to understand the nature of the abstract representations and computational processes responsible for our ability to reason, speak, perceive, and interact with the world. In addition, a commitment to a materialist resolution of the mind–body problem requires that we search for the manner in which these representations and processes are neurally instantiated in the brain. Given this dual aim, one might proceed in one of two ways: (1) from the bottom up, commencing with the study of how low-level information and computations are encoded in the neuroanatomy of the brain, in the hope of working upwards toward an understanding of the properties of higher-level cognitive processes; (2) from the top down, using behavioral data from a variety of sources to provide an abstract characterization of cognitive processes and then utilizing these results to guide our search for the neural mechanisms of cognition. At present, our understanding of the neural basis of higher-level cognition is virtually negligible. Thus, it is probably fair to say that a contemporary study of any of the domains discussed in part II of this volume may be carried out only in a top-down fashion.

It is sometimes claimed that top-down research into cognition is incapable of producing anything but *castles in the sky* having little relevance to human thought. However, a brief look at the history of science shows us that the study of a phenomenon in the absence of an understanding of its underlying mechanisms is by no means novel and has, moreover, sometimes had significant success. One prominent example is Gregor Mendel's study of heredity. By observing the external patterns in which properties of pea plants such as color and height were transmitted from one generation to the next, Mendel was able to deduce the existence of genes and the fundamental laws by which they combine. Of course, Mendel had no conception of the biological character of genes, nor any understanding of the reasons why his combinatorial laws held. Yet, his results have been substantially vindicated and formed the impetus for research into the biological basis of genetic material, culminating in Watson and Crick's discovery of DNA.

How might we mimic Mendel's methodology (and success) in the study of the mind? That is, how do we go about building an abstract theory of a cognitive process which can form the basis for subsequent neurological investigations? One route starts from the assumption that there are specialized representational structures which underlie mental processing within each cognitive domain. Just as Mendel's experience with pea plants led him to propose that heredity is best explained by positing an abstract representation, the gene, along with laws governing its behavior, so too can we use data from human behavior to lead us to the discovery of the abstract structures and laws governing cognition.

Within the cognitive sciences, this type of research has its roots within the domain of LINGUISTIC THEORY, specifically within the area of syntax, the study of sentence structure. Any study of humans' capacity for natural language syntax must face the age-old challenge of deriving infinite capacity (i.e., we can understand and produce arbitrarily long sentences, including those we have never before heard) from the finite resources provided by our physical endowment. During the first half of this century, work in the foundations of mathematics and computer science led to the development of a variety of mathematical and logical systems that fortuitously provided a formal means for answering this challenge. This led to innovative work by structuralist linguists such as Zelig Harris, which for the first time provided a set of mathematically precise representations and rules characterizing the well-formed utterances in a particular corpus of language use. Such precision represented a great advance, since it became possible for the first time to see the exact implications of analyses. Yet, structuralists largely adhered to a behaviorist stance on human psychology and hence took their formal representations to be merely descriptive devices and of no psychological import. In the mid-1950s, Noam Chomsky broke with this behaviorist view and suggested that the results of linguistic analysis should indeed be understood as the mental representations that comprise an individual's linguistic competence – that is, the knowledge which underlies the ability to speak or understand a language. In this framework of generative grammar, Chomsky maintained the focus on providing a mathematically precise characterization of grammar, keeping some of the formal apparatus advocated by Harris, while adding a new range of mathematical devices. However, he understood these devices as providing abstract descriptions of the mental computations underlying language (Chomsky, 1986).

In the remainder of this article, I will have much to say on precisely how this use of mathematical structures has played a beneficial role in our understanding of human grammar. We will look at two case studies, each demonstrating that the computations involved in linguistic cognition attend to remarkably abstract representational structures, quite far removed from the stimuli of the external world. We will then briefly consider the mathematical foundations which support these linguistic investigations and discuss the resultant productive interactions between studies of the properties of mathematical and grammatical structures. Finally, I will mention some other areas of cognitive science in which there have been attempts to make similar use of such abstract structures in modeling cognition.

Finding structure in language

Let us begin to investigate what kind of structures have been found to underlie human language. Consider the following English sentence:

(1) The student has finished her homework.

In order to make a question from this statement, we must change the order of the words, moving the auxiliary verb *has* to the front of the sentence:

(2) Has the student finished her homework?

We can ask the following question: what was the nature of the computation which affected this change in ordering? The simplest answer might go something like this:

(3) To make an English question, move the auxiliary verb to the front of the sentence.

This rule makes very few commitments about what structures underlie English sentences. It requires only that our linguistic computations recognize the notion *front of the sentence* and have the ability to identify elements in the category of auxiliary verbs – *have, be* – or modal elements like *should, could,* etc. This simple formulation is insufficient, however. In sentences involving two auxiliary verbs like (4), it does not tell us which auxiliary, *has* or *was,* to move.

(4) The student has finished her homework which was assigned today.

To address this problem, we need only complicate the rule slightly:

(5) Move the first auxiliary verb to the front of the sentence.

This rule makes only one additional ontological commitment: namely, the notion of *first*. This notion has a straightforward translation into sensory terms: that is, temporally earliest in the speech stream. Thus, it introduces no controversial assumptions about linguistic computation. Unfortunately, even this more complicated rule remains inadequate, as demonstrated by cases like the following:

(6) The student who is eating has finished her homework.

If we follow (5) and front the (temporally) first auxiliary verb, the element *is*, the result is a severely ill-formed string (where the asterisk indicates ungrammaticality):

(7) *Is the student who eating has finished her homework?

To produce the well-formed version of this question, we must instead move the temporally second auxiliary, *has*:

(8) Has the student who is eating finished her homework?

It turns out that we must complicate our rule still further to achieve the desired result:

(9) Move the first auxiliary verb which follows the subject to the front of the sentence.

Since the string "the student who is eating" constitutes the subject, we move the next auxiliary verb in the sentence – that is, *has* – so producing (8). In the previous cases, the temporally first auxiliary is also the one which follows the subject. Observe that this rule differs from the previous one in that it makes reference to the abstract notion of *subject*, one which does not have any direct sensory characterization. Thus, we must assume that grammatical representations include a certain amount of structural analysis so as to allow detection of the subject.

Recall that we take these grammatical rules and representations to form part of a speaker's knowledge of his or her language. Consequently, we are obliged to face the problem of how such rules and representations arise in the mind of a child during the process of language acquisition. From this perspective, it is interesting to note that in Crain and Nakayama's (1987) experiment, which elicited sentences of the relevant type, none of the 30 three- to five-year-old children learning English whom they tested ever gave a response like (7), as would be expected if they had fixed on the simpler, but incorrect, rule for question formation in (5). This is quite puzzling in face of the observation that questions of the type in (8), which are necessary to distinguish this possibility from the correct one, are ordinarily (i.e., outside the experimental context) quite rare, so much so that they may never occur in a child's linguistic input. Why, then, do children uniformly ignore the simpler possibility during the process of inducing the grammar of English, and instead proceed to hypothesize the more complex rule? Chomsky suggests that the resolution of this puzzle (often referred to as the "poverty of the stimulus") lies in the recognition of a certain amount of innate grammatical knowledge, what he calls Universal Grammar (UG) (see Article 45, INNATE KNOWLEDGE).

Chomsky argues that UG predisposes children to learn grammars of a certain sort: in the case at hand, ones that utilize *structure-dependent* rules – that is, making reference to notions of abstract grammatical structure, such as subject, and not to other notions, even if definable directly in sensory terms. This structural dependence property of UG also explains why many apparently *simple* grammatical rules are absent from the languages of the world. For example, there is no language which forms its questions using rule (5), or by reversing the order of the words in the sentence, or by switching the second and fourth words. It is hard to find an independent reason why such things should not be possible. Such rules are all simple to state and would be easy to compute, and a language which used them would be no worse off in terms of communicative possibilities. They simply do not seem to be part of any known human language. Reasons such as these make it quite difficult to provide functional explanations for the precise character of grammatical structures (see Article 31, FUNCTIONAL ANALYSIS).

Let us turn to another type of evidence which brings the detail of grammatical representations into greater focus. Consider the following pair of sentences:

(10) (a) John thinks he should eat spinach.
(b) He thinks John should eat spinach.

The first of these has two possible interpretations: it is either a statement of John's beliefs about his own dietary requirements or, alternatively, John's beliefs about the dietary requirements of some other unidentified male. By contrast, the second may only be interpreted as a statement about the beliefs of someone other than John about John's dietary needs. Similar contrasts hold in the following pairs:

(11) (a) John's belief that he should eat spinach is unshakable.
(b) His belief that John should eat spinach is unshakable.

(12) (a) I gave John his spinach.
(b) I gave him John's spinach.

In each pair, the pronoun may be understood as referring to, or co-referential with, John only in the first example, not in the second.

What is the explanation of these contrasts? As before, the simplest grammatical constraint ignores structural considerations and refers to the ordering among elements in the sentence:

(13) A pronoun may not be interpreted as identical in reference with a proper name if the pronoun precedes the proper name in the sentence.

Once again, however, the simple formulation is inadequate. In all of the following sentences, the pronoun precedes the proper name, yet they may both refer to the same individual: namely, John.

(14) (a) His mother thinks John should eat spinach.
(b) After talking with his mother, John concluded I should eat spinach.
(c) People who meet him in the morning find John irritable.

To understand the contrast between these cases and the (b) examples in (10)–(12), we must consider some details of underlying grammatical structure. It is usually assumed that a sentence like (10b) has the following treelike representation:

(15)

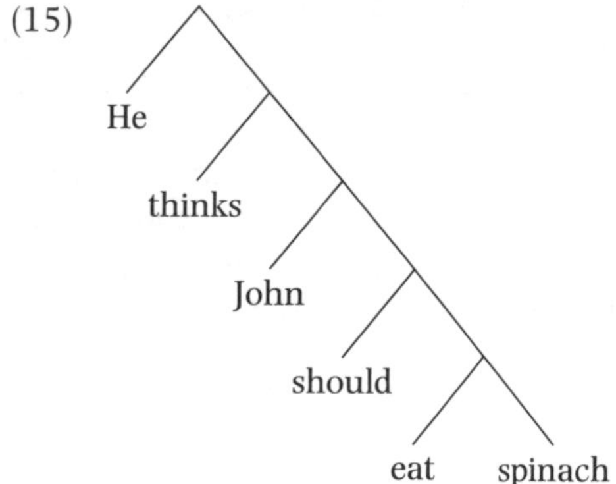

This structure represents the hierarchical groupings of the individual words into phrases, and these phrases into larger structures. For example, in this structure the verb *eat* forms a unit, or *constituent*, with the noun *spinach*, since there is a single node in the tree which dominates them both but nothing else. By contrast, the pair *should eat* does not form a constituent, since the only node which dominates them also includes the noun *spinach* (these three words together do, therefore, form a constituent). Constituency is explicable in both semantic and syntactic terms. We say the phrase *eat spinach* is a constituent since it performs a semantic function, one which is predicated of the embedded subject *John*. From the syntactic side, we can now interpret our previous conclusion that grammatical operations are structure-dependent as meaning that they may manipulate only constituents. From this, we may again conclude that the sequence *eat spinach* is a constituent, since it may be moved to the front of the sentence (16) or deleted entirely (17):

(16) (a) Eat spinach, he thinks John should.
(b) *Should eat, he thinks John spinach.

(17) (a) He thinks John should.
(b) *He thinks John spinach.

What, then, is the structural relationship in (10b) which renders the co-reference between *he* and *John* impossible? To help us see this, consider the structure of sentence (14a):

(18)

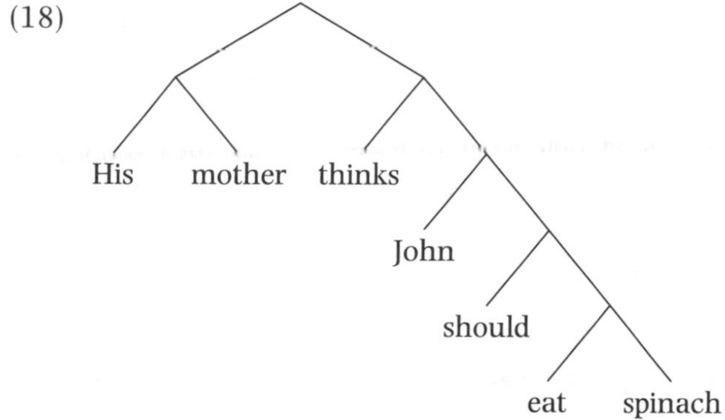

The sole difference between this structure and that above lies in the form of the subject constituent. In (15), this consists of the pronoun *He* alone. Here, however, the pronoun *His* is first combined with the noun *mother*, and this unit is then combined with the remainder of the structure. We can exploit this small structural difference to account for the interpretive differences we have already seen.

(19) A pronoun may not be interpreted as identical in reference with a proper name if the smallest (nontrivial) constituent containing the pronoun also includes the proper name.

By "nontrivial constituent" we mean one which includes something other than the pronoun itself. This constraint is quite natural from a structural perspective. It expresses in the simplest manner possible the structural relationship between the pronoun and the name *John* in (10b). What is striking is that our constraint in (19) accounts for a wide range of examples including all those in (11) and (12) as well as those in (14). For example, given the following structure for example (14b), we can see why the pronoun may be co-referent with *John*:

(20)

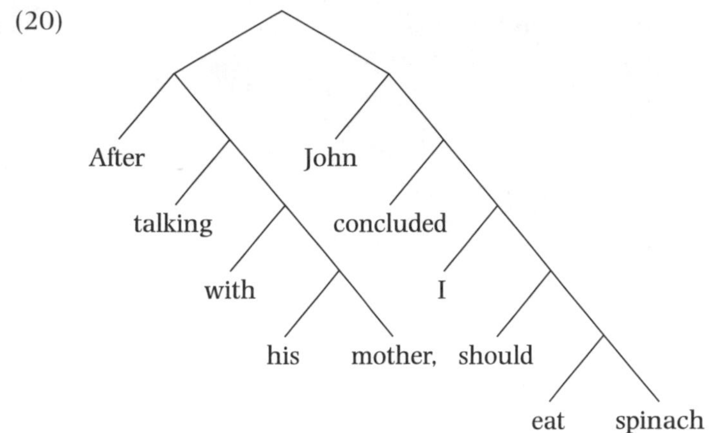

Since the smallest constituent containing the pronoun *his* is *his mother*, a constituent which does not include *John*, the constraint in (19) does not restrict the reference of the pronoun. Our success in providing an explanation for these facts about possible co-reference provides compelling evidence both for the constraint in (19) as well as for the fine structure of our abstract grammatical representations to which it makes crucial reference. This constraint tells us something even more surprising about the structures underlying our grammars when we consider the following variation on (14b):

(21) After talking with John's mother, he concluded I should eat spinach.

Once again, we find that co-reference between *he* and *John* is blocked. Yet, our constraint does not seem to help us. The structure in this case is identical to that in (20), so that the smallest constituent containing the pronoun is *he concluded I should eat spinach*, which does not include *John*. Before despairing, let us take a look at the verb *talking* in this sentence. Although there is no visible (or audible) expression of the subject of this verb, our intuitions tell us that the subject – that is, the *talker* – is necessarily co-referent with the pronoun *he*. Traditional grammar accounts for this intuition by positing an invisible pronoun in this position, much like that which occurs in the subject of imperatives. If we take this traditional idea seriously and suppose that there is actually an invisible pronoun in the structure precisely where the subject would ordinarily appear, we can understand this case in terms of our constraint: since the smallest nontrivial constituent containing the invisible pronoun also includes *John*, *John* may not be co-referential with the invisible pronoun. As we observed above, this invisible pronoun is necessarily co-referential with the subject pronoun *he*. By transitivity, then, we derive the result that *John* may not be co-referential with *he*.

The morals of this second investigation are similar to those we drew above. We see again that it is necessary for grammatical computations to be sensitive to abstract properties of grammatical structure, including hierarchical relations and the presence of invisible pronouns. Furthermore, the constraint in (19) has interesting implications for the problem of the acquisition of linguistic knowledge. It has been shown experimentally by Crain and McKee (1985) that children as young as three years of age do not make errors of the sort that would result from ignorance of constraint (19), and it seems quite clear that they are never taught this constraint, either explicitly or implicitly. Furthermore, every known human language abides by constraint (19). This cross-linguistic invariance, coupled with the poverty of the stimulus, suggests that UG innately specifies not only the format for the linguistic rules which the child acquires, but also in some cases the rules themselves.

Choosing structural representations

In the last section, we saw how our understanding of human grammar has profited from structural analysis: that is, the assumption that underlying our linguistic competence there are abstract rules and representations of a particular sort. Yet, our discussion of the properties of the grammatical rules and representations and why they were selected has been rather vague and informal. It is clear, however, that our success in applying structural analysis depends crucially upon choosing the appropriate underlying structures. Let us therefore turn briefly to the question of how this choice has been made in the case of grammatical analysis.

When applying structural analysis to some domain, we must make at least two choices. First, we must make the sometimes difficult decision as to what objects are to serve as the atomic elements manipulated and combined in our abstract structures. Indeed, it is here that Mendel's contribution to the study of heredity lies: that is, in the recognition of the atomic notion of the gene. In the case of syntax, the unit that we have assumed, and that has proved appropriate for analysis, is the word. It is important to observe that the recognition of the word represents a significant abstraction from the observable data. No direct analog exists in spoken language: the speech stream contains no reliable indications of word boundaries, and two instances of what we might call the same word can vary significantly in how they sound. Nor is the notion of word reducible to anything related to units of meaning. Individual words can be composed of separate meaningful subparts (called *morphemes*) which are combined together in different words:

(22) (a) help-less-ness
(b) help-ful
(c) care-less-ness
(d) care-ful

Instead, the notion of *word* comes to us through the human development of writing systems, many of which segment sentences into medium-size chunks above the level of individual sounds. Note that our choice of words as the atomic unit of syntax is crucial: Our constraint on co-reference, for example, would be obscured significantly if we were dealing with combinations of units smaller or larger than words.

Having identified our atomic units, we must address the second question of structural analysis: What kinds of rules and representations govern the combinations and manipulations of these atomic elements? In our two case studies, we concluded that words are grouped into treelike structures, parts of which may be transposed and/or deleted. Yet, our introduction of these abstract representations may have seemed a bit out of the blue. Why were such representations chosen? The answer to this question lies in a fortunate synergy in the first half of this century between studies of language and research into the foundations of mathematics and computer science. Around this time, formal models were being studied that could characterize the behavior of recently developed computing devices (see Partee et al., 1993, for an overview). These models provided a means of specifying an infinite number of objects/computational sequences using only a finite amount of information, much as a computer program can specify an infinite number of different outputs depending upon the input. One such abstract model that was developed was that of string rewriting systems. In such a system, we start out with some basic string, usually the symbol *S*, and then iteratively rewrite portions of this string according to what the rules specify. Thus, given the system of rules in (23), we can carry out the sequence of rewriting steps in (24).

(23) S → *frog* S
S → *pond*

(24) S ⇒ *frog* S ⇒ *frog frog* S ⇒ *frog frog frog* S ⇒ *frog frog frog pond*

Each rule tells us that the symbol to the left of the arrow (*S* in both rules in (23)) may be rewritten as the sequence of symbols to the right of the arrow. We continue with this process until there are no longer any symbols left which can be rewritten according to the rules. Notice that both of the rules in this system have an *S* on their left, and therefore either may be chosen at any point in the process of rewriting. However, as soon as we choose the second rule, our rewriting terminates, since there will be no remaining occurrences of the symbol *S*. It is for this reason that this sequence of rules allows us to generate all and only the strings which start off with a sequence of zero or more occurrences of *frog* followed by a single instance of *pond*.

As rewriting systems were developed, it became clear that they would also be useful for characterizing the sentences of human languages. Moreover, it was noticed that sequences of rewriting steps like that in (24) could be taken to generate a treelike structure: at each point of rewriting, we add a layer in the tree, placing the newly inserted material below the rewritten symbol. Thus, the sequence of rewriting in (24) corresponds to the following tree:

(25)

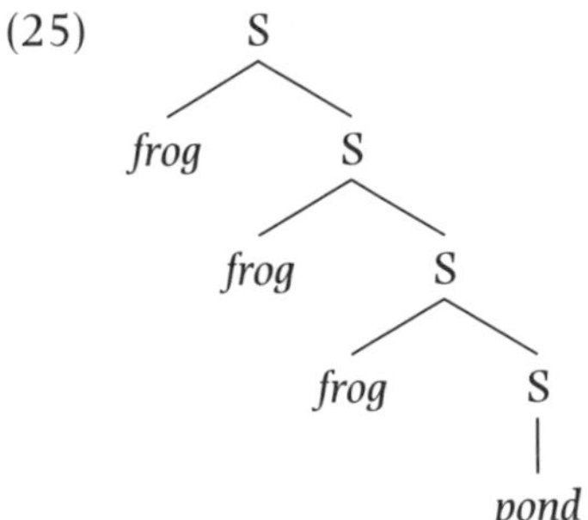

To produce the structures of natural languages, we can use systems of the same character. Thus, the following system of rules generates an infinite set of English sentences (assuming we replace the lower-case part of speech symbols with an appropriate word):

(26) S → noun-phrase verb-phrase
verb-phrase → auxiliary verb-phrase
verb-phrase → verb noun-phrase
verb-phrase → verb S
noun-phrase → name
noun-phrase → pronoun
noun-phrase → article noun

Interestingly, the sequences of rewriting steps which may be carried out using this system of rules correspond precisely to tree structures with the constituency we argued for above. The following is an example of one such structure:

(27)

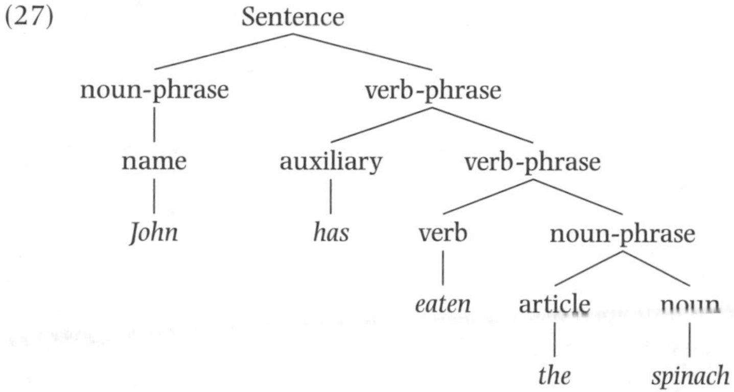

As the formal work produced a range of models which differed in their expressive power, linguists began to raise the question of whether the grammars of natural languages correspond in any interesting way to a particular class of mathematical devices. If this were true, it might allow us to understand restrictions on possible human grammars (embodied in UG) in formal terms. Chomsky (1957) argued that a certain class of devices, known as *finite state automata,* were too weak to serve as a formal basis for human grammars, since they cannot generate the recursive structural representations appropriate for certain sentences. Instead, he adopted the use of rewriting systems of the sort just discussed, called *context-free grammars* (CFGs), to generate constituent structure representations. Chomsky also adopted certain operations, called *transformations,* which would manipulate the tree structures produced by the CFG, in order to handle cases like English question formation. It turned out, however, that the introduction of transformations expanded the expressive power of the system considerably. In response to this, a number of proposals have been made which aim to reduce the expressiveness of grammars by eliminating transformations from them. It has been shown that CFGs alone are insufficient for this, because of the existence of certain constructions in Swiss German (Shieber, 1985). This leaves open the question of the extent to which human grammatical systems may be formally characterized, though a number of promising possibilities exist (cf. Partee et al., 1993, for further discussion).

Abstract structures in other domains

The methodology of structural analysis discussed in this article has been applied beyond the narrow realm of natural language syntax that we have considered here. Within the study of language, similar methods of analysis have been pervasively applied to the study of sounds (phonology), words (morphology), and meanings (semantics), yielding a range of abstract structural representations whose properties bear considerable explanatory burden. There are a wealth of cases in each of these domains analogous to those discussed here, though space prevents us from going in to them (see Akmajian et al., 1995, for a traditional overview, and Jackendoff, 1994, for one more focused on connections with cognitive science). Additionally, these representations have shed substantial light on the processes of language acquisition and language change. It has been found that variation in the types of sentences that are used, whether during the course of children's acquisition of their native languages (see Article 6, COGNITIVE AND LINGUISTIC DEVELOPMENT) or in the centuries-long periods of linguistic change, are best characterized not as superficial, haphazard alterations, but rather in terms of parametric modifications to the fundamental underlying grammatical rules and constraints.

Moving outside the domain of language, one application of these same methods has been in the study of music cognition. Just as the representations of linguistic theory arise out of an attempt to model speakers' intuitions about well-formedness and possible meanings of sentences in their language, Lerdahl and Jackendoff (1983) present a cognitive theory of musical representations in an attempt to model the intuitions that listeners have concerning properties like meter and grouping of notes in a piece of music. Their representations have in common with the syntactic structures discussed above their hierarchical nature, but differ in at least two respects. First, as Lerdahl and Jackendoff note, the constituents of syntactic structures are assigned a meaningful category, be it sentence or noun-phrase. Thus, what the first rule in (26) tells us is that the combination of a noun-phrase and a verb-phrase forms a sentential constituent. In Lerdahl and Jackendoff's musical structures, the combination of two constituents X and Y signifies only that one, say X, is the elaboration of the other, say Y. In this case, the resulting combination of X and Y is not essentially different in nature from the element Y occurring on its own. The second difference between the syntactic and musical representations lies in the mode in which their well-formedness is specified. We suggested above that syntactic representations are well formed exactly when they accord with all the rules and constraints of the grammar, like those in (9), (19), and (26). Lerdahl and Jackendoff propose that well-formedness for musical structure is determined instead by a set of preference rules. These preference rules may at times conflict with one another, and in such cases the conflict is resolved on the basis of the strength of the relative preferences. Consequently, there will be some musical structures that violate some of the preference rules yet are well formed. Lerdahl and Jackendoff point out that although their musical structures differ from syntactic representations, they are closely analogous to those exploited in another linguistic domain: namely, that of phonology, particularly in the areas of stress and prosody, which are the rhythmic and melodic properties of spoken language. This is suggestive of a more general cognitive capacity for dealing with temporal organization of various sorts, which could underlie both musical and linguistic cognition. Concerning the different mode of specifying well-formedness, it is intriguing to note that the recently proposed *optimality theory* adopts the view that preference rules (or violable constraints) are indeed the appropriate means for characterizing the well-formedness of linguistic structure (Prince and Smolensky, forthcoming). If this view is on the right track, it suggests once again that the apparent differences across these domains may not be as large as they first seemed.

We also find work in visual PERCEPTION that exploits complex representations and constraints on them to explain behavioral data from early vision (Marr, 1982), such as the *misperceptions* involved in visual illusions and subjective contours, as well as from high-level vision (Biederman, 1987), like the ability to recognize objects in spite of variations in perspective or the presence of visual noise. To give one example, Irving Biederman proposes that the basic three-dimensional appearance of objects is mentally represented as the combination of simple volumes called "geons." The nature of the combinatorial relationships determines why some views of objects, such as rotations in the plane, are less easily recognized than others, like rotations in depth, in that only the former induce a change in the TOP-OF relation and thereby present a visual stimulus which does not exactly match the mental representation.

Finally, results from a number of studies of REASONING and PROBLEM SOLVING have been used to argue that humans use structured representations in carrying out these tasks (Johnson-Laird, 1983; Newell and Simon, 1972). Philip Johnson-Laird, for instance, attempts to explain the difficulties which people experience in performing certain types of deductive inference. He suggests that people do make inferences from statements like "All scientists are skeptics" and "Anne is a scientist" to "Anne is a skeptic" not by manipulating the standard sorts of logical inference rules, but instead by constructing mental models of the situations which are described. Some collections of statements do not uniquely describe one single mental model, and an individual who is to make a correct inference must simultaneously consider multiple models. Johnson-Laird demonstrates that the level of difficulty for inferences is predicted by the number of mental models that must be considered (as well as the order in which they must be examined), rather than by the character or length of steps of application of logical inference rules.

In these latter domains, abstract mathematical structures have been important in certain lines of research, particularly research focused on computational issues. Yet, in work tightly linked to empirical investigations, structures of the type brought up in our discussion of syntax have not played the significant role in shaping research directions and theories that they have in the study of language. One might interpret this state of affairs as demonstrating that the abstract structures appropriate for these other domains differ quite radically from those used in linguistic cognition. Indeed, it may well be that such structures are of an as yet undiscovered character. If this is true, this might be an indication that different cognitive domains are processed in distinct modules and hence do not share the same types of representations (see Article 49, MODULARITY). An alternative explanation of the difference in the success of structural analysis across different domains might tie it to our not yet having found the appropriate atomic elements which the abstract representations combine and manipulate. In the case of high-level vision, for example, determining the right set of atomic units from which object representations are composed remains a matter of significant debate. Note that the problem of atomic units is significantly more difficult in visual perception and reasoning than language, largely due to the relative paucity of external evidence for the nature of internal representations. If this is the correct diagnosis, we can hope that it is simply a matter of time before more of the cognitive sciences are able to unlock the underlying laws and principles of cognition that the mind still holds secret.

References and recommended reading

*Akmajian, A., Demers, R. A., Farmer, A. K. and Harnish, R. M. 1995: *Linguistics: An Introduction to Language and Communication*. Cambridge, Mass.: MIT Press.

Biederman, I. 1987: Visual object recognition. In S. M. Kosslyn and D. N. Osherson (eds), *Visual Cognition: An Introduction to Cognitive Science*, Cambridge, Mass.: MIT Press, 121–65.

Chomsky, N. 1957: *Syntactic Structure*. The Hague: Mouton.

—— 1986: *Knowledge of Language: Its Nature, Origin, and Use*. New York: Praeger.

Crain, S. and McKee, C. 1985: Acquisition of structural restrictions on anaphora. In S. Berman, J. W. Choe, and J. McDonough (eds), *Proceedings of the 16th North Eastern Linguistics Society Meeting*, Amherst, Mass.: Graduate Linguistics Student Association, 94–110.

Crain, S. and Nakayama, M. 1987: Structure dependence in grammar formation. *Language*, 63, 522–43.

*Jackendoff, R. 1994: *Patterns in the Mind: Language and Human Nature*. New York: Basic Books.

Johnson-Laird, P. 1983: *Mental Models*. Cambridge, Mass.: Harvard University Press.

Lerdahl, F. and Jackendoff, R. 1983: *A Generative Grammar of Tonal Music*. Cambridge, Mass.: MIT Press.

Marr, D. 1982: *Vision*. San Francisco: Freeman.

Newell, A. and Simon, H. A. 1972: *Human Problem Solving*. Englewood Cliffs, NJ: Prentice-Hall.

Partee, B., ter Meulen, A. and Wall, R. E. 1993: *Mathematical Methods in Linguistics*. Dordrecht: Kluwer.

Prince, A. and Smolensky, P. Forthcoming: *Optimality Theory: Constraint Interaction in Generative Grammar*. Cambridge, Mass.: MIT Press.

Shieber, S. 1985: Evidence against the context-freeness of natural language. *Linguistics and Philosophy*, 8, 333–43.

PART IV

STANCES IN COGNITIVE SCIENCE

36

Case-based reasoning

DAVID B. LEAKE

Introduction

In some views of cognitive science, episodic memory is simply one of many pieces of the puzzle of cognition: MEMORY, REASONING, and learning can be studied largely independently of one another. The case-based reasoning (CBR) stance takes the contrasting position that human reasoning, memory, and learning are inextricably bound together. In this view, reasoning is primarily based on remembering and reapplying the lessons of prior episodes. A case-based problem-solver, for example, solves new problems by retrieving traces of relevant prior problems from memory, establishing correspondences between those problems and the new situation, and adapting the prior solutions to fit the problem at hand. Learning is a natural side effect of this process, as traces of new episodes are stored for future use.

The case-based model of reasoning and learning provides an account of how people can reason effectively even in complex situations and using imperfect information. For example, consider the problem faced by a teacher deciding how to help a troubled student. The decision must be made despite a lack of guaranteed rules for successful intervention, and despite imperfect information on the student's state of mind and the myriad outside influences on the student. Nevertheless, if the teacher is experienced, he or she can make a decision based on prior cases of students with similar problems. By remembering what did – and did not – work with them, and why, the teacher can reuse successful strategies and avoid repeating past mistakes. The teacher can also learn, by remembering the current situation, decision, and results, how to handle similar problems in the future.

Observations of human reasoning from experience have prompted interest in the CBR process both from artificial intelligence practitioners aiming to develop computer systems that reason and learn (see Article 16, MACHINE LEARNING) and from cognitive scientists aiming to understand human reasoning and learning. Understanding human case-based reasoning requires addressing questions about multiple research areas of cognitive science. For example, case-based reasoning is fundamentally analogical: it solves new problems and interprets new situations by analogy to prior episodes (see Article 1, ANALOGY). Likewise, case-based reasoning is intimately connected to areas such as memory and similarity. However, the case-based reasoning stance involves investigating these areas from a new perspective. First, it examines each process as part of an integrated whole, in the context of the others. Second, it calls for studying them in the pragmatic context of large-scale, real-world tasks, such as planning and explanation. This requires analyzing not only the reasoning processes involved but also the structure, content, and organization of the knowledge required for those processes to succeed for real-world tasks.

The case-based reasoning stance points to three central research questions for cognitive science. The first concerns *knowledge access*, or *indexing* – how memory is organized and accessed to retrieve relevant information. The second concerns *knowledge adaptation* – how relevant episodes are modified to apply to a new situation. The third concerns *learning* – how lessons of new experiences are extracted, represented, and organized.

Each of the three questions addresses how people perform complex cognition for real-world tasks, using their real-world knowledge. This makes them very difficult to answer with laboratory experiments. However, case-based reasoning research is beginning to provide answers to each one, through a methodology of developing computer models whose designs are strongly influenced by observations and studies of human reasoning and learning processes, but whose development is largely guided by examining functional requirements for the processes and knowledge involved.

For example, one way in which case-based reasoning research attacks the indexing problem is to analyze the types of reasoning tasks that cases can serve, the circumstances under which cases are likely to be useful, and the indexes required for them to be retrieved at the right times. The resulting theories of indexing are tested and refined by being implemented in computer models whose behavior can be examined. Empirical results from the computer models provide information about the theories' functional sufficiency and about gaps in the theories that must be addressed. Computer models of human case-based reasoning also suggest testable hypotheses about how people learn and how to structure EDUCATION to aid human learning. For example, empirical evaluations of the learning process of Michael Redmond's CELIA (Redmond, 1992) – a computer model of how problem solving changes with experience as cases and other types of knowledge are acquired – predict the relative role of cases and domain models in learning, the relative speed with which each body of knowledge grows, and the value of presenting a range of examples early on in the learning process.

Taking case-based reasoning seriously as a human reasoning process suggests that education should be aimed at facilitating students' natural learning through case-based reasoning, as well as at teaching students to perform case-based reasoning as effectively as possible. A strong new direction in case-based reasoning research focuses on developing innovative curricula and classroom activities built on the lessons of case-based reasoning research.

This chapter provides a brief overview of the case-based reasoning stance towards cognitive science, its development, and some of the research directions it has engendered. Janet Kolodner (1994) provides an extensive discussion of case-based reasoning as a cognitive model and presents important computer models of human case-based reasoning. Kolodner's (1993) textbook provides a thorough examination of case-based reasoning, both as a cognitive model and as an artificial intelligence technology inspired by observations of human reasoning. Case studies of selected computer models for multiple tasks are presented in a volume edited by Christopher Riesbeck and Roger Schank (1989). A volume edited by David Leake (1996) provides an overview of the principles of case-based reasoning research, a tutorial, and case studies addressing key issues.

Reasoning from remindings

Remindings are commonplace in everyday human reasoning. In children and adults, and in novices and experts, across a broad spectrum of cognitive tasks, memories of previous episodes provide useful information to guide understanding, problem solving, and prediction.

Previous episodes may be used in a straightforward way for routine processing of simple situations. For example, once when taking my two-year-old for a walk, I asked him where he would like us to turn at an intersection. He picked the direction we had turned the day before to go to the supermarket and explained his decision by his memory of the previous trip: "I have a memory: Buy donut."

Memories of previous episodes can also guide the reasoning of experts performing difficult tasks. Radiologists, for example, use records of previous cases as a basis for choosing X-ray treatments; lawyers argue and judges reach verdicts by comparing current circumstances with those of precedent cases (see Article 58, LEGAL REASONING); politicians shape their strategies according to memories of their prior successes and failures; governmental policymakers weigh new plans in terms of the experiences of the past. When people are confronted with complex situations, similar previous episodes provide a basis for new reasoning. When the stock market crashed in 1987, the *New York Times*'s analysis of the crash was framed by the question, "Is 1987 like 1929?" When Americans debate the wisdom of intervening in a foreign country, the debate is often framed by the question, "Will this be another Vietnam?"

Reasoning from prior episodes has a number of functional benefits compared to traditional rule-based models of reasoning. First, it helps to generate effective solutions in situations whose causal structure is not completely understood. Rule-based models assume that conclusions are drawn by chaining together generalized rules, and that learning involves deriving and storing new generalized rules for future use. However, real-world events are often too complex and too imperfectly understood to immediately distill them into generalized rules. The case-based reasoning process avoids this problem by deriving conclusions only when they are needed, directly from the episodes themselves, in the context of the new situation. The Vietnam War provides an example of the benefit of the case-based approach. The Vietnam War is a potential source of countless useful lessons for the future, but attempting to anticipate all of them a priori, or to determine the precise circumstances in which a given lesson should be applied in the future, would be an impossible task. Decisions about whether Vietnam is a relevant precedent in a new situation, about which of its lessons apply to that situation, and about how they apply, are best made in the context of the new situation.

Second, even when general rules are available, cases can usefully augment generalized knowledge. Rules intended to describe real-world phenomena are often unreliable, because they cannot enumerate the infinite set of factors that might affect their validity in a particular situation. Consequently, even when there exist rules from which a solution could be derived, it may be valuable to favor solutions that have proved successful in similar prior situations. Likewise, cases can capture and warn of exceptions to the rules.

Reasoning from cases rather than rules can also lead to more efficient processing. Consider, for example, the task of filling out an income tax form. Although it is possible to categorize types of income using general tax rules, it is much easier to simply reuse the categories selected and applied the year before. Empirical tests comparing artificial intelligence systems using case-based reasoning to rule-based systems show that case-based reasoning can provide a significant increase in speed.

Another benefit of case-based reasoning is that cases retain sufficient detail to allow them to be used for novel purposes, even if those purposes were not anticipated at storage time. This provides a basis for creative problem solving. For example, the chief engineer at Honda reported that one of their technological innovations was based on applying a specific observation in a new context:

> We could not think of a way to smoothly change the camshaft profile [for the VTEC-E engine]. It was a severe problem which we finally solved by inserting and removing a pin from the valve rocker gear. The person who developed this apparently got the idea by taking a serious look at a chicken skewer in a yakitori bar. (*Automobile Magazine*, January 1996)

Learning from experiences

After a case-based reasoner – human or machine – has applied a prior case to a new situation, it learns by storing the results in memory as a new case. If the results were successful, remembering that case in the future can provide suggestions and shortcuts for future problem solving. If the results were unsuccessful, remembering the case can warn about potential problems. In addition, explanations of current problems can help to avoid future difficulties, by enabling reindexing of erroneously retrieved cases to make sure that they are not mistakenly retrieved in the future. Thus in case-based reasoning every processing episode provides an opportunity for learning.

The learning of case-based reasoners combines aspects of both inductive and theory-driven learning. Adding cases to a case library is a form of inductive learning, but one that differs markedly from models of inductive learning (such as neural network models) that abstract away from specific examples to learn only generalizations. Models of case-based reasoning also differ in taking seriously the role of prior knowledge in learning. Selection and generalization of indices, for example, are often based on explanations of why particular features of a situation are relevant in a particular task context. Nevertheless, large chunks of knowledge from specific experiences temper the domain theory and supplant general knowledge. As in the chunking process of SOAR and the knowledge compilation process of ACT* (see Article 42, PRODUCTION SYSTEMS), the CBR learning process increases the availability of specific knowledge, and later processing favors that specific knowledge over initial rules. Case-based reasoning, by learning and reusing specific cases, reapplies the most specific knowledge possible.

The case-based reasoning process

Cases record specific, operational knowledge tied to a larger context. They encapsulate both the circumstances of that context and the outcomes or lessons learned. For example, a case might represent an anomaly and its explanation, a plan and its outcome, or a legal argument and its results. Cases can be used for two types of purposes: interpretation and problem solving (see Article 20, PROBLEM SOLVING). Problem-solving CBR adapts prior solutions to solve new problems. For example, an architect designing a building may base his or her design on memories of successful designs for similar buildings, adapting them to fit new requirements and to avoid pitfalls encountered in the past.

Interpretive CBR uses prior cases as the basis for forming a judgment about or a classification of a new situation, by comparing and contrasting a new example with cases that have already been classified. This is in the same spirit as exemplar-based models of categorization (see Article 8, CONCEPTUAL ORGANIZATION). Exemplar-based models posit that category information is stored in the form of ungeneralized instances of a category, and that each new example is categorized by assigning it to the category of the *closest* prior exemplar. For example, it is natural to predict whether someone will enjoy a movie by remembering the reaction to the most similar movie he or she saw before. Interpretive CBR may also involve sophisticated reasoning about the relevance of particular cases. For example, when cases are used to make legal decisions, a reasoner must make arguments comparing and contrasting a potentially large set of precedent cases to the new situation (Ashley and Rissland, 1987).

Case-based reasoning is made up of a number of sub-processes, with specific sub-processes depending on whether case-based reasoning is being used for problem solving or for interpretation. Both problem solving and interpretive case-based reasoning begin with a process called "situation assessment," which elaborates the problem situation and characterizes it in a form compatible with the indexes used for the case library. Situation assessment is needed because it is impossible to anticipate all the contexts in which a case may be useful. As a result, the indexes used when storing a case may not reflect all the situations under which that case should be applied. Consequently, in novel situations, it may be necessary to redescribe the current situation in order to generate indexes that will lead to useful retrievals.

An example of the redescription process is provided by SWALE, a model of case-based reasoning for generating creative explanations of anomalous events (see Schank et al., 1994). One of the anomalies that the system explains is the sudden death of a star racehorse named Swale. To explain the death, the system attempts to retrieve prior explanations indexed under deaths of racehorses, but finds none in its memory; it then generalizes the index to deaths of animals and succeeds in retrieving prior cases with candidate explanations. However, none of those explanations applies to Swale, so it must generate other indices to find additional explanations. To do so, the system recharacterizes Swale's death in other ways (e.g., as destruction of valuable property, as the death of a young performer, and as death despite peak physical condition). Using the index "death despite peak physical condition," it retrieves a case involving the death of the runner Jim Fixx, who died when exertion overtaxed a hidden hereditary heart defect. This case suggests a plausible explanation for Swale's death.

Either during or after retrieval, analogical *mapping processes* establish correspondences between the current situation and the retrieved case, and *similarity assessment processes* identify relevant similarities and differences. (Some CBR systems reason from multiple cases, but this discussion simplifies the process by assuming that only a single case has been retrieved.) In problem-solving CBR, problematic differences are repaired by *adaptation processes* that fit the retrieved case to the new situation; in interpretive CBR, problematic differences may be accounted for by explaining why those differences are unimportant in the current context. *Evaluation processes* identify problems in the result, guiding incremental adaptation or further justification. The results of the reasoning process, combined with feedback about the success of the solution, are the raw material for *learning processes* that install new episodes in memory for future use and refine indexes. This learning provides the basis for better future reasoning.

An example processed by the computer model CHEF illustrates the major processes used in problem-solving CBR. CHEF was developed to investigate planning as a case-based reasoning process (Hammond, 1989a). CHEF built recipes – cooking plans – by recalling prior recipes and adapting them to fit current needs. The system remembered its failures, so that they could be avoided in the future; its successes, so that they could be reused; and the repairs it made to failed recipes, so that they could be reapplied. One of the planning problems given to CHEF involved building a recipe for the stir-fry dish "beef and broccoli." To generate a new recipe, CHEF begins by retrieving from memory the recipe whose goals most closely match the set of goals for the current task (in this example, the goals are to make a stir-fry dish, to include beef, and to include broccoli). During similarity assessment, differences between the goals of old and new dishes are weighted by a metric for comparing the importance of particular types of differences, to select the *most similar* prior recipe – the one expected to provide the best starting point. For example, it is harder to adapt a stir-fry dish to make a dessert than to adapt a beef dish to make a pork dish, so differences involving the type of dish are weighted more heavily than goals involving different meats. Initially, the recipe in CHEF's memory most similar to beef and broccoli is one for beef and green beans. CHEF generates the desired recipe with a simple adaptation, substituting broccoli for green beans and adding a new step required by broccoli but not green beans: chopping the broccoli into pieces. Based on its knowledge of the recipe taken as a starting point, CHEF predicts the characteristics of the new dish, such as tender beef, a savory taste, and crisp broccoli.

When CHEF tests the new recipe in its simulated world, one of its expectations fails: the broccoli is soggy. CHEF diagnoses the problem as involving a bad interaction between two concurrent steps in the plan. Because the beef and the broccoli were stir-fried simultaneously, liquid released by the beef during stir-frying caused the broccoli to be cooked in a pan that was wet rather than dry. As a result, the broccoli steamed rather than fried. (This was not a problem in the recipe taken as the starting point, because green beens are more robust than broccoli.) Based on its explanation, CHEF repairs the recipe by applying the strategy *split and reform* – changing the recipe to stir-fry each ingredient separately and combine their results. This produces a successful recipe which CHEF stores for future use. In addition, based on its explanation of the problem, CHEF refines its indexes. First, it forms a generalization about when to anticipate the problem in the future: whenever meat is stir-fried with a crisp vegetable. From this generalization CHEF learns that when it needs to generate a stir-fry recipe involving any meat and crisp vegetable, it should try to retrieve recipes designed to deal with the liquid problem.

When CHEF is later called upon to generate a recipe for chicken and snow peas, it anticipates the potential problem of the snow peas becoming soggy and searches memory for a recipe for a similar dish that addresses the problem. As a result, CHEF retrieves the beef and broccoli recipe as its starting point, in preference to other recipes that (before its learning) would have appeared to have more in common with the requested one (e.g., chicken and green beans).

Development of the case-based reasoning stance

The case-based reasoning stance arose from a number of currents of research. One important component was computer modeling of experts' use of cases in task areas such as legal and medical reasoning. Many other contributions to development of the case-based reasoning stance came from studies of the role of memory in story understanding. In the 1970s, considerable research was directed towards schema-based models of understanding. According to such models, the understanding process is based on reapplication of existing structured knowledge describing normative expectations about a situation. For example, Roger Schank and Robert Abelson (1977) proposed *scripts* as a type of event schema. Each script describes a stereotyped sequence of events that take place in a particular context, such as the standard events in going to a restaurant (wait for a table, sit down, order, etc.). After a script has been identified as relevant to a story, the script is used to guide inferencing about the reasons for the events stated in the story and to fill in unspecified details.

Psychological experiments by Gordon Bower, John Black, and Terrence Turner (1979) supported the role of scriptal knowledge in human understanding but also pointed to an important discrepancy between the predictions of script-based computer models and human memory. In script-based computer models, all information about a particular script was localized in a single package, independent of other scripts. If this were taken as a model of human memory, it would suggest that people who process a story with a script will store it in a localized form as well. To examine how script-based information was stored, Bower, Black, and Turner had subjects read stories involving scripts that were distinct, but related to each other, such as scripts for visiting a doctor and for visiting a dentist. The stories that subjects read did not include all the events in each script; parallel script actions were selected to be omitted in one story but included in the other. For example, subjects might be asked to read a story about a visit to a doctor that explicitly included the event "John arrived at the doctor's office," and to read a story about a visit to a dentist that did not make an explicit mention of arriving at the dentist's office. After an intervening task, the subjects were asked to recall the stories that they had read. It was found that subjects were more likely to report script actions that had been left unstated if they had read stories which stated the parallel actions. Such cross-schema influences are not accounted for by models in which memory is organized by schemas that are independent of each other.

Such models also failed to account for cross-contextual learning. When people learn about how to pay for a visit to a doctor's office, what they learn appears to be naturally transferred to other contexts, such as paying in a dentist's office. However, models based on fully independent schemas do not support this type of learning, nor do they account for how spontaneous remindings can occur across contexts (e.g., being reminded, when someone takes out a credit card to pay in a dentist's office, of a problem when a doctor's office refused to accept one). Thus three phenomena needed to be explained: cross-schema confusions during retrieval, cross-contextual learning, and cross-contextual remindings.

Roger Schank (1982) responded by proposing dynamic memory theory. Dynamic memory theory postulated a new type of knowledge structure, memory organization packages (MOPs). Like scripts, MOPs characterize event sequences; unlike scripts, MOPs are hierarchical. They are composed of other MOPs and lower-level components, called "scenes," which can be shared with other MOPs. For example, the act of payment after a professional office visit tends to be standard regardless of the particular profession involved. Although M-doctor-visit and M-dentist-visit are distinct MOPs, both share a single PAY scene. Consequently, what is learned about paying in one context, changing the PAY scene in memory, is automatically reflected in all MOPs that share the PAY scene. In addition, each processing structure serves as a memory structure. When episodes are retrieved, they must be reconstructed from the information stored under the MOPs and scenes used to process their constituent parts.

Thus in dynamic memory theory, processing structures, memory structures, and episodic memory are inextricably bound together. Spontaneous remindings occur during understanding because prior episodes – the cases of a case-based reasoner – are stored in memory under the structures used to process them and are accessible to provide guidance when those processing structures are used in the future. One of the first cognitive models investigating this integrated view of memory was Janet Kolodner's CYRUS, developed in the early 1980s. CYRUS examined the relationship between encoding, storage, and retrieval and implemented a model of the retrieval process based on constraints implied by human retrieval. For example, because people have difficulty enumerating long lists of items (e.g., enumerating the 50 states of the USA), the CYRUS computer model does not allow the members of categories in its memory to be enumerated. Instead, it models retrieval as a two-step cycle which first formulates a sufficiently detailed description of the desired item to distinguish it from other similar items and then discovers whether that item is in fact in memory. This research developed criteria for the types of schemas useful for storing episodes, how schemas change over time, and how schemas organize specific episodes.

Likewise, Christopher Riesbeck developed the ALFRED system to investigate the ramifications of failure-driven reminding for incremental learning. ALFRED used rules that reflect everyday knowledge to understand newspaper articles in the domain of economics. If one of the system's rules of thumb failed, an exception episode describing the failure and the recovery procedure was stored in memory, indexed under the faulty rule. If the rule failed again, the exception episode was remembered, and the rule was revised, using the exception episode to provide data about the types of situations under which the rule failed to apply.

By the mid-1980s, computer models of case-based reasoning were being investigated in multiple task domains, for tasks such as planning, diagnosis, interpretation during legal reasoning, criminal sentencing, story understanding, creative explanation of anomalous events, and design. This research resulted in functionally motivated models of memory organization, retrieval, similarity assessment, adaptation, and storage.

At the same time, studies of human learning and problem solving supported the role of remindings in human reasoning in many tasks. For example, it was shown that people learning to use a word processor reason about how to perform new tasks by remembering how they carried out old ones, that both novice and expert mechanics use prior experiences to help generate hypotheses about problems and to help select appropriate diagnostic tests, that expert decision-makers use analogs to suggest starting points for problem solving and to help evaluate candidate solutions, and that human explainers favor explanations based on prior experiences as they explain anomalies. These results are surveyed in Kolodner, 1993, 1994.

Some research directions

Memory organization

How human memory is organized, how memory organization is refined, and how retrieval reflects changing task contexts are central questions for case-based reasoning. One source of clues to human memory is to consider the remindings that occur in people and ask what types of features are shared by the old and new situations. Analyzing why a particular episode is retrieved in a given situation – and, equally important, why other episodes in memory that may share features with the new situation are not retrieved – provides information about the content of the indexes involved.

The case-based reasoning literature presents many examples of remindings based on abstract, thematic features that are shared by episodes in very different contexts. For example, Roger Schank (1982) described being reminded of *Romeo and Juliet* when seeing *West Side Story*. Accounting for such remindings depends on identifying the fundamental similarity that exists between the two stories, despite their many differences. Schank proposed that these types of similarity can be described by the type of goal being pursued – here, the mutual goal of the members of each couple to act together – and the conditions under which the goal is being pursued – here, outside opposition. To account for such remindings, Schank proposed that both stories are organized by a single memory structure that captures their abstract similarity, one organizing instances of "mutual goal; outside opposition." Schank called the abstract memory structures *thematic organization packages* (TOPs).

Human cross-contextual remindings support the idea that thematic indexes play a role in human memory organization. However, an important research question concerns the circumstances under which thematic remindings occur. Experiments in the 1980s on analogical reasoning showed that people may not be reminded of the most appropriate cases during problem solving and suggested the importance of *surface features* in guiding remindings. On the other hand, a series of more recent studies demonstrates the ability of human subjects to achieve remindings based on important features and to avoid being misled by superficial similarities. For a sampling of perspectives framing the issues involved and the relationship between human analogical reasoning and CBR, see Hammond, 1989b, pp. 125–52.

At the same time that psychological studies continue to examine human retrieval, computer models are being used to examine the functional requirements for memory processes and the content of indices in order for useful remindings to occur. This is a particularly active area of current case-based reasoning research.

Similarity assessment and knowledge adaptation

Modeling the similarity assessment and knowledge adaptation processes is a continuing challenge for case-based reasoning research. Deciding whether a previous case should be considered similar to a new situation, for example, depends on addressing the problem of identifying relevant features in particular task contexts. Similarity assessment is closely tied to the knowledge adaptation problem: Similarity assessment identifies the differences between old and new situations that need to be repaired by case adaptation.

Models of the case adaptation process are often rule-based, relying on fixed sets of rules, raising questions about how adaptation rules are acquired and how the cost of case adaptation is controlled. A number of early models of case-based reasoning proposed applying case-based reasoning to the adaptation process itself, relying on prior examples of adaptations to guide the adaptation process, rather than reasoning from scratch. Later models, such as CHEF, stressed the role of causal explanations of problems in determining how cases should be adapted. The mid-1990s saw a resurgence of research on case adaptation, including research by David Leake and colleagues on methods in which general adaptation principles are gradually augmented with learned domain-specific cases to guide adaptation and memory search. Attention also began to be directed towards the role of meta-reasoning in case-based reasoning to refine the case-based reasoning process. This work is surveyed in Leake, 1996.

The integration of multiple knowledge sources and reasoning processes

A natural question is how CBR relates to other forms of reasoning. Case-based reasoning can be viewed as one component of an integrated problem-solving architecture (e.g., see Veloso in Leake, 1996, pp. 137–49), or within an architecture for which CBR is the primary reasoning process, supported by other reasoning mechanisms (e.g., Hinrichs, 1992). Another question is how multiple cases and other knowledge sources are brought to bear on complex problems. Solving complex problems depends on the ability to reason from multiple remindings and generalizations throughout the reasoning process.

CBR and creativity

A common misconception about case-based reasoning is that it applies only if new problems are very similar to those solved in the past. However, the case-based reasoning architecture provides a framework for understanding creative reasoning: creativity can enter into case-based reasoning through flexible retrieval processes which result in novel starting points for solving new problems, through mapping processes that form novel correspondences, and through flexible case adaptation to generate novel solutions. These types of processes have been studied as a basis for modeling creativity for tasks such as explanation (Schank et al., 1994) and design (e.g., see Wills and Kolodner, in Leake, 1996, pp. 81–91).

CBR and education

Case studies have long been the basis of legal and medical education, and the *learning by doing* framework that case-based reasoning suggests is at the heart of problem-based learning methods in education. However, the lessons from computational models of CBR go further, providing new concrete suggestions about what makes a good problem, the range of problems that students should solve, and the kinds of resources that should be made available to student learners (Kolodner et al., 1996). Case-based reasoning highlights, for example, the importance of providing problems that present difficulties for students, enabling them to identify and repair gaps in their knowledge. It also suggests supporting students' learning by presenting them with cases chosen to help point out potential issues for them to address, to guide their learning; to suggest potential solutions or parts of solutions to consider; and to help in anticipating the effects of their decisions.

In addition, the case-based reasoning stance – that human reasoning is largely case-based – suggests that an important aspect of education is developing students' abilities as case-based reasoners. For example, when students reason from prior cases, they may be misled if they blindly apply the first case they retrieve: it is crucial for students to learn to evaluate the relevance of prior cases and the solutions those cases suggest.

Research groups led by Roger Schank and by Janet Kolodner are applying principles from the case-based reasoning cognitive model to develop a new generation of curricula, learning activities, and computer systems to support the learning process. (See e.g., Schank, in Leake, 1996, pp. 295–347.) The resulting instructional systems provide students with opportunities for learning by doing in rich task environments, driven by compelling goals, and also support students' learning by presenting them with useful information about others' experiences, in the form of relevant cases at appropriate times.

Summary

The case-based reasoning stance treats experiencing, remembering, and learning as inextricably bound together. The key questions it proposes for cognitive science concern the structure, content, and organization of knowledge in memory, the knowledge and processes required to retrieve relevant information, how to adapt retrieved information to fit new situations, and how to extract, represent, and organize the lessons of new experiences. Case-based reasoning emphasizes the role of concrete, operational knowledge. Rather than focusing on how basic knowledge can be composed to generate new solutions, case-based reasoning focuses on how large structures – cases – can be modified to fit new situations and views the learning of new cases as an integral part of the reasoning process.

A series of computer models, inspired by observations of human behavior, have built up understanding of the functional requirements for processes involved in case-based reasoning and the knowledge they require. These models provide computational definitions of central cognitive science concepts such as similarity and relevance. In addition, examining case-based reasoning as a cognitive architecture suggests concrete answers to classic questions ranging from the sources of creativity to how to aid students' learning. At the same time, CBR research is identifying further questions to be answered by a combination of computer modeling and psychological experimentation. Thus the case-based reasoning stance suggests not only central issues for cognitive science but also how to address them with a research methodology built on the interplay between studies of human behavior and analysis of functional constraints on the processes and knowledge needed to account for that behavior.

References and recommended reading

Ashley, K. and Rissland, E. 1987: Compare and contrast, a test of expertise. In *Proceedings of the Sixth Annual National Conference on Artificial Intelligence*, San Mateo, Calif.: AAAI, Morgan Kaufmann, 273–84.

Bower, G., Black, J. and Turner, T. 1979: Scripts in memory for text. *Cognitive Psychology*, 11, 177–220.

Hammond, K. 1989a: *Case-Based Planning: Viewing Planning as a Memory Task.* San Diego, Calif.: Academic Press.

Hammond, K. (ed.) 1989b: *Proceedings of the DARPA Case-Based Reasoning Workshop.* San Mateo, Calif.: Morgan Kaufmann.

Hinrichs, T. 1992: *Problem Solving in Open Worlds: A Case Study in Design.* Hillsdale, NJ: Erlbaum.

*Kolodner, J. 1993: *Case-Based Reasoning.* San Mateo, Calif.: Morgan Kaufmann.

—— 1994: From natural language understanding to case-based reasoning and beyond: a perspective on the cognitive model that ties it all together. In R. Schank, and E. Langer (eds), *Beliefs, Reasoning, and Decision Making: Psycho-Logic in Honor of Bob Abelson,* Hillsdale, NJ: Erlbaum, 55–110.

Kolodner, J., Hmelo, C. and Narayanan, N. 1996: Problem-based learning meets case-based reasoning. In *Proceedings of the Second International Conference on the Learning Sciences,* Charlottesville, Va.: AACE Press, 188–95.

*Leake, D. (ed.) 1996: *Case-Based Reasoning: Experiences, Lessons, and Future Directions.* Menlo Park, Calif.: AAAI Press.

Redmond, M. 1992: *Learning by Observing and Understanding Expert Problem Solving* (Ph.D. thesis, College of Computing, Georgia Institute of Technology). Technical report GIT-CC-92/43.

*Riesbeck, C. and Schank, R. 1989: *Inside Case-Based Reasoning.* Hillsdale, NJ: Erlbaum.

Schank, R. 1982: *Dynamic Memory: A Theory of Learning in Computers and People.* Cambridge: Cambridge University Press.

Schank, R. and Abelson, R. 1977: *Scripts, Plans, Goals and Understanding.* Hillsdale, NJ: Erlbaum.

Schank, R., Riesbeck, C., and Kass, A. (eds) 1994: *Inside Case-Based Explanation.* Hillsdale, NJ: Erlbaum.

37

Cognitive linguistics

MICHAEL TOMASELLO

A central goal of cognitive science is to understand how human beings comprehend, produce, and acquire natural languages. Throughout the brief history of modern cognitive science, the linguistic theory that has been most prominent in this endeavor is *generative grammar* as espoused by Noam Chomsky and colleagues. Generative grammar is a theoretical approach that seeks to describe and explain natural language in terms of its mathematical form, using formal languages such as propositional logic and automata theory. The most fundamental distinction in generative grammar is therefore the formal distinction between semantics and syntax. The semantics of a linguistic proposition are the objective conditions under which it may truthfully be stated, and the syntax of that proposition is the mathematical structure of its linguistic elements and relations irrespective of their semantics.

Recently, however, a new class of linguistic theories has emerged. These theories seek to analyze natural languages not in terms of their mathematical form, but rather in terms of their psychological functions. The focus is therefore on the cognitive and social processes of which natural languages are constituted, including such psychological phenomena as perception, attention, conceptualization, meaning, symbols, categories, schemas, perspectives, discourse context, social interaction, and communicative goals. The broadest term to cover all these theories is *functional linguistics* (see Article 31, FUNCTIONAL ANALYSIS). At the most fundamental level of analysis, functional linguistics rejects the generative grammar analogy between natural and formal languages, along with its concomitant distinction between semantics and syntax. In functional linguistics natural languages, like biological organisms, are composed most fundamentally of structures with functions. Linguistic structures vary from relatively simple entities such as words and grammatical morphemes to more complex entities such as phrases and linguistic constructions. All linguistic structures have functions, and in all cases this function concerns communication, including such things as reporting an event, identifying the roles played by participants in an event, asking a question, establishing a topic of discourse, and taking a particular perspective on a scene. For functional linguists, therefore, the most fundamental distinction in natural languages is *not* between meaningful linguistic elements and their algorithmic combination irrespective of meaning (i.e., mathematical semantics and syntax), but rather between structure and function, symbol and meaning, signifier and signified.

Within functional linguistics, *cognitive linguistics* refers to the set of theories that are primarily concerned with the cognitive dimensions of linguistic communication. Although there were important precursors in the work of linguists such as Charles Fillmore and Leonard Talmy, cognitive linguistics had its clear origins as a scientific paradigm in 1987 with the publication of George Lakoff's *Women, Fire, and Dangerous*

Things: What Categories Reveal about the Mind and the first volume of Ronald Langacker's *Foundations of Cognitive Grammar* – followed immediately by the founding of the International Cognitive Linguistics Association and its official journal *Cognitive Linguistics*. The fundamental stance of cognitive linguistics may best be summarized in terms of two key issues: the nature of linguistic meaning and the nature of grammar. In the view of some cognitive scientists, the cognitive linguistics approach to these two issues constitutes a revolution in our understanding of how human language and cognition operate. In this chapter I focus on these two issues, concluding with a brief discussion of the significance of the cognitive linguistics stance for cognitive science.

The nature of linguistic meaning

According to the cognitive linguistics perspective, natural languages are composed of social conventions (linguistic symbols) that specific groups of people have created for purposes of communicating with one another. In the process of linguistic communication the speakers of a language employ particular conventions/symbols to exhort their listeners to conceptualize particular events and situations in particular ways. It is therefore misleading to say that language *depends on* cognition, as if they were two separate entities. Rather, the more accurate characterization is that natural languages are nothing more or less than ways of symbolizing cognition for purposes of communication. This cognitive linguistics view of language as one particular manifestation of human cognition is best illustrated by three phenomena: (1) the dependence of word meanings on surrounding cognitive frames, (2) the myriad ways in which a single referential situation may be linguistically construed, and (3) the ever-changing meanings for which particular linguistic symbols are used historically, including metaphorical meanings. I treat each of these in turn.

First, in many linguistic theories the semantics of a language is viewed in the manner of a dictionary. That is, speakers are seen to possess cognitively distinct mental lexicons, within which there is a list of linguistic items, each of which has a meaning that may be described independently with something like a list of semantic features. The problem with this view is that many linguistic items take their meaning from the role they play in larger *forms of life*, and thus they require a description more encyclopedic in nature. For example, the word *bachelor* – which is formalized in some semantic theories as something like *adult* + *male* + *unmarried* – does not apply easily to such unmarried adult males as Tarzan, the Pope, and eunuchs. These individuals meet the formal criteria for *bachelor*, but they are not good exemplars, because they do not participate in the cultural setting from which the word takes its meaning: a cultural setting in which men of a certain age are expected to marry (Lakoff, 1987). Other examples of words whose significance is embedded in larger cultural frames include *trump* (which requires the game of bridge), *pedestrian* (which requires traffic), *change* (which requires the cultural convention of money), and *clue* (which requires a mystery). Even though it is clearest with highly culturally bound words such as these, the same basic principle applies as well to many other words that seem initially to be more context-independent; for example, a *leaf* can only be understood in the context of a tree, and a *knuckle* can only be understood in the context of a finger (which requires a hand, and so on; Langacker, 1987). In general, the meaning of many, perhaps most, linguistic expressions can be adequately characterized only with respect to some larger conceptual domain that is not, strictly speaking, a part of its meaning, but only provides a frame for that meaning.

Second, many linguists and cognitive scientists have implicitly operated with an objectivist view of linguistic semantics. On this view, a linguistic entity *stands for* things and situations in the world, so that entity's semantics comprises those things and situations in the world for which it stands. But this view of linguistic meaning basically ignores semantic differences that depend on the different perspectives that may be taken on one and the same objective situation. Clear examples of this more subjectivist view of linguistic meaning are provided by the following alternative descriptions of single situations:

The roof slopes upward. / The roof slopes downward.
John kissed Mary. / Mary was kissed by John.
The glass is half empty. / The glass is half full.
He has a few friends in high places. / He has few friends in high places.

In each case one and the same situation is described differently, depending on the point of view the speaker wishes to communicate (Langacker, 1987). People may also use different formulations to describe a single situation at different levels of detail. For example:

This is a triangle. / This is a three-sided polygon.
This vehicle is in my way. / This blue van is blocking my way into the driveway.
Susan managed to open the door with Jim's key. / Jim's key opened the door.
Bill flew to New York. / Bill bought a ticket, drove to the airport, boarded an airplane, etc.

One and the same referential situation may also be described in different words depending on the background frame of the communicative situation. Thus, the exact same piece of real estate might be described thus:

Hiker on a hilltop: "There's the coast."
Sailor at sea: "There's the shore."
Skydiver from the air: "There's the ground."
Child on vacation: "There's the beach."

The main point of all these examples is that human languages provide their speakers with a whole battery of symbolic resources with which they may induce other people to construe a particular situation or event in a particular way. The ways in which a situation or event may be construed linguistically are myriad, depending *inter alia* on the communicative intentions of the speaker, the canonical background frame of the expression, and the knowledge the listener may be assumed to possess in the communicative interaction.

Finally, there is the fact that the meanings of particular linguistic symbols in particular languages are constantly changing as their speakers put them to new uses, including metaphorical uses. These changes of meaning are not rare events, and the use of metaphors is not a specialized, atypical use of language. Lakoff and Johnson (1980) argue and present evidence that most everyday language includes the use of linguistic items originally conventionalized for other semantic purposes. These range from fairly subtle semantic extensions, such as *running* for political office and being *in* an organization, to more obviously metaphorical extensions, such as being *out of one's mind* or being a *lost soul*. Moreover, what Lakoff and Johnson discovered was that in human linguistic communication people do not just use isolated semantic extensions and metaphors in sporadic, unsystematic ways; but rather, they often structure whole experiential domains metaphorically. For example, following the metaphor that "Time is money," people say such things as:

I *spend* too much time watching TV.
That detour *cost* me 2 hours.
The delaying tactics *bought* them more time.

But time may also be seen in terms of space:

I don't know what lies *ahead* for me.
His youth is *behind* him now.
I'll be there *at* 5:00 *on* the 11th of July.

An especially powerful discovery about the metaphorical dimensions of language is that people often use more concrete domains of knowledge to structure and comprehend more abstract ones. This is manifest in people's frequent use of terms for very basic aspects of experience, such as bodily actions and simple perceptual transformations of objects, to structure more abstract domains. For example, we understand the English expressions *in* and *out* most fundamentally for such things as putting objects into and taking them out of containers; but we also put arguments in and take arguments out of our speeches. We use *off* and *on* most basically for putting clothes on and taking them off our bodies or putting objects on and taking them off tables; but we also say that a tennis player is on her game or off her game. Lakoff and Johnson's claim is that there are certain fundamental domains of human experience – constituted by what they call *image schemas* – that serve as prototypes of some very general referential situations, and thus as especially powerful source domains for metaphorical construals (Johnson, 1987). Overall, it may be said that semantic extensions and metaphorical construals pervade human language use, and their existence demonstrates that linguistic meaning is part and parcel of a process in which people continually adapt their existing means of linguistic expression for particular communicative goals.

The most general point to be made from all three of these considerations is that it is basically impossible to isolate linguistic meaning from cognition in general in the manner of a mental lexicon divorced from other aspects of human cognition and communication. Cognitive linguistics therefore adopts an encyclopedic, subjectivist approach to linguistic meaning in which human beings create and use linguistic conventions in order to symbolize their shared experiences in various ways for specific communicative purposes. These different experiences and purposes are always changing; so they can never be captured by an itemized, objectivist description of linguistic elements and their associated truth conditions. For an adequate description of linguistic semantics from the cognitive linguistics point of view, what is needed is a psychology of language in terms of such things as cognitive structures, the manipulation of attention, alternative construals of situations, and changing communicative goals.

The nature of grammar

In the cognitive linguistics view, the grammar of a language is best characterized as "a structured inventory of symbolic units" each with its own structure and function (Langacker, 1987). These units may vary in both their complexity and generality, with words being only one type of symbolic unit. At the simplest level of analysis, all the structures of a language are composed of some combination of four types of symbolic element: words, markers on words (e.g., the English plural *-s*), word order, and intonation (Bates and MacWhinney, 1989). Each of the several thousand languages of the world uses these four elements, but in different ways. In English, for example, word order is most typically used for the basic syntactic function of indicating who did what to whom; intonation is used mainly to highlight or background certain information in the utterance; and markers on words serve to indicate such things as tense and plurality. In Russian, on the other hand, who did what to whom is indicated by case-markers on words, and word order is used mostly for highlighting and backgrounding information. In some tone languages (e.g., Masai), who did what to whom is indicated via special intonational contours. In general, any of these four symbolic devices may serve virtually any semantic or pragmatic function in a particular language. Moreover, these structure–function relationships may change over time within a language, as in the English change from case marking to word order for indicating who did what to whom several hundreds of years ago.

These four types of symbolic elements do not occur in isolation, but in each language there are a variety of linguistic constructions composed of unique configurations of these elements (Goldberg, 1995). Linguistic constructions are basically cognitive schemas of the same type that exist in other domains of cognition. These schemas/constructions may vary from specific to general. For example, the one word utterance "Fore!" is a very simple, concrete construction used for a specific function in the game of golf. "Thank you" and "Don't mention it" are multi-word constructions used for relatively specific social functions. Some other constructions are composed of specific words along with *slots* into which whole classes of items may fit: for example, "Down with ____ !" and "Hooray for ____ !" Two other constructions of this type that have more general applications are:

the *way* construction:	She made her way through the crowd. I paid my way through college. He smiled his way into the meeting.
the *let alone* construction:	I wouldn't go to New York, let alone Boston. I'm too tired to get up, let alone go running around with you. I wouldn't read an article about, let alone a book written by, that swine.

Each of these constructions is defined by its use of certain specific words (*way*, *let alone*), and each thus conveys a certain relatively specific relational meaning, but is also general in its application to different specific content (Fillmore et al., 1988).

There are also constructions that are extremely general in the sense that they are not defined by any words in particular, but rather by categories of words and their relations. Thus, the ditransitive construction in English prototypically indicates transfer of possession and is represented by utterances such as "He gave the doctor money." No particular words are a part of this construction; it is characterized totally schematically by means of certain categories of words in a particular order: noun-phrase + verb + noun-phrase + noun-phrase. No construction is fully general, however; so in the ditransitive construction the verb must involve at the very least some form of motion (as in "He threw Susan money," but not *"He stayed Susan money"). Other examples of very general English constructions are the various resultative constructions (e.g., "She knocked him silly," "He cleaned the table off"), constituted by a particular ordering of particular categories of words, and the various passive constructions (e.g., "She is loved by Harry," "She got kissed"), which provide a unique perspective on scenes and are constituted by a particular ordering of word categories as well as some specific words (e.g., *by*) and markers (e.g., *-ed*). All these more general constructions are defined by general categories of words and their interrelations; so each may be applied quite widely for many referential situations of a certain type. These abstract linguistic constructions may be thought of as cognitive schemas of the same type found in other cognitive skills: that is, as relatively automatized procedures that operate on a categorical level.

An important point is that each of these abstract linguistic schemas has a meaning of its own, in relative independence of the lexical items involved (Goldberg, 1995). Indeed, much of the creativity of language comes from fitting specific words into linguistic constructions that are nonprototypical for that word. For example, the verb *kick* is not typically used for transfer of possession, and so it is not prototypically used with the ditransitive construction. But it may be construed in that way in utterances such as "Mary kicked John the football," because kicking can be seen as imparting directed motion to an object with another person as terminus. This process may extend even further to such things as "Mary sneezed John the football," which requires an imaginative interpretation in which the verb *sneeze* is not used in its more typical intransitive sense (as in "Mary sneezed"), but rather as a verb in which the sneezing causes directed motion in the football. If the process is extended far enough, to verbs for which it is difficult to imagine directed motion, the process begins to break down – as in "Mary smiled John the football." The important point is that in all these examples the transfer of possession meaning (that the football goes from Mary to John) comes from the construction itself, not from the specific words of which it is constituted. Linguistic constructions are thus an important part of the inventory of symbolic resources that language-users control, and they create an important *top-down* component of the process of linguistic communication – in keeping with the role of abstract schemas in many other domains of human cognition.

All constructions, whether composed of one word or many categories of words in specific orders with specific markers and intonations, derive from recurrent events, or types of events, with respect to which the people of a culture have recurrent communicative goals. This means that a major function of all linguistic constructions is attentional – for instance, to take one or another point of view on a situation or event – with particular participants either in the foreground or the background (the other major class of functions being the expression of the speaker's speech-act goal, as in asking a question). For example, the same event may be depicted as:

Fred broke the window with a rock.
Fred broke the window.
The rock broke the window.
The window got broken.
It was Fred who broke the window.
It was the window that Fred broke.
What Fred did was break the window.

In each of these construals of the event, the perspective is slightly different, and Fred's and the rock's roles in the process are made attentionally salient to different degrees (Croft, 1991), with each construal being most felicitously used for a particular communicative purpose in a particular discourse circumstance. Many of the more abstract and complex constructions of a language are created for precisely these types of attentional functions.

Different languages are constituted by different specific symbols and constructions, of course. In some cases these differences have become relatively conventionalized across different linguistic structures within a language, so that we may speak of different *types* of languages with regard to how they symbolize certain recurrent events or states of affairs. An important area of research in cognitive linguistics, therefore, concerns the different resources that different languages provide for symbolizing certain universal events and situations (van Valin and LaPolla, 1997). For example, almost all people speaking almost all languages have general constructions for talking about someone causing something to happen, someone experiencing something, someone giving someone something, an object moving along a path, and an object changing state. As one instance, there are two very common ways in which languages depict motion events, as characterized by Talmy (1988):

English: The bottle floated into the cave.
Spanish: La botella entró la cueva flotando ("The bottle entered the cave floating").

In English the path of the bottle is expressed by the preposition *into*, and the manner of motion is expressed by the verb *float*; whereas in Spanish the path is expressed by the verb *entrada*, and the manner of motion is expressed by the modifier *flotando*. Because this difference is pervasive and consistent in the two languages, we may say that in depicting motion events Spanish is a verb-framed language (because the path of motion is typically expressed by the verb), whereas English is a satellite-framed language (because the path of motion is typically expressed by the preposition). There are other typological differences among languages as well.

The cognitive bases of linguistic constructions have been most thoroughly investigated by Langacker (1987, 1991). Most importantly, Langacker has provided an account of the different cognitive operations that characterize the two categories of word that form the heart of the most general constructions in most of the world's languages: verbs and nouns. Verbs form the relational backbone of linguistic expressions and have to do with processes that unfold over time or else states that remain stable over some period of time. Thus, to be able to say that something has moved or changed, there must have been at least two *moments of attention*: one in which an entity was in one location or state and another in which it was in another location or state. For example, we cannot make the judgment that "She crossed the river" on the basis of a single snapshot of a woman at any location in or near a river, but rather we must have something like a first snapshot in which she is at a location on one bank of the river, a temporally subsequent snapshot in which she is in the river, and another in which she is on the opposite riverbank. We can also say "She has crossed the river" as a description of a completed event, in which case the woman is simply standing on one bank. But implicit in this expression is some previous process by means of which she traversed the width of the river. Finally, we can also say "She is across the river" for this same situation (woman standing on one bank), but in this case there is no implication that a process of crossing ever occurred. Note that the description of states, as in "She remains across the river," also requires at least two moments of attention in which the woman stays in the same location on the other side of the river (in one snapshot she might be engaged in initiating an activity). Interestingly, most languages allow their speakers to use some nouns as verbs in certain situations, in which case some kind of process interpretation is required, as in "brush with a brush" and "hammer with a hammer," "dock the boat," and "table the motion" (typically an action closely associated with the object).

Nouns are words used to indicate the participants in events or situations. Most prototypically these are spatially bounded entities such as people or trees or bicycles. But nouns may also be used to designate temporally bounded entities such as flashes and blips, and even more abstractly bounded entities such as Tuesdays or corporations or virtues. For Langacker, the key cognitive operation involved is the *bounding* of a portion of experience so as to create a *thing* as distinct from its surroundings. This process is conceptual in nature; so it can be applied to literally any experience, as illustrated by the fact that nouns may be used to talk about what are clearly events in nature (e.g., the parade, the party). Indeed, in most languages there are processes by means of which a verb form like *to swim* may be turned into a noun like *swimming* if it is thought of as a participant in an event or state of affairs, as in "This swimming strengthens my leg muscles." The bounding process that creates nouns thus reflects not the independent structure of the world, but rather the fact that an important communicative function in linguistic communication is the identification of *things* to be talked about.

This view of linguistic communication and the cognitive processes on which it depends is obviously very different from that of generative grammar and other formalistic approaches. But cognitive linguistics can nevertheless account for all the major phenomena of generative grammar. For example, on the generative grammar view, natural language structures may be used creatively, because speakers possess a syntax divorced from semantics. On the cognitive linguistics view, on the other hand, linguistic creativity results quite simply from the fact that speakers have formed highly general linguistic constructions composed of word categories and abstract schemas that operate on the categorical level. That linguistic categories and schemas are formed in the same basic way as other categories and schemas is evidenced by the fact that they show the same kinds of prototypicality effects and metaphorical extensions as other categories and schemas (Lakoff, 1987; Taylor, 1996). Also, generative grammar analyses depend crucially on hierarchically organized tree structures that are seen as unique to language (see Article 35, STRUCTURAL ANALYSIS). But cognitive linguists see this hierarchical structure as a straightforward result of the *chunking* processes characteristic of skill formation in many other cognitive domains. What is unique to language is the way this process is guided by communicative function, so that, for example, a noun phrase as a coherent unit of multiple words gains its coherence by virtue of the fact that the referential event being spoken about must be *grounded* in the actual interactional event of speaking – so that in attempting to indicate a particular object for a listener in a particular communicative context a speaker may say most appropriately either "the bike in the garage," "that bike," or even "it," in all cases fulfilling the same communicative function of helping the listener to identify the referent (Langacker, 1991). Finally, the traditional syntactic function of subject of a sentence receives a cognitive-functional treatment in cognitive linguistics in terms of the participant in an event or state of affairs on which the speaker is focused or that the speaker chooses to make most prominent attentionally for the listener (Tomlin, 1995). For example, we may say either "The tree is next to the house" or "The house is next to the tree," the only difference being which entity, the tree or the house, is taken as the attentional reference point.

The overall point of most interest and importance to cognitive science is that in the cognitive linguistics framework, grammatical skills may be explained in fundamentally the same terms as other complex cognitive skills; there are no hidden principles, parameters, constraints, or deep structures that operate in language but not in other cognitive domains. A language consists of nothing other than its inventory of symbolic structures and constructions, each adapted to a delimited set of communicative functions. There are quite concrete linguistic structures that speakers conventionally use to induce others to construe or attend to a situation in a particular way (specific words and combinations of words), and there are also more abstract and schematic constructions that are composed of combinations of linguistic categories, and so indicate relatively abstract meanings. Although the way in which cognition is manifest in language may have some of its own peculiarities (just as the way in which cognition is manifest in other cognitive activities displays its own peculiarities), in general it is accurate to say that the structures and functions of language are taken directly from human cognition and social interaction, and so linguistic communication, including its grammatical structure, should be studied in the same basic manner, using the same basic theoretical constructs, as all other cognitive skills.

Significance

To conclude, two revolutionary implications of cognitive linguistics for cognitive science should be highlighted. The first concerns the ease with which cognitive linguistics captures the dynamic nature of linguistic communication relative to more formalistic theories (which are always static, due to their essentialistic categories and rules). For example, cognitive linguistics may be used to provide a biologically realistic account of how language arose in human evolution from processes of nonlinguistic communication, as human beings acquired the capacity to use intersubjective symbols where before they had used only ritualized signals. Similarly, cognitive linguistics may also be used to analyze language change in human history, including the processes of grammaticalization by means of which different societies have created typologically different forms of language for their own collective purposes. Finally, cognitive linguistics may also provide for a psychologically realistic account of how human children acquire the specific structures and constructions of their native languages as they draw on more general skills of event cognition, categorization, joint attention, and cultural learning (Tomasello, 1992). All three of these dynamic accounts are made possible by the fact that cognitive linguistics explicitly recognizes that speakers adapt their linguistic structures over time to changes in their cognitive resources and communicative goals.

The second revolutionary implication concerns the nature of cognitive representation in general (Mandler, 1997). Formalistic views of human cognitive functioning and representation (e.g., those represented by generative grammar and other *language of thought* positions) suffer from such fundamental defects as the symbol-grounding problem: mathematical algorithms and computer programs manipulate what to humans are symbols, but they do not ground those symbols meaningfully in experience (suggesting that they are symbols only for the creators and users of the programs). Cognitive linguistics provides an alternative to this view that has no grounding problems. It has no grounding problems because natural languages are viewed as nothing other than sets of social conventions by means of which human beings communicate with one another about their experience. Human linguistic competence is thus composed of the same basic elements as many other cognitive skills, but it has unique properties as well, which arise as those elements are used in the process of communication. This view of human cognition and linguistic communication lends itself readily to all kinds of empirical investigations of the type common in cognitive science.

Overall, cognitive linguistics is an approach to language and cognition that should be very congenial to cognitive scientists, since it investigates the cognitive operations by means of which human beings use language to manipulate the attention of other human beings with respect to events and situations about which they wish to communicate. Importantly, in most cases it tries to do this while taking into account research on human cognition in other areas of cognitive science (e.g., ATTENTION and CONCEPTUAL ORGANIZATION). Cognitive linguists typically pursue their investigations by looking at how people use and understand particular types of linguistic expressions in particular types of naturalistic situations, a method not especially familiar to mainstream cognitive scientists. But the goals of cognitive linguistics may also be pursued by means of traditional psycholinguistic methods such as learning experiments, error analysis, and reaction time studies, and indeed the use of multiple methods toward the same scientific goal would seem to be a desirable direction for future research.

References and recommended reading

Bates, E., and MacWhinney, B. (eds) 1989: *The Cross-Linguistic Study of Sentence Processing.* New York: Cambridge University Press.

Croft, W. 1991: *Syntactic Categories and Grammatical Relations: The Cognitive Organization of Information.* Chicago: University of Chicago Press.

Fillmore, C. J., Kay, P. and O'Conner M. C. 1988: Regularity and idiomaticity in grammatical constructions: the case of *let alone. Language,* 64, 501–38.

*Goldberg, A. 1995: *Constructions: A Construction Grammar Approach to Argument Structure.* Chicago: University of Chicago Press.

Johnson, M. 1987: *The Body in the Mind: The Bodily Basis of Meaning, Imagination, and Reason.* Chicago: University of Chicago Press.

*Lakoff, G. 1987: *Women, Fire, and Dangerous Things: What Categories Reveal about the Mind.* Chicago: University of Chicago Press.

Lakoff, G. and Johnson, M. 1980: *Metaphors We Live By*. Chicago: University of Chicago Press.

Langacker, R. 1987: *Foundations of Cognitive Grammar*, vol. 1. Stanford, Calif.: Stanford University Press.

—— 1991: *Foundations of Cognitive Grammar*, vol. 2. Stanford, Calif.: Stanford University Press.

Mandler, J. 1997: Representation. In D. Kuhn and R. Siegler (eds), *Handbook of Child Psychology*, vol. 2: *Cognition, Perception, and Language*, New York: Wiley.

Talmy, L. 1988: The relation of grammar to cognition. In B. Rudzka-Ostyn (ed.), *Topics in Cognitive Linguistics*, Amsterdam: John Benjamins.

Taylor, J. 1996: *Linguistic Categorization*, 2nd edn. New York: Oxford University Press.

Tomasello, M. 1992: *First Verbs: A Case Study of Early Grammatical Development*. Cambridge: Cambridge University Press.

Tomlin, R. 1995: Focal attention, voice, and word order. In P. Downing and M. Noonan (eds), *Word Order in Discourse*, Amsterdam: John Benjamins.

*van Valin, R. and LaPolla, R. 1997: *Syntax: Structure, Meaning, and Function*. Cambridge: Cambridge University Press.

38

Connectionism, artificial life, and dynamical systems

JEFFREY L. ELMAN

Introduction

Periodically in science there arrive on the scene what appear to be dramatically new theoretical frameworks (what the philosopher of science Thomas Kuhn has called *paradigm shifts*). Characteristic of such changes in perspective is the recasting of old problems in new terms. By altering the conceptual vocabulary we use to think about problems, we may discover solutions which were obscured by prior ways of thinking about things. Connectionism, artificial life, and dynamical systems are all approaches to cognition which are relatively new and have been claimed to represent such paradigm shifts. Just how deep the shifts are remains an open question, but each of these approaches certainly seems to offer novel ways of dealing with basic questions in cognitive science.

While there are significant differences among these three approaches and some complementarity, they also have a great deal in common, and there are many researchers who work simultaneously in all three. The goal of this chapter will be, first, to trace the historical roots of these approaches, in order to understand what motivates them; and second, to consider the ways in which they may either offer novel solutions to old problems or even redefine what the problems in cognition are.

Historical context

When researchers first began to think about cognition in computational terms, it seemed natural and reasonable to use the digital computer as a framework for understanding cognition. This led to cognitive models which had a number of characteristics which were shared with digital computers: processing was carried out by discrete operations executed in serial order; the memory component was distinct from the processor; and processor operations could be described in terms of rules of the sort found in programming languages.

These assumptions underlay almost all of the most important cognitive theories and frameworks up through the 1970s, as well as a large number of contemporary approaches. These include the physical symbol system hypothesis of Alan Newell and Herbert Simon, the human information processing approach popularized by Peter Lindsay's and Donald Norman's text of the same name, and the generative linguistics theory developed by Noam Chomsky.

The metaphor of the brain as a digital computer was enormously important in these theories. Among other things, as the cognitive psychologist Ulric Neisser has pointed out, the computer helped to rationalize the study of cognition itself, by demonstrating that it was possible to study cognition in an explicit, formal manner (as opposed to the claim of the behavioral psychologists of the earlier era, who argued that internal processes of thought were not proper objects of study).

But as research within this framework progressed, the advances also revealed shortcomings. By the late 1970s, a number of people interested in human cognition began to take a closer look at some of the basic assumptions of the current theories. In particular, some people began to worry that the differences between brains and digital computers might be more important than hitherto recognized. In 1981, Geoffrey Hinton and James Anderson put together a collection of papers (*Parallel Models of Associative Memory*) which presented an alternative computational framework for understanding cognitive processes. This collection marked a sort of watershed. Brain-style approaches were hardly new. Psychologists such as Donald Hebb, Frank Rosenblatt, and Oliver Selfridge in the late 1940s and 1950s, mathematicians such as Jack Cowan in the 1960s, and computer scientists such as Teuvo Kohonen in the 1970s (to name but a small number of influential researchers) had made important advances in brain-style computation. But it was not until the early 1980s that connectionist approaches made significant forays into mainstream cognitive psychology. The word-perception model of James McClelland and David Rumelhart published in 1981, had a dramatic impact; not only did it present a compelling and comprehensive account of a large body of empirical data, but it laid out a conceptual framework for thinking about a number of problems which had seemed not to find ready explanation in the human information processing approach. In 1986, the publication of a two-volume collection, by Rumelhart and McClelland and the PDP Research Group, called *Parallel Distributed Processing: Explorations in the Microstructure of Cognition*, served to consolidate and flesh out many details of the new approach (variously called *PDP, neural networks*, or *connectionism*). Before discussing the sorts of issues which motivated this approach, let us briefly define what we mean by connectionism.

What is connectionism?

The class of models which fall under the connectionist umbrella is large and diverse. But almost all models share certain characteristics.

Processing is carried out by a (usually large) number of (usually very simple) processing elements. These elements, called *nodes* or *units*, have a dynamics which is roughly analogous to simple neurons. Each node receives input (which may be excitatory or inhibitory) from some number of other nodes, responds to that input according to a simple activation function, and in turn excites or inhibits other nodes to which it is connected. Details vary across models, but most adhere to this general scheme. One connectionist network is shown in figure 38.1. In this network, the task is to take visual input and recognize words – in other words, to read.

There are several key characteristics worth noting. First, the response (or activation) function of the units is often nonlinear, and this nonlinearity has very important consequences for processing. Among other things, the nonlinearity allows the systems to respond under certain circumstances in a discrete, binary-like manner, yielding crisp categorical behavior. In other circumstances, the system is capable of graded, continuous responses. Second, what the system *knows* is, to a large extent, captured by

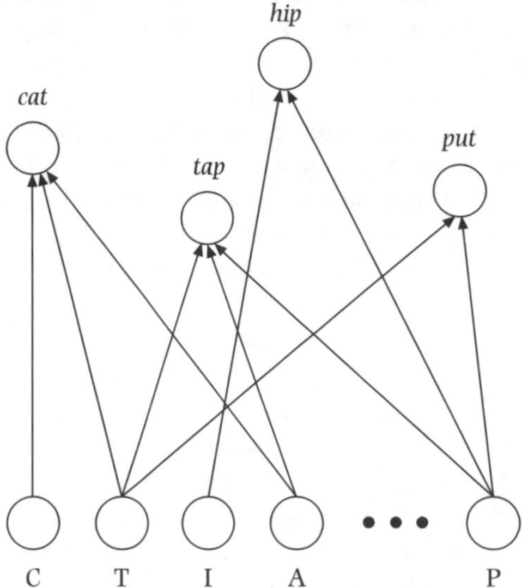

Figure 38.1 A simplified neural network for reading. The bottom layer of units consists of letter detectors, and the top layer of word detectors (not all letters or words are shown). "Excitatory" connections (lines with arrows) connect letters to the words containing them; "inhibitory" connections (lines with filled circles) between words cause them to compete (since only one word can be present at a time). More realistic networks take into account the fact that the same letter may appear in different positions in the word; they also provide for other types of information (sound, meaning) to influence reading.

the pattern of connections – who talks to whom – as well as the weights associated with each connection (weights serve as multipliers). Third, rather than using symbolic representations, the vocabulary of connectionist systems consists of patterns of activation across different units. For example, to present a word as a stimulus to a network, we would represent it as a pattern of activations across a set of input units. The exact choice of representation might vary dramatically; at one extreme, a word could be represented by a single, specialized input unit (thus acting very much like an atomic symbol); at the other extreme, the entire ensemble of input units might participate in the representation, with different words having different patterns of activation across a shared set of units.

Given the importance of the weighted connections in these models, a key question is: Who determines the weights? Put in more traditional terms: Who programs the networks? In early models, the connectivity was done by hand (and this remains the case for what are sometimes called *structured connectionist models*). However, of the exciting developments which made connectionism so attractive to many was the development of learning algorithms by which the networks could be *programmed* on their own. In other words, the networks could themselves learn the values for the weights. Moreover, the style of learning was inductive: Networks would be exposed to examples of a target behavior (e.g., the appropriate responses to a set of varied stimuli). Through learning, the network would adjust the weights in small incremental steps in such a way that, over time, the network's response accuracy would improve. Ideally, the network would also be able to generalize its performance to novel stimuli, thereby demonstrating that it had learned the underlying function which related outputs to inputs (as opposed to merely memorizing the training examples).

(It should be noted that the type of learning described above – so-called supervised learning – is but one of a number of different types of learning that are possible in connectionist networks. Other learning procedures do not involve any prior notion of *correct behavior* at all. The network might learn instead, for example, the correlational structure underlying a set of patterns.)

Now let us turn to some of the interesting properties of these networks, as well as the controversies which surround them.

Issues and controversies

Context and top-down processing

Early models of human information processing typically utilized a kind of *template-matching* approach in many tasks, especially those for which pattern recognition played an important role. With the advent of the computer metaphor, this approach gave way to the use of rules for analysis. A central feature of such models was the notion of a strict flow of information processing, which proceeded in what was called a *bottom-up* manner (e.g., perceptual features of a stimulus were processed first, yielding a representation which was then passed on to successively higher stages of processing). These assumptions were challenged by a number of findings. For example, it was found that subjects' ability to perceive a single letter (presented very rapidly on a computer screen) was influenced by the context in which it occurred. A letter could be identified better if it appeared in a real word, rather than appearing in isolation or embedded in a nonword letter string. In the domain of speech, it was discovered that the perception of ambiguous sounds was also affected by their context. Thus a sound which was perceptually midway between a *k* and a *g* would be heard by listeners alternatively as a *k* if what followed was *-iss* or as a *g* if followed by *-ift*. This seemed to be an example of top-down processing (since it was assumed that word knowledge came later in processing – so was *higher* – than perceptual processing, which was *lower*). In other types of experiments (e.g., in which subjects listened to sentences containing potentially ambiguous words such as *bank*) it appeared that prior context often plays a powerful role in biasing comprehension. This was called a *context effect*.

These results suggested that so-called higher processing (in the above examples, knowledge of the lexicon or contextually established information) might influence supposedly lower processes (such as visual or auditory perception). Indeed, a burgeoning experimental literature soon suggested that the degree of context effects and top-down influences might be very extensive. Furthermore, there was growing evidence that the human cognitive system was able to process at multiple levels in parallel, rather than being restricted (as was the digital computer) to executing a single instruction at a time.

The word-perception model of McClelland and Rumelhart (1981) was one of the first attempts to account for these sorts of data in a model which was frankly parallel, highly interactive, and departed significantly from the old-style digital framework. McClelland and Rumelhart called this the *interactive activation model*.

How to account for regularity? Rules or associations?

One of the basic challenges in cognitive science is how to account for behavior. If human behavior were either entirely random or limited to a fixed repertoire of actions which could be memorized, then there would be little to explain. What makes the problem so interesting is that behavior is patterned, and is often productive (we generalize these patterns to novel circumstances).

A natural way to account for the patterned nature of cognition is to assume that underlying these behaviors is a set of rules. Rules provide a compact and elegant way to account for the abstract, productive nature of behavior. Rules also offer a way to capture system-level properties: that is, there exists the possibility that rules can interact in complex ways. The problem that can arise, however, is that some human behaviors are often only partially general and productive. A good example of this (and one which has been well studied and the topic of considerable debate) is the formation of the past tense in English verbs.

The majority of English verbs form the past by adding the suffix *-ed* to the verb stem (e.g., *walk* + *ed, plant* + *ed,* etc.). This pattern might be captured by positing a rule for the regular past-tense formation. At the same time, there are also other verbs whose past-tense forms seem idiosyncratic: *go* → *went, sing* → *sang, hit* → *hit.* One way of dealing with such apparently irregular forms is to suppose that they are simply exceptions which must be memorized by rote and *listed* in some mental dictionary.

One strong piece of evidence in favor of this account was the observation that many children seem to go through several phases when they learn the past tense. Initially, at the stage where they know only a small number of verbs, some children begin by producing past-tense forms correctly for both regular and irregular verbs. Later, these children begin to make mistakes on the irregular verbs, treating them as if they were regular (*go* → *goed*). Ultimately, these children learn which verbs should be treated as regular – obey the rule – and which as irregular – must be memorized. Thus, performance has a kind of U-shape to it, starting off relatively good, getting worse, and finally becoming good again. The rule-based account explains this phenomenon by supposing that in early development children are simply memorizing all verbs. At some point, they discover the regularity in past-tense formation and start using the rule – only they have not yet learned that the rule is not fully applicable to all verbs and overgeneralize to irregular ones. The final stage is achieved when they learn which verbs are regular and which are irregular.

A difficulty with this account is that although some irregular verbs appear to be truly exceptional, in the sense that they are unique (e.g., *is* → *was, go* → *went*), others clump together in groups (e.g., *sing* → *sang, ring* → *rang*; or *catch* → *caught, teach* → *taught*; or *hit* → *hit, cut* → *cut*). These groups can not only be defined in terms of phonological similarity, but there is evidence that if confronted with a novel word which closely resembles one of the groups and asked to produce a past-tense form, native speakers will sometimes produce an irregular one: *pling* → *plang* (presumably, on analogy with *ring* → *rang*).

In 1986, David Rumelhart and James McClelland published the results of a connectionist simulation in which they trained a network to produce the past-tense forms of English present-tense verbs. Learning involved gradually changing the weights in the network so that the network's overall performance improved. Rumelhart and

McClelland reported that the network was not only able to master the task (though not perfectly), but that it also exhibited the same U-shaped performance found in children. They suggested that this demonstrated that language performance which could be described as rule-governed might in fact not arise from explicit rules.

This paper generated considerable controversy, which continues to the present time. Steven Pinker and Alan Prince (1988) wrote a detailed and highly critical response, in which they questioned many of the methodological assumptions made by Rumelhart and McClelland in their simulation and challenged their conclusions. In fact, many of Pinker and Prince's criticisms are probably correct and were addressed in subsequent connectionist models of past-tense formation.

A great many subsequent simulations have been carried out which correct problems in the Rumelhart and McClelland model. These simulations have in turn generated an ongoing debate about new issues. In recent writings, Pinker and Prince have suggested that although a connectionist-like system might be responsible for producing the irregulars, there are qualitative differences in the way in which regular morphology is processed which can only be explained in terms of rules. This has become known as the *dual mechanism* account. Proponents of a single mechanism approach argue that a network can in fact produce the full range of behaviors which characterize regular and irregular verbs.

Recursion and compositionality

Linguists have long noted that human languages have a curious property. Consider the grouping of words that is called a noun-phrase. Noun-phrases are things such as "John," "the old man," "the curious yellow cat," or "the mouse under the chair." Notice in this last example, "the chair" is itself a noun-phrase. Thus, noun-phrases may contain other noun-phrases; in fact, there is no principled limit to the degree of such self-embedding (e.g., "the mouse under the chair in the house that Jack built"). Such self-embedding is called *recursion* and refers to the possibility that a category may be defined in terms of itself, even if indirectly.

(Another way of thinking about recursion is in terms of circular definitions. We might define the category noun-phrase as comprising many things: e.g., *Adj N, det N, N PP*. This means that possible noun-phrases might be an adjective followed by a noun ("old car"), or a determiner followed by a noun ("the table"), or a noun followed by a prepositional phrase ("book on the table"). Suppose prepositional phrases, in turn, are defined as made up of a preposition ("on") followed by a noun-phrase ("the table"). This circular definition gives us recursion, since the definition of a noun-phrase allows the possibility that it might be made up of other things, including noun-phrases! This may seem strange, but it actually makes a lot of sense in language, because it explains why we can not only say things like "book on the table" but "book on the table in the room," or "book on the table in the room in the house," etc. More importantly, for many purposes grammatical rules treat all these things as the same kind of entity – which they are: noun-phrases.)

The notion of *compositionality* is closely related, and refers to the structural relationship between different elements which can arise when recursion occurs (as well as in other circumstances). Thus, one might say that the noun-phrase "spider on the chair" is composed of a noun which is modified by a prepositional phrase, which in turn is composed of a preposition followed by a noun phrase, etc. The tree diagram in figure 38.2

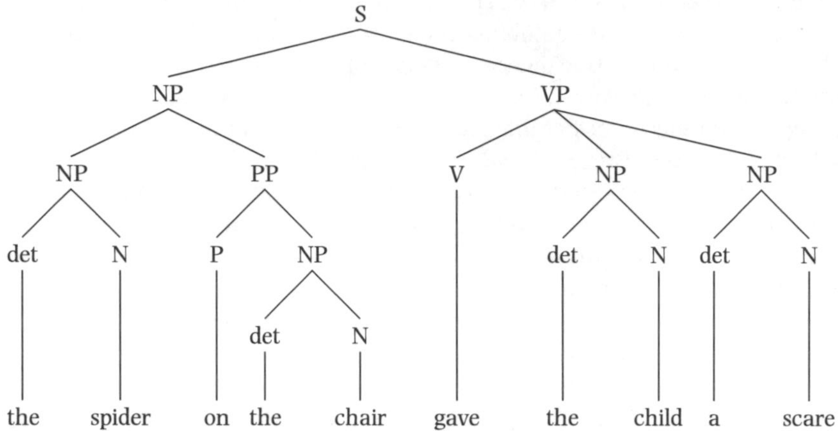

Figure 38.2 A "phrase structure" tree diagram representing the assumed grammatical structure of the sentence "The spider on the chair gave the child a scare." Lower elements in the tree are said to be "constituents" (or parts) of higher elements. There is considerable debate in the linguistic literature about the correct form for such tree structure representations (this is a very simplified version), and many modern theories have alternative ways of capturing constituent (part–whole) relationships.

is a typical notation used by linguists to capture such structural relationships (see Article 35, STRUCTURAL ANALYSIS).

The Rumelhart and McClelland verb-learning simulation discussed above dealt with issues in morphology (e.g., verb inflections), but soon other connectionist simulations were developed which modeled syntactic and semantic phenomena. All of those simulations, however, involved sentences of pre-specified (and limited) complexity. In 1988, Jerry Fodor and Zenon Pylyshyn wrote a paper in which they called attention to this shortcoming and argued that the deficiency was not accidental. They claimed that connectionist models were in principle unable to deal with unbounded recursion or represent complex structural relationships (constituent structure) in an open-ended manner. Fodor and Pylyshyn argued that since these phenomena are hallmarks of human cognition, connectionism was doomed and that only what they termed *classical* approaches (e.g., those based on the digital computer metaphor of the mind) would work.

The issues raised by Fodor and Pylyshyn have generated a large body of responses from connectionists, as well as further criticisms by proponents of the classical approach. Paul Smolensky provided one possible solution to Fodor and Pylyshyn's arguments (see Article 48, LEVELS OF EXPLANATION AND COGNITIVE ARCHITECTURES and Article 52, RULES). Another response involves using what are called *recurrent networks* (one version of a recurrent network is shown in figure 38.3).

In a recurrent network, the internal (or *hidden*) units feed back on themselves. This provides the network with a kind of memory. The form of the memory is not like a tape recorder, however; it does not literally record prior inputs. Instead, the network has to itself learn how to encode the inputs in the internal state, such that when the state is fed back it will provide the necessary information to carry out whatever task is being learned.

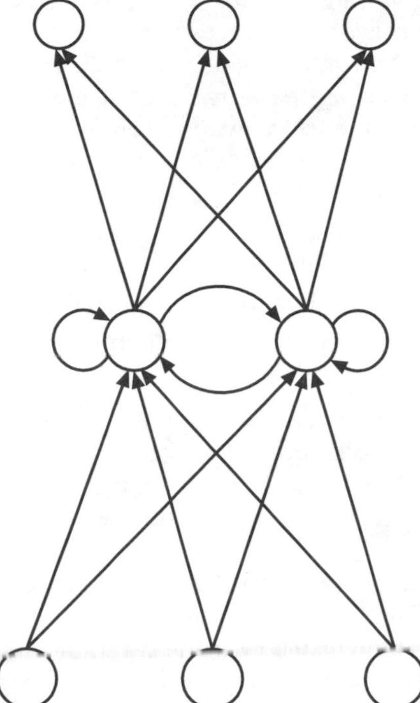

Figure 38.3 A recurrent network. The flow of information (activation) goes from the bottom of the network to the top, as represented by the direction of arrows connecting units. The middle units, however, also have recurrent connections, so that those units activate themselves in addition. (Recurrent connections are shown in bold.)

In 1991, David Servan-Schreiber, Axel Cleeremans, and Jay McClelland demonstrated that such a network could be trained on an artificial language generated by something called a finite state automaton (FSA; an FSA is the simplest possible digital computer; the most powerful type of computer is a Turing machine). The recurrent network's task is simply to listen to each incoming symbol, one at a time, and to predict what will come next. In the course of doing this, the network infers the more abstract structure of the underlying artificial language.

The FSA language, however, lacked the hierarchical structure shown in figure 38.2. Using simple recurrent networks that he introduced in 1990, Jeff Elman (1991) reported a simulation that could also process sentences which contained relative clauses – which involve hierarchical/compositional relations among different sentence elements. Later, Jill Weckerly and Elman reported that these networks show the same performance asymmetry in processing different types of embedded sentences as do humans. Their networks, like people, find sentences such as (1) more difficult than sentences such as (2).

(1) The mouse that the cat that the dog scared chased ran away.
(2) Do you believe the report that the stuff they put in coke causes cancer is true?

(Both sentences are *hard* in the sense that they have complicated structures and are somewhat unnatural. However, both are grammatically legitimate. The question is why most people find the first sentence much harder to understand than the second. It can't be just their structure, since they have nearly identical grammatical structures.)

Understanding exactly how such networks operate is an interesting problem in its own right, and there are strong connections between their solutions and the dynamical systems we discuss below.

Other phenomena – and shortcomings

Since the early 1980s, the connectionist paradigm has grown dramatically, and there are now models which attempt to answer questions in a wide range of areas. A few of these are:

- brain damage – if one *lesions* networks, does this have similar effects to those observed when humans suffer brain damage?
- reading – can networks be taught to read languages such as English, in which the mapping from letter to sound is not straightforwardly captured by a simple set of rules?
- development – can networks model the developmental process which occurs as children grow into adults?
- pattern completion – humans show an uncanny ability to fill in or reconstruct information which is missing in many patterns (such as partially obscured faces); do networks have similar capabilities?
- philosophy – do connectionist models offer new ways of understanding philosophical concepts such as representation, information, belief, etc.?

Although the explosive interest in connectionism has resulted in what seem like genuinely new ways of dealing with long-standing problems in cognition, there are a number of problems for which connectionist models seem not to offer any direct solution. This is not to say that connectionism is wrong, simply that even if it is the right approach, it is not the whole story. Two other recent approaches may help in this regard: artificial life and dynamical systems. We turn first to artificial life.

Artificial life

Connectionist models are powerful induction engines. That is, they learn by example and use the statistics of those examples to drive learning. The attraction of the approach is that although learning is statistically driven, the outcome of the learning process is a system whose knowledge is generalizable to novel instances.

But there are several respects in which connectionist models seem deficient. Furthermore, these deficiencies are similar to those found in almost all artificial intelligence models. Here are three examples of the most striking shortcomings.

Disembodied intelligence An enormous amount of the cognitive behavior of biological organisms is tightly coupled to the bodies in which the behavior is manifest. The way we think about space (for example) is highly dependent on properties of our visual system. The way we think about ourselves and the world depends on how we experience it, and our experience is vastly different from that of a fish or a bird or a cat (see

Article 39, EMBODIED, SITUATED, AND DISTRIBUTED COGNITION). Both traditional AI models and connectionist models tend to ignore the role of bodies; these models are in fact disembodied from the start.

Passive versus active – the importance of goals A neural network is a passive thing. Before it learns anything, left to its own devices, it does not do anything very interesting. Even after learning, few connectionist models display any behavior which is internally generated. In general, most connectionist and AI systems are reactive. Or if not, their goals are preprogrammed and determined by an outside agency (their programmer).

Yet even a very primitive biological organism displays goal-directed behavior. A snail, left alone on a table, will wander around in search of food. A baby, left to play in its crib, will spontaneously make noises, move around, and find things to amuse itself. Perhaps more importantly, when an external stimulus does impinge, the baby's reaction – and how much it processes that stimulus – depends on whether the object is interesting (a mother's face is vastly more interesting than a book). Put simply, biological organisms have an agenda.

Social versus asocial cognition Almost all AI and connectionist models view cognition as an essentially individual phenomenon: it occurs within the skull. Thus, these models focus on competencies such as chess, problem solving, pattern recognition, etc. But, as Edwin Hutchins and many other culturally oriented cognitive scientists have pointed out, in humans in particular, cognition is a social phenomenon. Placed alone in the Sahara (or even a more hospitable environment), an individual human would not display any of the behaviors we take as characteristically human (building computers, traveling to the stars, creating skyscrapers, etc.). A tremendous amount of our cognitive capacity depends on external artifacts: the physical and social structures we create in order to help us solve problems which could not be solved by one person alone (see Article 40, MEDIATED ACTION).

Early artificial life: vehicles

In 1984, the biologist Valentino Braitenberg published a short monograph called *Vehicles: Experiments in Synthetic Psychology*. The book consisted of 12 short *thought-experiments*, in which Braitenberg invited the reader to imagine different primitive vehicles. Each vehicle was simply a block of wood with a pair of wheels in the rear, sensors (where headlights would be), and connections from sensors to the motors which drove each wheel. The exact nature of the sensors and their connection to the motors varied with each vehicle.

Braitenberg then considered how the different vehicles might behave when placed on a surface, possibly with other vehicles, and exposed to a stimulus, such as a light source. Some of the vehicles moved toward the light and then, at the last minute, veered away. Others sped aggressively toward it and smashed into it. Others circled it, warily.

The chapters describing the various vehicles bore names such as "Love," "Hate," "Values," "Logic." And indeed, observed from outside, it was not difficult to imagine these vehicles as animate creatures, motivated by anger or affection or even complex reasoning processes. Of course, the circuits inside were actually quite simple. Braitenberg's point – one which was of interest to many connectionists as well – is that simple systems often give rise to complex behaviors. His monograph is a powerful, graphic warning about the *attribution problem*: It is easy to attribute more than is warranted to a mechanism, especially if we already have preconceived notions about what mechanisms must underlie a given behavior.

The hardest kind of intelligence: staying alive!

In 1987, a workshop (the first of what would become a series) was held at Los Alamos National Laboratory. Researchers from a wide range of disciplines met to exchange views on what was becoming a theme of growing interest in a number of different scientific communities: Artificial life, or Alife, as it is more popularly known.

Although the methods and specific goals of the different subcommunities varied, there were a number of perspectives which were shared. One idea was captured in Alife researcher Rik Belew's comment that "The smartest dumb thing anything can do is to stay alive." This accorded with ideas that had been developed by MIT roboticist Rodney Brooks. Brooks pointed out that the bulk of evolution had been spent getting organisms to the stage where they had useful sensory and motor systems; phenomena such as tool use, agriculture, literature, and calculus represent only the most recent few *seconds* on the evolutionary clock. Brooks inferred from this that one should therefore concentrate on the hard job of building systems which have sound basic sensorimotor capacities; the rest, he suggested, would come quickly after that.

Emergentism

Another central insight which underlies much of the work in Alife is the notion of emergentism: many systems have behaviors – *emergent properties* – which result from the collective behavior of the system's components rather than from the action of any single component. Furthermore, these behaviors are often unanticipated (and in the case of artificial systems, unplanned or unprogrammed).

Examples of emergentism abound in nature. Indeed, our very bodies are a compelling example. Our 100 trillion or so cells interact in complex ways to produce coherent activity; no single cell – or even single group of cells – predicts or accounts for the highest-level behavior. Social organizations are another example. No matter how autocratic the social structure involved, complex interpersonal dynamics usually give rise to group behaviors which could not have been predicted in advance. Many Alife researchers have come to the conclusion that emergentism is a hallmark of life. Artificial systems which exhibit emergentism (particularly behaviors which in some way resemble those of biological life forms) are especially interesting.

One example of what seems like a very simple system that displays interesting emergentism comes from what are called cellular automata. These are systems which are built out of (usually) two-dimensional grids. At any given point in time, each cell in the grid can assume one of a small number of states; most simply, ON (*alive*) or OFF (*dead*). At each tick of the clock, cells may change their state, according to a simple set of rules which usually depend on the states of a cell's eight immediate neighbors. A simple rule set which is the basis for a popular computer game (the *Game of life*) is the rule of 23/3: If a cell which is already alive has exactly two or three neighbors which are also alive, it survives to the next cycle; if a cell is not alive but has exactly three living neighbors, then it is *born*; in all other cases, a cell dies (or remains dead). If one

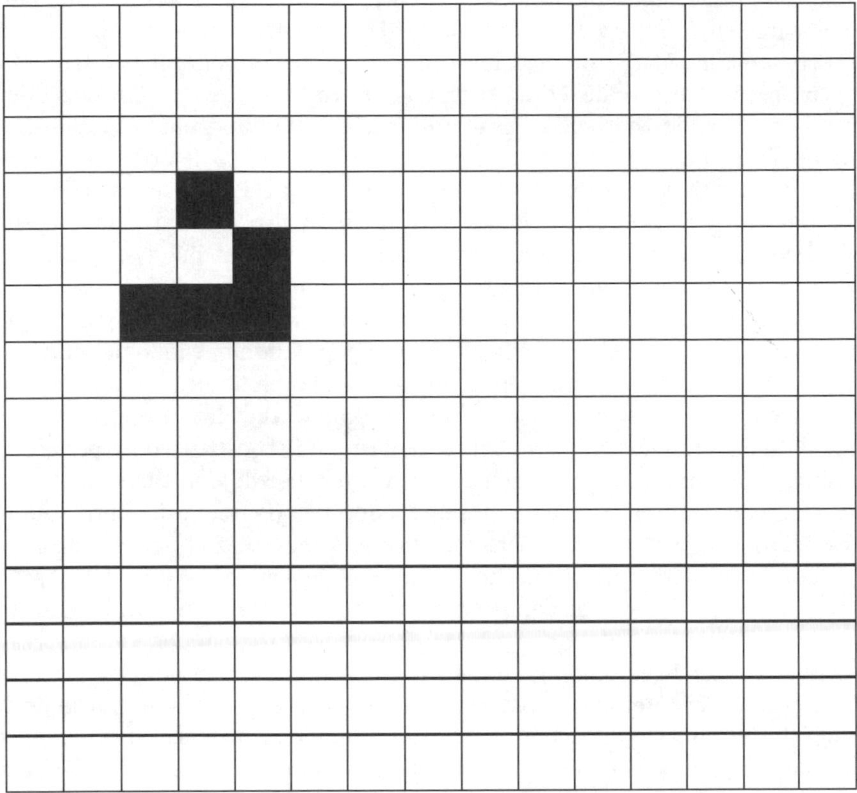

Figure 38.4 A grid for a cellular automaton. The cells in black are in an "on" state. Each tick of the clock, every cell updates – and possibly changes – its state according to some simple rule (the rule usually involves checking to see how many of the immediately neighboring cells are "on" or "off"; depending on the rule chosen and the state of its neighbors, a cell may change state at the next time cycle). The pattern shown here is called a "glider," because over time, the pattern appears to tumble and move slowly across the grid.

seeds the initial population of cells with the pattern shown in figure 38.4, something very striking happens. Over time, the pattern changes in a way which looks as if it is tumbling and deforming and in the process glides down and to the right. This is called a *glider*.

As well as providing additional examples of what seems like biological behavior (e.g., some patterns reproduce copies of themselves), cellular automata can be viewed as complex mathematical objects, and their properties have been extensively studied. More recent work by Melanie Mitchell and Randy Beer has also investigated ways in which these systems can solve computational problems.

Evolution

Most artificial systems are built by some external being. Biological systems, however, evolve. Furthermore, biological change has both a random element, in the form of random genetic variations, and a quasi-directed element, insofar as variants which are better adapted to their environment often produce more offspring, thus altering the genetic makeup of succeeding generations. This insight prompted computer scientist John Holland, in a 1975 monograph called *Adaptation in Natural and Artificial Systems*, to propose what he called the "genetic algorithm," or GA. The GA was intended to capture some of the characteristics of natural evolution in artificial systems.

Imagine, for example, that one has a problem which can be described in terms of a set of yes/no questions, and the goal is to find the right set of answers. The solution may be hard if there are interactions between the answers to particular questions; indeed, there may be multiple solutions, depending on how different questions are answered. The GA would model this by constructing an artificial *chromosome*; this is really just a vector of 1's and 0's, each bit position standing for the answer (1 = yes, 0 = no) to a different question. We begin with a population of randomly constructed chromosomes. Each chromosome represents a possible solution to the problem, so we can evaluate it to see how well it does. This determines the chromosome's *fitness* (on analogy with evolutionary fitness). A new generation is constructed by preferentially replicating those chromosomes which performed better, while at the same time randomly switching a small percentage of the 1's and 0's. (Better results are obtained if one also allows *crossover* to occur between chromosomes, so that part of one chromosome is combined with part of another.) The new generation is then tested, and its fitness is used to determine the composition of a third generation – and so on until a best solution is found.

The GA has much in common with natural evolution. It is especially powerful when there are high-order interactions between the many different sub-parts to a problem. Although the original GA makes simplifying assumptions which are questionable (e.g., there is not the genotype–phenotype distinction found in nature), it is widely used in Alife, sometimes in conjunction with neural networks. The next simulation gives an example of this.

Goals

We noted above that connectionist models have a disturbing passivity which is quite unlike biological organisms. One could construct a network in advance in order to endow it with what look like internally generated behaviors, but this is hardly the solution found in nature. Instead, many of the behaviors we think of as goal-directed – such as the search for food, the desire to mate and bear offspring, the responses to danger – are usually adaptive, and they are the product of the evolution of our species and not taught to us. Thus, it seems more appropriate to model such behaviors using an evolutionary approach such as the GA. One can even do this with neural networks.

Nolfi, Elman, and Parisi (1994) demonstrated the evolution of a simple kind of goal-directed behavior in neural networks in the following way. They constructed an artificial two-dimensional environment consisting of a 10 by 10 grid of cells. A small number of cells contained food. A population of artificial organisms were then *tested* in this environment. Each organism was a simple neural network. Two input units provided the animal with information about the smell (direction and strength) of nearby food; input units were connected to a bank of seven hidden units (the animal's *brain*), which projected to two motor output units. These two motor units allowed the animal to move forward, turn left or right, or pause. At the outset, each of the 100 animals in the first generation had random weights on the connection strengths, so the animals' behavior was disorganized. In some cases, animals would simply stand still for their whole lifetime. In other cases, animals would march forward until they fell off the edge of the world. This behavior was hardly surprising, given the random nature of the initial connections. Occasionally, an animal would stumble on some food, in which case it would eat it. At the end of an animal's life, it died. However, if by chance it had wandered across some food (most did not), then it was able to give birth to offspring. These offspring were copies of the parent, except that in a few cases random weight changes were introduced.

After 50 generations, very different-looking individuals had evolved. Placed in a world, an animal would head directly for the nearest food, ingest it, and then proceed to the next food; in short order, the animal gathered up all the food in its world. Viewed from above, the apparently deliberate and very efficient hunting for food certainly looked goal-driven. (But, mindful of Braitenberg's vehicles, we must remember that these behaviors were generated by relatively small neural circuits.)

What are the limits?

The Alife approach is still fairly young, and much of the work has a preliminary character. As a corrective to previous modeling approaches, there is no question that the Alife perspective is valuable. The emphasis on emergentism, the role of the environment, the importance of an organism's body, the social nature of intelligence, and the perspective offered by evolution are notions which go well beyond Alife. Further, by trying to understand (as Chris Langton has put it) life, not necessarily as it is, but as it *could be*, we broaden our notion of what counts as intelligent behavior. This expanded definition may in turn give us fresh ways of thinking about the behavior of more traditional (biological) life forms. But it is also clear that Alife has a long way to go. As is true of many modeling frameworks, the bridge between the initial *toy* models and more complete and realistic models is a difficult one to cross.

Cognition as a dynamical system

At the outset, it was pointed out that the digital framework, *mind as computer*, has permeated work in cognition until recent times, and that connectionism can be understood at least in part as an alternative which views *mind as brain*. The digital framework has also had a profound impact on the way we think about computation and information processing. Much of the formal work in learning theory, for example, draws heavily on results from computer science.

Interestingly, there is one subset of researchers who have not adopted the digital framework; these are people who study motor activity. Researchers such as Michael Turvey and Scott Kelso (to name only two prominent scientists from a large, active community) have instead used the tools of dynamical systems to try to understand how motor activity is planned and executed. This seems natural, given that motor activity has a dynamical quality which is difficult to ignore. For example, when we walk or run or ski, our limbs move in a rhythmic but complex manner and involve behaviors which change over time (and hence are dynamic). More recently, scientists from various other domains in cognitive science have also begun to explore the dynamical systems framework as an alternative to thinking about cognition in terms of digital computers.

What is a dynamical system? Most simply, it is a system which changes over time according to some lawful rule. Put this way, there is very little which is *not* a dynamical system, including digital computers! In practice, dynamical systems must also be characterized in terms of some set of components which have states, and the components must somehow belong together. (In other words, my left foot and the Coliseum in Rome do not constitute a natural system – unless perhaps my left foot happens to be kicking stones in the Coliseum.) The goal of dynamical systems is to provide a mathematical formalism which can usefully characterize the kinds of changes which occur in such systems.

There are a number of important constructs which are important in dynamical systems theory. For instance, having identified the parts of a system which are of interest to us (e.g., the position of the jaw, tongue, and lower lip), we can assign numerical values to these entities' current state. We can then use (in this example) a three-dimensional graph (one axis each for jaw, tongue, and lower lip position) to visualize the way in which all these components change their state over time. This three-dimensional representation is often called the *state space*. If we are interested in a formal characterization of how the system changes over time, then this leads us to use differential equations. These capture the way in which the variables' values evolve over time and in relation to one another.

A final example of a construct used by dynamical systems theory is the notion of an *attractor*. An attractor is a state toward which, under normal conditions, a dynamical system will tend to move (although it may not actually get there). A child on a playground swing constitutes a dynamical system with an attractor that has the child and the swing at rest in the bottom vertical position. The swing may oscillate back and forth if the child is pushed or pumps her legs, but there is an attracting force which draws the child back toward the rest position. The goal of a dynamical systems analysis of this situation would be to describe the behavior of the system using mathematical equations which tell us how the state of the system (e.g., the position of the child at any given moment) changes over time.

The dynamical hypothesis for cognition

Swings and springs and fluids in motion are obvious domains in which a dynamical perspective applies. How might it apply to cognition?

This question was raised at a conference at Indiana University in 1991; the topic of the conference was *Mind as Motion*, which is also the title of the book (edited by Robert Port and Timothy van Gelder (1995)) that was subsequently produced from the conference. In their introduction to the book, Port and van Gelder list several reasons why one should think of cognition as a dynamical system. They include the following.

Cognition and time Cognitive behaviors are not atemporal; they exist and unfold over time. Dynamical models take as their goal the specification of how changes in a system's states occur. Thus, any useful account of cognitive behavior must necessarily explain such temporal changes; and this is precisely what dynamical models take as their goal.

Continuity in state Although computational models can be formulated which model change as a serial movement from one discrete state to another, natural cognitive systems often change in a continuous manner such that there is never any state which is discretely separable from the next. As we listen to a sentence, for example, our understanding of the words we hear builds up gradually (sometimes, with expectations about an upcoming word even before we hear it, or other times delayed until the word is past). The meaning of the sentence as a whole unfolds gradually, rather than occurring all at once at the end. (Henry James is known for sentences which go on interminably, sometimes over pages; yet a reader need not wait till the very end till she understands the sentence.) It is a natural property of dynamical systems that they model the continuously changing nature of states; indeed, sometimes there never is an *end* state.

Multiple simultaneous interactions One of the problems which confronted the human information processing framework, described in the earlier section on connectionism, was how to deal with situations in which many things interacted in complex ways. The problem was not only conceptual (how to think about such systems) but also computational: the digital computer carries out only one (or at best, a very few) instructions at any one time – very fast, perhaps, but only one at a time. As a system grows in complexity, the interactions between the system's parts can grow exponentially, rapidly outstripping the possibility of modeling behavior using a digital machine. Dynamical systems, on the other hand, allow us to focus (i.e., model and formalize) precisely what is otherwise difficult: the many simultaneous interactions in a system which affect its overall behavior.

Self-organization and the emergence of structure We have already talked about emergentism; self-organization is the ability of a system to develop structure on its own, simply through its own natural behavior. Dynamical systems provide a good framework for understanding how and why such characteristics emerge.

For example, neither the stripes on a zebra nor the patch-like stripes of visual cortex (these are the ocular dominance columns which separately process inputs from the left and the right eye) are likely to be programmed in from birth. Instead, these stripes emerge out of initially relatively uniform patterns, which, over time, involve dynamic interactions (between substances controlling skin pigmentation in the zebra and neurons which are stimulated by left and right eyes in the case of the visual cortex). The underlying dynamics were first described by the famous mathematician and cryptographer Alan Turing and are known as reaction–diffusion equations – and they are dynamical equations.

The dynamical hypothesis for cognition is quite new. The body of work relating to motor behavior is the most substantial; far less has been done in realms of higher cognition. One example of how it might be applied to the case of language comes from work by Janet Wiles, Paul Rodriguez, and Jeff Elman (1995), who used a dynamical systems analysis to analyze a recurrent neural network that was trained on a counting task.

In this task, a recurrent network was trained on an artificial language in which every *sentence* consisted of some number of *a*'s and *b*'s; grammatical sentences had the form a^nb^n. In other words, legal sentences had some number (n) of *a*'s, followed by exactly the same number of *b*'s; for example: *aaabbb, ab, aaaaaabbbbbb* are all legal, whereas *abb* and *aaaaabbbb* are not. What makes this language at all of interest is that, although it is very simple, it has certain properties which resemble human language and are known to be hard (specifically, it requires some sort of *stack*, or memory device, for remembering how many *a*'s there were, in order to known how many *b's* to expect). After training, the network was able not only to process strings it had seen, but to generalize its knowledge to strings that were longer than any encountered during training. Given that the network does not possess a stack of the sort familiar to computers, how did it solve this problem?

The network had only two hidden units, so these two units together define the relevant state for the network. During processing, each unit's activation is represented by some number between 0 and 1; together, we can treat the two numbers as corresponding to the x and y axes of a two-dimensional state space. Thus, over time, the network's internal space will *move* through this x–y plane. Using dynamical systems analysis, Rodriguez and his colleagues discovered that, like the child on the swing, the network had attractors. Metaphorically (and simplifying a bit), one can think of the dynamics as follows, imagining that instead of a network we are working with our child on the swing. When the network is listening to the initial (*a*) portion of a sentence, it is as if the child were being pulled back; the more *a*'s, the further back the child is drawn. When the *b*'s start coming in, the child is released and allowed to swing, eventually coming to rest. Just how long it takes the swing to return to the rest position depends on how far back the child was pulled. The network operated in a similar fashion. This is an interesting example, because it suggests a way to use dynamical systems to tackle problems (such as processing language) which are central to cognition.

Conclusion

These three approaches represent attempts to deal with the shortcomings of cognitive models. The connectionist framework is primarily concerned with biological implementation and with problems in learning and representation. Can an inductive learning procedure discover abstract generalizations, using only examples, rather than explicitly formulated instructions? And how do the resulting knowledge structures capture the generalizations? Are there important differences between the ways in which networks represent generalizations and traditional rule systems?

The focus of Alife is different. Alife rejects the view that cognition consists only of highly developed mental activities such as chess, and emphasizes the intelligence which is required simply to survive. The role of adaptation and the role of evolution as achieving adaptation are valued; and a theme common to much of the work in Alife is the emergence of structure and behaviors which are not designed, but rather are the outgrowth of complex interactions.

The dynamical systems approach is also concerned with interaction and emergentism; more generally, it can be viewed as a mathematical framework for understanding the sort of emergentism and high-order interactions which are found in both connectionist and artificial life models. Dynamical systems also reflects a deeper commitment to the importance of incorporating time into our models.

The three approaches have much in common. They all reflect an increased interest in the ways in which paying closer attention to natural systems (nervous systems, evolution, physics) might elucidate cognition. None of the approaches by itself is likely to be complete; but taken together, they complement one another in a way which we can only hope presages exciting discoveries yet to come.

References and recommended reading

*Bechtel, W. and Abrahamsen, A. 1991: *Connectionism and the Mind: An Introduction to Parallel Processing in Networks*. Cambridge, Mass.: Basil Blackwell.

Braitenberg, V. 1984: *Vehicles: Experiments in Synthetic Psychology*. Cambridge, Mass.: MIT Press.

Elman, J. L. 1990: Finding structure in time. *Cognitive Science*, 14, 179–211.

—— 1991: Distributed representations, simple recurrent networks, and grammatical structure. *Machine Learning*, 7, 195–225.

Fodor, J. A. and Pylyshyn, Z. W. 1988: Connectionism and cognitive architecture: a critical analysis. In S. Pinker and J. Mehler (eds), *Connections and Symbols*, Cambridge, Mass.: MIT Press, 3–71.

Hinton, G. E. and Anderson, J. A. 1981: *Parallel Models of Associative Memory*. Hillsdale, NJ: Erlbaum.

Holland, J. H. 1975: *Adaptation in Natural and Artificial Systems*. Ann Arbor: University of Michigan Press.

*Langton, C. G. 1989: *Artificial Life*, Santa Fe Institute Studies in the Sciences of Complexity Proceedings, vol. 6. Redwood City, Calif.: Addison-Wesley.

McClelland, J. L. and Rumelhart, D. E. 1981: An interactive activation model of context effects in letter perception: part 1. An account of the basic findings. *Psychological Review*, 86, 287–330.

*McClelland, J. L., Rumelhart, D. E. and the PDP Research Group 1986: *Parallel Distributed Processing: Explorations in the Microstructure of Cognition*, vol. 2: *Psychological and Biological Models* Cambridge, Mass.: MIT Press

Nolfi, S., Elman, J. L. and Parisi, D. 1994: Learning and evolution in neural networks. *Adaptive Behavior*, 3, 5–28.

Pinker, S. and Prince, A. 1988: On language and connectionism: analysis of a parallel distributed processing model of language acquisition. *Cognition*, 28, 73–193.

*Port, R. and van Gelder, T. 1995: *Mind as Motion: Dynamical Perspectives on Behavior and Cognition*. Cambridge, Mass.: MIT Press.

Rumelhart, D. E. and McClelland, J. L. 1986: On learning the past tenses of English verbs. In D. E. Rumelhart and J. L. McClelland (eds), *Parallel Distributed Processing: Explorations in the Microstructure of Cognition*, vol. 2: *Psychological and Biological Models*, Cambridge, Mass.: MIT Press, 216–71.

*Rumelhart, D. E., McClelland, J. L. and the PDP Research Group 1986: *Parallel Distributed Processing: Explorations in the Microstructure of Cognition*, vol. 1: *Foundations*. Cambridge, Mass.: MIT Press.

Servan-Schreiber, D., Cleeremans, A. and McClelland, J. L. 1991: Graded state machines: the representation of temporal contingencies in simple recurrent networks. *Machine Learning*, 7, 161–93.

Wiles, J., Rodriguez, P. and Elman, J. L. 1995: Learning to count without a counter: a case study of dynamics and activation landscapes in recurrent networks. In *Proceedings of the Seventeenth Annual Conference of the Cognitive Science Society*, Hillsdale, NJ: Erlbaum, 482–7.

39

Embodied, situated, and distributed cognition

ANDY CLARK

Wild brains

Biological brains are first and foremost the control systems for biological bodies. Biological bodies move and act in rich real-world surroundings. These apparently mundane facts are amongst the main driving forces behind a growing movement within cognitive science – a movement that seeks to reorient the scientific study of mind so as to better accommodate the roles of embodiment and environmental embedding. Two claims characterize the common core of this emerging approach:

(1) That attention to the roles of body and world can often transform our image of both the problems and the solution spaces for biological cognition.

(2) That understanding the complex and temporally rich interplay of body, brain, and world requires some new concepts, tools, and methods – ones suited to the study of emergent, decentralized, self-organizing phenomena.

These core claims are sometimes joined by some further, more radical speculations, namely:

(3) That these new concepts, tools, and methods will perhaps displace (not simply augment) the old explanatory tools of computational and representational analysis.

(4) That the familiar distinctions between perception, cognition, and action, and indeed, between mind, body, and world, may themselves need to be rethought and possibly abandoned.

In this brief treatment I aim to lay out and discuss some of the concrete results that have inspired this complex of claims and speculations and then to suggest a more conciliatory image of their consequences.

Human infants

Consider a familiar class of real-world agents – human infants. A newborn infant, held suspended off the ground, will perform a recognizable stepping motion. After a few months, this response disappears, only to reappear at about 8–10 months old (the age at which infants begin to support their weight on their feet). Independent walking cuts in at about a year. One explanation of these changes involves what the development psychologists Esther Thelen and Linda Smith have dubbed a "grand plan, single factor view." Such a view depicts the emergence of independent walking as the expression of a kind of detailed genetic blueprint. Thelen and Smith, however, reject all versions of the grand plan, single factor view and instead depict learning to walk in terms of the complex interplay of multiple factors involving body, brain, and world in essentially equal terms.

Evidence for the multifactor view comes from a series of elegant micro-developmental studies involving bodily and environmental manipulations (Thelen and Smith, 1994). For example, the reflex stepping action seen to disappear at about two months can be restored by holding the baby upright in water. The key parameter that causes the early disappearance of the stepping reflex is thus, it seems, just leg mass! The period during which the stepping reflex is not normally seen corresponds to the period during which the sheer mass of the infant's legs defeats an inherent spring-like action in the muscles. Immersion in water reduces the effective mass and thus allows stepping to reappear. This story contrasts strikingly with a variety of accounts that seek to explain the disappearance by invoking maturational processes in the brain or the predetermined phases of a *locomotion program*.

Further support for the alternative, multifactor view comes from a series of environmental manipulations in which nonstepping infants (aged between one and seven months) are held on a moving treadmill. Under such conditions coordinated stepping is again observed. Moreover, this stepping shows adaptation to variations in treadmill speed and even compensates for asymmetric belt speeds (the treadmill comprised two independent belts capable of driving each leg at a different speed). These results demonstrate a major role for a kind of mechanical patterning caused by the backward stretching of the legs brought about by the action of the treadmill. This component of the stepping response is clearly in place even while other factors (such as leg mass) prevent its expression under ecologically normal conditions. Thelen and Smith conclude that the developmental pattern of infant stepping does not, after all, reflect the predetermined stages of a stored program or the influence of any single variable or parameter. Instead, it comes about as a result of the interplay of a variety of forces spread across brain, body, and world. Such forces include bodily features (such as leg mass), mechanical effects (stretch-and-spring), ecological influences (water, treadmills), and cognitive impetus (the will to move). The upshot is an image of the human infant as a system in which brain, body, and world conspire to yield the complex behavioral profiles that cognitive science seeks to explain and understand.

Autonomous agents

An autonomous agent is a creature (real, artificial, or simulated) that must act and survive in a reasonably demanding, not too heavily regimented environment. Such creatures must often act quickly, and on the basis of noisy, incomplete, or conflicting information. Humans, mice, and cockroaches are all examples of natural autonomous agents. Recent years, however, have seen an explosion of research on so-called artificial autonomous agents – robots (or simulated robots) that must act and survive in a variety of difficult settings such as the flow of a crowded office, the crater of a volcano, or the surface of the moon.

A famous early example of autonomous agent research is the robot Herbert, built in the 1980s in the Mobile Robot (Mobot) Laboratory at the Massachusetts Institute of Technology. Herbert's job was to collect empty soft drink cans left around the laboratory. In this real-world environment, Herbert had to co-exist with the human researchers — he had to avoid bumping into them and avoid disrupting delicate ongoing construction work. More positively, Herbert needed to move around, identify, and acquire abandoned cans.

One kind of solution to such a problem would be to exploit sophisticated scanning devices tied to a complex image processing and planning system. The goal of such a setup would be to generate a detailed internal model of the surroundings, to identify cans, and then to plot an efficient collection route. Given the current state of the computational art, however, such a solution would be costly, imperfect, and fragile. The system would spend long periods lost in thought, contemplating its future plans of action. And once in action, it would be easily upset by new events, people moving around, and so on. Naturally, there are ways to begin to address all these problems. But instead of tackling a long list of such problems, the designers of Herbert decided to try a fundamentally different approach. For all these problems and pitfalls, they began to suspect, were really artifacts of an overtly centralized and intellectualist approach to the generation of real-world, real-time behavior: an approach that leans too heavily on detailed inner models and long chains of inference, and that makes too little use of the simplifications and shortcuts afforded by simple environmental cues and the robot's own capacities for action, motion, and intervention.

An alternative approach, pioneered by Rodney Brooks (1991), exploits such cues and shortcuts for all they are worth. Brooks advocates the use of a *subsumption* architecture – a design that incorporates a number of quasi-independent subsystems, each of which is responsible for one self-contained aspect of the creature's activity, and which are not coordinated by any central system. Instead, the various subsystems are capable only of sending simple signals that bypass, override, or occasionally modify the response of other subsystems. The activity of one subsystem (or layer) will thus often subsume that of another.

Herbert, built by Brooks's graduate student, Jonathan Connell, in the 1980s, was designed as just such a decentralized bag of tricks agent. The basic layers of the subsumption architecture generated simple locomotor routines, and a ring of ultrasonic sonar sensors supported a behavior of stopping and reorienting if an obstacle was spotted immediately ahead. These routines were, however, overridden as soon as the simple visual systems detected a rough table-like outline. At this point the surface of the table would be scanned, using a laser and a video camera. If the basic outline of a can was detected, the whole robot would rotate on its wheels until the can was fixed in its center of vision. At that point the wheel motion ceased, and a simple signal activated a robot arm. The arm moved blindly, aided by simple touch sensors that lightly skimmed the surface of the table. On encountering a can shape, a grasping behavior was activated, and the target object was acquired.

Herbert is just one example of many, and the full subsumption architecture is not universally in vogue. But it provides an illustration of a number of features characteristic of a wide range of recent endeavors in autonomous agent research. The robot does not generate or exploit any detailed internal models of its surroundings (and hence need not constantly update such models over time); it does not engage in complex inferences and planning; the various sensory systems do not feed data to a central intelligent system, but instead participate rather directly in the generation and control of particular behaviors. The robot uses its own motion in the world to aid its sensory searches and to simplify behavioral routines, and it allows its activity to be sculpted and coordinated by the flux of local environmental events. In short, Herbert, like the human infants discussed above, is a good example of a system in which body, brain, and world all cooperate to yield a computationally cheap, robust, and flexible variety of adaptive success.

Large-scale systems

A major theme of embodied cognitive science is, we have seen, the emergence of adaptively valuable patterns of activity out of the blind (noncentrally orchestrated) interactions of multiple factors and components. Such a perspective lends itself naturally to the consideration of patterns that characterize the activity of larger groups of individuals, such as ant colonies, ship's crews, and social, political, or commercial organizations. Certain ants, to take a well-worn example, forage by a process known as mass recruitment. If an ant finds food, it leaves a chemical trail as it returns to the nest. Other ants will follow the trail. If they too find food, they will add to the chemical concentration. A process of positive feedback thus leads rapidly to a high concentration of chemical signals that in turn orchestrate the activity of many hundreds of ants. The strikingly organized apparition of a steady flow of ants systematically dismantling a recently discovered food source is thus the outcome of a few simple rules that sculpt the first responses of the individual ants and that lead the ants to alter their local environment in ways that lead to further alterations (the increasing concentration of the trail) and to a distinctive and adaptively valuable system of group foraging (for details and a computer simulation, see Resnick, 1994, pp. 60–7). Ants, termites, and other social insects use a variety of such strategies to build complex structures (such as termite nests) without leaders, plans, or complete inbuilt programs. The individuals know only to respond in specific ways to specific environmental patternings that they encounter. But those responses in turn alter the environment and call forth other responses from other individuals, and so on.

Consider now the operation of a human collective – the navigation team of a seafaring vessel. The cognitive scientist, anthropologist, and mariner Edwin Hutchins (1995) describes in exquisite detail the way in which a typical navigation team succeeds without benefit of a detailed controlling plan and the complex ways in which representations and computations are propagated through an extended web of crew and artifacts. Individual crew members operate within this web in ways that display more than a passing similarity to the strategies of the social insects! Many ship duties consist in being alert for a specific environmental change (e.g., waiting for a sounding) and then taking some action (e.g., recording the time and sending the sounding to the bridge). These actions in turn alter the local environment for other crew members, calling forth further responses, and so on. In addition, a vast amount of work involves the use of external structures and artifacts, such as alidades, maps, charts, and nautical slide rules. Some of these devices function so as to re-represent acquired information in forms which make it easier for us to use or transform it. The nautical slide rule, for example, turns complex mathematical operations into perceptual scale-alignment tasks of a kind that human operators usually find easier to perform. In fact, the entire work space is structured so as to simplify the computations and operations falling to individual biological brains.

None of these observations is in itself terribly surprising. But Hutchins shows, in convincing detail, just how genuinely distributed (between agents) and reshaped (by the use of artifacts, spatial layout, and simple event-response routines) the ship navigation task has become. The captain may set the goals. But the mode of execution, by means of a complex web of information gatherings, transformations, and propagations, is nowhere explicitly laid out or encoded. Instead, success flows from the well-tuned interaction of multiple heterogeneous factors encompassing biological brains, social practices, and a variety of external props and artifacts.

One striking difference, of course, separates the case of human collective success from that of, for example, social insects. In our case, the necessary harmonization of brains, bodies, and environmental structure is itself, in part, an achievement of human learning and cognition. We create designer environments that alter and simplify the computational tasks which our brains must perform in order to solve complex problems. The key to understanding this extra dimension (one deep root of our unusual type of adaptive success) lies surely in our ability to use and profit from the very special tool of human public language. Language is perhaps the ultimate artifact: the one responsible for our apparently unique ability to think about the nature of our own thoughts and cognitive capacities and hence to seek to deliberately alter our world in ways that allow us to press maximum effect from the basic computational capacities of our biological brains. That, however, is a story for another day (Clark, 1997, ch. 10).

Complementary virtues?

Having sampled some real research in embodied, embedded, and distributed cognition, we can now ask: What significance (if any) does all this have for the philosophy and practice of the sciences of the mind? It is probably too soon to make a solid assessment, but we can make some progress by reviewing the four claims laid out in the first section, in the light of the examples just sketched. In order, then:

> Claim 1: that attention to the roles of body and world can often transform our image of both the problems and the solution spaces for biological cognition.

This claim is surely correct. We saw, for example, how the robot Herbert uses physical motion to simplify on-board computation; how the problem of learning to walk, for real infants, is defined against the mechanical backdrop of spring-like muscles; and how the physical work space, instrumental supports, and division of human labor all play a major role in enabling ship navigation. The morals that I would draw are twofold.

First, we should be aware of the importance of what I shall call *action-oriented* inner states in the control of much everyday, on-line behavior. By action-oriented states, I mean states that are especially well geared to the computationally cheap production of appropriate responses in ecologically normal conditions. A recent research program that stresses such states is the so-called animate vision paradigm (Ballard, 1991), with its emphasis on the use of indexical and/or locally effective personalized representations and the use of real-world motion and action to defray computational costs. It is also possible (with a little poetic license) to read J. J. Gibson and the ecological psychologists as likewise stressing the close fit between certain inner states and the specific needs and potentialities of action determined by a given animal/environment pairing. Second, we should acknowledge an important role for what Kirsh and Maglio (1995) nicely term "epistemic actions": namely, actions whose purpose is not to alter the world so as to advance physically toward some goal (e.g., laying a brick for a walk), but rather to alter the world so as to help to make available information required as part of a problem-solving routine. Examples of epistemic actions include looking at a chessboard from different angles, organizing the spatial layout of a hand of cards so as to encode a record of known current high cards in each suit, laying out our mechanical parts in the order required for correct assembly, and so on. Epistemic actions, it should be clear, build designer environments – local structures that transform, reduce, or simplify the operations that fall to the biological brain in the performance of a task.

A cognitive science that takes seriously the twin notions of action-oriented inner states and epistemic actions will, I believe, have gone a long way towards remedying the kind of disembodied intellectualist bias identified at the start of this essay.

> Claim 2: that understanding the complex and temporally rich interplay of body, brain, and world requires some new concepts, tools, and methods – ones suited to the study of emergent, decentralized, self organizing phenomena.

This is a potentially more challenging claim, suggesting as it does that taking brain–body–world interactions seriously requires us to develop whole new ways of thinking about cognitive phenomena. In this vein, a number of cognitive scientists have recently begun to investigate the potential use of the concepts and tools of dynamical systems theory as a means of understanding and analyzing the kinds of cases described above. Dynamical systems theory is a well-established framework for understanding certain kinds of complex physical phenomena (see Article 38, CONNECTIONISM, ARTIFICIAL LIFE, AND DYNAMICAL SYSTEMS). What is novel, then, is the attempt to apply this framework to the class of phenomena standardly investigated by the sciences of the mind. The object of a dynamical systems analysis is the way the states of a system change and evolve over time. Many of the guiding images are geometric, since the system is pictured as evolving within a certain *state space* (a set of dimensions of variation discovered or defined by the theorist). Given a state space, the task of a dynamical analysis is to present a picture of how the values of the state space variables evolve through time. The product is thus an understanding of the system's space of potential behaviors in terms of a set of possible trajectories through the space. This understanding can have both a quantitative and a qualitative dimension. Quantitatively, it may consist in the description of a dynamical law that strictly governs the evolution of the state variables (often, such a law will amount to a set of differential equations). Qualitatively, it may consist in an understanding of the properties of regions of the state space. For example, some regions will be such that any trajectory heading close by becomes, as it were, diverted so as to pass through the region. Such regions (or points) are known as *attractors*. Conversely, some regions or points will be such that trajectories heading in that direction get deflected. These are *repellors*. One especially powerful construct, as far as the understanding of embodied, embedded systems is concerned, is the idea of a *collective variable*. This is a state variable that folds together a number of forces acting on the system and plots only the overall effect of these forces. A very simple example, used by the philosopher Timothy van Gelder, would be tracking the behavior of a car engine by plotting engine temperature. Such a strategy abstracts away from the details of specific components and focuses attention on the overall results of a multitude of internal interactions (between parts) and external influences (such as temperature, humidity, and road conditions). It thus reduces a potentially very complex (*high-dimensional*) description (one involving distinct variables for each separate influence) to a simpler, more tractable (*low-dimensional*) one. If it is well chosen, such a low-dimensional description can nonetheless be of great explanatory and predictive value. In the case of the car, for example, it can help predict and explain low fuel economy and misfiring. To get more of the flavor of dynamical tools in use, the interested reader should consult, for example, the extended treatment in Kelso, 1995.

Dynamical systems approaches do, I conclude, offer concepts, tools, and methods well suited to the study of emergent, decentralized, self-organizing phenomena. Such approaches are attractive because of their intrinsic temporality (the focus on change over time) and their easy capacity to criss-cross brain–body–environment boundaries by using constructs (such as collective variables) that merge many influences into a single factor. As a result, dynamical analyses are tailor-made for tracking emergent phenomena – ones that depend on the complex and often temporally rich interplay between multiple factors and forces spread across brain, body, and world. It is thus no surprise that Thelen and Smith, for example, go on to offer dynamical analyses of the various infant behaviors described earlier – behaviors which, we saw, were indeed determined by a subtle interplay of neural, bodily, and environmental factors. The precise status of these new tools, however, remains unresolved, and is the subject of the third – and more problematic – claim, namely:

> Claim 3: that these new concepts, tools, and methods will perhaps displace (not simply augment) the old explanatory tools of computational and representational analysis.

Well, perhaps. But this claim strikes me as somewhat premature. There are two main reasons for caution. First, there is an issue about scaling and tractability. As the number of parameters increases, our intuitive geometric understanding breaks down. A lot therefore turns on the availability of simpler low-dimensional descriptions arrived at by the canny use of collective variables and the like. This strategy, however, quickly gives rise to a further worry: namely, whether useful low-dimensional dynamical descriptions will always be available and what is lost (as something inevitably must be) by their use. Second, there is an issue concerning the type of understanding that attends even a successful dynamical account. For it is a type of understanding that sits uncomfortably close to description, as opposed to functional explanation. We learn what patterns characterize the system's evolution over time (and hence can make useful predictions). We learn, for example, that the overall system state A must give way to overall system state B, and so on. But we do not thereby understand the actual role of specific components or the detailed organization of the system. A useful illustration of this problem, drawn from Sloman (1993, p. 81), concerns the attempt to understand contemporary digital computers. One possible strategy would be to treat all the computer's sub-states (the states of the arrays, registers, etc.) as dimensions and to display the activity of the device as a set of transitions between such globally defined overall states. Such an approach nicely displays the system's overall evolution over time. But it hides away the details concerning what is going on in the individual registers and arrays. In particular, it hides away the facts concerning which sub-states can vary independently of one another, which sub-states vary with which external parameters, etc. In short, it highlights the global behavior at the expense of local structural and functional detail. Which approach works best for any given system is a complex empirical question (sometimes we really do need both). It is also a question that remains unresolved in the case at hand: namely, the explanation of intelligent behavior in biological organisms.

We should also be careful to distinguish two issues that are easily run together in any assessment of dynamical modeling. One, as we just saw, concerns the grain of such analyses – their tendency to invoke abstract collective variables so as to provide tractable descriptions of complex emergent patterns. Another concerns the type of vocabulary used to characterize the states of the systems. Some dynamicists (such as Thelen and Smith) do indeed eschew the use of representation talk – talk of the knowledge encoded in specific inner states – as a means of understanding systemic organization. But others (see, e.g., several treatments in Port and van Gelder, 1995) seek to combine the use of dynamical tools with talk of inner states as representations. Such accounts associate representational roles with specific dynamical features. Attractors, trajectories, and collective variables may thus be interpreted as the systemic bearers of specific contents or the players of specific semantic roles. These hybrid dynamical models constitute, I believe, the most promising way in which dynamical theory may illuminate genuinely cognitive phenomena. The value of such approaches will, however, depend on the extent to which genuinely cognitive functions turn out to be tied to genuinely high-level emergent systemic features: the kind that might be constituted by the complex feedback and feed-forward interactions of multiple neuronal populations and circuits, or by temporally rich interactions involving bodily and environmental factors. If, by contrast, typical cognitive functions depend on much more local, component-based systemic features, the scope and value of even hybrid dynamical approaches will be diminished.

Border disputes

And so to metaphysics and the fourth and final claim to be examined:

> Claim 4: that the familiar distinctions between perception, cognition, and action, and indeed, between mind, body, and world, may themselves need to be rethought and possibly abandoned.

Three broad classes of argument are invoked in support of such claims. The first, and least persuasive, is based on an idea of equal distribution of causal influence. Thus, Thelen and Smith showed (see above) that infant stepping is in some sense equally dependent upon bodily, neural, and environmental factors: it is not specified by a full neural program but emerges from the interaction of bodily factors (leg mass), environmental context (the treadmill and water-immersion experiments), and neural and anatomical structure. This leads them to comment, for example, that "it would be equally credible to assign the essence of walking to the treadmill than to a neural structure" (1994, p. 17).

The authors are surely correct to stress the way in which attention to bodily and environmental factors can transform our image of the neural contributions to behavioral success. But it would be a mistake to infer from the (let us grant) equal causal roles of various factors to any lack of demarcation between them. It is not obvious in what sense, then, we really do here confront an "inextricable causal web of perception, action and cognition" (1994, p. xxii). It is a complex web, to be sure. But genuine inextricability would seem to require something more. The smooth running of a car engine reflects a complex interplay of inner factors and environmental conditions (heat, humidity, and so on). But the complexity of the interplay in no way motivates us to abandon the idea of a real distinction between the various types of factors involved. Likewise, the demarcations between perception, cognition, and action, or between mind, body, and world, do not seem unduly threatened by demonstrations of complex causal interplay, subtle trade-offs, and the like.

There is, however, one special type of causal interplay that may indeed raise tricky questions concerning systemic and functional boundaries. This type of case has been discussed by the cognitive scientists Francisco Varela, Evan Thompson, and Eleanor Rosch (1991) and centers on a special class of phenomena that I dub "processes of continuous reciprocal causation" (CRC for short). CRC is a close relative of the idea, popularized by the cybernetics movement of the late 1940s and 1950s, of *circular causation*. To get the idea, think of the way a crowd sometimes starts to move in a unified direction. The motion of the crowd is nothing but the sum of the motions of the individuals. Yet a given individual both contributes to, and is affected by, the overall motion. As soon as (by chance or by some external push) a degree of consensus about direction begins to emerge, weak or undecided individuals are sucked into the flow. This generates an even more powerful consensus, which in turn sucks in more individuals. There is thus a kind of circular positive feedback loop going from individual to crowd and back again.

Ordinary circular causation, however, is compatible with discrete temporal stages of influence, in which one system (e.g., a hi-fi component) sends a signal to another that then waits and subsequently sends a signal back, and so on. The notion of continuous reciprocal causation is meant to pick out the special subset of cases in which no discrete temporal staging interrupts the coupled dynamics of the linked systems. For example, consider a radio receiver: the input signal here acts, as Tim van Gelder (personal communication) has pointed out, as a continuous modulator of the radio's behavior (its sound output). Now imagine that the radio's output is also continuously transmitted back so as to modulate the other transmitter, thus creating a circle of continuous reciprocal causal influence. In such a case there is a complex, temporally dense interaction between the two devices – an interplay that could easily lead to rich coupled dynamics of positive feedback, mutual damping, or stable equilibrium, depending on the precise details of the signal processing. Given the continuous nature of the mutually modulatory influences, the usual analytic strategy of divide and conquer yields scant rewards. We can easily identify the two components; but we would be ill-advised to try to understand the behavioral unfolding of either individual device by treating it as a unit insulated from its local environment by significant boundaries of transduction and action. For the spatial envelope demarcating each component is of little significance if our object is to understand the evolution of the real-world behaviors. There are genuinely distinct components here. But the correct unit of analysis, if we seek to understand the behavioral unfolding of either component, is the larger coupled system which they jointly constitute. To whatever extent brain, body, and world can, at times, be joint participants in episodes of continuous reciprocal causal influence, we may indeed confront behavioral unfoldings that resist explanation in terms of inputs to and outputs from some supposedly insulated cognitive engine. (Conversely, when we confront episodes of genuinely *decoupled* information processing and problem solving (as in the rotation of a mental image or off-line planning and cogitation), the classical image of an insulated cognitive engine will again be appropriate – although even in these cases, episodes of dense, continuous, mutual modulation between neural circuits may impede attempts to assign specific information-processing roles to component neural areas and pathways.) These issues are discussed at length in Clark, 1997.

A third class of argument capable of putting pressure on the intuitive divisions between mind, body, and world turns on the functional role of certain bodily or environmental resources. Thus recall Kirsh and Maglio's useful notion of epistemic actions: actions whose purpose is not to affect the world per se but to change the world *so as* to alter or reduce the computational loads on inner mechanisms as we try to solve a problem. The contrast is thus between an action like building a wall to keep out intruders (nonepistemic: the goal is merely to alter the world in an immediately useful way) and an action like shuffling Scrabble pieces in the hope of presenting visual cues better suited to prompt the recall of candidate words for use in the game. This is an epistemic action insofar as we don't generate the spatial reshuffling for its own sake, but rather for its value as part of the problem-solving process. In such cases it is as if the in-the-head computational process incorporated a sub-routine that happens to work via a call to the world. The sub-routine is, however, a bona fide part of the overall computational process – a process that just happens to be distributed across neural and environmental structures. Similar stories could be told about the use of nautical artifacts, the spatial arrangement of maps on the desk of a navigation team (Hutchins, 1995), the use of fingers in counting, and so on. Body and world can thus function as much more than the mechanism and arena of simple practical actions. They can provide a valuable extension of cognitive and computational space: a work space whose information-storing and information-transforming properties may complement those of the on-board organ of computation and reason, the biological brain. (Notice that the putative fact that we sometimes have a special kind of introspective access to what is going on in our own heads is not fatal to such claims, for it is agreed on (nearly) all sides that there are at least some inner and genuinely cognitive processes to which we have no such access. The outer events that form proper parts of extended computational processes would naturally fall into this latter category.) The question of whether such extended computational processes can ever constitute a genuine extension of the *mind* into the local environment is, however, both complex and vexed. Such extensions do not, in any case, seem to undermine the general distinctions between perception and action or between world and mind. At most, they temporarily shift the boundaries of the underlying computational states and processes.

Claim 4, I conclude, remains unproved. The distinctions between perception, cognition, and action and between mind, body, and world seem real enough and not at all threatened by our increasing recognition of the complex interplays and trade-offs that bind them together. Behavioral unfoldings involving the continuous, mutually modulatory interplay of brain, body, and world may, however, present more of a problem. This is a ripe topic for future research.

Conclusions: minds and their place in nature

The insights emerging from work on embodied, situated, and distributed cognition are of major importance for the development of a more balanced science of the mind. This new wave of research is a much-needed antidote to the heavily intellectualist tradition that treated the mind as a privileged and insulated inner arena and that cast body and world as mere bit-players on the cognitive stage.

No one, to be sure, ever thought that body and world had no role to play in the orchestration of successful action! But actual cognitive scientific practice has shown an unmistakable tendency to focus on inner complexity at the expense of attention to other factors. (Notable exceptions to this trend include J. J. Gibson and the ecological psychology movement and work in animate vision (Ballard, 1991).) This inner orientation shows itself in a wide variety of ways, including the (now decreasing) use of anesthetized animals for neural recording (see Churchland and Sejnowski, 1992, p. 440), the development in artificial intelligence of complex planning algorithms that make no use of real-world action except to carry out explicitly formulated plans and to check that the actions have the desired effects (see the discussion in Steels, 1994), and the fundamental image of symbolic reason whereby problems are first rendered as strings of inner symbols, which then form the domain of all subsequent problem-solving activity (Newell and Simon, 1981). In addition, the long accepted notion that the task of the visual system is to reconstruct a detailed three-dimensional image of the current scene (Marr, 1982) may also be seen as a consequence of this inner orientation – a consequence strongly challenged by recent work (such as Ballard, 1991) in which the visual system constructs only a sequence of partial and task-specific representations and relies heavily on our active capacity to repeatedly direct visual attention (via saccadic eye movements, etc.) to different aspects of the scene.

It is increasingly clear, then, that natural solutions to real-world problems frequently reveal brain, body, and world as locked in a much more complex conspiracy to promote adaptive success. It also seems likely that, to do justice to this complexity, we will indeed need to exploit new tools, concepts, and methods – perhaps importing apparatus from the established framework of dynamical systems theory so as to better track and explain emergent, decentralized, self-organizing phenomena. Nonetheless, it is not yet clear that such new ideas and methods should be seen as replacing, rather than augmenting, more traditional approaches. Nor does it seem likely that any of these developments will undermine the most basic distinctions between cognition and action or between mind, body, and world.

Mind, I conclude, has indeed too often been treated as an essentially passive item: an organ for recognition, classification, and problem solving, but one not intrinsically and fundamentally geared to the control of real-time action in a rich, highly exploitable real-world setting. As a result, perception, motion, and action have been seen as strangely marginal affairs: practical stuff to be somehow glued on to the real cognitive powerhouse, the engine of disembodied reason. It is this methodological separation of the task of explaining mind and reason (on the one hand) and real-world, real-time action (on the other) that recent developments in the study of embodied cognition call into serious question. This is a worthy challenge indeed.

References and recommended reading

Ballard, D. 1991: Animate vision. *Artificial Intelligence*, 48, 57–86.

Brooks, R. 1991: Intelligence without representation. *Artificial Intelligence*, 47, 139–59.

Churchland, P. and Sejnowski, T. 1992: *The Computational Brain*. Cambridge, Mass.: MIT Press.

*Clark, A. 1997: *Being There: Putting Brain, Body and World Together Again*. Cambridge, Mass.: MIT Press.

Hutchins, E. 1995: *Cognition in the Wild*. Cambridge, Mass.: MIT Press.

Kelso, S. 1995: *Dynamic Patterns*. Cambridge, Mass.: MIT Press.

Kirsh, D. and Maglio, P. 1995: On distinguishing epistemic from pragmatic action. *Cognitive Science*, 18, 513–49.

*Levy, S. 1992: *Artificial Life*. London: Jonathan Cape.

Marr, D. 1982: *Vision*. San Francisco: W. H. Freeman and Co.

Newell, A. and Simon, H. 1981: Computer science as empirical enquiry. In J. Haugeland (ed.), *Mind Design*, Cambridge, Mass.: MIT Press, 35–66.

Port, R. and van Gelder, T. (eds) 1995: *Mind as Motion: Dynamics, Behavior, and Cognition*. Cambridge, Mass.: MIT Press.

Resnick, M. 1994: *Turtles, Termites and Traffic Jams: Explorations in Massively Parallel Microworlds*. Cambridge, Mass.: MIT Press.

Sloman, A. 1993: The mind as a control system. In C. Hookway and D. Petersen (eds), *Philosophy and Cognitive Science*, Cambridge: Cambridge University Press, 69–110.

Steels, L. 1994: The artificial life roots of artificial intelligence. *Artificial Life*, 1 (1 and 2), 75–110.

Thelen, E. and Smith, L. 1994: *A Dynamic Systems Approach to the Development of Cognition and Action*. Cambridge, Mass.: MIT Press.

*Varela, F., Thompson, E. and Rosch, E. 1991: *The Embodied Mind*. Cambridge, Mass.: MIT Press.

40

Mediated action

JAMES V. WERTSCH

The study of mediated action focuses on how humans use *cultural tools*, or *mediational means* (terms used interchangeably), when engaging in various forms of action. The cultural tools involved may range from simple mnemonic devices, such as marks on a stone, to natural language and computers, and the kind of action involved may be socially distributed or carried out by individuals.

At the heart of analyses of mediated action is an irreducible tension between cultural tools, on the one hand, and agents' active uses of them, on the other. A focus either on cultural tools or on agents in isolation fails to come to terms with this defining property of mediated action. Focusing on agents in isolation often amounts to accepting the tenets of *methodological individualism*: that is, assuming that "no purported explanations of social (or individual) phenomena are to count as explanations, or . . . as rock-bottom explanations, unless they are couched wholly in terms of facts about individuals" (Lukes, 1977, p. 180). This assumption, which underlies many studies of psychological processes, fails to take into account the ways in which cultural tools shape human action. Conversely, focusing on cultural tools in isolation all too easily leads to a kind of social or instrumental reductionism in which active agents disappear and human action is viewed as being mechanistically determined by mediational means. Both forms of reductionism can be avoided by maintaining a focus on mediated action. From this perspective, what is to be described, interpreted, or explained is mediated action, a unit of analysis whose properties span the gap between traits of individuals and properties of instruments or contexts.

There are striking similarities between many studies of mediated action and research in other areas of cognitive science. For example, analyses of distributed cognition focus on how humans working with instruments such as computers and humans working in groups form integrated cognitive systems that cannot be understood by examining the elements of such systems taken separately (e.g., Hutchins, 1995; Norman, 1988). Analyses of mediated action pursue this line of reasoning by assuming that virtually any human mental process is distributed. Even an individual thinking in seeming isolation typically employs one or another set of linguistic or other semiotic tools, the result being that mediational means shape the performance at hand.

Authors who start from perspectives other than distributed cognition or mediated action are often tempted to place quotation marks around words such as *remember* and *think* when these words are predicated of cockpits, human machine systems, groups, and other such nonindividual agents. From this perspective, these terms are assumed to apply to individuals, and the inclusion of cultural tools or groups in the picture is taken to involve metaphorical extensions of core, unmarked meanings of the terms.

The fact that modifiers such as *distributed, socially distributed,* and *socially shared* are often placed before a term like *cognition* is a reflection of this assumption.

By contrast, researchers concerned with mediated action and distributed cognition view such assumptions as limiting their efforts to examine complex phenomena. While recognizing the importance of individual psychological and neural levels of analysis, these researchers view such levels as moments in a more inclusive picture. They do not believe that their basic unit of analysis can be reduced to these moments, or that these moments can be examined in isolation and subsequently combined into a more complex picture.

An understanding of mediated action must involve an analysis of the settings, or contexts, that embed the actions. Some actions are loosely structured, such as when two people who do not know each other improvise a conversation in a novel environment. At the other extreme would be a thoroughly domesticated, formalized setting that embeds a system of activity with a historically entrenched set of roles and tool-using practices. Examples here would include navigational procedures followed on a navy ship (Hutchins, 1995), bureaucratic dynamics in a business office, and experimental practice and theory construction in scientific laboratories.

The study of mediated action often traces its intellectual roots to the Russian psychologist and semiotician Lev Semënovich Vygotsky (1987) and his colleagues and followers, such as Aleksandr Romanovich Luria and Aleksei Nikolaevich Leont'ev. Vygotsky's influence continues to be felt in at least two respects that distinguish the study of mediated action from most other lines of inquiry in cognitive science. First, Vygotsky was fundamentally concerned with how human consciousness is socially, culturally, and historically situated. While recognizing the role of universals in human mental functioning, his focus was on how specific forms of mental functioning reflect and reproduce concrete social, cultural, and historical settings. This line of reasoning has given rise to contemporary sociocultural studies (Wertsch, 1991) in which the key to understanding the relationship between sociocultural settings and human action is the cultural tools that are employed in mediated action.

A second way in which the Vygotskian heritage is in evidence in contemporary sociocultural studies of mediated action is in the focus on language and speech. Although Vygotsky formulated his position on mediation in terms of general claims about the tools which humans employ, language occupied center stage in his empirical analyses. For example, it lay at the center of his analyses of concept development and of the self-regulative potential of *egocentric* and inner speech. Some striking parallels to this line of reasoning can be found in recent discussions in philosophy and cognitive science. For example, building on the notion that language is "the ultimate artifact," Andy Clark (1997) argues that it serves as "a tool that alters the nature of the computational tasks involved in various kinds of problem solving" (p. 193), and Daniel Dennett (1996) has pursued similar lines of reasoning in arguing that language provides an essential key to understanding human consciousness.

In contrast to many contemporary linguistic analyses that focus on the structure of language, Vygotsky's primary concern was with the instrumental role which language plays in social and individual functioning. He was particularly concerned with the form–function relationships that characterize human social and individual functioning, examining such issues as the use of language in *complexive* and conceptual reasoning and the emergence of inner speech. In outlining his claims about such issues, Vygotsky relied on genetic, or developmental analyses, arguing that an understanding of mental functioning must be derived from the study of its origins and the transformations it has undergone in various *genetic domains* (Wertsch, 1991). In this latter regard, Vygotsky investigated changes that occur over sociocultural history, as well as microgenetic transformations that occur during the performance of a single mental act; but he paid particular attention to ontogenetic progressions and regressions, especially as they occur during childhood.

An essential part of Vygotsky's line of reasoning about the development of human consciousness concerns the social origins of individual mental functioning. His claims about this issue were much stronger than simply that individual processes are influenced by the social setting in which they develop. Instead, he argued that human mental processes such as memory and reasoning make their first appearance on the *intermental* plane and then appear on the *intramental* plane. Hence, like contemporary scholars of distributed, socially distributed, and socially shared cognition, Vygotsky began with the assumption that *thinking, memory,* and other such terms apply to social as well as individual phenomena. Indeed, he argued that intramental functioning largely derives from intermental functioning. In his view, intramental functioning emerges in individuals as a result of taking over the various aspects of the intermental functioning in which they have participated. From this perspective, the organization of intermental functioning is of crucial interest, since it is this organization that shapes what will emerge on the intramental plane.

The notion of mediated action provides crucial underpinning for this line of reasoning. It is by mastering the cultural tools used to participate in mediated action on the intermental plane that intramental functioning emerges. Language plays a particularly significant role in this regard. Specifically, speech (i.e., a form of mediated action using language as a mediational means) is viewed as one of the principal means for incorporating children and other apprentices or tutees into intermental functioning, and this speech is then mastered by these apprentices or tutees to form the intramental plane of functioning. For example, speech may involve a dialogue organized around questions posed by a tutor and responses provided by a tutee in a problem-solving setting.

Vygotsky (1987) examined such interchanges between tutors and tutees from several perspectives. For example, in his account of the *zone of proximal development* he outlined the relationship between intermental and intramental functioning in terms of implications for mental testing and productive forms of instruction. The zone of proximal development is defined as the distance between the *level of actual development*, which is what a tutee can do independently, and the *level of potential development*, which is what a tutee can do in collaboration with a more capable peer or tutor. In Vygotsky's view, a narrow focus on the level of actual development results in assessment and instructional strategies that examine only where individuals have been and fails to provide crucial information about whether and how they might progress to the next level of development. By contrast, his focus was on intermental functioning, with its attendant use of language and other mediational means, and how such intermental functioning can generate new forms of individual functioning.

Building on the foundations provided by Vygotsky and his followers, contemporary analysts have extended the notion of mediated action by outlining several of its specific properties. The beginning assumption for this effort is again the basic claim that mediated action is defined by the irreducible tension between cultural tools and the active use of these cultural tools by individuals or groups. In what follows, three basic properties of mediated action are outlined: (1) mediational means and hence mediated action are socioculturally situated; (2) mediational means are associated with constraints as well as affordances; and (3) the relationship between agents and mediational means can be characterized in terms of *appropriation* as well as *knowing how*.

One of the principal reflections of Vygotsky's influence on accounts of mediated action and one of the points at which these accounts come into contact with findings from disciplines such as sociology, history, and anthropology concerns the sociocultural situatedness of mediational means. One and the same set of cultural tools is not provided by all sociocultural settings, and cultural tools are not invented *ex nihilo* by individuals. Instead, they are fundamentally shaped by the institutional, cultural, and historical forces that characterize a particular sociocultural setting. Among other things, this is manifested in the fact that cultural tools are associated with particular forms of power and authority in a particular setting. Just as Donald Norman (1993) has asserted that "technology is not neutral," studies of mediated action recognize that certain cultural tools, such as patterns of speaking, may be *privileged* (Wertsch, 1991) in one sociocultural setting but not in others.

Such observations suggest that mediated action is a good candidate for a unit of analysis in interdisciplinary research. Scholars in disciplines such as anthropology, sociology, and history often provide valuable insights into what cultural tools are available in a particular sociocultural setting, insights that scholars in psychology, cognitive science, and other disciplines can employ when trying to understand human action. Representatives from these latter disciplines, in turn, can provide valuable insight into how a sociocultural setting is produced and reproduced. From this perspective, mediated action provides a kind of meeting point where representatives of various disciplines may be able to coordinate their efforts.

Study of the forces that give rise to cultural tools has not usually been the main focus of analyses of mediated action, but there are a few general points that can nevertheless be made. Perhaps the most interesting claim to explore in this connection is that many of the cultural tools employed in mediated action were not designed for the role they have come to play. An illustration of this can be found in the keyboards used to type in English. (For a detailed case study of this example, see Norman, 1988.) Almost all users of such keyboards use the so-called QWERTY version, named after the fact that these letters are located at the upper left-hand quadrant of the array. Unless otherwise informed, most users of this keyboard assume that it was designed to facilitate their typing. As James Wertsch (1991) has noted, however, just the opposite is the case from today's perspective. The QWERTY keyboard was designed in an era of mechanical typewriters, when the biggest impediment to efficient typing was having two or more keys jam. As a result, the designers of the QWERTY keyboard specifically devised it to slow typists down.

With the appearance of electric typewriters and word processors, there is obviously no such need to slow typists down. Nevertheless, the vast majority of individuals who type in English continue to use the QWERTY keyboard, something that is made all the more striking by the fact that there is a readily available alternative keyboard design that is superior for most typists in terms of speed and accuracy. This Dvorak keyboard is relatively easy to master, and most computer keyboards can easily be converted to its configuration.

The fact that the vast majority of individuals typing in English continue to use the QWERTY keyboard speaks to the power of historical, economic, and other forces in shaping the cultural tools we employ and hence the mediated action we carry out. It is an illustration of the fact that many cultural tools may not be designed, or may not have evolved, to facilitate the forms of mediated action in which they are currently employed. The particular case of the QWERTY keyboard is sometimes viewed as an isolated, quaint illustration of how technological and economic forces can go wrong. As authors such as Norman (1993) have argued, however, institutional, cultural, and historical forces often result in technology that is far from ideally designed from the perspective of the user, and this raises the question of whether similar issues might not be involved for all sorts of mediational means.

Natural language presents an intriguing set of problems from this perspective. For the most part, language is not consciously planned or designed, a point that makes it somewhat different from the QWERTY keyboard example. However, many of the lessons of this illustration apply to language as well. For example, literacy and its impact on social and individual action raise several interesting questions. Literacy skills acquired in formal educational settings are associated with a specific set of cognitive skills, and the kind of language use required in formal literacy training is related to a willingness and ability to engage in tasks such as syllogistic reasoning.

However, it is generally accepted that literacy did not emerge as part of an effort to facilitate skills such as those required in syllogistic or other abstract reasoning tasks. Instead, literacy emerged in response to needs such as keeping records and conducting communication about commercial transactions. Furthermore, specific writing systems have often emerged when speakers of one language have borrowed the script used for another. Such facts serve to reinforce the claim that many cultural tools arise in response to forces that have little to do with the range of functions they are eventually required to serve.

These points about the forces that go into the production of cultural tools are implicitly tied to assumptions about their use, or consumption. In this connection the second basic property of mediated action to be outlined here is that mediational means always involve constraints as well as affordances. Building on ideas proposed by James Gibson, Norman (1988) has defined affordances as the "perceived and actual properties of the thing, primarily those fundamental properties that determine just how the thing could possibly be used" (p. 9). In the context of mediated action, the notion of affordance concerns the fact that specific mediational means facilitate certain patterns of action. This can be taken to be one of the implications of the illustration about typewriter keyboards. It is no accident that all recent world records for typing speed have been set with the Dvorak keyboard. At a general level this layout clearly has greater affordances for rapid typing in English than the QWERTY keyboard. This is not to say, however, that the Dvorak keyboard somehow causes or mechanistically determines performance at a certain level. There are still individual differences in the use of it, or any other cultural tool, and indeed there are undoubtedly some individuals for whom the QWERTY keyboard will always be a better cultural tool than the Dvorak keyboard. In the context of typing English, however, the general superiority of affordances provided by the Dvorak keyboard for most typists is clear.

Most analyses of the impact of introducing a new cultural tool into mediated action focus on the enhanced performance or affordances that are anticipated. For example, the Dvorak keyboard is typically described in terms of its superiority to the QWERTY keyboard. When discussing differences between using, say, calculators as opposed to slide rules, the emphasis is similarly on the new affordances provided by the former. And in the case of language the focus has usually been on how new forms and functions such as literacy provide additional power to human reasoning. This sort of benign, if not optimistic view of how the appearance of new cultural tools is associated with progress was clearly a part of the perspective Vygotsky (1987) outlined in connection with abstract rationality and conceptual reasoning.

An issue that is beginning to receive wider attention among analysts of mediated action has to do with how cultural tools involve constraints, as well as affordances. In many cases these constraints come to be recognized only through cultural or historical comparison. For example, it was only after the introduction of the Dvorak keyboard that the constraints of the QWERTY keyboard came to be widely appreciated, and after the appearance of a new word processing program, the constraints of previous ones often appear to be so serious that it is hard to remember how we could have lived with them. The fact that such constraints in cultural tools are routinely recognized with the appearance of new ones suggests the possibility that *any* cultural tool brings constraints as well as affordances with it.

This point applies to language as much as to any other mediational means. For example, building on the ideas of Edward Sapir and Benjamin Lee Whorf, John Lucy (1992) has outlined an argument consistent with claims about constraints as well as affordances associated with the grammatical structure of various languages. His argument is not that particular grammatical categories and distinctions mechanistically enhance or constrain the way humans think or remember. Instead, his point is that cognitive tendencies, or patterns of *habitual thought* of particular sorts, are afforded and constrained by the grammatical structures of particular languages.

Similarly, it has been argued that the rise of literacy on a mass scale restructured consciousness and, in so doing, has exacted a high price on certain forms of human mental functioning. From this perspective, literacy curtails or constrains several desirable attributes of human consciousness found in societies that rely primarily on oral traditions. For example, orality is associated with aesthetic and rhetorical activities that have often been found difficult to maintain in literate societies. Although analyses of these issues have not been specifically formulated in terms of mediated action, they are quite consistent with the claim that forms of semiotic mediation should be viewed in terms of the constraints, as well as the affordances, they provide.

The third general property of mediated action to be reviewed here concerns issues that are often discussed under the heading of "internalization." This is a term that is widely used in psychology and cognitive science, but a consistent interpretation of what it means is noticeably lacking. Indeed, it often carries unwanted conceptual baggage because of the spatial metaphors it employs. These metaphors often entail internal–external divisions and forms of methodological individualism that are considered to be untenable when made explicit. A fundamental methodological tenet of research on mediated action is that the boundaries of the unit of analysis are flexible and fluid. Hence there is no need to differentiate between internal and external spaces in some rigid way. For this reason, a distinction between *knowing how* and *appropriation* will be used, instead of *internalization*, in what follows. In the context of the discussion of mediated action, the expressions *knowing how* and *appropriation* are concerned with two forms of relationships that may exist between agents and the cultural tools they employ.

Knowing how to use a cultural tool means being able to use it in a socioculturally appropriate manner. In this sense, knowing how is associated with the skills entailed in the use of cultural tools, as opposed to the *knowing that* associated with propositional knowledge (Ryle, 1949). Knowing how is a notion that applies to cultural tools ranging from language to computers to bicycles. There are clearly some individuals who are better than others at using such cultural tools, pointing to the fact that development and individual differences are important topics to be examined under this heading. Unlike the term *internalized*, the term *knowing how* does not entail the idea that the processes involved must be understood as existing on an internal plane, something that is clear in the case of bicycle riding, but also needs to be recognized when considering the use of computers, or even language. The focus is instead on how the psychological and neural processes required to use a cultural tool emerge, rather than on how external processes are somehow transferred to a preexisting internal plane.

Discussions of internalization, at least in cognitive psychology and cognitive science, are primarily concerned with processes of knowing how. While an analysis of knowing how to use cultural tools is obviously essential for any adequate account of mediated action, studies of mediated action also frequently touch on a second phenomenon sometimes lumped under the heading of internalization. This second phenomenon concerns *appropriation*, a process that may operate independently of knowing how to use a cultural tool. In contrast to knowing how, where the issue is whether individuals can use a cultural tool appropriately, the issue of appropriation concerns individuals' willingness to use the tool and their sense of ownership of it.

The notion of appropriation derives largely from the ideas of the Russian literary scholar and semiotician Mikhail Mikhailovich Bakhtin (see Wertsch, 1991) about the ways in which linguistic forms and meanings *belong* to speakers. From this perspective, the process of speaking is always one of taking words that belong to others and making them one's own. In some cases, this is relatively straightforward, and the two *voices* involved operate in relative harmony. In other cases, however, tensions may arise over the *ownership* of forms and meanings, and there may be resistance on the part of the speaker to using the words or, more generally, the cultural tools of others. The dynamics of such relationships have major implications for whether and how various forms of mediated action will be carried out.

In general, appropriation is a less well-developed notion than that of knowing how in studies of mediated action, but it will undoubtedly play an increasingly important role in such studies as they delve further into issues of identity, EMOTIONS, and other little-studied dimensions of human action. In this connection, it will be particularly interesting to examine cases in which individuals who demonstrate a facility in using a mediational means exhibit an unwillingness or resistance to doing so. Such resistance may surface either in the form of explicit reflection or through other aspects of practice.

This kind of phenomenon is particularly striking in the case of language use. It is not uncommon for individuals who have obvious facility in employing a language, dialect, expression, narrative, or other linguistic form to exhibit a clear resistance to using it. In such cases, they may state or otherwise suggest that a particular way of speaking is *not theirs* or that it *belongs to others*. For example, a bilingual individual may go to great lengths to avoid speaking one of the languages she knows because of such attitudes. In such cases, the distinction between knowing how to use a cultural tool and appropriating it is clear and can be quite important when trying to interpret mediated action. While such cases stand in contrast to most, where knowing how to use a cultural tool is closely connected with appropriating it, they provide a reminder that these two processes need to be differentiated in the study of mediated action.

In sum, the analysis of mediated action is concerned with how humans employ cultural tools in social and individual processes. Because of its focus on the irreducible tension between agents and cultural tools which defines mediated action, this analysis stands in contrast with others that focus on individuals or on instruments in isolation. Many studies of mediated action are grounded in the ideas of Vygotsky, with the result that analyses of sociocultural situatedness and of language as a cultural tool have been particularly important. Contemporary analyses of mediated action are beginning to go beyond Vygotsky's formulation, to examine issues such as the conditions that have given rise to cultural tools, the constraints as well as affordances associated with them, and the distinction between knowing how and appropriation.

References and recommended reading

Clark, A. 1997: *Being There: Putting Brain, Body, and World Together Again.* Cambridge, Mass.: MIT Press.

Dennett, D. C. 1996: *Kinds of Minds: Toward an Understanding of Consciousness.* New York: Basic Books.

*Hutchins, E. 1995: *Cognition in the Wild.* Cambridge, Mass.: MIT Press.

Lucy, J. A. 1992: *Language Diversity and Thought: A Reformulation of the Linguistic Relativity Hypothesis.* Cambridge: Cambridge University Press.

Lukes, S. 1977: Methodological individualism reconsidered. In S. Lukes (ed.), *Essays in Social Theory*, New York: Columbia University Press, 177–86.

Norman, D. A. 1988: *The Psychology of Everyday Things.* New York: Basic Books.

*—— 1993: *Things that Make us Smart: Defending Human Attributes in the Age of the Machine.* Reading, Mass.: Addison-Wesley.

Ryle, G. 1949: *The Concept of Mind.* London: Barnes and Noble Books/Harper Row.

*Vygotsky, L. S. 1987: *The Collected Works of L. S. Vygotsky*, vol. 1: *Problems of General Psychology. Including the Volume Thinking and Speech*, trans. N. Minick. New York: Plenum.

*Wertsch, J. V. 1991: *Voices of the Mind: A Sociocultural Approach to Mediated Action.* Cambridge, Mass.: Harvard University Press.

41

Neurobiological modeling

P. READ MONTAGUE AND PETER DAYAN

Introduction

A cartoon description of the goals of cognitive science and neuroscience might read respectively "How the mind works" and "How the brain works." In this caricature, there would seem to be little overlap in the vocabularies employed by each domain. The cartoon cognitive scientist could speak at length about decision making and short-term memory in a relatively self-consistent manner, without any need to make reference to the language of neuroscience. Likewise, the cartoon neuroscientist could provide an immense body of physical detail about the function of neurons, synapses, and their component parts. She could even build models about how collections of neurons work together, or even how they might have developed.

In both the cognitive and the neural cases, such descriptions are inadequate; some phenomena will appear enduringly complicated, admitting no simple theory at a single level. In the cartoon scenarios, it is possible that many mindlike phenomena are not reducible in any strong way to descriptions enlisting interactions among components in the brain. There are a number of sophisticated arguments suggesting why such a reduction is or is not possible, likely, or fruitful – we do not enter into this debate here. Instead, we will illustrate with practical examples the mileage that can be obtained from a kind of *squeeze* approach to the problem of relating cognitive and neural descriptions. Complexities found by the cognitive scientist find natural explanation in the neural substrate; equivalently, the natural theoretical context which is necessary for interpreting the neuroscientist's results comes from hypothesized purposes.

This approach is fairly straightforward: take a consistent description of some behavioral or cognitive phenomenon, such as decision making on simple choice tasks, along with a description of what may be relevant neural constraints, and squeeze. The squeeze amounts to building a connection from the vocabulary in one domain to the vocabulary in the other. To the extent that reality and practicality permit, this seat-of-the-pants heuristic for theory construction pushes top-down and bottom-up constraints toward one another – hence the title of this chapter. The stated approach clearly exposes our bias: we assume that concepts like attention, reward, memory, decision, etc. will find some mapping onto the descriptions of brain function.

The squeeze was first formalized by David Marr, using ideas about computational equivalence. He showed how different notions of computation can be used in modeling cognitive and neural phenomena. In his view, cognitive tasks have to be specified precisely enough that one can write a computer program that demonstrably performs them. The action of collections of neurons that are believed to be necessary for this task must also be specified precisely enough that one can write another computer program that captures their behavior. Crudely speaking, the squeeze is successful if the two programs are computationally equivalent.

We illustrate the general approach using two examples. The first concerns how animals learn to predict events in the world that have rewarding consequences and how they can use the predictions to control their actions in order to optimize their rewards. Here, there is an enormous amount of neural, behavioral, and computational data that can be used to constrain the model at multiple levels. The second example concerns attention. Attention is less well understood than prediction at almost any level. We therefore explore at some depth just one aspect: namely, what it could mean for different sets of neural inputs to control the output of a neuron according to different contents of attention. Before discussing these examples, we describe classes of neurobiological models.

Methods of neurobiological modeling

There exist two classes of computational models in neurobiology. One concentrates on capturing closely the substantial information that modern neurobiology has garnered on the processes operating within and between single neurons – for example, the way that current flows through dendritic or axonal arbors, the effects of the many different sorts of ion channels, the ways that receptor molecules are influenced by neurotransmitters, etc. These models have been extremely important in understanding certain phenomena, including the origin of action potentials, oscillations in membrane potentials, and the integrative function of dendrites. They have been less illuminating at a systems level, because neurobiological models are intrinsically so complicated. Moreover, even these detailed models omit large numbers of phenomena, and there is no guarantee that the details on which they focus are the appropriate ones.

Computational models in the other class operate at the level of whole neural systems. In the best examples, the focus is on how collections of neurons cooperate to implement appropriate computations. The neural substrate is represented using artificial neurons that influence one another through modifiable synapses (see figure 41.1). The neural units are typically extremely simple representations of real neurons and ignore many biophysical details. These representations are then analyzed using mathematical techniques or are simulated on digital computers or both.

The models differ in the kinds and number of details incorporated, depending on the problem at hand. For us, the squeeze is key; the models have to represent some known feature(s) of the neural substrate, but also have to be simple enough to admit computational analysis at another level. We shall see that the resulting models make both biological and behavioral predictions that can be separately tested.

Prediction of reward

The ability of an animal to anticipate future salient stimuli is a form of prediction: representations of sensory events must reliably reflect the likelihood, time, and magnitude of future important events such as food and danger. Experiments have established that both vertebrates and invertebrates are capable of making and using such predictions to modify their internal models of the world and to choose actions appropriately. The concept of prediction is a computational one: a system uses its current state and

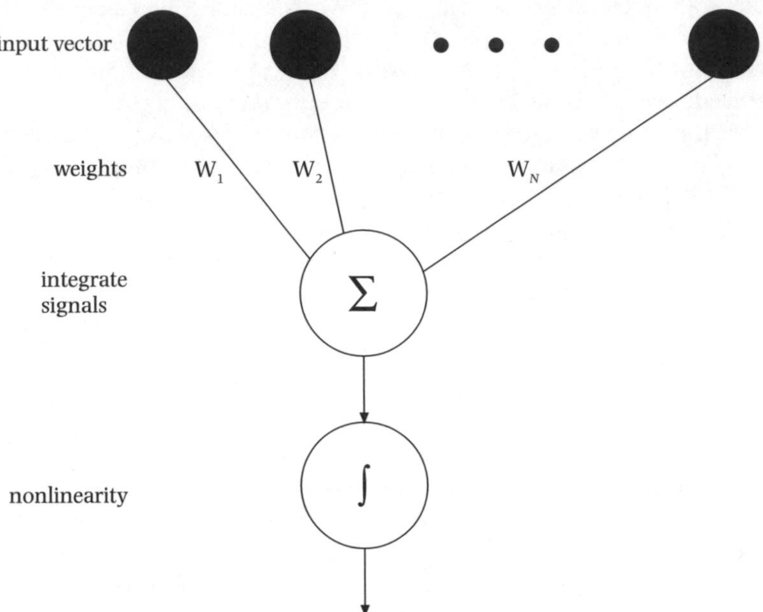

Figure 41.1 Simulated neural units. Artificial neural units take on a variety of forms, but the general scheme is shown here. A neural unit collects information from other neural units along connections (synapses). Each connection is associated with a weight ($w_1, w_2, \ldots, w_n$) that can be changed according to preset rules. The unit typically integrates its inputs (here shown as a summation) and some kind of nonlinearity before producing an output.

its past to anticipate the likely future state of itself and the world. Hence, prediction can be defined outside the realm of a behaving animal. Nevertheless, as we have described, behavioral experiments are used to assay this capacity in animals.

Unlike prediction, reward is a concept that is defined, not just assayed, by how an animal behaves. An organism, given multiple behavioral choices, will allocate some portion of its time or visits to each. Reward is assumed to be a latent quality of each behavioral choice. The magnitude of the reward content is defined by the relative proportion of time or visits allocated to the choice. This is a very behaviorist notion, but it permits easy quantification of many types of behavioral experiments, especially those involving decisions among alternatives. In past behaviorist traditions, these kinds of operational definitions became prohibitively restrictive in the classes of mechanistic explanations that they would permit.

We have been interested in understanding how animals learn to predict and also how they use rewards to (learn to) choose between actions. Substantial constraints are available from the theory of adaptive and optimal control (how systems of any sort can make predictions and choose appropriate actions), animal learning theory (the performance of animals in classical and instrumental conditioning tasks), and the neurobiology of reward (studies in electrical self-stimulation and the neuropharmacology of drugs of addiction). We begin by examining a simple neurobiological model which draws on both behavioral data and physiological data to hypothesize that a simple, direct representation of reward exists in the brains of honeybees and humans.

These models suggest one way in which predictions are constructed and used by real brains.

Learning decision making in honeybees, humans, and networks

We have previously proposed that the representation of reward and expectations about future reward are constructed in part through systems of neurons situated in the midbrain and basal forebrain of mammals and in analogous structures in invertebrates. As discussed below, these groups of neurons are known to be associated with the processing of information about affective values, send extremely long axons to widespread regions of the brain, and deliver neuromodulatory substances, including dopamine, serotonin, acetylcholine, and norepinephrine. By appealing to an established body of computational theory called the *method of temporal differences* (TD) (Barto, 1994), we have constructed a theory in which activity patterns in the cerebral cortex make predictions about future rewarding events through their connections onto subcortical nuclei (figure 41.2; a *nucleus* is a group of neurons in the mammalian central nervous system; the analogous structures in the peripheral nervous system or in invertebrates are called *ganglia*).

In this TD model, the world for an animal in a conditioning paradigm is cast in simplified form as a Markov decision problem. Here, there are states, which are signaled by stimuli such as lights and tones or positions in a maze; transitions between states, such as transitions in a maze; and actions, such as lever presses, which can affect rewards directly and can also affect transitions between states, as in a maze. The task for the animal is to choose appropriate actions that maximize rewards over the long term. The challenge is that actions can have delayed affective consequences; for example, in a maze, a poor early move can make for very long paths, but this fact may only be apparent when the animal finally reaches the goal. A standard way to solve such decision problems is called *dynamic programming* and comes from the field of optimal control. In dynamic programming, a system adopts a policy, which is just some consistent way of choosing actions at states, learns to evaluate states under this policy, and then improves the policy on the basis of these evaluations. The value of a state under a policy is the average amount of reward that the system can predict it will receive if it starts in that state and follows its policy. The policy can be improved by choosing actions that lead to high- rather than low-value states.

TD offers a method of performing dynamic programming in an approximate manner. The values of states are estimated using weights. These estimates can be improved by measuring the degree to which they are inconsistent with each other and with the delivery of rewards and punishments from the world – for example, if two places in a maze are one step apart, then the estimate of the distance to the goal from one should be one more than the estimate from the other. Any inconsistency is called a *prediction error* and is what is used to drive learning. In the model, the fluctuating outputs of neurons in subcortical nuclei (denoted P in figure 41.2) represent this prediction error. Hence, the fluctuating delivery of neuromodulator carries information about these errors to widespread target regions. If the inconsistency is in the estimation of reward, then increases in neuromodulator delivery literally mean "Things are better than expected," and decreases in neuromodulator delivery mean "Things are worse than expected." In fact, this same error signal can have two roles: (1) training the predictions

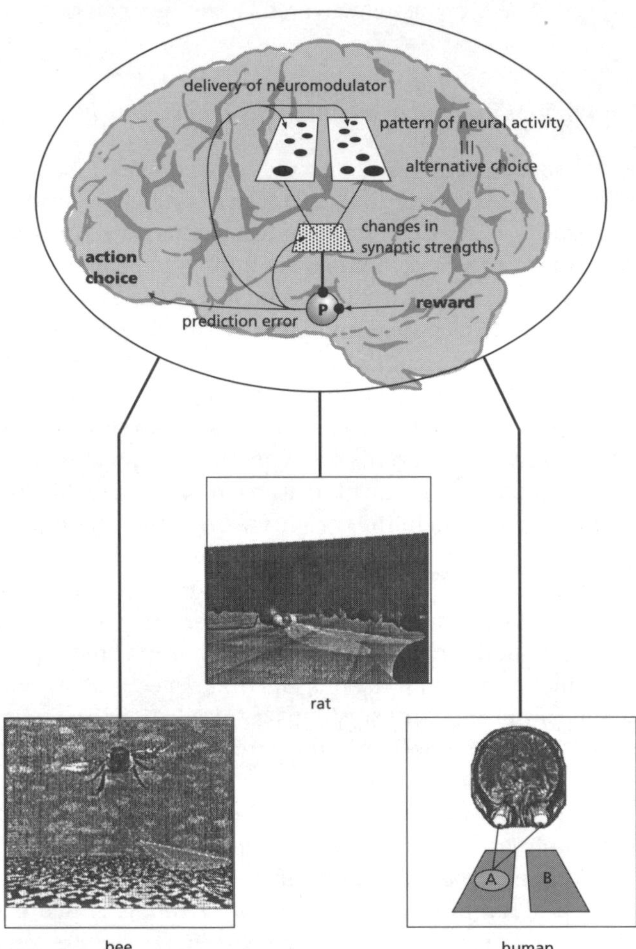

Figure 41.2 Neural representation of reward and expected reward. Activity patterns in the cortex construct expectations of the time and magnitude of future rewarding events and send this information through highly convergent connections to a group of modulatory neurons (labeled P). Also converging at neuron P is direct information about rewarding events such as the intake of food or pleasurable sensations. Neuron P is a linear unit, and, on the assumption that the cortical activity arrives at P in the form of a temporal derivative, the fluctuating output of P represents ongoing errors in the predicted amount of total future reward and the amount actually received. In the absence of direct reward input, the output of P is used to bias actions. In the presence of direct reward input, the strength of synaptic contacts is modified, and this updates the organism's model of the world. This arrangement has been used to model the choices made by flying bees, rats moving in a two-dimensional arena, and humans.

to be correct – any net bias in the fluctuations indicates an error in the expectations; and (2) choosing and training the choice of good actions – if the expectations are correct on average, then a positive fluctuation indicates that the associated action may be better than average. This second role for the signal implements the approximation to the way that policies are improved in dynamic programming.

As indicated in figure 41.2, this basic theory has been applied to bees foraging over fields of flowers, rats foraging in a two-dimensional arena, and human decision making on a simple card-choice task. In the case of the bees and the rats, a virtual world was constructed using computers, and a virtual rat and a virtual bee were permitted to move about in these worlds. This methodology provides a fruitful experimental testing ground for the behaviors that result from the operation of biological learning rules under the influence of some representation of the environment and sensory apparatus of the animal.

These models make testable predictions about the behavior expected from the bees, rats, and humans. The models incorporate biological assumptions; hence, they also offer predictions about the behavior of neurons. In these examples, the unifying factor is a well-understood computational theory that permits us to assign computational functions to specific biological constraints. The fact that the behavior of the models on foraging, learning, and decision-making tasks matches the behavior of the appropriate animal provides further support for the approach. As computing technology evolves, it may become possible to use large-scale simulations to make testable predictions about the interaction of multiple organisms in a simulated world.

These experiments primarily addressed the issue of behavioral choice. We have used exactly the same model to study the physiological behavior of neurons that deliver to their targets a neuromodulator called dopamine during the course of experiments that probe the way that animals come to predict events with rewarding consequences. We will describe this in some detail below.

Dopamine and reward

Dopamine (DA) is a neuromodulator that has long been associated with reward processing. In the mammalian brain, dopamine is produced and distributed by nuclei located in the midbrain. One of the dopamine nuclei is called the ventral tegmental area (VTA). The neurons in the VTA send axons to brain structures known to be involved in reward processing (one important site is the nucleus accumbens). As outlined below, three major lines of experimental evidence suggest that the VTA and dopamine in general are involved in reward processing.

First, drugs like amphetamines, which are known to be addictive, are dopamine re-uptake inhibitors (i.e., they prolong the action of dopamine near the sites where it is released). Second, neural pathways connected to the VTA are among the best targets for electrical self-stimulation experiments. In these experiments, rats press bars which deliver an electrical current at the site of the electrode. The rats choose this self-stimulation over food and sex. Third, agents which block the action of dopamine on dopamine receptors lead to extinction behavior in instrumental conditioning: animals that press a bar to get a reward will stop pressing the bar when given the full reward under measured doses of haloperidol (a dopamine receptor blocker), just as if they were no longer being rewarded. In spite of these very concrete results suggesting a role for dopamine in reward processing, the actual relationship between dopamine release and reward delivery is complicated. In many cases, for example, the delivery of reward to an animal is not followed by any increase in the delivery of dopamine. Also, dopamine plays many other important roles.

Dopamine delivery and prediction

Another main source of neurobiological constraints on this issue is a series of experiments performed by Wolfram Schultz and his colleagues (Schultz, 1992). They characterized the electrophysiological properties of dopamine neurons in the monkey VTA and substantia nigra (another dopamine nucleus involved in motor acts). These workers recorded from dopamine neurons while animals were learning and performing simple behavioral tasks for reward (apple juice). They found a subset of dopamine neurons in the VTA whose activity clearly relates to reward processing, but not in a simple fashion.

In one task, monkeys were presented with a light which signaled the delivery of reward (apple juice) provided that the animal performed an action within a pre-specified amount of time. In the context of this simple task, the light consistently predicts that reward will be delivered consequent on the action. Through training, the animals' reaction times for the action decrease, and they clearly use the onset of the light stimulus as a cue that reward will follow if they act correctly. These statements all rely on behavioral assessments; however, Schultz and his colleagues found that the dopamine neurons changed their firing rates in ways that consistently related to the learning displayed by the animals. A number of consistent features emerged in these studies:

- Early in training (naive animal), most dopamine neurons increased their firing rate when reward was delivered and showed no change in firing rate upon presentation of the light.
- Later in training (trained animal), most dopamine neurons increased their firing rate when the light came on and showed no change in firing rate upon delivery of reward.
- If two sensory cues consistently precede delivery of reward, then changes in the dopamine neurons' firing rate shift from the reward to the earliest consistent predictor of reward.

Remarkably, these neurobiological data mirror the computational requirements of a prediction error signal as specified in a theory based on the method of temporal differences (TD). As we described briefly above, our model for neuromodulatory control of learning and action choice fits into a temporal difference framework. The striking fact is that the prediction error signal in TD has precisely many of the characteristics listed above for the dopamine neurons:

- Early in learning, when the computational agent does not know that a cue predicts the delivery of reward, it is surprised by the delivery of reward; that is, there is a substantial inconsistency between its predictions and the outcome, so there is a substantial positive prediction error (the increase in firing upon delivery of reward).
- Once the agent knows that the cue predicts the reward, then the reward itself is expected and leads to no prediction error (after learning, there is no change in firing upon delivery of reward).
- When the predictive sensory cue appears, it was not predicted, and there is a prediction error consequent on the cue (after learning, there is an increase in firing after the onset of the predictive sensory cue).
- For two sensory cues that both predict reward, a TD model learns that the earliest cue can itself predict the reward that the later cue predicts and so itself attracts all the net prediction error.

Of course, there are a number of problems with the model and the areas in which we made arbitrary choices, went beyond the available data, or brushed aside genuine complexities that might be inconvenient for the model. It is from these problems that behavioral and neurobiological experiments naturally arise. Notable concerns are:

- The basal firing rates of the dopamine cells are low (around 1 Hz), suggesting that increases and decreases in firing rate cannot carry the same amount of detailed information about prediction errors. This fact, along with other theoretical and experimental observations, suggests that there may be an opponent system to dopamine that constructs and delivers information about punishments and withheld rewards. There is reason to believe that this might be one of the roles of the serotonin system.
- Through their widespread axons, the dopamine cells distribute information about prediction error to widespread structures. In its simplest form, the model requires dopamine to control synaptic plasticity for synapses that construct the predictions. The location(s) of the memories are also unclear. We have suggested the amygdala as one likely site, based on its pivotal position in the limbic system and evidence that interfering with the amygdala interferes with forms of secondary conditioning. This phenomenon probes the affective values associated with stimuli.
- Some simple learning paradigms are best described in terms of attention: the animal allocates more or less attention to particular stimuli based on its experience with them. This differential allocation results in more or less learning accruing to those stimuli during learning. Possible mechanisms behind selective attention are discussed in the next section but have not been incorporated in the models described above.

These inadequacies notwithstanding, this model operates at four different levels of description. The temporal difference model matches animal-learning data; however, it also implements known techniques of optimal control, most notably the engineering technique of dynamic programming. This link enables the squeeze – any system implementing an algorithm like temporal differences can reliably learn to perform appropriate actions that can even require complex sequences of choices. The key signal for temporal differences is the prediction error. Schultz's data strongly support the hypothesis that this error is being carried by the fluctuating output of dopamine neurons. Assessing the appropriateness of behavior lies in the realm of ethologists, providing the fourth descriptive level.

Attention

One critical element missing from the above discussion is attention: how various sensory cues and prediction errors are marked as being more or less salient. Our examples above have not provided for such effects. The concept of ATTENTION originated in the vocabulary of psychology, but it has eluded being made computationally, psychologically, or neurobiologically crisp. One category of experimental observation is that, on presenting the same set of stimuli to an animal on different occasions, different stimuli seem to be favored in terms of reaching consciousness, attracting learning, controlling behavior, and even determining the activities of neurons. Other attentional phenomena such as orienting behavior are not well characterized by this description, but there is no reason to expect that everything we call attention should comprise a natural class.

Unlike the case of prediction, it is hard to specify a precise computational problem that attention is solving. One popular possibility is that attention is important because the way in which the neural substrate performs computations is such that it gets confused if multiple cues are processed simultaneously. From the experimental end, attentional effects have been probed in various ways. One is assessing changes in neuronal activity and local blood-flow changes in identifiable brain regions. As described and probed by Michael Posner, his colleagues, and many other groups, attentional effects, as measured by NEUROIMAGING technology, are not associated with a single brain area or with the whole brain. These workers have sought to determine how attention influences brain activity: either increasing or decreasing it, depending on the tasks to which a subject is put.

In addition to neuroimaging experiments, detailed electrical recordings from the brains of alert primates indicate that attentional effects, as assayed by changes in behavior or in perceptual thresholds, correlate with dramatic changes in the electrical activity of identified neurons in areas of the cerebral cortex devoted to vision (see Article 34, SINGLE NEURON ELECTROPHYSIOLOGY). The mechanisms and constraints that permit this kind of control of neural activity are in general unknown. But these experiments show clearly the net effect: namely, that certain synapses become ineffective and/or others become augmented. But they do not distinguish between radically different types of mechanisms.

Generally, theories of attention are significantly underconstrained by the available data at any level of description. Attention therefore constitutes a class of phenomena in which interaction between models and experiment can be particularly crucial in going beyond the phenomenology to the neural mechanisms – they solicit the very evidence that would make squeezing effective. Below, we suggest both the spatial scale and the neural loci through which attentional effects could emerge in a working brain. In all our discussion, we have in mind effects that most likely occur at the level of the cerebral cortex and basal ganglia of mammals, particularly primates.

Spatial scale of attentional mechanisms in the brain

In order to investigate how neural tissue in the cerebral cortex could implement constraints that result in attentional effects, a decision about scale must be made. Do attentional effects emerge at the level of brain regions, neural circuits, single neurons, groups of synapses, single synapses, or perhaps at an even smaller scale?

We first inquire about the smallest scale at which neural activity could be modulated. It is already clear from neuroimaging and electrophysiological studies that changes in the activity of groups of neurons correlate with behaviorally assessed attentional effects. It is therefore reasonable to assume that the activities of single neurons are similarly affected. We suggest here that the physical substrates of attentional effects in the brain could exist at a smaller scale still: the single synapse.

In the mammalian cortex, synapses average about 1 micron in diameter, and their density falls somewhere between 0.7 and 1.2 billion synapses per cubic millimeter, which amounts to about 1 synapse in every cubic micron of tissue – that is, a billion connections in a region about the size of a match head. Since synaptic densities are so high, most notions of *nearby* include a large number of synapses. It is not a sufficient framing of the problem to assert simply that the function of single synapses is modulated during attentional effects. A number of important points remain: (1) Where does the information originate that modulates the function of a synapse that receives this information? (2) How is the information delivered to the synapse in question? (3) What is the postulated effect on the synapse?

We describe a set of answers to these questions below. We do not delve deeply into question (1); rather, we assume that some region or regions of the cortex become specialized to construct and distribute the salience of various sensory events. One might expect prefrontal regions to be particularly involved in the voluntary control of attention. However, one can also imagine that more local, automatic mechanisms of suppression and enhancement of neural activity are just as important for attentional processing – for example, automatic segmentation of a visual scene that permits recognition of parts of the scene. In this latter case, it is unlikely that modular regions of prefrontal cortex would be the final arbiters.

Embedded in question (2) are subtle issues of how the information is coded and the physical substrates used to transmit it to the synapse. There is a wealth of possibilities here, ranging from changes in synaptic function caused by the release and vascular distribution of some humoral factor, to the rapid delivery of a neuromodulator like norepinephrine through activity in axons originating in the locus coruleus (a midbrain nucleus that distributes norepinephrine to the cerebral cortex and other structures). In both cases, synapses can be affected according to a volume effect: hormones and neuromodulators both act at a distance from their sites of initial distribution.

These mechanisms for distributing attentional information share the problem that the delivery mechanism communicates information to very many synapses; therefore, synapses which receive the information need some mechanism to assess whether they should be suppressed or enhanced on its basis. Temporal correlation between the electrical activity of the synapse and delivery of the *attentional signal* is one way to make this assessment: short-term fluctuations in the *bottom-up* synaptic activity associated with the object of attention would filter through to the areas controlling attention and would then be reflected in commensurate short-term fluctuations in the *top-down* attentional signal they broadcast. The resulting correlations between these two signals are straightforward to measure. The neuromodulators could also be targeted more precisely to particular synapses: the synapses may possess the right combination of receptors, making them more sensitive to a particular neuromodulator, or the attentional signal could be targeted through a fixed anatomical connection. In any case, comparatively few cells deliver neuromodulators to enormous numbers of synapses; hence, precision in the delivery of the information is lost and must be recovered by some other mechanistic trick. We have identified two possible general schemes.

Below, we outline in detail how specific synapses and/or cells could be selected by some attentional signal. By making the assumption that the synapse is the smallest scale at which control of neural function can be exercised, we arrive at two different physical schemes for how an attentional signal could be constructed and used in real brains.

The resource consumption principle: attentional selection in volumes of neural tissue

It is well accepted that the synapse is a junction that passes information from one neuron to another neuron or to volumes of neural tissue (white structures in figure 41.3B). In a real nervous system, information travels from one neuron to another in the form of electrical impulses called *action potentials*. Action potentials travel along thin branched fibers called *axons*, which terminate on other neurons through enlarged endings called *synaptic terminals*, or *boutons*. Information about action potential arrival at a synaptic terminal is passed to the recipient neuron through the rapid release and diffusion of chemicals from the synaptic terminal. This transfer of information is rapid, because the gap between the end of the axon and the next neuron is very small (20 billionths of a meter); hence, diffusion of the chemical rapidly influences the next neuron. There is one drawback: the arrival of the impulse causes calcium to flow very rapidly into the synaptic terminal, and, without calcium entry, the terminal will not release its neurotransmitter (i.e., no information is transmitted to the receiving neuron). Normally, the level of calcium inside the terminal is extremely low, but when the action potential (electrical impulse) arrives, the calcium flows in through channels, and the levels inside the terminal increase rapidly. This flow of calcium into the terminal is absolutely necessary for the terminal to function: impulses invading a terminal in a region of tissue without calcium will not be transmitted to the next neuron. In this context, the calcium present outside synapses acts like a limited, shared resource that synapses must obtain in order to operate. One of us (Montague) has suggested that the above facts about calcium and neural transmission amount to an abstract processing principle that permits volumes of brain tissue to select a set of functioning synapses. This idea is called the *resource consumption principle* (RCP) (see figure 41.3).

The resource consumption principle appeals to a fluid metaphor when treating the function of synapses. In this theory, there are two classes of fluids. The first class, called the *resource*, must be moved from outside to inside a synapse, in order for the synapse to function. The second class of fluid is envisioned as a composite of many separate fluids, each representing different kinds of information delivered throughout a volume of neural tissue through the release of different types of neurotransmitters. The collection of input signals is treated as a vector and is called a *key*. As stated, each component of the key is treated as a fluid available homogeneously throughout a volume of neural tissue.

Synapses are pre-equipped with receptors that recognize different combinations of neurotransmitter fluids. The combination of receptors on each synapse is also envisioned as a vector and is called a *combination lock*. Each time a molecule of neurotransmitter binds a receptor, a quantity of the resource is moved from the outside to the inside of the synapse where the binding takes place. The scheme is: (1) the key is presented to a volume of neural tissue; (2) the key matches some synapses' locks better than others and causes the matching synapses to consume more of the shared resource; (3) those synapses that have consumed the most resource tend to function (transmit); (4) at a slower time scale, the resource is replenished in the surrounding volume of tissue (see figure 41.4).

Since the resource is in limited supply, synapses that consume it do so at the expense of neighboring synapses in the local volume of tissue. In this fashion, there is enough resource for only a subset of synapses to function (transmit), and so a fierce competition for resource is set up because of the way that volumes of tissue are organized and the dependence on resource. In a direct sense, the volume of neural tissue *attends* to those synapses (locks) that have successfully consumed the resource. This description omits detailed considerations about the dynamics of these processes, but the general idea is communicated.

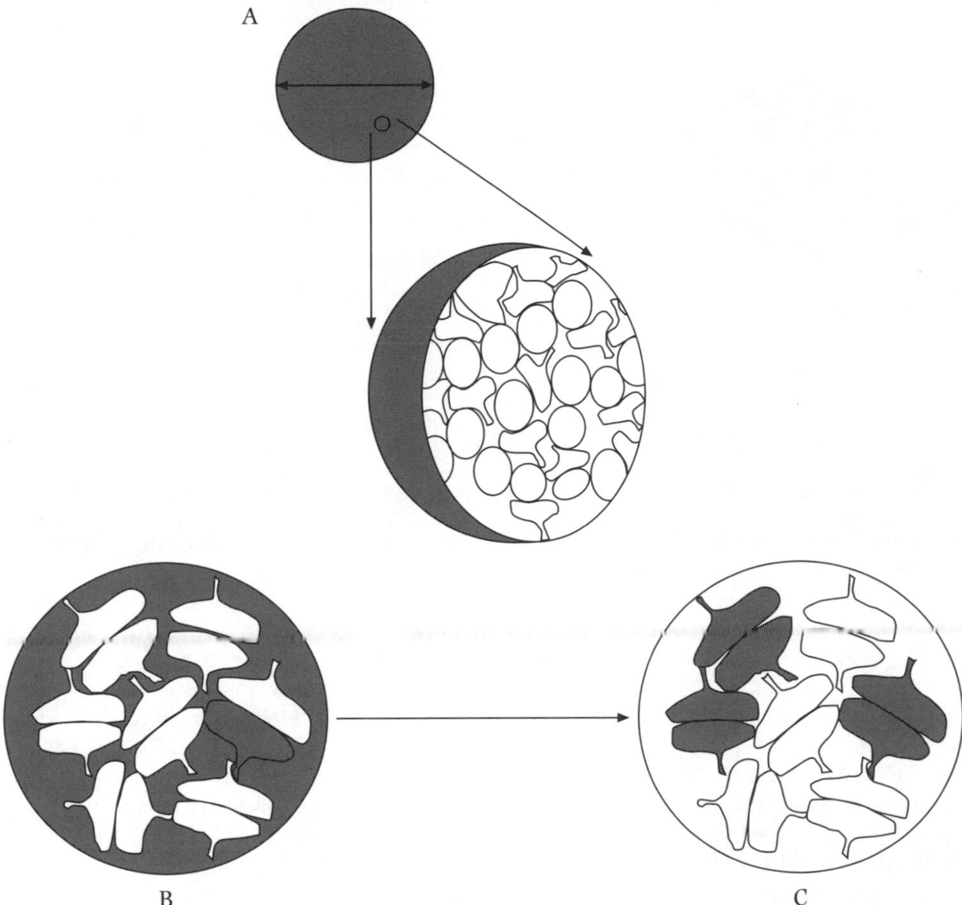

Figure 41.3 The resource consumption principle (RCP). A: Synapses are immersed in a homogeneously distributed resource fluid. A sub-volume of this region is shown as inset B. In order to function, the resource (gray) must be moved into a synapse (white structures). In this particular example, the space enclosed by the synapses represents over 85 percent of the volume; therefore, the resource is in limited supply and is consumable by a small fraction of synapses in the volume, as shown in C.

In these proposed mechanisms, attention to one set of synapses rather than another is granted as a result of a competition directly through the tissue space. Volumes of tissue that are far apart interact in the same manner through longer-range axonal connections that contribute synaptic terminals to common volumes. The difficult question not answered by this description is how an attentional signal is broadcast widely enough so that this local competition for resource can automatically decide on a set of working synapses.

The capacity for some set of synapses to match a particular key must preexist within the tissue before the key is experienced. It is intriguing to ask whether basic organizational properties of neural tissue define the primitives out of which various forms of attention are constructed. In the case of the resource consumption principle, attentional

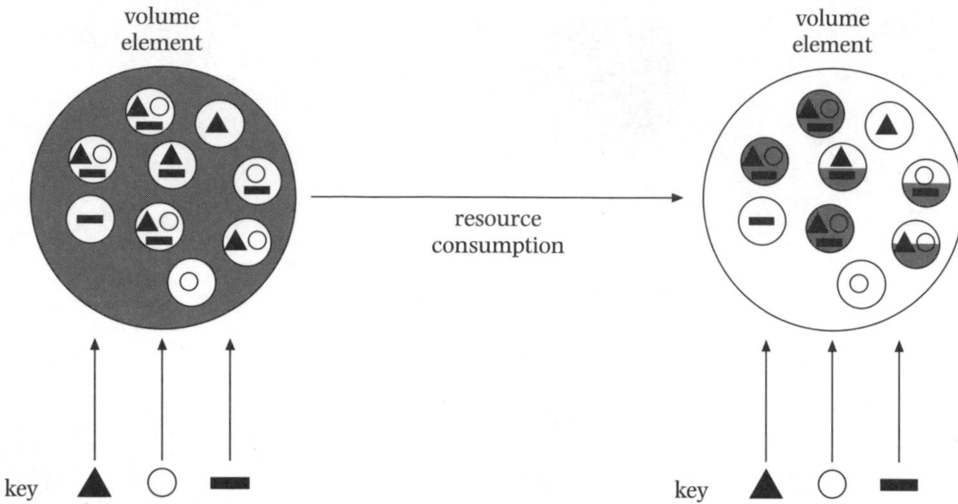

Figure 41.4 Matching of locks and keys through a fluidlike connectivity. The collection of signals that impinge on the volume is represented as a vector $v = (v_1, v_2, v_3 \ldots)$, where each component of the vector v_i represents a different signal type. We call this collection of signals the "key." Each component v_i of the key is presented homogeneously throughout the volume. In this fashion, each component of the key acts like a separate fluid homogeneously distributed throughout the volume. Different synapses are sensitive to different keys, because they possess various combinations of *receptors* each sensitive to one type of fluid – that is, one component of v. The collection of receptors on each synapse can also be envisioned as a vector and is called a combination *lock*. As shown, presentation of a particular key matches the combination locks of some synapses better than those of others, causing the matching synapses to consume resource (gray) it has captured. The resource allows the probability that a synapse is activated (ON) to be a function of the distribution of this required resource "fluid." The main idea is that those synapses marked as ON define the attention of the engine – that is, those synapses that work under the influence of the current key.

limitations result first and foremost from a commodity that is literally in limited supply and shared throughout a volume of tissue. There are certainly other strong possibilities for attentional effects that would exist at the level of entire circuits of neurons. We discuss one possibility below.

Attentional selection in recurrent reverberatory loops

The resource consumption principle locates attentional suppression of synapses directly as a property of the synapses themselves in conjunction with their three-dimensional neighborhood. However, there is a completely different way in which suppression or enhancement could take effect. The key is to consider the net influences of synapses on the postsynaptic cells. Individual synapses that are directly suppressed under the RCP could actually be functioning; however, the postsynaptic cells might not provide a faithful report of their activity (i.e., the cells are effectively inhibited or just not enhanced). It is unlikely that this inhibition is direct; for instance, there are no long-range GABA-ergic inhibitory axons. The inhibition could, however, be indirect.

One model for this indirect inhibition or excitation comes from the notion of self-excitatory loops through the striatum and the thalamus. These are brain regions involved in motor control that have a critical dependence on dopamine delivery for their proper function. Anatomical data suggest that neurons in somatosensory and motor regions of the cerebral cortex project to small clusters of neurons in the striatum (i.e., there is divergence in the cortex to striatum connections, since neurons from one area of the cortex project to many striatal clusters, each of which also receives information from other cortical areas). Conversely, these clusters send information indirectly back to those same areas of the cortex that generated them (i.e., there is convergence in the striatum to cortex connections). The main neurons in the striatum are inhibitory: they may inhibit each other to a degree partly controlled by dopamine levels (although this is somewhat controversial). The primary source of this dopamine is the substantia nigra (mentioned above).

Based on this suggestive anatomy and the physiology of the inhibitory cells in the striatum, it has been proposed that different motor actions are represented in the cortex and that these compete for control of the motor effectors (e.g., the limbs) at the level of the striatum. The mutual inhibition in the striatum provides a competitive mechanism whereby actions that are incompatible with each other compete – for example, extensor and flexor muscles for the same movement are not activated simultaneously. The effect of winning the competition is that the excitatory loop involving the neurons representing that action in the cortex and the striatum is *opened*, thereby encouraging the relevant action to be performed. Cells in the motor cortex representing actions that do not win the competition are not boosted in this manner, and so their actions are not performed.

As pointed out by Ann Graybiel (1995), the anatomy of this system forces relevant information from the cerebral cortex to compete for control of motor output at the level of the striatum. Moreover, Graybiel notes that the striatum is the target of a large number of neuromodulatory systems, allowing many kinds of information to influence the competition for motor output.

One can conceive of attention directed to one stimulus out of a collection in the same way as attention directed to one action out of a collection. Indeed, Neil Swerdlow and George Koob (1987) have suggested that the nucleus accumbens or ventral striatum might play the same role for stimuli that the dorsal striatum plays for actions. Cells representing different stimuli in the cortex would therefore compete not at the level of their synapses or even directly with each other, but rather at the level of the striatum, competing to gain access to an excitatory loop. Synapses within the cortex would then appear to be more or less effective according to the victory or defeat of their postsynaptic cells in the striatal competition.

There is a diverse cocktail of neuromodulator receptors in the nucleus accumbens. These molecular signaling pathways would then be available to bias the competition in light of predictions of the presence or absence of rewards. There is already direct evidence that dopamine and serotonin exert influence over phenomena in conditioning that have attracted explanations in terms of attention. Also, patients with schizophrenia, a disease that is believed to involve misfeasance in the dopamine system, show symptoms suggesting deficits in their attention.

Similar issues about other means of controlling the focus of attention apply to this model as to the RCP, except that there is now a defined structure on which any manipulations must take their effect. The connections between prefrontal areas and the basal ganglia could mediate explicit attentional control – we imagine the system learning how to pay attention to particular stimuli by learning what outputs to provide to which cell clusters in the striatum.

Summary

We have described both the intents and the processes of neurobiological modeling. The best neurobiological models are computationally explicit enough to perform complex tasks that animals can clearly perform themselves and close enough to slight abstractions of the neurobiological data that they can be constrained (and therefore falsified) by neural findings. We have described the *squeeze* that results in the best circumstances. All models of interesting behaviors are radically underconstrained at any given level. By taking on results at multiple different levels, even if they are couched in different vocabularies, the models become much better specified.

We showed two examples of modeling: one rather better worked out than the other. Learning to predict rewards and act appropriately on the basis of those predictions is highly adaptive. Although there are many ways that a vertebrate might go about making these predictions, we pointed to the evidence that they do it in a particular way, by showing the relationship between the firing of dopamine cells in reward-related areas of the brain and a key signal in one class of prediction algorithms called *temporal difference algorithms*. This link not only provided a rationale to the otherwise rather perplexing behavior of the dopamine cells in response to rewards, but also suggested why cells in two different dopamine-projecting regions might fire the same way, and it has led to suggestions for a number of other experiments. The same model applies to the selection and learned selection of good actions in behavioral choice tasks that have been applied from honeybees to humans. Of course, the model is quite abstract and quite incomplete in various important ways. However, it shows how one can take advantage of the squeeze.

The other topic for modeling is attention. We were not able to progress so far in this direction, mostly because the phenomena are not delimited so well in any of the different vocabularies – it is just not clear exactly what features a model should have. To show that modeling is still possible even under these circumstances, we focused on a particular notion, which is common across all models of attention: namely, that synapses change their net efficacies as a result of attentional processing. According to one model, these efficacies change directly. Synapses are constantly competing with their literal neighbors for a shared resource, and the competition can be biased through a number of different message systems in the brain which carry information relevant to the focus of attention. According to the other model, the relevant synapses do not change their efficacies directly; rather, the cells that they influence are part of self-excitatory loops through the striatum. In this latter model, competition happens in the basal ganglia, rather than locally in the cortex, and biasing happens there too. The models are not mutually exclusive.

None of these models is complete. They pose more questions than they answer. Nevertheless, neurobiological modeling offers a powerful complement to existing experimental techniques in neurobiology. Our capacity to collect data has far outstripped our capacity to construct appropriate theoretical contexts for them. Experiments are in any case intrinsically theory-driven. Neurobiological modeling amounts to specifying those theories precisely enough that they can be programmed or analyzed, and specifying them so that they respect the data not only from one level of inquiry but from them all.

References and recommended reading

Barto, A. G. 1994: Reinforcement learning control. *Current Opinion in Neurobiology*, 4, 888–93.

*Churchland, P. S. and Sejnowski, T. J. 1992: *The Computational Brain*. Cambridge, Mass.: MIT Press.

*Dayan, P. 1994: Computational modelling. *Current Opinion in Neurobiology*, 4, 212–17.

Graybiel, A. M. 1995: The basal ganglia. *Trends in Neurosciences*, 18, 60–2.

*Kandel, E. R. and Hawkins, R. D. 1992: The biological basis of learning and individuality. *Scientific American*, 267, 78–86.

*Montague, P. R. 1996: The resource consumption principle: attention and memory in volumes of neural tissue. *Proceedings of the National Academy of Sciences*, USA, 93, 3619–23.

Montague, P. R., Dayan, P., Person, C. and Sejnowski, T. J. 1995: Bee foraging in uncertain environments using predictive hebbian learning [see comments]. *Nature*, 377, 725–8.

Montague, P. R., Dayan, P. and Sejnowski, T. J. 1996: A framework for mesencephalic dopamine systems based on predictive hebbian learning. *Journal of Neuroscience*, 16, 1936–47.

Pulvirenti, L. and Koob, G. F. 1994: Dopamine receptor agonists, partial agonists and psychostimulant addiction. *Trends in Pharmacological Sciences*, 15, 371–9.

Schultz, W. 1992: Activity of dopamine neurons in the behaving primate. *Seminars in the Neurosciences*, 4, 129–38.

Swerdlow, N. R. and Koob, G. F. 1987: Dopamine, schizophrenia, mania, and depression: toward a unified hypothesis of cortico-striato-pallido-thalamic function. *Behavioral and Brain Sciences*, 10, 197–245.

42

Production systems

CHRISTIAN D. SCHUNN AND DAVID KLAHR

Cognitive universals

"Fiiirre!" If someone were to shout that while you were in the midst of reading this essay, you would, like most people, stop reading and look around the room for the source of the shout, or the fire itself. You would also consider whether the likelihood of a fire was sufficiently high to cause you to take appropriate action – for example, locate a fire extinguisher, call the fire department, or leave the room. Of course, you have not been sitting around waiting for someone to yell "Fire!" You were engaged in some task (unrelated to fires), yet were able to react to that stimulus. By contrast, if you were a police officer who was being trained at a firing range, you would be doing exactly that (i.e., waiting for someone to yell "Fire!"), and when you heard "Fire," you would pull the trigger on your weapon, rather than look around for the source of the cry.

In both situations, two kinds of knowledge would be brought to bear in responding. On the one hand, you would have facts (e.g., that someone yelled "Fire," that you are in the library or on a firing range, etc.), and on the other, you would have skills regarding how to respond (e.g., how to find an exit, how to fire a gun, etc.). Furthermore, responding could involve both preexisting knowledge of both types (e.g., that you are on a firing range, or how to fire a gun) and newly acquired knowledge of both types (e.g., that someone yelled "Fire!" or where to look for the exit in this particular room).

This example illustrates the cognitive universals that builders of production system models believe to be fundamentally important aspects of intelligent behavior (both human and artificial). First, at any point in time there are always many possible actions from which one must be selected, and the selection happens very quickly. For example, the reaction to the shout "Fire!" involves quickly selecting among many possible responses. Second, actions are taken sequentially. For example, leaving the room involves a series of actions (locating the door, moving towards the door, opening the door, etc.). Third, there are two very different kinds of knowledge: *knowing that* and *knowing how*. In our example, you could know that you are currently on a firing range or know that someone yelled "Fire!" By contrast, you know how to fire a weapon, and you know how to exit a room. Your knowing that would determine which of your knowings how would be appropriate in this instance. Fourth, the process of behaving is fully interleaved with the process of learning. In our example, you would learn both things about the current situation and how to act while you were acting, and this learning would influence subsequent behavior. These cognitive universals are summarized in table 42.1.

These universal aspects of behavior led Allen Newell, one of the founding fathers of cognitive science, to propose production systems as an information processing theory

Table 42.1 Cognitive universals deemed important by production system modelers

1	Parallel consideration of possible actions
2	Serial execution of actions
3	Distinction between *knowing that* and *knowing how*
4	Simultaneous performance and learning

of human learning 25 years ago (Newell, 1973). In this chapter, we will present an overview of what production systems are and how they achieve these cognitive universals. Then we will consider what the basic theoretical assertions of production systems are, illustrating these assertions with more detailed discussions of production system components.

What is a production system?

Production systems consist of two basic entities: one for representing objects and features in the world and one for representing skills or procedures for interacting with the world. Mostly for historical reasons, the set of entities used to represent objects and features is usually called *working memory*, but it would be more accurate to call it *declarative memory* (since what psychologists call "working memory" is typically much more limited in size). Declarative memory contains statements and beliefs about the world. These include things known for a long time (e.g., that rooms have red fire exit signs) and more recent knowledge such as assertions about the immediate context, both *external* (e.g., someone just yelled "Fire!") and *internal* (e.g., my subgoal to get to the exit).

By contrast, productions are rules stated in an *if–then* form (where the *if* side is called the *condition*, and the *then* side is called the *action*). For example, one might have the production, "If you hear the word 'Fire!,' then look to find the source of the yell." The condition side of the production is a list of entities that must appear in declarative memory (e.g., hearing the scream "Fire!," being in certain place, etc.). When the conditions of a production are true of the current state of declarative memory, then the production is said to *match*. The action side of a production can refer to either behavioral actions (e.g., leaving a room) or new declarative memory elements (e.g., a new current location or a new goal). Thus, declarative memory is used both to decide which productions to apply and to store the results of production actions.

Cognitive universals and the components of production systems that implement them

How do production systems attain the cognitive universals described in the preceding section? For each of the four universals, there is a corresponding major component of production systems. In the following section, we will discuss those components and show how they attain each of the cognitive universals.

Parallel selection among options – conflict resolution

It is the productions that give a production system the power to be reactive to changes in the environment and to consider large numbers of responses simultaneously. At any given time, all productions are considered simultaneously, to determine which ones are currently applicable (i.e., which productions match the items currently in declarative memory).

What happens when more than one production's conditions are currently satisfied? For example, in the case of someone yelling "Fire!" while a person is trying to read an essay, how does the person know whether to continue reading or to pay attention to the shout? The productions relevant to reading are still applicable, even though the productions relevant to dealing with the shout have also become applicable. The process by which a production system decides among potentially applicable productions is called *conflict resolution*. An important thing to note about conflict resolution is that it involves a theory of decision making: how does one decide among different potential actions? For production systems, these decisions are viewed as an integral part of the computational framework.

As production systems have evolved over the past 25 years, a number of conflict resolution schemes have been used. The early schemes used various combinations of the following approaches: (1) favoring productions whose conditions refer to declarative memory elements that have been most recently added or changed; (2) favoring productions with many conditions (i.e., more specific ones) over productions with few conditions (i.e., more general ones); or (3) simply setting a rank ordering among productions (e.g., fire-attending productions are always considered first). These schemes are difficult to implement, because the ordering of productions frequently needs to change from one task to another, and this change in ordering is not well predicted by recency or specificity. Nonetheless, these are usually the approaches that are tried first in building a production system model, because they are fairly simple.

Alternative schemes order productions according to how often they have been used in the past (with preference given to those that have been used most often) and how often the productions have been used successfully. In this way, production order can be adapted to different tasks via experience with each task. We will return to this issue later, when we discuss learning in production systems.

A more radical approach to conflict resolution is to not do it at all, but instead apply all rules that could be applicable to the current situation in parallel. In one such scheme, the productions only make suggestions about what to do (applying knowledge from past experiences), and then some other conflict resolution scheme has to decide among those suggestions. In another, all productions fire and reallocate a limited pool of activation *resources* to the declarative memory elements on their action sides. But, in effect, all such schemes merely defer the conflict resolution decision and delegate it to some other part of the architecture. The basic issue of conflict resolution and the ultimate sequential selection of motor actions cannot be avoided in a production system framework.

Serial execution of actions – subgoaling

How are production systems able to follow a sequence of steps in service of some goal? There are two answers to this question. First, productions can refer to declarative memory items that are the result of a previous production action. For example, in adding 15 and 19, the production for setting the carry in the tens column can only occur after the production that retrieves the sum of nine and five.

The other answer is that the conditions of productions can also refer to goals and subgoals. For example, one might have the production, "If my goal is to leave the room, then set a subgoal to locate the exit." With a number of such productions, a production system can thus follow a long sequence of steps by setting a sequence of subgoals.

Knowing that versus knowing how: declarative memory versus productions

Production systems easily capture the distinction between *knowing how* and *knowing that*: the distinction between items in declarative memory and productions corresponds to the distinction between *knowing that* and *knowing how*. There is considerable psychological and neurophysiological evidence suggesting that the human brain treats these kinds of knowledge separately. For example, some brain-damaged patients can learn new cognitive skills yet are completely unable to learn new facts (see also Article 29, DEFICITS AND PATHOLOGIES). This distinction also explains how people can behave in a rulelike fashion without explicitly knowing the rules. For example, people usually have little difficulty applying the grammatical rules of their native language (procedural knowledge), without being able to explain that grammar (declarative knowledge).

Simultaneous performance and learning – production rule learning and tuning

A fundamental challenge for both builders of AI systems and cognitive modelers is to produce a model that can both perform some task and learn (i.e., improve with experience). Production systems provide important insights into, and strong theoretical assertions about, how learning might occur in a performance system; so we shall explore this issue in some depth.

Learning in production systems Most of the production systems of the early 1970s were incapable of learning. In the few systems that did learn, the mechanisms for controlling when learning occurred were not especially powerful and rarely produced any serious accounts of human learning. At that time, cognitive scientists built models of a particular task at a particular level of competence. That is, they built models of how a person (or machine) might perform some task but did not address the question of how the person (or machine) might acquire those abilities. The modelers simply assumed that some learning mechanism was capable of producing the models that they built, and that the (eventual) addition of the learning mechanism would not interfere or interact with the performance of their models.

When cognitive scientists first became interested in learning and developmental issues in the late 1970s, they built a series of models, each of which described performance at a particular level of competence (or developmental stage). However, they did not offer computational accounts of how a model of one stage might evolve into the next stage. The belief at that time was that the learning mechanisms could be induced from a solid understanding of performance at the different stages. Further, it was assumed that the learning mechanisms, once developed, could simply be applied to a model of one performance level to produce a model of the next performance level. However, the efficacy of this approach has not yet been demonstrated, and the emergent view is that the early assumptions were probably incorrect.

As workable mechanisms of learning were developed in the 1980s, modelers discovered that there was a problem: under some circumstances, the addition of learning mechanisms radically altered the functioning of the performance model. That is, when learning was *turned on*, some models either ceased to function, because learning immediately moved them into an unworkable state, or they had to be rewritten from scratch to produce learning that noticeably improved system performance. Moreover, the workable learning mechanisms produced a more limited set of kinds of productions and memory elements. That is, many of the productions that were built into the early, static models were simply not learnable by the learning mechanisms. Of course, one might always believe that learning algorithms developed in the future might make these unlearnable productions learnable. However, at the very least, these problems undermined the early notion that one could build a competence model first and the learning model second. Nonetheless, this remains an unresolved issue, and many production system models are still being built without an active learning mechanism.

Mechanisms for learning new productions Production systems are based on the assumption that the fundamental unit of thought is the production, and that competence undergoes discrete jumps as new productions are learned. Most importantly, the learning is assumed to be specific to a particular production, rather than general improvement across numerous productions at once. Therefore, specifying how new productions are learned is a central theoretical and practical feature of a production system.

One production-learning mechanism is *compilation*, in which a new production is produced that performs in one step the action of several productions. An important variant of compilation is the chunking algorithm used in the Soar production system (Newell, 1990). The *chunking* algorithm determines which pieces of declarative knowledge were used by a recently successful sequence of productions and then creates a new production that looks for those declarative memory elements and directly produces the desired conclusion. For example, if the system determines through trial and error (i.e., many production firings) that a certain key opens a particular door, then it will create a production that recognizes the relevant features of the key and the door and directly selects the correct key.

Another mechanism is ANALOGY. This mechanism, used in the ACT-R production system (Anderson, 1993), converts examples in declarative memory into productions. In particular, the system tries to draw an analogy between the current goal and the corresponding goal in an example and creates productions that achieve the current goal using steps analogous to the ones used to achieve the example goal. For example, if the system was trying to solve an addition problem and had recently processed an example, it would try to draw analogies between the corresponding steps and states to get the answer to the current problem. As a result of this analogy process, new productions are created that then can be used the next time in lieu of the analogy process.

Other production-learning mechanisms create productions by combining or mutating existing productions. For example, one might add or delete conditions to a production, thereby making it more or less situation-specific. Classifiers are a special case of this process (Holland, 1986). They consist of a set of rules for classifying instances into different categories. New rules are created by randomly mutating some of the conditions of existing rules. The new rules are then strengthened or weakened according to their effectiveness in classifying instances.

Although there are several different mechanisms for learning new productions, current production systems that learn use only one production-creation mechanism, and each uses a different one. Why is this? It is not that the learning schemes are incompatible. Indeed, one could easily imagine a system that involves more than one of the production-creation mechanisms.

Several factors were instrumental in the decision to construct each of the production system architectures with only one production-creation mechanism. First, using only one mechanism is parsimonious. This feature is desirable in a scientific theory of human behavior, because the more components a theory has, the more difficult it is to test it. Second, having only one mechanism provides a strong test of the power of that mechanism. At this point, the exact limitations and potential of each of the learning mechanisms have not yet been fully explored. In particular, it is not clear that more than one mechanism is required (which is surprising, given how different the production-creation schemes seem, as we have seen). Finally, for computer scientists interested in performance factors, using only one mechanism reduces the computational demands: the learning algorithms tend to be very computationally intensive, and using more than one learning mechanism would severely affect system performance.

Mechanisms for tuning productions A frequent finding in studies of learning is that performance continues to improve with practice. That is, despite the fact that the learning is specific to a particular sub-skill, one usually finds continual speedup on that sub-skill with practice. How might a production system achieve this gradualness in learning if learning new productions creates discrete jumps in performance? Two solutions to this dilemma have been incorporated into production system architectures. One is to formulate productions at a very fine-grained level of detail, such that many productions are required to produce each external action. In such a scheme, the addition of each production would produce only a minor improvement in performance. Another solution is to associate with each production performance parameters that cause productions to perform slowly, suboptimally, or infrequently when they are first created, and then gradually become faster, more efficient, or more frequent. It is important to note that even with the addition of these more graded features, the fundamental unit of learning and transfer in a production system architecture is the production.

There are two primary mechanisms for tuning production-performance parameters. First, productions can be strengthened each time they are used and weakened over time with disuse. This production strength can be used to determine how fast a production will be selected or how likely that production is to be selected. Alternatively, productions can be tuned by keeping track of how often they lead to problem-solving success or failure. Under such a scheme, productions are more likely to be selected if they usually lead to success than if they usually lead to failure.

Meta-theoretical assumptions

Although particular production systems involve many different theoretical assumptions, there are some core assumptions that are common across all production systems. They are listed in table 42.2. In this section we will discuss meta-theoretical assumptions involved in the claim that there is a correspondence between the cognitive universals in table 42.1 and the production system commonalities in table 42.2.

The first three assumptions have already been treated in some depth. It is worth noting that psychologists using production systems differ in terms of how strongly

Table 42.2 Core theoretical assumptions of production systems

1	Procedural knowledge is stored as if–then rules (called productions).
2	The execution of a production is the fundamental unit of thought.
3	The acquisition of a production is the fundamental unit of learning.
4	Declarative memory is represented by framelike propositions.
5	The results of computation are stored in a (potentially temporary) declarative memory.
6	Knowledge is (mostly) modular.
7	All intelligent behavior can be captured in one cognitive architecture.

they hold these three assumptions. Some view production systems as a loose metaphor for understanding the activities of the mind (i.e., the mind generally acts *as if* it were a production system). Others use production systems to make an unambiguous theoretical assertion about the nature of the mind: the mind *is* a production system. The latter position is stated forcefully by John Anderson in his book on the ACT-R theory of human thought: "What is happening in the human head to produce human cognition?," he asks rhetorically. His answer is unequivocal: "Cognitive skills are realized by production rules. This is one of the most astounding and important discoveries in psychology" (Anderson, 1993, p. 1). In the rest of this section, we explain the other four core theoretical assumptions.

The framelike structure of declarative memory elements

Production systems typically represent declarative memory items in terms of entities called *frames* or *schemas*. Each frame is simply a list of attribute–value pairs in which attributes represent dimensions (e.g., color, size, location, etc.) that take on the values of the entity that the memory item denotes. For example, a declarative memory item representing some visual object might have a slot for the object's color, another slot for the object's shape, and yet another slot for the object's position. Different kinds of items can have a different set of slots. One can think of the different combinations of slots as representing different object categories, as well as relations between objects.

Such framelike memory structures provide precise, powerful representations of things in the world, including objects, relations between objects, and relations between relations. This representational power is especially important when one tries to build systems that do complex problem solving. However, this form of memory representation often has difficulty in situations where the knowledge is more continuous and less hierarchical (e.g., low-level vision).

Interestingly, the particular organization of declarative knowledge in a production system usually does not have immediate consequences for the system's performance. That is, one can get similar behavior from very different organizations of memory items. For example, one can use a single declarative memory element with many slots representing all that one knows about some individual, or one can have a large number of declarative elements each representing individual facts about that individual. A production system can function equally well with either representation scheme. The reason is that what matters is primarily whether information is contained somewhere in memory, not so much which information is stored together. If a different organization is selected, the productions are rewritten to accommodate the new structure. It is important to note, however, that in the production systems that learn, the organization of declarative memory can have a strong influence on performance.

Results of computation are stored in a (potentially temporary) declarative memory

As noted earlier, declarative memory does more than represent objects and features in the environment; it also represents the intermediate results for tasks that cannot be solved all in one step. For example, when mentally multiplying two two-digit numbers, you must mentally store the intermediate products. Thus, a production system for doing this would contain some declarative memory elements that represent the (external) multiplicands, as well as other declarative memory elements to represent the (internal) partial products. Another way in which declarative memory serves this function is in storing goals and subgoals.

This function of declarative memory raises another important and related question: Are these declarative memory elements permanent? In particular, are all the intermediate products of complex tasks erased after the task is complete, or do they leave long-lasting declarative memory elements? The basic problem is that the more information there is sitting around in declarative memory, the more likely it is that many productions will be satisfied simultaneously. This, in turn, complicates the process of conflict resolution. Moreover, this issue relates to a common psychological finding considered to be a basic feature of human cognition: the limited nature of short-term or working memory.

Production system designers have proposed a wide range of answers to these questions. At one extreme are systems in which items stay around forever once they are created. At the other extreme are systems in which items are deleted once the system moves onto the next task. The only way in which such systems can remember facts over long time spans is to have productions that re-create the facts in declarative memory when they are required. Intermediate between these two extreme approaches to dealing with the duration of declarative memory elements are those systems in which the elements vary in activation (which in turn determines how available or easily retrieved they are). The activation increases each time the represented facts or items are encountered and decays with time after each encounter. At first blush, it would seem obvious that the vast body of empirical evidence from experimental studies of human MEMORY could be used to select among these approaches. However, it turns out that one can produce the effect of a limited working memory using any of these schemes, and the ultimate answer will require both further experimental evidence and detailed modeling of those experimental results.

Knowledge is (mostly) modular

How does knowledge interact, and how does learning become generalized? Production systems provide strong answers to these fundamental questions: (1) learning occurs at the unit of productions; (2) transfer from one situation to another occurs to the extent that the same productions are applicable in both situations. This assumption about the modularity of productions allows production system designers to determine, via a detailed analysis of a task domain or a careful encoding of the verbal and behavioral protocols of human problem-solvers or both, what the individual productions are and simply add them to the system. One does not have to decide *where* to put a production; its conditions define when it will be used.

Because of their modularity, production systems scale up well to complex tasks. That is, not only do production systems function well on small, simple tasks; they also function well in more realistic environments involving many sub-tasks and thousands (or more) of bits of knowledge. For example, there is a production system called TacAir Soar which has tens of thousands of productions, can fly a simulated plane in a dogfight (in real time) while doing language comprehension and production, and is capable of providing a verbal summary of the mission afterwards (Tambe et al., 1995).

It is important to note that this modularity is not perfect. In particular, one often finds that some productions are not entirely independent of all other productions. As noted earlier, this can be particularly troublesome in production systems that learn, because whenever one has a sequence of productions that are needed to perform one task, then the removal or addition of a production can affect the function of the other same-task productions. For example, if a newly added production changes the internal goal in the beginning or middle of a sequence of steps, then the production which is to compute the next step may be affected. Moreover, many productions rely on other productions to set things up in a very particular form so that they can be accessed in just the right way. Thus, in many complex situations, the modularity of the production system is only partial, and groups of productions must be constructed with careful consideration given to their potential interactions. The way in which cognitive scientists deal with this complication is to do careful task analyses of the steps required by a task and of the relationship between them. This is one of the most difficult aspects of building production system models.

For computer scientists, whose goal is to build intelligent systems, rather than explain human intelligence, this partial nonindependence is usually a nuisance. For psychologists, however, this aspect reflects known properties of human learning. For example, learning how to play racket ball can interfere with one's tennis abilities (see also Article 49, MODULARITY).

All intelligent behavior can be captured in one cognitive architecture

From the perspective of a psychologist trying to develop a scientific theory in the form of a computational model, production systems have a very important advantage: it is possible to see how one underlying computational framework (called a *cognitive architecture*) could give rise to all cognitive activities. The advantage of using such a unified architecture was clear even in the early days of cognitive science. Without a unified architecture, a modeler had to develop a theory from scratch for each new task. Models developed in such a fashion had a very ad hoc flavor, and it was never clear how such models could be combined with models of other cognitive activities to produce overall behavior. Production systems were first proposed just as this issue of avoiding a plethora of ad hoc models was raised, and, even today, proponents of the unified architecture idea are for the most part production system modelers.

Conclusion

In this chapter, we have covered the set of cognitive universals that production systems were designed to capture, the basic components of production systems, and their core meta-theoretical assertions. In addition to those common features, there are components unique to particular production systems, including spreading activation between declarative memory elements via associative links, capacity limitations on activation, and learning in declarative memory. These design parameters represent different theoretical assumptions and lead to important performance differences. Although we do not have space to discuss them here, we mention them to convey the sense in which production systems are to be conceived as tools under development, rather than fixed theoretical statements.

However, there are a number of areas in which production system models have already done very well, and are arguably the strongest (and occasionally only) models in those areas. These areas are almost exclusively instances of higher-level cognition and generally require the coordination of many kinds of knowledge. They include learning mathematical skills such as algebra and geometry, learning computer-programming skills, language comprehension, scientific discovery, and many other forms of high-level, complex action and reasoning. For more detailed descriptions of these examples, we refer interested readers to Anderson, 1993; Halford and Simon, 1995; Just and Carpenter, 1987; and Steier and Mitchell, 1996.

References and recommended reading

*Anderson, J. R. 1993: *Rules of the Mind.* Hillsdale, NJ: Erlbaum.

*Halford, G. and Simon, T. (eds) 1995: *Developing Cognitive Competence: New Approaches to Process Modeling.* New York: Academic Press.

Holland, J. H. 1986: Escaping brittleness: the possibilities of general-purpose learning algorithms applied to parallel rule-based systems. In R. S. Michalski, J. C. Carbonell, and T. M. Mitchell (eds), *Machine Learning: An Artificial Intelligence Approach,* Los Altos, Calif.: Morgan Kaufmann, vol. 2, 593–623.

*Just, M. A. and Carpenter, P. A. 1987: *The Psychology of Reading and Language Comprehension.* Boston, Mass.: Allyn and Bacon.

Newell, A. 1973: Production systems: models of control structures. In W. G. Chase (ed.), *Visual Information Processing,* New York: Academic Press, 463–526.

—— 1990: *Unified Theories of Cognition.* Cambridge, Mass.: Harvard University Press.

*Steier, D. and Mitchell, T. (eds) 1996: *Mind Matters: A Tribute to Allen Newell.* Mahwah, NJ: Erlbaum.

Tambe, M., Johnson, W. L., Jones, R. M., Koss, F., Laird, J. E., Rosenbloom, P. S. and Schwamb, K. 1995: Intelligent agents for interactive simulation environments. *AI Magazine,* 16.

PART V

CONTROVERSIES IN COGNITIVE SCIENCE

43

The binding problem

VALERIE GRAY HARDCASTLE

Our brains process visual data in segregated, specialized cortical areas. As is commonly remarked, the brain processes the *what* and the *where* of its environment in separate, distal locations. Indeed, regarding the *what* information that the brain computes, it responds to edges, colors, and movements using different neuronal pathways. Moreover, so far as we can tell, there are no true association areas in our cortices. There are no *convergence zones* where information is pooled and united; there are no central neural areas dedicated to information exchange. Still, the visual features that we extract separately have to come together in some way, since our experiences are of these features united together into a single unit. The *binding problem* is explaining how our brains do that, given the serial, distributed nature of our visual processing. How do our minds know to join the perception of a shape with the perception of its color to give us the single, unified experience of a colored object?

This problem has a venerable history in philosophy, first appearing in its modern guise in David Hume, as he, following John Locke, speculated on the rules that our minds must follow in uniting simple impressions into more complex ideas. He recognized that the rules of association alone could not be enough: incoming stimuli are always changing, yet we manage to experience ideas as constant across time. Somehow our *faculties of imagination* step in and fill the gap between stimulus impressions and later memories and ideas. Immanuel Kant too recognized that mere spatial contiguity and temporal conjunction would not unite certain incoming stimuli into bound impressions at the exclusion of others. Both Hume and Kant concluded that our minds must add something to our perceptions so that our experiences are of a three-dimensional, object-filled world.

This history – and its solution – recapitulates itself in contemporary cognitive science. Like Hume and Kant, cognitive scientists recognize that the story of visual perception told thus far is incomplete. The brain must rely on something besides physical connectedness among cortical areas to generate united percepts. But what? Association, even in the head, is not enough. What would be?

Below I outline a *top-down* and a *bottom-up* approach to the problem of perceptual binding and examine the more popular hypotheses now being explored. None of these solutions can be correct as it stands, though, given what we know already about how brains work. Perhaps, as I suggest in the second half of this chapter, these failures indicate that binding is a pseudo-problem that could be dissolved by looking at perception from a different level of analysis.

The top-down dialectic

One way in which we gain empirical access to our brains' binding procedure is through the study of illusory conjunctions. We occasionally make mistakes when we combine shape and color features in creating a representation if the stimuli appear only briefly and our attention is elsewhere. For instance, if subjects are flashed a stimulus consisting of a red X, a green T, and a blue O flanked on both sides by numbers and are asked to notice and pay attention to the numbers, they may see (and later report) a blue X, a green T, and a red O. Indeed, some are so convinced of their perception that they do not believe that the actual stimuli were otherwise. The obvious hypothesis is that their brains, taxed by the difficult task, bound the separately processed colors and shapes together incorrectly.

Anne Treisman takes evidence along these lines to suggest that we have neural signals which convey the presence of distinctive features; we have *feature maps* for length, color, cardinality, orientation, curvature, movement, depth, and various topological properties (Treisman and Schmidt, 1982). Later stages of processing integrate the information from the early feature extraction phase into object-specific representations by focusing attention on the feature maps. Treisman hypothesizes that focused attention operates by means of a *master map* of locations that registers the presence and discontinuities of color. It selects all the features present at the attended location, binds them into a temporary object schema, and then compares that schema with stored descriptions (see figure 43.1). Illusory conjunctions occur as selective attention misbinds groups of spatially contiguous features together into units.

So, the illusory conjunctions at least stem from poor location information in some aspects of mental processing. But not everyone shares Treisman's conclusion that lack of attention is also necessary. For example, William Prinzmetal believes that poor spatial resolution is responsible for the illusory conjunctions (Prinzmetal and Keysar, 1989). Against Treisman's master map theory of feature integration, Prinzmetal argues that while diverting attention may indeed decrease spatial resolution (and so indirectly increase illusory conjunctions by further limiting location information), it also affects feature identification and perceptual organization. In support of his position, he points out that illusory conjunctions are more likely to occur when features are part of the same perceptual *unit*, and that we are more uncertain of where things are within a group than across groups.

Like the Gestaltists, Prinzmetal holds that external organizational principles, such as contiguous areas, common surfaces, subjectively defined groups, and syllable-like units in words, work to constrain what little spatial information we have. We have a set of processing functions that determine color, texture, and so on, and that information is *integrated* over previously defined objects or surfaces. He suggests that the algorithms we use in visual processing filter out noise and fill in absent stimuli using the most economical coding possible. One possibility is that the algorithms compute borders accurately but hues much more coarsely. In addition, visually sensitive neurons integrating information over receptive fields, which omit spatial information, would compound relatively poor color resolution in any algorithm. However, the ubiquitous feedback connections in our visual system allow information to flow back through the various stages of processing to make up for this deficit. We use object representation *templates* to constrain feature processing and to determine which processed bits should

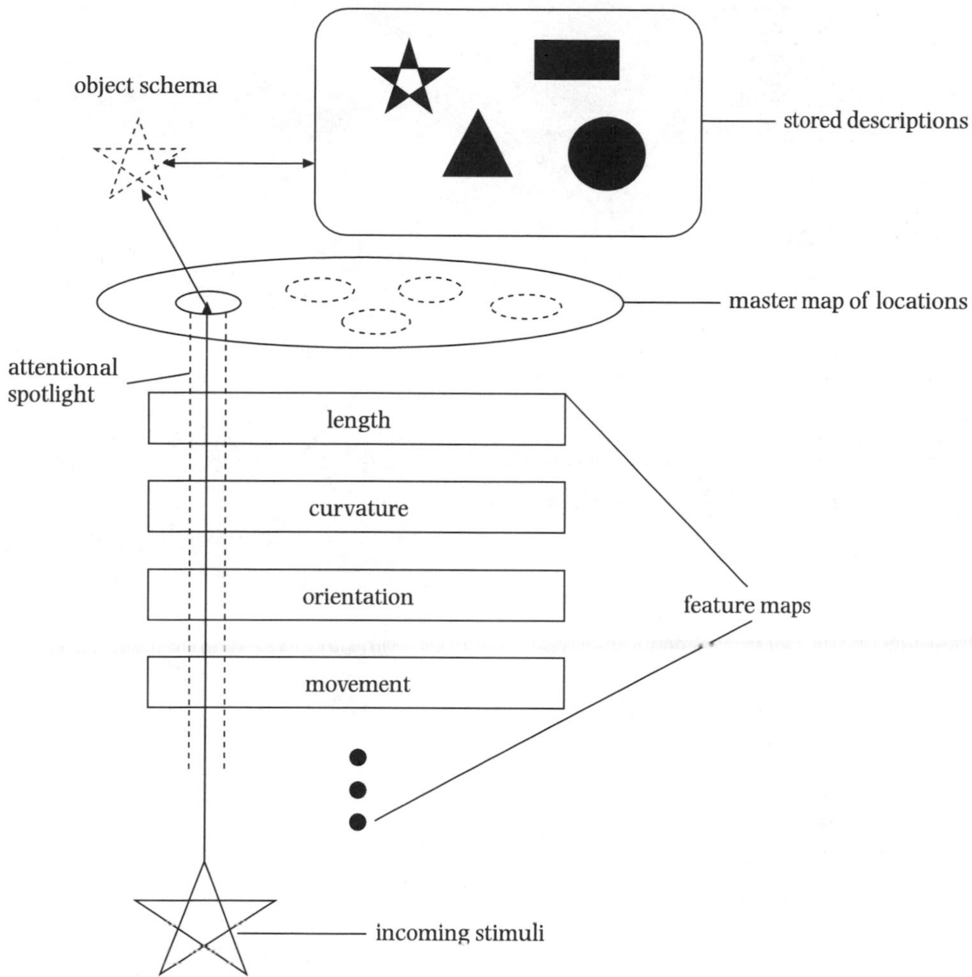

Figure 43.1 Treisman's model of perception proposes that early vision encodes some simple and useful properties of a science in a number of feature maps. Focused attention, using a master map of locations, integrates the features present at particular locations. The integrated information is then compared with descriptions of objects in a recognition store.

be united together into one percept (see figure 43.2). In essence, our memories of past object organization influence our present color and shape judgments.

A bottom-up contribution

It is not obvious how either Treisman's or Prinzmetal's (or any other top-down) solution to the binding problem might be implemented in the brain, however; for they all assume that various features of different objects are active simultaneously. Of course, the different neuronal patterns we use to represent objects or aspects of objects have to coexist within the same physical wetware. At the same time, we accept parallel

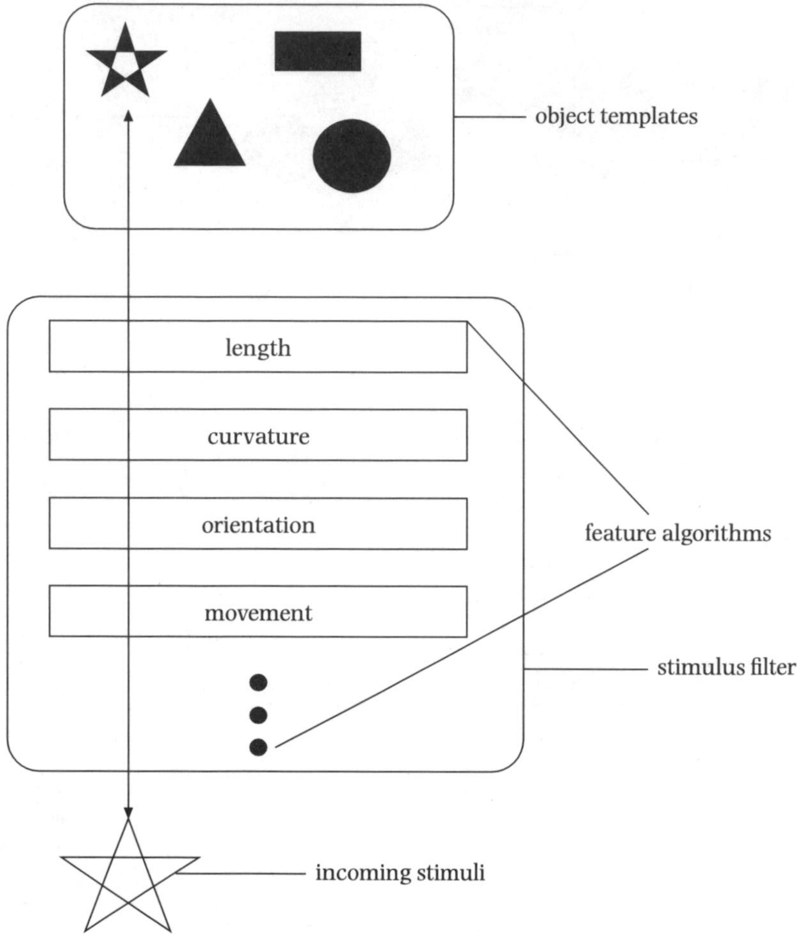

Figure 43.2 Prinzmetal suggests that we have object representation "templates" which constrain feature processing and help determine which features should be united together to form an object hypothesis. Reciprocal connections in our visual system feed information back among the stages of processing, so that our final perceptions can influence feature judgments.

distributed processing as the fundamental computational mechanism of the brain. But, if the brain is in fact fundamentally and massively parallel, we are left with what Christoph von der Malsburg (1995) calls the "superposition catastrophe." If we take mental *symbols* to be different subsets of coactive neurons within the same brain structure, as they are in the classical framework of neurobiology, then if more than one symbol becomes active at a time (as they surely must, given what we know about feature extraction), their coactivation superimposes them, and we get an undifferentiated mess. So, for example, if we are confronted with a red square and a blue triangle, we will not be able to tell that the red belongs with the square and the blue with the triangle, since our neuronal *symbols* for red, blue, square, and triangle will all be active together in a single, complex firing pattern.

If using the physical connections among neurons to tie different subsets of firing patterns together cannot work, the brain must still find a solution somewhere, for it does solve the binding problem. Time is a likely candidate, for mental and biological functions operate on two different time scales. There is the psychological time scale, which characterizes mental processing. This is ordered on tenths of a second; that is, we clock information processing in psychology in 10 msec units. There is also a faster, neuronal time scale, which characterizes cellular fluctuations and is ordered on thousandths of a second; we clock the activity of individual neurons in 1 msec units. The mean activity of a set or assembly of neurons evolves on the longer, psychological time scale, but the activity of the individual cells that comprise the assembly operates on the faster time scale. Correlation of the single cell activity fluctuations within a group of neurons could be what ties together the distributed firing patterns into a single unit. Moreover, several patterns of correlated activity could coexist if the patterns are desynchronized relative to one another, thus alleviating the superposition catastrophe. In other words, single cell activity (which would underlie and support higher-level, averaged activity) may be the appropriate place to look for the mechanisms that account for *higher-level* perceptual phenomena.

However, it is difficult to translate questions concerning how a single cell behaves into questions dealing with how populations of neurons interact. Since each area of the brain is composed of a staggeringly enormous number of neurons, recording the firing pattern of a single cell may not capture any relevant higher-level patterns or processes. Indeed, since there are tens of thousands of cells in each visual column, any one cell from which a scientist records could have a function orthogonal to the task at hand. And until neuroscientists could make the leap from studying single cells to studying populations of cells, many interesting questions concerning visual perceptual experience in the nervous system were simply unapproachable.

But within the last ten years or so, two sorts of techniques have emerged as ways to start studying the behavior of larger numbers of neurons: multiple simultaneous recordings from single cells and multiple-channel electroencephalograph recordings from relatively small populations of cells. As a means whereby to connect neurophysiological structures with larger neurophysiological or psychological functional units, these procedures allow scientists to examine the intrinsic dynamic operations of circumscribed cortical areas and to start assessing their relevance for visual perception. In particular, when coupled with computer technology (which permits simultaneous visual stimulation of different areas in the visual field keyed to the cells whose spike trains are being recorded), they gave scientists a way to start determining the principles of visual processing within single cortical areas. The function of the distributed systems connecting the various visual subsystems can then be studied using mathematical correlation methods to assess the cooperative firing across groups of neurons in different areas.

Though it is still largely unknown how the visual system links features of a visual scene together, neuroscientists are beginning to give a few answers to some of the smaller questions and are starting to speculate on what principles might underwrite psychology's larger story. For example, Reinhard Eckhorn and his research group at Philipps University in Germany took simultaneous multiple recordings of single unit activity, multiple unit activity, and local slow-wave field potentials from areas 17 and 18 in lightly anesthetized cats with independently drivable microelectrodes (Eckhorn et al., 1988). They evaluated their data for receptive field properties, orientation and direction tuning, and short-epoch cross-correlations between various combinations of the different types of recordings and discovered that the neurons of an assembly partly synchronize their outputs through a transition to a phase-locked oscillatory state. Assemblies relatively far from one another which have similar receptive field properties – including assemblies in different areas of the brain – also synchronize their activities if they are stimulated at the same time. Operating at a frequency between 40 and 80 Hz (the gamma range), these oscillations occur in all three types of recordings. Eckhorn speculates that this sort of oscillatory pattern may serve as a general mechanism for the binding of stimuli features in the primary visual cortex by linking excitatorially connected neurons with similar receptive fields. In this way, the dynamic assemblies would *define* an object in virtue of resemblance among features.

There are three possible sources for the oscillations in area 17: intrinsic membrane properties in presynaptic cells (e.g., the interneurons or the pyramidal cells), oscillations in thalamic input, or intracortical feedback pathways. Very recent data indicate that the synchronous oscillations are probably cortical phenomena alone (Schechter, 1996). These results suggest that the brain uses these temporal patterns for some peculiarly cortical activity.

Moreover, Charles Gray and Wolf Singer have shown that phase-locked synchronization across spatially separate columns for cells with similar orientation preferences are influenced by global properties of the stimulus (Gray et al., 1989). (Indeed, such synchronization has been found between neurons in area 17 of both left and right cerebral hemispheres.) When two short light bars move in opposite directions over two similar receptive fields, the responses of the cells show no phase locking. When they move in the same direction, the cells are only marginally synchronized. When the cells are shown a single long bar of light, their responses are strongly phase-locked. These results indicate that phase-locked oscillations depend on various large-scale aspects of the stimuli, such as form or motion, which local responses cannot reflect when taken individually. Therefore, the synchronization may also serve to represent the higher-order features in a pattern. For example, it may help in figure–ground segregation.

Although the phase-locked oscillations may yet turn out to be an artifact of some other, more fundamental process, it may also be that stimulus-evoked resonances provide a code to correlate information within and between different sensory systems. And as modelers in artificial intelligence pick up on the mechanism of phase-locked oscillations and explore the principles underlying this sort of binding solution, it does appear that this hypothesis provides the ideal sort of answer to psychology's binding problem. In fact, it is congenial to both Treisman's hypothesis of a master map of locations, in which attention is used to tie features together, and Prinzmetal's notion that organizational assumptions in perception plus feedback from higher-level processes account for perceptual binding. For example, computer modeling has demonstrated that some nonlocal feedback must play a fundamental role in the initial synchronization and the dynamic stability of an oscillation. And this feedback mechanism could exist either as Treisman's central feature-locator attentional mechanism or as Prinzmetal's lower-level, distributed recurrent feedback loops. It is no wonder, then, that phase-locked oscillations are being taken seriously as a new, exciting, possible solution to the superposition catastrophe and the binding problem.

Difficulties with the 40 Hz solution

However, the neurophysiologists who suggest that synchronized oscillations are the key to solving the binding problem confuse the different types of joining together a brain must do. *Binding* generally refers to the joining together of the individually processed features at the *psychological* level; what we experience subjectively are colored, shaped objects. Unfortunately, sometimes *binding* is also used to refer to the joining together of the component pieces of some features – grouping together the neurons that each signal a bit of an object's shape into a single assembly, for example. A better term for this process is *segmentation*. Segmentation operates prior to psychological binding, for psychology (generally speaking) takes a unified feature as the basic unit in perceptual theories.

The early stages of the visual system parse incoming two-dimensional patterns of retinal stimulation into cohesive features. They must somehow figure out how to indicate that this neuron over here with this particular receptive field is signaling a bit of the same feature as that neuron over there with that particular receptive field. This can be handled purely topographically until brain areas MT or V4 are reached. There the receptive fields are large enough that spatial confusions could arise, and hence some other mechanism is required to individuate object features. It is then a separate step to bind these segmented features into unified percepts. The question before us is how to explain this latter sort of perceptual binding.

What Eckhorn and his group and Gray and his colleagues have indexed is segmentation, not binding. They presented cats with very simple stimuli, a bar or a grid pattern, and then tested the response of cells with similar receptive fields *for single features*. If two cells are sensitive to similar orientations, for example, then their experiments predict that the cells will fire in phase-locked synchrony if the visual system is shown a bar in that orientation. This is no trivial result, but it does not show what would happen if two cells, one color-responsive and one orientation-responsive, were both activated by an appropriately colored and oriented bar. In fact, in cells they tested, Gray and Singer found that if neurons in different hypercolumns were sensitive to different features but still responded to the same particular input, then these cells were *not* phase-locked. Though some sort of oscillations may be the correct answer to the problem of perceptual binding, we do not have anything near conclusive evidence linking phase-locked oscillations with the subjective experience of unified objects. Those who see phase-locked oscillations in individual neurons as the additional temporal process required to overcome superposition difficulties celebrate prematurely.

Binding as a pseudo-problem: a systems-dynamical approach

I suggest that we should approach the binding problem from a different perspective. It is my contention that the alleged difficulty we have in figuring out how the individual features in our complex perceptions become stuck together into identifiable objects is a pseudo-problem; it is the product of looking at that aspect of perception at the wrong level of organization in the brain (see Article 48, LEVELS OF EXPLANATION AND COGNITIVE ARCHITECTURES). To see what I am talking about, let us begin with some analogous situations.

Consider the rolling pattern of boiling water. Suppose we want to explain how that pattern comes to exist. If we trace the activity of a single water molecule as it breaks into its component pieces, we don't find a rolling pattern; we find random motions. If we trace the individual activity of many water molecules, we still only find lots of individual random patterns. So where does the rolling pattern come from? Answer: tracing the motion of individual atoms and molecules is the wrong level of organization for discerning rolling. We need to look at the cumulative effects of all the water at the same time. The only way to do this is to look at the water at a higher level of organization – the level of water patterns – and describe the motion at that level. However, the problem of where the rolling patterns come from disappears at the higher level, for rolling is just what happens when water boils. The problem of how boiling water rolls is a pseudo-problem. The higher-level organization we see in boiling water is the product of the individual motions of individual molecules and atoms, to be sure, but we don't have a way to generalize from the sum of the individual motions to the overall pattern. The rolling pattern simply emerges when the water is heated and its molecules start to break down.

Now consider the Texas shoreline along the Gulf of Mexico. It is depicted on maps as a gentle concave curve, and it certainly appears that way when photographed from a satellite or high-flying airplane. But if we walk along the shore, we can't see the gentle curve; we can't even see part of it. We see grains of sand and pounding waves, and the shore *line* – if there is one at all – changes with each tide, with each wave. Even if we could freeze the ocean and the sand at some instant in time, the line that we could trace along the beach where the water meets the land would be very jagged as we move along and around the particles of sand. So are the maps and the pictures wrong? No; they are merely depicting the beach as it appears from a different perspective, from the perspective of taking in all the sand and all the water at once. From far away, the beach has a definite shape and is smooth; at close range, the beach's shape changes continuously and is quite irregular.

Perhaps we should understand our unified percepts in a similar manner. Looking at the activity of single cells or of several cells taken at once is looking at the wrong level of organization to understand our complex perceptions taken in their entirety. If we could look at the cumulative effects of all the cells in an area taken at once, then we would see a pattern that is the unified perception of some object, just as we see the boiling pattern when we look at heated water all at once and the curve of the beach when we look at the sand all at once.

This alternative perspective means that we should tie *bound* perceptions to synchronized activity over larger areas of the cortex than individual cells or small groups of neurons. We can see these patterns of perception most strikingly in the electrical activity we record using electrodes over groups of neurons. (The recordings are known as electroencephalograph waves, or EEG waves.) But at this higher level, we cannot distinguish any separate information-processing streams that somehow need to come together. Consequently, the problem of perceptual binding – how the brain ties together individual features into representations of single objects – fails to appear here, for neural activity already exists in unified patterns.

The spontaneous firing of neurons seems to be random. This activity is largely incoherent, and we can think of it as producing a background tapestry of noise against which we receive sensory inputs. Stimuli input disturbs this noisy spontaneous behavior

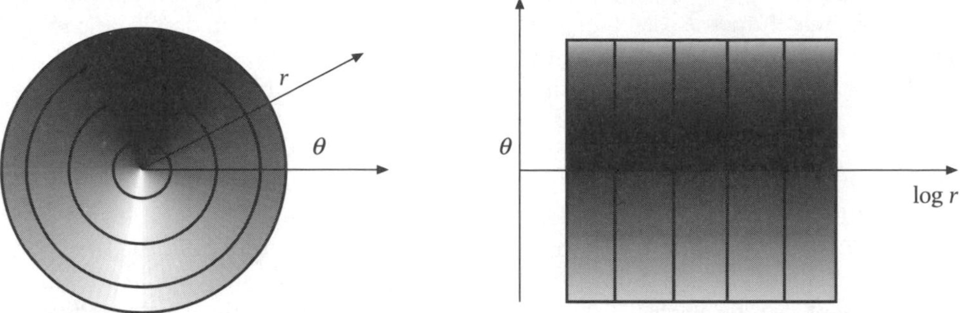

Figure 43.3 On the left is a typical "funnel" pattern in an early hallucinatory experience. On the right is the funnel pattern as it appears in EEG recordings of the cortex: as a set of bars equal in size, arranged parallel to the y-axis. Notice that the cortical pattern is a simple transformation of a set of concentric circles.

by forcing various neuronal groups to fire in stable, rhythmic patterns. Adopting this perspective means that our bound perceptions are mapped to the transient firing patterns of large groups of neurons whose behavior we hope has a particular mathematical description. If this is correct, we should find a higher-level structural pattern emerging from lower-level neuronal behavior that indexes our perceptual experiences. Although I grant that this story can seem quite fanciful, remarkably enough, we do seem to find exactly the correspondences I suggest. Let me give an example using the very simple visual patterns seen in early drug-induced hallucinations.

Most visual hallucinations begin with the experience of simple geometric shapes, such as gratings, lattices, checkerboards, cobwebs, funnels, tunnels, or spirals. If we could isolate some pattern of activity occurring in hallucinating cortices that is absent from normal controls, then we would have picked out a neural correlate to the hallucinatory perceptual pattern. There are few studies that record the firing patterns of humans during hallucinatory experiences. However, EEG waves have been recorded over the cortices of hallucinating cats. These show a higher-order, synchronous, oscillatory pattern of the sort I have been discussing. What is even more interesting, as Bard Ermentrout (1979) shows, is that the patterns recorded over cat cortex are *geometric transformations* of the patterns which the cats are presumably seeing. Most significant is that the transformations follow the topographically organized path which we can trace from the retina to the cortex. Though the neuronal projections from the retina to the cortex are organized topographically, the path twists and turns along the way. If we were to untwist and unturn the cortical pathway, we would find that the patterns recorded moving over the cortex (rolls and waves) correspond exactly to what we hallucinate (spirals and grids) (see figure 43.3).

Of course, most of our visual experiences are much more complicated than these simple experiences, so we would not expect an EEG recording of normal perception to show such obvious regularities. Instead, we would expect to find more complicated oscillatory patterns. And such patterns are beginning to be uncovered. For example, areas in which stimulus-evoked oscillations have been found include the rabbit olfactory cortex; cat, hedgehog, and rat olfactory systems; the auditory cortex and the hippocampal areas in cats; the hippocampus of guinea pigs; and monkey and cat visual cortices. Similar higher-order resonances have also been found in humans. EEG and ERP recordings show the oscillatory patterns above our association, limbic, and motor areas, as well as the auditory pathway, cerebellar cortex, and neocortex. Preliminary results using magnetic field tomography in primary motor and sensory cortices also corroborate the brain-wave data.

Walter Freeman, among others, maps these sorts of patterns into an *attractor phase space*, which suggests that they are the same sorts of pattern as those found over the hallucinating cat cortices (Skarda and Freeman, 1987). He records the higher-level neuronal oscillations using 64 electrodes over the rabbit olfactory bulb. He then plots the EEG activity of electrodes on a 64-dimensional graph, with each axis representing the activity of a single electrode. This graph represents the *phase space* for the olfactory bulb. It includes all the possible activity that the 64 electrodes could pick up when recording over the olfactory system. A 64-dimensional space is very complex, but Freeman discovered that when the rabbits were smelling something that was meaningful to them, then the activity of the bulb seemed to be *attracted* to a small corner of the phase space. Different attractor areas are keyed to different meaningful smells. If we think of each axis as representing the values that a feature or a sub-feature could take in perception (these probably would not correspond to the features which psychology has identified as salient), then the various attractor regions would correspond to sets of activated features grouped together into meaningful units. In other words, the attractor regions in some multidimensional phase space correspond to bound perceptions.

One advantage of using this sort of analysis for understanding perception is that we can describe the sorts of paths which the oscillating neuronal areas follow in the multidimensional phase space (as recorded with the electroencephalogram) using relatively simple nonlinear equations. Moreover, since all neurons have preferred ways of firing, there are only a few places in the multidimensional space that attractors end up residing in. So, out of the complex multidimensional phase space we discover a simple order to our perceptions.

Adopting this perspective means that things like EEG patterns are the relevant information-bearing states in complex neuronal systems. Using attractor dynamics to describe our perceptions gives rise to an entirely different conception of information processing than what we traditionally find, which are descriptions of single cell interactions and individual connectivities. The sequence of patterns that the oscillatory network passes through over time – the movement from one attractor to another – constitutes the psychologically relevant *computations* of our brain. Consequently, the details of the individual neuronal events – for example, what Eckhorn, Gray, and Singer are charting – are not important in explaining psychological phenomena. From a dynamical systems perspective, the macroscopic patterns we record of the activity of thousands of neurons at a time are where the information processing in an organism actually takes place.

Furthermore, and more important for our purposes, if we consider entire cortical regions as single systems each acting as one resonating unit, then a separate mechanism for uniting disparate feature bits is unnecessary. In other words, from this perspective, binding does not require a process to tie bits of perceptual information together. The basic stuff of cognition consists in the oscillatory patterns which emerge as we learn to recognize and categorize stimuli, and our perceptual states just are the movements that the network goes through in an abstract phase space. Perhaps, as important as lower-level single cell research is in understanding how neurons behave and how neural networks are constructed, it is not the appropriate level of analysis for perceptual experiences. Instead, the true story may be found in system dynamics.

This is not to say that as a result of taking this higher-level view, we do not need feature-detectors or any of the other cells hypothesized in the lower-level hierarchical theories. The lower level is still there and is still useful for explaining many phenomena, feature detection included. There is nothing wrong with being a pluralist about levels. We explain potassium influx and efflux at one level of analysis, and aphasia and agraphia at another. Similarly, we might explain feature detection at one level and the unified nature of our perceptions at another.

That aside, from the perspective of system dynamics, the puzzle of how to explain binding is laid to rest or, at least, is set aside. Our inability to point to some place in our information-processing stream that unites our feature bits could be an artifact due solely to adopting a lower-level, serial, distributed model of perception. If we think in terms of larger processing areas in a continually active brain, perceptions become dynamic, oscillatory patterns – without a binding problem. But only further research will tell us whether this perspective is correct or whether binding really is a problem which the brain needs to solve.

References and recommended reading

Eckhorn, R., Bauer, R., Jordan, W., Brosch, M., Kruse, W. and Munk, M. 1988: Coherent oscillations: a mechanism of feature linking in the visual cortex. *Biological Cybernetics*, 60, 121–30.

Ermentrout, B. 1979: A mathematical theory of visual hallucination patterns. *Biological Cybernetics*, 34, 137–50.

*Freeman, W. J. 1991: The physiology of perception. *Scientific American*, 264, 78–85.

Gray, C. M., Konig, P., Engel, A. K. and Singer, W. 1989: Oscillatory responses in cat visual cortex exhibit inter-columnar synchronization which reflects global stimulus properties. *Nature*, 338, 334–7.

*Hardcastle, V. G. 1994: Psychology's binding problem and possible neurobiological solution. *Journal of Consciousness Studies*, 1, 66–90.

Prinzmetal, W. and Keysar, B. 1989: Functional theory of illusory conjunctions and neon colors. *Journal of Experimental Psychology: General*, 118, 165–90.

Schechter, B. 1996: How the brain gets rhythm. *Science*, 274, 339–40.

Skarda, C. A. and Freeman, W. J. 1987: How brains make chaos in order to make sense of the world. *Behavioral and Brain Sciences*, 10, 161–95.

*Treisman, A. 1986: Features and objects in visual processing. *Scientific American*, 254, 114–25.

Treisman, A. and Schmidt, H. 1982: Illusory conjunctions in the perception of objects. *Cognitive Psychology*, 14, 107–41.

*von der Malsburg, C. 1995: Binding in models of perception and brain function. *Current Opinion in Neurobiology*, 5, 520–6.

44

Heuristics and satisficing

ROBERT C. RICHARDSON

Introduction: bounded rationality

Bounded rationality is a fundamental feature of cognition. We make choices between alternatives in light of our goals, relying on incomplete information and limited resources. As a consequence, PROBLEM SOLVING cannot be exhaustive: we cannot explore all the possibilities which confront us, and search must be constrained in ways that facilitate search efficiency even at the expense of search effectiveness. If we think of problem solving as a search through the space of possibilities as it was conceptualized by Allen Newell and Herbert A. Simon, limitations on search entail that for all but the simplest problems we can investigate at most a small proportion of the possibilities. Moreover, our evaluation of these possibilities will be uncertain and incomplete. We must rely on heuristic methods for pruning the tree of possibilities and for evaluating the possibilities we consider. In medical diagnosis, for example, the problem is one of inferring what disease is present from a given set of symptoms, when no one symptom is invariably present with a given disease, and the same symptom can have various causes. Moreover, patients can be suffering from more than one disease, a fact which compounds the difficulties of accurate diagnosis. Given that symptoms cannot generally be assumed to be independent and are noisy indicators of the underlying disease, the size of the search space turns out to be large, and inference concerning underlying diseases is inevitably uncertain. One solution is to limit the range of the search, eliminating some candidates as implausible; that is, we may adopt assumptions limiting the range of diseases considered to a narrow range of those antecedently likely. Good diagnosticians evidently narrow the range to no more than a handful of candidates. Search must be limited in order to be efficient.

Bounded rationality contrasts sharply with what Simon (1983) calls *Olympian rationality*, embodied most clearly in the model of (subjective) expected utility and Bayesian models of human judgment. Fundamentally, on the theory of (subjective) expected utility, problem solving or decision making requires a strategy set, a set of outcomes associated with each alternative in that set, a utility function defined over the outcomes, and a rule for maximizing expected utility (see Article 10, DECISION MAKING). The agent must know the alternative strategies available and be able to predict their consequences or likely consequences. There must be a utility function ordering possible futures and a decision rule based on the array of futures associated with the various alternatives. The model has proved to be remarkably resilient, and over the 50 years following John von Neumann and Oskar Morgenstern's (1947) revival of it, the theory has been elaborated in a variety of ways which enhance its descriptive adequacy: it has been extended to incorporate subjective weighting of alternatives, to take into account various attitudes toward risk, and to allow for nonlinear preference functions. A Bayesian model of human reasoning, similarly, requires an exhaustive set of hypotheses, a probability distribution on those hypotheses which is coherent, and conditional probability assignments for evidence on the various alternative hypotheses (see Article 21, REASONING). In order to assess the impact of some piece of evidence, an individual must know the theoretical alternatives and how likely each is on prior evidence. The probability assignments to the various hypotheses must meet the conditions on coherence laid down in the probability calculus. To then determine the impact of evidence on a given hypothesis, the individual must know how likely that evidence is on each of the competing alternative hypotheses. Both accounts of rationality have some critical features in common. They require an exhaustive knowledge of the possible alternatives, an assessment of the value of each alternative, and a decision rule for ordering them. These are daunting requirements, especially when there are many alternatives or when they are not independent; they are not generally likely to be met by search with limited resources, much less in human decision making.

The model of (subjective) expected utility and Bayesian models of human judgment are generally conceived of as normative models of decision making and are contrasted with descriptive models garnered from empirical studies of human behavior. Economists and philosophers line up defending a theory of rational choice with a normative orientation, while psychologists often adopt a descriptive project. A model of bounded rationality does not respect a sharp divide between the normative and the descriptive; instead, it seeks an account of rationality responsive to human limitations and aspirations. The result is not the elimination of a distinction between normative and descriptive projects, but an ambivalence towards any theory which divorces the two.

Satisficing and human reasoning

It was reflection on the factors which make Olympian rationality unrealistic as a description of the procedures of human rationality that led Simon to defend a *satisficing* approach for human decision making, beginning in the mid-1950s (Simon, 1979, 1983). Simon recognized that, in actual decision making, humans are rarely presented simultaneously with all the available alternative modes of action. Even if we were thus informed, or if we simply restricted our decision procedure to those options known to be available, we humans are not generally sufficiently aware of the set of likely outcomes attendant on those actions, much less the probabilistic array associated with each choice. Even given such comprehensive knowledge, we do not generally consider detailed scenarios for the future, with exhaustive and exclusive alternatives in place and conditional probabilities neatly assigned to them. Lastly, even granting this wealth of information and the will to use it, our utility functions are notoriously ill defined, in being intransitive, nonlinear, and encompassing incommensurable values. As Simon said:

> The classical theory is a theory of a man choosing among fixed and known alternatives, to each of which is attached known consequences. But when perception and cognition intervene between the decision-maker and his objective environment, this model no longer proves adequate. We need a description of the choice process that recognizes that alternatives are not given but must be sought; and a description that takes into account the

> arduous task of determining what consequences will follow on each alternative. (1979, p. 272)

Satisficing essentially substitutes a stopping rule for one that would maximize utility. To see what this means intuitively, imagine that we search through alternatives until we find some choice which has outcomes at or above a specified threshold. Once an option passing muster is found, it is then embraced *without any further search or evaluation*. The procedure can be refined to allow for adjustment of the threshold value either up or down, depending on the ease with which we find options at or above the threshold value, thus allowing a satisficing rule to be adjusted sufficiently to reach a decision even in an environment which is fundamentally inhospitable.

Satisficing was meant to provide an improved, more realistic decision procedure within the constraints of bounded rationality. We are not required to make an exhaustive search through the available options. We simply look at enough options to find one which has consequences which are minimally acceptable. A satisficing rule also relieves us from the need for a utility function strong enough to allow for systematic comparisons and rank orderings of outcomes. With a satisficing rule, we need to find an alternative whose consequences are generally acceptable, but we will not need to make systematic and exhaustive comparisons with the outcomes of other behavioral alternatives. Simon writes:

> An earmark of all these situations where we satisfice for inability to optimize is that, although the set of available alternatives is "given" in a certain abstract sense (we can define a generator guaranteed to generate all of them eventually), it is not "given" in the only sense that is practically relevant. We cannot within practicable computational limits generate all the admissible alternatives and compare their respective merits. Nor can we recognize the best alternative, even if we are fortunate enough to generate it early, until we have seen all of them. We satisfice by looking for alternatives in such a way that we can generally find an acceptable one after only moderate search. (1981, p. 139; cf. Simon, 1983)

Task complexity does affect actual decision strategies and outcomes. In decision problems, when there is an increase in the number of alternatives (with either risky or nonrisky choice), in the number of dimensions, or in the time pressure, then choice strategies change, the quality of choices decreases, response variability increases, and, paradoxically, the confidence of subjects increases.

A. L. Samuel's (1959) landmark studies in machine learning exhibit bounded rationality as clearly as does human cognition. Checkers is a reasonably simple game, yet complicated enough to tax time and search resources even for large computers. It is a well-defined game in the sense that what constitutes a win, a lose, or a draw is sharply defined, as is the space of board positions. Samuel observes that there is no known algorithm which will guarantee a win or a draw in checkers. Moreover, since an exhaustive search would require exploring something on the order of 10^{40} possible games, there is no real alternative but to engage in a selective search. That is, the only tractable procedures are ones which do not guarantee ideal solutions and may not guarantee solutions at all. In the case of checkers, it is possible to enumerate the possible moves from a given board position and to evaluate board positions at various depths. Samuel's program evaluated board positions in terms of the ability of the players to move and the relative number of pieces. The evaluation of moves from an evaluation of the associated board positions followed a minimax procedure, choosing the move which maximizes the machine score, assuming that the opponent will respond in order to minimize it. In other words, a move was selected provided that, among the possible moves, its worst outcome was better than the worst outcome associated with alternatives to it. Samuel programmed the machine to search for appropriate heuristic procedures for evaluating moves in terms of subsequent board positions – in effect, to learn checkers. Yet the machine was inevitably limited in resources, and its search reflected this fact. Samuel's machine, for example, limited how many moves it looked ahead, depending on a number of variables, including whether a jump is available at various depths. In the limit, the program stopped at twenty moves ahead. That is more than humans generally pursue, and Samuel's machine came to play a solid game of checkers; but it is nonetheless a real limitation on search. Samuel's machine suffers from bounded rationality no less than its human opponents. The significance of bounded rationality only increases when we turn to problems that are ill defined, or in which the search space is ill structured.

In spite of these *prima facie* advantages to the analysis of decision making in terms of satisfaction rather than optimization, it has become common to maintain that satisficing is merely a case of optimizing under side constraints. In order to reconstrue a merely satisfactory decision in such a way that it is optimal, it is necessary only to constrain the problem space so that additional alternatives are excluded. Alternatively, we may restructure the problem space, assigning additional costs to exploring alternatives; for example, it is common to appeal to the associated costs of additional search or lost opportunities. We could as well add limitations on memory, attention, or motivation. This will allow us to redescribe *any* satisfactory solution as optimal under some set of constraints, particularly when the *constraints* are themselves unconstrained.

Nonetheless, there are two fundamental reasons for maintaining a contrast between optimizing and satisficing. Both are sufficient to undermine the conclusion that satisficing *generally* reduces to optimizing under side constraints. The first depends on differences in the procedures and in the formal requirements for reaching a decision. Though there may be some set of constraints under which a solution is optimal considered from an Olympian standpoint, differences in terms of the kind or character of the information available may guarantee that the procedures used in finding solutions do not utilize those constraints. For example, a reduction of satisficing to optimizing requires knowledge of the existence of the side constraints, their likely effects, and the ability to take them into account in a systematic way. This is as unrealistic as the original problem it was designed to replace. They are psychologically unrealistic. Perhaps we *might* have been creatures for which these would have been realistic procedures. Perhaps we *should* have been creatures which followed such procedures. However, these are not realistic expectations for humans as we are in fact constituted. Satisficing is thus intended to provide a theory of bounded rationality adapted to the actual limitations of human cognition. The issues of what information is in fact available, what information is used, and what capacities the agent has for processing the information are central (Simon, 1979, pt 1, ch. 1).

The second reason for maintaining the contrast concerns the limitations imposed by the structure of the environment (cf. Simon, 1979, pt 1, ch. 2, and Simon, 1981). Models optimizing choice require that the organism consider all the possible alternatives and evaluate them in such a way as to gain a complete comparative ranking. One factor distinguishing satisficing is that the decision making in which real organisms engage does not leave them in a position even to generate the behavioral options, much less evaluate the likely consequences. The simple fact is that there is generally no procedure for doing so, or no procedure which is available in real time with finite resources. In many cases, perhaps most, the options to be considered are presented sequentially rather than simultaneously, and we need to make a decision to accept or reject the options *as presented*. We may think of the options as presented with a window of opportunity. When that window is open, it is open for a limited time, and once the opportunity to take it is past, the option can no longer be recovered. Even if there is an abstract sense in which the option might remain an open one – for example, it may be true that if we were again in the same place, we could still capitalize on the option – this may provide little real opportunity; the option may in fact not be available, or it may simply be too costly to make it worth our time and energy to return to that place. Thus, we have a picture of an organism presented sequentially with a series of options. At each step, the organism must decide whether to accept or reject a given option. Once rejected, an option cannot be recovered efficiently. In such a situation, optimal choice is a matter of luck at best; for optimization requires that we be able to choose between a set of options in such a way as to maximize utility. Even if we do make the optimal choice, by luck perhaps, the procedure which such an environment requires is hardly an optimizing one.

A satisficing procedure, by contrast, is tailor-made for such circumstances. We take the options as they come, settling for the first one which passes muster. It is worth underscoring that these differences are not peripheral or incidental. The assumptions of transitivity and completeness inherent in Olympian rationality imply that the choice made is independent of any questions of, for example, the order in which alternatives are considered. History does not matter. Satisficing, by contrast, is not committed to either transitivity or completeness. Satisficing acknowledges the contingencies of history. Thus, there are reasons to maintain that, though theories of utility maximization are capable of describing the outcomes of choices, they offer little insight into the procedures or mechanisms effecting behavior.

Heuristics and decision making

Instead of thinking of search as concerned with specific decisions, we may also think of it as governed by methods or procedures selected from among a myriad possible methods or procedures. Procedures, or rules, involved in reasoning can themselves be satisfactory or optimal. In practice, there is little alternative but to forgo optimal procedures; indeed, there are even very few games for which there are known to be procedures guaranteeing solutions. Even in relatively well-defined domains, search resources limit us to satisfactory procedures. The procedures which humans actually use in making decisions are also limited in application and effectiveness. If an *algorithm* is a procedure that will guarantee a solution to a problem if the problem has a solution, then a *heuristic* is a procedure that may not reach a solution at all, even if there is one, or may not even provide an answer (cf. Newell et al., 1958).

Heuristic procedures have a number of additional characteristics. Heuristics reduce the complexity of a computational task and in that sense reflect bounded rationality. By reducing the complexity of the computation, they reduce the demands on resources such as time, memory, and attention. Often, in fact, heuristic procedures are the only procedures which are realistic. Heuristics are more or less reliable as guides to good answers, and their reliability varies substantially between domains. As a result, heuristics may be satisfactory procedures, even if they are not optimal ones. Finally, heuristic procedures will sometimes fail to reach solutions or will generate incorrect solutions, and these failures will be systematic. Given a heuristic procedure, it should be possible to predict its biases, and given information about the biases a system manifests, it should be possible to induce something about system design. Embracing heuristic procedures is essentially a matter of embracing decision procedures which are satisfactory but not optimal. In this sense, heuristic procedures can be understood as the consequence of satisficing in the choice of procedures. That is, they involve satisficing at a higher level.

For illustration, we can turn to another artificial example. Chess provides an interesting, historically important case. There are now chess programs which compete favorably with human experts, though they generally rely on strategies not available to human subjects; for example, the best programs may examine as many as a million possibilities before selecting a move, whereas a human expert will generally consider at most a dozen or so. As a result, they are of limited interest in assessing human problem solving even if they are feats of technological prowess. Insofar as we hope to glean some information concerning *human* problem solving from simulations, we are faced with the problem of evaluating whether machines and humans use similar heuristics. Chess is a finite game, with a finite number of possible positions, but it is not one for which there is any known algorithm, or brute force solution. Computers and humans alike rely on tractable methods for evaluating moves short of terminal positions. The *only* feasible procedures in this case are heuristic ones. In Newell, Shaw, and Simon's work, alternative moves were evaluated in terms of goals such as safety of the king, material balance, and center control. They were careful to emphasize that the heuristic procedures implemented were supposed to be "principles of play similar to those used by human players" (1958, p. 65). The feature which they found distinguishes expert players from novices is their ability to encode strategic positions efficiently. Simon and his collaborators subsequently offered evidence to show that human memory for board positions is organized around defensive patterns (Simon, 1979, pt 6, chs. 2, 4, and 5), and simulations based on these principles subsequently showed at least a qualitative similarity to human behavior (Simon, 1979, pt 6, ch. 3). The methods for evaluating intermediate positions, and thus the procedures for determining moves, are heuristic procedures, reliable but limited in application and success. Search is selective. Representation of board positions is keyed to defensive structures. Evaluation of intermediate board positions depends on a few significant cues to game quality. The procedures are limited in application and dependent on knowledge of the domain. There is no guarantee that there will be a solution, where a solution would be a good move or the best move. There is no guarantee that the alternatives chosen will be optimal. The procedures are nonetheless satisfactory for competent chess, even if they are less than optimal.

The seminal work of Amos Tversky and Daniel Kahneman on human judgment, beginning in the early 1970s, illustrates a similar moral and supports the view that human reasoning systematically violates the standards for statistical reasoning, ignoring, among other things, base rates, samples size, and correlations (cf. Kahneman et al., 1982). They take a Bayesian account of probabilistic reasoning as a normative standard, establishing a standard for an ideally rational agent. They found that humans systematically fall short of the standard, concluding that "man is apparently not a conservative Bayesian: he is not a Bayesian at all" (ibid., p. 46). They provide many familiar examples of our judgmental infelicities. Here is a well-known one, which they provided to subjects (ibid., pp. 91ff.):

> Linda is 31 years old, outspoken and very bright. She majored in philosophy. As a student, she was deeply concerned with issues of discrimination and social justice, and also participated in antinuclear demonstrations.

Subjects were then asked to rank eight statements in terms of how probable they are, including, importantly, these three:

(1) Linda is active in the feminist movement.
(2) Linda is a bank teller.
(3) Linda is a bank teller and is active in the feminist movement.

This is an instance of what they call a *conjunction fallacy*: since (3) is a conjunction of (1) and (2), it cannot be more probable than either; yet subjects typically rated (3) as more likely than (2). Tversky and Kahneman argue in light of such experimental results that humans operate with judgmental heuristics rather than with anything approximating the formal rules of Bayesian decision theory. Results such as these indicate that people evaluate probability in terms of a heuristic they call *representativeness*, which they describe as the degree to which an individual, or event, resembles a prototypical member of a category. To return to the example, the description of Linda supposedly would make her typical, or representative, of feminists or of feminist bank tellers, but not of bank tellers. Tversky and Kahneman allow that these heuristics do produce reasonable judgments in a wide range of cases but hold that they lead to errors in comparison with the Bayesian standard. Moreover, they claim that the representativeness heuristic and what they call an *availability heuristic* are biased and subject to systematic errors. The experimental support for the existence of these heuristics depends on producing cases in which subjects deviate from the normative standard, since otherwise it would be possible to hold that we *are* intuitive Bayesians; it is important that the deviations observed are precisely what is indicated by the heuristics. Given an array of reliable procedures, there should be general agreement on some prescribed judgments insofar as they are reliable; the differences between the procedures will be more visible in the pattern of their failures.

The representativeness heuristic does capture a number of phenomena which have been reliably produced in experimental settings, including not only the conjunction fallacy but also neglect of base rates, insensitivity to sample size, and nonregression to the mean. There have been a variety of criticisms of Tversky and Kahneman, and there is considerable uncertainty over how useful the approach is in understanding human judgment. It is clear, at the very least, that there is considerable need to specify more clearly the conditions under which representativeness or availability effects occur and to elaborate and refine the heuristic models. There have also been more radical challenges. Some have rejected the normative standard provided by Bayesian decision theory (e.g., Cohen, 1981). Others have challenged the general appeal to heuristics and biases as an approach to understanding human judgment. Gerd Gigerenzer, in particular, has presented what he says is a comprehensive alternative to the program of Tversky and Kahneman (see especially Gigerenzer, 1991; also Lopes and Oden, 1991). Gigerenzer has argued that many of the effects ascribed to a representativeness heuristic disappear when problems are posed in terms of frequencies rather than the likelihood of individual events. Thus, in an experiment paralleling the case above, Gigerenzer (1991) provided a problem with the same preface concerning Linda, but added:

> There are 100 people who fit the description above. How many of them are:
> bank tellers?
> bank tellers and active in the feminist movement?

The *conjunction fallacy* is significantly reduced in this format, though it does not disappear. Tversky and Kahneman had previously noted the difference between frequency and single-case formats on probability judgments, and Gigerenzer has shown that in frequency formats, not only the conjunction fallacy but also base rate neglect and overconfidence effects are reduced. There is some ambiguity in the results, in part because of unclarity about the magnitude of this effect. (Gigerenzer reports a reduction in conjunction errors from roughly 85 percent to between 10 and 20 percent. Gigerenzer and Hoffrage (1995) found the frequency of the normatively appropriate Bayesian answers to diagnostic problems to be in the 50 percent range. Replications of Gigerenzer's methodology have found less striking reductions.) Nonetheless, shifting to a frequency format can improve judgmental performance.

The differences between the two programs of research are not as pervasive as might be thought. It is important to notice in particular that sensitivity to the way a question is framed is exactly what would be predicted by a heuristic approach. Gigerenzer emphasizes that the rules applied depend on the way in which the information is presented, an observation which is both true and important. He urges in addition that when the problems are posed in a way that reflects frequencies rather than single-case probabilities, there is greater reliability. Gigerenzer claims that Tversky and Kahneman's methodology misleads us in relying on single-case probabilities, and that once set in the proper frequentist format, human judgmental capacities improve:

> the mind acts as if it were a frequentist; it distinguishes between single case events and frequencies in the long run – just as probabilists and statisticians do. Despite the fact that researchers in the "heuristics and biases" program routinely ignore this distinction fundamental to probability theory, when they claim to have identified "errors," it would be foolish to label these judgments "fallacies". (Gigerenzer, 1991, p. 95)

The *natural* format for the human mind, Gigerenzer claims, is frequentist. Gigerenzer certainly overstates the case in claiming that the mind acts as if it were a frequentist, just as Tversky and Kahneman overstate the case in claiming that human judgment is essentially non-Bayesian. The experimental results at best support the claim that given a frequentist formulation of the problems, humans are less likely to commit the fallacies of reasoning which lead Tversky and Kahneman to describe humans as irrational. This suggests that although we can follow the prescriptions of standard Bayesian reasoning – whether by invoking a mentalistic version of Bayes's rule or not is left open – this does nothing to undercut the significance of Tversky and Kahneman's positive proposals or the claim that, for example, representativeness plays a significant role in our reasoning about chance. Tversky and Kahneman's work does, in any case, show that given a *standard* single-case format, we are prone to commit errors in reasoning. If we do not deploy some heuristic such as representativeness, then why are we prone to increased errors in the *standard* nonfrequentist formats? The presence of the errors and the robustness of Tversky and Kahneman's results are not undercut by Gigerenzer's attempts to make cognitive illusions *disappear*. However, Gigerenzer's results *do* support the claim that we are capable of more nearly proper probabilistic reasoning than Tversky and Kahneman have been inclined to recognize. Gigerenzer's own alternative to Tversky and Kahneman's explanation of the experimental results, which he terms *probabilistic mental models* (Gigerenzer et al., 1991), is itself a form of heuristic search. The experimental evidence seems to support a variety of mechanisms, depending on context and format.

Conclusion

The perspective offered by satisficing and the attendant empirically driven work on the heuristics of human judgment might seem to widen the gulf between normative and descriptive enterprises. Rejecting the ideal in favor of the satisfactory would mean moving away from the normative standard for judgment; and heuristics gain significance by contrast with the ideals provided by Bayesian and expected utility models. Nonetheless, the distinction between normative and descriptive enterprises was not always as sharp as it now appears (see Gigerenzer et al., 1989). Classical probabilists recognized no significant distinction between objective and subjective probabilities, much less between normative and descriptive problems. They sought a theory of *reasonable judgment*, with all the ambiguity that suggests. We are now accustomed to a sharper distinction between questions concerning how humans do reason and how they should. I think that we should be wary of any sharp distinction between questions about how humans *do* come to decisions and how humans *should* come to them. The relationship of these studies is at this point unresolved; in any case, when we are concerned with a theory of rational choice, the limitations on human reasoning cannot be neglected for a theory of human rationality.

References and recommended reading

Cohen, L. J. 1981: Can human irrationality be experimentally demonstrated? *Behavioral and Brain Sciences*, 4, 317–31.

Gigerenzer, G. 1991: How to make cognitive illusions disappear: beyond heuristics and biases. *European Review of Social Psychology*, 2, 83–115.

Gigerenzer, G. and Hoffrage, U. 1995: How to improve Bayesian reasoning without instruction. *Psychological Review*, 102, 684–704.

Gigerenzer, G., Hoffrage, U. and Kleinbölting, H. 1991: Probabilistic mental models: a Brunswikian theory of confidence. *Psychological Review*, 98, 506–28.

*Gigerenzer, G. et al. 1989: *The Empire of Chance: How Probability Changed Science and Everyday Life*. Cambridge: Cambridge University Press.

*Kahneman, D., Slovic P. and Tversky, A. (eds) 1982: *Judgment under Uncertainty: Heuristics and Biases.* Cambridge: Cambridge University Press.

Lopes, L. L. and Oden, G. C. 1991: The rationality of intelligence. In E. Eells and T. Maruszewski (eds), *Rationality and Reasoning: Essays in Honor of L. J. Cohen*, Amsterdam: Rodolpi, 199–223.

*Newell, A. and Simon, H. A. 1972: *Human Problem Solving.* Englewood Cliffs, NJ: Prentice-Hall.

Newell, A., Shaw, J. C. and Simon, H. A. 1958: Chess playing programs and the problem of complexity. *IBM Journal of Research and Development*, 2, 320–35. Repr. in E. A. Feigenbaum and J. Feldman (eds), *Computers and Thought*, New York: McGraw-Hill, 1963, 39–70.

Samuel, A. L. 1959: Some studies in machine learning using the game of checkers. *IBM Journal of Research and Development*, 3, 211–29. Repr. in E. A. Feigenbaum and J. Feldman (eds), *Computers and Thought*, New York: McGraw-Hill, 1963, 71–108.

Simon, H. A. 1979: *Models of Thought.* New Haven and London: Yale University Press.

—— 1981: *The Sciences of the Artificial*, 2nd edn. Cambridge, Mass.: MIT Press.

—— 1983: *Reason in Human Affairs.* Stanford, Calif.: Stanford University Press.

von Neumann, J. and Morgenstern, O. 1947: *Theory of Games and Economic Behavior.* Princeton: Princeton University Press.

45

Innate knowledge

BARBARA LANDAU

At the heart of cognitive science lie two problems: the nature of our knowledge and how it emerges. For many centuries, these issues were the province of philosophers only. Nativists such as René Descartes argued that much of our knowledge was innate, driven by the character of the human mind and only indirectly by the nature of the particular events we might experience. By contrast, empiricists such as John Locke argued that very little of our knowledge was innate; rather, he argued that knowledge was acquired through a lengthy process of learning in which the putative primitive elements of thought – the sensations – came to be associated with each other, yielding higher-level concepts. While no longer the sole province of philosophy, the nativist and empiricist frameworks have come to organize some of the most important issues in cognitive science today: How can we characterize human knowledge? How does it emerge? What are the relative contributions of innate structure and learning through experience?

From the nativist perspective, the general answer to these questions is that the human mind is predisposed to interpret the world in terms of certain categories: for example, language, space, and number. Carrying out the nativist program requires a full understanding of the specific structure of these categories and the way they emerge in early development. Cognitive science provides a unique means for triangulating these problems via different disciplines. For example, the nature of linguistic structure is the focus of intense inquiry by linguists; the question of how language and space might be instantiated in human minds and in machines is the focus of inquiry for perceptual and cognitive psychologists and computer scientists; and the question of how knowledge of language, space, and number emerge in young learners is a primary focus of cognitive and developmental psychologists.

In this essay, the nativist framework will be presented as an important perspective on these questions, and the perspective will be defended by showing how the nativist framework has borne fruit in three specific domains: language, space, and number. Within these domains, a number of empirical hallmarks of innate endowment can be found. These hallmarks include the early and rapid emergence of competence despite very different environmental conditions for learning, the domain-specific nature of this competence, and the disparity between possible models in the environment and actual acquisitions by the learner. As a whole, these hallmarks suggest that humans are born with predispositions to create skeletal knowledge systems particular to specific domains. These skeletal systems are nourished by information from the environment, but their character is not *caused* by it.

Before turning to these empirical observations, however, it is important to note that the nativist perspective does not rule out learning or developmental change. Nativists and empiricists alike acknowledge that there may be definite observable changes as knowledge emerges. Often these changes are clear evidence of learning. In some cases, such learning can be more salient to the naked eye than what remains constant, as when we observe mere three-year-olds fluently speaking their native tongue, whether it be English, French, or Farsi. From the nativist perspective, however, innate endowment need not be anathema to learning and developmental change. Just as the biologist must explain the change from tadpole to frog but must also explain why all offspring of frogs first develop into tadpoles, the nativist can embrace change together with innate guiding principles. Although we may be struck by the extent of learning apparently required for a three-year-old to master a given language, we must also account for why this learning takes place so rapidly and so uniformly, under a wide variety of environmental conditions. Those aspects of domain-specific knowledge that emerge early and rapidly under quite different conditions of learning can be argued to constitute candidate cognitive universals.

Emergence of competence under varying conditions of learning

Development normally occurs as children look, listen, and move about their environments. One might assume, as did the classical empiricist philosophers, that input from the sensory systems is necessary for acquiring the fundamental systems of knowledge. Cases of natural accidents, where children are born either deaf or blind, have recently shown this assumption to be false. Rather, the acquisition of knowledge is surprisingly robust and impervious to a great deal of variation in experience.

Language

The linguist Noam Chomsky (1980) has argued that the structure and functioning of human language are as specialized as those of any other biological organ, say, the heart or the kidneys. From this perspective, the environment – one's native language – provides the conditions under which the innate principles of language can emerge in behavior. Thus it is a specific environment that causes children to learn English if they are raised hearing English, but French if raised hearing French. However, regardless of the specific language learned, acquisition emerges on a relatively constant developmental timetable, even in cases where the normal conditions for learning are not met. As Eric Lenneberg (1967) argued, language development is surprisingly impervious to a great deal of variation in experience, as one would expect of any biologically important system.

A number of recent studies illustrate such robust emergence of language in cases where children are deprived of important aspects of normal experience. Susan Goldin-Meadow and her colleagues (Feldman et al., 1978; Goldin-Meadow et al., 1994) have studied the case of deaf children who have not been exposed to any formal language. The children were not exposed to a structured spoken language, because they were congenitally and profoundly deaf, and therefore could not hear the language spoken by their parents. Neither were they exposed to a structured sign language such as American Sign Language, because their parents had made the decision not to expose them to formal signed languages. Although the parents did gesture while speaking to the children, there was no evidence that these gestures had truly linguistic form or structure.

Despite this lack of linguistic input, these deaf children did produce gestures and gestural combinations as they communicated with their parents. These gestures were used to communicate the same kinds of meanings that are conveyed universally by children learning a spoken language: for example, comments on the objects and actions in their environment. Because the nature, range, and structure of the gestures did not bear a close resemblance to those gestures produced by their mothers as they spoke to the children, the investigators concluded that the children had invented a language of their own. The key question then became whether these invented languages exhibited the same kinds of fundamental properties as are found in other natural human languages, whether spoken or signed.

A series of detailed studies on one of the deaf children, David, showed that his invented language did indeed show many of the hallmarks of natural languages. For example, David's early vocabulary included gestures that expressed meanings conveyed in many languages by nouns and also gestures expressing meanings normally conveyed by verbs. David combined these gestures into signed utterances, just as hearing children learning spoken languages produce early combinations of words that make up short, simple sentences. Furthermore, David used systematic ordering for the gestures in order to convey different meanings, in the same way that hearing children use word order to convey meaning (e.g., in English, "Baby drink" versus "Drink juice").

Importantly, a close examination of David's gestures revealed that his early corpus of noun and verb gestures fell into two distinct sets. He reserved certain gestures for the noun meanings, and others for the verb meanings. The noun–verb distinction is a hallmark of natural languages, found in all languages. Apparently, David's invented system preserved this distinction, even though he had available no external model by which to organize his linguistic system. By the age of about three years, David began to use some of the noun gestures as verbs and some of the verb gestures as nouns. This would be comparable to our use in English of the noun *fax* as a verb, as in "I am faxing the paper to you," or the verb *run* as a noun, as in "I took my dog for a run." English-speaking children are known to engage in such creative exchange of words between word classes by the age of about three as well. In these cases, children are careful to distinguish the nouns from the verbs by placing each in its correct sentential position and using morphology appropriate to each class (e.g., placing the article *a* before the word *run* to make it a noun, or placing the tense-marker *-ing* after the word *fax* to make it a verb). David, too, was careful to distinguish between uses of a gesture as a noun or a verb. Just like children learning a spoken language, David systematically marked the gestures as noun or verb by appropriately changing their position in his signed utterances and altering the actual gesture somewhat for its use as noun or verb, akin to the morphological changes between noun and verb in English. The emergence of such fundamental distinctions as that between nouns and verbs in David's invented language strongly suggests that he sought some formal linguistic means to distinguish between the two word classes. He chose hand shape and sentential position to do so. This recruiting of resources to mark an important distinction suggests that the learner possesses innate predispositions to structure human language in certain ways, independent of exposure to a formal linguistic system.

Another case that illustrates the robust nature of language learning can be found in studies of language acquisition by children who are blind from birth. Like the deaf child, the blind child lacks a dimension of experience that might be critical for learning language. Although the blind child can hear spoken language, it is difficult for her to interpret what she hears, for much of the information needed to interpret ongoing events comes to us through vision. In the extreme, one might expect a blind child to be completely incapable of learning those parts of the vocabulary that encode visual experience. According to classical empiricist philosophers John Locke and David Hume, the blind could never come to have concepts that encode visual experience; hence they could never attach meanings to visual terms such as "look" and "see." Locke's argument was simple: If sensory experience is necessary for the formation of concepts and the acquisition of words for those concepts, then the absence of visual experience will inevitably lead to the absence of words reflecting that experience.

Lila Gleitman and I (Landau and Gleitman, 1985) studied language learning in several congenitally blind children, asking whether blindness would generally impede language learning and, in particular, whether the blind child would show an absence or distortion of words that encode visual experience, such as *look* and *see*. An overall analysis of language learning in three children blind from birth showed that, although there was an initial mild delay, the children acquired a vocabulary quite similar in size, content, and rate of acquisition to normal, hearing children. Similarly, the syntactic properties of their utterances – the number of words per sentence, the internal structure, and the morphology – were comparable to normal, hearing children. The rate of language learning for the blind child was, overall, quite similar to that of the sighted child, proving that blindness need not impede the fundamental characteristics of language acquisition. This is consistent with Lenneberg's observation that language learning appears to proceed normally even in cases where there is rather gross experiential deprivation. Even though the blind child cannot use rich information afforded through vision to interpret the utterances she hears, language learning appears to take place rather normally.

A more direct test of the empiricist hypothesis was made by examining whether and how the blind child could learn the visual vocabulary. This series of studies was initiated by a striking event. At the age of two, one of the blind children – Kelli – produced the word *see* among her earliest verbs. Kelli's spontaneous uses of *see* and *look* were then examined, and her comprehension of the terms in different contexts was assessed. Most of her production of *see* occurred when she wanted to come into contact and explore an object: for example, when she requested to "See camera!" and then explored it with her hands. The idea that she might be using the visual terms *see* and *look* to mean "explore with the hands" was tested using a series of comprehension tasks, comparing her performance to that of sighted children of the same age who were either permitted to use their vision or blindfolded. For example, when sighted children were told, "Look up!," they typically turned their heads and eyes upwards, to visually apprehend a target; they did this even when they were tested blindfolded. In comparison, when Kelli was told, "Look up!," she raised her hands upwards, just as the sighted child did with her eyes. This suggested that Kelli understood the command "Look!" to mean "Apprehend with the hands." Further probes confirmed this interpretation. For example, if told, "Look at the book," Kelli would explore the book thoroughly with her hands; but if told, "Touch the book," she merely contacted the book with her hands but did not explore it. If told, "Look at the toy real hard," she explored the toy most thoroughly with her hands; but if told, "Touch the toy real hard," she struck it vigorously just once.

From what experiential base did Kelli decide that *look* and *see* meant "explore with the hands"? Kelli's mother used these terms in everyday discourse, but an examination of her usage showed that she did not use them consistently and exclusively under conditions in which Kelli was actually exploring objects with her hands. However, Kelli must have needed a term exclusively dedicated to describing her haptic exploration of objects; so it seems likely that, lacking any other clear-cut use for the terms, Kelli commandeered the terms *look* and *see* for just this use. As a testimony to the naturalness of this meaning, many languages of the world (though not English) do have single words that mean "explore with the hands." One example is the French word *tâter*, which, significantly, is used by the blind and of the blind in describing their haptic exploration of objects.

Thus Kelli apparently constructed meanings for the words *look* and *see* that did not exist in the environmental and linguistic input around her and were not identical to the meanings that sighted children have for these terms. However, Kelli's invented meanings were related to those of sighted children in an interesting way. Because the tactile and haptic systems are the major spatial exploratory organs for the blind, Kelli was in a deep sense interpreting the word *look* to mean "explore." The only difference between Kelli's meaning for the term and that of sighted children was that different organs of exploration served as the primary means of *seeing*.

Space

One of the touchstones of the classical nativist and empiricist debate among philosophers concerned the question of whether our knowledge of space depends on the nature of experience. The case of the blind man was considered critical: Given the very different nature of experience for the blind and the sighted, could it ever be possible for the blind man to possess the same knowledge as the sighted? According to Descartes, the shape of our knowledge is determined by innate structures of the mind; hence the modality of experience would have no effect on the fundamental character of this knowledge. By contrast, Locke argued that the blind man and the sighted man could never come to possess the same knowledge; because the mode of experience critically affects the shape of our knowledge, a system built primarily on the basis of vision could never be equivalent to one built primarily on the basis of touch.

Recent studies of the nature of spatial knowledge in the blind child suggest that Descartes' conclusion was probably more valid than Locke's. My colleagues and I (Landau, et al., 1984) studied a young child who was totally blind from birth. Informal observations suggested that this child navigated competently in familiar environments. The question then arose whether she possessed a system of spatial knowledge that could allow her to construct a spatial representation – a kind of mental map – of a novel layout. In order to answer this question, we brought the $2\frac{1}{2}$-year-old child into a novel 8 foot by 10 foot room in which four landmarks had been set (see figure 45.1). The child was first guided to a starting point (M). She was then walked to T, shown the landmark there, and walked directly back to M. Then she was walked from M to P, shown the landmark there, and walked directly back to M again. Finally, she was walked from M to B and back to M again.

Following this introduction to the layout, the child was given a set of spatial inference problems, in which she was asked to move between landmarks on her own. She was led from M to T, was asked to find P, and was allowed to move completely on her own. When she reached P, she was asked to find T, again on her own. These trials were then repeated, followed by requests to go from T to B and back again and from B to P

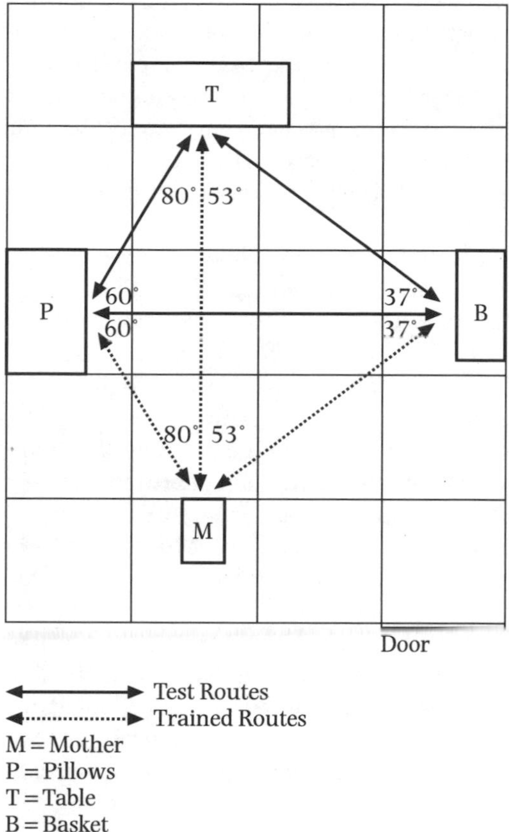

Figure 45.1 Spatial layout used for testing spatial inferences in the blind or sighted child (Landau et al., 1984). The child started at M and was guided from M to T and back again, then from M to P and back again, and then from M to B and back again. Then she was led from M to T and asked to find P on her own. When she reached P, she was asked to find T, again on her own. Following this, she was also asked to find B from T and P from B, all on her own. Finding these new (non-trained) paths required that she make spatial inferences using information gained during training.

and back again. The question was whether she would be able to find these landmarks on her own and, more importantly, what paths of movement she would follow. If she was only able to retrace paths she had already traveled, then she would be restricted to moving along known paths. For example, she might move from T to P by tracing the path of T to M, followed by the path of M to P, both of which she had followed before. Alternatively, she might be capable of making spatial inferences to construct *novel* paths – perhaps straight-line paths – between the landmarks. The latter would show that she had in mind a complete, unified representation of the spatial relationships among the four landmarks, and that she was able to use this representation to make spatial inferences – that is, to mentally generate novel paths among the landmarks.

It was found that this child did, indeed, move along roughly straight-line paths between pairs of landmarks. A range of measurements showed that, when first given the instruction to move to a designated landmark, she oriented towards that landmark and then moved along a path that took her more or less directly to it. Various control experiments with the same child showed that her performance was not guided by any cues emanating directly from the landmarks (such as odors) or from the room in general (such as echoes); furthermore, her performance was roughly equal in accuracy to that of sighted blindfolded subjects of the same age, suggesting that her level of accuracy (which was not perfect) was roughly what one could expect from a two- to three-year-old child.

These findings suggest that the blind child possesses spatial knowledge that is apparently quite similar to that of the sighted child. What is the nature of this knowledge? Landau and colleagues argued that the navigation task described above requires a spatial representation based on metric information: that is, one that includes information about angles and distances among landmarks (see Article 12, IMAGERY AND SPATIAL REPRESENTATION). Such information is geometrically necessary for computing the shortest straight-line paths among landmarks in the task described above. (Try for yourself to solve the inference problem without using such information!) C. Randy Gallistel (1990) has argued that such representations are common to many species, ranging from humans to rats, honeybees, and sand scorpions. The case of the blind child shows that visual experience is not necessary for the emergence of mental representations that preserve the relative metric locations of objects in space. These representations allow the child to move independently among objects along paths she has already traveled and along paths she has never traveled before. Such a system of knowledge arises whether the child has experienced the rich spatial information afforded by vision or the less rich information about the larger environment that is afforded by the other senses, making plausible Descartes' proposal that certain properties of space are innate.

Domain-specific competence

Arguments about nativism and empiricism often engage the question of whether the structure of knowledge in a given domain is specific to that domain or shared by other domains. For example, cognitive scientists have asked whether the basic units of language are specific to language only, or whether they are common to both linguistic and nonlinguistic processes. If each domain is qualitatively different from the next, then this increases the likelihood that learning must be constrained from the start, possibly by special-purpose learning mechanisms designed for each domain. The qualitative differences across domains such as language, space, and number suggest that learning within each might be guided by domain-specific organizations, a second hallmark of innate endowment.

Language

One example of domain-specific learning can be found in the study of how people use sounds to make meaningful contrasts among words. Every language possesses a set of *phonemes* – sound contrasts that result in a difference in meaning for a native speaker. For example, in English, the acoustic contrast between *p* and *b* is *phonemic*, because the difference between the sounds can be used to signal a difference of meaning: for example, in the pair of words *bat* and *pat*. Pioneering work by Alvin Liberman and colleagues (1967) showed that adult speakers perceive these distinctions *categorically*. Even though there is considerable variation in the way that the *p* sound itself is acoustically realized in speech (depending on what sounds precede and follow it), speakers of English treat all *p* sounds as roughly equivalent and as different from all *b* sounds. Although such phonemic differences are quite salient to native speakers of English, the same distinctions do not necessarily signal differences of meaning in other languages; for example, the /p/ versus /b/ distinction is not phonemic in Kikuyu. Similarly, there are phonemic distinctions that do not appear in English but do appear in other languages of the world. For example, in English, the sound *t* is produced by placing the tongue just behind the teeth. However, in Hindi, there are two different *t* sounds, which produce a difference in meaning. One is produced by placing the tongue on the teeth (called a *dental t*), the other by placing the tip of the tongue back against the roof of the mouth (called a *retroflex t*). Although such a difference in the articulation of *t* would not create two words with different meanings for adult speakers of English, such a difference does so for adult speakers of Hindi.

Given such cross-linguistic differences in what sound contrasts are phonemic, the question arises as to how differential sensitivity to these contrasts develops, and whether this sensitivity is specific to the task of learning language. Janet Werker (1989) has conducted substantial research on this topic. In her studies, infants listen to repeated presentations of a given syllable and are trained to turn their heads when they hear something *different*. For example, infants might hear repeated presentations of the syllable /ba/, followed at some point by the syllable /pa/. If they are sensitive to the phonemic difference between these two syllables, they should turn their heads when they hear /pa/; if they are not sensitive to this difference, they should not turn their heads. The questions here are what set of distinctions infants are sensitive to, when and how this changes over development, and whether the sensitivity is specific to the context of language.

Using this method, Werker discovered that infants between the ages of about six and eight months are sensitive both to those contrasts that are phonemic in their language *and* to those that are not phonemic in their language but are phonemic in other languages. Using the examples above, a six-month-old growing up in an English-speaking household will show sensitivity to the contrast between /pa/ and /ba/ (relevant for English) and also to the contrast between the two *t* sounds (dental and retroflex) which is relevant in Hindi but not in English. By around ten months, infants begin to show narrower sensitivity, converging on the contrasts that are specifically relevant to their native language and becoming less sensitive to contrasts that are not phonemic in their language. Thus, between six and ten months of age, infants change from being sensitive to a universal inventory of phonemic contrasts – those contrasts that occur across the entire range of human languages – to being sensitive only to those contrasts relevant to their native language. This narrowing of sensitivity follows us into adulthood, when we are quite poor at discriminating pairs of syllables that differ only in some nonnative contrast.

These findings raise the question of whether this learning – and *unlearning* – is domain-specific: that is, whether this decrease in sensitivity to nonnative contrasts amounts to complete loss of sensitivity to the relevant acoustic distinctions, or whether the loss is specific to tasks engaging knowledge of language. Recent evidence suggests the latter to be probably true. As one example, Werker and colleagues gave English-speaking adults a discrimination task in which the stimuli were either (a) full syllables exhibiting nonnative contrasts such as the Hindi contrast discussed above, or (b) shortened syllables in which the critical acoustic properties of the contrasts were preserved, but enough other material was deleted so that the stimuli did not sound like speech at all. The adults were not able to discriminate the nonnative contrasts in the first condition but were able to do so in the second. This suggests that the adults' difficulties in discriminating nonnative contrasts were not due to absolute lack of sensitivity to the relevant acoustic properties but, rather, to the context in which the information was embedded. The acoustic information relevant to discriminating Hindi syllables was simply not engaged by English speakers when syllables (linguistic units) were to be discriminated. This suggests that the reorganization of sensitivity that occurs during infancy is linked specifically to the domain of language and, presumably, to the task of learning language.

Number

A recent line of investigation by Karen Wynn (1992) provides compelling evidence that infants possess principles of number that enable them to perform simple mathematical computations. Additional studies indicate that these computations are domain-specific, in that they *care about* numerical properties of objects but not nonnumerical ones.

Wynn used a method that is now quite standard in infant cognition: that of measuring infants' responses to violations of expectation. Infants tend to look longer at surprising events than at events they do not find surprising. In order to determine whether four-month-olds possess expectations about certain numerical principles, Wynn asked whether infants would be surprised if they observed events in which there were apparent violations of simple addition or subtraction.

Infants in an *addition* condition spent several minutes observing one small interesting object placed on a small stage in front of them. Then, a screen appeared, hiding the object from the infants' view. The infants then watched as they saw the experimenter's hand place a second object behind the screen. They then saw the experimenter's hand retreat from the screen, empty. From the adult's perspective, watching this event gives the distinct impression that the experimenter has placed a second object behind the screen to join the first. If the infants remembered that one object was on the stage when the screen hid it from view, if they also observed the second object being placed behind the screen, *and* if they were capable of determining that one object added to another yields two objects, then the infants should expect there to be two objects behind the screen. Because infants are known to look longer at surprising events, Wynn predicted that if infants were able to make this numerical inference, they should not be surprised to observe two objects, but they should be surprised to observe one object, since this outcome would violate the principle of addition.

Wynn tested this expectation by dropping the screen and letting the infants observe two different events, one after the other. In one event, there were two objects visible when the screen was dropped, just the number one would anticipate from a computation of 1+1. In the other event, there was only one object visible when the screen was dropped, just the number the infants had originally seen, but certainly not the number predicted from the correct mathematical computation. Infants looked longer at the numerical violation: that is, the event showing a single object. To confirm that the infants really were responding to specific number (rather than, say, overall larger quantity), Wynn conducted another experiment, in which infants were shown two objects to which one was added in the same manner. The correct numerical answer to this addition problem is different from that in the previous experiment: that is, it is 3, not 2. Wynn found that infants looked longer at the result of 2: that is, a violation of addition for the numerical problem 2+1. Wynn also included a *subtraction* condition, in which infants were shown two objects, the screen was placed in front of them, and the experimenter visibly removed one. For adults, this sequence of events leads to the conclusion that there is just one object remaining behind the screen. The test situation is then exactly analogous to the *addition* condition, except that the predictions differ. If infants can *subtract*, then they should not be surprised to observe just one object when the screen is removed, but should be surprised to observe two objects. These predictions were confirmed.

A recent replication and extension of Wynn's experiments shows that the infants' expectations in the addition task were truly numerical, hence domain-specific. That is, number mattered to the infants in this task, but other properties of objects not relevant to number did not matter. Simon, Hespos, and Rochat (1995) repeated Wynn's experiment with one important difference. Instead of just adding one object to another and then testing to see whether infants expected a sum of 1 or 2, they also varied whether the objects being added were identical to or different from the objects already hidden behind the screen. For example, if the first object that infants observed standing on the stage was an *Ernie* doll, then the second (added) object might be either a second Ernie doll or a quite different looking *Elmo* doll. Infants' expectations of both number and identity were then assessed. In the testing sequence, infants observed events that were either (a) violations or nonviolations of numerical outcome and (b) violations or nonviolations of identity outcome. For instance, if an Elmo had been *added* to an Ernie, the proper outcome would be just two dolls, one Ernie and one Elmo. Improper outcomes could be produced by either numerical violations (just one doll, be it Ernie or Elmo) or identity violations (two Ernies or two Elmos). The results showed that infants looked longer at violations of *numerosity* but not at violations of specific identity. In a control experiment, Simon and colleagues showed that the infants' disregard for the objects' identities was not due to poor discriminability or lack of salience.

This result is important, because it shows that when the infants' knowledge of number was engaged, the specific identities of the objects did not matter at all. It is a foundational principle of number that numerosity is independent of identity. For example, numbers can range over any set of objects, regardless of their identity, their homogeneity, their location, etc. The domain-specific nature of infants' number knowledge allowed them to compute the addition problem without engaging their knowledge of specific object identities.

Contribution by the learner

The third hallmark of innate endowment is the child's *creativity*. When children say or do things that have not been directly modeled by their care-givers, these acts constitute strong evidence for the independent contribution of the learner. Among the most salient examples of such creativity are the errors produced by young children learning language. Two- and three-year-olds may say "I eated all my dinner" or "My foots have ten toes," even though they have never heard these forms of the verbs. Examination of such errors shows that children make mistakes because they are following rules they have learned from hearing English – in this case, that formation of the past tense requires adding /ed/, and formation of the plural requires adding /s/. Children then apply the rules even in *irregular* cases, despite the fact that they have surely heard only *ate* (not *eated*) and *feet* (not *foots*).

Such errors are one example of the more general case in which the child's performance does not match up precisely with the input in the environment. In such cases, one can observe quite clearly the contribution of the learner's own internal schemas – plans for knowledge of language, number, or space. Learners use these schemas as the overall outlines which are filled out by relevant bits of information from the environment.

Language

Recall that certain properties of language appear to emerge in young children even if there is no linguistic model for them in the environment. The case of deaf children who invent a sign language tells us that certain fundamental distinctions (such as that between nouns and verbs) will appear, even if the relevant linguistic input is completely absent from the environment. Additional studies by Elissa Newport and colleagues show that even when young learners are exposed to *imperfect* input, they will reorganize and systematize these data, creating a language that is more like other natural languages than the language they are actually exposed to.

Singleton and Newport (1994) studied the case of acquisition of American Sign Language by a congenitally deaf child named Simon, whose congenitally deaf parents were not native signers themselves. Simon's parents had learned ASL somewhat late in life, after age 15, and were not fluent in spoken English, due to their profound hearing losses. As is typical of late learners of a language, Simon's parents did not possess full native control over the structure of the language. In particular, their control over the structure and combinatorial rules for morphology (units of meaning equal to or smaller than whole words) in ASL was significantly worse than that of native learners of ASL. Because Simon's only exposure to ASL was through his parents, his language input was not always fully grammatical and often lacked units of meaning that are required in ASL. For example, Simon's parents often omitted obligatory morphemes or combined them improperly, providing input that one could describe as degraded relative to a fully grammatical corpus.

Singleton and Newport studied Simon's acquisition of ASL, asking whether he would learn a language that was closely modeled on his parents' input (complete with its lack of well-formedness) or whether his own system would show modifications more in line with a fully grammatical system. In one test, they gave Simon, his parents, and appropriate control subjects (native deaf signers) sets of action scenes which were to be described in sign. Each of the scenes depicted moderately complex actions (e.g., a doll jumping through a hoop), whose description would require a specific set of ASL morphemes representing the objects, their motion, path, location, etc., properly ordered in the sentence.

An overall assessment of accuracy (i.e., proper inclusion of morphemes) showed that Simon's parents performed considerably worse than deaf adult native signers – about 70 percent correct overall, compared to 90 percent. However, Simon performed as well as deaf native child signers of the same age – about 90 percent correct on morphemes expressing motion and location. Thus, despite Simon's relatively impoverished input, his own knowledge of language seemed to be the same as that of children learning from fully grammatical input.

Even more striking, the results showed that Simon's language was more systematically grammatical than that of his parents. For example, although his parents frequently produced the proper morpheme representing an object's path, they did not do so all the time (as would a native signer). Rather, they typically used it about 70 percent of the time, making a variety of errors on the remaining 30 percent of occasions. Simon, however, produced this same response much more frequently, as if he took the most predominant morpheme in that context and *boosted* its frequency so as to use it in a rulelike fashion – more systematically and more regularly than his parents, and more consistent with a fully grammatical ASL. Frequency alone, however, was not sufficient for learning. For example, Simon's mother very frequently used a particular hand shape for a wide variety of objects – in essence, this was a sign that did not pick out any special semantic distinction. But despite the fact that this sign was very frequent in the input, Simon did not boost its frequency – in fact, he rarely used it at all.

These findings suggest that Simon built his modified language on moderately frequent signs that systematically represented some meaningful distinction. In this way, his language was more systematic than that of the input he received from his parents. Simon's *creation* of a more grammatical linguistic system using somewhat degenerate input shows how learners impose their own organization on the input in their environment.

Number

As with language, there is evidence from the domain of number that children bring to the learning task certain principles or structures that serve to organize incoming information. These principles appear quite strikingly in cases where children's counting behavior contains errors, but is nevertheless rule-governed.

Rochel Gelman and C. Randy Gallistel (1978) first showed the power of such organizing principles in young children's spontaneous counting. Children typically begin counting around the age of two, and when they do, the nature of their counting behavior reveals that they are using principles whose application takes them well beyond the data that have been presented to them.

According to Gelman and Gallistel, children's counting behavior follows a set of principles that dictate what can be counted and how to count those things. For example, the one–one principle states that one and only one number tag may be assigned to each item to be counted; one cannot use the same number tag to count two different objects in a given set. The stable order principle states that number tags must be used in the same order on every counting occasion; if ordering were not preserved, then one would not be able to ascertain equivalent numerosity on two different occasions. The abstraction principle states that anything is a potentially countable *object*; hence same or different objects can be counted as a single set, objects can be grouped and regrouped and counted as part of a new set, etc. One might think that children learn to count simply by memorizing the list of count tags (number names) and repeating them when asked to count. But Gelman and Gallistel argue that counting is much more complex and subtle: that, in fact, the counting principles serve as the skeleton into which learners fit the actual names of numbers.

The most striking evidence for this idea comes from Gelman and Gallistel's observations of young children's *errorful* counting – in particular, idiosyncratic number lists. For example, two-year-olds often count sets inaccurately, saying, for example, "two, six" for an array of two objects. An older child might count a two-object array correctly but a four-object array saying, for example, "one, six, ten, fourteen." In neither case do the number tags correctly apply to the object sets; hence these tag lists could never have been part of the input to the child. However, the children's errorful counting is systematic. First, the same child will tend to use the same idiosyncratic list over and over again. Second, the children's lists observe the counting principles. For example, the one-one principle is observed, because children use individual, different tags for each object; the stable order principle is observed, because the children use these same tags in just the same numerical order from trial to trial, preserving the ordinal character of the count tags. Gelman and Gallistel report that some children even recruit tags from a wholly separate domain, using the alphabetic letters as count tags – for example, counting a set of objects as "a, b, c, x." Even in these cases, the counting principles are observed. Just one tag is used for each object, the order of the tags is preserved, and the ordering within the alphabetic tag list maps accurately onto the quantities being counted, with the earlier letters mapping onto smaller numbers and the later letters onto larger numbers. The use of the alphabet to fuel the counting principles is especially striking, since the particular tags used are clearly not a part of the language's natural set of count tags.

The counting principles, and their application by children to ascertain the numerosity of an object set, illustrate once more how internal mental structures guide the behavior of young children. The counting principles are surely embodied in the counting activity of parents; but nowhere in the input are they explicitly described to the learner. Given that the principles themselves, and not the exact characteristics of the input, guide children's counting, it must be concluded that the learner's internal schemas for number contribute as much to the process of learning as the environmental events that provide grist for the mental mill.

Summary and conclusions

The notion that the mind comes prepared to interpret the world in terms of certain categories such as language, space, and number is not new. However, since the time of Descartes – and, most notably, over the past 20 years – there has been a burgeoning of empirical evidence from cognitive science that testifies to the importance of native endowment in guiding the acquisition of knowledge in these domains. This chapter has described a variety of such findings, each of which is consistent with Descartes' rationalist program.

In particular, three hallmarks of native endowment have been discussed: early, rapid emergence of competence over very different environmental conditions for learning; the domain-specific nature of this learning; and the contribution by the learner, as evidenced by qualitative differences between the environmental input for learning and the learning that actually occurs. Each of these is consistent with the idea that learners come to the task of learning prepared with some initial biases, schemas, and structures which will be used to filter, analyze, and reorganize data provided by the environment. These predispositions appear to be domain-specific, with each engaging highly specific structures, requiring highly specific sensitivities to different kinds of information, and perhaps even recruiting different mechanisms for obtaining, processing, and organizing this information.

References and recommended reading

Chomsky, N. 1980: *Rules and Representations.* New York: Columbia University Press.

Feldman, H., Goldin-Meadow, S. and Gleitman, L. R. 1978: Beyond Herodotus: the creation of language by linguistically deprived deaf children. In A. Lock (ed.), *Action, Symbol, and Gesture: The Emergence of Language,* New York: Academic Press, 351–414.

*Gallistel, C. R. 1990: *The Organization of Learning.* Cambridge, Mass.: MIT Press.

Gelman, R. and Gallistel, C. R. 1978: *The Child's Understanding of Number.* Cambridge, Mass.: Harvard University Press.

Goldin-Meadow, S., Butcher, C., Mylander, C. and Dodge, M. 1994: Nouns and verbs in a self-styled gesture system: what's in a name? *Cognitive Psychology,* 27(3), 259–319.

*Landau, B. and Gleitman, L. R. 1985: *Language and Experience.* Cambridge, Mass.: Harvard University Press.

Landau, B., Spelke, E. and Gleitman, H. 1984: Spatial knowledge in a young blind child. *Cognition,* 16(3), 225–60.

Lenneberg, E. H. 1967: *Biological Foundations of Language.* New York: Wiley.

Liberman, A., Cooper, F. S., Shankweiler, D. P. and Studdert-Kennedy, M. 1967: Perception of the speech code. *Psychological Review,* 74, 431–61.

Simon, T. J., Hespos, S. and Rochat, P. 1995: Do infants understand simple arithmetic? A replication of Wynn (1992). *Cognitive Development,* 10(2), 253–70.

Singleton, J. and Newport, E. 1994: When learners surpass their models: the acquisition of American Sign Language. MS, University of Rochester.

Werker, J. 1989: Becoming a native listener. *American Scientist,* 77, 54–9.

Wynn, K. 1992: Addition and subtraction by human infants. *Nature,* 358, 749–50, 792.

46

Innateness and emergentism

ELIZABETH BATES, JEFFREY L. ELMAN, MARK H. JOHNSON, ANNETTE KARMILOFF-SMITH, DOMENICO PARISI, AND KIM PLUNKETT

The nature–nurture controversy has been with us since it was first outlined by Plato and Aristotle. Nobody likes it anymore. All reasonable scholars today agree that genes and environment interact to determine complex cognitive outcomes. So why does the controversy persist? First, it persists because it has practical implications that cannot be postponed (i.e., what can we do to avoid bad outcomes and insure better ones?), a state of emergency that sometimes tempts scholars to stake out claims they cannot defend. Second, the controversy persists because we lack a precise, testable theory of the process by which genes and the environment interact. In the absence of a better theory, innateness is often confused with (1) *domain specificity* (outcome X is so peculiar that it must be innate), (2) *species specificity* (we are the only species who do X, so X must lie in the human genome), (3) *localization* (outcome X is mediated by a particular part of the brain, so X must be innate), and (4) *learnability* (we cannot figure out how X could be learned, so X must be innate). We believe that an explicit, plausible theory of interaction is now around the corner, and that many of the classic maneuvers to defend or attack innateness will soon disappear. In the interim, some serious errors can be avoided if we keep these confounded issues apart. That is the major goal of this chapter: not to attack innateness but to clarify what claims about innateness are (and are not) about.

What will a good theory of interaction look like when it arrives? It is useful here to distinguish between two kinds of interactionism: *simple interactions* (black and white make gray) and *emergent form* (black and white get together, and something altogether new and different happens). In an emergentist theory, outcomes can arise for reasons that are not predictable from any of the individual inputs to the problem. Soap bubbles are round because a sphere is the only possible solution to achieving maximum volume with minimum surface (i.e., their spherical form is not explained by the soap, the water, or the little boy who blows the bubble). Beehives take a hexagonal form because that is the stable solution to the problem of packing circles together (i.e., the hexagon is not predictable from the wax, the honey it contains, or from the packing behavior of an individual bee). D'Arcy Thompson offered hundreds of examples like these to explain the emergence of different bodily forms, up and down the phylogenetic scale. Jean Piaget argued that logic and knowledge emerge in just such a fashion, from successive interactions between sensorimotor activity and a structured world. In the same vein, it has been argued that grammars represent the class of possible solutions to the problem of mapping hyperdimensional meanings onto a low-dimensional channel, heavily constrained by the limits of human information processing (see Article 31, FUNCTIONAL ANALYSIS). Logic, knowledge, and grammar are not given in the world, but neither are they given in the genes.

Emergentist solutions of this kind have been proposed again and again in the developmental literature, as a way out of the nature–nurture controversy ("That which is inevitable does not have to be innate"). Unfortunately, the metaphors invoked by proponents of emergentism do not constitute a convincing theory of complex cognition, and the detailed descriptions of behavioral change offered by Piagetian scholars have never yielded the formal theory of development that Piaget sought for more than six decades. As a result, ardent nativists such as Noam Chomsky (1980) have viewed Piaget as a radical empiricist indistinguishable from Skinner in his reliance on environment as the ultimate cause of development. A similar fate has befallen those who propose an emergentist account of language (Gibson, 1992).

We believe that a more convincing emergentist account of development is now possible, for three reasons. First, developmentalists have begun to make use of insights from nonlinear dynamics (see Article 38, CONNECTIONISM, ARTIFICIAL LIFE, AND DYNAMICAL SYSTEMS). This is the latest, and perhaps the last, frontier of theoretical physics, offering insights into the processes whereby complex, surprising, and apparently discontinuous outcomes can arise from small quantitative changes along a single dimension. Beehive metaphors have thus given way to an explicit, formal account of emergent form. Second, it is now possible to simulate behavioral change in multilayered neural networks, systems that embody the nonlinear dynamical principles required to explain the emergence of complex solutions from simpler inputs. Third, students of behavioral development are becoming aware of some remarkable breakthroughs in developmental neurobiology. As we shall see, today's neurobiological results are very bad news for yesterday's nativists, because they underscore the extraordinarily plastic and activity-dependent nature of cortical specialization and buttress the case for an emergentist approach to the development of higher cognitive functions.

Even within an interactionist view of this kind, one has to start somewhere. The constraints on emergent form offered by genes and environment must be specified. What do we mean when we say that a given outcome is innately constrained? As a first approximation, we can define *innateness* as a claim about the amount of information in a complex outcome that was contributed by the genes (keeping in mind, of course, that genes do not act independently, and that they can be turned on and off by environmental signals throughout the lifetime of the organism). Elsewhere we have proposed a three-level taxonomy of claims about innateness, ordered from strong to weak with regard to the amount of information that must be contributed by the genes for this claim to work (Elman et al., 1996). Each level is operationally defined in terms that correspond to real brains and to artificial neural networks, as follows:

Representational constraints refer to direct innate structuring of the mental/neural representations that underlie and constitute *knowledge*. Synaptic connectivity at the cortical level is the most likely candidate for the implementation of detailed knowledge in real live brains, because that is the only level that has the coding power for higher-order cognitive outcomes. In artificial neural networks, this level is operationalized in the weighted connections between processing units.

Architectural constraints refer to innate structuring of the information-processing system that must acquire and/or contain these representations. Although representation and architecture are not the same thing, there is no question that the range of representations which a system can take is strongly constrained at the architectural level. In traditional serial digital computers, some programs can only run on a machine with the right size, speed, and power. In neural networks, some forms of knowledge can be realized or acquired only in a system with the right structure (the right number of units, number of layers, types of connectivity between layers, etc.). In fact, there is now a whole subfield of neural network research in which genetic algorithms are applied to uncover the class of architectures that are best suited to a given class of learning problems (see Article 16, MACHINE LEARNING).

To operationalize architectural constraints in real brains and in neural nets, we break things down into three sublevels:

> *Basic computing units.* In real brains, this sublevel refers to neuronal types, their firing thresholds, neurotransmitters, excitatory/inhibitory properties, etc. In neural networks, it refers to computing elements with their activation function, learning algorithm, temperature, momentum, and learning rate, etc.
>
> *Local architecture.* In real brains, this sublevel refers to regional factors like the number and thickness of layers, density of different cell types within layers, type of neural circuitry (e.g., with or without recurrence). In neural networks, it refers to factors like the number of layers, density of units within layers, presence/absence of recurrent feedback units, and so forth.
>
> *Global architecture.* In real brains, this sublevel includes gross architectural facts like the characteristic sources of input (afferent pathways) and patterns of output (efferent pathways) that connect brain regions to the outside world and to one another. In many neural network models, the size of the system is so small that the distinction between local and global architecture is not useful. However, in so-called modular networks or expert networks, it is often useful to talk about distinct subnets and their interconnectivity.

Chronotopic constraints refer to innate constraints on the timing of developmental events, including spatiotemporal interactions. In real brains, this would include constraints on the number of cell divisions that take place in neurogenesis, spatiotemporal waves of synaptic growth and pruning, and relative differences in timing between subsystems (e.g., differences among vision, audition, etc. in the timing of thalamic innervation of the cortex). The same level is captured in neural networks by incremental presentation of data, cell division schedules in growing networks, adaptive learning rates, and intrinsic changes in learning that come about because of node saturation.

See Elman et al., 1996, for detailed examples at all these levels. For our purposes here, the point is that strong nativist claims about language (Fodor, 1983; Pinker, 1994), physics (Spelke, 1991), or social reasoning (Leslie, 1994) have to assume representational nativism, implicitly or explicitly, because that is the only level with the required coding power for the implementation of knowledge that is independent of experience. For example, Noam Chomsky (1975) has proposed that "Linguistic theory, the theory of UG (Universal Grammar) . . . is an innate property of the human mind" (p. 34), and that we should conceive of "the growth of language as analogous to the development of a bodily organ" (p. 11). The mental organ metaphor leaves little room for learning. Indeed, Chomsky has argued that "a general learning theory. . . . seems to me dubious, unargued, and without any empirical support" (1980, p. 110). Piatelli-Palmarini (1989, p. 2) echoes this theme, stating that "I . . . see no advantage in the preservation of the term *learning*. I agree with those who maintain that we would gain in clarity if the scientific use of the term were simply discontinued." Where would such rich innate structure reside? Pinker suggests that this innate knowledge must lie in the *microcircuitry* of the brain. We think that he is absolutely right: If the notion of a language instinct means anything at all, it must refer to a claim about cortical microcircuitry, because this is (to the best of our knowledge) the only way that detailed information can be laid out in the brain.

This kind of representational nativism is theoretically plausible and attractive, but it has proved hard to defend on both mathematical and empirical grounds. On mathematical grounds, it is difficult to understand how 10^{14} synaptic connections in the human brain could be controlled by a genome with approximately 10^{6} genes, particularly when (a) 20–30 percent of these genes at most go into the construction of a nervous system (Wills, 1991), and (b) humans share approximately 98 percent of their genes with their nearest primate neighbors. But the problem is even worse than that. Paul Churchland (1995) reminds us that each synaptic connection can take multiple values. If we assume conservatively that each connection can take ten values, Churchland calculates that the synaptic coding power of the human brain contains more potential states of connectivity than there are particles in the universe! Genes would need a lot of information to orchestrate a system of this size. Of course, a detailed mapping from genes to cortex would still be possible if genes behaved like letters in the alphabet, yielding up an indefinite set of combinations. But this is not the case; instead, genes operate within a highly constrained spatiotemporal and chemical matrix (Edelman, 1987; Wills, 1991), using and reusing topological principles that have been conserved over millions of years and thousands of species.

One could argue that the innate component of knowledge occupies only a fraction of this massive state space, despite its richness. However, the past two decades of research on vertebrate brain development suggest that fine-grained patterns of cortical connectivity are largely determined by cortical input. For example, we know that the auditory cortex will take on retinotopic maps if input from the eye is diverted there from its normal visual target, that plugs of cortex taken from one cortical area and transplanted in another will take on the representations that are appropriate for the input they receive in their new home, that alterations in the body surface of an infant rat lead to corresponding alterations in the cortical map, that the *where is it?* system will take over the *what is it?* function in infant monkeys with bilateral lesions to inferior temporal cortex, and that human infants with left hemisphere lesions that would lead to irreversible aphasia in an adult go on to attain language abilities that are well within the normal range. (For reviews and references, see Elman et al., 1996, ch. 5; Johnson, 1997). In short, there is very little evidence today in support of the idea that genes code for synaptic connectivity at the cortical level. Instead, brain development in higher vertebrates appears to involve massive overproduction of elements early in life (neurons, axons, and synapses), followed by a competitive process through which successful elements are kept, and those that fail are eliminated (Edelman, 1987). Pasko Rakic refers to this competition as the process by which experience literally *sculpts* the brain.

In addition to this sculpting through regression, experience may also add structure across the course of life, inducing synaptic sprouting in just those areas that are challenged by a brand-new task (Merzenich and Jenkins, 1995).

Although there is surprisingly little evidence for innate representations at the cortical level, there is substantial evidence for innate architectures and innate variations in timing. This includes evidence that neurons *know* where they are supposed to go during cell migration (Rakic, 1988) and evidence that axons prefer particular targets during their long voyage from one region to another (Niederer et al., 1995). Could this kind of innateness provide the basis for an innate Universal Grammar? Probably not, because (a) these gross architectural biases do not contain the coding power required for something as detailed and specific as grammatical knowledge, and (b) the rules of growth at this level appear to operate across species to a remarkable degree. For example, Deacon (1997) describes evidence for lawful axon growth in the brain of the adult rat, from cortical transplants taken from fetal pigs!

There are also regional variations in the neurochemical substrate (e.g., somatosensory cortex transplanted to a visual region will take on visual maps but still expresses the neurochemicals appropriate for a somatosensory zone (Cohen-Tannoudji et al., 1994)), and regional variations in cell density (e.g., primary visual cortex is exceptionally dense, a characteristic that seems to be determined during neurogenesis, before any information is received (Kennedy et al., 1990)). Hence cortical regions are likely to differ from the outset in style of computation, which means that they will also vary in the kinds of tasks they can perform best. In other words, the competition that characterizes brain development does *not* take place on an even playing field. The game is rigged from the beginning, to privilege some overall *brain plans* over others. However, it is also clear that many alternative brain plans are available if the optimal form is precluded for some reason.

Bates and colleagues (1997) have argued that left hemisphere specialization for language in humans depends on indirect, architectural constraints like these. The temporal and frontal regions of the left hemisphere play a major role in the mediation of language production in over 95 percent of normal adults, leading to irreversible aphasia if specific left hemisphere sites are damaged. Yet, as noted above, infants with homologous injuries do not grow up to be aphasic. How can that be? If left perisylvian cortex isn't necessary for normal language, where does the typical *adult brain plan* come from? Studies of infants with focal brain injury show that the temporal (but not the frontal) region of the left hemisphere is indeed specialized at birth, because children with left temporal lesions are significantly slower in the development of expressive (but not receptive) vocabulary and grammar. However, this regional difference is no longer detectable by the time children with the same early injuries are seven years old, which means that a great deal of reorganization must have taken place across the first years of life. Evidently other regions of the brain are capable of taking on the representations required for normal language. Bates and colleagues suggest that left temporal cortex is initially specialized not for language itself, but for the extraction of perceptual detail (e.g., damage to the same regions has specific effects on the extraction of detail from a visual-spatial pattern). Under normal conditions, this indirect bias in computing style leads to left hemisphere specialization for language. But the representations required for language are not (and apparently need not be) present from the beginning, because the same cat can be skinned in a number of alternative ways.

Because the evidence is not good for strong, representational forms of nativism, the differences that we observe from one species to another must be captured primarily by architectural and chronotopic facts. The final product emerges from the interaction between these constraints and the specific problems that an organism with such structure encounters in the world. Within this framework, let us reconsider some of the classical arguments for innate knowledge, arguments which, we believe, confuse levels of analysis that should be kept separate.

Innateness and domain specificity

It has been argued that language is so peculiar, so specific to the domain in question, that it could not possibly be learned or processed by a domain-general system. Similar claims have been made about face perception, music, mathematics, and social reasoning. We argue that claims about domain specificity must (like innateness) be broken down into different levels before we can approach the issue empirically. Using language as our test domain, here is a brief overview.

Behavioral specificity

Languages represent a class of solutions to a problem that is undeniably unique in its scope and nature: the problem of mapping a hyperdimensional meaning space onto a low-dimensional channel (see Article 31, FUNCTIONAL ANALYSIS). There may be a casual resemblance to domains like bird song (learning in the vocal channel), chess (a complex set of solutions to a game that only humans play), or music (rule-governed transitions in sound), but these similarities are largely superficial. Turkish case inflections do not *look like* chess, bird song, or music; but they do look a lot like case inflections in Hungarian. That is, languages have very little in common with other cognitive systems, but they do have a lot in common with each other. Where do these commonalities come from? The meaning space involved in the language-mapping problem includes experiences that are shared by all normal members of the species, and the channels used by human language are subject to universal constraints on information processing (e.g., perception, memory, articulatory planning). Under these circumstances, we should not be surprised to find that the class of solutions to the problem is quite limited, constituting a set of alternatives that might be referred to as "Universal Grammar." We will stipulate that domain-specific behaviors have emerged in response to this mapping problem, and that natural languages draw from a common set of domain-specific solutions. But such facts do not constitute *ipso facto* evidence for innateness, because the same solutions could have emerged on an emergentist scenario.

Representational specificity

If an individual reliably produces the behaviors required to solve a domain-specific problem, it follows that he or she must possess a set of domain-specific mental/neural representations that support the behavior. That is, every representation must be implemented in a form that is somehow distinguishable from other aspects of knowledge (see section on innateness and localization below). This generalization holds whether the representations in question are innate or learned; hence the specificity of a representation is simply not relevant to the innateness debate.

Specificity of mental/neural processes

This is the level at which innateness and domain specificity finally cross. Is it possible for a domain-general architecture to acquire and/or process domain-specific representations? Notice that *domain-general* need not mean *a device that can learn and do anything.* We have already stipulated the need for a good match between problems and architectures in neural network research. There is no device that can learn and do everything. The debate is more specific. Can a domain like language be learned and/or processed by *any* system that is not specifically tailored for and dedicated to linguistic events? This is an empirical question, and the answer is not yet in. However, evidence in support of the domain-general view is available from simulations of domain-specific learning in general-purpose neural networks (see section on innateness and learnability below) and from the plastic reorganization of language and other higher cognitive functions observed in children with focal brain injury.

But what if the representation at issue is bizarre and not at all predictable (as far as we can see) from the problem to be solved? How could a general architecture ever acquire such a thing? Our belief that a structure is inexplicable may be nothing more than a comment on our ignorance. For example, Hubel and Wiesel discovered that the visual cortex of the cat contains odd little neurons that only fire to lines at a particular orientation. Why should such a peculiar structure emerge? Yet we know that such structures do emerge reliably every time a multilayered neural network is forced to extract three-dimensional information from a two-dimensional array (Shatz, 1996). We cannot predict the line-orientation solution just by looking at the inputs to the problem of mapping three dimensions onto two, but apparently just such a solution is required, whether or not it is built in. In other words, although the line-orientation detectors in visual cortex *could* be innate (phylogeny insuring a useful solution), these simulations show that they do not *have* to be innate. The same may be true for the odd-looking structures that comprise human grammars and human social reasoning.

Genetic specificity

Skipping over the intervening levels, arguments linking domain specificity and innateness are sometimes based on the specific patterns of impairment observed in individuals with genetic damage. Suppose, for example, that we uncover a form of language impairment that is associated with a genetically transmitted disorder. Doesn't this constitute direct evidence for the innateness of a domain-specific ability? Not necessarily. After all, language is entirely absent in cases of cerebral agenesis (where no brain grows at all above the brain stem level), but no one would argue that the absence of a brain provides interesting evidence for a domain-specific language faculty. Genetically based language disorders provide evidence for the innateness of a domain-specific faculty if and only if we can show that the genetic defect affects language *in isolation* (see Article 29, DEFICITS AND PATHOLOGIES).

Specific language impairment, or SLI, is defined as a significant disorder in which language falls well below that appropriate to mental age, in the absence of mental retardation, frank neurological impairment, hearing loss, severe social-emotional distress, or environmental pathology. It has been shown that SLI tends to run in families, and some have argued that this disorder constitutes the required evidence for a genetic defect that affects only grammar. A celebrated case in point is the London family in which (it was reported) a genetically transmitted impairment was observed that affects only regular grammatical morphemes (e.g., walk → walked), with no other effect on any other aspect of cognition or language, including irregular grammatical morphemes (e.g., give → gave). The initial report generated a great deal of excitement (Gopnik, 1990; Pinker, 1994), but it was ultimately shown to be premature and largely incorrect. More comprehensive studies show that the affected members of this family suffer from a host of deficits within and outside language, and the putative dissociation between regular and irregular morphemes does not replicate (Vargha-Khadem et al., 1995).

Thirty years of research on other children and adults with SLI yield a similar conclusion. Specific language impairment correlates with a range of relatively subtle deficits outside the boundaries of language proper, including aspects of attention, symbolic play, mental imagery, and the detection of rapid sequences of sounds. In short, specific language impairment is a genetically transmitted disorder, but it is no longer clear (despite the name) that it is specific to language, much less to some peculiarity of grammar.

The converse is also true: Deficits specific to language have turned out not to be innate, at least not in any interesting sense. For example, we now know that grammatical morphology (including all those *little words* and endings) is an especially vulnerable domain; whether or not it is impaired in isolation, morphology shows up as a major problem area in SLI, Down syndrome, and other disorders where a genetic base is known or suspected. However, specific problems with grammatical morphology have also been shown in many different forms of acquired brain injury (with little respect for lesion site), in neurologically intact individuals with hearing impairment, and in college students forced to process sentences under adverse conditions (e.g., with perceptually degraded stimuli, or with reduced attention and memory due to a competing task). Grammatical morphemes tend to be low in perceptual salience and imageability, and, perhaps for this reason, they constitute a *weak link in the processing chain*. The fact that they are preferentially disrupted in genetically based syndromes does not necessarily constitute evidence that they are innate in any domain-specific way. Damage to the human elbow has a very specific effect on tennis, but that does not mean that the elbow is a tennis processor, or that the genes that participate in elbow construction do so for the good of tennis. We have already stipulated that language is not tennis, but the metaphor is appropriate on this particular point.

To summarize, innateness and domain specificity are not the same thing, and the case for innateness can never be made simply by listing strange phenomena (i.e., the Madame Tussaud strategy). We turn now to species specificity, localization, and learnability, special cases of the effort to prove innateness by showing that a domain is *special*.

Innateness and species specificity

In this variant of the domain-specificity approach, it is argued that a domain must be innate because only humans do it – or, at least, only humans do it very well. This would include language, but it also includes music, politics, religion, international finance, and ice hockey. To be sure, there are rudimentary variants of human skills in other species, including language. The identification of such infrahuman precursors is useful, because it can tell us something about the evolution of language and other uniquely human functions. But the case for an innate, domain-specific system cannot be made simply by pointing out that nobody else has what we have. Although we are the only species that plays chess, no one wants to argue that we start out with a chess faculty in any interesting sense.

Of course, many of us don't play chess, but all normal humans use language. Do species specificity and universality together constitute evidence for an innate faculty? Possibly, but both facts could be explained by factors that are related only indirectly to the domain in question. To date, no one has ever identified a neural structure that is unique to humans – that is, a human-specific neuronal type, neurotransmitter, or pattern of cortical layering – or even (depending on how we define *area*) a human-specific area of the brain. Our undeniably unique array of skills appears to be built out of quantitative variations in the primate brain plan: for example, expansion of frontal cortex relative to other regions, proportional enlargement of secondary areas within visual cortex, more direct cortical control over the mouth and fingers. The latter innovation sounds like it might have emerged especially for language (or, more generally, for culture), and at some level that may be the case. It is interesting to note, however, that the same direct connections from cortex to the periphery are present in the embryonic rat but are eliminated before they have a chance to become functional (Deacon, 1997). Although we do not belong to the school of evolution that explains everything through brain size, species-specific abilities could be an unintended by-product of a much more general change in computing power (Wills, 1991). Species specificity alone does not constitute evidence for a specific mental organ.

Innateness and localization

This is also a form of the specificity argument: Mental organs are special because they take place in their own part of the brain, what Fodor (1983, p. 99) calls a "specialized neural architecture" (a term that conflates the representational and architectural levels we laid out above). If we could show, for example, that the brain handles regular and irregular grammatical morphemes differently, wouldn't that constitute evidence for two innately specialized, domain-specific processors? Not necessarily.

First, everything that we know is mediated by the brain. If we experience two stimuli in exactly the same way, then (by definition) we do not *know* that they are different. If we do experience them differently, then that difference must be reflected somewhere in the brain. Every new piece of learning changes the structure of the brain in some fashion, however minor. Consider, for example, a recent demonstration that chess experts show different patterns of cortical activity at different points in the game (Nichelli et al., 1994). This does not mean that we have an end game organ, not even in the adult state. And it certainly does not mean that we were born with one. All knowledge presupposes localization in some form (compact and local, or broadly distributed), and hence demonstrations of localization do not constitute evidence for innateness. This is true whether the localization is universal (all humans show the same pattern) or variable (some people handle the same content in different places).

However, the converse is not true: If a cognitive ability is innate, then it must be realized in some topographically specifiable way. That is how genes work: that is, by coding proteins in a spatially, temporally, and chemically defined matrix (Edelman,

1987; Wills, 1991). That is precisely why the evidence for cortical plasticity is so devastating to representational nativism. To fend off this evidence, one might envision a scenario in which the genes that set up the nervous system travel around in the bloodstream looking for a friendly environment in which a specific mental organ can be built. After all, every cell in the body contains the entire genome. Perhaps the language genes are wandering about, waiting for a signal that says "Start building a language organ now, here." There are certainly examples in the literature where the right thing does get built in the wrong place, or at least in an atypical place (e.g., the master gene for the eye, which can be multiplied in various places). But this kind of evidence appears to be the exception. And at least in that case (in contrast to higher cognitive functions like language), there is a specifiable set of physical constants involved which dictate the shape of the thing to be built. For the same reason that we cannot really build a dinosaur out of genes in a piece of amber (the Jurassic Park scenario), genes for language or music or face perception do not travel around in a lifeboat looking for a place to land. Localization does not presuppose innateness, but claims about innateness do presuppose a physical base. That is why nativists are wise to look for the neural correlates of the system they are interested in.

Innateness and learnability

Within linguistics, claims about innateness have been made that bypass all these lines of empirical evidence. The ultimate form of the eccentricity argument goes like this: X (usually language) is so peculiar, so unlike anything else that we do, that it cannot be learned by garden-variety learning mechanisms. Children (it is claimed) must acquire a grammar that is more powerful than his/her degenerate input can support. They are able to go beyond their data and zero in on the right grammatical target only because they already know a great deal about the class of possible grammars. The most principled form of this argument is based on a formal proof of learnability in computer science called "Gold's theorem" (Gold, 1967), which showed that grammars of a particular class cannot be induced or *guessed* from a finite base of positive evidence (i.e., examples of sentences in the language) in the absence of negative evidence (i.e., examples of sentences that do not belong in the language).

A thorough, or even a superficial, treatment of this argument goes far beyond our purview here, except to note that all learnability proofs rest upon at least four kinds of assumptions: a definition of the grammar to be acquired (e.g., grammars defined as strings of symbols generated by one or more recursive rules), a characterization of the data available to the learner, a specification of the learning device that goes to work on these data, and a criterion that defines successful learning. If the grammar to be acquired is very abstract, if our criterion for success is very high, and/or if the learning device is weak and the data are degenerate, then it follows incontrovertibly that the grammar cannot be learned without a great deal of innate knowledge to make up for those weaknesses. Does any of this apply to human language? As it turns out, Gold's theorem applies only if we make assumptions about the learning device that are wildly unlike any known nervous system. And that is the only formal proof around at this writing. No one has done the work to find out whether grammars of a different kind are learnable (e.g., probabilistic mappings from meanings onto sound), or whether a learning device with vastly different properties could acquire such a grammar (e.g., a multilayered neural network).

In the interim, there are now simulations of grammatical learning in neural networks that could be viewed as learnability proofs of a sort (for a review, see Elman et al., 1996). For example, Elman (1993) has shown that an artificial grammar with center embeddings and long-distance dependencies (i.e., agreement marking) can be learned by a simple recurrent network that lives in time and tries to guess what is coming next, one word at a time. The system accomplishes this with a positive database only (i.e., feedback comes in the form of guesses that are confirmed or disconfirmed when the next word comes in). It also makes errors and then recovers from those errors, in the absence of negative evidence, providing *prima facie* evidence against the generality of Gold's theorem for language learning in a different kind of system. However, it is not the case that any neural network can accomplish this task. Elman discovered that his network could only learn the grammar if it was first exposed to simple sentences, with complex sentences introduced later. But of course, this is not true for human children, who hear at least a few embedded sentences from the very beginning. Elman found that he could obtain the same result by a simple trick: Start the system out with a rapidly fading memory (instantiated in the units that copy the system's internal state on a previous trial) and gradually increase that memory (independent of learning itself) up to the adult form. As a result, the network could learn off short strings only in the early stages of learning – even though simple and complex strings are both available in the input from the very beginning. This single example illustrates our earlier point about different levels of innateness: A grammar that is unlearnable under one set of timing conditions becomes learnable in a recurrent network when the timing conditions change – all of this accomplished without building innate representations into the architecture before learning begins.

In short, we cannot conclude from the presence of eccentric structures that those structures are innate – not even if they are unique to our species, universal among all normal members of that species, localized in particular parts of the system, and learnable only under specific conditions. The same facts can be explained by replacing innate knowledge (i.e., representations) with architectural and temporal constraints that require much less genetically specified information. This kind of emergentist solution to the nature–nurture controversy has been around for many years, but it has only become a scientifically viable alternative in the last decade. As a result, the long-awaited reconciliation between Plato and Aristotle may be at hand.

References and recommended reading

Bates, E., Thal, D., Trauner, D., Aram, D., Eisele, J. and Nass, R. 1997: From first words to grammar in children with focal brain injury. In D. Thal and J. Reilly (eds), special issue on the origins of communication disorders. *Developmental Neuropsychology*, 13(3), 275–343.

Chomsky, N. 1975: *Reflections on Language*. New York: Parthenon Press.

—— 1980: On cognitive structures and their development: a reply to Piaget. In M. Piatelli-Palmarini (ed.), *Language and Learning: The Debate between Jean Piaget and Noam Chomsky*, Cambridge, Mass.: Harvard University Press, 35–54.

Churchland, P. M. 1995: *The Engine of Reason, the Seat of the Soul: A Philosophical Journey into the Brain*. Cambridge, Mass.: MIT Press.

Cohen-Tannoudji, M., Babinet, C. and Wassef, M. 1994: Early determination of a mouse somatosensory cortex marker. *Nature*, 368, 460–3.
*Deacon, T. W. 1997: *The Symbolic Species: The Coevolution of Language and the Brain.* New York: Norton.
Edelman, G. M. 1987: *Neural Darwinism: The Theory of Neuronal Group Selection.* New York: Basic Books.
Elman, J. L. 1993: Learning and development in neural networks: the importance of starting small. *Cognition*, 48, 71–99.
*Elman, J., Bates, E., Johnson, M., Karmiloff-Smith, A., Parisi, D. and Plunkett, K. 1996: *Rethinking Innateness: A Connectionist Perspective on Development*, Cambridge, Mass.: MIT Press.
Fodor, J. A. 1983: *The Modularity of Mind: An Essay on Faculty Psychology*. Cambridge, Mass.: MIT Press.
Gibson, E. 1992: On the adequacy of the Competition Model. *Language*, 68, 812–30.
Gold, E. M. 1967: Language identification in the limit. *Information and Control*, 16, 447–74.
Gopnik, M. 1990: Feature-blind grammar and dysphasia. *Nature*, 344 (6268), 715.
*Johnson, M. H. 1997: *Developmental Cognitive Neuroscience: An Introduction.* Cambridge, Mass.: Blackwell.
Kennedy, H., Dehay, C. and Horsburg, G. 1990: Striate cortex periodicity. *Nature*, 348, 494.
Leslie, A. M. 1994: Pretending and believing – issues in the theory of Tomm. *Cognition*, 50, 211–38.
Merzenich, M. M. and Jenkins, W. M. 1995: Cortical plasticity, learning, and learning dysfunction. In B. Julesz and I. Kovacs (eds), *Maturational Windows and Adult Cortical Plasticity*, Proceedings of the Santa Fe Institute Studies in the Sciences of Complexity, vol. 23. Reading, Mass.: Addison-Wesley, 247–72.
Nichelli, P., Grafman, J., Pietrini, P., Alway, D., Carton, J. C. and Miletich, R. 1994: Brain activity in chess playing. *Nature*, 369, 191.
Niederer, J., Maimon, G. and Finlay, B. 1995: Failure to reroute or compress thalamocortical projection after prenatal posterior cortex lesions. *Society for Neuroscience Abstracts*, 21.
Piatelli-Palmarini, M. 1989: Evolution, selection and cognition: from "learning" to parameter setting in biology and the study of language. *Cognition*, 31, 1–44.
Pinker, S. 1994: *The Language Instinct: How the Mind Creates Language.* New York: William Morrow.
Rakic, P. 1988: Specification of cerebral cortical areas. *Science*, 241, 170–6.
Shatz, C. 1996: The emergence of order in visual system development. *Proceedings of the National Academy of Sciences, USA*, 93, 602–8.
Spelke, E. S. 1991: Physical knowledge in infancy: reflections on Piaget's theory. In S. Carey and R. Gelman (eds), *Epigenesis of the Mind: Essays in Biology and Knowledge*, Hillsdale, NJ: Erlbaum, 133–69.
Vargha-Khadem, F., Watkins, K., Alcock, K., Fletcher, P. and Passingham, R. 1995: Praxic and nonverbal cognitive deficits in a large family with a genetically transmitted speech and language disorder. *Proceedings of the National Academy of Sciences, USA*, 92, 930–3.
Wills, C. 1991: *Exons, Introns, and Talking Genes: The Science behind the Human Genome Project.* New York: Basic Books.

47

Intentionality

GILBERT HARMAN

What is intentionality?

A proper understanding of intentionality is crucial to the study of a number of topics in cognitive science, including perception, imagery, and consciousness. The term itself, *intentionality*, can be misleading, in suggesting intentional action, doing something intentionally, with a certain aim or purpose. In cognitive science, the term is used in a different, more technical sense. Intentionality involves reference or aboutness or some similar *relation* to something having what the scholastics of the Middle Ages called *intentional inexistence* (Brentano, 1874).

When Mary thinks of George Miller as a cognitive scientist, the *intentional object* of her thought is George Miller, and the *intentional content* of her thought is that George Miller is a cognitive scientist. She has a mental representation of him as a cognitive scientist. What Mary thinks about has *intentional inexistence* in the sense that her thoughts may be wrong and she can have thoughts about things that do not even exist. She may think incorrectly that George Miller is a computer scientist or even that Santa Claus is a computer scientist.

If you treat intentionality as a *relation* to an intentional object, you must remember that it is not a real relation in the way that kissing or touching is. A real relation holds between two existing things independently of how they are conceived. When a woman kisses a man and the man she kisses is bald, the woman kisses a bald man. But Mary can think about a man who happens to be bald without thinking of him as bald; she may represent him as hairy. Similarly, Mary can think of someone who does not exist but cannot kiss or touch someone who does not exist.

Looking for something is an example of an intentional activity in this technical sense of *intentional* as well as in the more ordinary sense having to do with what you are aiming at. You sometimes look for things that turn out not to exist. Ponce de Leon searched in Florida for the fountain of youth. Alas, there was no such thing to be found.

There can be intentionality without representation. For example, needing something is an intentional phenomenon. The grass in my lawn can need water even though it is not going to get any and even if there is no water to give it. But the grass does not represent the water it needs.

Other examples of intentional phenomena include spoken and written language, gestures, representational paintings, photographs, films, road maps, and traffic lights. It is controversial how these last instances of intentionality are related to the intentionality of thoughts and other cognitive states.

Logical worries about intentional objects

Nonexistent intentional objects like Santa Claus and the fountain of youth raise difficult logical puzzles if taken seriously as objects. What properties do they have? What sorts of properties does Santa Claus have, as he is conceived by a certain child? Perhaps he is fat, lives at the North Pole, dresses in red, drives a sleigh, brings presents to children at Christmas time, and has at least eight reindeer. But intentional objects cannot always have all the properties which they are envisioned as having, because, as in the case of the child's conception of Santa Claus, a nonexistent intentional object may be envisioned as existent, and it is inconsistent to suppose that something could be both existent and nonexistent (Parsons, 1980).

You must resist the temptation to try to resolve such problems by identifying intentional objects with mental objects such as ideas or mental representations. That identification does not work. The child does indeed have an idea of Santa Claus, and Ponce de Leon had an idea of the fountain of youth. But the child does not believe that his idea of Santa Claus lives at the North Pole. Nor was Ponce de Leon looking for a mental representation of the fountain of youth. He already had a mental representation; he was looking for the (intentional) object of that representation.

Is it enough to say that a nonexistent intentional object is a merely possible object – an object that exists in some possible world or other, but not in the actual world? That is not a completely general account, because some intentional objects are not even possible. Someone may try to find the greatest prime number without realizing that there is no such thing. The intentional object of the attempt – the greatest prime number – is not a possible object. There is no possible world in which it exists.

One controversy concerning intentionality concerns how to provide a logically adequate account of talk of intentional objects. That is a controversy in philosophical logic (Parsons, 1980) and may not be especially important to the rest of cognitive science.

The moral is that, on the one hand, you have to take talk of nonexistent intentional objects with a grain of salt, without being too serious about the notion that there really are such things. On the other hand, you have to be careful not to conclude that the child pondering Santa Claus isn't really thinking about anything or that Ponce de Leon wasn't really looking for anything as he wandered through Florida.

The intentionality of feelings

To what extent does cognition involve intentionality? In one view, everything cognitive is intentional: intentional inexistence is the mark of the mental, according to Franz Brentano. Another view allows for nonintentional aspects of cognitive states, *raw feels*.

Clearly, many feelings recognized in *folk psychology* have intentionality and are not simply *raw feels*. A child hopes that Santa Claus will bring a big red fire truck and fears that Santa Claus will bring a lump of coal instead. The child is happy that Christmas is tomorrow and unhappy that he hasn't been a good little boy for the last few weeks. A child's hopes, fears, happiness, and unhappiness have intentional objects and intentional content.

It is unclear whether all feelings or emotions have intentional content in this way. Do feelings of *free-floating* anxiety and depression have no intentional content, so that you are not anxious about anything or depressed about anything, but just depressed? Or do such states have a very general, nonspecific content, so that you are anxious about things in general or depressed *about things in general*, just not anxious or depressed about something specific? It is hard to say what turns on the answer to this question.

Perception

Perceptual experience has intentionality inasmuch as it presents or represents a certain environment. How perceptual experience presents or represents things may be accurate or inaccurate. Things may or may not be as they seem to be. Sometimes what you see or seem to see doesn't really exist, as when Macbeth hallucinated a bloody dagger.

The intentional content of perceptual experience is perspectival, representing how things are from here, or even representing how things are as perceived from here. The content of the experience may even be in part about the experience itself; what is perceived is perhaps seen as causing that very experience.

The dagger is an intentional object of Macbeth's perceptual experience. That's what he is or seems to be aware of. You may be tempted to think that Macbeth must be aware of a mental image of a dagger; but that is like thinking that Ponce de Leon must have been trying to find an idea of the fountain of youth.

Sense data

The controversial *sense datum theory* holds that perceptual experience is most directly the experience of mental images or representations, and only indirectly the experience of things represented by those mental images. Notice that the sense datum theory goes beyond the hypothesis that perception involves the construction of various inner representations of the environment; the sense datum theory holds in addition that the perceiver is in the first instance aware of such inner mental representations.

The sense datum theory is sometimes defended by appeal to the following, fallacious *sense datum argument*:

1. Suppose Fred hallucinates a pink elephant.
2. Suppose also that nothing pink is in Fred's environment.
3. Nevertheless, Fred is certainly aware of something pink.
4. What Fred is aware of is not in his environment (from 2).
5. If Fred is aware of something that is not in his environment, it must be something mental.
6. So, Fred must be aware of something mental that is pink, a pink mental representation (from 3, 4, and 5).
7. On another occasion, Fred might have exactly the same visual experience when he was really seeing a pink elephant and not hallucinating.
8. So, on that occasion too, Fred must be aware of a pink mental representation (from 6 and 7).
9. Whenever Fred perceives something, he could conceivably have had exactly the same experience as a result of hallucination.
10. So, whenever Fred perceives something, in the first instance he must be aware of an inner mental representation (from 8 and 9).

This surprisingly tempting argument is fallacious in steps 4–6, through ignoring the intentionality of perception. It is like arguing that Ponce de Leon was looking for the idea of a fountain of youth rather than a real fountain whose waters really keep you young.

Mental imagery

Mental imagery has intentionality. What you image or imagine is the intentional object of your imagining or imaging. When you picture Lucy's smile, her smile is what you imagine. Theories of imagery offer accounts of the structure of the inner representation involved in one's imagery and the processes that operate on that structure. But what you imagine is not that inner mental representation; you imagine Lucy's smile.

The term *mental image* is ambiguous. Sometimes it refers to the thing imagined, the intentional object, Lucy smiling. Sometimes it refers to the imagining of that thing, picturing Lucy smiling. Sometimes it refers to the hypothetical inner representation formed when something is imagined, an inner mental picture or description of Lucy smiling. It is important not to confuse these things. Otherwise, the substantial claim that imagination involves the construction of inner pictures or other sorts of mental representations with specified structures will be conflated with the obvious fact that you are capable of imagining various things.

Similarly, it is important to distinguish imagining something revolving from actually revolving a mental representation in your mind or head; and it is important to distinguish imagining scanning a scene from scanning an inner mental representation (Block, 1981).

It is controversial what sort of *introspective* awareness you have of your inner mental representations. Matters are only confused through failure to distinguish the various senses of *mental image*. You have something that might be called *introspective* awareness of mental images in the first sense: namely, the intentional objects of your thoughts. You often know what you are thinking about, imagining, perceiving, and so forth. It is unclear whether you have any corresponding access to the mental representations, if any, underlying your thinking, imagining, perceiving, and so forth (see Article 12, IMAGERY AND SPATIAL REPRESENTATION).

The intentionality of bodily sensations

Some feelings – namely, emotions – have intentionality. What about the simplest kinds of feelings: bodily sensations like pains? Do these involve mental representation? Do they have intentionality, or are they mere raw feels?

Many pains have the sort of signaling function that implies intentionality, indicating distress at a certain place in the body. The content of this signal may be incorrect, as in the case of referred pain, where an irritation in your back can cause a pain in your foot. The part of your body in which a pain appears to be located is the relevant intentional object. That bodily part is presented as distressed. Sometimes that relevant part of the body doesn't even exist, as in the phantom limb phenomenon. Feeling pain in a phantom limb is like thinking of Santa Claus. In either case, the intentional object of the pain does not exist.

As in other cases, so too in the case of pain, it is important to distinguish the intentional object from the experience of that intentional object. Ordinary use of the term *pain* seems to mix these up. You count pain as an experience. But you locate it at its intentional object. The pain, you say, is located in your foot.

Do you have introspective access to anything beyond the intentional content of the experience of pain? It is very hard to say. When you attend to my pain, you attend to features of its location: that is, to the location of its intentional object. When you attend to a pain in your foot, you seem to attend to apparent events in your foot, not to anything in your head (which is presumably where the event of pain is located).

The intentionality of language, pictures, and symbols

Mental states and events are not the only things with intentionality. There is the intentionality of grass's need for water. There are also stop signs, electrical diagrams, portrait paintings, novels, and computer programs, all with some sort of intentional content. Language can be used to describe something correctly or incorrectly. What is described may or may not exist. The thing described is the intentional object of the description. What the description says is its intentional content. A painting of Lucy represents her as smiling. She is the intentional object, and the way she is represented is the intentional content of the painting. A painting of a unicorn is a representation of something that does not really exist. The *relation* between the painting and what it is a painting of is an intentional relation.

Many theorists follow Locke (1690) in supposing that the intentionality of many of these external representations – language, pictures, and diagrams – is derivative from the intentionality of mental states and events. In this view, language is used to express thoughts, and the content of linguistic expression derives from the intentional content of the corresponding thoughts. Similarly, perhaps, in this view, what pictures represent is determined by the content of perceptions which people have on viewing the pictures.

Stalnaker (1984) and other theorists think that the explanation goes in the other direction. In their view, people are capable of having thoughts with any sort of fine-grained intentional content only because they make use of a language which allows subtle differences in meaning; the intentional content of thoughts, in this view, derives from the linguistic content of the words or other representations that would be used to express the thoughts. Creatures without language have states with the sort of intentional content that needs have. Needs are insensitive to the difference between necessarily equivalent intentional contents. Water is H_2O, so whatever needs water, needs H_2O. Creatures with language can have states with more discriminating intentional content. So, a person who does not realize that water is H_2O may want water without wanting H_2O, or vice versa. In this view, such discriminative wanting is possible only because the person's desire can be in part a desire about language. One intentional object of the wanting is a linguistic expression, the word *water* or the expression *H_2O*.

Other theories of mental representation suppose that thought has intentionality because it involves mental representation with intentionality, whereas mental representations may be, but need not be, expressions in a language like English. Where in such views does the intentionality of the mental representations come from? If mental representations have intentional content in the way that outer representations like sentences and paintings have intentional content, and these outer representations have content because of the way they are related to intentional mental states, the account leads to a circle or an infinite regress: a mental state has intentional content because it involves a mental representation with that content, whereas the content of the mental representation derives from the content of the mental states that it is used to express!

Communication and calculation

One way to avoid such circularity is to distinguish the use of symbols in communication from their use in calculation or thought. Normally, when you use symbols to calculate something – for example, when you use numerical symbols to calculate the balance in your checkbook – you are not using those symbols to communicate with anyone, even yourself. Similarly, you may use a map not to show someone something, but to figure out how to get from where you are to where you want to go. You reason something out in language, in order to see what follows from certain assumptions, not in order to tell someone something you already know. In another case, you use a diagram to figure something out. You use models, including mental models, in order to solve problems.

In one view (Sellars, 1963), the intentionality of symbols used to think with derives from the way these symbols are used. Numerals acquire their meaning from their use in counting. The symbol for plus (+) has the meaning it has because of the way it is used in addition. Similarly, for other mathematical symbols. These are implicitly defined by their use in calculation. Similarly for logical concepts like *if, and,* and *not.* The content which these symbols have arises from the way they are used to construct complex representations and the implications recognized among representations so constructed. Other concepts have content because of their role in perception. The normal concept of *pink* has content because it plays a role in the visual experience of pink things.

Sellars's theory is a *use* theory of mental content, also called a *conceptual role* theory (Harman, 1987). The content of a symbol used in thought derives from the way that symbol is used in thought in relation to other symbols and in relation to things in the perceived environment.

Words and other symbols are also used in communication. In this view, the content of such communications derives from the thoughts expressed, thoughts that may well be thought *in* the very language used to communicate them. The content of the language used in thought, however, derives in this view from its calculative use, its use in thought, rather than its use to communicate thoughts.

First- versus third-person accounts of intentionality

Any attempt to explain intentional content in terms of use or conceptual role faces the following difficulty. Understanding the intentional content of a concept (i.e., understanding the concept) and understanding the conceptual role of the concept (i.e., understanding what the conceptual role of the concept is) are very different things. You can have a detailed understanding of the conceptual role or use of a concept without understanding the concept, and you can understand a concept perfectly without being able to specify exactly how the concept is used. For example, you might know exactly how a particular symbol is used in relation to other symbols and the environment without realizing that the symbol means *plus*. Similarly, you can fully understand addition and the concept of *plus* without being able to describe exactly how that concept is used in relation to other concepts and the environment.

To have a concept is automatically to understand the concept, whether or not you know how the concept is used. Furthermore, to understand another person's thoughts, it is not enough (and not required) that you understand how the concepts involved in those thoughts function. You need an understanding of the other person's thoughts *from the inside*. You need to *know what it is like* to have such thoughts. You need to relate the other person's thoughts to equivalent thoughts of your own that you understand.

Some theorists put the point like this: you need a *first-person* understanding of intentionality, an understanding from the point of view of the thinker. It is not enough to have a *third-person* understanding from the point of view of an observer of the thinker (Nagel, 1974).

This need not mean that a conceptual role or use theory is incorrect. Perhaps intentionality is a matter of use or conceptual role. But you have to distinguish two sorts of understanding of intentionality: the internal first-person understanding you have by virtue of being the person who uses representations in a certain way and the third-person understanding you have in terms of rules of use. Compare two different ways in which a person can know how to swim: being able to swim and being able to describe what is done when someone swims.

The Chinese room argument

Searle (1980) does not agree that the use or conceptual rule account can be correct. He argues that in giving an account of intentionality, you must begin from the first-person case, starting with your own thoughts. These thoughts have their intentional content essentially and intrinsically, by contrast with the intentionality of sentences, pictures, and diagrams, which lack intrinsic intentionality and possess intentionality only because of their relations to mental states with intrinsic intentionality. Intentionality originates in thinking; thoughts have *original intentionality*. Other things have only derivative intentionality.

Searle's *Chinese room argument* claims to show that a purely use or conceptual role theory of original intentionality cannot be correct. Suppose you had a specification of the roles of concepts used by a native speaker of some dialect of Chinese. Such a specification could be converted into rules for using the concepts, rules that could be followed by an appropriately programmed computer, or, for that matter, by a native speaker of English who does not know Chinese. Imagine such an English speaker with a list of the relevant rules as a subject of an experiment. The subject is placed alone in a room. Chinese sentences are slipped under the door, where they are received by the subject as marks on paper. The subject looks up and follows various rules and eventually puts marks on another piece of paper, which he slips back under the door. This process continues as long as is desired. To a Chinese-speaking observer, the pieces of paper passed back and forth seem to represent an intelligent conversation in Chinese. Such an observer might be convinced that the person in the room knows Chinese. But in fact that conclusion is wrong. Despite following relevant rules, the person in the room has no idea what the *conversation* is about and knows nothing of Chinese. Therefore, according to Searle, the sort of intentional understanding involved in knowing Chinese is not just a matter of use and conceptual role.

Suppose the rules in question are used to build a robot Chinese speaker. The robot might be constructed in such a way as to appear to be a person. Given the right rules, the robot would interact with ordinary Chinese speakers in ways that were indistinguishable from the ways in which a real person might interact. Searle allows that you could attribute intentionality to the inner states and processes of this robot. However, he claims that the intentionality so attributed would not be intrinsic to the states of the robot; it would not be the sort of original intentionality that your own mental states and processes have. You would be attributing derivative *as if* intentionality to the robot, just as you now attribute derivative intentionality to a computer program that keeps track of your checking account balances. Your computer doesn't really have any views about your checking account balance, and the robot doesn't really have any views about the price of eggs, even though it may make sense to talk as if they had such views.

Others disagree, arguing, for example, that you would have the same reason to attribute original intentionality to the robot in question as you have to attribute original intentionality to other people. Searle claims that you do not have the same reason. In the case of other people, you can appeal to an argument by analogy to your own case: you know directly about the intentionality of your own mental states; you have reason to think that other people are relevantly similar to you; so you have reason to think that they have states with intrinsic intentionality. The robot is constructed differently from the way in which you are constructed, so you do not have the same reason to attribute intrinsic intentionality to its states, according to Searle.

Conclusion

Intentionality is an important characteristic of cognition. It is useful to think of cognitive states as involving *relations* to *intentional objects*, even though the notion of an intentional object raises deep questions in philosophical logic. It is unclear whether all mental life involves intentionality, whether there are raw feels. Certainly, many kinds of feelings involve intentionality: emotions, for example, and bodily feelings. Knowledge and perception have intentional content; appreciation of this fact undermines the standard sense datum argument and helps to avoid mistakes in studying imagery. Understanding the intentionality of language, pictures, and other symbols and representations requires a distinction between using symbols to communicate ideas and using symbols to calculate or think with. The intentionality of symbols used in communication may be derivative of the original intentionality of symbols used in thought and calculation. However, it is controversial whether the mere use of symbols in the right way is enough to give them original intentionality.

References and recommended reading

Brentano, F. 1874: *Psychologie vom empirischen Standpunkt.* Leipzig. Ed. and trans. as *Psychology from an Empirical Standpoint.* London: Routledge and Kegan Paul, 1973.

Block, N. (ed.) 1981: *Imagery*. Cambridge, Mass.: MIT Press.

*Dennett, D. 1971: Intentional systems. *Journal of Philosophy*, 68, 87–106. Repr. in *Brainstorms*, Cambridge, Mass.: MIT Press, 1981, 3–22.
Gauker, C. 1994: *Thinking Out Loud: An Essay on the Relation between Thought and Language*. Princeton: Princeton University Press.
Harman, G. 1987: (Nonsolipsistic) conceptual role semantics. In E. LePore (ed.), *New Directions in Semantics*, London: Academic Press, 55–81.
Locke, J. 1690: *An Essay concerning Human Understanding*. London.
Nagel, T. 1974: What is it like to be a bat? *Philosophical Review*, 83, 435–50.
*Parsons, T. 1980: *Nonexistent Objects*. New Haven: Yale University Press.
Searle, J. 1980: Minds, brains, and programs. *Behavioral and Brain Sciences*, 3, 417–57.
—— 1983: *Intentionality*. Cambridge: Cambridge University Press.
Sellars, W. 1963: Some reflections on language games. In *Science, Perception, and Reality*, New York: Humanities Press, 321–58.
Stalnaker, R. 1984: *Inquiry*. Cambridge, Mass.: MIT Press.

48

Levels of explanation and cognitive architectures

ROBERT N. MCCAULEY

Introduction

Some controversies in cognitive science, such as arguments about whether classical or distributed connectionist architectures best model the human cognitive system, reenact long-standing debates in the philosophy of science. For millennia, philosophers have pondered whether mentality can submit to scientific explanation generally and to physical explanation particularly. Recently, positive answers have gained popularity. The question remains, though, as to the analytical level at which mentality is best explained. Is there a level of analysis that is peculiarly appropriate to the explanation of either consciousness or mental contents? Are human consciousness, cognition, and conduct best understood in terms of talk about neurons and networks or schemas and scripts or intentions and inferences? If our best accounts make no appeal to our hopes or beliefs or desires, how do we square *those* views with our conception of ourselves as rational beings? Moreover, can models of *physical* processes explain our *mental* lives? Does mentality require a special level of rational or cognitive explanation, or is it best understood in terms of overall brain functioning or neuronal or molecular or even quantum activities – or any of a dozen levels of physical explanation in between? Also, regardless of how they compare with explanations cast at physical levels, what is the status of psychological explanations that appeal fundamentally to mental contents?

As a means for beginning to address such questions, proposals about cognitive architecture concern which kind of explanation best characterizes primitive psychological activities. Although, technically, approaches to modeling those activities are unlimited, two strategies have enjoyed most of the attention. The prominence of the classical account and the distributed connectionist (or parallel distributed processing (PDP)) account, notwithstanding, nothing bars the development of other proposals.

Classicism employs rules that apply to symbolic representations to explain cognitive processing. PDP systems propagate activation through networks of processing units from an input layer to an output layer, without appealing to either symbols or their (rule-governed) manipulation. Proponents of these views debate whether a PDP account of cognition characterizes the human cognitive architecture or merely supplies details concerning the implementation of a classical architecture. Their answers depend upon how they regard the notion of cognitive architecture, how they assess the adequacy and centrality of classical accounts, and how they interpret PDP models. Their answers also depend upon what they assume about the relationships between scientific inquiries aimed at explaining the same phenomena but proceeding at different explanatory levels.

The second section discusses analytical levels in science and surveys philosophical accounts of reductionism. The third section considers the question of our cognitive architecture, outlines the classical account and the challenges it poses to connectionism, and surveys various connectionist responses. The final section describes recent integrative models of cross-scientific relations and their implications for these discussions.

Levels of explanation and inter-theoretic reduction

Scientists' facility with the concept *explanatory level* notwithstanding, clear, unambiguous criteria exist neither for specifying the notion of an explanatory (or analytical) level nor, often, even for distinguishing particular levels. *Within the cognitive sciences* computer scientists use these terms to describe hierarchies of compiled programming languages. Philosophers of science, by contrast, use them to talk *about the cognitive sciences'* relations to one another. This second use (which, arguably, encompasses the first) is especially helpful when considering the bearing of levels of explanation on hypotheses about cognitive architecture arising from *multidisciplinary* enterprises in cognitive science.

Many criteria for locating levels of explanation *among the sciences* roughly converge – at least with respect to theorizing about the structural relations of systems. For example, analytical levels partially depend upon viewing nature as organized into *parts and wholes* and largely mimic *levels of aggregation* (as opposed to simple considerations of scale). If one entity contains others as its parts, and its description requires further organizing principles beyond those concerned with those parts, then it occurs at a higher level of aggregation. The *range* of the entities that constitute any science's primary objects of study and that science's principal units of analysis also track this arrangement of analytical levels. The lower a science's analytical level, the more ubiquitous the entities it studies. For example, subatomic particles, discussed in physics, are the building blocks of all other physical systems (molecules, biological systems, galaxies, social groups, and more). Although complexity has no simple or single measure, the relative complexity of the (aggregated) systems generates a similar picture. Sciences at lower analytical levels study (relatively) simpler systems, at least to the extent that increasingly higher-level sciences deal with increasingly restricted ranges of events concerning increasingly organized systems whose study requires additional explanatory principles. The order of analytical levels also corresponds to the chronological *order in natural history* of the evolution of systems. The lower a science's level, the longer the systems it specializes on have been around.

Presumably, our most successful theories provide important clues about the furniture of the universe. This suggests that levels of analysis in science correspond to levels of organization in nature. Typically, what counts as an entity depends on both the redundancy of spatially coincident boundaries for assorted properties and the common fate (under some *causal* description) of the phenomena within those boundaries. So, for example, both their input and output connections and their various susceptibilities to stains aid in identifying cortical layers in the brain (see Article 4, BRAIN MAPPING). Emphasizing causal relations insures that the sciences dominate such deliberations. The greater the number of theoretical quarters from which these ontological commitments receive empirical support, the less troublesome is the circularity underlying an appeal to *levels of organization* as criteria for their corresponding levels of analysis.

Methodological considerations also segregate analytical levels, but less systematically. Sciences at different analytical levels ask different questions, promote different theories, and employ different tools, methods, and standards. Theories at alternative explanatory levels embody disparate idealizations that highlight diverse features of the phenomena. Such criteria can serve to arrange the major scientific families into levels (physical, biological, psychological, and sociocultural sciences), but each of these families includes separate sciences that, in turn, contain sublevels. We can identify seven sublevels within neuroscience alone (molecules, synapses, neurons, networks, maps, subsystems, and the central nervous system overall).

When analyzing inter-level relations in science, philosophers have cut through these vaguenesses surrounding the identification and differentiation of explanatory levels. They have concentrated on only one relation (reduction) between only one component of levels (theories). Traditional reductionism conceives of all inter-theoretic relations as explanatory and of all explanations as deductive inferences in which at least one of a theory's laws serves as a premise. In the case of reductive explanation, the immediate goal is to show that, with the aid of bridge laws, the explanatory principles of successful upper-level theories follow as deductive consequences from the laws of lower-level theories. The bridge laws establish connections between the two theories' predicates, providing grounds for the explanation of the upper-level theory and for the revelation that its entities are *nothing but* combinations of lower-level entities. Thus, in principle at least, the lower-level theory can, allegedly, replace the upper-level theory without explanatory or ontological loss.

Arguments persist about every feature of this proposal, but those about the bridge laws' status matter most. Ambitious reductionists, who aspire to both explanatory consolidation and ontological economies, argue that only comprehensive inter-theoretic *identities* of the two theories' predicates will insure the desired results. Type-identity theorists find ambitious reductionism particularly congenial, since that view claims that a successful reduction of psychology to neuroscience will certify the identity of mind and brain.

Securing a reductive explanation on the traditional view, however, involves no more than the bridge laws specifying lower-level conditions *sufficient* for upper-level patterns. Under those circumstances, reductions prove domain-specific, limited in scope, and less sweeping ontologically, since any systematic bridge laws will probably apply only in circumscribed settings. For example, even the reductive explanation of the Boyle–Charles law by statistical mechanics reduces not the notion of temperature but only *temperature of a gas*. For a variety of reasons, most philosophers are not optimistic about the possibility of obtaining comprehensive identities between psychological and neuroscientific predicates. This less ambitious view of reductive explanation recommends detailed analyses of scientific research on the relevant systems. Inevitably, the complexities which such analyses reveal do not readily lend themselves to either easy or comprehensive ontological pronouncements.

Two other reactions to the projected failures of systematic inter-theoretic mappings between psychology and neuroscience have gained attention. The first is *eliminativism*, which seeks the same economies as do the ambitious reductionists but does so by exploiting another dimension of traditional reductionism. Defenders of the traditional account regarded it not merely as a model of inter-theoretic relations between different analytical levels, but also as an account of theoretical *progress* within a particular level. They especially emphasize the corrections which a successor theory offers its predecessor. Noting that in some cases of intra-level theoretical progress the requisite bridge principles – let alone inter-theoretic identities – were not even remotely plausible, critics of traditional reductionism spotlighted episodes in which victorious theories simply eliminated their predecessors. The oxygen theory of combustion did not reduce the phlogiston theory; it eradicated it. Such episodes illustrate the most extreme form of inter-theoretic correction. Eliminativists in the philosophy of psychology apply these lessons to the inter-level case of neuroscience and psychology, holding that if psychological theories do not map reasonably well onto neuroscientific theories, then they will undergo elimination, in light of the neurosciences' superior merits and promise (Churchland, 1989).

Traditional reductionists and eliminativists conflate disparate forms of inter-theoretic relations when applying the same model of reduction both to (1) relations between theories at different explanatory levels and (2) theoretical progress within a single explanatory level (McCauley, 1986). Inter-theoretic corrections can occur in both sorts of cases; they must in the second. Elimination often occurs in the second too. But at least in the science of the past century elimination is virtually nonexistent in the first – certainly *when both* the upper-level science (experimental psychology in this case) is institutionally well established and the elimination is alleged to span the divisions between the major families of levels listed on page 612 (as the elimination of psychology by neuroscience is). (This contrasts with merely *consolidating* a theoretical account of what had previously been regarded as diverse phenomena at various sublevels *within a single level* – in the way, for example, that Maxwell's theory of electromagnetism did.)

The other prominent response to reductionism also questions the availability of adequate inter-theoretic connections. Jerry Fodor defends the *irreducibility of psychology* by insisting on the letter of ambitious reductionism. If bridge laws must provide type identities between psychological and neural predicates, then, Fodor (1975) argues, successful reduction will prove virtually impossible. Fodor does not deny that psychological states are brain states. He just denies the availability of *systematic* connections capable of linking theoretical predicates. He does not repudiate the identity of psychological and neural *tokens*; he just rejects the identity of psychological and neural *types*. Each token of some psychological type is a token of some physical type; however, every token of that psychological type is not a token of *one* particular physical type.

Two general considerations encourage Fodor's skepticism. The first concerns the disparate explanatory tactics pursued at different analytical levels. Psychology and neuroscience manifest all the methodological dissimilarities between analytical levels outlined on page 612. They often study human cognition and behavior in radically different ways. To the extent that psychological – but *not* neuroscientific – investigations presuppose conceptions of rationality, these two disciplines address distinct concerns and utilize idealizations that diverge drastically sometimes. Consequently, they often spawn explanatory principles and predicates that interpret closely related phenomena in substantially different ways. Thus, Fodor argues that psychology and the other *special sciences* formulate generalizations concerning types whose tokens' physical descriptions frequently have little or nothing in common. Fodor notes, for example, the diversity of physical things that serve as money (let alone those that might instantiate some belief). Such diversity blocks inter-level identities, because types of states and processes construed psychologically correspond only to large disjunctions of types construed physically. On this view, psychological states are so multiply realizable in our neural substrate that any possible connecting principles would prove neither theoretically interesting nor heuristically useful, since they would contain unmanageably large disjunctions of predicates.

The second ground for skepticism about securing adequate bridge laws suggests an even broader sort of multiple realizability and directly links questions about how analytical levels relate to questions concerning our cognitive architecture. Workable bridge laws are particularly unlikely when, as in psychology (of both the commonsense and the scientific varieties), the complex systems under study demand theories that often characterize states and processes *functionally*. Psychological theorizing has bred far clearer accounts of what such things as desires, parsers, mental images, and episodic memories *do* than of what they *are*.

Our inclination to assign such capacities and states not only to other animals, but especially to computers, implies a recognition that systems quite unlike us superficially might, nevertheless, process information similarly. Hence, many psychological generalizations apply readily enough to them. This fact has two important consequences. First, if parts of our psychological theories usefully describe and explain the behavior and *cognitive lives* of other animals and computers, then the bridge principles necessary to connect some psychological predicates to physical predicates will prove both intractably complex and metaphysically diverse. They must encompass not just disjunctions of a particular human brain's states on different occasions, or even disjunctions of various human brains' and animal brains' states on various occasions. They must also encompass disjunctions of an indefinitely large assortment of machine states (that constitute such cognitive accomplishments as, for example, adding two plus two).

These considerations – conjoined with those arising from the cross-classification of psychological and neural types that results from the diverging agendas of those two sciences – suffice, according to Fodor, to rule out ambitious reductionists' attempts to reduce or dispense with either our psychological theories, the explanations they inform, or the predicates they employ. On this view, psychology is autonomous and unconstrained by neuroscience.

The fit between many of our psychological generalizations and the behavior of computers also has a convenient *strategic* consequence for psychological theorizing. Since, like us, computers can carry out all sorts of cognitive tasks, and since we know just about all there is to know about how computers work, a natural strategy (indeed, some have argued that for computationalists the only plausible strategy) for theorizing about how *we* work is to assume that we work like computers. This assumption introduces the issue of cognitive architecture.

Cognitive architecture: classicism and beyond

This analogy between humans and computers is cognitive science's preeminent source of theoretical inspiration. Indeed, nearly all theorists hold that at *some* level of abstraction the relation is not mere analogy but identity. This view permits cognitive scientists to explain human cognition by appealing to the concepts and principles of machine computation. Still, beyond a commitment to the notion that cognition involves computations over representations, the precise directions in which this relation should lead us remain controversial. The emergence of distributed connectionist models over the past decade or so has stimulated debates about the character of both the representations and the computations involved in cognitive processing.

All parties to these debates agree that the nature of the underlying mechanisms restrict the character of the representations and computations in any computational system (though not all agree that the system is computational in the first place). Those mechanisms comprise the *structural* constraints on a (programmable) system's cognitive processing (in contrast to programs' various orchestrations of cognitive functioning). In a computer, these basic mechanisms determine its *functional architecture.*

The concept of cognitive architecture results from applying this notion of the functional architecture of a computer to the human cognitive system. Any computational model of human cognition that aspires to exceed mere input–output equivalence inevitably embodies assumptions about cognitive architecture. The ultimate aim is to specify the constraints which brain mechanisms impose on human cognition. The goal is to provide at least a functional characterization of the basic principles and relations that shape how that neural hardware operates cognitively.

From a computational standpoint, a model of our cognitive architecture should prove *strongly equivalent* with this neural system. The model should not only be input–output equivalent, it should capture the system's primitive representational states *as both primitive and representational,* and it should portray our cognitive processing as transitions between such states (without appealing to other representational states). Although classical and connectionist proposals currently dominate discussions, the space of possible architectures is enormous, leaving room for plenty of new proposals in the future.

Demonstrating a model's empirical accountability requires specifying ways in which human performance can bear on its assessment. For example, architectural features should be cognitively impenetrable. Putative architectural constraints should remain impervious to changes in a person's beliefs; thus, nothing learned should alter architectural constraints. Cognitive scientists have suggested that other types of evidence are relevant as well, including relative sensitivity to damage and chronometric measures of performance.

Unfortunately, additional considerations complicate the evaluation of such evidence. The behavior of a computational system is not just a function of architectural constraints. Programs also play a decisive role. Without extensive knowledge of design, distinguishing those aspects of behavior that arise as a result of the architecture from those that arise as a result of the programs it supports is rarely an easy task – let alone when the systems in question are organic, and the designer is natural selection. When cognitive systems consist of neurons, rather than computer chips, and the designer is evolution, instead of engineers, it is a fairly safe bet that at least sometimes the architecture realizes cognitive functions differently from how digital computers do.

This does not, however, automatically favor distributed connectionist models over classical models of cognitive architecture – certainly not until we know more about the principles guiding neural functioning. Classicism holds that a model of our cognitive architecture provides only a *functional* characterization of the underlying mechanisms. A vast array of physical arrangements can implement the configuration of functional relations which these abstract models describe. On *any* computational view, distinguishing a *cognitive* level from the neuroscientific level of explanation depends precisely on the fact that models of cognitive architecture involve abstractions away from many of the brain's physical details. Computationalists of both the classical and the connectionist varieties assume that the neural level will *not* prove the best level for characterizing the cognitive architecture. On the classical account, the cognitive level, at which models of cognitive architecture are fashioned, is the lowest analytical level at which states of the system *represent* features of the world. Many connectionists (e.g., Smolensky (1988)) demur – arguably providing more fine-grained analyses of these issues in the process. Brief summaries of classicism and the connectionist alternatives follow.

For the purposes of *theorizing*, proponents of classical models insist on a *principled* subdivision of the cognitive level into a semantic (or knowledge) level and a symbol (or syntactic) level. Theoretical assertions at the semantic level describe human thought in terms of goals and knowledge. As with commonsense psychology, considerations of meaning and rationality order semantic materials. The pivotal assumptions in classical proposals, however, concern the symbol level:

1 Mental symbols are context-independent representational primitives that possess their representational contents by virtue of their *forms*.
2 A finite set of such symbols can represent distinct semantic contents uniquely, because these symbols are the fundamental constituents of a quasi-linguistic system that possesses a concatenative syntax and semantics (that comprehensively parallel one another).
3 The formal, syntactic features of these symbols correspond precisely to neural properties that are pivotal in the etiology of behavior.

The language of thought (LOT) hypothesis (Fodor, 1975) sketches *how* the forms of complex mental representations can coincide point by point with the contents they represent, insuring that no change in content occurs without some change in form. The hypothesis is that they do so roughly as sentences in a language seem to. The forms and the corresponding contents of complex symbolic structures are distinctive combinations of the forms and the corresponding contents of the primitive symbols that are their constituents. The syntactic principles of the brain's computational language are recursive. Because the forms of symbolic expressions uniquely code their representational contents, principles describing the transitions between mental states can be cast syntactically. This is, in effect, to appropriate proof theory from logic to model cognitive processing. Proof theory utilizes a system of syntactic rules for deriving sentences, without appealing to semantics.

Because our mental representations have internal structures, and because principled combinations of primitive mental symbols account for those structures, Fodor and Pylyshyn (1988) insist that thought is:

1 *compositional* – primitive mental symbols are the representational elements from which complex representations are composed;
2 *productive* – although finite in number, they can produce an infinite number of complex mental representations by recursive means; and
3 *systematic* – since, *ex hypothesi*, the forms of the cognitive system's primitive symbols singularly represent their contents, and since the roles they play in the constituent

> structures of complex representations turn completely on those forms, thought is systematic: that is, the ability to entertain some thoughts is intrinsically connected with the ability to entertain others involving the same representational contents.

Finally, classicism holds that the brain states that instantiate the primitive symbols play an essential role in causing our behavior. In addition to the syntactic principles that order them as (representational) primitives, these symbols also submit to neural descriptions that conform to the demands of some eminent, but yet to be imagined, theories in neuroscience. Those theories will identify particular brain states that both instantiate these symbols and exhibit causal relations that match this symbol system's concatenative character. Syntactic principles mediate between our mental states' representational contents and their causal roles, reassuring us that mental representations play just the parts they ought to causally. Thus, the classical account of our cognitive architecture not only provides a framework for preserving our commonsense psychology's explanatory powers; it also envisions a scientific psychology that relies fundamentally on the conceptual framework of that commonsense view.

Connectionist architectures seem to diverge from classical models on nearly every front. They typically consist of a network of simple units in which activation is propagated along connections from input units to one or more layers of hidden units to a set of output units. They do so with neither a program nor a central processor controlling their performance. Frequently, numerous excitatory and (sometimes) inhibitory connections link the units. The links, which are typically feed-forward but can also be feedback (or *recurrent*), have adjustable connection strengths (or *weights*) that influence the amount of excitation or inhibition transferred from one unit to another.

On the basis of excitatory stimuli impinging at the input layer, the units' current levels of activation, and the configuration of all the connections and their weights, connectionist networks produce a pattern of activation (or *activation vector*) at the output layer. Adjusting the connection strengths via feedback learning rules, on the basis of the output vector's divergence from some goal, gradually trains an adequately configured network to respond more appropriately, not only to familiar materials but also to novel materials that manifest similar patterns, thus exhibiting how the system's knowledge resides in its weights. A PDP network's representational capacities, as indicated by its ability to generate appropriate output vectors, regularly involve activity throughout the entire network. Thus, representations are *distributed*. (This is in contrast to *localist* versions of connectionism, which assign semantic contents to specific units.)

These models are frequently introduced by noting their apparent affinities with brain structures, on the assumption that on these fronts, at least, they have an automatic advantage over classical notions of cognitive architecture. Points of similarity between PDP networks and the brain include their parallel processing and distributed representations, as well as their analog capabilities, fault tolerance, and dynamic states. Although no principled barrier precludes either distributed representations or parallel processing in classical architectures, with the exception of the parallel processing in production systems, both are infrequent in classical models.

Perhaps most importantly, connectionist models, unlike classical ones, do not involve the manipulation of symbols according to stored rules. Whatever the preferred interpretations accorded the inputs and outputs of a PDP network, the fundamental principles that characterize the alterations in individual units' activation levels (which collectively determine networks' trajectories through their state spaces) are mathematical equations that make no appeal to quasi-linguistic forms or semantic contents.

The critical question concerns the relationship between accounts of PDP networks and classical accounts of cognitive processing. Virtually all commentators see important *discontinuities* between the principal analytic categories and explanatory principles that these two accounts employ. Commentators on cognitive architectures, just like commentators on inter-theoretic reduction, part company on the implications of such discontinuities.

Classicists have argued that if connectionist models account for cognitive architecture, then their explanatory principles must appeal to representational states capable of semantic evaluation, which for the classicist also means that they are capable of serving as the constituents of syntactically complex structures. Moreover, if connectionists accept the systematicity of thought, then they must either show that it need not turn on compositionality or demonstrate how connectionist architectures can accommodate that property too.

Fodor and his collaborators maintain that connectionist models don't measure up. Although distributed representations have parts, those parts neither support semantic evaluations nor exhibit the properties of classical constituents. Connectionism lacks the means even to express psychological generalizations that classical theories capture. Thus, they hold that if the explanatory principles of PDP models address brain processes and states, then they do so at an analytical level that is sub-representational and therefore, according to classicism, noncognitive. Instead of characterizing our cognitive architecture, connectionist models only offer information about how a classical architecture might be *implemented* in brains, though even that is only a conjecture.

Fodor and Pylyshyn offer two reasons why matters of implementation are comparatively unimportant. In accordance with the autonomy of psychology, they argue, first, that as a theory of neural implementation merely, connectionism no more constrains cognitive theorizing than do theories from even lower levels, since implementation is a transitive relation all the way down. The *representational* character of mental symbols constitutes a fundamental barrier to the reductionistic program for the theoretical unity of science.

Second, and more importantly, the computer analogy assumes that *cognitive-level* theorizing concerns functional architecture – not the details of its implementations, for the staggering range of physically possible implementations renders them comparatively uninteresting. Fodor and Pylyshyn (1988, p. 63) hold that, in modeling abstract cognitive processes, "there is simply no reason to expect isomorphisms between structure and function" and, more generally, that "the structure of 'higher levels' of a system are [sic] rarely isomorphic, or even similar, to the structure of 'lower levels' of a system." Although, in principle, physical explanations can be had, they supply no insight into *cognitive-level* generalizations; hence, they count only as matters of implementation. The primary charge that connectionism concerns only implementation, then, rests on the same concerns over multiple realizability that generally plague ambitious accounts of reduction.

Advocates of alternative approaches to these issues *acknowledge* the explanatory discontinuities that classicists emphasize; however, they draw quite different conclusions!

Proponents of the dynamical approach diverge most radically from classicism. Employing arguments that, at times, mimic those of eliminativists in the debates about reduction, they envision an account of cognition that *transcends* the notion of representation. They abandon the entire computational project, dispensing, in effect, with the cognitive level and, thereby, with worries about cognitive architecture. Like eliminativists, they are not at all sure that the psychology that classicists defend provides any systematic insights about either the brain or behavior. They focus on networks as dynamical systems, emphasizing the explanatory comprehensiveness and detail of the relevant differential equations. They suspect that the theoretical distance from computation and representation to the mathematics of dynamical systems is unbridgeable. Finally, on this view, even "classic PDP-style connectionism . . . is little more than an ill-fated attempt to find a halfway house between two worldviews" (Van Gelder and Port, 1995, p. 34).

Defenders of connectionism, who put greater stock in our commonsense psychology, still repudiate classicists' defenses of it. Accepting the terms of classicism's challenge, they maintain that even if the eliminativists are right and PDP networks do not instantiate computable functions, they still utilize *representations*. For example, Terrence Horgan and John Tienson (1996) accept LOT and systematicity but deny that either involve a classical, combinatorial syntax. They regard psychological generalizations as *ceteris paribus* laws only and thus as incompatible with the hard rules which classicism requires. Mental representations that classicists regard as *complex* are realized as *primitive* symbols, in the way that irregular past tenses in English (*went*, not *goed*) seem to be (see Article 52, RULES).

While also defending our commonsense framework, Andy Clark (1993) rejects LOT and its accompanying features – certainly as classically comprehended. He too questions classical syntax, noting that PDP researchers have methods for producing structure-sensitive processing without concatenative coding. More generally, if classicism's firm distinctions – for example, between data and processing – do not fit PDP networks, perhaps those distinctions should be *recast*, rather than find connectionism wanting. Considering a flourishing program of connectionist research, let alone a potentially vast collection of yet unanticipated computational devices, why should classical views of basic computational notions, and especially the notion of representation, remain unchallenged? Measures of similarity between representational vehicles in networks can model the semantic similarity of representational contents. The explicitness of a representation need not turn on the tokening of a symbol, but on the ease of use and the multiple exploitability of the information within the system.

Clark emphasizes how representations co-evolve with processing dynamics in human development. That approach renders representational contents dependent upon the processor's capacities and the environment it operates in. Thus, Clark salvages commonsense psychology by attributing to it agendas overwhelmingly inspired by social and cultural practices. For example, he regards the cognitive underpinnings of concepts not as occurrent brain states but as a body of knowledge and skills informing manifestations whose only underlying unity is sociocultural.

Although he stresses the *compatibility* of classical and connectionist insights, Paul Smolensky (1988, 1991, 1995) also contests whether classical conceptions constrain connectionist accounts. He questions the classical assumption that micro-level accounts of processing details cannot have consequences for conceptions of structural relationships at the level of symbols.

Smolensky notes that connectionist models would implement classical architectures (in the programming language sense) only if classicism provided a precise, comprehensive, algorithmic account of cognitive processing. Anything less means that classical and connectionist models only *approximate* one another. Thus, Smolensky explicitly casts these controversies within the framework of levels of explanation *in science* rather than in programming languages. PDP modeling constitutes a *sub-symbolic paradigm* operating at the *sub-conceptual level*, which falls between the conceptual and the neural and currently offers the best means for theoretically connecting symbolic computation to neural functioning.

Smolensky maintains that such inter-level interaction can lead to the improvement of higher-level theories, hence sub-symbolic research can refine the classical approach. Employing processing algorithms affording greater precision and detail, connectionists can relate activity patterns to conceptual-level descriptions. Such integrative research across explanatory levels results in the *reconceptualization* of the notion of cognitive architecture. Smolensky advocates an "intrinsically split-level cognitive architecture" (1991, p. 204). *Syntactic* relations are characterized in terms of algorithms that describe the alterations in individual processing units' activation levels, while semantic interpretation transpires in terms of larger activity patterns. Sub-symbolic analyses offer new formal instantiations of computational concepts. Smolensky emphasizes that explicating classical notions in the language of continuous computation relies on a semantic shift accompanying the shift to the sub-conceptual level.

Smolensky's claim that sub-symbols, as activity patterns in networks, correspond to symbolic constituents has stirred debate. His critics insist that this constituency is not classical. Smolensky responds that sub-symbolic accounts provide penetrating *approximations* of compositional structure and LOT. The pivotal question is how the variable activities in networks achieve symbols' representational stability. Smolensky replies by turning this problem on its head, noting how connectionist nets readily accommodate the *context sensitivity* of representations (for which considerable psychological evidence exists). Sufficient representational stability depends not on symbolic form but on a *family resemblance* (1988, p. 17) among those vectors that, in different contexts, carry out some functional, sub-symbolic role.

Smolensky (1995) has elaborated an integrated connectionist/symbolic (ICS) architecture with which he aims, ultimately, to surmount any simple distinction between classical and connectionist architectures. Smolensky's ICS architecture employs general PDP principles constrained by tensor product structures that insure that both the semantics and the functions to be computed can be managed symbolically, even though symbols play no causal role in the computations. Harmonic nets, which are structured to maximize parallel soft-constraint satisfaction or *harmony* gradually, realize the various higher cognitive processes that symbolic accounts describe. Smolensky argues (1995) that such a sub-symbolic reduction motivates revisions in symbolic accounts that enable a richer theoretical integration of the two levels, resulting in the *preservation* of classical insights. Smolensky offers harmonic grammars and optimality theory's contributions to syntactic studies and phonology as illustrations of revisions that enrich classical accounts and preserve their most important claims.

Integrative models of cross-scientific relations

Smolensky anticipates the same sort of approximate reduction arising from the co-evolution of theories at different levels that various philosophers have championed. Co-evolving theories often yield progressively better inter-theoretic mappings. The increased integration associated with such co-evolution will not eliminate or replace symbolic accounts, but rather improve research at the conceptual level.

Smolensky's comments that any definition of constituency that provides "explanatory leverage is . . . valid" and that classical architecture is a "scientifically important" approximation of the underlying dynamics at the sub-conceptual level (1991, pp. 210, 203) accord with the *pragmatism* of recent *integrative models* of cross-scientific relations. Increasingly, philosophers argue that the welcome simplicity associated with reductionism exacts too high a price. Reductionism neglects all relations between explanatory levels except those between theories, and it conceives all inter-theoretic relations in terms of reductive explanation. Compared to reductionist accounts, integrative models explore a wider range of just the sort of cross-scientific relations that are particularly prominent in interdisciplinary research typical in cognitive science. Examining issues of discovery, evidence, method, and more, advocates of integrative models foresee *many* illuminating relationships (besides possible reductions) between psychological, connectionist, and neuroscientific models.

William Bechtel and Robert Richardson (1993) argue that the chief goal of reductionistic research among practicing scientists is the discovery and explication of the *mechanisms* underlying the functioning of complex systems. Pursuing the strategies of structural decomposition and functional localization, scientists steadily unveil the various micro-level mechanisms realizing higher-level patterns. This activity neither eliminates nor replaces the complex system or macro-level theories.

Smolensky notes that considerations of mathematical modeling, more than neural considerations, drive developments in connectionist research. He also emphasizes the accuracy, precision, and comprehensiveness of dynamical systems theory as an account of connectionist processing. He nonetheless conceives of connectionist modeling as a kind of primordial *neuro*computational research. *Contra* Van Gelder and Port (1995), developing theories of a system's dynamical features at one analytical level does not usually warrant ignoring theories of that system's parts and structures at that or higher levels (Clark, 1997). Explanatory levels contain theories of a system's synchronic and diachronic dimensions. Integrative models propose that interactions between research on synchronic and diachronic matters at a single level and between research of either sort at different levels are *mutually* enriching.

Research at lower levels can refine and even correct higher-level approximations. But integrative models also show how upper-level research (e.g., in psychology) can play a significant role in *justifying* lower-level proposals and motivating innovative research at intermediate levels. Attention to the psychological evidence *enhances* the precision and plausibility of connectionist and neuroscientific models (McCauley, 1996).

Valerie Hardcastle (1996) stresses how these interdisciplinary endeavors stimulate research in *bridge sciences* (such as evoked-response potential studies) and contribute to the *explanatory extension* of the sciences involved, either by conceptual refinement or by the theoretical support of one science for an antecedently problematic assumption of an other. Hardcastle criticizes classicists' assumptions about the clarity of distinctions between structures and functions that are so pivotal to their sharp distinction between architecture and implementation. She argues that whether a description counts as structural or functional depends upon the analytical levels involved, the questions asked, the related explanations available, and the background knowledge at hand. What might look like implementational detail from a higher-level perspective (e.g., different measures of clustering in network activity yielding different accounts of conceptually interpretable patterns) may have architectural implications from a lower-level perspective (if, for example, the differences among these measures' accounts of such patterns are found to turn systematically on micro-level variables).

Multiple realizability does not necessarily present intractable problems for integrative models. Alternative realizations of psychological states raise neither barriers to cross-scientific connections nor grounds for declaring disciplinary autonomy, but opportunities for further empirical research about the complexity of the interface between the psychological and the neural. If something like the identity theory were to prove plausible even for some extremely limited cognitive domain, the possibility of alternative realizations will certainly not deter scientists from exploring and exploiting all the resulting cross-scientific connections! If multiple instantiation of psychological functions proves the rule, it does not follow that – and in many cases there is little reason to expect that – neuroscientists face an unmanageably large number of alternatives. *Even* if token physicalism is basically correct, the important question for integrative models is whether it might sustain some cross-scientific connections that advance research in cognitive science. Unlike most reductionists and many of their prominent critics, integrative modelers do not presume that the answer to that question can be determined on principled grounds.

References and recommended reading

*Bechtel, W. (ed.) 1986: *Integrating Scientific Disciplines*. The Hague: Martinus Nijhoff.

Bechtel, W. and Richardson, R. C. 1993: *Discovering Complexity*. Princeton: Princeton University Press.

*Bickle, J. 1995: Connectionism, reduction, and multiple realizability. *Behavior and Philosophy*, 23, 29–39.

Churchland, P. M. 1989: Eliminative materialism and the propositional attitudes (1981). Repr. in *A Neurocomputational Perspective*, Cambridge, Mass.: MIT Press, 1–22.

Clark, A. 1993: *Associative Engines: Connectionism, Concepts, and Representational Change*. Cambridge, Mass.: MIT Press.

—— 1997: *Being There: Putting Brain, Body, and World Together Again*. Cambridge, Mass.: MIT Press.

Fodor, J. A. 1975: *The Language of Thought*. New York: Thomas Y. Crowell Co.

Fodor, J. A. and Pylyshyn, Z. W. 1988: Connectionism and cognitive architecture: a critical analysis. *Cognition*, 28, 3–71.

Hardcastle, V. G. 1996: *How to Build a Theory in Cognitive Science*. Albany, NY: SUNY Press.

Horgan, T. and Tienson, J. 1996: *Connectionism and the Philosophy of Psychology*. Cambridge, Mass.: MIT Press.

McCauley, R. N. 1986: Intertheoretic relations and the future of psychology. *Philosophy of Science*, 53, 179–99.

—— 1996: Explanatory pluralism and the co-evolution of theories in science. In R. N. McCauley (ed.), *The Churchlands and their Critics*, Oxford: Blackwell, 17–47.

*McLaughlin, B. P. 1997: Classical constituents in Smolensky's ics architecture. In M. L. Dalla Chiara et al. (eds), *Structures and Norms in Science*, Dordrecht: Kluwer Academic Publishers, 331–43.

Smolensky, P. 1988: On the proper treatment of connectionism. *Behavioral and Brain Sciences*, 11, 1–74.

—— 1991: Connectionism, constituency, and the language of thought. In B. Loewer and G. Rey (eds), *Meaning in Mind: Fodor and his Critics*, Cambridge, Mass.: Blackwell, 201–27.

—— 1995: Reply: constituent structure and explanation in an integrated connectionist/symbolic cognitive architecture. In C. MacDonald and G. MacDonald (eds), *The Philosophy of Psychology: Debates on Psychological Explanation*, Oxford: Blackwell, 223–90.

*Van Gelder, T. 1995: What might cognition be if not computation? *Journal of Philosophy*, 91, 345–81.

Van Gelder, T. and Port, R. F. 1995: It's about time: an overview of the dynamical approach to cognition. In R. F. Port and T. van Gelder (eds), *Mind as Motion*, Cambridge, Mass.: MIT Press, 1–43.

49

Modularity

IRENE APPELBAUM

Introduction

Imagine walking into Starbuck's, ordering a double latte, meeting a friend, drinking up, and leaving. In the course of this simple event, you would engage in a wide variety of cognitive activities, among them problem solving, face recognition, speech production and perception, memory, and motor control. How does the mind – an apparently unitary entity – accomplish such a diversity of tasks? Is the mind partitioned into diverse mechanisms, each responsible for a different job? Or are more uniform, general-purpose mechanisms deployed for different cognitive purposes? Which tasks even count as the same, and which as different? Is visual recognition a single task, or are the mechanisms that recognize objects fundamentally distinct from those that recognize faces? Is speech produced and perceived by similar processes or by different ones? More generally, how, and how much, do such different processes interact?

It is to these and related questions that the debate over the modularity of mind is addressed. Because the issue is not the character of cognitive capacities *per se*, but the organization and distribution of the systems that underlie these capacities, the issue of modularity is often described as concerning the *architecture*, or *design principles*, of the mind (see Article 48, LEVELS OF EXPLANATION AND COGNITIVE ARCHITECTURES).

Proponents of modularity argue that the mind comprises separate subsystems carrying out relatively specific functions, relatively automatically and autonomously. Theories differ as to how isolated, automatic, and specific these modules are claimed to be, and as to which cognitive processes are thought to be modular. Theories of modularity may be distinguished, in other words, in terms of their answers to the conceptual question, What *makes* something a module?, and the empirical question, Which cognitive processes *are* modules, so described?

Although largely unpopular earlier in this century, some form of the modularity thesis is now a prominent, even dominant, view. One reason for this change in the intellectual tide concerns the role of empirical evidence in this debate. Current defenses of modularity theory are distinguished by the fact that experimental data are marshaled in support of the view.

The appeal to empirical evidence does not easily resolve the debate, however, because there is wide disagreement over how this evidence should be interpreted. Questions remain as to how and how much interaction there is, both among modules and between modules and nonmodular systems. There are also questions about the internal structure of modules themselves: Are they further decomposable into sub-modules, and if so, how, and how much, do sub-modules interact with each other and with their parents? Do the properties associated with modules constitute necessary and sufficient criteria for being a module, or are they merely generally characteristic properties? Are some properties more essential than others? If so, which ones?

In addition to the conceptual question (What makes something a module?) and the empirical one (Which specific processes are in fact modular?), a third, more methodological dimension cuts across the debate. The modularity thesis is not just a descriptive claim about the internal organization of the mind, but a normative claim about how the mind ought to be studied.

In what follows, I will investigate these questions as they arise for several subdisciplines of cognitive science: philosophy of psychology, linguistics, neuropsychology, and developmental psychology. I will be especially concerned to emphasize the possibilities, as well as the difficulties, involved in bringing empirical evidence to bear on the debate over the modularity of the mind.

Philosophy of psychology: Fodor and the modularity of mind

Jerry Fodor's book *The Modularity of Mind* (1983) has become a central reference point for discussions about modularity. At the time of its publication, however, a modular approach had already been defended in a number of domains. Such an approach is to be found, for example, in David Marr's *principle of modular design*, in Kenneth Forster's *autonomous* model of lexical access, in Noam Chomsky's notion of a *language organ*, in Michael Posner's distinction between *automatic* and *strategic* processing, and in Herbert Simon's concept of a *nearly decomposable system*. Fodor's contribution was thus less to initiate discussion about modularity than to systematize and promote it.

We can understand Fodor's central claims about modularity in terms of the three dimensions enumerated above: conceptual, empirical, and methodological. At the conceptual level, Fodor claims that modular systems are distinguished by their characteristic properties and function. Functionally, he distinguishes three kinds of mechanisms: (1) transducers, (2) modules, and (3) central systems. The function of transducers is to receive energy impinging at the organism's surfaces and translate it into a representational form accessible by other psychological systems. The function of central systems is that of inference and belief fixation. The function of modules is to mediate between transducers and central systems. Although this mediation may operate in either direction, Fodor discusses almost exclusively modules which take transduced representations and infer hypotheses about their distal sources which then become available for use by central systems. More generally, Fodor (1983, p. 40) says, the function of such modules is "to present the world to thought."

Modules are intermediate between transducers and central systems not only in terms of the order of processing but in terms of the complexity of processing as well. Like central cognitive mechanisms, modular mechanisms are supposed to be inferential and computational; but, like transducers, they are assumed to be reflexive and automatic.

In *The Modularity of Mind* Fodor identifies nine properties that are claimed to be responsible for the automatic, autonomous nature of modular processing. Modules, Fodor says, (1) are *domain-specific*, (2) operate in a *mandatory* fashion, (3) allow only *limited central access* to the computations of the module, (4) are *fast*, (5) are *information-*

ally encapsulated, (6) have *shallow outputs,* (7) are associated with *fixed neural architecture,* (8) exhibit *characteristic and specific breakdown patterns,* and (9) exhibit a *characteristic pace and sequencing* in their development.

In later essays, however, Fodor emphasizes informational encapsulation to the exclusion of the others as the single defining characteristic of a module. An informationally encapsulated system operates largely in isolation from the background information at the organism's disposal. Informational encapsulation constrains a priori the amount and type of data available for consideration in projecting hypotheses about the distal layout. Moreover, this constraint on information is achieved architecturally rather than substantively. That is, in solving a particular computational task, the modular mechanism can only make use of information within the module; it has no capacity to bring even relevant information to bear if it happens to lie beyond the module's boundaries.

It is important to distinguish informational encapsulation from domain specificity, which some other writers take to be the defining feature of a module. To say that modules are domain-specific is to say that they operate on distinct classes of stimuli: only a specific stimulus domain will trigger the operation of any given module. Fodor (1983, p. 103) describes the difference between informational encapsulation and domain specificity as follows: "Roughly, domain specificity has to do with the range of questions for which a device provides answers (the range of inputs for which it computes analyses); whereas encapsulation has to do with the range of information that the device consults in deciding what answers to provide."

Central systems – those responsible for inference and belief fixation – are, according to Fodor, nonmodular and hence unencapsulated. Such systems are characterized by the absence of antecedently established constraints on the information which they can recruit in the course of their operation. More positively, in an analogy to the process of confirmation in science, Fodor describes central systems as *isotropic* and *Quinean*. Isotropic processes are those in which information from arbitrary knowledge domains may be relevant to the confirmation of a given hypothesis. "Everything the scientist knows," Fodor explains (1983, p. 105), "is, in principle, relevant to determining what else he [she] ought to believe." By a *Quinean* system, Fodor means one in which the degree of confirmation of a hypothesis depends not only on its intrinsic features but also on its relation to all other system beliefs.

At the empirical level, Fodor's principal claim is that perception is modular but higher-order cognition is not. Perceptual, but not cognitive, processing is accomplished by encapsulated mechanisms which operate independently of the rest of the organism's knowledge. In Fodor's usage, therefore, the phrase "modularity of mind" implies only that some processes (the perceptual ones) are accomplished by encapsulated mechanisms, not that the mind in general is modular.

The example that Fodor most often invokes to illustrate this view is the Müller–Lyer visual illusion, in which two parallel lines are flanked by arrows, pointing inward in one case and outward in the other. Although the lines are objectively of equal length, the one with outward-pointing arrows appears shorter. Even when one knows that the two lines are of the same length, they continue to look as if they are of different lengths. It is this persistence of the illusion and the discrepancy between how the lines look and what is believed about them that Fodor cites to support the claim that (visual) perception is modular. Even when the organism knows that the two lines are of the same length, it cannot use this knowledge to affect its perception, suggesting that the visual processes are encapsulated from such (module-external) information.

A second empirical claim that Fodor makes is that language is like perception in being modular, rather than central, like cognition. Because perception and language are not usually classified as being of a common type, Fodor coins the term *input system* for what he claims is the (natural) kind of mental system comprising perception and language (though strictly this kind includes both input and output systems).

Note, in passing, that the term *cognitive* is commonly used in two different senses: as a general, neutral term for all mental capacities, including perception, in which case it contrasts roughly with *bodily*, and in a narrower, more restricted sense, in contrast with *perceptual*. It is this latter usage that Fodor has in mind when identifying as cognitive such nonmodular, central systems as attention, memory, inductive reasoning, problem solving, and general knowledge.

Finally, at the methodological level, Fodor argues that the distinction between modular and nonmodular psychological systems is coextensive with the distinction between those psychological systems that can be fruitfully studied scientifically and those that cannot. Modular systems are good candidates for scientific investigation; central systems are not, because they are Quinean and isotropic. In particular, central, or unencapsulated, systems are subject to unconstrained data search. This at once makes such systems rational – they can take into account anything the organism knows or believes – but it also makes them susceptible to what is known as *the frame problem*: the difficulty of finding a nonarbitrary strategy for restricting the evidence that should be searched and the hypotheses that should be contemplated in the course of rational belief fixation (Fodor, 1987, p. 26).

The frame problem is something inherently faced by any unencapsulated, rational system. On the one hand, the lack of constraint on potentially relevant evidence implies that there is no natural end to deliberation. On the other hand, evidence must be constrained if a system is to function at all, and it must be constrained nonarbitrarily if it is to function rationally. (Modular processing is not rational processing precisely because its data base of information is constrained arbitrarily – i.e., architecturally.) Because the identity and degree of relevant considerations change from situation to situation, Fodor believes that relevance cannot be formalized in a theory, and therefore that central systems cannot be the object of fruitful scientific investigation.

Fodor's view implies rather dire consequences for the future of cognitive science. Although cognitive science has been concerned to explain the processes of perception (especially vision), the centerpiece of the project has been the dream of explaining more general cognitive abilities such as thought, memory, and problem solving. Fodor's claim is that these processes, being quintessentially unencapsulated, are ones that we have little hope of understanding and hence are ones that we should, as a matter of research strategy, abandon. Bold by intellectual temperament, Fodor dubs this methodological point "Fodor's First Law of the Nonexistence of Cognitive Science" (1983, p. 107).

In sum, Fodor makes three principal claims about modularity: the empirical claim that perception, but not cognition, is modular; the conceptual claim that modules, but not central systems, are informationally encapsulated; and the methodological claim that encapsulated processes, but not unencapsulated ones, are amenable to scientific study. Taken together, these three claims form an argument against the possibility of doing cognitive (as opposed to perceptual) science.

Language and modularity

The modularity thesis has been investigated in most detail in the domain of language. In the dominant tradition of generative grammar, a tradition initiated by Chomsky in the 1950s, a core assumption has been that the processes responsible for language production and perception are largely innate and modular. To emphasize the functional independence of linguistic from other cognitive processes, Chomsky has described the language module as an independent "mental organ."

Nevertheless, because generative linguistics (see Article 15, LINGUISTIC THEORY) concentrates on explaining linguistic competence (the tacit knowledge that is said to underly our ability to use language) rather than linguistic performance (the actual use of language in concrete circumstances), debates about modularity, which concern performance issues of how language is processed, have most often taken place in psychology and psycholinguistics, rather than in linguistics proper.

More recently, the field of COGNITIVE LINGUISTICS has attempted to offer an alternative to the Chomskyan modular approach by arguing that linguistic abilities are derived from more general, cognitive ones.

The lexicon

Speech perception is usually thought to involve a distinct stage, between the perception of phonemes and the recognition of syntactic structure, which is responsible for word recognition. A word-recognition model must explain how, given some particular acoustic and phonetic input, a single item is selected from the mental lexicon. The issue that arises for modularity is whether syntactic and semantic knowledge influence this selection process. A modular approach argues that such contextual knowledge does not exert any influence; a nonmodular approach, that it does.

More specifically, the debate concerns the point during processing at which word recognition interacts with higher levels of linguistic knowledge. On a modular approach, such as that advocated by Forster (1976), interaction occurs only after an initial word has been chosen solely on the basis of acoustic and phonetic information. On the nonmodular approach defended by Marslen-Wilson and Welsh (1978), syntactic and semantic knowledge are claimed to interact with acoustic and phonetic input in helping to determine the initial choice of lexical percept.

Closely related to the question of whether or not word recognition is modular is the question of whether it is a (purely) *bottom-up* or a (partly) *top-down* process. "Top" and "bottom" here refer to the relative abstractness of the information being processed. Information to be processed may be thought of as hierarchically organized, with more concrete information related to the proximal sensory input at the *bottom* and increasingly abstract, conceptually determined information at the *top*.

In a purely bottom-up system, processing decisions are based only on information from below. In a top-down system, information from higher levels of conceptual processing can also influence lower-level processing. Thus Forster's *autonomous* model is strictly bottom-up, whereas Marslen-Wilson's and Welsh's *interactive approach* allows for both bottom-up and top-down processing.

Forster's autonomous model is described as "indirect," because the input from the lower-level stimulus does not directly access the mental lexicon, but only an intermediate *access file*. The access file contains *bins* with access codes for individual words in the lexicon. Because of the process whereby a single candidate is selected from among the words in the chosen bin, the model is considered *active*. While the bin itself is identified on the basis of its acoustic-phonetic pattern, words inside the bin are organized simply by frequency, so the appropriate word code cannot be selected in advance. Instead, the system must actively search through the bin to find the word which matches the input signal. What is crucial about this model – what makes it modular – is that this search takes place on the basis of acoustic and phonetic information alone. Not until a word has been recognized is syntactic and semantic information brought to bear.

Marslen-Wilson and Welsh object that Forster's model provides insufficient basis for selecting a single entry from among the words in the bin. The bin, they charge, is accessed too early in processing – after only a couple of phonetic segments have been heard – for the available lower-level stimulus information to uniquely specify the appropriate word. Yet higher-level information, which together with the partial acoustic-phonetic information *would* be sufficient, is ruled out by the strictly bottom-up nature of the theory.

The interactive model which Marslen-Wilson and Welsh propose instead is also active, but, unlike Forster's model, it is direct. What makes it a direct access model is that the acoustic-phonetic representation triggers word candidates in the lexicon directly; there is no intermediate access-file stage. This is also what makes it a partly bottom-up model: the initial group of word candidates, called a *cohort*, is chosen on the basis of lower-level stimulus information alone. The model is active because the process of selecting and recognizing a single word involves candidate words actively taking themselves out of competition as they become disqualified, until only a single candidate remains.

Marslen-Wilson and Welsh cite evidence for this interactive approach from several experiments, including some involving shadowing tasks, in which subjects are asked to repeat speech as it is spoken. Subjects frequently and fluently restore distorted words to their correct form. For example, when the speech signal contains the nonword *compsiny* in the appropriate sentential and semantic context, subjects will repeat *company*. It is assumed that subjects actually *hear* the word *company*, rather than simply *infer* that that is what they should have heard. Without the appropriate context, however, or in isolation, subjects quite clearly hear (and repeat) the distorted *compsiny*.

Norris (1982) has defended a version of Forster's autonomous model by arguing that autonomy is compatible with the model having multiple outputs among which contextual knowledge then selects. Since there is no point at which contextual information influences decisions internal to the module, Norris argues, the strictly bottom-up, or autonomous, character of the model is preserved.

Tyler and Marslen-Wilson (1982) counter by pointing out that Norris's view preserves autonomy only by undermining the work it is supposed to do. The module's job, they argue, is now redefined as that of making continuous, multiple, candidate suggestions, while the actual job of selecting a single word occurs at a later, post-modular stage of processing.

Although Marslen-Wilson and Welsh's *cohort theory* has subsequently undergone considerable modification, later versions retain the core interactionist principles of this original, influential version.

The phonetic module

In the debate over the modularity of the lexicon, both sides presuppose that the derivation of a phonetic representation from an acoustic speech signal is straightforward and unproblematic. On inspection, however, it turns out that the process of phonetic perception is anything but straightforward. Indeed, nearly half a century of effort has been devoted to trying to explain it satisfactorily. The difficulty arises, in large measure, because the acoustic properties of a given phonetic segment (e.g., a consonant or a vowel) vary widely, depending on the surrounding phonetic context, and the same acoustic property may signal different phonetic features in different contexts. Instead of a one-to-one relationship, in other words, the mapping between acoustic segment and phonetic segment is many-to-many.

In the face of this "lack of invariance," as the problem is known, the motor theory of speech perception (Liberman and Mattingly, 1985) has adopted a modular approach which claims that invariant properties are to be found in the mechanisms of production. That is, the immediate objects of speech perception on this view are not to be found in the proximal stimulus (the acoustic signal), or in the peripheral articulatory movements of the vocal tract, but in the still more distal objects variously identified as the speaker's *phonetic intentions* or *phonetic gestures*.

According to this view, speech is special among perceptual phenomena because, in addition to perceiving it, we also produce it. Phonetic gestures are thought to be the primitives of both speech production and perception. And the perception of phonetic gestures is thought to be effected in a module which maps the variable acoustic speech signal to the postulated invariant phonetic gestures. The principles which govern speech perception, on this view, are thus not general auditory ones, but the specialized phonetic principles of the phonetic module. Recent versions of the theory appeal explicitly to Fodor's conception of a module to underwrite the claims for the specialized nature of phonetic perception.

One sort of empirical evidence frequently cited in support of the modular nature of phonetic perception, both by proponents of the motor theory and by Fodor, derives from experiments showing that visual information about the speaker's articulatory movements can influence phonetic perception. In experiments of this sort (McGurk and MacDonald, 1976), subjects are presented with conflicting visual and auditory information. For example, a subject views a film of a person making the articulations appropriate for a syllable such as [ba] but, synchronized with these lip movements, hears a tape of someone saying [na]. What subjects report hearing in such cases is [ma], which contains elements of both the auditory and the visual information: like [ba], it is a bilabial consonant and, like [na], it is a nasal consonant. Moreover, subjects are unaware that different information has been contributed by different sensory channels; they simply hear an integrated speech percept.

Proponents of the motor theory argue that this result, referred to as the "McGurk effect," supports a modular approach to phonetic perception, because it seems to show that the objects of speech perception are neither auditory nor visual, but more abstract, modality-independent constructs such as the phonetic gestures proposed by their theory and processed by a module. It is important to note, however, that the question of whether the objects of perception are distal and modality-independent is distinct from whether they are perceived in a module. The evidence from bimodal speech perception seems to provide evidence for the former, but not necessarily for the latter. Moreover, the claim that a module is needed to explain how invariant phonetic objects are recovered from the variable acoustic signal is, without an account of how the module is supposed to work, not very helpful. But the motor theory provides no such account; it simply asserts that the module *does all the hard work* and accomplishes its task *effortlessly* and *automatically*.

Dominic Massaro (1987) provides an opposing interpretation of the McGurk effect. He argues that the processes invoked in auditory-visual speech perception are not specific to this domain but are similar to those invoked in other domains such as person impression, learning of arbitrary categories, and sentence interpretation. The problem which this raises for a modular interpretation is that it seems to show that the principles governing speech perception are not domain-specific and, more generally, that speech perception and domains generally thought to be nonmodular operate according to the same kinds of principles. Both sorts of processes, Massaro argues, involve integrating multiple sources of top-down and bottom-up information and classifying the resulting pattern in terms of the relevant categories.

Thus, the McGurk effect is invoked as evidence both that speech perception is modular and that it is not.

Neuropsychology

Neuropsychology is the study of patients with abnormal brain functions (see Article 29, DEFICITS AND PATHOLOGIES). Because such patients often exhibit highly selective impairments – for example, short-term memory may be impaired, while long-term memory remains intact – they are thought to provide evidence that the mind is modular. The assumption here is that if the mind was, instead, a uniform general-purpose system, damage to the brain would have more uniform effects on function.

An alternative interpretation, which seems to challenge this modular conclusion, claims that the selective nature of impairments reflects a hierarchical, rather than an independent, relationship among cognitive functions. On such a view, more complex tasks are simply ones that require a greater quantity of largely homogeneous cognitive resources. By decreasing the overall amount of brain power available, the claim is, brain damage has a disproportionate impact on cognitive functions demanding the most resources.

However, this nonmodular view is incompatible with an important source of evidence arising from *double dissociations*. In double dissociations, two patients or groups of patients exhibit complementary deficits. According to Timothy Shallice (1988, p. 34), "a dissociation occurs when a patient performs extremely poorly on one task . . . and at a normal level or at least at a very much better level on another task." In a double dissociation a second patient shows the reverse performance pattern on the same two tasks. If the selective impairment of brain functions reflected merely the quantity of cognitive resources required for different functions, one would expect all patients to exhibit the same pattern of cognitive deficits.

Yet, as Shallice argues, double dissociations do not provide decisive evidence in favor of modularity. Although, *assuming* the mind to be modular, we would expect double dissociations, Shallice cautions that we can draw no definite conclusion about *whether* mind is modular simply because we find such dissociations.

The assumption that double dissociations imply modularity is nevertheless pervasive in neuropsychology. Shallice and his colleagues attribute this view to two related factors. First, David Plaut (1995) points to a failure to distinguish the claim that double dissociations and modularity *fit together so naturally* from the claim that the former genuinely *imply* the latter. Plaut acknowledges that modularity may be seen as a natural interpretation of the evidence from double dissociations in the sense that, on such an interpretation, the taxonomy of cognitive abilities mirrors that of processing mechanisms. The idea here is that this isomorphic relationship is the simplest and so, intuitively, the most natural interpretation. But, Plaut claims, to move from a claim about what seems intuitively natural to an assertion about how nature actually works – or *has* to work – is illegitimate.

To legitimize this move, one would need additionally to assume that general-purpose and modular systems exhaust the types of possible cognitive systems. The acceptance of this claim is the second factor that, in Shallice and Plaut's view, accounts for the widespread belief that modularity can be inferred simply from the existence of double dissociations. Since, as we noted above, a general-purpose architecture is incompatible with the existence of double dissociations, the inference *would* be justified if these *were* the only two types of cognitive architectures available. But Shallice's point is precisely that they are not. Shallice identifies a number of nonmodular (or only partly modular) processing systems that might give rise to the pattern of deficits exhibited in double dissociations, including overlapping processing regions, coupled systems, semi-modules, and multilevel systems. Once such a repertoire of alternative explanations is recognized, the mere elimination of general-purpose systems does not license the conclusion that the mind is modular.

Another type of nonmodular system that has seemed especially unlikely to generate double dissociations is a connectionist network (see Article 38, CONNECTIONISM, ARTIFICIAL LIFE, AND DYNAMICAL SYSTEMS). Yet, in recent work by Plaut and Shallice, a connectionist network lesioned to simulate brain damage does just that. The network, which was trained to pronounce written words based on their meanings, exhibited a double dissociation between abstract and concrete word meanings following damage.

Although questions remain regarding the relevance of connectionist models for understanding human cognitive processes, this result is important, because it provides an example of distinct behavioral deficits that do not correspond to distinct structures in the system. That is, although a lesion in one location generates a concrete, but not abstract, word-reading deficit, and a lesion in a different location generates an abstract but not concrete word-reading deficit, the system does not contain distinct components for pronouncing concrete words and abstract ones.

Rather, the complementary deficits arise from the differential contribution that different components of the system make to the pronunciation of these two word classes. The *direct* pathway from orthography to meaning is relatively more involved in the pronunciation of abstract words than concrete words. Thus, damage to this pathway has a disproportionately strong effect on abstract words. The pronunciation of concrete words, by contrast, relies more on what is known as the *cleanup* pathway. Thus, the ability to pronounce concrete words correctly is more impaired by a lesion to this area than the ability to correctly pronounce abstract words. It is important to emphasize that in the normal functioning network, both pathways are involved in the pronunciation of both classes of words. As Plaut points out, "it would be a mistake to claim that the direct pathway is specialized for abstract words while the clean-up pathway is specialized for concrete words."

The observation that cognitive functions can be damaged selectively is intimately connected with the claim that, in such cases, regions of the brain are damaged selectively as well. The discipline of neuropsychology only gets off the ground once a correlation between cognitive activity and brain damage is identified. While there is much hesitation regarding how narrowly this correlation should be specified, it is a commonplace assumption in neuropsychology that functionally independent cognitive systems are physically localizable. Indeed, one reason why information-processing models are attractive to neuropsychologists is that, in addition to providing a much-needed intermediate level between brain and behavior, they can, as Shallice (1988, p. 15) puts it, "be easily 'lesioned' conceptually."

Developmental psychology: beyond modularity

Endorsing the modularity of cognitive processes tends to go hand in hand with endorsing their innateness, as Fodor's position illustrates. Similarly, a nonmodular approach is closely aligned with a developmental approach to cognition, as exemplified in the views of the pioneering developmental psychologist Jean Piaget. In her book *Beyond Modularity*, Annette Karmiloff-Smith (1992) attempts to synthesize these two perspectives. Like Fodor, she acknowledges the existence of modules; but, unlike him, she denies that they are innately specified in detail. Like Piaget, she advocates a strong developmental contribution to cognitive processing; but, unlike him, she denies that development can be characterized in domain-general terms. Instead, domain-specific predispositions which selectively focus the infant's attention are claimed to give development "a small but significant kickstart" (Karmiloff-Smith, 1994, p. 693).

While the defining property of Fodor's modules is, as we have seen, informational encapsulation, Karmiloff-Smith's modules are characterized as domain-specific systems of knowledge which need not be informationally encapsulated. In addition, the distinction between modules and central systems is less rigid for Karmiloff-Smith than it is for Fodor.

But what is most significant about Karmiloff-Smith's approach is not her claim that modules arise as the result of a developmental process, but her claim that modules may not represent the final phase of this process. The specialized, automatic store of knowledge characteristic of modules is not, in Karmiloff-Smith's view, the most advanced form of cognitive development. Rather, she claims, human beings are capable of more creative cognitive processing which goes *beyond modularity*.

In such processes, modular knowledge is used for cognitive purposes other than its application in its own domain. More specifically, knowledge – in the form of internally represented rules and representations – itself becomes the object of knowledge for other cognitive processes. Instead of manipulating some range of phenomena in the world, the phenomena being manipulated, in such cases, are representations in the head. Unlike modular knowledge, which is fixed and always applied in the same way, in the present cases, cognitive processes can manipulate representations in a variety of ways for a variety of purposes. For example, in the case of knowledge of music, Karmiloff-Smith claims, this process allows different notes and chords to be manipulated independently instead of being accessible only in their *run-off sequence*.

Because each such use of modular knowledge involves representing mental representations in a different way, Karmiloff-Smith describes this process as one of "*representational redescription.*" In curiously Hegelian language, she notes that representational redescription turns "information that is *in* the mind into progressively more explicit knowledge *to* the mind" (1994, p. 698). It is this open-ended ability to represent the fixed store of modular knowledge explicitly that Karmiloff-Smith identifies as the creative and peculiarly human aspect of cognitive processing.

Conclusion

Some form of the claim that the mind contains independent functional subsystems is accepted as a competitive hypothesis in most disciplines bordering on cognitive science – philosophy, linguistics, psychology, and neuropsychology. In some – theoretical linguistics and cognitive neuropsychology – it is the dominant view. Although much of the evidence in support of a modular approach is empirical, this evidence is often equivocal, being invoked simultaneously by proponents and opponents of modularity. Thus, the debate over the modularity of mind is likely to persist for some time. We may, however, expect the principal alternative to modularity to shift from general-purpose structures to other nonmodular types of specialized structures.

References and recommended reading

Fodor, J. A. 1983: *The Modularity of Mind.* Cambridge, Mass.: MIT Press.

*—— 1985: Precis of modularity of mind. *Behavioral and Brain Sciences*, 8, 1–42 (with commentaries).

—— 1987: Modules, frames, fridgeons, sleeping dogs, and the music of the spheres. In J. L. Garfield (ed.), *Modularity in Knowledge Representation and Natural-Language Understanding*, Cambridge, Mass.: MIT Press, 25–36.

Forster, K. I. 1976: Accessing the mental lexicon. In R. J. Walker and E. C. T. Walker (eds), *New Approaches to Language Mechanisms*, Amsterdam: North Holland, 257–87.

*Garfield, J. L. 1987: *Modularity in Knowledge Representation and Natural-Language Understanding.* Cambridge, Mass.: MIT Press.

Karmiloff-Smith, A. 1992: *Beyond Modularity.* Cambridge, Mass.: MIT Press.

—— 1994: Precis of *Beyond Modularity*: a developmental perspective on cognitive science. *Behavioral and Brain Sciences*, 17, 693–745.

Liberman, A. M., and Mattingly, I. 1985: The motor theory of speech perception revised. *Cognition*, 2, 1–36.

Marslen-Wilson, W. D. and Welsh, A. 1978: Processing interactions and lexical access during word recognition in continuous speech. *Cognitive Psychology*, 10, 29–63.

Massaro, D. W. 1987: *Speech Perception By Ear and Eye: A Paradigm for Psychological Inquiry.* Hillsdale, NJ: Erlbaum.

McGurk, H. and MacDonald, J. 1976: Hearing lips and seeing voices. *Nature*, 264, 746–8.

Norris, D. 1982: Autonomous processes in comprehension: a reply to Marslen-Wilson and Tyler. *Cognition*, 11, 97–101.

Plaut, D. C. 1995: Double dissociation without modularity: evidence from connectionist neuropsychology. *Journal of Clinical and Experimental Neuropsychology*, 17, 291–321.

Shallice, T. 1988: *From Neuropsychology to Mental Structure.* Cambridge: Cambridge University Press.

Tyler, L. K. and Marslen-Wilson, W. D. 1982: Conjectures and refutations: a reply to Norris. *Cognition*, 11, 103–7.

50

Representation and computation

ROBERT S. STUFFLEBEAM

Most cognitive scientists believe that cognitive processing (e.g., thought, speech, perception, and sensori-motor processing) is the hallmark of intelligent systems. Aside from *modeling* such processes, cognitive science is in the business of mechanistically *explaining* how minds and other intelligent systems work. As one might expect, mechanistic explanations appeal to the causal-functional interactions among a system's component structures. Good explanations are the ones that get the causal story right. But getting the causal story right requires positing structures that are really in the system. After all, within the context of a mechanistic explanation, to posit X is to claim not only that X exists but that X is doing some causal labor. Because most cognitive scientists believe that cognitive processing requires the use, manipulation, and storage of internal representations, they characteristically posit internal representations to explain how intelligent systems work.

Representations, to put it crudely, are *things* that stand for something. A thing *stands for* something just in case it bears content – a relation that is satisfied whenever some X is *about*, *symbolizes*, *depicts*, or otherwise stands for some Y. Any entity that bears content *within* a system is an *internal representation*. Philosophers have long posited such entities in their explanations of cognitive processing. For instance, when one visually perceives a cat, one does not have the cat in one's head, but rather a mental image or percept of the cat. To utter "cat" when one has this percept requires that one have some *idea* of what a cat is – a *cat concept*. Since percepts, ideas, and concepts are necessarily *entities* that stand for something, common sense dictates that explaining perception, thought, and other cognitive processes requires positing internal representations.

Sometimes what is commonsensical is wrong. For instance, the earth is not flat; nor does brain size correspond to intelligence. Especially concerning questions of what exists, the nod generally goes to science over common sense. *Naturalism* is the reason why. It names the view that there is no standpoint outside science from which to explore questions concerning *what there is* (matters of ontology) and *how we know* what there is (matters of epistemology). In recent years, naturalism has been driving attempts of philosophers to make their theories of mind more scientifically respectable – that is, more biologically plausible, testable, and falsifiable. Philosophers are not the only ones who bear the burden for offering naturalized theories and explanations. Cognitive scientists, like other scientists, bear that burden too. After all, the ontology of minds and other intelligent systems gets fixed by the *scientific* character of cognitive scientific explanations. This is one reason why philosophers with an investment in naturalizing theories of mind typically put great stock in representation-laden cognitive *scientific* explanations.

But naturalizing the mind and other intelligent systems comes at a price. On the one hand, it is true, as W. V. O. Quine says (1960, p. 22): "everything to which we concede existence is a posit from the standpoint of a description or the theory-building process, and simultaneously real from the standpoint of the theory that is being built." And nothing so binds the methodologically diverse research fields named by *cognitive science* as does their conviction that without internal representations over which to operate (e.g., to compute, to use, to manipulate, or to store), no intelligent system could do what it does. Consequently, cognitive scientists of nearly all stripes and theoretical persuasions tend to posit internal representations to explain how intelligent systems work. On the other hand, it is also true that it does us no explanatory good to base our posits on notions of representation that are so ill defined, trivial, or imprecise, that our posits lack any empirical import.

What is at issue is not that scientific posits are sometimes wrong (e.g., *phlogiston* is not the *medium* of fire). Nor is it at issue that central to our reliance upon science to further empirical knowledge is the notion that no empirical hypothesis is immune from revision. On the contrary, if we take fallibility and revisability seriously, serious questions arise about the ontological status of internal representations. Such is the case because, first, what *representation* means varies from discipline to discipline and from theory to theory. Aside from complicating matters of interdisciplinary discourse in a discipline that is *inherently* interdisciplinary, *representational pluralism* in cognitive science ensures that much of what get posited as internal representations are representations just in virtue of the description *we* put on cognitive processing. Second, cognitive scientists use *representation* to refer to a wide range of phenomena (e.g., processes, mappings, rules, theories, information-bearing states, causally co-varying structures, etc.). As such, it is not obvious that everything that gets called a representation warrants the name. What is worse, many things that get called representations are representations in virtue of notions of representation that are so trivial and uninteresting that cognitive scientists are *guaranteed* to find them. That is not good science. Third, although it is almost universally assumed that all cognitive processes are computational processes, and all computational processes require internal representations as *the medium* of computation, an anti-representationalist challenge has arisen from discussions of several computation-related issues: *Are* intelligent systems computational systems? Is a *symbolic* computational framework a plausible framework for explaining biological cognitive processing? Do computational *simulations* explain how the mind/brain works? Thus, for a variety of reasons, some cognitive scientists contend that the status of internal representations may be as problematic as that of phlogiston. I happen to be one of them.

My purpose in this essay is to explore some of the issues that make the practice of positing internal representations controversial – indeed, problematic. Two justifications for the practice will be discussed: folk psychology and computationalism. The former has not only dominated common sense-related discussions of mental causation in the philosophy of mind; it is the sort of explanation that issues from the knowledge-level descriptions championed by Alan Newell and other computer scientists. The latter has dominated mechanism-related discussions of intelligent systems in cognitive science. Both justifications are controversial. But so too are most attacks on computationalism. Exploring and resolving some of these controversies requires much in the way of reexamination, including (1) what makes something a representation; (2) the relation between computationalism, representationalism, and the *medium* of computation; (3) the widespread notion of synonymy between *computation* and *symbolic-digital processing*; and (4) anti-computationalism.

Mental causation: folk psychology

For most philosophers of mind, what justifies the practice of positing internal representations is the view, famously defended by Jerry Fodor (1987), that there can be no intentional causation without explicit representation. One way this notion gets fleshed out is in terms of folk-psychological explanations.

Folk-psychological explanations purport to explain an agent's behavior by appealing to the mental causes for the agent doing thus-and-so. For example, my wife is about to go to the video store. A folk-psychological explanation of Beth's going-to-the-video-store behavior would look like this:

1 Beth believes that the tape she rented yesterday is due today.
2 Beth believes that by going to the video store, she can return the tape on time.
3 Beth desires that the tape be returned on time.
4 *Ceteris paribus*, agents act on their beliefs so as to get what they desire.

∴ Beth will go to the video store.

The mental causes here, as in all explanations of this sort, are an agent's *propositional attitudes*. A propositional attitude is a psychological disposition (e.g., desire, belief, fear) towards a specific content-bearing state (e.g., that the tape be returned on time, that elephants are noble creatures). Because what one can desire now can later dispose one toward, say, fear or hope, one can be disposed towards a given content-bearing state in different ways. In any event, all propositional attitudes bear content. The content of a propositional attitude is expressed through the *that* clause of a proposition. Given that anything which bears content is a representation, it is not hard to see how the use of folk-psychological explanations justifies positing internal representations to explain cognitive processing. Indeed, without a commitment to some or other *theory of (mental) representation*, folk psychology would be beyond the pale of scientific respectability. That folk psychology should prove scientifically unrespectable would be nothing less than "the greatest intellectual catastrophe in the history of our species" (Fodor, 1987, p. xii). While avoiding such apocalyptic consequences requires positing internal representations, it also requires a commitment to *representationalism* – the view that without a *medium* of internal representations, intelligent systems could not do what they do (see Article 47, INTENTIONALITY).

The adequacy of any theory of representation will be measured by (1) how well it addresses the ontological issue of what it takes for something to be a bearer of content. It must also explain (2) how internal representations are produced (e.g., by the environment or by interaction with other internal representations); (3) how internal representations figure causally in the production of behavior; and (4) how to individuate the content of representations (i.e., what it is that makes representation X have content Y).

Many philosophers consider the last desideratum to be the most important. In fact, how content *ought* to be individuated is generally *the* problem when philosophers speak of "the problem of representation." The problem, in short, is discovering a *naturalized theory of content* – a scientifically plausible, empirically constrained account of how representations have the content they do. Such a theory is supposed to answer not only the ontological question, "How can anything manage to be about anything?"; it is also supposed to explain why it is "that only thoughts and symbols succeed" (Fodor, 1987, p. xi).

Naturalized theories of content are controversial. The chief source of conflict concerns whether content should be individuated in a *wide* fashion, sensitive to an individual's environment, or a *narrow* fashion, sensitive to, say, an individual's learning history. Proponents of the former are called "externalists." Proponents of the latter are called "individualists." Individualists and externalists themselves further divide regarding whether a *causal* (information-theoretic) or *function-based* approach works best. While all causal approaches individuate content in terms of how a representation was *produced*, some champions of this approach are externalists (Fred Dretske), others are individualists (Jerry Fodor). For those who champion a *function-based* semantic theory, content gets individuated in terms of a representation's *purpose* or *function* – that is, in terms of how it gets *used*. Some proponents of this approach are externalists (Ruth Millikan), others are individualists (Brian Loar).

While it is true that nothing would *be* a representation unless it bore content, we need not get further entangled in how to individuate content. After all, whether it is appropriate to posit internal representations is an issue that primarily concerns the ontological status of internal representations, not what they are about – a task that occurs only *after* one has identified something as a representation. The issue at stake here, however, is the *prior* question concerning what counts as a representation. About *that*, this much is clear: if to bear content is what makes something a representation, then because all photos bear content, not all representations are either symbols or thoughts. Indeed, even a rock can be a representation (cf. Fodor, 1987, p. xi). To illustrate, grab four rocks, toss them to the ground, then arrange them to depict the location of St Louis relative to, say, Atlanta, Boston, and Chicago. Because each of these rocks now stands for a particular city, they are all representations. Hence, even if symbols and thoughts are the only things that can be *internal* representations, clearly they are not the only things that can *be* representations.

So, are all internal representations either symbols or thoughts? Putting aside for the moment the issue of what counts as a symbol, cognitive scientists consider many sorts of things to be internal representations that are not thoughts: for example, states of sensorimotor processing, somatosensory maps, schemas, prototypes, rules, theories, computational states, and states that causally co-vary with one another (see Article 51, REPRESENTATIONS). That representations are posited to fill a host of explanatory gaps is another feature of representation-talk that ought to give us pause. So too should this: neuroscientists consider causal co-variation to be a sufficient stands-for relation. For instance, many neuroscientific *explanations* make an ontological appeal along one (or both) of the following lines:

> The neural activity of structure B (or population of neurons B) co-varies with the neural activity of structure A (or population of neurons A). Thus, B is a representation of A.

> Structure B (or population of neurons B) carries information of type F (or function F). Thus B is a representation of F.

Neither of these uses of representation-talk is appropriate. If they were, then *every* biological state would be a representation, rendering representation-talk vacuous. Here is why. First, not only do all *biological* states causally co-vary with other states, *everything physical* causally co-varies with something. Strictly speaking, therefore, *everything physical* would be a representation. That is absurd. Second, *all* biological states are put to use toward *some* function (possibly multiple ones). So, if being used by the system toward some function was a sufficient way for an object to be in a stands-for relation, again, we would lose whatever mechanistic gain such talk contributes to explanations of intelligent systems. So much the worse for representational pluralism. At the very least, it is not obvious that everything that gets called a representation merits the name. This too should give us pause.

We need not delve further into folk psychology either. After all, more evidence than the prevalent use of folk psychology is needed to secure the *scientific* status of its explanations. Consider the folk-psychological explanation that I offered earlier. Because Beth *really* went to the video store because I asked her to return the video, premises 1–4, while all true, appear not to constitute the *real* explanation. Explanatory failures of this sort are more the norm than the exception. And since *any* action by an agent can be explained in a folk-psychological manner, such explanations are far too easy to come by. Assuming that folk psychology is even a theory, that we can always posit a set of propositional attitudes to *explain* even habitual actions (e.g., walking) does not bode well for a theory that purports to explain *purposive* behavior.

Cognitive science is no more driven by folk psychology than astronomy is driven by astrology. Rather, as is often noted, the hypothesis driving cognitive science is that the mind is a computer. Consequently, even someone who denies the existence of beliefs and desires can *still* be committed to the notion that representations play a fundamental role in cognitive processing. Such is the view championed by Paul and Patricia Churchland. Not only do they deny the legitimacy of folk psychology; they justify their representationalism on computational grounds. So too does almost everyone in cognitive science. As such, let us move on to the major reason why most cognitive scientists posit internal representations – *computationalism*.

Computationalism

Computational processes are always implemented within a computational framework – classicist, connectionist, Bayesian, etc. (see Article 48, LEVELS OF EXPLANATION AND COGNITIVE ARCHITECTURES). Since cognitive processes are assumed to be computational processes, explaining how an intelligent system works requires positing some computational framework. The link between computationalism and representationalism appears to be direct, for without a *medium* of internal representations, computational systems could not compute. Thus, it is claimed, all computational systems require internal representations as a *medium* of computation. The received computationalist gloss on the nature of this medium is that internal representations are, in some sense, *symbols*.

Linking computationalism with *symbolic* representations is problematic. But so too is putting a representationalist gloss on *nonsymbolic* computational processing. To see why, we need to delve into the nature of computational systems. But let me first say something about the term *computation*; as is often noted in the literature, its meaning is hard to pin down. This much is clear: *computation* admits of two senses; one is a mathematical *function*, the other is a *process*, generally mechanical. In the *function* sense, a computation is a mathematical abstraction that accounts for a mapping between elements of two classes, usually inputs and outputs of a system. The mapping function is the algorithm, or rule, specifying how the first element is related to the second. For instance, *multiply the first by 3* is the rule or algorithm mapping the elements of the ordered pairs $\{(1,3), (2,6), (3,9)\}$. Expressed algebraically, the rule is $y = 3x$. In the *process* sense, *computation* refers to the act of computing, implementing, or satisfying some function f.

To avoid confusion, I use *function* when talking about the mathematical abstraction (or rule). When I use "computation," it is the process sense that I have in mind. Now to the task of characterizing computational systems.

Computational systems

We do not find computational systems *in the wild*, because whether something is a computational system is not *just* an empirical matter. Rather, something is a computational system always relative to a computational *interpretation* – a description of a system's behavior in terms of some function f. In other words, something warrants the name "computational system" only *after* its behavior gets interpreted as satisfying, implementing, or computing some specific function. Thus, individuating computational systems is always interpretation-dependent. Not only is there the subjective matter of whether we care about individuating an object as a computational system, there are certain pragmatic considerations that figure in determining which function among equivalent functions is being computed.

Two senses of "behavior" underlie the practice of individuating computational systems: one is the inward sense of *how* a system does what it does – the internal processing; the other is the outward sense of *what* a system does (e.g., fall to the ground, produce speech, etc.). Consequently, there are two sorts of computational interpretations. A computational interpretation *in either sense* justifies calling an object a "computational system." Thus, there are two general classes of computational systems: O-computational systems and I-computational systems. Why this distinction is important will become clear presently.

O-computational systems (OCSs) When an object's *outward behavior* receives a computational interpretation, it is an OCS. Since "every physical system in the universe" implements *some* function or other (Dietrich, 1994, p. 13), computational interpretations of outward behavior are easy to come by. This can be shown with the aid of Sophie, my cat, who I am holding about two feet off the floor. As I release her, she falls to the ground. Because her falling-to-the-ground behavior satisfies the distance function, $D(t) = gt^2/2$, Sophie is an OCS. She might *also* be an ICS (see below).

Not all outward computational interpretations are so vacuous. For example, all the major bodies in our solar system move in predictable orbits. The orbit, or outward behavior, of each of these bodies has been interpreted in terms of differential equations. Because differential equations are mathematical functions *par excellence*, our solar system is a paradigmatic OCS. This does not entail that OCSs are ubiquitous. After all, even if the outward behavior of "every physical system in the universe" *can* receive an outward computational interpretation, not every one of them *does*. Hence, not everything is an OCS. And since "computer" names systems whose *inward behavior* (e.g., input-to-output transformations) receives a computational interpretation, unless a given OCS also receives an inward computational interpretation, computer-talk is inappropriate.

Since our solar system is not an *intelligent* system, this aside about a paradigmatic OCS might seem beside the point. It is not. The solar system and prototypical intelligent systems – namely, (human) persons – both share some important features: (1) their behavior emerges from the tightly coupled interaction with simpler systems; and (2) their behavior can be interpreted in terms of differential equations. As these are the features of *dynamical systems*, it follows that some intelligent systems are dynamical systems. Although it might seem obvious that persons, like planets, are embedded within an environment rich in other dynamical systems that shape their behavior, explaining *intelligent* systems within the framework of dynamic systems theory (DST) is *very* controversial. This is so, in part, because DST forces cognitive scientists to reexamine the practice of wedding intelligent systems to *internal* computational processing. Aside from being motivated to dispel problematic ontological commitments, DST (among other approaches) aims to reconnect cognitive processing with the world. The issue of *re*-connection arises because in *behaviorism*, which was the former received scientific view of cognition, internal processing paled in causal significance when compared to the environment. But in the current received scientific view, *computationalism*, the environment is all but ignored in favor of internal processing. *Cognitivism*, which is the view that the mind is to the brain as a program is to a symbolic-digital computer, carries this computational solipsism to its logical extreme. But most proponents of DST see cognition as the product of an agent who is closely coupled with her environment. On this view of intelligent systems, not only does cognitive processing get *extended* out into the environment, but the boundaries for the intelligent system do so as well (see Article 38, CONNECTIONISM, ARTIFICIAL LIFE, AND DYNAMICAL SYSTEMS, and Article 39, EMBODIED, SITUATED, AND DISTRIBUTED COGNITION).

Given that outward computational interpretations are interpretations of *outward* behavior, it should be obvious that not all computational explanations license a commitment to internal representations. Such is the case independent of DST-related reasons for reexamining the practice of wedding intelligent systems to *internal* representational processing. The task now is to determine whether all *inward* computational explanations license a commitment to internal representations.

I-computational systems (ICSs) When an object's *inward behavior* receives a computational interpretation, it is an ICS. An inward computational interpretation is warranted when a system has the following features: input states, output states, and transitions between those states. When one specifies the function (rule or algorithm) that describes what the system must do to the input(s) to get to output(s), one has specified the mapping function f. In so doing, one has rendered an inward computational interpretation.

Consider, say, the floor beneath you. Floors contain many molecules. In a sufficiently large floor, the movement of these molecules could be *described* as satisfying just about any function. Indeed, since computer programs are functions *par excellence*, the movement of molecules within your floor could even be described as satisfying the function underlying Word®, which is currently running on my Mackintosh PC. So, if we treat the movement of the floor's molecules as inputs, outputs, and transitions between them, your floor can receive an inward computational interpretation. And if it does, your floor would be an ICS. Now consider my Mac. On its hard drive are stored scores of programs. As you might well imagine, since programs are functions, inward computational interpretations of my Mac are easy to come by. Unsurprisingly, PCs are paradigmatic ICSs.

I have chosen your floor and my Mac as exemplars of ICSs because I want to underscore two important but often overlooked truths about such systems. First, rendering an inward computational interpretation does *not* entail that the system in question is actually computing – that is, mechanically following the algorithm. Second, no ontological commitments about internal representations follow from merely describing an object as an ICS. Not only are these points lost on a majority of computationalists, they are also lost on many anti-computationalists, most notably John Searle (1990). In abridged form, Searle argues as follows:

1 Every physical system in the universe can receive an inward computational interpretation. (His example: The movement of molecules within a sufficiently large wall can be *described* as implementing, say, the WordStar® program on his PC.)
2 Thus, everything *is* an ICS – indeed, "everything is a digital computer" (p. 26).
3 Digital computers actually compute, and *anything* can be described as a digital computer (e.g., floors, rocks).

∴ Computationalism is explanatorily bankrupt.

This sort of anti-computationalism is ill founded. Though not all the reasons why need detain us, identifying some of them will help to explain the above truths about ICSs.

First, even if the inward behavior of every physical system in the universe *can* receive an inward computational interpretation, not every one of them *does*. So, if receiving an inward computational interpretation is necessary for something *to be* an ICS, not everything *is* an ICS.

Second, though almost anything can receive an inward computational interpretation, it does not follow that doing so in the case of just *any* object does us any explanatory good. While all such individuations depend on interpretations of input-to-output transformations, surely not just *any* internal goings-on count as such activity – for example, the movement of molecules within a wall.

Third, *descriptions* of mere satisfaction are mere descriptions. While they may be useful, mere descriptions, computational or otherwise, carry little explanatory weight. This is why treating an object *as if* its inward behavior consisted in input-to-output transformations fails to illuminate our understanding of how it works (assuming it even makes sense to say that it *has* inward behavior). Because only *explanations* license ontological commitments, the difference between a mere computational interpretation and a mechanistic explanation is a difference that makes a difference when the task becomes fixing the ontology of a given intelligent system.

Note that mere inward computational interpretations do not apply only to ordinary objects. Far from it. Consider the amount of labor that cognitive scientists have expended on attempting to *simulate* some aspect or another of human cognitive processing – for example, playing chess. Although many programs *simulate* human chess-playing behavior, it does not follow that the existence of any such program explains how *brains* work when humans play the game. Many computationalists – especially cognitivists – appear to believe otherwise. For cognitivists, understanding how the mind works just is a matter of discovering the right program. And *any* program that simulates cognitive behavior appears to fill the bill. Enter Searle. When Searle and other anti-computationalists attack computationalism, most of their effort is directed at undermining cognitivism and dispelling unlicensed inferences from simulations to explanations. The latter, especially, is one reason why anti-computationalists are so skeptical about internal representations.

Last, whereas Searle's wall and your floor, at best, *merely satisfy* their interpreted functions, Searle's PC and my Mac *actually* compute. The difference between *mere* and *actual* ICSs (or computers) is another difference that makes an ontological difference.

Actual computation?

Here is the problem: Since the individuation of computational systems *is* always interpretation-dependent, would not anything that we call an ICS be a *real* (*actual* or *true*) ICS? Insofar as we do not find computational systems in the wild, because whether something is a computational system is not *just* an empirical matter, yes. In much the same way that both a rock and a photo can be *real* representations, all ICSs are *real* computational systems, albeit only in a weak sense.

Although individuating computational systems is not *just* an empirical matter, it does not follow that *no* empirical matter matters. Many computationalists used to believe otherwise. They argued that because any function can be implemented in a variety of systems, what matters is the computational interpretation, not the empirical specifics concerning how the function is implemented. While some computationalists stubbornly cling to this aprioristic notion (believing that how minds work can be *explained* independent of how brains work), many now realize that one stands little chance of explaining how an intelligent system works unless one's computational interpretation is wedded to, and constrained by, certain (empirical) matters of implementation. This *naturalistic* approach presupposes that some systems mechanically implement their function(s). For systems that *actually compute*, the label "computational system" has no bearing on what they do, because their computational nature does not depend *solely* on someone's computational interpretation. Were this not the case, there would be no computational basis for a distinction between, say, your floor and my Mac. Since not all ICSs actually compute, not all ICSs are *actual* computational systems, however *real* in the weak sense they might be.

So, what empirical matters bear upon whether a system actually computes? According to contemporary orthodoxy, actual computation requires "the rule-governed manipulation of internal symbolic representations" (van Gelder, 1995, p. 345). This explains (1) why many cognitive scientists treat "computation" as shorthand for symbolic-digital processing; (2) why "computer" is treated as synonymous with digital computer; and (3) why Searle and other anti-computationalists focus their critical attention on computational *explanations* based on either the *architecture* of digital computers or *simulations* on digital computers.

Nevertheless, there seem to be at least two paradigms of actual computational processing (see Article 26, ARTIFICIAL INTELLIGENCE):

> In symbolic-digital processing, computation involves simple mechanical operations applied to discrete and explicit symbol tokens – determinable quasi-linguistic structures that mediate the combinatorial, rule-following production of a system's output. It is the sort of processing occurring in classicist systems.
>
> In nonsymbolic-analog processing, computational operations are defined over continuous sets involving direct, quantitative input–output relations (i.e., as the input varies continuously, so do the outputs). It is the sort of processing occurring in systems implementing parallel distributed processing (PDP). The computational primitives here are analog quantities or distributed representations – that is, distributed patterns of activation among the processing units (see Article 38, CONNECTIONISM, ARTIFICIAL LIFE, AND DYNAMICAL SYSTEMS).

If we assume that brains are *actual* ICSs, then, ignoring hybrids, they must implement one or the other of these two types of actual computation. Not only are both sorts of computational primitives called *internal representations*, but each is considered the *medium* of its respective style of computation. Consequently, the ontological status of internal representations in biological cognitive processing would seem to be assured. Do not hold your breath.

Cognition without representation?

Let us cut to chase. To explain how a given system does what it does, positing internal representations would be required just in case the system trafficked in *entities* whose content-bearing status does *not* depend just on our descriptions or interpretations. Adopting the less than ideal vocabulary found in the literature on intentionality, I call such content-bearers "intrinsic representations." Take your average photograph. Not only would it bear content even if no one were to see it; it will bear content for as long as it exists. Such is the case because photos are ontologically dependent on being content-bearers. They have this feature because unlike, say, rocks, photos are produced by a process designed to produce content-bearers. Not everything has this feature. The contrast class is *extrinsic representations* – content-bearing *entities* whose status as representations *does* depend just on our descriptions or interpretations. Anything can be described *as if* it bore content; anything can be an extrinsic representation. But not everything is an *intrinsic* representation.

Do brains produce (internal) intrinsic representations? I think they do. Whereas photos are produced by a mechanistic process *designed* to produce entities that are ontologically dependent on being content-bearers, a plausible *evolutionary* analog would be the products of mental imagery. Surely *mental images* of one's past experiences are intrinsic *re-presentations* if anything is. Linguistic tokens are another candidate. After all, once the content of a linguistic item is fixed, at least for a particular linguistic community during such-and-such time, tokens of that type will always bear content. So, if either mental images or linguistic utterances are intrinsic representations, *some* intrinsic representations are products of biological cognitive processing. What is at issue is this: Do internal representations mediate the processes underlying the production of such representations? As these processes are supposed to be *computational* processes, the answer should be yes.

If brains are *actual* ICSs because they implement classicist computation, positing internal representations to explain biological cognitive processing would *seem* to be both necessary and uncontroversial. After all, symbolic-digital processing is *necessarily* processing over discrete, explicit symbols and rules. While the stands-for relation between a symbol and its content is arbitrary, nothing would *be* a symbol unless it were a content-bearer. Consequently, all symbols are intrinsic representations.

When we peer into the innards of a PC, we find a rule-based production system – a program (written in, say, LISP) with discrete (and explicit) rules that operate over discrete syntactic structures in the production of new syntactic structures. If we assume that syntactic structures are symbols (which not everyone is willing to do), we will be sure to find entities that compel us to say "Lo, a symbol; lo, a rule." If brains implement symbolic-digital processing, we *should* be able to structurally decompose brains into their discrete rules and symbols. But we cannot. While we can *describe* the brain *as if* it implemented symbolic-digital processing, that is a far cry from the brain actually doing so. Connectionists and anti-computationalists take the absence of discrete, explicit symbols and rules as evidence that brains are not symbolic-digital computers. Most classicists do not, however, for they are *not* using "symbol" in the sense of a discrete (and explicit) sign. Rather, they consider a symbol to be *any* internal pattern. But since every structure in the body generates *some* pattern or other (be it chemical, electrical, or metabolic), classicists *guarantee* that one will *always* be able to put a symbolic gloss on *any* biological process, even digestion. That is not good. For if *any* internal pattern is a symbol, no amount of empirical evidence could count *against* the notion that all intelligent systems operate over symbols (or internal representations). This violates one of the cardinal tenets of empiricism, *falsifiability*.

If brains do not implement symbolic-digital processing, does it follow that brains are not *actual* ICSs? No. Brains could implement nonsymbolic-analog processing – a possibility about which traditional anti-computationalists are curiously silent. Such processing is radically different from symbolic-digital processing (e.g., successful neural networks are not programmed to do what they do; PDP involves statistical operations defined over continuous sets, etc.). While connectionists *simulate* PDP on digital computers, there are important differences between *artificial* neural networks and *real* ones – populations of neurons in the brain. For instance, whereas analog quantities do the computational work in brains, in artificial neural nets it is the patterns of activation among the processing units, so-called distributed representations. Both of these *mediums* are poor candidates for being representations. Let us ignore worries about causal co-variance, multiple ends, and whether nondiscrete entities or processes make intrinsic content-bearers. Instead, note that neither analog quantities nor distributed patterns of activation *mediate* nonsymbolic-analog processing; rather, they *are* the processing.

Given the right sort of interpretation, analog quantities or distributed patterns of activation, like anything else, can *be* representations. But since it is the interpretational process alone that makes them representations, at best, they are *extrinsic* ones. While such constructs are descriptively useful, trying to pass them off as *internal* representations trivializes whatever gain representation-talk is supposed to contribute to our understanding of nonsymbolic-analog processing (Stufflebeam, 1995). It also immunizes representationalism from being falsified. So much the worse for representationalism.

Conclusion

Representation-talk is wrought with controversy. This is so, in part, because cognitive scientists posit representations while remaining ambivalent, at the very least, about the ontological problems associated with the practice. Also, it is far from obvious that everything that gets called a *representation* merits the name, much less whether they are *internal* representations. Resolving these and related tensions requires much in the way of reexamination and includes asking the following questions:

- Why should representation-laden computational descriptions qualify as mechanistic explanations?
- To what extent are *internal representations* artifacts of the interpretation *we* put on cognitive processing?
- To what extent do our commonsense intuitions about vision predispose us to *find* representations in perceptual processing, even though representation-talk seems appropriate only when the system needs to keep track of external objects that are *not* immediately present?
- If any internal pattern of activation counts as a symbol (or an internal representation), what possible empirical evidence would count against the notion that all intelligent systems operate over symbols (or internal representations)?
- Is there any level of complexity below which one *would not* posit internal representations to explain how a system works? If there is, why are mechanistic explanations of the simplest biological processes representation-laden?
- How much computational labor do biological intelligent systems off-load to their environment, thus minimizing the need for *internal* representations?

This is only the short list. I do not pretend to have resolved all these issues here.

The upshot of what soul-searching we *have* done is, I believe, a fracture in the supposed link between computationalism and representationalism. And aside from sensitizing ourselves to the unconsidered use of representation-talk, another result is that we can be full-blooded computationalists without committing ourselves to the view that the brain processes information in the same way as do our representation-laden computer simulations. Where the ontology of biological intelligent systems is concerned, representation-related conservatism is a small price to pay for a commitment to naturalism, hallowed be its name.

References and recommended reading

Churchland, P. S. and Sejnowski, T. J. 1992: *The Computational Brain.* Cambridge, Mass.: MIT Press.

Dietrich, E. (ed.) 1994: *Thinking Computers and Virtual Persons.* San Diego, Calif.: Academic Press.

Fodor, J. A. 1987: *Psychosemantics.* Cambridge, Mass.: MIT Press.

Quine, W. V. O. 1960: *Word and Object.* Cambridge, Mass.: MIT Press.
*Searle, J. R. 1990: Is the brain a digital computer? *APA Proceedings*, 64, 21–37.
Stufflebeam, R. S. 1995: Representations, explanations, and PDP: is representation-talk really necessary? *Informatica*, 19, 599–613.
van Gelder, T. 1995: What might cognition be if not computation? *Journal of Philosophy*, 92, 345–81.

51

Representations

DORRIT BILLMAN

Cognition is the flexible coupling of perception and action. Whether direct or complex, this coupling depends on representing information and operating upon it. Thus, representation and its partner, processing, are the most fundamental of ideas in cognitive science. *Representations* are the bundles of information on which processes operate. Cognitive processes such as perception and attention encode information from the world, thus creating or changing our representations. Processes of reasoning and decision making operate on representations to form new beliefs and to specify particular actions. *Process* refers to the dynamic use of information. *Representation* refers to the information available for use. Loosely speaking, *representations* include the ideas, sights, images, and beliefs that fill our thoughts and also the sensations and dispositions which may fall outside our awareness. Because representation is such a central concept in cognitive science, the term is used in a number of related senses, and I will note a couple of these more specialized uses as the need arises.

We have many intuitions about the information that is part of our own thinking or that is needed for the operation of an artificial system, and these intuitions are often a valuable starting point and source of hypotheses about representation. However, it is also frequently the case that our intuitions are incorrect or lacking altogether. This leaves a large set of problems regarding representation open for study, and cognitive scientists investigate many of them. What are the representational components of visual perception? What representations does an infant have that aid initial language learning? What representations will allow a computer system to diagnose blood diseases or a robot to navigate in unfamiliar territory? Different research goals emphasized by different disciplines within the cognitive sciences motivate different types of questions about representation: what people use, what a computer application needs, or what the nature of logic, language, or imagery might be.

Sometimes it is useful to separate questions about representation from those about processing. Consider a psychological example. An air traffic controller might err because she did not represent critical information about loss of altitude or because of a processing slip due to overloaded attention at the critical moment; identifying which was the case might be important both theoretically and practically. The difference between representation and processing is often a useful contrast.

The most fundamental contrast in understanding representation, however, is the contrast between the representation and the thing represented. All representation systems involve a relation between a represented world and a representing world (Palmer, 1978). A represented world provides the content that the representations are *about*, and the representing world is the system for preserving some information from the represented world (see Article 47, INTENTIONALITY).

Our mental representation of some event does not contain the same information as did the event itself. This difference shows up when two people reminisce about the same conversation and discover that their memories are very different; of course, if each mental representation had the same information as the event itself, then two mental representations of a given event would be the same. Even the simplest percept is not the same as the stimulus which triggered it. Our perception selects, organizes, and sometimes distorts information from the perceived world. The perception of one individual differs from that of another, and differences among species are even greater.

Because this separation between a represented and a representing world is so fundamental, we will begin by considering five types of representation system, which differ, not in what is being represented, but in the status of the representing world. In particular, we will introduce more carefully the idea of mental representation, which is at the heart of cognitive science. Then we consider three aspects of representation – content, organization, and format – and related controversies.

Five realms of representation

Several very different kinds of representation systems are addressed in cognitive science. Sometimes cognitive scientists are not explicit about the differences, and this can lead to confusion. There are at least five different *realms* of representation. I'll call them *external, mental, computational, theoretical,* and *physiological,* as they do not all have standardized labels. For each I sketch the relation between the represented world and the representing world.

External representations are systems such as writing, pictures, or maps. For maps, the representing world is the particular map, with the system of conventions for how elements of the map correspond to elements of the territory, and the represented world is the territory. For language the story is complex: loosely, the representing world is an utterance with the system for interpreting it, and the represented world is some combination of (a) states of affairs in the world to which the speaker wishes to draw attention and (b) aspects of the speaker's beliefs and intentions. The study of external representation systems is an important part of cognitive science. For example, cognitive engineering addresses the design of good texts, diagrams, or interfaces, and cognitive anthropology studies external representation systems as important products of a culture. But the focus of cognitive science is the mental representations that fill our thoughts, encode our perceptions, and envision our goals and plans. (See Articles 5, 54, 56, and 60: COGNITIVE ANTHROPOLOGY, EDUCATION, EVERYDAY LIFE ENVIRONMENTS, and SCIENCE.)

Mental representations are the internal systems of information used in perception, language, reasoning, problem solving, and other cognitive activities. Mental representations cannot be observed directly. Their nature is inferred from observing the information to which a person is sensitive and the distinctions a person uses. As with external representations, there may be different kinds of mental representation systems, such as kinesthetic, linguistic, and visual. What is the represented world when the representing world is mental representation? Most simply, mental representations represent information about the external world – the perception of a face or memory of a conversation. Further, some of these external things are themselves representations: photos, textbooks, menus, and so on. In addition, mental representations can be about internally generated information, such as remembering a past thought or considering a newly generated idea or goal. Something is a mental representation because of its role in a person's (or animal's) cognitive system, not because it is about one thing versus another. (Some researchers, perhaps following Piaget, restrict the term *mental representation* to presentation again of information from long-term memory, unavailable from perception; but this restricted use is not the dominant one.)

Computational representations are used by a computer system, just as mental representations are used by a human cognitive system (see Article 26, ARTIFICIAL INTELLIGENCE). Just as people have mental representations for carrying on the internal business of cognition, computer programs have internal, computational representations for carrying out computations. The represented world can include the external situation, such as the temperature in Timbuktu or the payroll deductions of the president, and also the internal state of the program, such as a counter in a loop or a location in memory. Of particular interest to cognitive science, some computational representations are also intended as models of human cognition, and thus they are also special cases of *theoretical representations.*

Theoretical representations are part of a theory about something. They provide an abstract model of the target domain, be it movement of beach sand, economic growth, or human cognition. My theory of perception might claim that people represent rectangles in terms of size and shape; my theory about stereotyping might claim that nongroup members represent the social group African-Americans with an average of media presentations; my theory about decision making might claim that people represent choices in terms of worst envisionable outcomes. Representations in a theory of cognition often have two layers of correspondence. First and foremost, the representations in the theory are taken to correspond to the mental representations in people's minds: that is, the represented world of the theory. If the theory is a better one, it will represent more distinctions that are actually important to human cognition and will not introduce distinctions which do not matter. Second, many of these theoretical representations of mental representations indirectly correspond to things in the world, such as an actual rectangle, US media presentations in the last year, or the Chernobyl explosion.

Finally, *physiological representations* are invoked when physiological structures are treated as elements of a representing world. A variety of methods, most dramatically functional magnetic resonance imaging (fMRI) but also positron emission tomography (PET), allow identification of certain areas of the brain that are active in certain kinds of tasks (see Article 32, NEUROIMAGING). Sometimes a certain area of the brain is active for a variety of tasks using the same kind of information, and processing of that kind of information is selectively disrupted by damage to that area of the brain. This sort of evidence has been found for a variety of types of information, including human faces, aspects of syntax, music, and recent personal experiences. Based on this, researchers may make claims that a particular brain structure represents a particular kind of information. At a very different level of analysis, some simpler aspects of *cognition*, such as low-level vision or classical conditioning, can be studied and modeled at the level of detailed neural circuits. Here a researcher might propose that the activation level of a particular neuron represents the contrast in a part of the visual field. (See Articles 4, 29, and 34: BRAIN MAPPING, DEFICITS AND PATHOLOGIES, and SINGLE NEURON ELECTROPHYSIOLOGY.)

How are these five realms of representation related? As we have seen, elements of the representing world in one realm may be part of the represented world of another. Cognitive science deals with some particularly complicated relations. First, mental representations can be about external representations. Second, in a theory about cognition, unlike a theory about earthquakes, say, the *theoretical representations* have *representations* (specifically, mental representations) in their represented world. Third, the representations in a computational model of cognition are simultaneously computational representations and theoretical representations. Further, only some aspects of the computational representation have anything to do with the theory, while other aspects are implementation details. Sometimes it is hard to identify just what aspects of a computational model are important theoretically and what aspects may be important pragmatically but have no consequence for the theoretical claims. Fourth, the relation between mental and physiological representations is a long-standing puzzle. It figures importantly in philosophical debates about reductionism and materialism (see Articles 47 and 48: INTENTIONALITY and LEVELS OF EXPLANATION AND COGNITIVE ARCHITECTURES). Historically, a strength of cognitive science and cognitive psychology was the discovery that a great deal could be learned about particular mental operations without appeal to particular brain locations or physiological processes. However, new brain imaging technology is providing a wealth of information that will help us connect our knowledge about mental representation to the physiological.

Evaluating any representation system requires an ability to *stand outside* it and view it using the tools provided by another representational system. The primary basis for evaluating a representation system in any realm is its content. A good representation is one that captures the important distinctions in the represented world and does not introduce spurious distinctions. For example, when comparing mental representations, that of a novice versus that of an expert, we may conclude that the expert's mental representation includes distinctions which the novice's lacks. In evaluating theoretical representations about cognition, we may criticize one theory for failing to include information which is in fact part of the mental representation being modeled. Comparing systems, or identifying strengths and weaknesses, requires this ability to stand outside the particular representation being evaluated.

In addition to capturing the right information, a good representation should use an organization and format that allow appropriate processing. Organization and format are important, because they influence how processes can encode, compare, and transform information.

People are quite flexible in using multiple representation systems. This ability is not just important to us as cognitive scientists evaluating cognitive theories; it is a fundamental aspect of human cognition. People in their everyday roles use multiple representation systems, switch among them, and supplement the weaknesses of one with the strengths of another. This flexibility is the source of much power in human cognition.

Content

Much cognitive science investigates representation content. Linguistics asks about the content of mental representations of language: what does someone know which enables them to master any language? In artificial intelligence, content of both computational and mental representation is a central issue: what information does the system need to represent in order to suggest good investment portfolios? Sometimes AI research looks to human performance to identify what information might be sufficient to do the job; other times the goal is specifically to develop a computational model of mental representations.

Questions about representation content pervade every part of psychology, from physiological to social. Within cognitive psychology, content questions are of paramount interest in cognitive development, instructional psychology, and study of expert–novice differences. By contrast, much work from the information-processing tradition focuses on the processes whereby information is attended to, remembered, or used. Here, researchers usually use material such that the representations can be assumed (e.g., that an A will be represented as the alphabetic character, not as an unorganized collection of lines, a house, or something else entirely).

Questions about the content of mental representations can be answered more directly by empirical inquiry than can questions about format or organization. The content of many mental representations is accessible for conscious report, and here much information can be had for the asking: questionnaires or self-report methods investigate beliefs about everything from consumer preferences to knowledge of physics. However, many other representations are not reportable, such as those in perception, action, and language systems and those of infants or nonhumans. In these cases the content of mental representations must be inferred more indirectly from behavior, such as preferential looking, ability to learn a discrimination, or patterns of errors. Often converging evidence from several tasks is needed.

Cognitive theories differ in the content attributed to mental representations: do infants have an understanding of ordinal relations between numbers, or do they just understand that a 2 is different from a 3? Do high school students have an integrated theory of physics which is Aristotelian in character, or do they have a piecemeal collection of inconsistent beliefs? I believe that cognitive theories can make successful predictions about content in domains where people have an integrated system of knowledge with constraints among aspects of that knowledge. For example, the explanations that an elementary school child will offer of basic astronomical phenomena can be predicted from a cognitive theory about the sequence of mental models that children develop; and performance on grammaticality judgments is well predicted by the content theories of syntax which linguistics provides.

Organization

Organization addresses the way in which units within a representation system are related to each other. Organization is important, because much information can be represented indirectly, in the way local units are organized or related to each other. Two examples: much information in a connectionist network is carried by the way the nodes are linked together; for concepts organized in a set inclusion, or taxonomic, hierarchy, information about more general concepts (such as *bird*) can be applied to more specific concepts (such as *canary*).

A few types of information seem particularly important in organizing human knowledge and also in many computational systems. These include set inclusion hierarchies, part–whole hierarchies, causal schemas, and associations based on frequency and recency of use. Human knowledge is often organized consistently within a quite local domain, but rarely does this consistency hold up more globally.

There are two aspects of organization that have been investigated in studies of people: organization for access and organization for inference. Much research on memory is research about access, asking how knowledge is organized such that one particular slice of knowledge quickly triggers access to another. (See Article 17, MEMORY.) For example, autobiographical memory is not organized sequentially like a video recording but has strong thematic structure rather than a simple temporal ordering. Research on inference and reasoning addresses organization for inference. This concerns not the speed with which one piece of information is accessed from another, but rather the organization or structure of beliefs. (See Articles 21 and 8 REASONING and CONCEPTUAL ORGANIZATION.) For example, verifying the truth of "Marie Antoinette had a spleen" involves inferring that this property which is true of people in general is true of this queen; inferences about the spread of a new disease are influenced by causal schemas about infection.

Format

A representation format is the system of conventions for expressing the content. The format of a representation is an important matter and has a long history in computer science; any time you program anything, you must commit yourself to one format or another. What formats are the best, the most efficient, the easiest to use? The term *representation* is sometimes used in artificial intelligence to refer to *format*, particularly to the nature of individual, localized units, and to contrast with *organization*.

Questions about format entered cognitive science from computer science because of their role in computational modeling. While it is easy to tell what format a system is using from *the inside*, by reading or writing some code, it is hard to tell from *the outside*, by observing behavior. Differences in format are not linked directly to differences in behavior, and for people it is the behavior, not the code, to which we have access. Despite, or perhaps because of, the difficulty of telling one format from another on the basis of behavior, the debates about representation format have been some of the most polemic of any within cognitive science.

Learning to understand representation format is really best done by example, and we will consider three choices about format which have been particularly controversial: Are analog, as well as propositional, formats needed to model cognition? Is connectionist or symbolic representation format the correct one? Is knowledge represented as instances or as rules?

Analog versus propositional representation format formed the center of a heated debate in the 1970s (Kosslyn, 1981; Pylyshyn, 1981). Propositional format is a discrete *language of thought* from which propositions can be constructed. Analog format is continuous with spatial and quantitative relations in the represented world, which are preserved in the very format of the representing world. Both parties in the debate agreed that propositional systems could represent any information that analog systems could. The debate was whether two formats are used in human thought, or whether a propositional format is sufficient. Single-format proponents argued that, given the power of propositional representations, we should not posit two formats and so raise problems of parsimony and difficulties of translation between them. Much of the debate turned on the nature of imagery. A large number of studies showed that imagery has many perception-like, continuous properties of the sort that naturally fall out from analog representations. Thus, either a representation *really* is analog, or it is propositional but constructed so as to mimic analog properties. The choice between these was certainly to favor the format that easily generates the desired properties. (See Article 12, IMAGERY AND SPATIAL REPRESENTATION.) This debate was also a lesson that representation formats which seem profoundly different may in the limit be indistinguishable (unless processing assumptions are included).

The choice between connectionist and symbolic formats was the burning issue about format of the 1980s (Fodor and Pylyshyn, 1988; Smolensky, 1988). Symbolic representation is based on generative combination rules operating on a set of primitive and of composed elements, the symbols. Throughout the early decades of cognitive science, cognitive models were primarily symbolic, with major successes in problem solving, decision making, concept learning, and language. Connectionist computation is based on spreading activation among nodes in a network. Motivating topics for connectionist architectures were perception, recognition, classification, and the learning of these tasks from exposure to examples.

Initially it appeared that the choice of connectionist versus symbolic format made fundamentally different claims about the nature of human cognition. However, with the development of connectionism and of comparisons between formats, it became clear (1) that connectionist as well as symbolic format allowed extremely powerful computations; (2) that both formats allowed computations too powerful to model people and were in need of restriction, not extension; and (3) that because of their power, symbolic and connectionist systems could mimic each other, as well as people.

What, then, is the significance of a connectionist versus a symbolic representation format? I see at least two substantial issues. First, while it is formally possible to do a given computation with connectionist or with symbolic computation, it may be very much easier with one than with the other. Different sets of problems have been solved by researchers in each community, and different sets of well-developed tools and known solutions are available. Thus building on one or the other foundation can radically change the ease of constructing a model of a particular task or phenomenon. This is a pragmatic issue rather than a theoretical one, but an important one, nevertheless. Second, building and comparing models with very different formats may aid in analyzing what principles are truly responsible for success. It is easy to erroneously attribute a model's successes to one component or choice when the predictive work is really being done elsewhere, or to believe that one component or choice is the best when many alternatives would be equally successful. For example, pushing the connectionist framework to model parsing of sequentially presented sentences may help identify properties which a successful connectionist *or* symbolic implementation share and which may thus really be necessary for a successful solution. A third issue is sometimes advanced in favor of a connectionist format. Connectionist systems, with their nodes and links, are said to resemble the brain, with its neurons and synapses, and hence to yield *neurally plausible* models. For the most part this suggestion is a promissory note, but further work might provide the wherewithal to cash it.

Since the general contrast between connectionist and symbolic representation formats does not produce contrasting empirical predictions, it is moot to argue whether mental representations really are connectionist or symbolic: we lack tests that would distinguish. Of course, two individual models, one connectionist and one symbolic, might make different predictions and be empirically distinguishable as better or worse models of some aspect of human cognition. The difference in prediction would not result directly from the difference in representation format, but rather from the particular processing and content claims together.

A third, current debate about representation may also prove to be just about format and the pragmatics of modeling, not about testable differences in mental representation. In a variety of domains, from concept learning to language, theories differ in whether they assume that knowledge is represented as rules (Nosofsky et al., 1994; Pinker, 1991) or as instances (Plunkett and Marchman, 1991; Redington and Chater, 1996). While these seem to be very different and to be empirically distinguishable, there is a notorious degree of mutual mimicry here too (Barsalou, 1990). Initially, rules were often considered to be quite abstract and exceptionless, and learning exceptions of any sort was taken to be a problem for rule models. However, many rule models now allow multiple, partially accurate, conflicting rules of varying specificity. Initially, instance models assumed specific, unselective, and veridical encoding of instances. However, many instance models now strategically allow selective and partial encoding of instances. Thus, as flexibility is added to both sorts of representation format, the information available in a *rule* and that available in an *instance* may become indistinguishable. We have learned something about format, however. For many tasks, a good model must allow variation in the specificity of its representations; using a format that allows only the most general (strict rules) or only the most specific (simple instances) representations won't do the job.

The moral so far is that many distinctions in representation format may be important as pragmatic choices about building computational models. However, many such differences in computational format do not imply any behavioral differences. When this is so, choice of format should be treated as part of only the *computational* representation, not the *theoretical* one.

Finally, I would like to suggest an example of one format distinction that may capture empirical, identifiable differences in mental representation: symmetrical versus asymmetrical access. Representational units may be organized symmetrically or asymmetrically. When information is organized asymmetrically, a fixed subset of information acts as a key to access the remaining information. Intuitively, this seems appropriate when one set of information is reliably available first, and this predicts some other information. This might characterize classification when a system needs to recognize category membership from a fixed set of properties or symptoms, or to select an action in a problem-solving sequence. A condition–action production rule, or procedural knowledge specifying what to do in specified circumstances, is an example of asymmetrical representation within the symbolic approach. Information specified in the condition (e.g., that the game is tic-tac-toe, that it is your move, that the grid is blank) is used to access information in the action part of the production (e.g., that you cannot be beaten, that you should place an X in the middle). However, information from the action part cannot activate or access information in the condition. Connectionist networks, where the input nodes represent one set of information (e.g., the state of the board) and the output conditions represent another kind of information (e.g., moves to make), are also examples of asymmetrical access.

When information is organized symmetrically, any subset of the information may be used to access the remaining information. Intuitively, this seems appropriate when there is a set of associated, interpredictive data, different subsets of which may be available or needed on different occasions. This might characterize knowledge about concepts such as *dog, bike, Yuppie,* or *wedding.* Declarative representation in symbolic systems is an example of symmetrical representation. Auto-correlational networks, in which the input nodes and the output nodes represent the identical set of distinctions, are an example of symmetrical representation in a connectionist framework, because any element can contribute to the activation of any other. Such representation formats are often used for pattern recognition.

The distinction of symmetry is an important one, because different, complementary tasks are easily accomplished with the different formats. Asymmetrical, if–then representations, with their more rigid access keys, are particularly felicitous for encoding information for processes which are efficient and rigid. Symmetrical, schema-type representations, with their more flexible access patterns, are particularly felicitous when information is flexibly used across contexts, when the task as well as the information available from the world can vary.

Does this internal difference between symmetrical and asymmetrical representation produce any observable difference in behavior? This difference in representation will not directly and inevitably lead to different behavior in all tasks. For example, a system using asymmetrical, rule-type representations can build two rules, one predicting that if something has feathers, it has a beak, the other predicting that if something has a beak, it has feathers. Conversely, a system built from symmetrical representations may still have asymmetries of association if *absent-minded* is more strongly associated with *professor* than *professor* is with *absent-minded.* Further, one system may include both types of representation (e.g., procedural and declarative) or mixed representation (e.g., a connectionist system in which input and output nodes represent an overlapping but not identical set of distinctions). Nevertheless, paired with reasonable processing assumptions, the two types of representations are likely to produce different behavior, particularly with respect to learning, transfer, and new uses of old information.

The general distinction between symmetrical and asymmetrical representation formats is not a standard one in cognitive science. Discussing it here serves three purposes. First, it is a difference in representation format which is general enough to allow comparison and analysis across very different systems. Often issues regarding representation are couched in terms specific to one framework, such that it is hard to tell whether the issue arises in other formats. For example, symbolic systems distinguish between productions, which typically encode if–then, procedural knowledge used to generate actions, and declarative knowledge, which typically encodes networks of knowledge about the world usable for many purposes but not directly linked to actions. This distinction within symbolic systems between productions and declarative knowledge is a special case of the contrast between asymmetrical and symmetrical representations. But, described in these terms, one would not notice that the same symmetric–asymmetric distinction is found in connectionist systems. Second, it illustrates the complexity of going from a difference among representational formats – even a very fundamental distinction – to a difference in behavior. Strengths or weaknesses in representation format can often be compensated for by processing, either to build or to use representations in flexible ways. Third, even though a representational distinction does not project directly to a difference in behavior, different kinds of information uses are most easily supported by different representational formats. Thus, in conjunction with modest processing claims, choice of representation format can have empirical consequences, as well as make modeling much harder or much easier.

Summary

Representation is one of the most central concepts in cognitive science: there is no cognition without representation, and no cognitive science either. Controversies about content and format of representation pervade cognitive science. Many are resolved empirically by investigation of what people do, others pragmatically by what works for a computational system. But even the unresolvable can advance the field by providing new ways of thinking about cognition and new tools for theorizing.

Analysis of representations is complicated by the fact that the representing world of one representational system (say, a picture) can be part of the represented world in another system (say, a mental image). Understanding this layered structure is important to understanding the complexities of everyday cognition, where people traffic in multiple external and mental representation systems. It is also important for understanding the activities we engage in as cognitive scientists, such as modeling mental representations with the computational representations in a simulation. The ability to work in and coordinate multiple representation systems is the root of much power in human cognition: amazingly, people can talk about what they see, sketch what is described, and invent new types of external representation systems from writing to computational modeling itself.

References and recommended reading

*Barsalou, L. W. 1990: On the indistinguishability of exemplar memory and abstraction in category representation. In T. K. Srull and R. S. Wyer, Jr. (eds), *Advances in Social Cognition*, vol. 3, Hillsdale, NJ: Erlbaum, 61–88.

Fodor, J. A. and Pylyshyn, Z. W. 1988: Connectionism and cognitive architecture: a critical analysis. Special issue on connectionism and symbol systems. *Cognition*, 28, 3–71.

Kosslyn, S. M. 1981: The medium and the message in mental imagery: a theory. *Psychological Review*, 88, 46–66.

Nosofsky, R. M., Palmeri, R. J. and McKinley, S. C. 1994: Rule-plus-exception model of classification learning. *Psychological Review*, 101, 53–79.

Palmer, S. E. 1978: Fundamental aspects of cognitive representation. In E. Rosch and B. B. Lloyd (eds), *Cognition and Categorization*, Hillsdale, NJ: Erlbaum, 259–303.

Pinker, S. 1991: Rules of language. *Science*, 253, 530–5.

Plunkett, K. and Marchman, V. 1991: U-shaped learning and frequency effects in a multilayered perceptron: implications for child language acquisition. *Cognition*, 38, 43–102.

*Pylyshyn, Z. W. 1981: The imagery debate: analog media versus tacit knowledge. *Psychological Review*, 88, 16–45.

Redington, M. and Chater, N. 1996: Transfer in artificial grammar learning: a reevaluation. *Journal of Experimental Psychology: General*, 125, 123–38.

Smolensky, P. 1988: On the proper treatment of connectionism. *Behavioral and Brain Sciences*, 11, 1–74.

52

Rules

TERENCE HORGAN AND JOHN TIENSON

Contemporary cognitive science has two principal branches: the classical computational approach (sometimes called *classicism*) and connectionism. Rules are fundamental to theorizing about the basic units of processing in both classicism and connectionism. But we will be concerned primarily with rules that apply to representations and that determine transitions from one cognitive/representational state to the next. Such rules are fundamental to classicism, since, according to classicism, cognitive processes simply are rule-governed cognitive state transitions. Rules that apply to representations are not a built-in feature of connectionist architecture, so the role of such rules in connectionism is less clear. We will argue that they are neither necessary nor (in general) desirable in connectionism. We discuss the status of rules first in classicism, then in connectionism.

Classical cognitive science

The interdisciplinary field known as cognitive science grew up with the development of the modern digital computer. Researchers from such diverse fields as computer science, neurophysiology, psychology, linguistics, and philosophy were brought together, not just by the understanding that the computer is a powerful tool for studying cognition, but, more importantly, by the conviction that the digital computer is the best model of cognition in general, and consequently of human cognition in particular. The picture of the mind as a computer dominated both theory and methodology in cognitive science and artificial intelligence for over a quarter of a century and is deservedly called the *classical* view in cognitive science.

The distinctive feature of the modern digital computer is that its processes are determined by a program – a system of rules that determine transitions from one state to the next. Computers process information represented in *data structures* or symbols, and it is these symbols to which the rules refer (see Article 26, ARTIFICIAL INTELLIGENCE). Thus, the classical approach gives a central and fundamental role to rules – specifically, rules for the manipulation and transformation of symbol structures. In the case of cognition, the symbol structures are the cognitive system's representations; their representational contents are the contents of thought. Thus, classicism is often called the *rules and representations* conception of mentality.

It is clear that there are at least two *levels* at which the processes of a computer or of a brain can be described: the physical, or neurobiological, level and the level of informational or mental content. (It is typical to refer to the level of content as a *higher* level; similarly, biology is a higher level of description than chemistry, chemistry higher than physics.) According to classicism, in order to tie these two levels together, and thus to understand how mental or informational processes can be realized in a physical medium, it is necessary to consider a third level of description between the top, cognitive level and the bottom, physical level. (In describing these three levels, we follow an influential account provided by David Marr (1982).)

At the top level one specifies a function or mapping from input to output, corresponding to the cognitive capacity being modeled. In modeling human vision, for example, the system takes representations of retinal stimuli as input and yields representations of objects in three-dimensional space as output. At this level one also justifies this input–output function in terms of the modeling task. One desideratum in modeling human vision, for example, would be for the system to be subject to the same visual illusions that human subjects experience, rather than producing correct representations of objects in three-dimensional space in all cases.

At the middle level one specifies a system of representations for the input, a system of representations for the output, and a system of rules – that is, a program or algorithm – for computing the output from the input. At the lowest level, the actual physical processes by which the computation is carried out are determined. (In practice, there are often several intervening levels between the top and middle levels and between the middle and bottom levels. A flow chart, for instance, amounts to a level of description below the top which specifies the steps by which the cognitive function is computed. Flow charts are elaborated by adding additional boxes between or within boxes (yielding a lower level), eventually *bottoming out* at the middle level. Impressive cognitive functions are thus seen as ultimately composed of numerous cognitive *baby steps.*)

Any variety of cognitive science must specify what cognitive transitions take place in performing the cognitive tasks it aims to model. Also, all varieties agree that cognitive systems are physically realized. And it is very natural (although perhaps not necessary for any possible flavor of cognitive science) to hold that there is a middle level of description between the cognitive and physical levels, and that this middle level is essentially a level of mathematical design. What is distinctive of classicism is the idea that the middle level involves computation: that is, rule-governed symbol manipulation. Thus, a basic contention of classicism is that the appropriate mathematical framework for cognition is the theory of computation. As we will see, the classical picture has significant (and possibly limiting) implications for the top level.

We can better understand the role of the middle level in showing how cognitive processes can be physically realized by considering the relationship between the top and middle levels of description as understood by classical cognitive science. When information processing is accomplished by computation, the rules executed by the computational system can be viewed in two different ways. From one perspective the rules refer explicitly to the task domain. For instance, one can formulate a set of rules for assigning classes to classrooms on a university campus, taking into account such constraints as the location of each classroom, the number of seats in each classroom, the number of seats requested for each class, and the time of day each class will be offered. These rules can be made precise, completely explicit, and exceptionless, so that a human being who had no understanding of the task domain could still determine room assignments by following the rules. To do this, one would need only the ability to follow simple instructions and perform elementary logical and mathematical operations. Terms for classes, rooms, etc. could be replaced by nonsense syllables or schematic letters. When rules about the task domain have been made precise, explicit, and exceptionless in this way, they can be viewed from another perspective, not as having representational content, but as purely formal symbol-manipulating rules that determine processing by reference to nothing other than the syntactic form of the representations. The elementary operations specified in such rules can then be mirrored by simple physical devices. Hence, such rules can be put in the form of a program that will run on a conventional computer. Both the rules constituting the program and the representations to which they apply thus have two guises, corresponding to the middle and top levels of description. The rules are purely formal, applying to representations solely on the basis of their structural-syntactic features; but the representations, and hence the rules, are also appropriately interpretable as being about objects and facts in the problem domain – classes, classrooms, class times, and so on.

We call such rules *programmable, representation-level rules*, although the designation may not be entirely happy because of the dual nature of the rules. In a conventional computer, and perhaps also in natural cognizers as classical cognitive science would understand them, the *mirroring* of representation-level processes in the physical device is quite complex. Thus representation-level rules are quite different from the rules that determine the basic physical processes of the device.

Classicism maintains that the rules that determine cognitive processes in (natural and artificial) cognitive systems also have these two guises. On the one hand, they must be purely formal, referring solely to symbolic states and processes that can be mirrored in the causal states and processes of a physical system. On the other hand, since the symbolic states are supposed to be implementations of *cognitive* states, the rules must also be interpretable as cognitive-level laws governing transitions among mental states – that is, as psychological laws. The two guises of these representation-level rules are the reason why rule-governed processing of formal/syntactic states directly implements cognitive state transitions; psychological laws at the cognitive level of description are directly mirrored by corresponding formal/syntactic rules at the mathematical level of description. And these, because of their formal nature, can be physically realized.

Cognitive processes are thus understood as the result of the system's conformity to physically implemented representation-level rules. A program that is intended to model a human cognitive capacity – say, visual perception or parsing sentences or getting about in crowded shopping areas – is a hypothesis about the states and processes that occur when a person exercises that capacity. The explanation of the capacity itself, according to classicism, is that a person has the capacity to perform that cognitive task by virtue of having a (possibly hard-wired) system of representation-level rules for doing so.

Many domains are not amenable to rules that will yield a correct result in every instance. In some cases, no such rules are possible; in other cases, rules of this kind are not known, or they are too complex and unwieldy to be of any use (or both). Computers have been usefully employed for information processing in such domains, because computers can be provided with programs for searching for a solution in an efficient manner – that is, *heuristic* programs, programs that do not guarantee a correct or optimal outcome in every case, but that employ reasonable strategies that will yield solutions in a large range of cases. Chess-playing programs, for instance, have heuristic rules to eliminate most possible moves and consider only the most plausible ones. The best move will sometimes be missed, but a good program will make a good move in almost every situation. One thing that can make one chess-playing program better than another is better heuristics (see Article 44, HEURISTICS AND SATISFICING).

Much research in classicism has been on heuristic programs, and for the most part, the representation-level rules posited by classicism are heuristic rules. It is a reasonable contention, within the classical paradigm, that most of human cognition is determined by heuristic programs.

The classical contention that cognition is mathematically realized by an algorithm involving symbolic representations – that is, by rule-governed symbol manipulation – involves three basic assumptions:

(1) Intelligent cognition employs structurally complex mental representations.
(2) Cognitive processing is sensitive to the structure of these representations (and thereby to their content).
(3) Cognitive processing conforms to precise, exceptionless rules, statable over the representations themselves and articulable in the format of a computer program.

Classicists also maintain that the systematic semantic coherence of human thought rests largely and essentially on the manipulation of representations that encode propositional content via language-like *syntactic* structure. So classicism makes an additional foundational assumption that is more specific than (1):

(4) Many mental representations have syntactic structure.

(Classicists can, and often do, allow that *some* of the computational processes of human cognition operate on representations without language-like syntactic structure (e.g., imagistic representations).)

Assumptions (1)–(4) pertain to the middle, mathematical level of description as conceived by classicism. But to maintain, as classicism does, that cognitive transitions are implemented by an algorithm that computes those transitions is to presuppose something about the top level too, something so basic that its status as an assumption is often not even noticed. For cognitive transitions can be computed by a physical system only if

(5) Human cognitive transitions conform to a *tractably computable* cognitive transition function.

A function is *computable* in the mathematical sense if there is (in principle) a means of computing, for any given input, the output of the function in a finite number of elementary operations. To say that a function is *tractably* computable adds that it is possible for these operations to be carried out by an appropriate kind of physical device. It is clear that classical cognitive science is committed to *tractable* computability, not just computability in the mathematical sense, since cognitive science is concerned with how human cognitive systems work. But even computability in the mathematical sense is a limitation that one would probably not think of imposing on cognitive processes unless one already thought that they were determined by cognitive-level rules.

We close this section by mentioning two things to which classicism is sometimes taken to be committed, but to which it is not in fact committed. Classicism is not committed to cognitive transitions that are deterministic. Classicists can, and often do, provide for nondeterministic cognitive transitions, by building some kind of randomizing element into the underlying algorithms – such as making the act performed depend upon the results of some entry in a random number table.

Also, classicism does not assert that the rules of cognitive processing must be represented by (or within) the cognitive system itself. Although programs are explicitly represented as stored *data structures* in the ubiquitous general-purpose computer, stored programs are not an essential feature of the classical point of view; a classical system can conform to representation-level rules simply because it is hard-wired to do so. It is plausible from the classical point of view, for example, to regard some innate processes as hard-wired. (Thus we have characterized the rules of classical cognitive science as programm*able*.)

Connectionist cognitive science

By the early 1980s, certain kinds of difficulties were arising quite persistently and quite systematically within classicism. Examination of these difficulties makes it seem likely that they are not mere temporary setbacks, but difficulties in principle, stemming from fundamental assumptions of the classical framework. The difficulties centered largely around what has come to be called the *frame problem*. In its original form, the frame problem was concerned with the task of updating one's system of beliefs in light of newly acquired information. If you learn that Mary has left the room, you will stop believing that Mary is in the room, and also stop believing, for example, that someone is sitting on the sofa and that there are four people in the room. You will also make some obvious inferences from the new information: for example, that the clothes Mary was wearing and the package she was carrying are no longer in the room. But most of your beliefs will not be affected by the new information. Human beings adjust their beliefs in response to new information so naturally that it is surprising to find that it is a problem. But it has proved quite difficult for classical cognitive science.

For a belief system of any size, obviously, it is not possible to examine each of the system's beliefs to see if it needs to be changed. Thus, Jerry Fodor (1983) characterizes the frame problem as "the problem of putting a *frame* around the set of beliefs that *may* need to be revised in light of specified newly available information" (pp. 112–13, emphasis added). Seen this way, the problem is fundamentally one of relevance: to provide an effective, general procedure that will determine the beliefs to which any particular new belief is at all *relevant*. Those are the beliefs that get *framed*. Which of these relevant old beliefs actually need to be revised in a given case is then a further question.

There are several other cognitive activities that pose similar problems of relevance: belief fixation (arriving at a new belief on the basis of diverse and perhaps conflicting evidence), retrieving from memory information that is relevant to solving a current problem or carrying out a current task, and forward-looking tasks such as deciding what to do next, deciding what is morally permissible or obligatory, and making plans.

Apparently, the classical approach in all these areas must be, as Fodor suggests, to attempt to put a *frame* around what is relevant: that is, to try to introduce rules which determine, for any given item of information, what is relevant to that item of information and what is not. Call such solutions to problems of relevance "*frame solutions.*"

Frame solutions appear to be doomed to failure. Human cognitive systems are open-ended. There is no limit to the things a human being can represent. And anything one can represent is potentially relevant to anything else one can represent. Relevance depends upon the question, topic, or problem at hand – in a word, upon context. For virtually any pair of items of information you pick, there will be some context in which one is relevant to the other. (It has been suggested that the price of tea in India is not relevant to the question of whether Fred has had breakfast by 8:30. The obvious reply is that it is relevant "if Fred happens to be heavily invested in Indian tea and the market has just fallen savagely" (Copeland, 1993, p. 115).)

Our suggestion, then, is that there are no such *relevance frames* in human cognition. But what other kind of solution is possible within the classical framework? Cognitive science lacks the slightest clue as to how representation-level rules could update memory appropriately or find relevant information efficiently for open-ended belief systems of the kind possessed by humans. Indeed, it seems entirely likely that it can't be done by systems of rules at all. As Fodor (one of the staunchest defenders of classicism) has written:

> The problem . . . is to get the structure of the entire belief system to bear on individual occasions of belief fixation. We have, to put it bluntly, no computational formalisms that show us how to do this, and we have no idea how such formalisms might be developed . . . In this respect, cognitive science hasn't even *started*: we are literally no farther advanced than we were in the darkest days of behaviorism. (Fodor, 1983, pp. 128–9)

The reemergence of connectionism in the 1980s was in large part a response to the problems in classical cognitive science. As problems persisted, many researchers looked elsewhere for a better prospect of positive results, and the only other game in town was parallel distributed processing – connectionism. But this raises a foundational question that has received surprisingly little discussion: Does connectionism have features (fundamentally different from those of classicism) that suggest that it can make progress, not just on other problems, but on the very problems that slowed progress in classical cognitive science?

Classical systems, by their very nature, involve both representation-level rule execution and representations with language-like syntactic structure. Neither of these features is required by the inherent nature of connectionist systems. Thus, syntactic structure and cognitive-level rules are two places to look for fundamental differences between connectionism and classicism.

Certain kinds of rules are very prominent in connectionist theory, but they are not representation-level rules. Activation updating within individual nodes and local activation passing from one node to another occur in accordance with rules. (In current connectionist modeling, these are programmable rules. This is why connectionist networks can be simulated with standard computers, as they are in virtually all connectionist modeling. But it is not part of connectionist theory that node-level rules must be programmable.) However, the processing that takes place locally between nodes and within individual nodes is not in general representational. Not all local node activations in a network model have representational content, and in some models the activation of a single node never has representational content – all representations, even the most basic or atomic, consist of activation patterns over a whole set of nodes.

Thus, the fact that individual nodes are rule-governed leaves open the question of whether the processes that representations undergo in connectionist models must conform to rules.

There is an important sense in which even node-governing rules are absent from connectionist systems: networks do not contain explicitly represented rules of any kind. It is sometimes thought that the absence of explicit rules constitutes a watershed difference between connectionism and classicism. But this is a mistake. As explained earlier, the rules posited by classicism can be hard-wired into a computational system rather than being encoded as representations. (Indeed, at least some rules executed by a classical computational system *must* be hard-wired. The node-governing activation-update rules of a connectionist network are analogous to basic hard-wired rules of classical systems.)

It is more common to focus on lack of syntactic structure as an alleged difference between connectionism and classicism (see, e.g., Churchland 1989, 1995, who sees this as a virtue, and Fodor and Pylyshyn, 1988, who see it as a deficiency). Such authors claim that the activation vectors that constitute representations in connectionist systems lack syntactic structure. (A vector is essentially an ordered *n*-tuple of items; an activation vector is an ordered *n*-tuple of activation values of specific nodes in a neural network.) This means that the processing of representations in connectionist systems is fundamentally different from the largely syntax-driven processing of representations in classical systems. These writers do not raise the question of whether connectionist processing conforms to programmable rules; implicitly, at least, they evidently suppose that it does, but they would suppose that the rules at work in connectionist systems apply to some kind of nonsyntactic formal/mathematical structures. Churchland describes processing as effecting *vector-to-vector transformations*, and he suggests that such transformations conform to rules that are sensitive to the vectorial structure of the representations. This approach, which we call *nonsentential computationalism*, repudiates assumption (4) of the five foundational assumptions of classicism mentioned earlier, while retaining the other four.

Nonsentential computationalism is not obviously a correct interpretation of all extant connectionist models. On the contrary, there are certain models that are naturally interpreted as involving both (1) representations that have syntactic structure and (2) processing that is sensitive to this structure. (Two important kinds of such models, due respectively to Jordan Pollack (1990) and Paul Smolensky (1990), are discussed in Horgan and Tienson, 1996, chs 4 and 5). Nor is nonsentential computationalism obviously the most natural or most attractive foundational framework for connectionist cognitive science. One serious reason for doubt is that nonsentential computationalism in effect offers just a seriously limited variant of classicism. It is a variant because it continues to hold that cognition is implemented by processes that conform to programmable rules (so it can be no more powerful than classical cognitive science). It is limited because it eschews an extremely powerful way of introducing semantic coherence into the computational manipulation of representations: the syntactic encoding of propositional information.

A fan of nonsentential computationalism might be expected to reply that connectionist models get by without any explicit stored memories, with lots of information *in the weights*, and that networks are not programmed. However, to the extent that connectionist processing conforms to representation-level rules, we could get these same features in a classical system in which (1) all the rules are hard-wired rather than explicitly represented, and (2) lots of information is implicitly accommodated in the (hard-wired) rules rather than being explicitly stored in memory.

We remarked earlier that the inherent nature of connectionist systems does not guarantee either of two features that are intrinsic to classicism: representation-level rule execution and representations with syntactic structure. So another approach to cognition that might be joined to connectionism is to deny that human cognitive processes conform to representation-level rules. (Such an approach need not repudiate syntactic structure.)

But is it possible for a connectionist system that employs representations to fail to conform to rules that refer to these representations? Indeed it is. In the first place, it is not necessary for the temporal evolution of a connectionist network to be tractably computable. The natural mathematical framework for describing networks is the theory of dynamical systems (Horgan and Tienson, 1996, ch. 4), and the temporal evolution of a dynamical system need not be tractably computable. And if the temporal evolution of a network is not tractably computable, there is no reason to believe that the cognitive evolution of the cognitive system which the network realizes will be tractably computable via representation-level rules.

But in the second place, it is important to understand that a connectionist model may not conform to representation–manipulation rules even if it does conform to sub-representational programmable rules that govern individual nodes and local inter-node transactions. (As noted already, most current connectionist models conform to programmable node-governing rules; the networks are simulated on standard computers.) As a prelude to explaining why not, we begin with a preliminary point that is important and not widely recognized. It is possible for a connectionist system to be *nondeterministic* at the representational level of description, even if the system is fully deterministic at the sub-representational level of node activation updating and local inter-node activation passing. This is because the same connectionist representation can be realized by many different sub-representational states of the system, and the representation-level outcome of processing can depend upon the specific way that a representational state is realized sub-representationally.

One source of multiple realizability of representations is different degrees of activation of nodes. The realization of a particular cognitive state, say A, might consist in each of a given set of nodes being active to at least a certain degree, say 0.8. Then some realizations of this cognitive state will have node N more highly activated than node M; others will have node M more highly activated. It can then happen that from some activation states that realize A the system goes into activation states that realize cognitive state B, while from others it goes into activation states that realize a different cognitive state C; so there will be no way of knowing the cognitive-level outcome just from knowing its initial total cognitive state. Being nondeterministic at the cognitive level can be a valuable asset in many kinds of competitive activities, such as playing poker and fleeing for one's life. (Note that no randomizing dice-throw rules are involved at any level of description, either representational or sub-representational, as would be required to make a classical system nondeterministic.)

This preliminary point establishes an important moral: namely, that key features of a connectionist system at the sub-representational level of description need not *transmit upward* to higher levels of description, because inter-level realization relations can work in ways that block such transmission. As we next explain, tractable computability of state transitions can also fail to transmit upward in connectionist systems, so that a system can fail to conform to programmable, representation-level rules, even though it conforms to programmable sub-representational rules.

Given that the transitions of the underlying network are tractably computable, one might think that the cognitive transitions realized in that network could be computed like this. Starting from a cognitive state, (1) select an activation state that realizes this cognitive state; (2) compute the network's transitions from this activation state through subsequent activation states; and (3) for each subsequent activation state, compute the cognitive state (if any) realized by that state.

Although the assumption that the transitions of the network are tractably computable guarantees (2), there is no guarantee that step (3) – or even step (1) – will be possible. The function from activation states to cognitive states need not be tractably computable. It is possible, for example, that the simplest, most compact way to specify that function might be via an enormous (possibly infinite) *list* that pairs specific total activation states with specific total cognitive states – a list far too long to be written using all the matter in the universe, let alone to constitute a set of programmable rules.

If the cognitive transitions implemented by a network are not computable in the way just suggested, they need not be tractably computable in any other way either. Thus, one should not infer from the fact that a network's activation-state transitions are tractably computable that it implements a cognitive transition function that is tractably computable. Nor should one suppose that an algorithm for computing the network's behavior over time automatically determines an algorithm for computing its cognitive transitions.

The possibility that the realizing function may not be tractably computable is not a mere abstract possibility. Certain connectionist learning algorithms allow models to select their own representations (Pollack, 1990; Berg, 1992; discussed in Horgan and Tienson, 1996, ch. 4). Representations are modified, along with weights, as learning progresses; this allows for more efficient schemes of representation, with weights and representations ending up *made for each other*. It is easy to suppose that complex cognitive systems that worked in this way (as natural cognitive systems apparently do) would have very complex, rich, subtle realization relations that are not tractably computable.

Given that it is possible for a connectionist cognitive system to fail to conform to programmable, representation-level rules, several questions arise. First, if cognitive transitions are not effected by executing such rules, how are they brought about? Second, are there reasons to think that it is desirable for a system not to be rule-describable? Third, if a system does not conform to rules at the cognitive level, can it be coherent enough and systematic enough to be called a cognitive system at all?

A very natural way to think about cognitive transitions in connectionist systems is in terms of content-appropriate *cognitive forces*. Beliefs and desires work together to generate certain forces that tend to push the cognitive system toward output states that would result in particular actions. But those forces can be overcome by stronger forces pushing in different, incompatible directions. A single clue in a mystery might point to the guilt of some suspects and at the same time tend to clear certain other suspects to varying degrees. Thinking of the clue produces forces that tend to activate some possible beliefs about whodunit and inhibit others. The interaction of cognitive forces in a cognitive system can be very complex. Forces can compete, in that they tend toward incompatible cognitive states, or they can cooperate, tending toward the same or similar outcomes. There can be a large number of competing and cooperating factors at work in a system at once. Connectionist models that perform multiple, simultaneous, soft constraint satisfaction provide suggestive simple models of the interaction of cognitive forces.

In a connectionist network the interaction of cognitive forces is physically implemented by spreading activation. But when a representation is realized by activation of a large number of nodes, the cognitive forces generated by the overall representation are distinct from the local physical forces produced by the individual nodes implementing the representation. (The individual nodes need not be similar to one another in the kinds of weighted connections they have to other nodes or representations, so they might have different causal roles from one another.)

The possible value of such a picture for dealing with relevance phenomena – phenomena associated with the frame problem in classicism – should be evident. Any two cognitive states that put out forces tending to activate other cognitive states will be capable of interacting causally when co-present in a cognitive system. And any two or more states that are relevant to the same problem will interact with respect to that problem – at least to the extent of tending to move the system in the direction of conflicting or compatible solutions. Thus, certain kinds of content-relevant interaction are *automatic* for systems that have states with content relevant cognitive forces. Potential interactions do not have to be anticipated in advance in terms of form or content – a key difference from classicist systems, in which the operative representation-level rules must determine all such outcomes. Furthermore, forces interact with one another in a manner appropriate not only to the contents of all the cognitive states currently activated in the system, but also to much nonactivated information that is implicit in the system's structure – *in the weights*, as connectionists like to say.

In natural cognizers, there are many systematic patterns by which cognitive forces are generated (many of which correspond to the generalizations of commonsense psychology). Appropriately related beliefs and desires conspire to produce forces that tend toward certain choices. (It is arguable that this pattern depends upon syntactic or syntax-like structure of the belief and desire states.) Repeated observation of a pattern of events results in cognitive forces that tend to produce expectations of similar patterns. In such cases there is a causal tendency to make such choices, have such expectations, etc. But these are *defeasible* causal tendencies: that is, it is always possible that the tendency will be overridden by a stronger force or combination of forces. Thus, although there are generalizations about the cognitive transitions that correspond to these patterns of cognitive forces, there are no programmable rules corresponding to these generalizations, because they have exceptions.

Furthermore, these generalizations cannot be *refined* into programmable rules by specifying the possible exceptions. Because of the potential relevance of anything to anything, it is not possible to spell out all of the exceptions in a machine-determinable way (Horgan and Tienson, 1996, ch. 6). The defeasibility of causal tendencies poses a deep problem for classical cognitive science, since all potential exceptions need to be specified in just such a way: they need to be explicitly covered, for instance, by *unless* clauses within representation-level rules. In the cognitive forces picture, *nothing* has to be done to deal with exceptions. They arise naturally as a feature of the architecture.

Although cognitive state transitions do not conform to representation-level rules, according to the connectionist-inspired conception of cognition that we are suggesting, systematic patterns among cognitive processes (such as those mentioned above) do conform to psychological laws of a certain kind: *soft* laws, as we call them. Soft psychological laws have ineliminable *ceteris paribus* (all else equal) clauses, allowing for exceptions that are not specified in the laws themselves. It is important that the exceptions allowed by such laws include a virtually endless range of exceptions based in the psychology of the cognizer. They are not merely *lower-level* exceptions resulting from factors like physical breakdown (e.g., having a stroke) or external physical interference (e.g., being hit by a bus). It is also important that the psychology-level exceptions are not mistakes or errors, but the result of the proper functioning of the cognitive system. We believe that soft laws characterize the kind of consistency and systematicity that natural cognizers actually have. They support explanation and prediction in wide ranges of cases. And they are the kind of generalization sought and found in many branches of cognitive psychology (Horgan and Tienson, 1996, chs 7 and 8).

It remains to be seen whether this nonclassical view of the mind will gain empirical support from ongoing work in cognitive science. Meanwhile, however, it is well to keep in mind that connectionist modeling does not presuppose or imply that human cognition conforms to programmable, representation-level rules, and that there are serious reasons to believe that human cognitive capacities essentially outstrip the capacities of systems that execute representation-level rules.

References and recommended reading

Berg, G. 1992: A connectionist parser with recursive sentence structure and lexical disambiguation. In *AAAI-92: Proceedings of the Tenth National Conference on Artificial Intelligence.*

Churchland, P. M. 1989: *A Neurocomputational Perspective: The Nature of Mind and the Structure of Science.* Cambridge, Mass.: MIT Press.

*—— 1995: *The Engine of Reason, the Seat of the Soul: A Philosophical Journey into the Brain.* Cambridge, Mass.: MIT Press.

*Copeland, J. 1993: *Artificial Intelligence: A Philosophical Introduction.* Oxford: Blackwell.

*Fodor, J. 1983: *The Modularity of Mind: An Essay on Faculty Psychology.* Cambridge, Mass.: MIT Press.

Fodor, J. and Pylyshyn, Z. 1988: Connectionism and cognitive architecture: a critical analysis. In S. Pinker and J. Mehler (eds), *Cognition and Symbols*, Cambridge, Mass.: MIT Press, 3–71.

Horgan, T. and Tienson, J. 1996: *Connectionism and the Philosophy of Psychology.* Cambridge, Mass.: MIT Press.

Marr, D. 1982: *Vision.* New York: Freeman.

Pollack, J. 1990: Recursive distributed representations. *Artificial Intelligence*, 46, 77–105.

Smolensky, P. 1990: Tensor product variable binding and the representation of symbolic structures in connectionist systems. *Artificial Intelligence*, 46, 159–216. Repr. in G. Hinton (ed.), *Connectionist Symbol Processing*, Cambridge, Mass.: MIT Press, 1991.

53

Stage theories refuted

DONALD G. MACKAY

This chapter examines the stages of processing meta-theory (SPM) that has guided construction of theories in psychology during the past 350 years, from philosopher René Descartes in seventeenth-century France to neuropsychologists Carl Wernicke and Paul Broca in nineteenth-century Europe to psychologists Dominic Massaro and Alan Baddeley in late twentieth-century America and Britain. The most basic SPM assumptions are that processing and storage of information take place within a finite number of autonomous modules or stages, and that some stages are sequentially ordered with respect to others. Flowcharts typically summarize these assumptions, as in figure 53.1. The traditional stages of processing for verbal information are comprehension, storage, retrieval, and production. SPM flowcharts can add new stages and can alter old labels to represent new types of information processing, but whatever the labels, the stages must be finite in number, distinct, independent, and sequentially ordered between input and output. For example, the storage stage in figure 53.1 begins only after comprehension is complete, and production begins only after storage and retrieval are complete.

To illustrate SPM in a recent manifestation, I will describe a 1993 theory of memory and some of its seemingly minor crises related to its SPM assumptions. I then review a range of findings that directly contradict SPM and suggest a new shape for future theories in psychology.

Minor crises in current theories of memory

The minor crises concern relations between memory for lists and memory for sentences. By postulating two autonomous systems for processing and storing lists versus sentences, current multi-store theories of memory illustrate the SPM assumption that information processing and storage take place within autonomous modules, or stages. For example, Alan Baddeley, a leading British researcher investigating the psychology of memory, postulates a memory system known as the "phonological loop," which processes and stores word lists in *raw* phonological form for short periods of time and is separate and distinct from the system for processing and storing the syntax and meaning of sentences (the central executive).

Baddeley's multi-store account of memory currently faces two sorts of empirical crises. The first concerns cases where sentence variables influence list processing in ways that would not be expected if fundamentally autonomous memory systems process sentences versus lists. By way of illustration, consider a recently discovered effect in my lab whereby syntactic and semantic factors influenced immediate recall of words in rapidly presented lists. I and my colleague Lise Abrams (1996) compared immediate memory for identical words in *chunked* versus *unchunked* lists that were six to eight

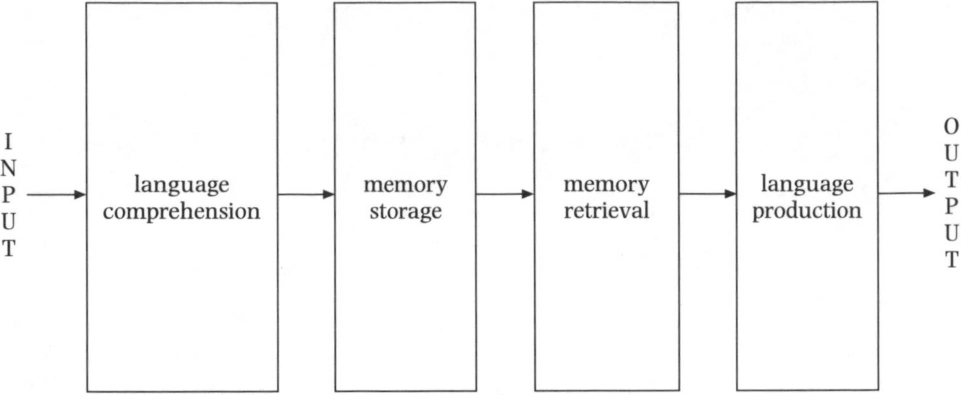

Figure 53.1 A standard stages of processing flowchart illustrating the stages for comprehending, storing, retrieving, and producing verbal materials.

words long and rapidly presented via computer so as to preclude rehearsal. Chunked lists, such as 1 (below), contained two familiar two-word phrases located at unpredictable positions in the strings, whereas unchunked lists, such as 2 (below), were identical except for the substitution of two unrelated words that destroyed the phrases. Results of this experiment showed that words in unchunked lists were more poorly recalled than identical words in the phrases of chunked lists. For example, *night* was recalled more poorly as an unrelated word in list 2 than in the phrase *night gown* in list 1, but the unrelated word *mind* was recalled equally poorly in both lists. Because phrases are fundamentally syntactic/semantic entities, these findings indicate that sentential factors (syntax/semantics) influence short-term memory within rapidly presented lists. To explain this result and meet this first crisis in general, multi-store theories must explain how semantic/syntactic factors influence a supposedly separate store traditionally viewed as purely phonological in nature.

1 Chunked list: phrase *good faith* mind *night gown* film (phrases italicized).
2 Unchunked list: phrase people faith mind night hose film (unrelated words).

The second crisis concerns phenomena in immediate recall of *sentences* that are attributable to factors that characterize lists. For example, my colleague Michelle Miller and I showed that a certain aspect of lists – namely, listlike prosody – when introduced into spoken sentences causes a short-term memory phenomenon (known as repetition deafness) that is otherwise observed only within lists. To explain this result and meet this crisis in general, multi-store theories must explain how a phenomenon can arise in the supposedly autonomous memory system for storing and processing sentences by introducing a characteristic of lists.

Problems with SPM

As the above crises illustrate, the problem with theories derived from SPM is not just the widely acknowledged crudeness or vagueness of SPM diagrams currently used to describe psychological findings. Rather, the widely unacknowledged problem is that available data directly contradict SPM assumptions. To make this point more generally and in greater detail, I next examine the traditional SPM stages in pairs, beginning with comprehension versus memory (i.e., storage and retrieval in figure 53.1), and review the original data favoring the hypothesis that textbook-title categories such as memory and comprehension constitute separate and sequentially ordered processing stages in the brain. I then show how more recent data have undermined both the original data and the basic SPM assumptions themselves. I conclude by showing why theories derived from SPM are also unlikely to support sophisticated applications of psychological knowledge.

Are comprehension and memory dissociable stages?

Nineteenth-century neuropsychological data suggested that comprehension and MEMORY are dissociable processes, such that comprehension disorders (e.g., Wernicke's aphasia) can occur without concomitant memory disorders (e.g., anterograde amnesia), and vice versa. However, neuropsychology moves on. More recent studies using highly sensitive implicit measures of on-line comprehension and memory have called this dissociation into question. The findings of Loraine Tyler, a neuropsychologist at Birkbeck College, London, illustrate some of these newer data. Tyler (1992) directly addressed the separability of comprehension and memory as distinct, sequentially ordered stages by testing whether the so-called comprehension deficit of Wernicke's aphasics is truly specific to comprehension or reflects both a comprehension deficit and a memory deficit that shows up only when tested via after-the-fact, explicit measures based on conscious judgments about prior comprehension (see Article 29, DEFICITS AND PATHOLOGIES).

Tyler's first step was the traditional demonstration that Wernicke's aphasics fail to respond accurately to questions such as "Is this sentence grammatical?," unlike normal controls presented with the same sentences. This frequently reported finding indicates that Wernicke's aphasics have a comprehension problem, or a memory problem, or both, when tested after the fact via explicit measures. Tyler's next step was to present similar sentences to her aphasics in an on-line priming task, where effects of ungrammaticality could be determined via indirect measures *at the time of processing.* These new data showed that her Wernicke's aphasics responded like normal controls, as if their comprehension deficit were indistinguishable from a memory deficit that can be overcome via tests that do not require conscious retrieval after on-line processing has occurred. This second finding undermines the first and most fundamental source of support for SPM and suggests that memory and comprehension may not be separate processing stages after all.

Tyler's experiments illustrate two widely used means of testing memory, known as direct and indirect tests. The latter show the effects of previously presenting a word (usually in an incidental context earlier in the experiment) on subsequent word perception (e.g., reduced recognition time) or on subsequent word production (e.g., reduced time to produce or name the word), without requiring conscious recollection of the prior experience with the word. By contrast, direct tests call for conscious recollection of the prior experience with the word(s) – for example, via cued recall, explicit recognition, or free recall of the previously presented word list. Within this larger context, Tyler's results comport with results from a wide range of direct and indirect tests of comprehension and memory *in normal people,* suggesting that identical mechanisms underlie everyday language comprehension on the one hand and the encoding and storage of verbal materials in laboratory studies of memory on the other.

The existence of identical mechanisms for comprehension, encoding, and storage of verbal information explains why nobody has ever been able to establish a detailed, convincing dividing line between where language comprehension leaves off and where storage begins, either empirically or theoretically, for either everyday behavior or experimental tasks. No such dividing lines can be established, because no memory-specific verbal processes exist independently of mechanisms that have evolved for learning, comprehending, and producing language.

Are memory retrieval and production dissociable stages?

Is memory retrieval (e.g., the process of retrieving words in verbal memory studies) distinct and separate from the everyday ability to produce words in sentences? No data directly support this basic SPM assumption, and considerable data contradict it. For example, data from the tip-of-the-tongue (TOT) phenomenon indicate that everyday language-production processes can be indistinguishable from word retrieval in verbal memory studies. TOTs normally occur when a speaker is unable to retrieve a familiar word such as *locust* or *Napoleon* during everyday speech production, even though they can often retrieve aspects of the word (e.g., its first letter, how many syllables it contains, its stress pattern, and other words similar in sound or meaning or both). Information about TOTs has come from three sources: diaries recorded at TOT onset during everyday speech production, questionnaires that assess a person's history of TOTs, and laboratory TOTs in response to questions such as "What do you call the leather band formerly used for sharpening an old-fashioned razor?" Interestingly, conclusions from all three sources of TOT data are indistinguishable, indicating that identical mechanisms underlie everyday speech production and memory retrieval in laboratory tasks.

Lise Abrams and I recently reviewed a wide range of other data that indicate not just close parallels, but identity between processes underlying everyday language production and laboratory word-retrieval tasks (see Article 14, LANGUAGE PROCESSING). For example, the time required to begin pronouncing a visually presented word is reduced if participants have previously encountered the same word in an incidental auditory context, and these response times are strongly correlated with cued recall of the corresponding word as having occurred earlier in the experiment. In short, pronunciation onset time indirectly measures memory for prior occurrence of the word and the ability to explicitly retrieve the word, but at the same time reflects a word *production* process. These and other results thus indicate that identical mechanisms are used for language production on the one hand and retrieval of verbal materials in laboratory studies of memory on the other. Moreover, far from being separate processing stages, memory-retrieval mechanisms and language-production mechanisms are unitary and inseparable, even though different types of tasks can tap into these unitary language abilities in different ways.

Are comprehension and production dissociable stages?

Comprehension and production are the most widely separated stages in SPM, occupying opposite ends of the processing spectrum in figure 53.1. They are also the oldest stages, first postulated by René Descartes in 1637, and were the first to receive empirical support, in work of Carl Wernicke published in 1874. Wernicke argued that comprehension and production constitute separate processing stages, because comprehension disorders in Wernicke's aphasics can occur without the concomitant production disorders seen in Broca's aphasics, and vice versa. However, more recent studies, using a variety of new and highly sophisticated techniques, suggest that aphasias are more complicated than was originally thought. With appropriate controls for lesion size, as well as for pragmatic (nonlanguage) aids to comprehension, Wernicke's and Broca's aphasics exhibit both receptive and expressive deficits that tend to be commensurate in nature and in extent. Moreover, microelectrode techniques have shown that stimulating one and the same cortical site can affect both production *and* perception of corresponding phonological units, suggesting that production and perception are inseparable at a neural level. Moreover, even more recent neuroimaging data support a similar conclusion (see Article 32, NEUROIMAGING).

Further support for common components underlying perception and production comes from a wide range of data from normal adults, including parallel empirical effects in production and perception, interactions between processes for production and those for perception, the nature of units for production and perception, the nature of errors in perception versus production, top-down effects in perception, bottom-up effects in production, effects of concurrent production on a perceptual illusion known as the "verbal transformation effect," and indirect tests of language perception and memory, showing, for example, reduced time to begin pronouncing a visually presented word due to prior auditory perception of the same word in an incidental context earlier in an experiment. In short, many different sorts of data indicate that perception and production cannot represent separate, independent stages of processing; and these include the original neurological data once thought to support their separation.

SPM and applications of psychological knowledge

One possible reaction to the empirical difficulties facing SPM is: "So what? Who cares about theory anyway? Isn't psychology primarily interested in experiments and practical applications, and haven't SPM flowcharts inspired lots of experiments over the past century?" Perhaps; but SPM experiments may have limited applicability. The problem can be illustrated via a *Gedanken* experiment involving a 400-year time warp: specifically, a test of the hypothesis that seventeenth-century physicists should adopt an SPM approach to ballistics that resembles twentieth-century SPM psychology. On this *Gedanken* hypothesis, seventeenth-century physicists should analyze ballistics into a sequence of ordered stages, beginning with construction of the projectile and launching device (input), through impact of the launched projectile (output). A typical flowchart in SPM ballistics might resemble figure 53.2a. Stage 1 (orientation) positions the gun relative to the earth. Stage 2 (ignition) inserts the projectile into the gun and fires it. Stage 3 (ascent) projects the missile to some determinable height (seventeenth-century physics has twentieth-century technology in this time warp) before it starts to plunge (stage 4, descent) and finally impact the earth (output).

With refined *observation* of these stages, one can imagine that SPM physicists might eventually be able to describe something like the actual path of a projectile, and to predict the path of identical projectiles fired under identical conditions. The problem is that SPM ballistics can describe only particular trajectories for particular projectiles

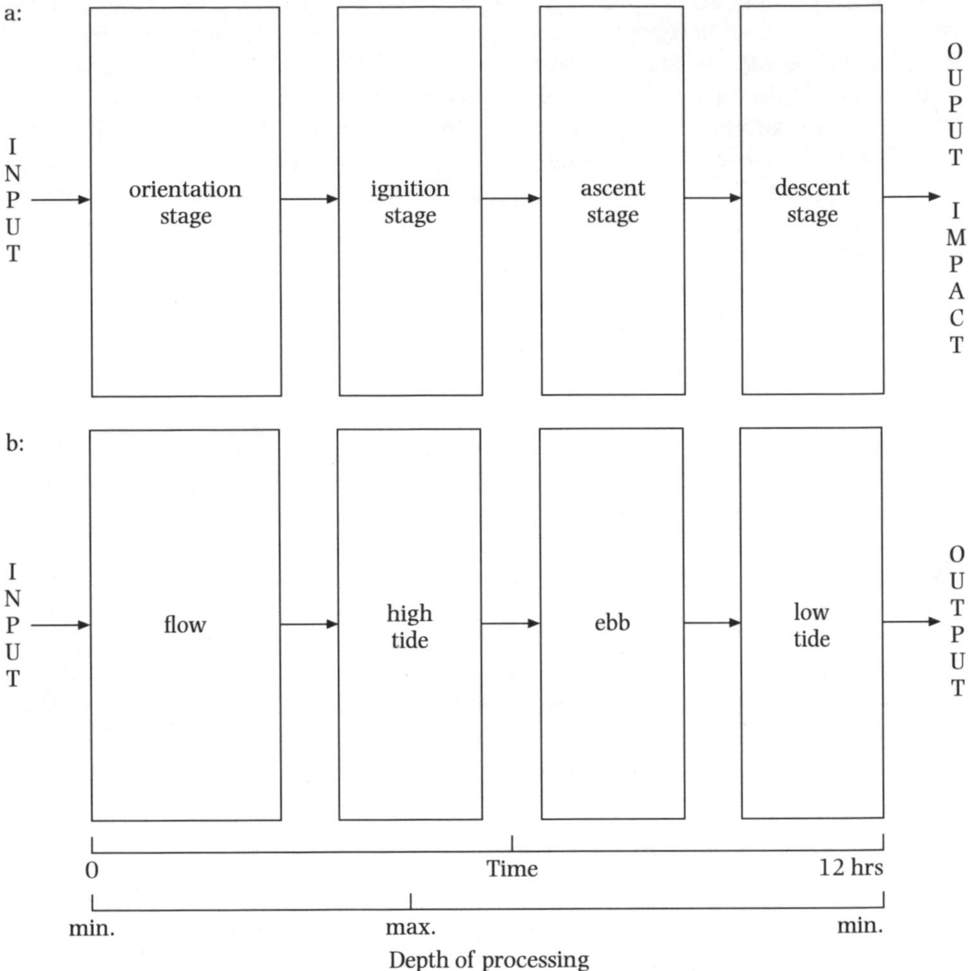

Figure 53.2 a: Hypothetical stages of processing in seventeenth-century information processing ballistics. b: Hypothetical stages of processing in sixteenth-century information processing oceanography.

fired under particular conditions, an achievement of dubious practical value, because launching a series of identical projectiles under identical conditions is not a reasonable goal. As a fundamentally descriptive approach, SPM can *in principle* provide no way of developing the explanatory or theoretical concepts that are needed for general applications in either physics or psychology. That is, predicting paths of different projectiles fired under variable (real-world) conditions requires theoretical concepts such as mass, gravitational force, inertia, kinetic energy, and centrifugal force that are unattainable within a descriptive approach such as SPM. In short, SPM would have constrained practical as well as theoretical developments in physics, just as in psychology.

This is not to say that SPM is incompatible with *all* possible applications. A *Gedanken* experiment with a 500-year time warp nicely illustrates this point. Hypothetical SPM oceanographers in sixteenth-century Europe might have developed flowcharts resembling figure 53.2b for predicting tide levels as a function of time in various European harbors. Stage 1 (flow) describes the water flow up to stage 2 (high tide), and stage 3 (ebb) describes the water flow down to stage 4 (low tide). These stages can be measured via *response times* and via *depth of processing* (i.e., water depth) at points of interest in the harbor. Clearly, *these* descriptive measurements do have *some* (restricted) practical value. Although general theoretical principles governing tidal phenomena have been understood for over a century, the details needed to apply these principles to particular harbors are so variable, so many, and so complex that pre-theoretical or description-based predictions are sometimes as effective for negotiating a particular harbor as more sophisticated, theory-based predictions employing the fastest state-of-the-art computers.

Nonetheless, pursuing SPM oceanography for an indefinite period of time will never give rise to explanatory or theoretical concepts such as mass, force, gravitational attraction between astronomical masses, centripetal acceleration, and friction, concepts that are needed to accurately predict tides *of any sort* – for example, *atmospheric tides, earth tides,* and *lunar tides* – *anywhere* in the universe. As a descriptive meta-theory, SPM precludes applications that are sophisticated and general.

Conclusions

This chapter has presented both empirical and meta-theoretical arguments against SPM. The meta-theoretical arguments focused on practical limitations of SPM as a descriptive meta-theory. The empirical arguments focused on evidence violating SPM assumptions – for example, evidence that lists and sentences are processed within the same memory system, and that neither storage and comprehension, nor memory retrieval and production, nor comprehension and production can be considered independent, sequentially ordered stages.

However, meta-theoretical dreams die hard. SPM theorists might hope that if traditional stages don't work, then perhaps other, more complex stages might work. Such hopes fly in the face of both experience and logic. More recent flowcharts have introduced labels such as "attention," "encoding," "recognition," and "identification," which lack precise and meaningful definition, and have further concealed our ignorance within a purely descriptive SPM framework rather than promote insight into underlying mechanisms.

What is needed instead is a clear, viable alternative to SPM that is explanatory or theoretical in nature, and over the past decade, the general shape of this new alternative has become increasingly clear. Common units and processes underlie perception, memory, and production of language, rather than separate units in separate processing modules that store lists versus sentences, or language comprehended versus produced versus learned. Support for this new view comes from a wide range of findings, including evolutionary analyses indicating that memory and production of language must have developed together, rather than separately, during human evolution. A set of mutations that enabled some humans to speak would improve their chances of survival only if they and others could understand, remember, and repeatedly use the same language code. And whatever abilities evolved to comprehend, store, and produce words in everyday life over the last million years must surely provide the basis for encoding, storing, and recalling words in twentieth-century experiments on verbal memory.

References and recommended reading

Burke, D. M., MacKay, D. G., Worthley, J. S. and Wade, E. 1991: On the tip of the tongue: What causes word finding failures in young and older adults? *Journal of Memory and Language*, 30, 542–79.

Gathercole, S. E. and Baddeley, A. D. 1993: *Working Memory and Language*. Hillsdale, NJ: Erlbaum.

MacKay, D. G. 1987: *The Organization of Perception and Action: A Theory for Language and other Cognitive Skills*. New York: Springer-Verlag.

—— 1993: The theoretical epistemology: a new perspective on some long-standing methodological issues in psychology. In G. Keren and C. Lewis (eds), *A Handbook for Data Analysis in the Behavioral Sciences: Methodological Issues*, Hillsdale, NJ: Erlbaum, 229–56.

MacKay, D. G. and Abrams, L. 1996: Language, memory and aging: distributed deficits and the structure of new versus old connections. In J. E. Birren and W. K. Schaie (eds), *Handbook of the Psychology of Aging*, 4th edn, San Diego, Calif.: Academic Press, 251–65.

Tyler, L. K. 1992: *Spoken Language Comprehension: An Experimental Approach to Disordered and Normal Processing*. Cambridge, Mass.: MIT Press.

PART VI

COGNITIVE SCIENCE IN THE REAL WORLD

54

Education

JOHN T. BRUER

Introduction

Since the mid-1950s, the various research programs within cognitive science have advanced our basic understanding of human mental function. Over the past 20 years, this basic science of mind has also contributed to the genesis of an applied science of learning and teaching that can powerfully inform educational practice and dramatically improve educational outcomes (Bruer, 1993). Classroom practice based on this applied science differs from traditional instruction in several ways. Instruction based on cognitive theory envisions learning as an active, strategic process. It assumes that learning follows developmental trajectories within subject-matter domains. It recognizes that learning is guided by the learners' introspective awareness and control of their mental processes. It emphasizes that learning is facilitated by social, collaborative settings that value self-directed student dialogue.

Each of these four instructional features has its roots in specific cognitive science research programs. Research on human memory has established that memory is an active, strategic process, supporting the contention that learning itself is active, strategic, and constructive. Research on problem solving within subject-matter domains has resulted in descriptions of domain-specific learning trajectories that specify in some detail the knowledge and skills required for expertise within domains and how knowledge and skills are best organized to enable expert performance. Research on metacognition has revealed the importance of self-awareness and self-control of cognitive processes and has shown how this awareness and control can guide understanding and learning. Research on sociocultural factors in cognition has provided significant new insights about the importance of language, collaboration, and social discourse in cognitive development and learning.

Memory: research and applications

Research on human MEMORY has been a central pursuit of experimental psychology since its inception a century ago. A claim fundamental to cognitive psychology, which distinguishes it from behaviorism, is that the mind is an active information processor, not a passive communication channel. Early on, cognitive psychologists argued that we overcome intrinsic limitations on our short-term and working memory capacity by actively recoding knowledge into more complex symbol structures, or *chunks*. This suggested that learning might involve active, strategic recoding of knowledge structures in an attempt to discover the most efficient chunks for any given task. Cognitive research elaborated on an earlier, 1932 insight of F. C. Bartlett about how long-term memory functions: Stimuli that cohere with prior existing memory schemata are better recalled than stimuli that fit poorly into prior schemata. The result was the development of the schema theory, which has contributed to how educational psychologists think about CONCEPTUAL CHANGE. Cognitive research on memory provides empirical support for constructivist approaches to learning and teaching.

One of the most educationally significant results arising out of this research program is the encoding specificity principle: To remember a percept, we perform specific encoding operations on it which determine what is stored in memory. In turn, what is stored in memory determines what cues will be effective in helping us retrieve that memory trace (Tulving, 1983). Success on a memory-retrieval task is not a function of *strength* of the mental representation alone. There is a strong interaction between memory's encoding and retrieval processes. In fact, the utility and efficacy of a particular encoding process depends on, and interacts with, eventual retrieval conditions. A more general, educationally salient formulation of this same basic result is Morris, Bransford, and Franks's (1977) *transfer- appropriate processing*: The value of particular types of acquisition activities can be assessed only in relation to the type of activities that subjects will be expected to perform at the time of retrieval or test. According to this principle, it is not possible to determine the value of learning activities in themselves. The value of a learning activity can be determined only relative to what one expects students to do with the material they are expected to learn.

Research on human memory tells educators two things. First, encoding interacts with retrieval, and acquisition conditions interact with recall performance. Thus, the nature of the learning activity itself in part determines the students' subsequent ability to transfer that learning to new situations. Second, the interaction between encoding and retrieval is mediated by, and depends upon, students' prior understanding, their preexisting knowledge, and their pre-instructional schemata. If memory is an active, constructive process, then students' prior knowledge structures, current learning conditions, and future application conditions are inextricably intertwined. Cognitively sound instruction should build on this architectural feature of human memory.

Recognizing that students' prior knowledge structures influence current learning has had a substantial impact on science instruction. Cognitive and educational researchers have documented numerous misconceptions that students, and indeed all of us, have about how the physical world operates. They have also found that these misconceptions are largely impervious to traditional science instruction. In physics, for example, misconceptions persist even after extensive, formal instruction through the college level. Traditional instruction does not correct students' prior misconceptions, because it ignores them. Ignoring students pre-instructional understanding allows students to interpret and encode traditional science instruction using these preexisting, naive memory schemata. The result is that students encode, or learn, schemata that are very different from those which teachers are attempting to impart.

Instructional approaches that attempt to assess students' pre-instructional knowledge and beliefs about scientific principles are significantly more successful than traditional science instruction in correcting misconceptions and imparting more expert-like understandings of science. Jim Minstrell and Earl Hunt, for example, developed a cognitive approach to high school physics instruction, a curriculum they called "Physics for Understanding" (see McGilly, 1994, pp. 51–74). Each instructional unit begins with a diagnostic test that allows the instructor to identify students' prior understandings and observe how students reason with them. Minstrell and Hunt call the pieces of science knowledge that students use in their reasoning *knowledge facets.* Among the knowledge facets which students bring to a specific problem, some are incorrect, but others are correct. Correct facets can be used as anchors for instruction, to help students construct more expert-like schemata; incorrect facets become targets for instructional change. Evaluations that have compared Minstrell and Hunt's approach to traditional instruction and to other experimental physics curricula show that students in the Minstrell–Hunt curriculum acquire significantly superior understandings of physics and scientific reasoning. Applying our understanding of memory in the design of science instruction can result in curricula which allow teachers to help students correct their naive understandings and misconceptions. Such instruction is significantly more effective than traditional approaches.

Transfer-appropriate processing provided the research base for an extended research program on memory and learning. This work extended beyond studying memory for facts to studying mnemonic and learning strategies and the interactions between facts and strategies. The work of the Cognition and Technology Group, Vanderbilt University (CTGV), illustrates how this research can contribute to the design of more effective instructional tools (see McGilly, 1994, pp. 157–200). In their research on school learning, the CTGV found that knowledge and skills acquired in the context of solving a problem were better retained and more readily transferred to novel problem-solving situations than information acquired through reading and study. The success of problem-based learning in experimental situations contributed to the design of an approach to classroom instruction which they called *anchored instruction.* This approach uses a video-disk adventure as an instructional anchor. The video anchor provides realistic, complex problems within which problem-oriented learning can occur. Learning while solving realistic, complex problems assures similarities between the learning, or acquisition, situation and the eventual retrieval, application, or test situation.

The CTGV's first anchored instructional tool, "The Adventures of Jasper Woodbury," was designed to improve middle-school mathematics instruction, particularly students' abilities to formulate and solve multi-step mathematical problems. Each Jasper adventure ends with a challenge that requires students to solve a complex, multi-step mathematical problem. All the information the students need is embedded in the video. Through student-directed discussion, this information is gradually extracted from the video, combined with mathematical principles, and organized into a problem solution.

The narrative, adventure structure of the anchor video itself provides students with a memory aid. Viewing and re-viewing the video helps the students gradually build deeper understandings of the adventure and the problem it poses. The video is accessible to students with vastly different levels of prior understanding, background knowledge, and abilities; yet the shared format helps the entire class rapidly develop common understandings and vocabulary. The coherence and complementarity of the learning and application tasks – of the encoding and retrieval situations – help students actively construct deep concepts and coherent mental models of the problem and mathematical reasoning.

Extensive classroom evaluations of the Jasper adventures have shown that Jasper students gain a slight advantage over students in traditional classrooms on standardized mathematics tests. However, Jasper students are significantly superior to students in traditional classrooms in their abilities to solve one-step, two-step, and multi-step mathematics problems. Jasper students' biggest gains over students in traditional classrooms lie in their abilities to formulate problems, plan problem-solving strategies, and identify the subproblems that must be addressed to reach a complete solution. These mathematical skills, skills that are rarely taught explicitly in middle-school curricula, are fundamental to applying quantitative thinking in everyday problem situations.

Problem solving: research and application

One of cognitive science's great contributions to our understanding of learning and instruction is that intelligence and expertise are domain-specific and develop along specific learning trajectories within each knowledge domain. Research on human PROBLEM SOLVING has been fundamental to this insight. In the 1960s, Herbert Simon and Allen Newell, who first formulated this research program, argued that if we want to understand learning within a knowledge domain, it is necessary to start with a detailed analysis of how people solve specific problems, over short periods of time, in that domain. Cognitive scientists within this research tradition attempt to analyze logically a problem-solving task to frame initial hypotheses about the knowledge and skills the task requires and about how the knowledge and skills might best be organized to solve the problem. Armed with these hypotheses, researchers observe subjects solving a problem, having the subjects *think aloud* while they do so. Researchers repeat this procedure using different problem types within a domain and collecting think-aloud protocols from subjects of varying skill levels, from novice to expert, in the domain. PROTOCOL ANALYSIS of subjects' think-aloud protocols leads to refined hypotheses and models of problem solving within the domain and eventually to the formulation of detailed learning and developmental trajectories within knowledge domains. In some cases these learning trajectories can be presented as production system models of problem-solving expertise within the domain (see Article 42, PRODUCTION SYSTEMS).

By the mid-1970s, research on problem solving began to specify in some detail portions of the learning trajectories within the traditional school subject domains of reading, writing, mathematics, and science. These learning trajectories, where they are available, have helped guide the design of improved instructional materials. The trajectories describe common bottlenecks to learning in specific domains and allow teachers to diagnose learning problems in terms of individual students missing particular understandings or skills, rather than attributing lack of progress to general cognitive deficits. This cognitive research program, when combined with an understanding of how human memory works, gives educators powerful tools. It provides educators with an empirically based technology for determining students' pre-instructional knowledge, for specifying the form of likely future knowledge states, and for choosing appropriate problems and learning activities to help students build the desired knowledge structures from their preexisting schemata (Siegler and Klahr, 1982).

Applications of this learning technology have occurred in almost every knowledge domain. However, the domain of number knowledge provides an excellent example of how this approach has helped some educators think about and provide mathematics instruction. Problem-solving research, combined with methods and theories of developmental psychology, have resulted in a profound advancement in our appreciation of how children naturally understand and learn about quantity and number, and how children use this natural understanding to learn first formal arithmetic in the elementary grades. Children are sensitive to number and quantity from infancy. Most arrive at school with a considerable command of informal number knowledge. This informal knowledge includes the ability to count, to compare numbers for size, and to invent strategies based on counting and comparing to solve simple arithmetic problems. From a cognitive perspective, then, given our understanding of memory, the challenge is to build on this informally acquired numerical knowledge to help children master elementary arithmetic and its formal symbolism of digits and equations.

Research has also revealed that not all children enter school with comparable informal understandings of number. Traditional arithmetic instruction often assumes that children possess the central conceptual structure of the number domain, which includes knowing the counting words and understanding principles of one-to-one correspondence, cardinality, and ordinality. Instruction also assumes that children know that arabic numerals and number words name the same things. However, children at risk for early mathematics failure often arrive at school with deficient understanding of ordinality; specifically, they are often unable to determine reliably which of two numbers is bigger or smaller. These children, too, are often unfamiliar with arabic numerals. For some reason, these knowledge gaps are particularly prevalent among children from low-income homes.

Using the learning trajectory in the number domain and its description of the central conceptual structure which children should have at school entry, it is possible to design diagnostic tests to identify the gaps in children's conceptual structures and to undertake specific curricular interventions to fill the gaps. One example of such a cognitive-developmental approach is the "RightStart" mathematics curriculum (see McGilly, 1994, pp. 25–50). This curriculum attempts to assure that all children leave kindergarten with the complete central conceptual structure needed to master first formal arithmetic. The curriculum explicitly teaches all the knowledge and skills contained in that structure. In extensive evaluations of this curriculum, kindergarten students identified as severely at risk for mathematics failure at school entry were, by the end of first grade, indistinguishable, and in some cases superior to, initially more capable middle-class students in their math performance and understanding. By the end of the first grade, these initially at-risk children performed at levels comparable to those achieved by Japanese students on computational tests used in international mathematics comparison studies.

Meta-cognition: research and applications

What most distinguished cognitive science from behaviorism was its recognition that a scientific psychology had to include hypotheses and theories about covert, internal mental processes. Psychology had to be a science of the mind if it was to be an adequate science of behavior. This expanded perspective contributed to the development of research programs on meta-cognition and meta-cognitive processes. Generally, *meta-cognition* refers to a person's knowledge and control of the cognitive domain.

Ideas about meta-cognition can be found in various cognitive research programs (Brown et al., 1983). Within the problem-solving tradition, reliance on verbal protocols – thinking aloud while engaged in problem solving – assumes that subjects have some awareness, access to, and control of the cognitive processes they invoke to solve specific problems. Within information processing psychology more broadly, the notion of executive function was often invoked to explain how we are able to allocate our limited attentional and working memory resources to carry out specific processes at the appropriate times when engaged in complex, recursive problem-solving routines. Piaget emphasized the importance of self-regulation in conceptual reorganization, where self-regulatory skills gradually evolved from autonomous or automatic, to active, to fully conscious. For Vygotsky, children's cognitive development involved moving from *other-regulation*, provided externally by more expert learners in the child's sociocultural community, to internalized self-regulation.

Meta-cognitive knowledge can be of two types. First, it can be knowledge about cognition and one's own cognitive processes. This type of meta-cognitive knowledge might best be described as declarative knowledge of how minds work. Such knowledge is explicit and usually appears relatively late in development. Of course, we can develop erroneous beliefs about how our minds work. Examples of such explicit meta-cognitive knowledge include awareness of memory limitations, of personal problem-solving skills, of a task's demands, and of the appropriate strategies to use on tasks, given one's limited capacities and skills. This kind of meta-cognitive knowledge can be acquired in one subject domain, then transferred to learning and problem solving in new domains. Research on meta-cognition has shown that the possession of, and ability to, transfer meta-cognitive skills among knowledge domains differentiate able from less able learners. In learning a new domain, able learners, or intelligent novices, possess and use meta-cognitive knowledge and skills, whereas less able learners, or unintelligent novices, do not. For example, intelligent novices appreciate the difference between memorization and comprehension, are aware that different learning tasks demand different learning strategies, and allocate study time accordingly. Unintelligent novices, or weak learners, do none of these things. Intelligent novices are likely to use self-tests and self-questioning as sources of feedback to correct their misconceptions and change inappropriate learning strategies, unlike unintelligent novices (Bransford et al., 1989). Some children acquire these skills naturally and informally; others fail to do so. Fortunately, the research also shows that instruction which is meta-cognitively aware can impart meta-cognitive abilities and thereby help unintelligent novices become more effective learners (see Article 29, DEFICITS AND PATHOLOGIES, and Article 59, MENTAL RETARDATION).

A second type of meta-cognitive knowledge includes the ability to regulate one's own cognitive processing using skills such as planning, monitoring, and checking. Some of these skills seem to be implicit, unconscious, procedural knowledge. Typically these skills are highly task- and situation-dependent. Some meta-cognitive skills of this kind might initially be explicit and subject to conscious control but become implicit and procedural through overlearning and automatization.

One of the best examples of instruction to teach skills of this kind is the "Reciprocal Teaching of Reading Comprehension," which applies ideas about meta-cognition to UNDERSTANDING TEXTS (Palincsar and Brown, 1984). Borrowing from the problem-solving tradition, "Reciprocal Teaching" tries to impart four monitoring and comprehension skills which expert readers use automatically and unconsciously, but which are noticeably lacking in unskilled readers. These skills are summarizing the gist of a text as one reads, predicting what might happen next in the text, clarifying difficult points in the text, and formulating questions to foster text comprehension. "Reciprocal

Teaching" helps unskilled readers, who can adequately decode written text, assimilate these meta-cognitive monitoring and control strategies.

In "Reciprocal Teaching" reading groups, students read a brief passage silently. One member of the group, initially the reading instructor, serves as leader. When everyone has finished the passage, the leader summarizes what has been read, clarifies difficult or unclear points for the group, formulates a question that might appear on a comprehension test about the passage, and makes a prediction about what might happen next in the story. For the next passage, another member of the group, this time a student, serves as the leader, until all students have taken their turn. Group members serve as supportive critics and aids in helping each of the leaders in turn to learn and apply these high-level comprehension strategies as they collaborate to construct meaning from the text.

Acquiring these meta-cognitive skills has a significant impact on students' reading comprehension. Working with seventh-grade remedial students who, before instruction, were reading at the third-grade level, "Reciprocal Teaching" raised student performance on reading comprehension tests from around 10 percent correct to 85 percent correct, improved classroom reading performance from the seventh to the fiftieth percentile, and dramatically improved students' classroom comprehension of science and social science texts. These results were attained after approximately 20 days of "Reciprocal Teaching" instruction, and students maintained these comprehension levels through months of follow-up studies. The "Reciprocal Teaching" approach has been extended by Ann Brown and Joe Campione into a middle-school instructional program called "Fostering Communities of Learners," which attempts to teach these meta-cognitive comprehension skills and other reasoning skills to students within the context of middle-school science instruction, again with considerable success (see McGilly, 1994, pp. 229–90).

Social cognition: research and applications

While each of the above examples has its dominant roots in a different cognitive research program, all of them exemplify the fourth feature of an applied science of learning. They all create and attempt to maintain social, collaborative classroom environments that allow and encourage self-directed student discussion. In Minstrell and Hunt's physics curriculum, each unit of the curriculum begins with a diagnostic quiz to identify students' existing knowledge facets, but students are then expected to state and defend their reasoning about how physical objects behave in the world. Throughout each curriculum unit, the teacher's role is to facilitate student thinking and to mediate a student-driven discussion of scientific reasoning and the application of physical principles. The CTGV's Jasper series uses video adventures to create socially shared, problem-solving contexts which again encourage student-directed discussion of problem formulation, problem solving, and quantitative reasoning. "RightStart" uses small group instruction, built around number games, to encourage student discussion of mathematical reasoning. The curriculum demands that students defend and explain answers and reasoning to their peers and encourages students to serve as helpful critics and external monitors of ongoing mathematical thinking. "Reciprocal Teaching" imparts high-level, expert comprehension strategies, but does so in a socially interactive learning environment, where students engage in dialogue about the strategies and how to use them to construct meaning from a text. Students collaborate and mutually support one another's learning.

All these instructional interventions, at least implicitly, recognize that learning occurs best in social, collaborative settings involving self-directed student dialogue. This feature of effective learning environments does not follow directly from any of the three previous research programs discussed here, but relies instead on research in SOCIAL COGNITION. Among the research programs mentioned above, the closest link would be to Lev Vygotsky's influence on meta-cognitive research. Vygotsky's more general views on the importance of sociocultural contexts and social mediation in learning and development, however, are fundamental to this fourth feature (see Article 40, MEDIATED ACTION). This specific connection is most explicit in the development of "Reciprocal Teaching," where early presentations of that work cited the impact of self-verbalization in helping children acquire social control skills. "Reciprocal Teaching" attempted to help students use self-verbalization to acquire meta-cognitive control in a cognitive domain such as reading comprehension. "Reciprocal Teaching" creates environments for self-verbalization, in which a more expert leader provides support and scaffolding for less able learners. The "Reciprocal Teaching" group provides a means of *other* or external regulation until the individual student can internalize the social process and become a self-regulated learner in the domain.

Thus, a major influence of Vygotsky on North American educational psychology has been an emphasis on the need to create environments where skilled *others* can socially mediate student learning, allowing the gradual internalization of knowledge and skills, leading to self-regulation and eventual mastery of higher cognitive skills. Why this approach works so well remains an open question. To say that simply learning is a social process, or that knowledge is a social construction, is to redescribe the phenomenon rather than explain it. Minimally, such learning environments make covert, implicit skills and strategies shared, public, and overt. They provide a structure to make implicit knowledge explicit. Furthermore, these learning contexts impart not only domain-specific knowledge and skills, but also illustrations and examples to students of how, within a society, culture, or disciplinary community, one is expected to think like an expert.

The full implications of sociocultural cognition for education remain largely unexplored, only because this is a relatively new, but rapidly expanding, research front. It is also an open question how research within this tradition might cohere with or complement research in other areas of cognitive science. Ideally, one might seek an intelligent division of labor. Research on memory, problem solving, and aspects of meta-cognition contribute to an empirical technology for discovering and describing the knowledge and skills which children need to become expert within knowledge domains. However, these approaches alone say relatively little about how to design learning environments that allow students to make optimal, self-conscious progress along a learning trajectory. A fundamental contribution of sociocultural research has been to call attention to the importance of the learning context itself and to stimulate careful investigation of learning environments, the use of language, the support provided by others, and the role of self-directed student discussion. If the other cognitive research programs described here can continue to provide detailed descriptions about *what* should be learned and in what sequence, sociocultural research might provide insights about *how* to structure environments to best impart that knowledge.

Conclusion

Forty years into the cognitive revolution, we have developed at least the rudiments of an applied science of learning. We are now poised to further exploit, develop, and assimilate this applied science to improve learning and teaching. The diverse methods, theories, and research programs underlying this applied science provide dangers if they are pursued in isolation, but opportunities if pursued collaboratively or in parallel. There is a danger that theoretical and methodological differences will needlessly divide the research community interested in education and cognition. There are genuine theoretical differences, and some of the differing positions are deeply held, but their ultimate implications for, and relevance to, better classroom practice may be minimal.

Those entering the field whose primary interest is in sociocultural cognition should not lose touch with previous and ongoing basic research on human memory. How memory works has been a pillar of cognitive science and one where continuing insights are accruing. Any applications of cognitive science to classroom practice must incorporate, or at least be consistent with, our understanding of human memory. On the other hand, some memory research must be pursued outside the laboratory and must consider the impact of sociocultural variables on both encoding and retrieval processes.

Problem-solving research provides a powerful technology with which to describe learning trajectories and design developmentally appropriate curricula. While many view this line of research as a mature one, we have, at best, only a meager understanding of learning trajectories within only a few disciplines. Much work that could have great educational significance remains to be done. Problem-solving research could achieve, and is achieving, more by expanding its notion of task environment and task analysis to include more social and cultural variables. The basic methods of this research program are as applicable to group, collaborative, and distributed cognitive processes as they are to individual problem solving and cognition.

Research on meta-cognition has provided some basic, powerful educational insights, however much remains to be done to refine, differentiate, and classify meta-cognitive processes. Although this concept has roots in diverse research programs, we may find that their respective characterizations of meta-cognition in fact describe sets of distinct skills, rather than varieties of a single cognitive phenomenon. As commentators and critics have pointed out, the meta-cognitive research area is characterized by a proliferation of terms, phenomena, and nuances that sometimes threaten coherence. A vague, diffuse construct – meta-cognition is all things to all people – is not useful and can provide little guidance to educators. Because meta-cognition is an important construct and one where all these various research programs have a tradition and common interest, it might be an ideal site on which to concentrate on complementary research activities.

All of the examples mentioned here, and indeed most of the effort in educationally relevant cognitive research, have addressed student learning. The challenge has been to apply our understanding of learning to instruction. In general, we have been slow to appreciate that teachers are also cognitive agents, and that teaching is a domain where knowledge and skills contribute to expertise. We have little understanding of the trajectory that leads from novice to expert teaching, of the meta-cognitive skills required for teaching, and of how to structure learning and professional development environments to impart these skills. Cognitive research on teaching can still be called the "missing research program." The various strains of cognitive research can all help fill out this seriously underdeveloped area of teacher cognition.

To advance an applied science of learning will require cooperation, flexibility, and some accommodation along all the various research fronts. Looking at memory only within the laboratory will make only minor contributions to an applied science of learning. Looking at small pieces of problem solving only within individuals will not significantly advance an applied science of learning. Claiming that all learning is by definition social and that there is no need to consider how the mind/brain works or how individuals differ in acquiring and organizing knowledge will not significantly advance this applied science either. A united front that recognizes the inherent strengths and limitations of the various research programs within cognitive science holds considerable promise for advancing an applied science of learning, a science that could have significant impact on learning and instruction within both formal and informal learning environments.

References and recommended reading

Bransford, J. D., Vye, N. J., Adams, L. T. and Perfetto, G. A. 1989: Learning skill and the acquisition of knowledge. In A. Lesgold and R. Glaser (eds), *Foundations for a Psychology of Education*, Hillsdale, NJ: Erlbaum, 199–250.

Brown, A. L., Bransford, J. D., Ferrara, R. A. and Campione, J. C. 1983: Learning remembering and understanding. In P. H. Mussen (ed.), *Handbook of Child Psychology*, vol. 3: *Child Development*, New York: Wiley, 77–166.

*Bruer, J. B. 1993: *Schools for Thought: A Science of Learning in the Classroom.* Cambridge, Mass.: MIT Press.

*McGilly, K. (ed.) 1994: *Classroom Lessons: Integrating Cognitive Theory.* Cambridge, Mass.: MIT Press.

Morris, C. D., Bransford, J. D. and Franks, J. J. 1977: Levels of processing versus transfer appropriate processing. *Journal of Verbal Learning and Verbal Behavior*, 16, 519–33.

Palincsar, A. S. and Brown, A. L. 1984: Reciprocal teaching of comprehension-fostering and comprehension-monitoring activities. *Cognition and Instruction*, 1, 117–75.

Siegler, R. S. and Klahr, D. 1982: When do children learn? In R. Glaser (ed.), *Advances in Instructional Psychology*, vol. 2, Hillsdale, NJ: Erlbaum, 121–211.

Tulving, E. 1983: *Elements of Episodic Memory*. New York: Oxford University Press.

55

Ethics

MARK L. JOHNSON

Every moral tradition and every moral theory necessarily presupposes some specific view of how the mind works and of what a person is. The cognitive sciences constitute our principal source of knowledge about human cognition and psychology. Consequently, the cognitive sciences are absolutely crucial to moral philosophy. They are crucial in two basic ways. First, any plausible moral system must be based on reasonable assumptions about the nature of concepts, reasoning, and moral psychology. Second, the more we know about such important issues as the role of emotion in moral deliberation, the nature of moral development, and the most realistic conceptions of human well-being, the more informed we will be in our moral thinking. Empirical investigations into mind thus provide a way of examining the presuppositional link between the *is* of mental functioning (e.g., how concepts are structured, the nature of rational inference, the role of emotion in reasoning) and the *ought* of morality – that is, the normative claims of our ethical system.

As an example of the critical function of cognitive science for ethics, consider its implications for two of our most prominent ethical systems: utilitarianism and Kantianism. Utilitarianism is the view that the morally correct action for a given situation is that action that would maximize the good, or well-being, that is realizable in that situation for the largest number of people affected by that action. Utilitarianism thus requires highly determinate concepts of *the good* (including concepts of self-interest and happiness), as well as a capacity to calculate rationally the probabilities governing the realization of these goals or states of affairs. However, empirical studies show that people are quite poor at performing the kinds of interest and welfare calculations required by classical utilitarianism, to such an extent that the theory becomes psychologically unrealistic.

Kantianism holds that morality is a system of absolute, universally binding moral laws that come from a universal reason (pure practical reason) and that can be applied to concrete cases to give us moral guidance. Kantianism thus requires the existence of a universal human reason and of moral principles that contain clearly defined literal concepts. But there is ample empirical evidence to suggest, first, that human reason is not transcendent, since it is inextricably bound up with our embodiment and emotions, and second, that our moral concepts are not literal, but rather metaphorical in nature. In this way, empirical results concerning cognition place major constraints on what an adequate moral theory would look like.

It is a shocking fact, therefore, that mainstream moral philosophy for the last two centuries has largely denied any significant role for empirical research on the mind. Cognitive science is typically dismissed as being irrelevant, on the basis of the mistaken assumption that there exists a rigid dichotomy between facts and values. The *irrelevance-of-cognitive-science* argument takes the following form: The cognitive sciences deal with empirical facts, seeking to describe and explain natural phenomena of the mind, such as why people think and behave as they do. It is then asserted that such facts about human thinking and behavior cannot give rise to normative claims about how people ought to behave. Therefore, it is claimed, cognitive science has little bearing on ethics.

The philosophical prejudice against cognitive science

The hostility of moral philosophy to cognitive science thus rests primarily on the notorious *is/ought* dichotomy – the erroneous view that facts and values are radically distinct and independent of each other. In his *Treatise of Human Nature* David Hume argued famously that he could see no justification for moral philosophers slipping imperceptibly from talk about what *is* (facts) to talk about *what ought to be* (normative moral judgments). Kant reinforced this purported split between descriptive empirical knowledge and normative moral judgments by claiming that the universality and necessity of absolute moral laws could never stem from merely empirical scientific generalizations but must come instead from the normative capacity of pure reason to issue universally binding moral principles. In the *Foundations of the Metaphysics of Morals*, he concluded that, "not only are moral laws together with their principles essentially different from every kind of practical cognition in which there is anything empirical, but all moral philosophy rests entirely on its pure part. When applied to man, it does not in the least borrow from acquaintance with him (anthropology) but gives a priori laws to him as a rational being."

This vehement denial of the relevance of empirical study for moral guidance was reinforced in a most fateful way at the beginning of the twentieth century by G. E. Moore's *Principia Ethica*, which had a profound influence on the direction taken by moral theory for many decades to follow. Moore asserted that the fundamental moral concept *good* denotes a simple, unanalyzable, *nonnatural* property of certain experiences or states of affairs. Although Moore never really explained what the nature of a *nonnatural* property was, he contrasted it with the *natural* properties – properties perceivable and measurable – that were the objects of study of the empirical sciences. Moore thus concluded that what he saw as the fundamental question of moral philosophy, the question of what *good* is, could never be answered by reference to the results of the empirical sciences. He coined the term *naturalistic fallacy* for what he regarded as the catastrophic mistake of attempting to define the concept *good* in terms of natural properties (or *any* properties, for that matter). The only role that Moore allowed for empirical inquiry in ethics was to supply causal knowledge of which actions were most likely to realize certain ends that were deemed to be good.

Moore influenced generations of subsequent moral theorists, who have simply assumed that there is a radical split between moral psychology (as an empirical, descriptive discipline) and moral philosophy (as rational analysis and normative prescription). They see no essential link between the *is* of moral psychology and the *ought* of moral philosophy.

Areas where cognitive science bears on ethics

From the perspective of the cognitive sciences, then, the most fundamental challenge is to show that the alleged *is/ought* split is mistaken, by showing how research in cognitive science is important for our moral understanding. We must examine the ways in which cognitive science constrains moral theory. Every moral theory must necessarily make assumptions about conceptual structure, reasoning, and the nature and limits of human understanding. It is the business of cognitive science to investigate just such issues, and in this way our empirical knowledge of the mind is directly relevant to moral philosophy.

Early in the twentieth century the idea that empirical science is not just a servant to moral philosophy, but rather is the core of ethics, was set forth cogently by John Dewey in *Ethics* and *Human Nature and Conduct*, where he argues that ethics is "ineradicably empirical, not theological nor metaphysical nor mathematical. Since it directly concerns human nature, everything that can be known of the human mind and body in physiology, medicine, anthropology, and psychology is pertinent to moral inquiry" (p. 204).

Dewey's argument that moral philosophy must incorporate the best empirical research on the mind was ignored by Anglo-American analytic moral theory. Only in the last few years have a handful of philosophers begun to take seriously the importance of cognitive science for moral theory. Most notably, Owen Flanagan (1991) has argued that a minimal psychological criterion of any adequate moral theory is that it be compatible with our most stable and reliable knowledge of human psychology. He formulates this requirement as a "Principle of Minimal Psychological Realism: Make sure when constructing a moral theory or projecting a moral ideal that the character, decision processing, and behavior prescribed are possible, or are perceived to be possible, for creatures like us" (p. 32).

However innocuous this principle might seem to most people, it has radical implications for the relation of cognitive science to moral theory. What it entails is that normative moral theory ought to be constrained by what the cognitive sciences are discovering about the mind and human psychology. It means that the normative dimension of moral thinking is not independent of the facts of moral psychology.

The chief problem for a more naturalistic view of ethics is to survey the kinds of empirical studies that are relevant to moral theory and to show how these results ought to influence our moral evaluations. For the most part, results from the cognitive sciences available to date function primarily to set basic constraints on the nature of a psychologically and cognitively realistic morality. These fundamental requirements for a tenable moral theory come from empirical research on the following major areas:

Concepts and rules

There is a dominant *moral law tradition* in the West that encompasses Stoicism, most of Judeo-Christian morality, Enlightenment rationalist ethics, and several contemporary moral theories, according to which morality is regarded as a system of universal moral laws or rules, discernible by human reason, and directly applicable to the kinds of concrete moral situations that people encounter in their lives (Johnson, 1993). Different traditions see these moral rules as having different sources – coming either from God (as in Judaism and Christianity), from a universal human reason (as in Kantianism), or as being socially constructed (as in Rawls's theory of justice) – but they all see morality as based on moral laws.

Since the cognitive sciences focus so heavily on the nature of conceptualization and reasoning, their results bear directly on the cognitive plausibility of moral law theories. For example, moral law theories typically require a set of universal, absolute, strict rules that specify morally correct behavior for the kinds of ethical situations that we are likely to encounter. To avoid relativism, these rules must apply for all times and cultures. Each rule must have a fixed, clear, literal meaning that is directly interpretable by ordinary people, so that they can decide how to act in a specific situation. Otherwise, there would be no proper or correct application of a given rule.

George Lakoff (1987) surveys a massive body of cognitive research showing that virtually none of our more abstract human concepts has this clear, literal, unequivocal internal structure. It is difficult to find an abstract natural concept that is literal, defined by necessary and sufficient conditions, and uniform in structure. On the contrary, many concepts are *fuzzy*, in the sense that they have shaded boundaries and a graded internal structure. There are degrees of category membership, rather than clear, distinct boundaries for what falls within the category. Members of a category often do not share any one set of essential features. Instead, they exhibit *prototype effects*, where one category member will be more prototypical – a more central member of the category – than other members. It is typical for people to conceptualize and reason by means of prototypes rather than by lists of features that are supposed to define a category. Paul Churchland (1995) has explored some of the ways in which our moral judgments are grounded in experientially developed moral prototypes. Many of our concepts thus have a complex *radial* structure (Lakoff, 1987), in which one or more prototypes are related to other category members by means of a variety of principles of extension, such as metaphor, metonymy, and functional relations (see Article 25, WORD MEANING).

Another important outcome of cognitive research that undermines moral law theories is the discovery that most of our abstract concepts are defined by sets of metaphors (Lakoff, 1987), some of which are typically inconsistent with one another. Johnson (1993) and Lakoff (1996) analyze a number of moral concepts, such as *rights, justice, revenge, well-being*, and *will*, to reveal underlying conceptual metaphors that jointly define our entire moral orientation: metaphors such as *morality is health, moral strength, moral boundaries, nurturance, light/dark, uprightness*, and *purity*. If our fundamental moral concepts are defined by *multiple* and possibly inconsistent conceptual metaphors, then the literalist picture of moral thinking (in which concepts are supposed to map directly and univocally onto situations) cannot be correct.

Reasoning

This large body of research on conceptual structure thus wreaks havoc on any traditional moral law theory, since it shows that morality is not principally a matter of learning and following univocal, literal rules. Because our moral concepts are structured by prototypes, metaphors, metonymies, and other imaginative devices, the moral reasoning we engage in with such concepts is almost never strictly deductive. Only in the most trivial cases – those in which the specific situation fits some prototype within a moral category – can we simply subsume a particular case under a general ethical rule. Most of our moral deliberation consists, instead, in exploring possible metaphorical and metonymic extensions from prototypical cases to nonprototypical cases. This does not mean that there is no place for moral principles, but rather that the principles we do have should be understood as idealized strategies, based on past experience and defined via prototypes, models, and metaphors.

The way we reason about a situation depends on the way we *frame* it. Consider, for instance, the abortion debate. How is it to be framed as a moral issue? Typically, it is framed as a matter of rights – the rights of the fetus versus the rights of the mother – and within this context the debate focuses almost exclusively on whose rights should take precedence. However, the abortion issue does not come with the label "rights" stamped on it. Framed instead as a question of social welfare, the debate would center around whether communal well-being is enhanced or diminished by the practice of abortion under certain specific constraints. Or, within the framework of political liberation, abortion might be regarded as justified only in situations where it is a means to realizing human freedom of oppressed communities.

Moreover, within various framings of the abortion issue, there is the further problem of deciding what counts as a *person*, given restrictions about how persons are to be treated. The concept *person* is a *contested concept*, one that has an underspecified core that is not contested but that can be extended in any number of highly controversial and contested ways. There is no absolute or frame-neutral stance from which to define or extend such terms; rather, social, historical, religious, philosophical, and political factors give rise to potentially incommensurable framings of these concepts.

In sum, this sort of complexity and frame dependence for our concepts and reasoning makes a traditional moral law account impossible. Moral law theories must presuppose precisely what cannot be taken for granted: namely, that our moral concepts are literal, internally homogeneous, and uncontested in any fundamental sense. Such cognitive results thus undermine any form of moral fundamentalism.

Another important source of converging empirical evidence concerning framing comes from psychological studies of the heuristics and models which people use in making probability determinations under conditions of risk (see Article 44, HEURISTICS AND SATISFICING). Much contemporary moral theory rests on a classic economic conception of rationality, according to which people are seen as being incentive-driven by what they perceive as in their self-interest and as basing their decisions, using logical reasoning, on all available information, including the impact that current choices will have on future choices (see Article 10, DECISION MAKING). Both classical utilitarianism, which determines the moral correctness of an action by how well it contributes to the overall well-being of all affected people, and egoism, which requires the calculation of self-interest, are clear expressions of economic rationality in ethics.

A host of studies growing out of the work of Amos Tversky and Daniel Kahneman (1974) suggest that a strict economic conception of rationality may not be cognitively realistic. People typically do not, and probably cannot, make good probabilistic judgments of the sort required by an economic conception of rationality. The way in which a decision under risk is framed makes a big difference to what people will decide is in their best interest in a particular situation. People are risk-averse when deciding about possible gains and risk-seeking when considering possible losses. The *endowment effect* makes people reluctant to risk assets that belong to their endowment, even when objective probabilities might suggest possible substantial gains. The extreme difficulty of making good probability judgments calls into question any simple economic model of rationality.

Critics of the Tversky–Kahneman studies have tried to defend the economic conception of rationality by arguing that the studies are flawed and that people are not really so bad at probabilistic reasoning. Still, no one disputes the fact that people employ a range of frames and heuristics for reasoning. Our reasoning, therefore, is not frame- or context-neutral, and this challenges both the absolutistic pretensions of classical moral law views and the classical economic model of reasoning. Simply put, the rational calculations of either self-interest or communal well-being, if not humanly impossible, are certainly extremely difficult to perform, especially since most of our moral goals require complex cooperative activity among many people over long periods of time. The idea that individuals can calculate what is in their own best interest appears to be an illusion, not to mention the impossibility of calculating what will produce the greatest communal good.

Emotion and moral deliberation

Most contemporary moral theory, as well as most of the Western moral tradition, assumes a fundamental ontological and epistemological gap between reason and emotion. On the basis of this alleged dichotomy, philosophers typically argue that morality stems either from reason alone or from feeling alone. In the first camp, Kant was the most notorious champion of *pure* reason, rejecting feeling as an inappropriate basis for moral judgment, since he thought that feeling was subjective and individual and was therefore incapable of supporting absolutely binding universal moral laws. In the second camp, Hume argued that moral judgments are based not on reason, but on moral sentiments. He saw reason as a calculating faculty that issues judgments about truth, whereas the passions alone move us to action: "Since morals, therefore, are meant to have an influence on the actions and affections, it follows, that they cannot be deriv'd from reason; . . . Morals excite passions, and produce or prevent actions; Reason of itself is utterly impotent in this particular. The rules of morality, therefore, are not conclusions of reason" (*Treatise of Human Nature*, book 3, part l, section l). Following Hume, twentieth-century emotivists, such as A. J. Ayer and C. L. Stevenson, claim that moral judgments are not cognitive or descriptive, but rather are expressions of emotion or attitude intended to influence action.

However much they appear to differ, both the champions of reason and the champions of emotion share the same mistaken assumptions that reason is radically separate from emotion and that moral judgment must be based on one side of this split, to the total exclusion of the other. Both these assumptions are false. Neither side in the debate has paid any significant attention to research on affect and reason, which is why most moral theories cannot provide an adequate account of moral motivation or of the role of emotion in moral deliberation (see Article 11, EMOTIONS).

Some of the most significant research on this topic to date is Antonio Damasio's (1994) investigation of brain damage that affects reasoning, including social and moral deliberation. Damasio argues that "certain aspects of the process of emotion and feeling are indispensable for rationality" (p. xiii). Especially when it comes to all forms of practical reasoning, reason cannot do its job independent of complex emotional processes that are connected to and monitor global states of the body. This does not mean that morality is not rational, but rather that our practical and moral reasoning are always oriented and guided by affective states. Thus, moral deliberation is neither purely rational nor purely a matter of feeling or emotion alone. Instead, reason can act only in concert with our emotions.

Another striking result of Damasio's research is that there is no such thing as a *pure* reason wholly independent of human embodiment. Bodily processes and experiences establish the patterns of our reasoning: "Our most refined thoughts and best actions, our greatest joys and deepest sorrows, use the body as a yardstick" (p. xvi). We are just beginning to understand the profoundly important role of embodiment in human moral understanding (Johnson, 1993; Lakoff, 1996), but the evidence indicates that the traditional conception of will and reason as disembodied, pure, and radically free is bankrupt.

Empathy and self-formation

Empirical studies suggest that, from a cognitive and developmental point of view, empathic feelings are the principal basis for our ability to care about other people and hence the source of morality (Goldman, 1993). Moreover, the self is an interpersonal self, existing in relation to others and by virtue of its ability to have empathic feelings for others. Stern (1985) presents a broad range of studies of infant development showing that, although there obviously are innate capacities, an infant's sense of self emerges only through interpersonal interactions in which he or she becomes cognitively and affectively attuned to parents and siblings. This *affect attunement* is a communicative process in which parent and child respond interactively to each other's moods, feelings, and attention, typically at a level beneath that of conscious awareness. In such an ongoing process of mutual coordination, parent and infant tend to match or loosely imitate each other's behavior, but often in different sensory modes, such as when a baby succeeds in grabbing a toy and lets out an exuberant "Aaaaah!," followed by the mother's scrunching up her shoulders and shimmying her upper body for a period approximating the duration of the "Aaaaah!"

Stern sees affect attunement as a precondition for true empathy, which involves additional cognitive processes, such as abstracting empathetic knowledge from the experience of emotional resonance and thereby understanding how another person feels and experiences a situation. In a series of studies over many years, Hoffman (1993) has assembled a massive case for the central role of empathy in any account of moral experience. He shows how empathy develops from birth, examines its crucial role as a moral motive, and studies sex differences and the contribution of sex-role socialization in empathy-based morality. Hoffman's results suggest five major ways in which empathy lies at the heart of moral motivation and judgment. (1) Early on, children who witness the distress of another person come to respond empathically with feelings more appropriate to the person in distress than to themselves. (2) This empathic distress scheme is readily extended to people who are not present and to imagined sufferings, because we have the imaginative capacity to represent to ourselves the experience of others. (3) Empathic affects are largely congruent with forms of caring and also with various forms of justice; in this way, our moral principles may be activated by basic empathy. (4) Because *cool* affective states can become *hot* cognitions, even abstract moral principles can acquire a motive force via the heating up of the empathic component of a principle. (5) Empathy plays a crucial role in moral judgment, such as when it supports moral impartiality.

This picture of the interpersonal self and of the crucial role of empathy and other emotions undercuts the traditional Enlightenment view of moral agents as autonomous, atomistic, rational egos with moral personality formed prior to their actions and social relations. Cognitive research supports the communitarian critiques set forth in works such as Alasdair MacIntyre's *After Virtue* and Michael Sandel's *Liberalism and the Limits of Justice*, which argue that contemporary versions of Enlightenment rationalism, such as Rawls's *A Theory of Justice* (1971) and Nozick's *Anarchy, State, and Utopia* (1974), assume a mistaken view of moral agents as existing prior to, and independent of, their actions and social interactions. The archetypal Kantian conception of the self as a radically free rational ego capable of deliberating and willing in accordance with pure practical reason is an illusion. We exist in and through others, and our moral concern depends on our ability to empathize with others. Cognitive studies of empathy and emotion thus promise to shed light on some of the most vexing problems of moral theory, such as how moral principles can have any motive force, why we care about the well-being of others, how impartiality is experientially grounded, and why there is no radical split between reason and emotion.

Moral development

Every moral theory presupposes some theory of moral development. Until quite recently, Kantians and other moral law theorists took great comfort from knowing that the two major studies of moral development, Jean Piaget's *The Moral Judgment of the Child* and Lawrence Kohlberg's *Essays on Moral Development*, seemed to support their general view of the structure of moral understanding. Kohlberg's view, in particular, recognized Kantian autonomy as the pinnacle of moral development. Kohlberg's six stages of possible moral development ranged from children following rules laid down by external parental authority (first stage), through intermediate stages of social constraint, and aiming toward autonomous guidance by universal ethical principles (final stage).

Over the last two decades this picture of mature morality as autonomous rule following in accordance with universal rational principles has come under severe criticism from further studies of cognitive development. Flanagan (1991) summarizes five major areas in which research has challenged the model of discrete, homogeneous stages through which children pass on their way to moral maturity. The main objections stem from evidence against the existence of holistic, unified, general-purpose stages and also from evidence that people seldom rise beyond a mixed blend of stages 2–3 and 3–4. Much has also been made of the problem of drawing any psychologically realistic claims from standard tests, especially those (like Kohlberg's) based on verbal reports about hypothetical situations. Moreover, when one begins to unpack some of the guiding assumptions built into the tests, such as that morality is a system of rules for behavior, it appears that many studies are begging key questions about the scope and nature of morality. For example, the assumption that *later* stages are *higher* stages automatically excludes the real possibility that mature moral understanding might integrate a number of different considerations and forms of judgment, instead of focusing on one privileged mode of thought.

What cognitive science discovers about moral development will place substantial constraints on the structure of a psychologically realistic moral theory. If, for instance, cross-cultural studies, such as Richard Shweder's *Thinking through Cultures*, reveal wide cultural differences concerning moral ideals and patterns of moral development, then claims about either absolute, universal values or universal moral rules become highly questionable.

Another important aspect of moral development concerns the role of the family as a basis for moral understanding and growth. Lakoff's (1996) analysis of the relation of morality and politics reveals the central role of family morality as a basis both for a person's general moral values and for their political orientation. Lakoff contrasts *strict father morality* (a morality that prizes authority, discipline, moral strength, and order) with *nurturant parent morality* (emphasizing empathy, care, nurturance, and compassion). He then argues that developmental studies in the area known as *attachment theory* uncover severe problems with strict father morality. For example, strict father morality prides itself on producing self-reliant, disciplined, autonomous people. In fact, studies show that it tends to produce people who lack self-confidence, cannot make decisions for themselves, are unable to criticize authority, and do not manifest great moral strength. In short, studies of moral development can sometimes lead directly to fundamental critiques of moral frameworks and can establish the range of psychologically plausible moral systems.

Gender

Especially since the Enlightenment, moral agency has typically been characterized by mainstream philosophers as gender-neutral, based on an allegedly universal rationality that is supposedly shared by all free and equal rational beings. Feminist philosophers have lately subjected this myth to scathing critique, pointing out the many ways in which assumptions based on traditional gender typing are built into the structure of the dominant universalist view. Only in the last 15 years has empirical research on gender differences been brought to bear on this important issue. Carol Gilligan (1982) challenged Kohlberg's studies (which were based on a small number of males) and opened the way for a more thorough and self-critical investigation of possible gender differences in moral understanding. While she distinguishes a morality of rights and justice from a morality of care and responsibility, she is careful not to draw this distinction strictly along gender lines. She suggests, however, that girls are, and boys are not, typically educated into the culture of care, thereby learning very different ways from boys to handle moral conflicts.

The main questions are: Does this gender-differentiated model really stand up under scrutiny? If there are such differences, are they based on genetic differences or on socialization? The evidence is not in yet to decide such issues. Flanagan (1991) offers an extensive survey of recent work on the issue of gender and moral orientation. He looks at studies of virtually every aspect of this question, from whether women score lower than men on Kohlberg-type tests (answer: there are no statistically significant differences) to whether there are only two basic gestalts for moral orientation (answer: no), to whether the justice and care gestalts appear to split neatly along gender lines (answer:

no). Flanagan concludes that, although such gender studies have been important in greatly enriching our moral understanding, there is not yet sufficient experimental evidence to draw strong conclusions about the role of gender in morality: "Moral personality is, in the end, too variegated and multipurpose to be analyzable in terms of a simple two-orientation scheme – even blended together" (p. 233).

Why cognitive science matters to morality

The above six areas mark out substantial domains of empirical research in the cognitive sciences that should directly constrain the form and content of any humanly realizable morality. Quite obviously, the joint constraints established by research in these areas leave open a wide range of possible moral orientations. There is no cognitively realistic way to avoid such pluralism. Nor would it be advisable to try to avoid it, even if we could, since it is the very existence of multiple conceptions of human well-being and modes of living that makes us aware of the limits of our own moral views and also opens up possibilities for moral growth.

Those who think that empirical research is irrelevant to moral theory often argue that empirical theories change over time, so that today's popular theory of cognition may be tomorrow's whipping boy, as new empirical studies raise questions and criticisms. But this is no argument against the relevance of cognitive science for morality, because morality, as Dewey said, is experimental – it is a massive ongoing communal experiment in solving the problems of human living and flourishing that confront us every day. A morality which cannot be revised as new discoveries about the mind are made known is a dead morality incapable of meeting the kinds of change that are part of human existence.

The cognitive sciences will not generally tell us how to act in particular situations. But they provide moral guidance because they contribute to our moral understanding – our understanding of human nature, of the workings of the mind, and of how we make sense of and reason about moral issues. Their normative significance comes from what they reveal about what it means to be human and about what is required for intelligent, critical, and constructive moral judgment and action.

References and recommended reading

Churchland, P. 1995: *The Engine of Reason, the Seat of the Soul: A Philosophical Journey into the Brain*. Cambridge, Mass.: MIT Press.

Damasio, A. 1994: *Descartes' Error: Emotion, Reason, and the Human Brain.* New York: Grosset/Putnam.

Dewey, J. 1922: *Human Nature and Conduct*. In *The Middle Works of John Dewey*, ed. J. Boydston. Carbondale, Ill.: Southern Illinois University Press.

*Flanagan, O. 1991: *Varieties of Moral Personality: Ethics and Psychological Realism*. Cambridge, Mass.: Harvard University Press.

Gilligan, C. 1982: *In a Different Voice: Psychological Theory and Women's Development*. Cambridge, Mass.: Harvard University Press.

Goldman, A. 1993: Ethics and cognitive science. *Ethics*, 103, 337–60.

Hoffman, M. 1993: The contribution of empathy to justice and moral judgment. In A. Goldman (ed.), *Readings in Philosophy and Cognitive Science*, Cambridge, Mass.: MIT Press, 647–80.

*Johnson, M. 1993: *Moral Imagination: Implications of Cognitive Science for Ethics.* Chicago: University of Chicago Press.

Lakoff, G. 1987: *Women, Fire, and Dangerous Things: What Categories Reveal about the Mind.* Chicago: University of Chicago Press.

—— 1996: *Moral Politics: What Conservatives Know that Liberals Don't.* Chicago: University of Chicago Press.

*May, L., Friedman, M. and Clark, A. (eds) 1996: *Mind and Morals: Essays on Ethics and Cognitive Science.* Cambridge, Mass.: MIT Press.

Stern, D. 1985: *The Interpersonal World of the Infant: A View from Psychoanalysis and Developmental Psychology.* New York: Basic Books.

Tversky, A. and Kahneman, D. 1974: Judgment under uncertainty: heuristics and biases. *Science,* 185, 1124–31.

56

Everyday life environments

ALEX KIRLIK

Introduction

Few scientific disciplines have the potential of cognitive science to speak to the problems which people face every day in dealing with an increasingly complex technological, social, and cultural environment. In principle, understanding activities such as perceiving, learning, reasoning, speaking, deciding, and acting should be relevant to improving either ourselves or our situations when cognition is necessary for achieving our goals. Whether by improving our inner world (e.g., through education) or our outer world (e.g., through design), cognitive science promises to provide resources to help people better adapt to their environments and better adapt their environments to themselves.

It is quite likely that you have engaged in a sort of informal, applied cognitive science yourself. Cognitive scientists often remark that people share a variety of beliefs and assumptions about how their own minds work – that is, a *folk psychology*. Not surprisingly, these folk theories also have their practical or applied counterparts: look around your home or office and observe the many ways you have organized your world to help you accomplish cognitive tasks. Creating a shopping list aids long-term memory, using a pen and paper while balancing a checkbook supports working memory, and highlighting important references in this book aids attention allocation. In addition, it is likely that you have purchased a variety of artifacts, such as calendars, calculators, or computers, that were created primarily to support cognitive activities (see Article 40, MEDIATED ACTION).

The vast majority of these cognitive artifacts were designed not by the application of cognitive theory but instead by appeal to folk-psychological intuitions, trial and error, and the forces of the marketplace. Only recently have researchers begun to consider how theories and findings from cognitive science can be systematically applied to the design of the cognitive environment. This design-relevant cognitive science research is characterized by a rich interplay between theory and design, in which basic science and application are intimately related and mutually informing. This mutual influence between theory and design in cognitive science has proved to be a healthy state of affairs, signaling some novel and valuable lines of attack for addressing some of our central theoretical problems. The very recent *opportunity* for an applied cognitive science, enabled by advances in information technology, may provide an important additional source of empirical constraint on theorizing needed to move toward a science of how people perform relevant cognitive tasks in actual situations. As I will discuss below, study of the role of cognitive artifacts in human activity has already provided valuable data challenging and informing theory in cognitive science.

In the following, I first provide a brief overview of the history of efforts to apply psychology to design. I then turn to the foundational task of laying out the logic underlying how inferences about cognition can reliably be made by observing instances of human interaction with both successfully and poorly designed artifacts. The empirical portion of the chapter consists of a discussion and analysis of a number of diverse, everyday activities in the home and the work place, with a focus on the role which designed artifacts play in influencing cognition and behavior. For each activity I also describe how an understanding of the artifact's role either has, or should, shed light on theoretical issues of current interest to cognitive science. Next, and in part to demonstrate that investigations of applied issues have been both fruitful and wide-ranging, I discuss a recent theoretical work inspired largely by studies of people interacting with designed environments. In closing, I offer comments on prospects for theories of everyday cognition, with an emphasis on how design-relevant research offers promise to shape these theories.

Designing with cognition in mind

Systematic application of psychological theories, methods, and findings to the design of artifacts has a much longer history than cognitive science itself. Around the time of World War II, the field of human factors emerged in response to practical design problems imposed by effectively coupling soldiers, sailors, and aviators with increasingly sophisticated technological systems such as artillery, naval ships, and aircraft. Most human factors research has dealt with performance-oriented issues such as visual ability, selective and divided attention, memory limitations, and motor control, with implications for training, personnel selection, and the design of interface displays and controls. The design of a modern automobile, for example, has been influenced in dozens of ways by human factors research: anthropometry (the study of human physical dimensions) has been used to assure that controls are within a reachable distance for the vast majority of the driving population; biomechanics (the study of human mechanical abilities) has been used to ensure that controls can be manipulated with acceptable force; and research on vision and perception has been used to improve the readability of dashboard displays. Much of this work applying the psychology of human performance to design is still relevant today and is required material for most industrial engineers and engineering psychologists.

In the past 20 years or so, research entailing the application of psychology to design has shifted its focus to more distinctly cognitive issues, such as the types of knowledge required to operate or diagnose a system, the acquisition of expertise, the processes of judgment, decision making, and problem solving, and the production of human error. Two factors have prompted this shift in emphasis. First, the increasing complexity of the systems controlled by humans (e.g., desktop computers, nuclear power plants, commercial aircraft) has created a need to understand how to design interfaces enhancing people's abilities to learn and reason using sophisticated knowledge. Second, the growth of cognitive science itself has provided the opportunity to begin to address these needs in a scientifically grounded manner.

Two new fields have recently emerged from the marriage of cognitive science and the design disciplines. Human–computer interaction (HCI) is an area of research focused on the problems of designing information technology to support the tasks that people perform using computer systems. Perhaps the moment defining the onset of this field was the publication of a book on the psychology of HCI by Stuart Card, Thomas Moran, and Alan Newell (1983). Based on the collaborative efforts of Xerox PARC and Carnegie–Mellon cognitive scientists, and strongly influenced by Alan Newell and Herbert Simon's seminal research on problem solving, this ground-breaking book provided a detailed model of the computer-user called the "model human processor," and demonstrated how the model could be used to predict durations for various computer-related tasks. This book also presented the GOMS (goals, operators, methods, selection of rules) model of user behavior, which has subsequently been extended and computationally implemented using the production system architecture by David Kieras of the University of Michigan and Peter Polson of the University of Colorado (1985). Applied mainly to procedural tasks such as using a word processor, the employment of computational models to describe user behavior, as pioneered by these researchers, has been extended in a variety of additional directions, such as predicting the ease of learning to use a computer application or predicting the effects of a user transferring from one application to another. This style of HCI research represents perhaps the most direct, detailed, and sustained attempt to apply existing cognitive science theory to interface design.

A second, closely related discipline that has recently emerged at the intersection of cognitive science and design is cognitive engineering. This term was popularized by Jens Rasmussen in a book on the cognitive engineering of complex, dynamic systems (Rasmussen, 1986), and by Donald Norman in a paper on the cognitive engineering of computer systems (Norman, 1986). Norman, for many years a cognitive scientist at the University of California, San Diego, and now working in the computer industry, is also well known for his innovative treatment of the subtleties of designing artifacts to support everyday cognition.

Cognitive engineering and HCI share many of the same concerns, but those researchers and practitioners most likely to identify themselves as cognitive engineers tend to be engaged in the design of complex, typically dynamic, human–machine systems. Rasmussen, formerly of Denmark's Risoe National Laboratory, spent many years studying nuclear power plant design and operator behavior, in order to improve system safety through interface design, training, and aiding. As this application domain would suggest, cognitive engineers are often concerned with questions associated with human–computer interaction; but cognitive engineers view information technology not as the target of human interaction per se, but rather as an intermediary between a person or team controlling a system and the system itself. Thus, cognitive engineers view their role as not only attempting to ensure the efficiency of human–computer interaction, but also to ensure that the type of interaction promoted by a particular interface is also appropriate to the demands of the larger system or context in which the interface is embedded. As such, cognitive engineers often spend as much time working to understand an application domain (e.g., commercial aviation, power plants) as they do in working to understand operator cognition and behavior in these domains, since interface design must be informed by both types of knowledge. A survey of the range of psychological issues of concern to cognitive engineering is provided by David Woods and Emile Roth in their paper on cognitive systems engineering (Woods and Roth, 1988).

There are clearly many areas of overlap between the disciplines of human factors, HCI, and cognitive engineering, and many researchers concerned with design identify themselves with more than one of these fields. All three disciplines touch on the relation between psychology and the design of artifacts in everyday life environments to some extent, and I will draw upon the work and opinions of a wide variety of researchers in the following overview and analysis.

Psychological theory and psychological artifact

John Carroll, for many years an HCI researcher at the IBM Watson Research Center who is now at Virginia Tech, has thought deeply about the relationship between theory and design in cognitive science. In a pair of papers (Carroll and Campbell, 1989; Carroll, 1991), Carroll persuasively argues that explicit extensions of academic cognitive science to HCI have had little actual impact on design practice. Moreover, he observes that the most influential HCI design work has been done without any guidance from cognitive science whatsoever. Carroll notes that task analysis techniques and modeling methods taken from cognitive science have been shown to have predictive value for describing only relatively simple, highly constrained cognitive tasks. Although some detailed aspects of a design might be informed by direct application of current cognitive theory, most (and all the molar) aspects of design must be created without scientific guidance. While there are undoubtedly those who would disagree with this view, few who have actually designed systems *from the ground up* would, and the more recent history of HCI research demonstrates a distinct shift of focus away from computational cognitive modeling as the dominant research methodology.

In noting that design often precedes theory in a variety of sciences (e.g., bridge building preceded physics), Campbell points out that HCI has a unique opportunity to contribute to cognitive theory by a process akin to reverse engineering. Each designed artifact embodies an implicit theory concerning the nature of human interaction. Uncovering the theories embodied in successful and unsuccessful designs can inform us about the functional properties of the environment that both foster and inhibit successful cognitive performance. An excellent example of the use of this methodology appears in an article by Edwin Hutchins, James Hollan, and Donald Norman (1986). These researchers observed that the *direct manipulation interface* (as opposed to typing verbal commands) was one of the most significant advances in interface design, yet its design was not informed by cognitive theory. They therefore reverse-engineered the direct manipulation interface to discover the environmental properties that give rise to direct engagement and fluent activity on the one hand and mediated and thus cognitively intensive activity on the other.

I invite you to consider a particularly troublesome interactive device (a VCR, perhaps) and identify the flawed cognitive theory implicit in its design. In my own case, I find that it would be relatively trivial to write a simple computer program that could unerringly operate my VCR, which speaks to the type of cognitive creature for whom it was optimally designed. However, my cognitive abilities are severely strained in trying to operate the VCR myself. I apparently have little difficulty interpreting the relatively simple procedures contained in the VCR's instruction manual, which is why I could easily write a computer program based upon these instructions. What I cannot do well, and what the computer excels at, is to recall and consistently execute these procedures at a later time when they are needed. Apparently, the implicit theory underlying the design of the VCR is that I have memory and execution abilities on a par with a computer system able to retrieve and run arbitrary procedures. That I cannot do so efficiently testifies to the fact that my memory and execution abilities are quite unlike those presumed by this theory. It is also revealing that exactly the opposite situation often holds in the case of my own interaction with either the natural environment or well-designed artifacts. In these environments, my behavior is often fluent and successful, but I would be hard pressed to write a procedural computer program that could succeed in these situations.

People's capacity for synthesis greatly exceeds their capacity for analysis, in that our creative products often reflect a degree of tacit knowledge that goes well beyond what we can explicitly state or formalize. Good designers can produce successful artifacts without explicit knowledge of the psychological theory that would justify why the design succeeds, just as one might be able to successfully stack a pile of books without explicit knowledge of the physics suggesting why the pile does not fall. The methodology of examining artifacts to uncover their implicit psychological theories has the potential to reveal information about how people interact with the world transcending the knowledge that could be gained through purely analytical methods.

Now is a time of high entropy in the field of HCI, as researchers are just as likely to be involved in the creation of their own theories of interaction as they are in creating design principles, frameworks, or techniques. As foreseen by Carroll, much of the literature in HCI currently consists of reporting the development of novel interface prototypes and studies of human interaction with various artifacts in an effort to determine what functional properties of the environment influence cognition and action (i.e., reverse engineering). As such, a good deal of modern HCI research has gained a distinctly ecological or functionalist focus, looking simultaneously inward toward the mind and outward toward the world for an understanding of cognition in context. In pursuit of this goal, recent years have seen a renewed interest in the ideas of earlier theorists such as Lev Vygotsky, James Gibson, and Egon Brunswik, who all believed that a theory of behavior must have the conceptual resources to span the boundary of the skin in order to describe both the internal and the environmental structures that influence cognition and action.

This renewed interest in functionalist or ecological approaches has been paralleled somewhat independently in the related field of cognitive engineering. Many researchers involved with the design of interfaces to complex, dynamic systems are also finding value in describing both the cognitive and the environmental determinants of behavior in order to guide design. As mentioned previously, many cognitive engineering researchers spend considerable time – in some cases years – achieving expert knowledge of the application domains in which they work. They do so because they understand that, in order to effectively support activity in these domains, they must have knowledge of both the internal, cognitive constraints upon behavior and the external, environmental constraints upon behavior. As a result, an increasing number of cognitive engineering researchers are taking an explicitly ecological approach to the analysis and design of human–machine systems, drawing upon the theories of James J. Gibson and Egon Brunswik.

What we may be witnessing in this recent return to an emphasis on environmental issues in HCI and cognitive engineering may be a backlash against what Donald Norman has described as cognitive science's original preoccupation with the *disembodied intellect* at the expense of seriously considering the external contributions to cognition and behavior. One should not be surprised that much of the impetus for this change comes from those concerned with design, since understanding interaction with artifacts demands consideration of environmental issues, whereas advances may perhaps be made on other cognitive fronts without meeting such a requirement head on. What kinds of evidence do studies concerned with the design of everyday life environments provide to suggest that a recentering of cognitive science to give equal attention to both internal and environmental issues will be fruitful?

Heading out for dinner

Hunger and thirst arrive, and you and some friends head to a restaurant for dinner. While waiting for a table, you move to the crowded bar and ask your friends what they would like to drink. Concentrating intently, perhaps by rehearsing their drink orders to yourself, you finally locate the bartender and shout out the list. The bartender prepares and serves the drinks flawlessly.

Why does it seem that the bartender does not have to concentrate as hard as you do in order to remember the list of drink orders? King Beach of the City University of New York conducted a study to shed light on this issue (Beach, 1988). Beach found ten novice and ten expert bartenders and had them perform a task modeled on a common bartending school activity called the "speed drill." In this task four drink orders are orally presented to a bartender, who then mixes the four drinks as quickly and as accurately as possible, and performance is measured for both speed and accuracy. In the first experimental manipulation, Beach required both groups of bartenders to count backward from 40 by threes on a subset of trials, in order to limit their ability to verbally rehearse the set of drink orders. As expected, he found that this manipulation greatly increased the number of drink errors made by novices. Perhaps surprisingly, however, the expert bartenders were unaffected by the distracting task of counting backward.

Beach's second experimental manipulation revealed the source of this expert–novice difference. Here, rather than allowing the two groups of bartenders to use standard glassware, he required them to use instead a set of identical, opaque black glasses for preparing the drinks on a subset of trials. Beach found that the novice bartenders were unaffected by this manipulation, whereas the number of errors committed by experts increased 17-fold! The expert bartenders were selecting glasses for drinks at the time of ordering and using these glasses as an external memory store. Joy Stevens, also of the City University of New York, has done a study on waitress memory with similar findings (Stevens, 1993). By contrast with the dominant theoretical view of skilled memory as entailing more numerous and more elaborate links in a semantic network in the head, Stevens concludes that her research instead provides evidence "for links between the person and the context in which she is working." Expertise in these everyday tasks clearly owes much to an ability to exploit regularities in the structure of the designed environment (see Article 39, EMBODIED, SITUATED, AND DISTRIBUTED COGNITION).

After drinks, your party is seated and orders steaks, some rare, some medium, and another well done. The cook grills your party's steaks, and all are served hot and according to specifications. I was recently interested in how cooks manage the apparently difficult task of simultaneously preparing multiple food orders and undertook an observational study of this situation at a local diner. In the situation studied, the cook sometimes had more than ten pieces of meat cooking on the grill simultaneously, and was also involved with a variety of other food-preparation activities. I observed three quite different strategies for managing the task of grilling meats. In the simplest cases, in which perhaps only a few meats were present on the grill, the cook would throw the meats on the grill in a haphazard fashion, and repeatedly check the underside of each piece of meat to ensure that it was cooked appropriately. Grilling behavior became much more interesting when a larger number of meats had to be prepared simultaneously.

In particular, one cook divided the grill into three sections, associated with meats to be prepared rare, medium, and well done. In this manner the cook was able to off-load the task of remembering how well each meat should be cooked by structuring the environment to contain this information. However, this cook still used a process of continually checking the underside of each piece to ensure that it was cooked to the degree called for by its position on the grill.

Even more interesting was the behavior of another cook, who managed to off-load the demands of even this latter task. This cook also divided the grill into regions associated with rare, medium, and well done, as three horizontal slices in the front, center, and rear of the grill surface. Meats to be cooked well were placed in the rearmost slice toward the far right of the grill, and meats to be cooked medium were placed in the center horizontal slice and also toward the right of the grill, but not as far to the right as the meats to be cooked well done. Finally, meats to be cooked rare were placed in the frontmost slice of the grill and toward the center of the grill looking left to right. After the *initial conditions* had been set in this manner, the cook then intermittently slid each piece of meat a small distance toward the left of the grill, inducing a slow leftward movement of the food over the course of cooking. In this way, each piece of meat could be flipped at the midpoint of its journey across the grill, and would signal its own completion as it reached the left barrier of the grill surface. By starting meats requiring more cooking time further toward the right side of the grill, and by inducing a relatively fixed velocity of meats to the left, this strategy ensures that each piece of meat is cooked to the correct doneness. This *trick* reduced the need for either checking the meat's underside or using an internal predictive model of the rate of meat cooking in lieu of perceptual information.

What lessons can we learn from these studies of humans interacting with artifacts in our visit to the restaurant? The skilled memory studies demonstrate that artifacts possess functional properties as components of a memory system crossing the boundary between person and environment. In addition, the cooking study demonstrates that when the given environmental design does not possess structure capable of playing a supportive functional role, people can and sometimes will create novel sources of structure that are capable of contributing a supportive functional role. In this case, the environmental structuring created by the cook's own activities resulted in reducing demands on internal memory.

These examples help explain why an increasing number of HCI and cognitive engineering researchers are entertaining functionalist or ecological perspectives. If we consider how we might construct a computational model to simulate the activities of either the bartender or the cook, it should be clear that these models would have to contain resources for describing both the internal and the environmental structures that participate in the execution of these skills. Models of the bartender using the opaque glassware, and of the cook using the haphazard strategy, would have to possess relatively sophisticated structure and function devoted to describing internal cognition, while the structure and function devoted to describing the external environment of these performers could be quite simple. On the other hand, models of the expert bartender using standard glassware, and of the cleverest cook, could have a comparatively simpler component describing internal cognition, while the structure and function devoted to describing the environments of these expert performers would have to be much more elaborate. The unit of analysis required to theorize about issues such as these is the functional system incorporating both person and environment. This systems perspective on the proper unit of analysis for theorizing and modeling behavior is the hallmark of functionalist or ecological approaches, representing the common ground of earlier theorists such as Vygotsky, Gibson, and Brunswik.

Dealing with machines

An increasing amount of everyday activity centers around interacting with machines. Studies of both successful and unsuccessful designs have provided important insights into the nature of human interaction with the world. The previously discussed case of reverse engineering the direct manipulation interface by Hutchins, Hollan, and Norman is a good example of learning from successful design. On the other hand, a failed design can also provide valuable information when the psychological theory underlying the design can be articulated. Lucy Suchman (1987) performed such a study from a distinctly anthropological perspective. An anthropologist at Xerox PARC, Suchman undertook to find out why a particular copier's *help* system was misnamed. The help system failed to provide timely, contextually appropriate assistance to novice users. This device, described as an "expert help system," was intended to guide copier-users through the actions necessary to perform a variety of copying tasks. Suchman observed that the psychological theory underlying the design of this system was the *planning model* taken from theoretical cognitive science.

In the planning model, human action is understood to flow from plans, which are represented sequences of intended actions. Generating a plan is accomplished by a process of problem solving, in order to choose a set of ordered actions that will change the current environmental state into a desired state. The expert help system in the copier which Suchman studied assumed that the human user could be described in such terms. Thus, the help system worked by using its knowledge of actions taken by the user to attribute a particular plan to the user. Given this knowledge of the user's intent, the system could then presumably provide the user with timely and contextually appropriate instructions on how the user's goal could be met by manipulating the copier's controls.

Suchman concluded that the failure of the expert help system's design was due to nothing less than the failure of central aspects of the planning model itself, as a theory of user behavior in this particular situation. Her conclusion was based on a fine-grained analysis of the detailed interaction between the help system and the users she studied. In short, Suchman argued that the planning model, as implemented in the help system, did not provide the resources to support the improvisational, sense-making activities that people engage in during conversation in order to achieve mutual understanding. She concluded that human action in many situations is more faithfully described as *situated* rather than planned. What this distinction amounts to precisely has been the topic of much recent debate in cognitive science, and a sufficient treatment of this issue cannot be provided here. Suffice it to say that a key element of the situated perspective is the view that the traditional planning model overestimates the amount of problem solving underlying behavior and instead views behavior as largely improvised by combining information from perhaps sketchy plans with information concerning many of the fine details of a person's environmental situation.

The design implications of this situated perspective are not yet clear. For example, it could be argued that the system which Suchman studied was both too smart and not smart enough to foster effective human activity. On the one hand, many of the so-called intelligent features of the help system clearly added a level of complexity and confusion to interaction that would otherwise not have been present (i.e., the device was too complex to be comprehensible and predictable to its users). User behavior would have been better supported perhaps with improved design of the apparently confusing copier interface itself, rather than by adding a sophisticated help system to assist users in coping with the deficiencies of the existing interface. An analogy can be given to the design of doors: the need for a *help* system on a door (a "Push" or "Pull" sign) typically indicates that the door itself provides a poor interface suggesting how it should be manipulated. The need for signs on doors can typically be eliminated by the design of improved interfaces (e.g., improved handles or knobs). On the other hand, an even more intelligent help system for Suchman's copier, with the full capabilities of a human expert, would also have better supported user activity than the minimally intelligent system that was in place. Technology is not presently available to build a help system with this level of intelligence. Although the design implications of the situated perspective are as yet unclear, this view highlights an essential truth about the often intimate relationship between people and the environment during activity, and thus the need for cognitive theory with resources rich enough to capture many of the fine details of a person's environmental situation.

Getting around and getting a head

I want to close this discussion of lessons learned through design and through studying the designed environment by mentioning a major theoretical work explicitly addressing the contribution of the artifact to cognition and behavior. Edwin Hutchins is a cognitive scientist at the University of California, San Diego, who brings an anthropologist's perspective to the study of cognitive science. During the 1980s and 1990s, Hutchins has devoted a considerable amount of time to observing operators of highly complex human–machine systems, such as naval ships and commercial aircraft. Among other things, these studies have revealed that artifacts play a crucial role in the successful functioning of these systems. In aviation, for example, Hutchins noticed that the cockpit contains a variety of devices that allow flight crews to essentially off-load cognitive demands to the world. Pilots use *speed bugs*, for example, which are adjustable devices that can be moved around the dial of cockpit speed indicators, to externalize memory of a number of critical airspeeds relevant to takeoff from the runway. Rather than having to remember these speeds, the pilots set these bugs prior to takeoff at predetermined critical speed values on the indicator's dial; they can thus literally see when the various threshold speeds are reached by watching the needle on the dial approach the positions of the bugs.

Hutchins has recently produced a ground-breaking theory of "cognition in the wild" based upon his various studies of marine navigation and the cultural, social, cognitive, and environmental contributions to this behavior. His view of cognition as distributed across people and artifacts and his analysis of how cognitive tasks are performed by sequentially re-representing information both internally and environmentally provide perhaps the best theoretical resources yet available for meeting the modern challenges of designing and understanding human interaction with cognitive artifacts. Like James Gibson, whose perceptual psychology was based largely on studying how people "get around in the world," Hutchins's studies of how people "get around" in a more high-tech sense promises to broaden the range of empirical data considered acceptable in cognitive science and perhaps also to prompt some theorists to reconsider the relationship between people and the world in the performance of cognitive tasks.

Conclusion

Application is truly where the rubber meets the road in cognitive science, and thus the place where current theoretical inadequacies are often felt first, and perhaps also most acutely. But application also has the beneficial quality of exposing flaws in theory where they might otherwise go undetected due to subtlety, overlooked due to wishful thinking, or uncorrected due to a lack of any hard proof that remedy is really necessary. The recent opportunity for an applied cognitive science, with its attendant design successes and failures, promises to provide an additional source of empirical constraint on theories of how people perform practically relevant cognitive tasks in everyday situations.

There will, of course, be no single theory of *everyday cognition*. Some everyday activities, of the type emphasized here, occur in intimate interaction with the external world. Others occur in more detached contemplation. I have naturally focused on the former type, because these are the activities that most concern design. Many cognitive scientists have serious doubts about the possibility of a theory of practical, everyday activity, noting that these activities are strongly shaped by accidental particulars of behavioral situations. Just as physics does not try to predict the exact course taken by a rolling stone, but instead tries to describe the underlying, so-called essential nature of the universe, in their view psychology should be similarly uninterested in the exact behavior of a bartender or a cook, and should instead aim to describe the underlying, presumably essential nature of cognition.

Theorists holding such a view may be unswayed by many of my preceding observations concerning the need for research aimed at discovering and describing the functional properties of the environment participating in human interaction with the world. I emphasize, however, that a functionalist or ecological view is no more interested in glassware, grills, copiers, or cockpits than a more traditional, cognitivist view is interested in any *particular* beliefs or representations of such. Computational modelers, for example, are interested not so much in the content of various kinds of knowledge, but rather in the functional properties of this knowledge. Similarly, the ecological or functionalist perspective also seeks abstract, generalizable theory of the functional properties of the environment involved in cognition, to complement abstract theory of internal functional properties. It is too early to know whether or not more comprehensive theories, describing both the internal and the external structures participating in human activity, will be forthcoming. Such theories, however, seem to be demanded by the empirical data, required for design, and can be significantly informed by studying people interacting with a designed environment.

References and recommended reading

Beach, K. D. 1988: The role of external mnemonic symbols in acquiring an occupation. In M. M. Gruneberg and R. N. Sykes (eds), *Practical Aspects of Memory: Current Research and Issues*, vol. 1, New York: Wiley, 342–6.

Card, S. K., Moran, T. P. and Newell, A. L. 1983: *The Psychology of Human-Computer Interaction*. Hillsdale, NJ: Erlbaum.

Carroll, J. M. 1991: Introduction: the Kittle House manifesto. In J. M. Carroll (ed.), *Designing Interaction: Psychology at the Human–Computer Interface*, New York: Cambridge University Press, 1–16.

Carroll, J. M. and Campbell, R. L. 1989: Artifacts as psychological theories: the case of human–computer interaction. *Behavior and Information Technology*, 8(4), 247–56.

*Flach, J., Hancock, P., Caird, J. and Vicente, K. 1995: *Global Perspectives on the Ecology of Human–Machine Systems*, vol. 1. Hillsdale, NJ: Erlbaum.

*Hutchins, E. 1995: *Cognition in the Wild*. Cambridge, Mass.: MIT Press.

Hutchins, E., Hollan, J. and Norman, D. A. 1986: Direct manipulation interfaces. In D. A. Norman and S. Draper (eds), *User-centered System Design: New Perspectives in Human–Computer Interaction*, Hillsdale, NJ: Erlbaum, 87–124.

Kieras, D. E. and Polson, P. 1985: An approach to the formal analysis of user complexity. *International Journal of Man–Machine Studies*, 22, 365–94.

Norman, D. A. 1986: Cognitive engineering. In D. A. Norman and S. Draper (eds), *User-centered System Design: New Perspectives in Human-Computer Interaction*, Hillsdale, NJ: Erlbaum, 31–62.

—— 1988: *The Psychology of Everyday Things*. New York: Basic Books.

Rasmussen, J. 1986: *Information Processing and Human–Machine Interaction: An Approach to Cognitive Engineering*. New York: North-Holland.

Stevens, J. 1993: An observational study of skilled memory in waitresses. *Applied Cognitive Psychology*, 7, 205–17.

*Suchman, L. 1987: *Plans and Situated Actions*. New York: Cambridge University Press.

Woods, D. D. and Roth, E. M. 1988: Cognitive systems engineering. In M. Helander (ed.), *Handbook of Human–Computer Interaction*, New York: North-Holland, 3–43.

57

Institutions and economics

DOUGLASS C. NORTH

Economic theory is built on assumptions about human behavior – assumptions which are embodied in rational choice theory. Underlying those assumptions are implicit notions about how the mind works. Until recently economists have not self-consciously examined those implicit notions, but recent work in economics and particularly game theory has forced economists to explore the sources of the beliefs that underlie economic choices and therefore to build a bridge between cognitive science and economics. In this essay I explore the path of economic reasoning that leads to engagement with cognitive science.

Introduction

The neoclassical approach to analyzing the performance of an economy assumes that in the face of pervasive scarcity individuals make choices reflecting their desires, wants, or preferences. Neoclassical theory is constructed by aggregating those preferences in the context of fixed resources, private goods, and given technologies. The result has been a powerful set of tools for analyzing resource allocation at a moment of time in developed economies under the assumption that the markets being modeled are governed by impersonal forces of supply and demand. The competitive model of neoclassical theory enshrined in general equilibrium theory makes a major contribution to economic understanding by demonstrating that a decentralized system of market forces would generate an efficient system of resource allocation. In this framework an individual agent's beliefs play no role in decision making.

Valuable as the neoclassical approach has been for the development of an elegant body of theory, it is a very imperfect tool for solving economic problems either at a moment in time or over time. Both information and the enforcement of agreements are imperfect, leading to transaction costs. Further, markets are the creatures of political forces. In the real world of imperfectly competitive markets, beliefs determine the choices of the actors. Agents' motivation is derived from their private information and expectations about price movements. Moreover, some goods and services are public – not only the traditional ones of national defense and public security, but also property rights and the rule of law. Since these goods and services are traditionally created through the political system, understanding them requires not only knowledge about people's preferences for such goods but their incentives to produce them, given people's beliefs about other people's willingness to pay for them. Preference-based models of either markets or elections are relatively simple. Beliefs, on the other hand, are anything but simple, because they involve some description of how people learn, how they update hypotheses and theories, and how they model the world they live in. And it is modeling beliefs that is at the heart of all theorizing in the social sciences.

Let us consider the implications of the economist's preference-oriented approach to modeling economic behavior. These models embody the *substantive rationality assumption*, which conceives of an actor as maximizing rewards subject to a complex and consistent ordering of preferences. Such models work well for competitive posted-price markets. The chooser need only choose the quantity to buy or sell, because the competitive environment so structures the situation that price can effectively be viewed as a parameter. If all choices were simple, made frequently with substantial and rapid feedback, and involved substantial motivation, then substantive rationality would suffice for all purposes. It would be both a predictive and a descriptive model of equilibrium settings, and learning models based upon it could be used to describe the dynamics out of equilibrium. But as soon as we move away from simple competition to situations in which the price depends on the behavior of other buyers and sellers, the complexity of the decision increases. Indeed, the interesting issues that require resolution come from the interaction of human beings in economic, political, and social markets. Knowledge of other people's actions and beliefs is an essential prerequisite to constructing useful models. But so too is a knowledge of their preferences, because it is the melding of preferences and beliefs that determines choices. The strategic interaction of human beings is the subject of game theory, and the vast literature on the subject that has evolved is a testimonial to its current appeal in the social sciences. But the current status of game theory itself makes clear that what has been missing in most game-theoretic models is "a description of the players' reasoning processes and capacities as well as a specification of their knowledge of the game situation" (Bicchieri, 1993, p. 127).

The puzzle I seek to unravel is still deeper than how people reason and learn. It is how do humans evolve and believe in hypotheses or theories in the face of uncertainty. Let me explain. Frank Knight (1921) made a fundamental distinction between risk and uncertainty. In the case of risk, probability distributions of outcomes could be derived with sufficient information, and therefore choices could be made on the basis of those probability distributions (the basis of insurance). In the case of uncertainty, no such probability distribution is possible; as a consequence, economists have held that it is impossible to develop a body of economic theory in such cases, and that economic reasoning per se will be of little value. However, human beings do construct theories all the time in conditions of pure uncertainty – and indeed act on them and sometimes die for them. Communism is the most famous modern secular example. Religions, too, are based on faith in a theory. But it is not just overall ideologies like communism or religions that should concern economic theory. It is the widespread existence of myths, taboos, prejudices, and simply half-baked ideas that serve as the basis of economic decision making. Indeed, most of the fundamental economic and political decisions that shape the direction of polities and economies are made in the face of uncertainty. You have only to open a newspaper and read the headlines to observe such decisions every day.

Cognitive science is potentially of great relevance to economics – not just insofar as it tries to explain how human beings learn and meld beliefs and preferences to reach decisions and hence the choices that underlie economic theory, but also as it tries to explain how and why they do develop theories in the face of pure uncertainty, what makes those theories spread amongst a population or die out, and why humans believe in them and act upon them. In the remainder of this essay I explore these issues in the expectation that, in the future, cognitive science may give us some definitive answers that can serve as the basis for major breakthroughs in economics and the social sciences generally.

Institutions

Neoclassical theory assumes that individuals' preferences are stable and that choices are made within a framework of constraints. The constraints include those imposed by income and technology, but not those imposed by the institutions of society. The reason for the absence of institutional constraints is that the chooser is assumed to have *perfect* information and therefore *certainty* about alternatives. Agents know what is in their self-interest, act in their self-interest, and are able to perform the calculations necessary to discriminate amongst alternative decisions. In such a world institutions are functionally unnecessary. Institutions exist to structure human interaction in a world of uncertainty, or, as Ronald Heiner put it in an article of fundamental importance, "The origins of predictable behavior" (1983), they arise from the effort of individuals in the face of pervasive uncertainty to reduce that uncertainty by limiting the choices available to the players and thereby making behavior predictable. Without institutions there would be no order, no society, no economy, and no polity. The construction of an institutional framework has been an essential building block of civilization.

Once we recognize the fundamental role of institutions in reducing uncertainty, we must restructure the theoretical framework used in economics and the other social sciences. Institutions not only provide the incentive structure of a society at a moment of time and thereby constrain the choice set, they are also the carriers of the process of change. Therefore, whether we are modeling economic performance at a moment of time or over time, institutions are central to the theoretical construct. But what are institutions, and where do they come from?

Institutions are formal rules (constitutions, statute and common laws, regulations), informal constraints (conventions, norms of behavior, and self-imposed codes of conduct), and their enforcement characteristics. Institutions reflect the beliefs of the players – or at least of those players able to shape the rules. Behind beliefs are language and the cultural heritage of the players – a subject discussed in the next section. Before turning to it, I wish to explore the way the institutional context influences choices. Let me return to the instance where a substantive rationality assumption works well. It works well when the perfectly competitive market constrains the choices of the players. Or, to put it directly, the rationality model works best when the institutional framework constrains the choice set and when what passes for rationality is in good part a function of the institutional framework. Note carefully, however, that this framework, the scaffolding, will not necessarily produce efficient economic results. Indeed, the scaffolding may so structure incentives that *rational* actors will make choices that produce inefficient economies. Indeed, sources of poor economic performance, as evidenced by poverty, low incomes, and stagnation, are a consequence of institutions that structure incentives that discourage productivity-improving activities. Because institutions are a creation of the belief systems of those players who can shape the rules of the game, we must examine the way in which diverse belief systems emerge; this requires a temporal perspective.

Development in time

Time, in this context, is the dimension in which human learning occurs; the collective learning of a society (to use Hayek's term) embodies its past learning. We can briefly characterize this historical process as follows: Given the genetic architecture of the brain with its proclivities for language (Pinker, 1994) and cooperative behavior (Barkow, et al., 1992), tribal groups evolved very differently in different physical environments. They developed diverse languages and, with different experiences, different mental models to explain the world around them. The language and mental models formed the informal constraints that defined the institutional framework of the tribe. They were passed down intergenerationally in the form of culture, taboos, and myths that provided cultural continuity. As specialization and division of labor developed, tribes evolved into polities and economies; the diversity of experience and learning produced increasingly different societies and civilizations, with different degrees of success in solving the fundamental economic problem of scarcity. As the complexity of the environment increased, human beings became increasingly interdependent. More complex institutional structures were essential to capture the potential gains from political and economic exchange.

Ever since Adam Smith, economists have recognized that the wealth of nations is a function of specialization, division of labor, and the size of the market. But what economists have only lately come to realize is that, as the market gets larger, more and more resources must be devoted to transacting – that is, to coordinating, integrating, and enforcing agreements. However, there is more to the process of market expansion than simply increasing resources devoted to transacting. With small, personal exchange it pays to cooperate, because the players interact repeatedly. But with impersonal exchange, to use the game theory analogy, it pays to defect. Historically it has been the creation of political and economic institutions that has altered the payoff to cooperation. Throughout most of history and in many contemporary societies, the institutions necessary to produce cooperation – particularly the political ones – have not been forthcoming. It entails a fundamental restructuring of a society to create a world of impersonal exchange. Because institutions reflect the belief system of a society, we must turn to the diverse cultural heritages of societies to see why the collective learning has not been conducive to creating the necessary institutions. The learning process appears to be a function of (1) the way in which the existing belief system filters the information derived from experiences, and (2) the different experiences confronting individuals and societies at different times. In some cases the initial belief system has not been congenial to institutional innovations that would permit impersonal exchange; in other cases the experiences were not those that would incrementally alter the belief system to create such institutions.

An historical illustration can help to bring the issues into focus. Avner Greif (1994) has explored the contrasting cultural background of Genoese and Maghribi traders in the late medieval Mediterranean trade and the different institutional frameworks they evolved to deal with the impersonal markets of long-distance trade. The Genoese evolved bilateral enforcement mechanisms, which entailed the creation of formal legal and political organizations for monitoring and enforcing agreements. This was an institutional framework that lent itself to further evolution of increasingly complex trade. The Maghribi, who had adopted the cultural and social attributes of Islamic society, developed in-group social communication networks to enforce collective action. These networks, while effective in relatively small, homogeneous ethnic groups, did not lend themselves to the impersonal exchange that arises from the growing size of markets and diverse ethnic traders. Greif describes the generality of these different belief systems for the Latin and Muslim worlds and then makes the connection between such belief structures in the European scene and the evolution of economic and political institutions.

The role of ideas and belief systems in shaping societies is not altogether new in the social sciences. Max Weber's celebrated *The Protestant Ethic and the Spirit of Capitalism* (1958) argued that Protestantism was the underlying source of capitalism. While his argument was flawed – it was broadly the Judeo-Christian tradition rather than Protestantism that was the source, and he failed to make the connection between beliefs and the consequent institutions – he at least recognized that ideas matter. Modern economic theory has no role for ideas. Preferences are assumed to be fixed, and, as noted above, beliefs are assumed to play no role in decision making. Only very gradually are economists coming to realize that they must model beliefs and the way they evolve if they are to make further progress in the discipline. What still requires explanation is the diversity of belief systems and their cognitive basis.

Ideologies

The pervasive human attempt to reduce uncertainty is the key to understanding the way in which humans process information and evolve belief systems. In order to make uncertain situations *comprehensible*, humans formulate explanations of different types. The pervasiveness of myths, taboos, and particularly religions throughout history (and prehistory, as well) suggests that humans have always felt a need to explain the unexplainable. A possible explanation for this need might be that the proclivity to adopt some explanation is more adaptive than a willingness to settle for no explanation.

Merlin Donald (1991) maintains that there have been two stages in the cultural development of human thought – the mythic and the theoretic. The former, which characterized thought before the ancient Greeks developed models of rational thought and argument, was characterized by the use of external formalism and was employed in the service of narratives. This earlier form of thought continues to play a critical role in ideologies. Indeed, the cultural heritage of a society provides the means to reduce the divergence of perceptions that arise from diverse experiences and constitutes the means for the intergenerational transfer of unifying perceptions. Cultural learning not only encapsulates learning from past experiences but also provides shared explanations for phenomena outside the immediate experiences of the members of the society in the form of myths, taboos, and dogmas.

Ideologies are organized belief systems, frequently having their origins in religions, which make both proscriptive and prescriptive demands on human behavior. They incorporate views about how the world does and should work. As such, they provide a ready guide to making choices. The ideologies that constrain choice making in political and other social contexts are often loose constructs that guide choices in the face of uncertainty just as surely as do more organized structures. But whether organized or *loose*, ideologies play a complementary role to institutions in making behavior predictable. While institutions structure the physical–social environment of human beings, ideologies structure the mental environment, thereby making predictable the choices of individuals over the range of issues relevant to the ideology. But what makes individuals susceptible to having their mental environment structured?

Institutions and choice

The quest for order leads humans to construct elaborate forms of social, political, and economic beliefs which then inform their collective problem-solving behavior. Such scaffolds consist of both the mental models they possess – that is, belief systems – and the external environment – that is, institutions. Part of the scaffolding is an evolutionary consequence of successful mutations and is therefore part of the genetic architecture of humans; part is a consequence of cultural evolution. Just what the mix is between the genetic architecture and the cultural heritage is in dispute. Evolutionary psychologists – psychologists who have donned the mantle of sociobiology – have stressed the genetic architecture in the scaffolding process at the expense of the role of the cultural heritage.

Recent research by experimental economists lends some support to the evolutionary psychologists' position. In a recent paper summarizing a large number of experimental game results, Elizabeth Hoffman, Kevin McCabe, and Vernon Smith (1995), report:

> people invoke reward/punishment strategies in a wide variety of small group interactive contexts. These strategies are generally inconsistent with, but more profitable than, the noncooperative strategies predicted by game theory. There is, however, consistency with the game theoretic folk theorem which asserts that repetition favors cooperation, although we observe a substantial use of reward/punishment strategies and some achievement of cooperative outcomes even in single play games.
>
> Non-cooperative outcomes are favored, however, where it is very costly to coordinate a cooperative outcome, in larger groups, and even in smaller groups under private information. In large groups interacting through markets using property rights and a medium of exchange, and with dispersed private information, non-cooperative interaction supports the achievement of socially desirable outcomes. Experimental studies have long supported this fundamental theorem of markets. This theorem does not generally fail, however, in small group interactions because people modify their strict self-interest behavior, using reward/punishment strategies that enable some approximation of surplus maximizing outcomes. Seen in the light of evolutionary psychology, such behavior is not a puzzle, but a natural product of our mental evolution and social adaptation.

Others, such as Stephen J. Gould, have suggested that there is a lot of slack in the genetic architecture which gives greater scope to cultural evolution. Gould has maintained not only that the selection environment changes, but that in many cases it is relatively *loose*, resulting in survival in which chance and breeding capabilities, rather than competitive pressures, may play a major role. Certainly many of our personal preferences – hunger, thirst, sex, and perhaps some of our beliefs – are genetically determined, but some preferences and most beliefs surely must be acquired.

Kenneth Binmore (1996), a leading game theorist, maintains that our genes probably do not insist that we prefer or believe certain things but are responsible for organizing our cognitive processes in terms of preferences and beliefs. He maintains that we come equipped with algorithms that not only interpret the behavioral patterns which we observe in ourselves and others in terms of preference-belief systems but actively build such models into our own operating systems. The evolutionary advantage of such an inductive process is that new behaviors are tested against past experience in our internal laboratory. Humans enjoy the benefits of having the potential to learn a second-best strategy in any game. Interactive learning is a two-stage affair in which we first receive a social signal that tells us how to behave and then test the behavior against our preferences, to see whether we wish to follow its recommendation.

One issue, then, is the extent to which the *mind is adapted* by several million years of genetic encoding, versus the influence of cultural evolution; another related issue is just how the mind works. Both are central to answering the questions posed at the end of the introduction, and we are far from settling them in either evolutionary psychology or cognitive science.

The foregoing summary of research results in experimental economics makes clear that there is evidence of an innate drive for cooperation; but the immense variation in the forms it takes and its varying degrees of success make the cultural component of cooperative behavior critical for the successful creation of impersonal political and economic markets. Creating cooperative frameworks of economic and political impersonal exchange is at the heart of problems of societal, political, and economic performance.

How does the mind work? There is an ongoing tension in the social sciences between evidence of human irrationality or limited rationality on the one hand (see Article 44, HEURISTICS AND SATISFICING) and the substantive rationality assumption of neoclassical economists on the other hand. "There is a large body of research that documents striking failures of naive humans when confronting relatively simple tasks in probability theory, decision making, and elementary logic. . . . There is [also] the continuing belief of psychologists and computer scientists that by understanding human problem solving performance we will be better able to build machines that are truly intelligent" (Lopes and Oden, 1991).

Related to this tension is another. Humans certainly reason, analyze, and deduce, and in doing so construct categories, representations, and models; and where deductive reasoning is not possible, humans reason by other, different means (see Article 21, REASONING). The use of ANALOGY, pattern recognition, and CASE-BASED REASONING suggests the widespread use of inductive reasoning. All this work focuses primarily on internal information processing operations, but there is a minority view, elegantly defended by Edwin Hutchins (1995), that emphasizes the cultural world in which cognitive behavior is embedded (see Article 39, EMBODIED, SITUATED, AND DISTRIBUTED COGNITION and Article 40, MEDIATED ACTION).

Such disagreements amongst cognitive scientists render tentative any conjectural applications to economics. Curiously enough, though, there are parallel sources of disagreement in economics. The rationality assumption in economics, in its pristine form, assumes that humans know what is in their self-interest and make choices accordingly. The implication is not only that humans are perfectly informed about all possible alternatives, but also that their choices are encumbered by the constraints imposed by technology and income (given fixed preferences) but unencumbered by the context within which humans make choices. By contrast, the new institutional economics argues that humans typically make choices under conditions of uncertainty, and that those choices are constrained by the institutional context.

It should be apparent from the previous sections of this essay that I believe that the most promising approach to applying cognitive science to economics is one which assumes that humans have a quite different kind of intelligence than is implied by the rationality postulate and its deductive corollary. The integration of institutional analysis within a culturally conditioned approach to cognitive science seems to me to be the most promising approach.

A connectionist framework, which I find most congenial, suggests that most learning comes from absorbing and adjusting to subtle events that have impacts on our lives and which incrementally modify our behavior ever so slightly (see Article 38, CONNECTIONISM, ARTIFICIAL LIFE, AND DYNAMICAL SYSTEMS). Implicit knowledge evolves without ever being reasoned out. In fact, we are relatively poor at reasoning compared to understanding and seeing the solutions to problems. We are good at comprehending and understanding if the issue is sufficiently similar to other events that have happened in our experience. We are good at pattern matching, and this is the key to the way we perceive, remember, and comprehend. This characteristic is the key to our ability to generalize and use analogy. Neural networks constitute fast but limited systems, which in effect substitute pattern recognition for reasoning. Pattern recognition provides just the right resources for motor control, face recognition, deciphering hand-written zip codes, but leaves much to be desired when it comes to providing a framework to build careful, sequential reasoning. Nevertheless, two factors have enabled humans to overcome this handicap: (1) scaffolding, in which the genetic and cultural heritage has already done most of the reasoning, and (2) representational redescription – the human capacity to continually reorganize knowledge via generalization and analogy to serve as the source of human creativity.

Here is the way a leading study on cognition characterizes the inductive process by which rule-based mental models are formed and revised in an ongoing process:

> The (cognitive) system is continually engaged in pursuing its goals, in the course of which problem elements are constantly being recategorized and predictions are constantly being generated. As part of this process, various triggering conditions initiate inductive changes in the system's rules. Unexpected outcomes provide problems that the system solves by creating new rules as hypotheses. Concepts with shared properties are activated, thus providing analogies for use in problem solving and rule generation. . . . The major task of the system may be described as reducing uncertainty about the environment. (Holland et al., 1986, p. 69)

But this ongoing inductive process by which rule-based models are formed must be put in the context of genetic encoding and the cultural scaffolding which imposes fundamental constraints on the choices of humans. Without drawing a neat division between the genetic and cultural components, it is still a powerful implication of this approach that the domain of choices available is radically limited by the preexisting scaffolding, as compared to rational choice models. We have a long way to go to answer the two questions posed at the beginning of this essay about how human beings learn and meld beliefs and preferences to make the choices that underlie economic theory and how and why they develop theories (and act on them) in the face of pure uncertainty.

References and recommended reading

Barkow, J. H., Cosmides, L. and Tooby, J. 1992: *The Adapted Mind: Evolutionary Psychology and the Generation of Culture.* Oxford: Oxford University Press.

*Bicchieri, C. 1993: *Rationality and Coordination.* Cambridge: Cambridge University Press.

*Binmore, K. 1996: "Evolution in Eden." In "Game Theory and the Social Contract II," draft MS.

Donald, M. 1991: *Origins of the Modern Mind: Three Stages in the Evolution of Culture and Cognition.* Cambridge, Mass.: Harvard University Press.

Greif, A. 1994: Cultural beliefs and the organization of society: a historical and theoretical reflection on collectivist and individualist societies. *Journal of Political Economy*, 102, 912–50.

Heiner, R. 1983: The origins of predictable behavior. *American Economic Review*, 73, 560–95.

Hoffman, E., McCabe, K. and Smith, V. 1995: *Behavioral Foundations of Reciprocity: Experimental Economics and Evolutionary Psychology*. University of Arizona Working Paper.

Holland, J. H., Holyoak, K. J., Nisbett, R. E. and Thagard, P. R. 1986: *Induction: Processes of Inference, Learning, and Discovery.* Cambridge, Mass.: MIT Press.

Hutchins, E. 1995: *Cognition in the Wild.* Cambridge, Mass.: MIT Press.

Knight, F. 1921: *Risk, Uncertainty, and Profit.* Boston: Houghton Mifflin Co.

*Lopes, L. and Oden, G. C. 1991: The rationality of intelligence. In E. Eells and T. Murazewski (eds), *Probability and Rationality*, Amsterdam: Rodopi.

Pinker, S. 1994: *The Language Instinct.* New York: W. Morrow.

Weber, M. 1958: *The Protestant Ethic and the Spirit of Capitalism.* New York: Scribner.

58

Legal reasoning

EDWINA L. RISSLAND

Legal reasoning is an engaging field for cognitive science, since it raises so many fundamental questions, such as the representation and evolution of complex concepts. This article focuses on aspects of legal reasoning that require reasoning with cases, often in concert with other modes of reasoning.

Research on computational models of legal reasoning began over 25 years ago with an article in the *Stanford Law Review* (Buchanan and Headrick, 1970). Motivated by early successes in ARTIFICIAL INTELLIGENCE (AI) with programs like DENDRAL, this article suggested the applicability of various AI methods and singled out certain thought processes, such as characterizing relevant facts, resolving rule conflicts, and finding analogies, as good research problems. This article was prescient in focusing on problems that have remained at the core of study and on the two major paradigms – rule-based reasoning and reasoning by ANALOGY (now largely the province of case-based reasoning) – that have dominated this field.

In the 1970s, many projects explored the use of logic and rule-based approaches. In the 1980s, significant new lines of research on open-textured concepts and case-based reasoning were begun. By the mid-1980s, the new paradigm of CASE-BASED REASONING (CBR) had emerged, in part because of work on modeling reasoning with legal precedents. Presently, researchers are pursuing a variety of approaches: rule-based (expert systems, logic programming), logic (classic, deontic, nonmonotonic), case-based reasoning, etc. Several have combined rule-based and case-based paradigms. Neural nets and other sub-symbolic approaches have been used very little. Advances in information-retrieval techniques and the explosion of information on the World Wide Web have revived interest in legal information retrieval.

The nature of legal reasoning

Legal reasoning has many characteristics that make it a rich field of study for cognitive scientists:

- Legal reasoning involves different modes of reasoning: reasoning with cases, rules, definitions, policies, and analogies.
- Law – particularly Anglo-American appellate law – is based on a specific mode of reasoning: *stare decisis*, the doctrine of precedent, that has as its heart reasoning from case to case and analogizing and distinguishing cases.
- There is great variety in the types of cases: hypotheticals and real precedents, negative and positive cases, easy-to-interpret and hard-to-interpret cases, prototypical cases, extreme cases, exceptional cases.

- There is great variety in the types of legal rules: statutory rules created by legislatures, doctrinal rules summarizing past cases, heuristic rules of individual practitioners, constitutional principles, *meta-rules* or maxims of how to reason.
- Legal concepts are typically *open-textured;* that is, they cannot be adequately defined using classical logic as universally valid, necessary and sufficient conditions that cover all cases. Legal rules are not inviolable; exceptions always emerge.
- Legal rules, concepts, and doctrines constantly evolve in response to new cases, changes in societal context, and in the legal reasoners themselves, such as judges. As Justice Benjamin Cardozo said, as soon as one creates rules and categories (to summarize the current state of the law), "the changing combinations of events will beat upon the walls of ancient categories." Change can be a dramatic *paradigm shift* or an incremental concept drift.
- Legal questions do not usually have one unassailable *correct* answer. For instance, in a legal dispute there is an answer proposed by each side. In the law, one is not merely allowed to, but is expected to argue, in the belief that the "truth will win out" by energetic advocacy and debate.
- Legal reasoners do more than follow existing rules and precedents; they argue, refine, and reformulate them. They engage in a kind of competitive theory formation, in which each side creates a legal theory to advance its goals in consonance with existing law.

Stare decisis, the doctrine of precedent

At the core of Anglo-American law is the doctrine of precedent, *stare decisis* (Levi, 1949). *Stare decisis* requires that similar cases be decided similarly. It places a high premium on reasoning with decided, past cases – *precedents* – to resolve new legal problems. It also employs reasoning with hypothetical cases – *hypos* – in interesting ways: for instance, to test the limits of an argument. The linchpin in *stare decisis* is the determination of *similarity* and *relevance* between cases, particularly a new problem case and the precedents. Determination of similarity and determination of relevance go hand in hand. Given that there are competing views of a problem – otherwise there would be no legal dispute – there are always competing proposals as to the set of most similar relevant cases that should control and determine its outcome. In addition, the march of time, with its attendant accretion of decided cases, can enormously affect the interpretation of a given fact situation. Thus the nature of case similarity and relevance is dependent on one's point of view and ultimate goals and is highly dynamic. This true dynamism contrasts strikingly with the popular caricatures of legal reasoning as a stodgy, mechanical process in which courts simply apply static rules to new facts.

The doctrine of precedent is used in all types of Anglo-American law: common law, which is based almost exclusively on reasoning with case precedents; statutory law, which involves reasoning with statutes (i.e., legislated legal rules); and constitutional law, which involves reasoning with broad legal principles. Statutory and constitutional law also involve reasoning with cases that have explicated, extended, and refined the law. We note that other legal traditions, such as Roman and Napoleonic law, often do not give cases such a central role. In particular, the civil code traditions used in much of Western Europe place a much higher emphasis on reasoning with rules.

Legal reasoning occurs in different contexts, including *advocacy, administration, advice giving,* and *adjudication.* Of these 4 A's, the first two have been the most thoroughly investigated. Advocacy, involving the creation of arguments, has largely been the focus of case-based reasoning. Administration, involving the application of administrative law (e.g., unemployment regulations), has been largely the province of rule-based approaches. Advice giving, which includes explicating the pros and cons of alternative courses of action as well as generating plans, is an underdeveloped area, which few projects have addressed. Very few projects have addressed adjudication; none has entertained the idea of an autonomous program acting as a judge. All realize that judging includes very sophisticated reasoning that must strike a delicate balance between precedent, societal needs, and individual equity.

Rules and logic

Early work led to many insights regarding the application of traditional expert systems, logic programming, and logic to all types of law. A striking result of this body of work was heightened awareness of the open-textured nature of legal concepts and difficulties with a strictly logical view of legal reasoning.

The project of Waterman and Peterson (1981) is illustrative. To study how experts, such as trial lawyers and insurance claim adjusters, assess the monetary settlement value, or *worth,* of product liability cases in tort law – the legal area that concentrates on issues of negligence and liability – they built an expert system, called "Legal Decision-making System." It also served as a test of the usefulness of the expert systems approach, which Waterman had helped pioneer, to law.

Even though tort law is primarily a common law area, rules were used to encode legal knowledge. For example, the following rule encodes the concept of *negligence*:

> IF (the product's user is not working in some area of some possible danger and the use of the product by the user does not involve ordinary-care)
> or (the product's user is working in some area of some possible danger (d) and the use of the product by the user does not involve ordinary-care and the danger (d) does not cause the lack of ordinary-care by the user)
> or the victim does not accept "the product is dangerous"
> THEN assert the use of the product by the user was negligent.
>
> (Waterman and Peterson, 1981, p. 390)

Waterman and Peterson concluded that the rule-based approach was successful since it came up with reasonable values for the worth of a case. However, they observed that it ran headlong into the problem of interpreting open-textured terms used in the rules. This is the central conundrum in this approach, since either the system must interpret terms by backchaining, using more rules and terms, or it must ask the user directly. Eventually, when the rules run out, the user may be forced to make difficult judgment calls.

Ideally, one would like the system to ask the user only about primitive, uncontroversial facts and to leave interpretation of them to the system. However, the line between fact and inference is hard to draw and maintain. For instance, one might be tempted to have the system ask the user "Did the user use ordinary care?" This sort of question might seem mundane to an expert systems builder, but to a legal expert it is very troublesome indeed, since it asks the user to make a profound legal judgment at the very heart of the case. The usual way around this in expert systems is to create rules that provide sufficient conditions for reaching such a conclusion. However, such a rule will itself be certain to contain troublesome terms. For instance, the rule defining *ordinary care* uses the predicate "The product is dangerous," and that term is not further unraveled but is asked directly of the user. The rules run out on a term whose interpretation is a key legal sub-issue in itself, and the user is asked to resolve it without any further support from the system.

Marek Sergot and Robert Kowalski, pioneers in logic programming, explored similar problems in statutory law. Their goal was to encode, or *formalize*, legal statutes both as an aid to the drafting or administration of them and as a test of logic programming techniques (Sergot et al., 1986). In one project, they focused on the British Nationality Act, a statute that exhibits all the complications and ambiguity inherent in statutory law. Since it was new, it involved virtually no case law. From the outset, they anticipated that there would be problems of vagueness with terms such as "being of good character" or "having sufficient knowledge of English." They believed that many such problems of interpretation could be reduced to questions of fact, askable of the user. However, like Waterman and Peterson, they found that the usual practice of unwinding such terms with more rules was not always satisfactory, since at the *bottom* one is still not at a level of nonjudgmental facts.

They also encountered fundamental difficulties with the PROLOG treatment of negation as failure, in which something is said to be false if it is not known or cannot be inferred to be true. For instance, if the program cannot show that an infant was born in the UK, it assumes it was not. This treatment of negation is one form of the so-called *closed world assumption.* They also encountered difficulties in the treatment of counterfactual conditionals. For instance, the statute uses the phrase "became a British citizen by descent or would have done so but for his having died." But it is not obvious what information should be considered in order to infer that a person would have been a citizen if the person had not died. Rules were used to grapple with such *but for* difficulties.

L. Thorne McCarty's TAXMAN project served a pivotal role in the development of the field as a whole. McCarty wanted to build a cognitive model of appellate argument as exemplified in the case of *Eisner v Macomber*, a very important, early (1920) United States Supreme Court case concerning the concept of *taxable income.* In TAXMAN I, McCarty tried traditional logical approaches. Like others, he was dissatisfied with its shortcomings, particularly in dealing with open-textured terms such as *taxable income* and *unrealized appreciation.*

McCarty's solution, embodied in his proposed TAXMAN II system, was to employ a *prototype and deformations* model of concepts (McCarty and Sridharan, 1982). The model is based upon the observation that a common rhetorical strategy in legal argument is to construct a sequence of mappings from a decided prototypical case to the disputed case, in order to argue why the disputed case should be resolved just as the prototype was.

McCarty's representation of a legal concept has three components: (1) an *invariant* description of the concept; (2) a set of exemplars, each of which matches some but not all aspects of the description; and (3) a set of transformations, called *deformations*, that describe how one exemplar can be mapped into another. The deformations give coherence to a concept and make it more than just a simple disjunctive definition of exemplars. For some concepts, it may be possible to link all the examples to a single exemplar – a *prototype* – and thus neatly represent the second and third components of the concept with one prototype plus a set of deformations. McCarty's representation framework relied heavily on the use of primitives, such as *permission* and *obligation*, from deontic logic. Only recently has McCarty (1995) implemented his model, although over the intervening years he developed a detailed representation scheme, called "Language for Legal Discourse," for concepts, such as *ownership* and *income*, needed for reasoning in areas like corporate tax law.

Using his prototypes-and-deformations framework, McCarty analyzed several important legal arguments: in particular, the main opinion by Justice Pitney and the dissent by Justice Brandeis in *Eisner v Macomber*. The Brandeis dissent connects various hypotheticals and two of the very few precedents (existing in 1920) concerning taxable distributions to argue that the distribution in *Eisner* should be taxable. McCarty represented these arguments by showing how to map the exemplars of taxable distributions (distribution of a corporation's cash and distribution of stock of an unrelated corporation) from the two Supreme Court precedents to the distribution of stock in this dispute. Justice Pitney's argument that the distribution in *Eisner* is not taxable is represented similarly in terms of transformations of exemplars of nontaxable distributions.

At about the same time as McCarty was developing his prototypes-and-deformations model, I was developing a *retrieval-plus-modifications* model of how one generates examples that satisfy given desiderata. Although my model, called *constrained example generation* or CEG, grew out of research on mathematical reasoning, I soon applied it to legal reasoning. There is a strong analogy between the way in which mathematicians create examples by modification of mathematically important aspects of existing examples and the way in which legal experts create hypotheticals by varying legally important aspects of actual and hypothetical legal cases. In the CEG model, a new example was generated by retrieving examples from an *examples-space* and then applying modifications to those that come closest to satisfying the new constraints. Examples were indexed by mathematical attributes as well as by a taxonomy of how examples are used in mathematical reasoning: easily accessible *start-up*, well-known *reference*, troublesome *counter* examples, etc. Modifications were indexed through a difference operator table linking attributes to modifications that affect them. CEG was an early precursor to what is now termed *adaptive CBR*.

The approaches of Rissland and McCarty share some striking similarities. Both highlight the importance of cases. Both focus on the manipulation and linking of cases, both real and hypothetical, by deforming or modifying them in legally important ways.

Open texture and the hard–easy distinction

Open-textured legal concepts and the key role played by exemplars in reasoning about them came under detailed scrutiny in Anne von der Leith Gardner's (1987) landmark research. She explored three closely related topics: (1) the nature of open-textured concepts; (2) the process of distinguishing between questions that a reasoner (person or machine) has enough information to resolve and those that can be argued either way; and (3) the use of legal knowledge expressed as examples as well as rules, particularly to mitigate insufficiencies of rule-based reasoning, such as resolving open-textured predicates when the rules run out or are otherwise inconclusive. She explored these issues in the context of so-called *issue-spotter* questions often asked of law students on law school and bar examinations.

Her work was motivated in part by the keen observations of legal scholars such as H. L. A. Hart, who applied the phrase "open texture" to legal concepts and described them as having a "core of settled meaning," where there is not much debate about how to interpret particular cases, and a *penumbra*, where interpretation is debatable. Hart and others developed a jurisprudence of *easy* and *hard* questions. Easy cases are those about which a body of legal experts agree on the interpretation; hard cases are those where there is dispute. Hard cases are associated with the penumbra, easy cases with the well-settled core.

On an issue-spotter question, one scrutinizes the facts, usually presented in a carefully crafted, elaborate hypothetical case, to determine which raise issues of debatable interpretation – that is, hard questions. One then discusses the pros and cons of the competing interpretations. Discussing easy questions is not required and is usually considered a waste of time, since it takes away from analysis of the hard issues. The goal is to produce justifiable explanations of the competing interpretations. This requires sifting what in search-based models of problem solving would be called the *plausible* from the merely *possible*. As the great American legal scholar Karl Llewellyn (1989) said, "[W]hile it is possible to build a number of divergent logical ladders up out of the same cases and down again to the dispute, there are not so many that can be built defensibly."

Gardner explored these questions in law concerning how contracts are formed by offer and acceptance, a well-honed topic in a standard first-year law school course in contract law. An easy offer-and-acceptance situation is one in which A makes an offer to B to sell him his car, "I'll sell you my car for $3,000," and B responds, "It's a deal; I'll buy it!" Because of the importance of what was said by the bargaining parties and the often everyday subject matter, Gardner made extensive use of the theory of speech acts and dealt with problems of representing commonsense knowledge.

Her program produced a two-tiered analysis. The top level is a decision tree, where each node corresponds to an event, such as the act of asking "Do you want to buy my car?", and each branch represents an interpretation, such as *offer* or *inquiry*. Leaf nodes represent the final interpretations, *contract* or *no contract*, that follow down from various paths. Associated with each node in the top-level tree is another tree representing the details of the analysis underlying the competing interpretations.

Gardner represented legal knowledge in three ways:

1 As an augmented transition network (ATN), which includes the states one can be in (e.g., an offer has been made) and transitions representing the allowed moves between states (e.g., for there to be a contract, there must be an offer followed by an acceptance);
2 As rules which model legal doctrine and technical definitions. These are used to determine if a state transition can be taken: that is, the *offer*, *rejection*, *counter-offer*, *acceptance*, etc., arcs of the ATN.
3 As a set of generalized fact patterns, each of which represents a stereotypical situation plus its interpretation and is linked to the legal term (e.g., *offer*) to which it speaks.

Hard questions can arise in three ways: (1) there are competing rules applicable in a given situation; (2) there are unresolved predicates in an attempted application of the rules; and (3) there is a lack of unanimity in the decided cases matching the situation.

Gardner's program begins by trying to apply the relevant rules needed to move the interpretation of the facts to the possible next states of analysis, as captured in the ATN. If no difficulties are encountered, then relevant cases – that is, the generalized fact patterns – are used as a check of the tentative interpretation derived by the rules. If there are no cases pointing to an *opposite* answer to that derived by the rules, then the question is easy, the tentative answer holds, and the analysis moves along. If there are both confirming and opposite cases, the interpretation is debatable, the question is considered hard, and a branch point in the analysis is created. If all the relevant cases are opposite, then the case-based answer overrides the rule-based answer, the question is considered easy, and the analysis is moved along.

The situation is similar when there are difficulties in applying the rules. If the cases point to more than one resolution, the issue is labeled "hard," and a branch point is created in the top-level output tree. If all the cases point to the same resolution, then it is considered to be the correct resolution, the question is labeled "easy," and the analysis proceeds without any branch point.

Gardner's work reifies the model of the hard–easy distinction put forth by legal philosophers like Hart: if there is disagreement in the case law, consider the question to be hard. It is consonant with the view of many famous jurists, including Oliver Wendell Holmes and Karl Llewellyn, that the essence of law is what it does with cases – as captured in the oft-quoted aphorism of Holmes that "The life of the law has not been logic: it has been experience" – and furthermore, that rules – what Llewellyn called "pretty playthings" – are secondary to cases. Of course, there are prominent legal scholars, like Ronald Dworkin, who take the position that hard questions may not be so hard.

Gardner's program performed well on test questions taken from a standard review book for contracts law. It even spotted issues that the authors had missed. More interestingly, it brought out weaknesses in this model of the hard–easy distinction. Using a diversity of interpretations in the set of relevant cases to label a question "hard" raises too many false alarms.

Her work added to the growing body of research on the importance of cases in modeling legal reasoning. However, in her program, cases were not the particular fact situations of actual legal cases. Rather, they were generalized patterns or templates. Further, there was no dynamic determination of relevancy, since cases were hard-linked to their concepts, and the same set of cases was retrieved regardless of problem situation. Such concerns became the focus in the next generation of systems addressing case-based reasoning.

Case-based reasoning

Kevin Ashley and I have pioneered a new model of legal reasoning called *interpretive* or *precedent-based case-based reasoning*. HYPO was the first AI and law system to explicitly model many key aspects of *stare decisis*, such as case similarity.

Using some ideas from my CEG, HYPO began as a program to model the creation of legal hypotheticals. In Ashley's doctoral research, it matured to a landmark program that dealt with the larger issue, precedent-based argument, and provided explicit models of how cases can be compared, analogized, and distinguished and how hypothetical cases can be created (Ashley, 1990). HYPO creates stylized arguments of the kind found throughout Anglo-American law, especially appellate law. In recent years, Ashley and his students have used HYPO to create intelligent tutoring systems, like CATO, which teach students how to make precedent-based legal and ethical arguments.

Appellate argument often involves domain-specific factors, each of which relies on a cluster of underlying facts, and each of which, all other things being equal, can be used to argue for a particular interpretation of the facts. In HYPO, such factors are called *dimensions*. Typically a line of cases – that is, a group of cases citing each other in support of a particular way of resolving an issue – employ a shared set of dimensions. For instance, in HYPO's trade secrets domain, the more people to whom a complaining company (i.e., the plaintiff) has disclosed its secret, the less persuasive it is to argue that any one particular party (e.g., a defendant) is responsible for misappropriating the company's trade secret. This way of arguing about a case is encoded in the dimension called *secrets voluntarily disclosed*. Another, *competitive advantage*, captures the idea that the greater the disparity in the time and resources (e.g., money) expended by two competing companies to bring a highly similar product to market, the more likely it is, *ceteris paribus*, that the company expending the fewer resources misappropriated a secret from the other one.

Dimensions can be thought of as *slices* through a high-dimensional case space or projections onto legally relevant subspaces of facts and considerations. Dimensions are encoded using intermediate-level features (e.g., *plaintiff has product information, plaintiff made disclosures*), which are computed from the entry-level facts. A dimension has one or two *focal slots*: features at the heart of arguments using the dimension. For instance, the number of disclosees is the focal slot of the secrets voluntarily disclosed dimension. By altering focal slot values, cases can be made stronger or weaker and moved along dimensions.

Dimensions are used to define relevance and to assess similarity. Any case in the case–knowledge base sharing at least one applicable dimension with a problem case is considered relevant. Relevant cases are sorted according to how on-point they are. Case A is considered *more on-point* than case B if the set of applicable dimensions which case A shares with the problem case properly contains those shared by case B and the problem case. Maximal cases in this (partial) ordering are called *most on-point cases*. The sorting is shown in a *claim lattice*, in which the problem case is shown as the root node and the most on-point cases in the next level of nodes. *Best* cases are most on-point cases that were decided for the interpretation being argued for.

For each best case, HYPO generated a *three-ply argument*, involving two sides, one for and one against an interpretation (i.e., plaintiff and defendant). HYPO can lead off with either. In a three-ply argument: (1) side 1 makes a *point* by drawing an analogy between the problem case and its best case; (2) side 2 makes a *response* by first pointing out distinctions between side 1's best case and the problem situation and, second, offering its own best case or a counterexample case; and (3) side 1 makes a *rebuttal* by first distinguishing away side 2's best case, and then offering a counterexample to it. For example:

> ⇒ Point For Defendant as Side 1:
> Where: Plaintiff disclosed its product information to outsiders. Defendant should win the claim for Trade Secrets Misappropriation.
> *Cite: Midland-Ross Corp. v. Sunbeam Equipment Corp.* 316 F. Supp. 171 (W.D.Pa., 1970)
> ⇐ Response for Plaintiff as Side 2:
> Midland-Ross Corp. v. Sunbeam Equipment Corp. is distinguishable because: In Midland-Ross, plaintiff disclosed its product information to more outsiders than in Crown Industries.
> Counter-examples:
> Data General Corp. v. Digital Computer Controls Inc. 357 A.2d 105 (Del. Ch. 1975), held for plaintiff even though in Data General plaintiff disclosed its product information to more outsiders than in Midland-Ross Corp. v. Sunbeam Equipment Corp.
> ⇒ Rebuttal for Defendant as Side 1:
> Data General Corp. v. Digital Computer Controls Inc. is distinguishable because: In Crown Industries, Plaintiff disclosed its product information in negotiations with defendant. Not so in Data General. In Data General, plaintiff's disclosures to outsiders were restricted. Not so in Crown Industries.
> Note:
> Plaintiff's response would be strengthened if: Plaintiff's disclosures to outsiders were restricted. Cf. Data General Corp. v. Digital Computer Controls Inc. A.2d 105 (Del. Ch. 1975)
>
> (Ashley, 1990, pp. 70–1)

HYPO also posed hypotheticals that show how to strengthen or stress aspects of the argument. HYPO created hypotheticals with dimension-based methods: make a nearly applicable dimension applicable by supplying missing facts, make a case weaker, stronger, or extreme by moving it along a dimension by manipulating focal slot values, etc. The HYPO model has also been used to analyze hypotheticals used in actual Supreme Court oral arguments (Rissland, 1989).

Karl Branting's GREBE developed a different model of case representation and precedent-based argument (Branting, 1991). GREBE represents cases in terms of their arguments as well as their facts. Arguments are represented using directed graphs, which show the relation between facts, intermediate conclusions, and final conclusions. GREBE reuses sub-pieces of old arguments, called *exemplar-based explanations* (*EBE*s), to generate new arguments. Case similarity is defined using the *structure-mapping* approach of Gentner and others, and best cases are selected using a form of classic search, called "A* search," used in AI.

Reasoning with cases and rules

Given the progress in rule-based and case-based reasoning and the deepened understanding of core problems like open texture, the natural next step is to study how rules and cases can be used together. Investigation of this issue began in the late 1980s, beginning with the CABARET project which I pursued with David Skalak at the University of Massachusetts and the PROLEX project led by Anja Oskamp and Rob Walker at the Computer/Law Institute at the Vrije Universiteit in Amsterdam. Closely following was the GREBE project of Karl Branting at the University of Texas and, more recently, the IKBALS project of John Zelenikow, George Vossos, and colleagues at LaTrobe and Melbourne universities in Australia.

Statutory law is the natural arena in which to study hybrid approaches. *Statutory interpretation* is the process of determining the meaning of a legal rule, including the meaning of constituent terms, and then applying it to a particular set of facts. Rule-based systems run into many problems of statutory interpretation, such as open texture. Other problems include:

- Rules have uncodified exceptions and unspoken requirements, making them impossible to apply in a mechanical way.
- Both terms and rules change in meaning and scope over time, for instance, before new rules can be drafted or old rules amended.
- Rules can use deliberately vague terms (e.g., *reasonable, good faith*) to allow for flexibility and discretion.
- Concepts can be formulated differently in different rules.

All statutory domains engender such problems, even those whose statutes have been written with the greatest care, such as tax and commercial law. Such difficulties are inevitable and not necessarily the result of faulty draftsmanship or legislative shortsightedness. No amount of fiddling with rules and terms will make them impervious to challenges posed by cases cast up from an ever-changing world, even if they are relentlessly rewritten to handle each new challenge.

For example, in the statute governing the taking of deductions for expenses incurred in the use of a home office, the US tax code says that a deduction may be taken for any portion of the dwelling that is *exclusively* used on a *regular basis*, as *the principal place of business*, as *a place of meeting or dealing* with patients, clients, or customers, or is in a *separate structure*. Nowhere in the code, however, are the key statutory predicates (italicized) defined. It is up to administrators and the courts to decide whether a particular set of facts meets the requirements. To do so, they rely on regulations, heuristic rules of practice, and precedents. Whereas in a common law area, cases carry the lion's share of the burden, in a statutory domain the burden is joint between rules and cases.

Of the hybrid systems, CABARET gives the most equal footing to case-based and rule-based reasoning (Rissland and Skalak, 1991). Its domain-independent architecture has the following major components:

1. Two primary, independent *co-reasoners*: a HYPO-style, case-based reasoner and a traditional (expert systems) rule-based reasoner.
2. For each reasoner, there is a dedicated *monitor* that makes observations on its processing, results and partial results and recasts these observations in a language (the *control description language*) understandable by the controller.
3. A *controller* uses the observations harvested by the monitors to propose and select tasks to be acted on by the individual co-reasoners.

The overall cycle of processing is that one of the co-reasoners works on a task, the monitor modules make observations, and, based on the observations, the controller posts new tasks and selects one to be worked on next by one of the co-reasoners. Upon conclusion, CABARET outputs a memo generated by filling in a stylized template.

CABARET uses four distinct sources of knowledge:

1 Knowledge needed for HYPO-style CBR: a case base, dimensions, similarity metrics, etc.
2 Knowledge needed for classic backward- and forward-chaining rule-based reasoning.
3 Control knowledge: domain-independent control rules, encoded in the control description language, to propose and rank tasks based on observations made by the monitors.
4 General domain knowledge: especially hierarchies available to all modules.

CABARET's control rules guide its problem solving and embody its theory of statutory interpretation. Examples are:

1 If one mode of reasoning fails, then switch to the other.
2 Once a conclusion is reached, switch the form of reasoning to check if it holds in the other mode.
3 If all but one antecedent of a rule can be established, then use CBR to show that the missed antecedent can be established using cases.
4 If all but one antecedent of a rule can be established, then use CBR to show that the missed antecedent is not necessary.
5 Use CBR on terms which are deliberately open-textured.

Such rules are part of a three-tiered model of statutory interpretation (Skalak and Rissland, 1992). It takes into account the point of view given to CABARET. The top tier contains four *argument stances*: (1) *broaden* a rule that on the face of it does not hold but which one wants to hold in the problem case; (2) *discredit* a rule that does seem to apply and that one doesn't want to; (3) *confirm* that a rule establishes a desired conclusion; (4) *confirm* that a rule fails to establish an undesired conclusion (see table 58.1):

Table 58.1

	Rule conditions met in problem case	Rule conditions not met in problem case
Pro	Confirm the hit	Broaden the rule
Con	Discredit the rule	Confirm the miss

Source: Skalak and Rissland, 1992, p. 82

The second tier contains tactics, called "argument moves" to carry out the stances. Four moves are possible for each stance. The cases available in the case base determine which ones are applicable. Note that argument stances are determined by the status of a rule *on the problem case*, and argument moves are determined by the status of the rule *on actual cases* in CABARET's case base. Argument moves are carried out with CBR *primitives* like *distinguishing* and *analogizing*, which form the third level.

Many other argument strategies and tactics are needed for a comprehensive theory of argument. Some are "slippery slope," "strawman," "turkey," "chicken and fish" (Skalak and Rissland, 1992). The last is a kind of double-negative argument in which one argues that the problem case is so *unlike* the cases in which an interpretation was held *not* to apply that the interpretation should apply in the problem case (Since turkeys are so unlike fish, which are not subject to the regulations for chickens, turkeys should be treated like chickens) (paraphrase of ibid., p. 19).

In summary, CABARET incorporates case-based and rule-based reasoning on an equal footing in an agenda-driven architecture and embodies a three-tiered theory of statutory argument. Just as HYPO makes computationally crisp ingredients of precedent-based argument, so CABARET does for statutory argument.

In conclusion, as this field has matured, it has deepened our understanding of precedent, the keystone of legal reasoning. That there has been a fruitful synergy between the study of legal reasoning and other disciplines like AI is evidenced by the number of truly ground-breaking systems that have been created by researchers working in this area. These have not only provided increasingly informed models of legal reasoning but have also contributed to fundamental progress in AI. For instance, the first case-based reasoning and hybrid case-rule systems were developed in the study of legal reasoning. Of course, the study of legal reasoning has also benefited from application of methods developed in AI. This fruitful synergy is bound to continue.

References and recommended reading

Ashley, K. D. 1990: *Modeling Legal Argument: Reasoning with Cases and Hypotheticals.* Cambridge, Mass.: MIT Press.

Branting, K.1991: Building explanations from rules and structured cases. *International Journal of Man–Machine Studies*, 34, 797–837.

Buchanan, B. G. and Headrick, T. E. 1970: Some speculation about artificial intelligence and legal reasoning. *Stanford Law Review*, 23, 40–62.

Gardner, A. von der Leith 1987: *An Artificial Intelligence Approach to Legal Reasoning.* Cambridge, Mass.: MIT Press.

*Levi, E. H. 1949: *An Introduction to Legal Reasoning.* Chicago: University of Chicago Press.

*Llewellyn, K. N. 1989: *The Case Law System in America*, trans. P. Gewirtz. Chicago: University of Chicago Press.

McCarty, L. T. 1995: An implementation of Eisner v. Macomber. In *Proceedings of the Fifth International Conference on Artificial Intelligence and Law* (ICAIL-95), New York: ACM Press, 276–86.

McCarty, L. T. and Sridharan, N. S. 1982: *A Computational Theory of Legal Argument.* Technical Report LRP-TR-13. New Brunswick, NJ: Laboratory for Computer Science Research, Rutgers University.

Rissland, E. L. 1989: Dimension-based analysis of supreme court hypotheticals. In *Proceedings of the Second International Conference on Artificial Intelligence and Law* (ICAIL-89), New York: ACM Press, 111–120.

—— 1990: Artificial intelligence and law: stepping stones to a model of legal reasoning. *Yale Law Journal*, 99, 1957–81.

Rissland, E. L. and Skalak, D. B. 1991: CABARET: rule interpretation in a hybrid architecture. *International Journal of Man–Machine Studies*, 34, 839–87.

Sergot, M. J., Sadri, F., Kowalski, R. A., Kriwaczek, F., Hammond, P. and Cory, H. T. 1986: The British nationality act as a logic program. *Communications of the ACM*, 29, 370–86.

Skalak, D. B. and Rissland, E. L. 1992: Arguments and cases: an inevitable intertwining. *Artificial Intelligence and Law: An International Journal*, 1, 3–48.

*Twining, W. and Miers, D. 1982: *How to Do Things with Rules*, 2nd edn. London: Weidenfeld and Nicolson.

Waterman, D. A. and Peterson, M. A. 1981: *Models of Legal Decisionmaking.* Report R-2717-ICJ. Santa Monica, Calif.: The Institute for Civil Justice, The Rand Corporation.

59

Mental retardation

NORMAN W. BRAY, KEVIN D. REILLY, LISA F. HUFFMAN, LISA A. GRUPE, MARK F. VILLA, KATHRYN L. FLETCHER, AND VIVEK ANUMOLU

One important problem in cognitive science is to understand the development of cognitive processes in children and to devise computer models to explore the mechanisms that underlie these changes (see Article 6, COGNITIVE AND LINGUISTIC DEVELOPMENT). Our research addresses these general goals. In particular, we are concerned with developmental changes in cognitive strategies in typical children and in children with mild mental retardation.

Before the late 1950s, psychological theories of mental retardation were very global, simply stating that individuals with mental retardation failed to learn because they had low intelligence. Philip Vernon, in the mid-1970s, pointed out that this view resulted in a circular explanation: to say that individuals with mental retardation have difficulty learning because they have low intelligence adds nothing to the understanding of the nature of mental retardation.

To break the circularity of this global approach, more specific theories of mental retardation began to evolve. Most either focused directly on the nature of memory deficiencies in individuals with mental retardation or made memory a central component. In the early 1960s, David Zeaman and Betty House developed an attention-deficit theory of mental retardation that localized the learning problem in attention. Norman Ellis identified the deficit in learning as a faulty short-term memory trace. Later, John Belmont and Earl Butterfield attributed the locus of the learning deficit to inadequate use of rehearsal strategies.

Though these more specific approaches attempted to break the circularity of the low-intelligence explanation of learning deficits in individuals with mental retardation, they created a deficit approach which has resulted in a rather unbalanced focus on limitations of individuals with mental retardation. A more comprehensive view of the cognition of children with mental retardation can be achieved by looking not only at ways they differ from, but also at ways they are similar to, children without mental retardation.

Our program of cognitive research started with the expectation that we would find deficiencies in the information processing of individuals with mental retardation. We did find some deficiencies, but the capabilities of the children in our studies were particularly salient against the zeitgeist which led us to expect only deficiencies. So, in devising a model of cognitive processes, we have taken a more balanced approach. Our first goal was to discover which processes in children with mental retardation might differ or be deficient and which might be similar or identical to those in children without mental retardation. We here provide a brief overview of what we found to be the main cognitive deficiencies and competencies of children with mental retardation.

Cognitive deficits

Research during the last 25 years has established deficiencies in three aspects of information processing in individuals with mental retardation. At the very early stages of information processing, individuals with mental retardation do not process some basic aspects of visual stimuli in the same way as individuals with average intelligence do. Robert Fox and Stephen Oross, at Vanderbilt University, conducted a series of experiments in the late 1980s showing that individuals with mild mental retardation do not process depth cues or movement cues with the same degree of accuracy as control individuals without mental retardation. In some of their studies, they used random dot stereograms, with one matrix of random dots presented to the right eye, the other to the left (dioptic presentation). Under some conditions, controls see a form (e.g., a square) immediately, whereas individuals with mental retardation do not form such percepts. The findings of Fox and Oross (1992) raise the possibility that some aspects of neural functioning involved in preattentive visual processing are deficient.

The second aspect of information processing that is deficient in individuals with mental retardation is encoding information under concepts. For example, the letters of the word *horn* are nothing more than a set of black lines placed on a page, but once encoded, these lines become recognized as a word corresponding to a musical instrument, a safety feature of an automobile, or a hard piece of cartilage extending from the snout of a rhinoceros, depending on the context. The decision that a word is *horn* and not another similar word such as *born* takes on the order of 0.4 seconds and may be made with varying degrees of accuracy.

There are many studies which suggest that individuals with mental retardation take longer and are less accurate in encoding information. For example, between 1970 and 1985, Norman Ellis, at the University of Alabama in Tuscaloosa, conducted a number of studies of short-term memory (see Article 17, MEMORY). In one prototypical study, participants were shown a single word and asked to recall it either immediately or after delays of up to 30 seconds. Individuals with mental retardation demonstrated consistently poorer recall than control subjects, even on the immediate recall test. Further, the magnitude of the difference remained the same with longer delays. These findings led to the conclusion that information is forgotten at the same rate in both groups, but that less information is encoded by the individuals with mental retardation (an encoding deficit).

Finally, the most firmly established finding in the study of individuals with mental retardation is that they have deficiencies in the use of cognitive strategies. In the 1970s, John Belmont and Earl Butterfield at the University of Kansas conducted a program of research on memory strategies used to remember a sequence of letters, words, and picture names. The participants were asked to recall the items in their order of presentation. During the study the participant controlled the exposure time of each item. In this type of task, individuals without mental retardation increased their study time as they progressed through the list, an indication that they were using a cumulative rehearsal strategy, reviewing the earlier items as the later items were exposed. Individuals with mental retardation, however, allowed the same exposure time for each item in the sequence. The failure to use a rehearsal strategy to keep track of the order of the items resulted in poor recall. Similar types of strategy deficiencies have been observed in a variety of memory tasks, and specific deficits have been found in memory-related processes such as the inhibition of irrelevant information (Merrill and Taube, 1996) and speed of processing (Kail, 1993).

Cognitive competencies

It is clear, then, that there are deficits in some aspects of information processing in individuals with mental retardation. However, in the 1970s, Ann Brown, at the University of Illinois, noted that not all aspects of information processing are deficient. She observed that nonstrategic processes, such as visual recognition memory, were equivalent in subjects with and without mental retardation. More recent work has greatly extended her results to phenomena such as automatic processing, spread of activation, short-term retention, stimulus organization, organization of semantic memory, and long-term retention; in these areas as well, no difference is found between individuals with and without mental retardation. Virtually all these areas of competency, however, were investigated in the context of the deficit approach. That is, the investigators did not discuss the implications of their findings for a more balanced view of both deficits and competencies.

Nevertheless, these results make a strong case that many structural features of memory in individuals with mental retardation are equivalent to those found in individuals without mental retardation, suggesting that the same information-processing architecture exists in children with and without mental retardation. The significance of these findings for a computational model is that these potential differences, having been ruled out, no longer serve as candidates to explain observed differences between the cognitive performance of individuals who differ in native intelligence.

Strategy competencies

Several investigators have found evidence for strategy competencies in children with mental retardation. Turner, Hale, and Borkowski (1996), in a longitudinal study of strategy development, found that the rate of increase in strategy use was the same for children with and without mental retardation. This finding suggests that the mechanisms of strategy change across childhood are similar, regardless of level of intelligence. Also, Baroody (1996) found that children with mental retardation are capable of *inventing* their own counting and addition strategies when learning simple facts of addition, a result that suggests strategy competencies beyond what would be expected by a deficit position.

Although others have obtained results showing strategy competencies, our research team is the only one which had as its goal the investigation of strategy competencies in children with mental retardation. In our studies, the importance of deficits in perception and encoding discussed previously has been minimized by using tasks in which the stimuli are easily perceived and in which the rate of presentation allows ample time for encoding. We designed our research to investigate whether a strategy deficit might account for the differences between individuals with and without mental retardation.

In one of our studies of strategy competency, Bray, Saarnio, Borges, and Hawk (Bray et al., 1994b) investigated external memory strategies in a task which allowed the use of both internal (e.g., verbal) and external (e.g., pointing at, orienting toward, and manipulating objects) memory strategies. The participants were shown movable objects and fixed targets and were asked to follow sequences of instructions, such as "Put the eraser on the chair; put the pencil on the table."

In a baseline condition, the children with mental retardation were *more* likely to use an object-oriented strategy (pointing to or holding the movable objects) than the controls. Although these strategies are not as efficient as target-oriented strategies (moving objects toward the targets), it is clear that children with mental retardation were actively attempting to remember the instructions. The groups differed in the likelihood that they would use an object-oriented strategy; but they choose similar tactics, suggesting common underlying cognitive abilities. Further, when children with mental retardation were given very minimal prompts, they devised more efficient, beneficial strategies that incorporate the same tactics as children without mental retardation. In contrast to a substantial literature showing deficiencies in the use of verbally based strategies, this study showed areas of overlap in strategy competency in children with and without mental retardation.

Similar results were obtained by Fletcher and Bray (1995) in a more complex task that allowed both verbal and external strategies. The memory task was embedded in a tape-recorded story in which the participant was guided through a *haunted house* by a *friendly ghost*. As the participant entered each imaginary room, (s)he heard a sequence of from one to seven statements such as "The broom is above the ghost" and "The lamp is on the blue side of the broom." The "blue side" referred to the right-hand side of the room, the "pink side" to the left-hand side. At the end of a sequence of such statements, the child was asked to recall the sequence by placing miniature objects on a matrix of velcro dots with a small, plastic ghost at the center of the matrix. The matrix represented the imaginary room.

Children with and without mental retardation all used a variety of external memory strategies, including pointing to the objects and holding and/or moving them on the table in front of them. In addition, many participants arranged the objects either in their hands or on the table. The arrangement strategies were the most interesting and the most effective. Using these strategies, the participants arranged the objects relative to a central point (ostensibly representing the ghost at the center of the matrix) either in their hand or on the table (e.g., for "The broom is above the ghost; the lamp is on the blue side of the broom," the broom was placed above the central point, and the lamp to the right of the broom).

In the external memory task, there was considerable similarity in the pattern of performance on the arrangement strategy and virtual overlap in performance on the pointing and holding strategies between the two intelligence groups. However, deficiencies appeared when no external strategies were permitted. In this task, the participants were given similar verbal instructions, but were asked to recall the instructions verbatim. This pattern of results suggested that individuals with mental retardation have strategy capabilities underestimated by verbally based tasks.

When the frequency of all observed external strategies and verbal strategies was combined on trials with three or fewer sentences to remember, there were *no differences* between the children with and without mental retardation. This suggests that the children with mental retardation were as likely to use a memory strategy as were children without mental retardation of the same chronological age. Although external memory tasks may offer more support than verbal memory tasks, additional supports such as verbal or physical cues may also increase external strategy use in children with and without mental retardation.

A study by Bray, Fletcher, Huffman, Hawk, and Ward (Bray et al., 1994a) described the use of external strategies in conditions that varied in the degree of situational support for strategy use. For 7-year-old children, the physical cues (a model of the computer screen), but not the verbal suggestion (to use any strategy they might choose), facilitated the use of external arrangement strategies which, in turn, aided recall. For 9- and 11-year-olds, both the verbal cues and the physical cues facilitated the use of arrangement strategies, whereas only the verbal cues did so in 17-year-olds. For 11-year-old children with mental retardation, neither the verbal nor the physical cues were sufficient to increase the use of arrangement strategies. For 17-year-old children with mental retardation, however, the combination of physical and verbal cues resulted in a level of strategy use equivalent to that of their chronological-age peers.

These results show that the strategy competency can be *activated* to the same level in children with and without mental retardation *without direct instruction*. Strategy competency in children with and without mental retardation is similar, but the cognitive potential of children with mental retardation requires more situational support before strategies are adopted.

Implications for computer models

If our team of cognitive scientists were to endeavor to write a computer program to simulate the performance of children with and without mental retardation, the deficit position would postulate that the strategy routines for children with mental retardation should contain deficits. Our empirical research does not support this type of model, however. It has shown that children with mental retardation *do* use strategies. The models would have to include some degree of situational support, not a lack of strategy capabilities.

Our team has approached the development of a cognitive model of strategy use in children with and without mental retardation by taking as a basic assumption that the structural features (architecture) are the same for children with and without mental retardation, but that the underlying competencies of children with mental retardation require more situational support before they become evident in the children's performance on cognitive tasks.

The utility of connectionist models for understanding strategies

We have employed connectionist ideas (see Article 38, CONNECTIONISM, ARTIFICIAL LIFE, AND DYNAMICAL SYSTEMS) to model strategy development in children. Perhaps the most compelling reason for using connectionist models (which use a neural network metaphor) is that these models, like the human brain, respond to multiple simultaneous constraints. Strategies develop in such models in nearly endless varieties in response to changes in context. Connectionist models also provide a way of looking at the development of rulelike behavior without assuming that the *rules* are in the child's head. Rather, rulelike behavior is generated in response to learning under multiple constraints and to being tested under conditions with the same or similar constraints (see Article 52, RULES).

Other research programs using connectionist models

Connectionist models have been successfully applied to a variety of problems in cognition, including problems of atypical development. For example, Ira Cohen in the early 1990s modeled the learning abilities of children with autism, using connectionist models to investigate the possible consequences of having too many or too few neuronal connections. His results indicate that models with too few connections lead to problems in discrimination learning and to poor generalization. Models with too many connections lead to good discrimination but poor generalization, the latter being the pattern typically observed in children with autism. This suggests the importance of investigating whether children with autism have an abnormally large number of neurons in their brains.

Hinton and Shallice (1991) provide a second encouraging example of the application of connectionist ideas to model atypical development. They introduced lesions into connectionist models trained to decode letter strings, by setting some connection weights to zero. The *damaged* networks exhibited a mixture of visual and semantic error patterns that were similar to those obtained in individuals with dyslexia. Virtually the same types of mixed error patterns were obtained, no matter where in the connection pool the *damage* was sustained. It is known that, along with other aspects of the syndrome, the visual and semantic error patterns observed with dyslexics vary widely. The qualitative correspondence between the simulations and the mixed error patterns obtained in dyslexia suggests that the lesions involved may not be specific to one site.

Description of a connectionist model of strategy development

We have used connectionist modeling in an effort to understand the nature of strategy deficiencies and competencies in children with mental retardation. Our generalized components model (Anumolu et al., 1996; Bray et al., in press) is modular in the sense that it consists of distinct, interrelated components, each designed to represent one aspect of strategy behavior. The development of each module was constrained by the tasks used in our empirical research to study external memory, prior empirical and theoretical concepts drawn from developmental psychology, prior connectionist research, and basic aspects of neurobiology.

The seven modules are shown in figure 59.1. The first is the sequencer module (which represents the sequence of sentences presented in our empirical research), and the second is the associative memory module (which learns and recalls representations of sentences like those used in our empirical work).

The strategy module consists of three nodes, each with selective connections to the entities in the object (e.g., eraser), target (e.g., table), and relation (e.g., on) pools of the associative memory module. This selective connectivity shown in figure 59.1 is crucial for understanding how the model generates different levels of recall, depending on the strategy used. Node 1 of the strategy module, representing an object-encoding strategy,

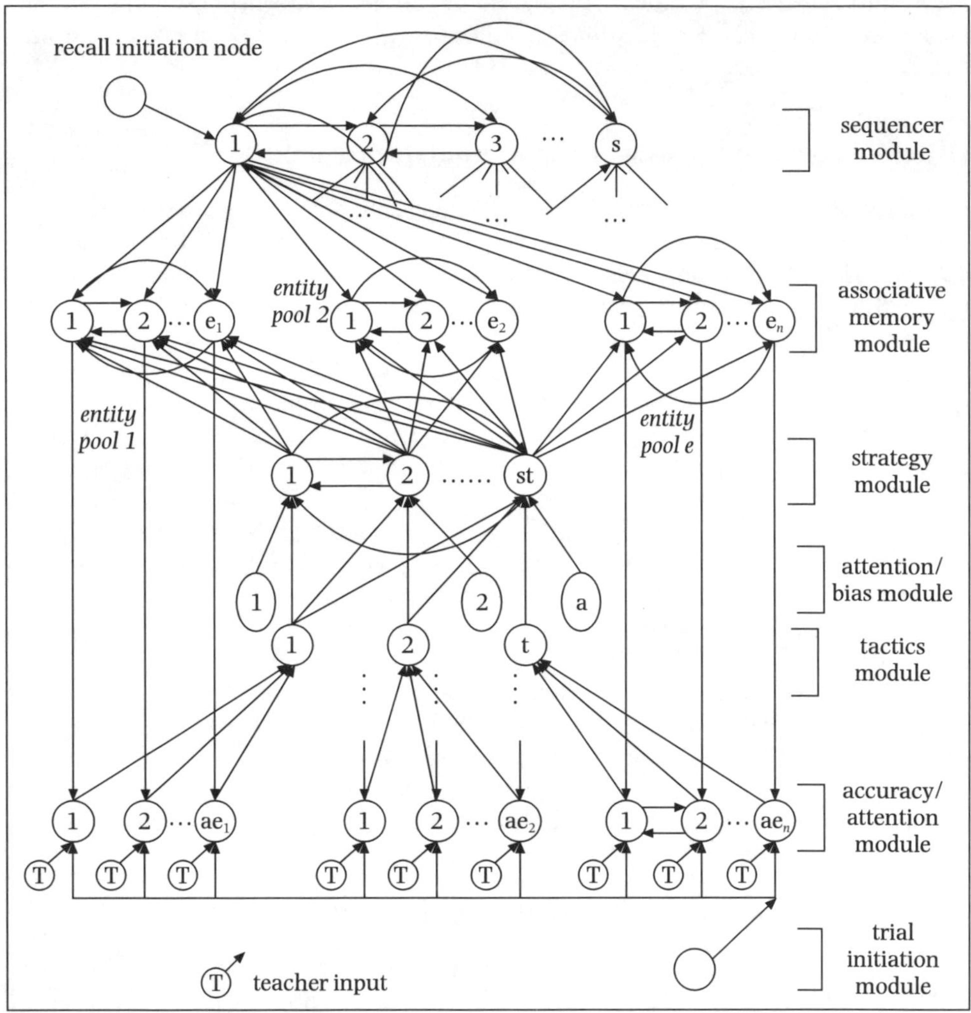

Figure 59.1 Generalized components model.

is connected only to nodes in the object pool, because only objects are involved in an object-encoding strategy observed in our empirical research. Node 2 of the strategy module is connected to the nodes of the object and target pools, because both objects and targets are encoded with this strategy. Node 3 of the strategy module is connected to the nodes of all three pools, because object–target–relation strategies involve encoding objects, targets, and relations. In this architecture, when a strategy is activated, it raises the activation of the corresponding nodes. Therefore, when an object-encoding strategy is activated, the activation value of the nodes of the object pool is raised.

The attention bias module represents different levels of attention to the components of the strategy used. For instance, a child might be biased to attend to the objects and give less attention to the target and the relations (where the objects are to be placed). In this case, the simulated child would probably use an object-oriented strategy rather than a more sophisticated strategy based on attention, where the object is to be placed in relation to the target.

The tactics module represents our theoretical construct that the tactics involved in strategies in general, and in external strategies in particular, have a hierarchical structure. The dimensional encoding mechanism was derived from theories that maintain that as children mature, they encode an increasing number of dimensions of tasks and events. In these theories, children begin by encoding information about only one dimension and move to encoding two and then three dimensions. In the accuracy-attention module, each node has one connection to an *external teacher*, which keeps track of whether recall was correct.

In our conceptual framework, the significance of the hierarchical nature of the tactics is that as children perform the external memory task, they perform actions very similar to those necessary to construct strategies. That is, in our external memory task, when responding to the sentence "Put the eraser on the chair," the child picks up the eraser (grasping tactic), moves it toward the chair (moving tactic), and places it on the chair (arrangement tactic). With experience, children parse the components of the response chain and, during the presentation of the sequence of sentences to be remembered, begin executing parts of the response chain in anticipation of the actual response. Young children begin executing the first component of this chain while listening to sentences by picking up a to-be-remembered object and holding it until the *bell* rings signaling the end of a sequence of sentences. Older children also execute the first and second tactics, moving the objects toward the target while listening to the sentences. With practice, older children and adolescents may execute all three tactics.

Our view is that the mechanism that underlies the *discovery* of these types of strategies is one in which the child parses the response chain required by the task. In doing so, he or she attends to an increasing number of elements of the response chain necessary for making a response. This is quite different from thinking of strategies as being *in the child's head*; rather, strategies evolve because the child attends to the appropriate aspects of the context provided by the task. In our view, strategy evolution is in response to the multiple constraints and resources provided by the context that directs the child's attention to the relevant aspects of the task.

The simulations, like the actual external memory task, involve two phases, a study phase and a recall phase. During the study phase, representations of the sentences are presented to the model, and a strategy is selected. The sentences are represented by the sequential activation of the appropriate nodes in the sequencer module and the corresponding entities in the entity pools. For instance, in simulating the presentation of the sentence "Put the eraser on the table," the first node of the sequencer module would be activated simultaneously with the *eraser*, *on*, and *table* nodes in the associative memory module. Using a Hebbian learning rule, the connections between the first sequencer node and the corresponding associative memory nodes would be increased. This procedure would continue for each sentence in a sequence.

During the recall phase, the model recalls the sentences presented in the last study phase. The recall interval begins with the activation of the recall initiation node, which in turn activates the nodes of the sequencer module. As each node is activated, the nodes of the associative memory module are activated in proportion to the previous learning during the study phase. In this way, it is likely that the weights connecting the first node of the sequencer and the items *eraser*, *on*, and *table* will be activated, resulting in the recall of the first sentence "Put the eraser on the table." Errors are made, however, because of the relative strength of weights for the connections to other entities in the associative memory module.

The components model generates several simulated behaviors that are similar to those observed in our empirical studies of external memory strategies and in Robert Siegler's studies of early addition strategies (Siegler and Shipley, 1995). In most simulation runs, the object, object–target, and object–target–relation strategies emerge in that order, as observed in our empirical research. Once a strategy *emerges*, it is likely that it will not be used exclusively, and the simulated child will occasionally use fewer, less sophisticated strategies, such as object encoding, after using an object–target encoding strategy. Accuracy of recall for our simulated children increases with the sophistication of the strategy, and there are primacy and recency effects in recall. Also, we are able to simulate the difference between children with and without mental retardation in terms of learning history, rather than specific process deficits. In both groups, we have been able to simulate different levels of situational support (e.g., verbal cues) by the manipulation of only one parameter.

These simulation results are important for several reasons. First, they show that individual differences can be modeled within the same architecture, consistent with the assumption that strategy potentials are the same in children with and without mental retardation. Second, these simulations illustrate that the model can simulate differences in experimental conditions with a minimum of parameter manipulation.

Conclusions

Our conceptual framework leads us to believe that a more balanced treatment of strategy deficiencies and competencies can provoke new research within the area of mental retardation. For example, future research on the parsing mechanism and on strategy use will focus on transfer of an uninstructed strategy. Devising a strategy by parsing the response requirements into tactics and assembling the tactics into a strategy results in a deeper level of processing than that provided by direct instruction. The exciting implication is that situations can be engineered so that children with mental retardation can *discover* new strategies without direct instruction. The self-organized knowledge so derived is more likely to be transferred to similar situations requiring similar strategies. It is well known that children with mental retardation have particular difficulty in generalizing strategies when trained directly. However, our research shows that children with mental retardation may devise effective external memory strategies with the same frequency as their chronological-age peers when given the appropriate physical and verbal prompts. This finding raises the possibility that these strategies devised without direct instruction will transfer to other tasks more readily than the same strategies taught directly to children with mental retardation.

Our view is that the future of research on strategy use in individuals with mental retardation will begin to focus more on strategy competencies than exclusively on strategy deficiencies. This will be a welcome change, since the deficiency approach has not led to the development of a framework for understanding the differences between individuals with and without mental retardation. Additionally, the connectionist modeling approach described herein may lead to a clearer understanding of the nature of strategy competencies and deficiencies in individuals with mental retardation and of mechanisms that may be responsible for the pattern of observed differences. Extant intervention techniques for the remediation of strategy deficiencies have met with limited success. It is our hope that the deeper understanding of the nature of mental retardation afforded by our approach will eventually lead to more effective educational training programs. These will be tailored to the strengths of individuals with mental retardation and will aid in the remediation of their deficiencies.

References and recommended reading

Anumolu, V., Bray, N. W. and Reilly, K. D. 1996: Neural network models of strategy development in children. *Neural Networks*, 10, 7–24.

Baroody, A. J. 1996: Self-invented addition strategies by children with mental retardation. *American Journal on Mental Retardation*, 101, 72–89.

*Bray, N. W., Fletcher, K. L. and Turner, L. A. 1996: Cognitive competencies and strategy use in individuals with mild retardation. In W. E. MacLean Jr. (ed.), *Handbook of Mental Deficiency, Psychological Theory and Research*, 3rd edn, Hillsdale, NJ: Erlbaum, 197–217.

Bray, N. W., Fletcher, K. L., Huffman, L. F., Hawk. T. M. and Ward, J. L. 1994a: Developmental differences in the use of models and verbal prompts in support of external strategies. Paper presented at the Conference on Human Development, Pittsburgh, Pa.

Bray, N. W., Saarnio, D., Borges, J. M. and Hawk, L. W. 1994b: Intellectual and developmental differences in external memory strategies. *American Journal on Mental Retardation*, 99, 19–31.

Bray, N. W., Reilly, K. D., Villa, M. F. and Grupe, L. A. In press: Neural network models and mechanisms of strategy development. *Developmental Review*.

Fletcher, K. L. and Bray, N. W. 1995: External and verbal strategies in children with and without mild mental retardation. *American Journal of Mental Retardation*, 99, 363–75.

*Fox, R. and Oross, S., III 1992: Perceptual deficits in mildly mentally retarded adults. In N. W. Bray (ed.), *International Review of Research in Mental Retardation*, vol. 18, San Diego, Calif.: Academic Press, 1–25.

Hinton, G. E. and Shallice, T. 1991: Lesioning an attractor network: investigations of acquired dyslexia. *Psychological Review*, 98, 74–95.

Kail, R. 1993: The role of global mechanisms in developmental change in speed of processing. In M. L. Howe and R. Pasnak (eds), *Emerging Themes in Cognitive Development*, vol. 1, New York: Springer-Verlag, 97–116.

*McClelland, J. L. 1989: Parallel distributed processing: implications for cognition and development. In R. G. M. Morris (ed.), *Parallel Distributed Processing: Implications for Psychology and Neurobiology*, New York: Oxford University Press, 8–45.

Merrill, E. C. and Taube, M. 1996: Negative priming and mental retardation: the processing of distractor information. *American Journal on Mental Retardation*, 101, 63–71.

*Siegler, R. S. and Shipley, C. 1995: Variation, selection, and cognitive change. In G. Halford and T. Simon (eds), *Developing Cognitive Competence: New Approaches to Process Modeling*, Hillsdale, NJ: Erlbaum, 31–76.

Turner, L. A., Hale, C. A. and Borkowski, J. G. 1996: Influence of intelligence on memory development. *American Journal on Mental Retardation*, 100, 468–80.

60

Science

WILLIAM F. BREWER AND PUNYASHLOKE MISHRA

The cognitive science of science studies the cognitive processes involved in carrying out science: How do scientists reason? How do scientists develop new theories? How do scientists deal with data that are inconsistent with their theories? How do scientists choose between competing theories? Research on these issues has been carried out by investigators in a number of cognitive science disciplines, particularly psychology, philosophy, and artificial intelligence. More detailed accounts of work in this area can be found in two recent conference volumes (Giere, 1992; Gholson et al., 1989).

At present the psychological and computational study of the scientific process is a relatively underdeveloped topic compared with the study of science by philosophers, historians, and sociologists. Nevertheless, the study of science and scientists within cognitive psychology and cognitive science has led to some important insights and is an exciting topic with enormous potential. Researchers in the cognitive science of science have used a wide range of methodologies. The most common approach has been laboratory studies of undergraduate participants. This type of research allows the use of laboratory control and can be used to develop the background knowledge needed to understand scientists; however, it suffers from not being directly about scientists. Actual laboratory research using scientists has been relatively rare. There have been some retrospective psychological studies of major figures in the history of science. These have the virtue of being about exemplars of exceptional scientific activity, but are limited to material that happened to find its way into the historical record. Recently there have been a few descriptive studies of working scientists. These have the virtue of ecological validity but do not allow for experimental control. There have also been a number of attempts to develop computational models of historical events in science; however, there have been few attempts to model working scientists.

Research in other areas of cognitive science informs and provides a background for the cognitive science of science. (See Articles 7, 20, and 21: CONCEPTUAL CHANGE, PROBLEM SOLVING, and REASONING.) The cognitive science perspective of this chapter means that we will not cover a number of topics that would be covered in a more inclusive discussion of the psychology of science (e.g., the personality and motivation of scientists, the social interactions of scientists, the institutions of science). We organize this chapter in terms of a simple heuristic: What do scientists do everyday in their capacity as scientists and what psychological processes are involved in those activities?

We propose that the activities of scientists can be grouped into three general categories:

1 Understanding and evaluating scientific information. Scientists spend much time reading the scientific literature and attending conferences. These activities suggest an investigation of how scientists understand and evaluate scientific data and theories.
2 Generating new scientific knowledge. Scientists design and carry out experiments and develop new theories. These activities suggest the study of scientific research strategies and the process of scientific discovery.
3 Disseminating scientific knowledge. Scientists invest considerable time writing up and giving talks about the results of their work. These activities suggest the study of the scientific writing process and the more general process of disseminating scientific information.

The above organizing heuristic provides an interesting perspective on the current literature on the cognitive psychology and cognitive science of science. Most research in this area has been directed at the issue of generating new scientific knowledge, with a particular focus on two subtopics: scientific discovery and data-gathering strategies in science. Our organizing heuristic suggests a wider range of topics for investigation.

In thinking about the topics to be covered in this chapter, we have adopted a second complementary heuristic: the focus of work in this area should be on issues relatively specific to scientists. For example, in the area of understanding scientific information, the study of the general processes involved in understanding difficult expository prose is not unique to scientists but could also be studied with other groups, such as historians or lawyers. However, understanding the role of data in theory evaluation, while not unique, seems much more specific to scientific thinking.

One final issue raised by our approach is the representativeness of the scientists discussed. Much of the thinking in the psychology of science has been driven by a small set of classic anecdotes about famous scientists: for example, the story of how Einstein was led to the development of the theory of special relativity by imagining himself traveling alongside a beam of light, or the story of how Kekulé discovered the ring structure of benzene by visualizing a snake grabbing its tail. These certainly are an interesting source of ideas; however, it seems to us that they ought to be used against a background of knowledge about how ordinary scientists work. We prefer the strategy used in areas such as the study of human memory, where individuals with exceptional memory are investigated to see how their memory performance relates to what is known about ordinary memory performance. Thus we think that the activities of exceptional scientists must be compared and contrasted with those of ordinary scientists. In the remainder of the chapter we will examine research on a number of selected topics in the cognitive science of science

Understanding and evaluating scientific information

When scientists read the work of other scientists or try to make sense of their own data, one important question that emerges is how the data are interpreted in the context of the theoretical beliefs of the scientist. One aspect of this issue is how scientists respond to anomalous data (data that go against their theories). Many popular and positivist accounts of science give primacy to data and assume that scientists rapidly change theories in the face of anomalous data. Other philosophers of science (Thomas Kuhn, Paul Feyerabend) focus on the powerful role of theory and argue that scientists often do not change their theories in the face of anomalous data.

Chinn and Brewer carried out a program of research on this issue (Brewer and Chinn, 1994). They used evidence from experimental psychology and from the history of science to argue that there are seven fundamental responses that scientists make to anomalous data. Scientists can *ignore* the data. They can *reject* the data (e.g., on methodological grounds). They can *exclude* the data from the domain of their theory by arguing that the theory is not intended to explain these particular data. They can hold the data in *abeyance* (i.e., concede that their theory cannot explain the data at present but that it will be able to in the future). They can accept the data but *reinterpret* it so that it is consistent with their theory. They can accept the data and make minor *peripheral changes* to their theory. And finally, they can accept the data and *change* their theory. Chinn and Brewer give a psychological analysis of these responses to anomalous data in terms of whether the scientist believes the data, whether the data can be explained, and whether the original theory is changed. They use this analysis to make the strong claim that these seven forms of response exhaust the psychologically plausible responses to anomalous data.

Chinn and Brewer also investigated how undergraduates respond to anomalous data. In these experiments, students read about a particular scientific theory (e.g., the meteor theory of mass extinctions in the Cretaceous period) and were then given a piece of anomalous data. Across a wide range of theories and experimental conditions, the students showed a strong tendency to hold on to their theories in the face of anomalous data by using one or more of the first five approaches listed above. Clearly these experimental studies need to be extended to examine the responses of scientists.

Thomas Kuhn's classic work on the nature of scientific revolutions has highlighted the issue of radical theory change in science. How can we understand the psychological processes involved when a scientist comes to reject one theory and replace it with another that is fundamentally different? Paul Thagard (1992) has recently developed a computational approach to these issues. He has formulated a theory of explanatory coherence which attempts to capture the relationships among concepts in a theory and the relationship between the theory and relevant data. He has implemented this theory in a computer program (ECHO) which represents theories in terms of nodes (concepts) and relations and uses a connectionist approach to calculate the degree of explanatory coherence for a theory.

Thagard has applied ECHO to a number of historical cases of scientific revolutions (e.g., Lavoisier, Darwin). He takes a cognitive approach to these historical cases and argues that a scientist choosing between two theories will choose the one with higher explanatory coherence. When presented with the historical facts in these cases of scientific revolutions, ECHO finds that the historically successful theory shows higher explanatory coherence. This is important work. However, if it is to serve as an adequate model of the psychology of the scientist, it needs further development. For example, Thagard's system currently compares two completed theories, whereas the historical record suggests that shifting from one theory to another is a slow, gradual process for the scientists involved. The system does not incorporate the psychologically important property of the theory-ladenness of data (i.e., that a scientist's theory influences his or her evaluation of the data). Finally, the system does not provide an account of what constitutes a psychologically appealing explanation for a particular datum. Although ECHO captures several important aspects of theory selection, clearly much more work is needed to provide a comprehensive model for theory choice.

Several investigators have attempted to take a cognitive science approach to the study of the work of a single famous scientist. Nancy Nersessian has studied Maxwell; Howard Gruber has investigated Darwin; and Ryan Tweney has examined Faraday. These accounts tend to be fairly detailed and complex. To give one brief example from this type of work, consider Tweney's (1992) discussion of Faraday's laboratory procedures. Faraday was an extremely prolific experimenter, and Tweney pointed out that Faraday kept extensive notebooks and developed a number of elaborate indexing schemes as memory aids, so that he could retrieve and understand the results of his own experiments.

Generating new scientific knowledge

Within the area of the cognitive study of science, the topic of research strategies for gathering data has received by far the most attention. Essentially all of this work derives from an experiment carried out by Peter Wason in 1960 (see Newstead and Evans, 1995, for reviews of this work). In Wason's experiment, undergraduate participants were told that the experimenter had a rule in mind involving three numbers; the participants were to generate number triples, and after each triple the experimenter would tell the participant if the number was correct or not. The experimenter then told the participants that the triple 2-4-6 was a correct instance of the rule. When the participants thought they knew the rule, they were to tell the experimenter, who would then inform them if they were right or not, and, if they were not right, the experiment continued. The rule used in the original and in most following experiments was: *numbers in increasing order of magnitude.* Wason intended this task to be a simulation of the scientific process of data gathering: the experimenter's rule corresponds to Nature, the undergraduates function as scientists, the generated triples correspond to designed experiments, and the feedback provided by the experimenter corresponds to data generated by the experiment.

Many of the participants in Wason's experiment never found the correct rule and instead focused on rules that were subsets of the correct rule (e.g., even numbers increasing by two). Following the work of the philosopher Karl Popper, Wason adopted the position that falsification is the normatively correct research strategy for data gathering. He argued that the students in this experiment tended to generate data consistent with their hypotheses and thus were using an unscientific research strategy.

Wason's experiment stands as a landmark in the cognitive study of science. It opened up the laboratory study of scientific research strategies. However, in retrospect, it is clear that in many important ways it was an inadequate experimental analog of scientific data gathering. For example, the *theories* in this task are analogs to scientific laws, not explanatory theories. The data never contain errors. The participants can directly ask *Nature* if their theory is correct. In this task (unlike real science), theories are easy to develop, and it is easy for participants to shift to a new theory. The correct theory, in combination with the initial seed (2-4-6), led the participants to develop theories that were subsets of the correct theory and so gave rise to a situation in which erroneous theories were highly confirmed by the data. Finally, it is not clear to us that the strong focus on disconfirmation in this literature makes sense when applied to the actual process of doing science. For example, in most real examples of science (e.g., the meteor theory of mass extinctions in the Cretaceous) the experiments that are carried out are derived from the theory (e.g., having found an iridium layer at a site in Italy, see if you can find an iridium layer in Denmark, or see if there was a large impact site that occurred at the beginning of the Cretaceous period) and have the potential to either support or disconfirm the theory. A number of these problems have been noted (and sometimes corrected) by other researchers in this area (Michael Gorman, Joshua Klayman, Michael Mahoney, Ryan Tweney).

Several investigators (Clifford Mynatt, David Klahr, Kevin Dunbar) have studied research strategies with task environments somewhat richer than the 2-4-6 task; however, it appears to us that additional work will be required to develop laboratory tasks that do a convincing job of capturing the research strategies which scientists actually use to gather data.

Quite a different approach to the study of scientific discovery has been taken by Langley, Simon, Bradshaw, and Zytkow (1987). These researchers developed computer simulations of various aspects of the process of scientific discovery. One of the most widely discussed programs that they developed is BACON, a program that carries out data-driven discovery. This program has been given data from a variety of historical cases (e.g., Kepler's third law, Black's law of temperature equilibrium) and has generated the appropriate mathematical functions.

BACON certainly provides a demonstration that it is possible to model part of the discovery process; however, one must note the limitations of this program. First, the discoveries made by the program are examples of scientific laws, not explanatory scientific theories such as the atomic theory or the theory of plate tectonics. Even with the discovery of laws, it would appear that in the historical cases the situation from the point of view of the scientist was less well structured than that given to the computer. In addition, the program has no motivated way to deal with experimental error. The occurrence of a single very erroneous data point would cause severe problems in BACON's ability to discover a relationship. Finally, it would appear that the background theoretical beliefs of scientists play a major role in most cases of scientific discovery, and this is missing from BACON. Langley and colleagues are aware of some of the limitations just discussed and have developed other programs (e.g., STAHL) which attempt to capture additional components of scientific discovery. Even though BACON and the other programs model only parts of the discovery process, they are still a very impressive achievement, since they demystify scientific discovery and show that at least some aspects of the process can be modeled by a computer.

Kevin Dunbar (1995) has recently carried out an observational study of scientists in several biological laboratories. He attempted to gather data from all the public aspects of research going on in these labs (including planning and execution of experiments, evaluation of results, group meetings, and public talks). This study provides a rich set of data on actual scientific practice. Among other things he has studied is the issue of how analogies are used in the scientific process. He finds that analogies to closely related problems were widely used to map the characteristics of previously successful experiments onto experiments with problems, in order to design new experiments.

Analogies to problems in different domains of inquiry were rarely used to drive scientific discovery, but were occasionally used to help someone understand an issue under discussion (see Article 1, ANALOGY).

Disseminating scientific knowledge

Scientists spend a significant amount of their time communicating their ideas (through journal articles, conference talks, e-mail, personal conversation, etc.). There has been relatively little research on this aspect of the scientific process. However, Greg Myers (1990) has studied the practices of working biologists engaged in different writing activities. He points out that the scientific journal article does not mirror the actual events in the laboratory that led to the article. Instead, the article is structured to give a series of logical arguments to support a theoretical position. He also notes that if the same scientist writes a popular article (e.g., for *Scientific American*) about the same experiments, the results take on, yet again, a very different textual form. In this case the findings tend to be written up as fact, without qualifiers and discussions of alternate theories. Clearly there is much additional work to be done in exploring the cognitive aspects of the rhetorical practices which scientists use in disseminating knowledge.

The research in the cognitive science of science reviewed here has already provided some interesting insights into the cognitive process underlying the activities of scientists. However, given the important role of science in modern society, it seems that the cognitive science of science has not received adequate attention. The task of understanding the cognitive processes involved in doing science is difficult, but the multiple methodologies used in cognitive science are particularly appropriate to this problem. This is an area with great potential.

References and recommended reading

Brewer, W. F. and Chinn, C. A. 1994: Scientists' responses to anomalous data: evidence from psychology, history, and philosophy of science. *PSA 1994*, 1, 304–13.

Dunbar, K. 1995: How scientists really reason: scientific reasoning in real-world laboratories. In R. J. Sternberg and J. D. Davidson (eds), *The Nature of Insight*, Cambridge, Mass.: MIT Press, 365–95.

*Gholson, B., Shadish, W. R. Jr., Neimeyer, R. A. and Houts, A. C. (eds) 1989: *Psychology of Science: Contributions to Metascience.* Cambridge: Cambridge University Press.

*Giere, R. N. (ed.) 1992: *Minnesota Studies in the Philosophy of Science*, vol. 15: *Cognitive Models of Science.* Minneapolis: University of Minnesota Press.

Langley, P., Simon, H. A., Bradshaw, G. L. and Zytkow, J. M. 1987: *Scientific Discovery: Computational Explorations of the Creative Process.* Cambridge, Mass.: MIT Press.

Myers, G. 1990: *Writing Biology: Texts in the Social Construction of Scientific Knowledge.* Madison: University of Wisconsin Press.

Newstead, S. E. and Evans, J. St B. T. (eds) 1995: *Perspectives on Thinking and Reasoning: Essays in Honour of Peter Wason.* Hove: Erlbaum.

Thagard, P. 1992: *Conceptual Revolutions.* Princeton: Princeton University Press.

Tweney, R. D. 1992: Serial and parallel processing in scientific discovery. In R. N. Giere (ed.), *Minnesota Studies in the Philosophy of Science*, vol. 15: *Cognitive Models of Science*, Minneapolis: University of Minnesota Press, 77–88.

Selective biographies of major contributors to cognitive science

WILLIAM BECHTEL AND TADEUSZ ZAWIDZKI

Hundreds of researchers have made significant contributions to cognitive science. What follows is a set of short academic biographies of people whom we believe should be counted on anyone's list of important contributors and whose work is discussed in the Companion. Not every important figure is included; and some people are included, especially from the history of cognitive science, who would not describe, or could not have described, themselves as cognitive scientists despite their considerable impact on the field. Because the number of important contributors expanded dramatically as cognitive science developed, we have not included any cognitive scientists who received the Ph.D. after 1970. A larger set of biographies, including many more contemporary researchers, can be found on the worldwide web at http://www.artsci.wustl.edu/~wbechtel/companion/biographies.html. We trust that the list will be useful to students doing research in cognitive science and to readers who wish to familiarize themselves with the work of specific contributors.

Abelson, Robert P. (b. 1928, Brooklyn, New York; Ph.D., Psychology, Princeton, 1953). Abelson has spent his career at Yale, where he has combined his interests in mathematical psychology and social psychology by focusing on phenomena such as attitude change. He developed an early interest in the use of computers to simulate both cognitive and social processes, including, for example, the American electorate in the 1960 and 1964 elections. He is best known in cognitive science for his work with Roger Schank on computational models of story comprehension. In *Scripts, Plans, Goals, and Understanding* (1977), they introduced the concepts of scripts, plans, and themes to handle story-level understanding. His most recent book is *Statistics as Principled Argument* (1995).

Anderson, James A. (b. 1940, Detroit; Ph.D., Physiology, MIT, 1967). Anderson has spent his career at Brown University, where he is currently professor and chair of cognitive and linguistic sciences. His work concentrates on applications of neural networks to such domains as human concept formation, speech perception, and models of low-level vision. His framework of a brain state in a box has been influential in neural network research. In 1995 he published *Introduction to Neural Networks.*

Babbage, Charles (b. 1791, Teignmouth, Devonshire; d. 1871, London). Elected to the Royal Society at age 24, Babbage became Lucasian Professor of Mathematics at Cambridge in 1827. He is often referred to as the "father of computing," in recognition of his design of two machines, the "difference engine" for calculating tables of logarithms by repeated additions performed by trains of gear wheels and the "analytical engine" designed to perform a variety of computations using punch cards. Babbage spent much of his life trying to build the difference engine, a prototype of which was not completed until long after his death. His inventions went beyond computing and included the speedometer and the train cow-catcher. Among his best-known writings are *Reflections on the Decline of Science in England* (1831) and *On the Economy of Machinery and Manufactures* (1833), the latter of which proposed an early form of operations research.

Baddeley, Alan D. (b. 1934, Leeds; Ph.D., Psychology, Cambridge, 1967). Baddeley, a leading researcher on memory, spent much of his career as director of the Applied Psychology Unit of the Medical Research Council in Cambridge, England. Among his important theoretical contributions has been the modeling of working memory in terms of an articulatory loop. His books include *The Psychology of Memory* (1976), *Your Memory: A User's Guide* (1982), *Working Memory* (1986), and *Working Memory and Language* (1993).

Bartlett, Sir Frederick Charles (b. 1886, Stow-on-the-Wold, Gloucestershire; d. 1969, Cambridge, England; M.A., Moral Sciences, University of Cambridge). Bartlett spent his professional career (1922–52) at Cambridge University, where he became the first professor of experimental psychology. He was editor of the *British Journal of Psychology* from 1924 to 1948, and was knighted in 1948. Bartlett is best known for his studies of memory using meaningful materials rather than nonsense syllables. In *Remembering* (1932) he showed how individuals, instead of merely reproducing the materials, organized them in terms of schemata.

Berlin, Brent (b. 1936, Pampa, Texas; Ph.D., Anthropology, Stanford University, 1964). After spending most of his career at the University of California, Berkeley, Berlin is now Graham Perdue Professor of Anthropology at the University of Georgia. He has conducted ethnobotanical studies among the Highland Maya of Chiapas, Mexico, and ethnobotanical and ethnozoological fieldwork among the Jívaro of Amazonas, Peru. In collaboration with Paul Kay, Berlin examined color terms in a wide number of languages, and established that there are eleven universal color categories and that these color terms enter a language in a strict order. Together they wrote *Basic Color Terms: Their Universality and Evolution* (1969).

Bobrow, Daniel G. (b. 1929, New York City; Ph.D., Mathematics, MIT, 1964). Bobrow has spent much of his career at Bolt Beranek and Newman and Xerox Parc. In addition to developing computer languages and programming systems, he has worked on such problems in artificial intelligence as natural language processing and knowledge representation. Through his active engagement in interdisciplinary collaboration, he helped to shape cognitive science in its early decades.

Boole, George (b. 1815, Lincoln, UK; d. 1864, Cork, Ireland). Mostly self-educated, Boole spent his academic career at Queen's College in Cork (1849–64). In his principal work, *An Investigation of the Laws of Thought on which are Founded the Mathematical Theories of Logic and Probabilities* (1854), Boole established a new branch of mathematics, symbolic logic, in which symbols are used to represent logical operations. In this book, Boole proposed a calculus (the Boolean algebra, in which symbols take one of only two values, 0 and 1) that he claimed was based on the nature of human logical thought. He saw his project as an attempt to translate thought into mathematical symbols.

Bower, Gordon Howard (b. 1932, Scio, Ohio; Ph.D., Experimental Psychology, Yale, 1959). Bower spent most of his career at Stanford University where he was mentor to many of the major contributors to cognitive psychology. His early work developed mathematical models of human learning, culminating in *Theories of Learning*, with Ernest Hilgard (1966), and *Attention to Learning* (1968), written with Thomas Trabasso. His work with John Anderson on human memory processes resulted in *Human Associative Memory* (1973). Other research of his deals with the influence of imagery, organizational factors, and emotions on memory and recall.

Brentano, Franz (b. 1838, Germany; d. 1917, Zurich, Switzerland; Ph.D., Philosophy, University of Tübingen, 1862). Brentano taught at the University of Würzburg (1866–74) and the University of Vienna (1874–95). An opponent of Wundt's emphasis on experimentation, Brentano is known especially for his appeal to intentionality, or internal object-directedness of thought, to mark the distinction between psychological and physical phenomena in *Psychologie vom empirischen Standpunkt* ("Psychology from an Empirical Standpoint") (1874).

Broadbent, Donald E. (b. 1926, Birmingham, England; d. 1993; Ph.D., Psychology, University of Cambridge). Broadbent was a researcher and director of the Applied Psychology Unit in Cambridge for a major portion of his career, before moving to Oxford in the 1970s. His research covered a wide variety of problems, from the design of postal zip codes to attention, in which he popularized the use of dichotic listening experiments. His books include *Perception and Communication* (1958), *Decision and Stress* (1971), and *In Defence of Empirical Psychology* (1973).

Broca, Paul (b. 1824, Sainte-Foy-la-Grande, France; d. 1880, Paris; M.D., University of Paris, 1849). Broca, who spent his career at the University of Paris (1853–80), had wide-ranging interests that included both neuroanatomy and anthropology. He is best known for his research on Laborgne, a patient who exhibited a severe deficit in articulate speech (the patient was known as "Tan," since that was the one expression he could utter). Broca argued that Tan's deficit originated from a lesion in the third frontal lobe, and that this was the center for articulate speech. The area came to be referred to as "Broca's area."

Brodmann, Korbinian (b. 1868, Liggersdorf, Germany; d. 1918, Munich; M.D., University of Leipzig, 1898). In research conducted between 1901 and 1910 at the Neurobiological Institute in Berlin, Brodmann argued that the human cortex is organized anatomically in the same way as the cortex of all other mammals. He showed that the cortex in animals and humans consisted of six layers, and, on the basis of anatomical differences in these layers, he developed a numbering system for areas of cortex which has become a standard basis for designating them. His work culminated in the publication of *Vergleichende Lokalisationslehre der Grosshirnrinde* in 1909.

Brown, Roger (b. 1925, Detroit; d. 1997, Boston; Ph.D., Psychology, University of Michigan, 1952). Brown spent most of his professional career at Harvard University and is celebrated for his longitudinal study of the development of language in three children, culminating in *A First Language* (1973). His other major books include *Words* *and Things* (1958), *Social Psychology* (1965), *Psycholinguistics* (1970), and *Psychology* (1975).

Bruner, Jerome (b. 1915, New York City; Ph.D., Psychology, Harvard, 1941). Having spent the major portion of his career at Harvard, Bruner later worked at Oxford and the New School for Social Research, and is currently research professor in psychology and senior research fellow in the School of Law at New York University. In the 1940s Bruner, together with Leo Postman, developed what came to be called the "new look" movement in perception, which emphasized the role of active psychological processes in perception. Together with George Miller, he founded the Harvard Center for Cognitive Studies in 1960. Following *A Study of Thinking* (1956), on concept acquisition, Bruner's interest turned increasingly to developmental psychology and the relation between culture and mental development. Most recently he has emphasized narrative and the nature of interpretive activity. His numerous books include *Actual Minds, Possible Worlds* (1987), *Acts of Meaning* (1991), and *The Culture of Education* (1996).

Buchanan, Bruce G. (b. 1940, St Louis, Missouri; Ph.D., Philosophy, Michigan State University, 1966). Buchanan spent much of his career in the Computer Science Department at Stanford University, where he was co-director of the Knowledge Systems Laboratory. He moved to the University of Pittsburgh in 1988 as co-director of the Center for Parallel, Distributed, and Intelligent Systems. Buchanan's research has focused on intelligent computer methods for knowledge acquisition and machine learning, scientific hypothesis formation, and construction of expert systems for scientific problems. He was one of the principal designers of a number of pioneering expert systems including DENDRAL, Meta-DENDRAL, MYCIN, E-MYCIN, and PROTEAN.

Chomsky, A. Noam (b. 1928, Philadelphia; Ph.D., Linguistics, University of Pennsylvania, 1955). Chomsky, who has spent his professional career at MIT, catalyzed a revolution in linguistics with his development of transformational grammars and arguments as to the shortcomings of statistical approaches to modeling grammatical knowledge. In 1959 he published a review of B. F. Skinner's *Verbal Behavior* in which he argued that language acquisition could not be explained with the resources of the classical theory of conditioning, but required the positing of representational structures governed by rules. Throughout his career Chomsky has regularly revised his accounts of grammar in an attempt to provide a more satisfactory account of acceptable linguistic structures. Chomsky has also been an ardent defender of a nativist account of our knowledge of grammar. Among his numerous publications are the following influential books: *Syntactic Structures* (1957), *Aspects of the Theory of Syntax* (1965), *Cartesian Linguistics* (1966), *Language and the Mind* (1968), *Rules and Representations* (1980), *Lectures on Government and Binding* (1981), and *The Minimalist Program* (1995).

Church, Alonzo (b. 1903, Washington, DC; d. 1995, Hudson, Ohio; Ph.D., Mathematics, Princeton, 1927). A faculty member at Princeton University from 1929 to 1967 and at the University of California at Los Angeles from 1967 until 1990, Church, one of the foremost logicians of the twentieth century, created the lambda calculus in the 1930s, which today is an invaluable tool for computer scientists. He is best remembered for Church's Thesis, that any effectively decidable function can be represented as a recursive function, and Church's Theorem (1936), which shows that there is no decision procedure for arithmetic.

Churchland, Patricia S. (b. 1943, Oliver, British Columbia, Canada; D. Phil., Philosophy, Oxford, 1969). After 25 years at the University of Manitoba, P. S. Churchland has been professor of philosophy at the University of California at San Diego since 1984. She has been an avid proponent of a reductionistic approach to mind; in *Neurophilosophy* (1986), a landmark book integrating philosophy with neuroscience, she defends a coevolution of mind and brain in which future psychological accounts will reduce to neuroscientific ones. She has collaborated in empirical work, in particular with T. Sejnowski, with whom she authored *The Computational Brain* (1992).

Churchland, Paul M. (b. 1942, Vancouver, British Columbia, Canada; Ph.D., Philosophy, University of Pittsburgh, 1969). After a lengthy stint at the University of Manitoba, P. M. Churchland became professor of philosophy at the University of California at San Diego in 1984. Churchland has been a major advocate of eliminativism, the doctrine that our everyday, commonsense, "folk" psychology, which seeks to explain human behavior in terms of the beliefs and desires of agents, is actually a deeply flawed theory that must be eliminated in favor of a mature cognitive neuroscience. Churchland first suggests this thesis in his 1979 book *Scientific Realism and the Plasticity of Mind*. In the 1980s, Churchland began to champion connectionism as a source of answers to traditional problems in the philosophy of mind and of science. His connectionist insights are presented in *A Neurocomputational Perspective* (1989); his latest book, *The Engine of Reason, The Seat of the Soul* (1995), extends his view to the social and moral dimensions of human life.

Cicourel, Aaron V. (b. 1928, Atlanta, Georgia; Ph.D., Anthropology and Sociology, Cornell University, 1957). Cicourel has been at the University of California at San Diego since 1970, where he is now professor of cognitive science, pediatrics, and sociology. A pioneer in relating cognitive science and sociology, Cicourel has concentrated on the local, ethnographically situated use of language and thought in natural settings. His 1974 book, *Cognitive Sociology: Language and Meaning in Social Interaction*, helped define this research area. His conviction that cognition is always embedded in cultural beliefs about the world and in local social practices has recently led him to explore the connections between neural development in children, human information processing, and the way in which socially organized ecologies influence the brain's internal organization and the child's capacity for normal problem solving, language use, and emotional behavior.

Clark, Eve V. (b. 1942, Camberley, UK; Ph.D., Linguistics, Edinburgh, 1969). E. V. Clark has been in the Department of Linguistics at Stanford University since 1969. Her research has focused on language acquisition, word formation, and lexical structure. For over 25 years she has directed the Child Language Research Forum at Stanford University and helped to edit its proceedings. Her books include *Psychology and Language* (1977), *The Ontogenesis of Meaning* (1979), *Meaning and Concepts* (1983), and *The Lexicon in Acquisition* (1993).

Clark, Herbert H. (b. 1940, Deadwood, South Dakota; Ph.D., Psychology, Johns Hopkins University, 1966). Clark has been in the Department of Psychology at Stanford University since 1969, where he has investigated cognitive and social processes in language use. He is especially interested in the interactive processes of conversation, which range from low-level disfluencies through acts of speaking and understanding to the emergence of discourse. He is also interested in word meaning and word use. His books include *Psychology and Language* (1977), *Arenas of Language Use* (1992), and *Using Language* (1996).

Collins, Allan (b. 1937, Orange, New Jersey; Ph.D., Cognitive Psychology, University of Wisconsin, 1970). Collins has spent most of his career as a research scientist at Bolt Beranek and Newman; since 1989 he has also been professor of education and social policy at Northwestern University. He is best known in psychology for his work on semantic memory and mental models, in artificial intelligence for his work on plausible reasoning and intelligent tutoring systems, and in education for his work on inquiry teaching, cognitive apprenticeship, situated learning, epistemic games, and systemic validity in educational testing. Collins was one of the founding editors of *Cognitive Science*.

Craik, Fergus I. M. (b. 1935, Edinburgh; Ph.D., Psychology, Liverpool, 1965). Craik has spent his professional career in the Psychology Department at the University of Toronto; he is now also affiliated with the Rotman Research Institute. Much of Craik's research has focused on memory, and he is perhaps best known for the concept of "levels of processing," which he and Robert Lockhart advanced in 1972 as an alternative to the hypothesis of separate stages for sensory, working, and long-term memory. According to the levels of processing framework, stimulus information is processed at multiple levels simultaneously, depending upon its characteristics, and the "deeper" the processing, the more will be remembered.

Damasio, Antonio R. (b. 1944, Lisbon, Portugal; M.D. and Ph.D., University of Lisbon, 1969, 1974). Currently, Damasio is M. W. Van Allen Professor and head of the Department of Neurology at the University of Iowa. His research in cognitive neuroscience has focused on large-scale neural systems and their role in mental function. Through his "semantic marker" hypothesis, he draws an intimate connection between emotion and cognition. His research on patients with frontal lobe damage, reviewed in his book *Descartes' Error* (1994), indicates that covert signals or overt feelings normally accompany response options and operate as a bias to assist knowledge and logic in the process of choice.

D'Andrade, Roy Goodwin (b. 1931, Brooklyn, New York; Ph.D., Social Relations, Social Anthropology, Harvard, 1962). Since 1970 D'Andrade has been in the Department of Anthropology at the University of California at San Diego. Early in his career he collaborated on a cognitively oriented computational analysis of kin terms and pioneered the use of scaling and other statistical techniques in anthropological linguistics. Later he used comparative cognitive studies to explore the cultural components of cognition. His books include *The Development of Cognitive Anthropology* (1995).

De Groot, Adriaan D. (b. 1914, Haarlem, The Netherlands; Ph.D., Psychology, University of Amsterdam, 1946). His early work on cognition, and in particular his thesis *Het denken van den schaker* (English translation, *Thought and Choice in Chess* (1965)), became an inspiration for later cognitive researchers such as Herbert Simon. De Groot's research, in particular, highlighted the importance, in order to achieve expert performance in chess, of the ability to organize or chunk information in strategic ways.

Dennett, Daniel (b. 1942, Boston; D.Phil., Philosophy, Oxford, 1965). Dennett has spent most of his career at Tufts University, where he is now director of the Center for Cognitive Studies. In his book *The Intentional Stance* (1987) he argues that intentionality can be explained in terms of a stance that we are forced to take toward complex, adaptive systems that behave rationally. His 1991 book *Consciousness Explained* argues that consciousness can be identified with a covert stream of internalized discourse. Other important works by Dennett include *Content and Consciousness* (1969), *Brainstorms* (1979), *Elbow Room* (1984), and *Darwin's Dangerous Idea* (1995).

Donders, Franciscus Cornelius (b. 1818, Tilburg, The Netherlands; d. 1889; M.D., Utrecht). Spending most of his professional career at the University of Utrecht, Donders developed the subtractive method whereby the time taken to complete one activity was subtracted from the time taken to complete another activity thought to differ in just one operation so as to determine the time required for the additional operation.

Dreyfus, Hubert L. (b. 1929, Terre Haute, Indiana; Ph.D., Philosophy, Harvard, 1964). Spending most of his professional career at the University of California at Berkeley, Dreyfus has for 30 years been a leading critic of AI. His critique draws upon continental philosophy and emphasizes embodiment of cognitive systems against an overly rationalist representational approach. His major books are *What Computers Can't Do: The Limits of Artificial Intelligence* (1972), *What Computers Still Can't Do: A Critique of Artificial Reason* (1992), and *Mind over Machine* (1986).

Ebbinghaus, Hermann (b. 1850, Wuppertal, Germany; d. 1909, Halle, Germany; Ph.D., Philosophy, University of Bonn, 1873). Ebbinghaus spent the major part of his career at the University of Berlin and the University of Breslau. In pursuit of his ambition to apply the scientific method to the study of "higher" cognitive processes, Ebbinghaus invented a new method for the study of memory. Using himself as sole subject, he learned lists of nonsense syllables to mastery and recorded the amounts retained, or the trials necessary for relearning, after a passage of time. His major work was *Über das Gedächtnis* (1885), (English translation, *Memory* (1913)). To publish work emanating from places other than Wundt's Leipzig laboratory, Ebbinghaus and König founded the *Zeitschrift für Psychologie und Physiologie der Sinnersorgane* in 1890.

Fechner, Gustav Theodor (b. 1801, Gross-Särchen, Prussia; d. 1887, Leipzig; M.D., University of Leipzig, 1822). Fechner spent much of his career at the University of Leipzig, first as professor of physics, then as professor of philosophy. Fechner was a pioneer in psychophysics, measuring sensation indirectly in units corresponding to the just noticeable differences between two sensations; the reports of his studies constitute a large part of the first of the two volumes of the *Elemente der Psychophysik*. Expanding on the earlier work of Ernst Weber, Fechner's Law holds that the intensity of a sensation increases as the log of the stimulus ($S = k \log R$).

Feigenbaum, Edward A. (b. 1936, Weehawken, New Jersey; Ph.D., Industrial Administration, Carnegie–Mellon University, 1960). Feigenbaum founded the Heuristic Programming Project in the Department of Computer Science at Stanford in 1965, where he continues research directed toward developing a general framework for modeling physical devices that supports reasoning about their designed structure, intended function, and actual behavior. He collaborated with Joshua Lederberg and Bruce Buchanan in developing DENDRAL, a pioneering expert system that could identify an organic compound by analyzing mass spectrography data.

Ferrier, Sir David (b. 1843, Aberdeen, Scotland; d. 1928, London; M.D., University of Edinburgh, 1868). A professor of medicine and neuropathology at King's College, London, Ferrier refined a technique for using mild electrical stimulation of brain areas to determine their function. In 1876 Ferrier published *The Functions of the Brain*, followed by *Cerebral Localization* in 1878.

Flourens, Marie-Jean-Pierre (b. 1794, Maureilhan, France; d. 1867, Montgeron, France; M.D., University of Montpellier, 1813). Flourens spent his career at the Collège de France (1828–67). A staunch opponent of neural localization of brain function, Flourens developed the technique of precise extirpation or ablation of cortical tissue. He found that the quantity of cerebral tissue removed was more important than its location. His work led to *Phrenology Examined*, which criticized the localization claims of Gall and Spurzheim and dealt their movement a major blow.

Fodor, Jerry A. (b. 1935, New York City; Ph.D., Philosophy, Princeton, 1960). Until 1986 Fodor was on the faculty at MIT, where he became an early expositor of Chomsky's program in linguistics, collaborated in psycholinguistic research, and developed his own strongly nativistic theory. Fodor defended the claim that thinking invokes a language-like medium in *The Language of Thought* (1975). His 1983 book, *The Modularity of Mind*, defends a strong version of faculty psychology, according to which the mind consists of informationally encapsulated, "low-level" perceptual modules which feed information to "higher-level" central cognitive processes which are non-modular. According to Fodor, only modular cognitive processes can be studied scientifically. Since moving to Rutgers University in 1988, Fodor has been an ardent critic of connectionist models of cognitive phenomena, arguing that they cannot account for the rationality of thought.

Freeman, Walter J. (b. 1927, Washington, DC; M.D., Yale, 1954). Since 1959 Freeman has been at the University of California at Berkeley, first in physiology and now in neurobiology. Freeman employs EEG and single unit recording to study cortical responses during goal-directed activity. His models of cortical activity using non-linear differential equations, especially his models of the olfactory bulb, have been influential in promoting the dynamical systems approach to cognition.

Frege, Gottlob (b. 1848, Wismar, Germany; d. 1925, Bad Kleinen, Germany; Ph.D., Mathematics, University of Göttingen, 1873). Frege taught at the University of Jena his entire career. A pioneer in modern logic, he constructed the first predicate calculus, developed a new analysis of basic propositions and quantification, formalized the notion of a proof in terms that are still accepted today, and demonstrated that one could resolve theoretical mathematical statements in terms of simpler logical and mathematical notions. To ground his views about the nature of logic, Frege conceived a comprehensive philosophy of language that introduced the important distinction between the sense and reference of linguistic terms.

Gall, Franz Joseph (b. 1759, Tiefenbrunn, Baden; d. 1828, Paris; M.D., University of Vienna, 1785). Gall practised medicine in Vienna from 1785 to 1807, and in Paris from 1807 to 1828. Gall developed a program correlating protrusions in the skull with psychological faculties, an approach he called "organology," but which his sometime collaborator Spurzheim called "phrenology." Between 1810 and 1819 Gall published *Anatomie et physiologie du système nerveux en général* in four volumes (the first two written with Spurzheim), of which the last three presented the phrenological doctrine. Between 1823 and 1825 Gall published his definitive statement on phrenology, a six-volume work entitled *Sur les fonctions du cerveau*.

Gazzaniga, Michael S. (b. 1939, Los Angeles; Ph.D., Psychobiology, California Institute of Technology, 1964). Having held faculty positions at Cornell Medical College and the University of California at Davis, Gazzaniga is currently David T. McLaughlin Distinguished Professor and director of the Program in Cognitive Neuroscience at Dartmouth College. Gazzaniga has been a leading investigator of cognitive disruptions arising in patients whose hemispheres have been disconnected through commissurotomy. He has also been one of the major institution builders in cognitive neuroscience, having founded the *Journal of Cognitive Neuroscience* and the Society for Cognitive Neuroscience, been chief organizer of the McDonnell Summer Institute in Cognitive Neuroscience, and edited *The Cognitive Neurosciences* (1995). Among his many books are *The Bisected Brain* (1970), *The Social Brain* (1985), *Mind Matters* (1988), and *Nature's Mind* (1992).

Gibson, Eleanor J. (b. 1910, Peoria, Illinois; Ph.D., Psychology, Yale University, 1938). Gibson, who spent most of her professional career at Cornell University, articulated an ecological perspective in developmental psychology: perceptual development was construed as a process of differentiation, and perceptual learning as an active process of information pickup. The perceptual world is not constructed by processes of association and inference; rather, the infant explores the array of stimulation, searching for invariants that reflect the permanent properties of the world and the persisting features of the layout and of the objects in it. What come to be perceived are the affordances for action made available by places, things, and events in the world. These ideas are expressed in *Principles of Perceptual Learning and Development* (1969) and *An Odyssey in Learning and Perception* (1991).

Gibson, James J. (b. 1904, McConnelsville, Ohio; d. 1979, Ithaca, New York; Ph.D., Psychology, Princeton, 1928). From 1928 to 1949 Gibson taught at Smith College; he then moved to Cornell University, where he remained for the rest of his career. Gibson is primarily known as the founder of the ecological approach to the study of perception, an approach which emphasizes the importance of rich, structured information that is already in the light reaching the retina. Subjects merely have to pick up this information, not construct it through information processing. Gibson laid out this approach in several books, including *The Perception of the Visual World* (1950), *The Senses Considered as Perceptual Systems* (1966), and *The Ecological Approach to Visual Perception* (1979).

Gilligan, Carol (b. 1936, New York City; Ph.D., Clinical Psychology, Harvard, 1964). Gilligan has spent her professional career at Harvard. She challenged Lawrence Kohlberg's pioneering studies of moral development, which focused exclusively on males. *In a Different Voice: Psychological Theory and Women's Development* (1982) argued that women often exhibit a different pattern of moral development, resulting in an emphasis on caring rather than adherence to rules. More generally, she has shown that mature moral development often involves an integration of both care and rule perspectives.

Gleitman, Henry (b. 1925, Leipzig; Ph.D., Psychology, University of California at Berkeley, 1947). Already legendary as a teacher and integrater of psychology at Swarthmore College, H. Gleitman then moved to the University of Pennsylvania 35 years ago. His research focused on cognitive processes, especially those involved in memory and language, social cognition, and the psychology of drama and humor. He wrote two of the most highly regarded introductory texts in psychology, entitled *Psychology* (1981/1995) and *Basic Psychology* (1983/1996), and (with L. Gleitman) wrote *Phrase and Paraphrase* (1971).

Gleitman, Lila R. (b. 1929, Brooklyn, New York; Ph.D., Linguistics, University of Pennsylvania, 1967). L. Gleitman has spent her career at the University of Pennsylvania, where she is currently professor of psychology and co-director of the Institute for Research in Cognitive Science. Noted for bringing a Chomskian perspective to bear on language acquisition research, she and her students have challenged conventional wisdom by showing that simplifications in speech to children do not enhance acquisition and that language development is robust in blind and deaf children. Most recently she has been a proponent of syntactic bootstrapping. She is the co-author of *Phrase and Paraphrase* (1971) and *Language and Experience: Evidence from the Blind Child* (1985).

Goldman-Rakic, Patricia (b. Salem, Massachusetts Ph.D., Experimental Psychology, UCLA). After more than a decade at the National Institutes of Mental Health, Goldman-Rakic moved to Yale School of Medicine in 1979. Her research focuses on the development, organization, and cognitive functions of the frontal lobe, especially the role of areas of prefrontal cortex in working memory.

Golgi, Camillo (b. 1843, Cortona, Tuscany; d. 1926, Pavia, Lombardy; M.D., University of Pavia, 1865). Golgi spent his entire career, from 1875 to 1926, at the University of Pavia. He received the Nobel Prize in physiology and medicine in 1906 for his work developing silver nitrate to stain nerve cells. Golgi rejected the claim that the nervous system is comprised of discrete cells, defending instead the reticular theory that posited continuity between nerve cells.

Goodman, Nelson (b. 1906, Summerville, Massachusetts; Ph.D., Philosophy, Harvard, 1941). After teaching at Tufts, the University of Pennsylvania, and Brandeis, Goodman completed his career at Harvard University. He argued that induction was governed by pragmatic considerations as to whether the predicates employed are sufficiently entrenched in our practice. He also contributed to aesthetics, arguing that the relation between pictures and what they picture is, like the relation between words and their referents, essentially arbitrary. His major works include *The Structure of Appearance* (1951), *Fact, Fiction, and Forecast* (1955), and *Ways of Worldmaking* (1978).

Greeno, James G. (b. 1935; Ph.D., Experimental Psychology, University of Minnesota, 1961). Greeno has been a professor of psychology at the universities of Michigan and Pittsburgh and professor of education at Berkeley and Stanford, where he is co-founder and senior research fellow in the Institute for Research on Learning. In his early career he was a mathematical psychologist working in such areas as learning and problem solving and wrote *Introduction to Mathematical Psychology* (1970). His interests moved towards educational applications of cognitive science, and he has especially focused on students' understanding and misunderstanding of mathematical concepts and ways to engage them in active learning. He has been at Stanford since 1988 and has been editor of *Cognitive Science* since 1993.

Gregory, Richard L. (b. 1923, London; MA, Experimental Psychology, Cambridge, 1950; Sc.D., University of Bristol, 1983). After reading philosophy and psychology at Cambridge under Sir Frederick Bartlett, Gregory pursued his research on perception at Cambridge (where he directed the Special Senses Laboratory), Edinburgh (where he helped found the Department of Machine Intelligence), and the University of Bristol (where he was director of the Brain and Perception Laboratory and founder of the Exploratory, a hands-on science center). Partly on the basis of responses to illusions, he defended the view that perception involves predictive hypotheses much like those employed in science. Gregory founded the international journal *Perception* in 1973; his books include *Eye and Brain* (1966), *The Intelligent Eye* (1970), *Illusion in Nature and Art* (1973), *Mind in Science* (1981), and *Odd Perceptions* (1986).

Griffin, Donald R. (b. 1915, Southampton, New York; Ph.D., Biology, Harvard, 1942). During his career, Griffin worked at Cornell, Harvard, and Rockefeller universities. As an undergraduate, he discovered that bats perceive the world using sonar, and much of his career was spent trying to understand their use of sonar to guide navigation. Later in his career Griffin argued for a mentalistic approach to animal cognition, defending the claim that other species experience the world in ways similar to our own. His major books include *The Question of Animal Awareness* (1976), *Animal Thinking* (1984), and *Animal Minds* (1992).

Grossberg, Stephen (b. 1939, New York City; Ph.D., Mathematics, Rockefeller, 1967). Much of Grossberg's career has been spent at Boston University, where he directs the Center for Adaptive Systems. Grossberg has been a pioneer in the development of neural networks and is especially known for his adaptive resonance theory (ART) models. His books include *Studies of Mind and Brain* (1982) and *Pattern Recognition by Self-Organizing Neural Networks* (1991).

Harlow, Harry F. (b. 1905, Fairfield, Iowa; d. 1981, Tucson, Arizona; Ph.D., Experimental Psychology, Stanford University, 1930). From 1930 to 1974, Harlow was director of the Primate Laboratory at the University of Wisconsin, which he founded. There, he conducted research on learning, motivation, the affectional systems, and the effects and treatment of social isolation in rhesus monkeys. Harlow's work on learning resulted in the development of the concept of learning set ("learning how to learn"), the Wisconsin General Test Apparatus for studying problem solving, and the error-factor learning theory, which postulates that learning takes place as various tendencies that make for errors are gradually eliminated. Harlow was also one of the first researchers to demonstrate that cognitive motives, such as exploration and curiosity, hitherto neglected by psychologists, were as important determiners of behavior as deficiency motives like hunger.

Harris, Zelig S. (b. 1909, Balta, Russia; Ph.D., Linguistics, University of Pennsylvania, 1934). Harris spent his academic career at the University of Pennsylvania, where he established the first department of linguistics in the United States. He was a major contributor to what is known as "structural linguistics," and is particularly known for his work on discovery procedures and the progress he made in extending the structural approach to syntax (a project carried further by his student, Noam Chomsky). His books include *Methods in Structural Linguistics* (1951), *Mathematical Structures of Language* (1968), *A Grammar of English on Mathematical Principles* (1982), and *Language and Information* (1988).

Hebb, Donald O. (b. 1904, Chester, Nova Scotia; d. 1985; Ph.D., Psychology, Harvard, 1936). Hebb spent most of his academic career at McGill University in Montreal, where he became an influential theorist concerned with the relation between the brain and behavior. His most important book, *The Organization of Behavior* (1949), emphasized the formation of cell assemblies in brain processing. He posited what is now known as Hebbian learning, in which the connections between neurons are strengthened if they are simultaneously active.

Hitzig, Eduard (b. 1838, Berlin; d. 1907, St Blasien, Baden; M.D., University of Berlin, 1862). Hitzig practised medicine in Berlin from 1862 to 1875; subsequently he directed asylums of the University of Zürich and the University of Halle. In 1870, with Gustav Fritsch, Hitzig established the electrical excitability of brain tissue by demonstrating that mild electric currents applied to specific brain locations in the dog resulted in particular muscular contractions.

Holland, John H. (b. 1929, Fort Wayne, Indiana; Ph.D., Computer Science, University of Michigan, 1959). Holland has spent his academic career at the University of Michigan, where he holds academic appointments in both computer science and psychology; he is also an external faculty member at the Santa Fe Institute. His research has addressed the adaptive character of cognitive processes, and he is the creator of the genetic algorithm, which uses processes comparable to those of biological evolution to arrive at computer programs that are well adapted to their tasks. His research on genetic algorithms is presented in *Adaptation in Natural and Artificial Systems: An Introductory Analysis with Applications to Biology, Control, and Artificial Intelligence* (1975/1992). Other books include *Induction: Processes of Inference, Learning, and Discovery* (1986), *Hidden Order: How Adaptation Builds Complexity* (1995), and *Emergence: From Chaos to Order* (1998).

Hopfield, John (b. 1933, Chicago; Ph.D., Physics, Cornell, 1958). A distinguished physicist who is currently professor of chemistry and biology at the California Institute of Technology, Hopfield previously held positions at Bell Laboratories, the University of California at Berkeley, and Princeton. Hopfield proposed a design for a neural network modeled on a spin glass, a type of physical system in which each atom in a matrix of atoms spins pointing up or down, influencing the spin of its neighbors until the matrix reaches a stable configuration. In the neural networks, known as "Hopfield nets," units influence the activation of their neighbors until a stable configuration is achieved.

Hubel, David (b. 1926, Windsor, Ontario; M.D., McGill, 1951). Hubel spent most of his career at Harvard, where he often collaborated with Torsten Wiesel. Together they showed that individual neurons in visual cortex responded optimally to specific stimuli: a line or edge at a particular orientation and, for the lowest-level neurons, particular locations. They also found that cells are organized in alternating columns, with each alternate column responsive primarily to the right or left eye. Other research focused on binocular vision and the importance of early visual stimulation to normal development. In 1981 they shared the Nobel Prize for medicine and physiology.

Jackendoff, Ray S. (b. 1945, Chicago; Ph.D., Linguistics, MIT, 1969). Jackendoff, professor of linguistics at Brandeis University, has used his studies of the semantics of natural languages to launch wide-ranging inquiries into the nature of mind and language. His recent books include *Consciousness and the Computational Mind* (1987), *Semantic Structures* (1990), *Languages of the Mind* (1992), and *Patterns in the Mind* (1994).

Jackson, John Hughlings (b. 1835, Providence Green, Hammerton, England; d. 1911, London; M.D., St Andrews University, 1860). Jackson spent most of his career at the National Hospital in Queen Square, London, where he conducted pioneering studies of epilepsy, aphasia, and paralysis. He rejected simple localizationist schemes in favor of a hierarchy of representations in cortex, with higher levels modulating the behavior of lower levels.

James, William (b. 1842, New York City; d. 1910, Chocorva, New Hampshire; M.D., Harvard University, 1871). James spent his career at Harvard, switching his appointment between physiology, philosophy, and psychology. He established the first American demonstration laboratory for psychology in 1875 and wrote *The Principles of Psychology* (1890). This landmark textbook for the new field of psychology covered such topics as brain function, habit formation, the stream of consciousness, the self, attention, association, the perception of time, memory, sensation, imagination, perception, reasoning, voluntary movement, instinct, the emotions, will, and hypnotism. His other classic psychological work, *The Varieties of Religious Experience* (1902), explored the relationships between religious experience and psychology.

Jenkins, James J. (b. 1929, St Louis; Ph.D., Psychology, University of Minnesota, 1950). Jenkins spent most of his career in the Department of Psychology and the Center for Research on Human Learning at the University of Minnesota before moving to the University of South Florida. Focusing his research on language processing, Jenkins was prominent among those of his generation who made the transition from behaviorism to information processing.

Johnson-Laird, Philip N. (b. 1936, Rothwell, Yorkshire; Ph.D., Psychology, University College, London, 1967). After positions at the University of Sussex and the Applied Psychology Unit of the Medical Research Council, Cambridge, Johnson-Laird became professor of psychology at Princeton University in 1989. His extensive research on language processing and reasoning, especially the use of mental models, is reported in *Language and Perception* (1976), *Mental Models: Towards a Cognitive Science of Language, Inference, and Consciousness* (1983), *The Computer and the Mind: An Introduction to Cognitive Science* (1988), *Deduction* (1991), and *Human and Machine Thinking* (1993).

Kahneman, Daniel (b. 1934, Tel Aviv; Ph.D., Psychology, University of California at Berkeley, 1961). After appointments at the University of British Columbia and the University of California at Berkeley, in 1993 Kahneman moved to Princeton University, where he has appointments in the department of psychology and the Woodrow Wilson School of Public Affairs. His research focuses on human judgment and decision making; in research done collaboratively with Amos Tversky, Kahneman attempted to show that human reasoning does not follow strict normative principles. Rather, people rely on heuristics that simplify problems, but sometimes lead to error. Many of the major results of this research were published in *Judgment under Uncertainty: Heuristics* *and Biases* (1982). More recently Kahneman's interests have turned to the social and affective determinants of belief and choice.

Katz, Jerrold J. (b. 1932, Washington, DC; Ph.D., Philosophy, Princeton, 1962). Katz, whose research focuses on semantics, the philosophy of language, and the foundations of linguistics, is professor of linguistics and philosophy at the Graduate Center of the City University of New York. With Jerry Fodor, he sketched "the structure of semantic theory" in generative grammar in 1963. Later he defended a Platonist view of language, treating it as an abstract structure which can be analyzed in the same manner as mathematical systems. His books include *The Philosophy of Language* (1966), *Semantic Theory* (1972), *Language and Other Abstract Objects* (1981), and *The Metaphysics of Meaning* (1990).

Kay, Paul (b. 1934, New York City; Ph.D., Social Anthropology, Harvard, 1963). Kay has spent his academic career at the University of California at Berkeley, first in anthropology and since 1982 in linguistics. Together with Brent Berlin, Kay conducted seminal research on color terms that resulted in *Basic Color Terms* (1969).

Kintsch, Walter (b. 1932, Temschwar, Romania; Ph.D., Psychology, University of Kansas, 1960). Since 1968, Kintsch has been on the psychology faculty at the University of Colorado in Boulder, where he is currently director of the Institute of Cognitive Science. Kintsch's primary interests are discourse, text comprehension, and meaning representation. His construction–integration model of text comprehension combines a construction process, in which a textbase is constructed from the linguistic input as well as from the comprehender's knowledge base, with an integration phase, in which this textbase is integrated into a coherent whole. Kintsch is a past editor of the *Psychological Review* and the *Journal of Verbal Learning and Verbal Behavior*; his books include *Learning, Memory, and Conceptual Processes* (1970) and *The Representation of Meaning in Memory* (1974).

Koffka, Kurt (b. 1886, Berlin; d. 1941, Northampton, Massachusetts; Ph.D., Psychology, University of Berlin, 1909). Koffka taught at the University of Giessen until 1927, when he moved to Smith College. Along with Wertheimer and Köhler, he was one of the founders of the Gestalt school of psychology and of its organ, the journal *Psychologische Forschung*. During his years at Giessen, Koffka wrote *The Growth of the Mind* (1927), in which he applied Gestalt notions to the problems of developmental psychology. Koffka did much to make Gestalt psychology more familiar to North Americans, especially through a long series of papers in the *Psychological Bulletin* in 1922. His later book, *The Principles of Gestalt Psychology* (1935), presented Gestalt psychology as a complete theory of behavior.

Köhler, Wolfgang (b. 1887, Revel, Estonia; d. 1967, Enfield, New Hampshire; Ph.D., Psychology, University of Berlin, 1909). Köhler taught at the University of Berlin until 1935 and then at Swarthmore College until 1955. Along with Wertheimer and Koffka, Köhler was one of the founders of the Gestalt school of psychology and of the journal *Psychologische Forschung*. Köhler wrote several books presenting aspects of the Gestalt school of psychology: *Gestalt Psychology* (1929), *The Place of Value in a World of Facts* (1938), and *Dynamics in Psychology* (1940). One of Köhler's major contributions was his work in the Canary Islands during World War I, where he demonstrated the perception of and response to relationships (rather than absolute stimulus values) in chicks (later known as transposition). There he also studied insight learning (closure over psychological gaps) in chimpanzees, as reported in *The Mentality of Apes* (1924). In his *Die physischen Gestalten in Ruhe und im stationären Zustand* (1920), Köhler argued that there is a correspondence in form between physical events in the brain and the subjective events caused by them.

Kuhn, Thomas S. (b. 1922, Cincinnati, Ohio; d. 1996, Cambridge, Massachusetts; Ph.D., Physics, Harvard, 1949). Most of Kuhn's career was spent in history of science at Harvard, Princeton, and MIT. His 1962 contribution to the *Encyclopedia of Unified Science*, also published separately as *The Structure of Scientific Revolutions*, posed a major challenge to the conception of unified science with which the editors of the series were working. Kuhn argued that scientific change often occurred not through integration of theories into broader theories, but through revolutions that would radically alter the prevailing paradigm that guided scientific research in a field. In advancing this view, Kuhn appealed both to work in Gestalt psychology and to Jerome Bruner's work on perception. Kuhn also challenged traditional views of the ordinary practice of science, arguing that during eras of normal science scientists were not testing theoretical perspectives, but trying to fit nature into an already accepted paradigm.

Lakoff, George (b. 1941, Bayonne, New Jersey; Ph.D., Linguistics, Indiana University, 1966). Beginning his career at the University of Michigan, Lakoff has been in the Linguistics Department at the University of California at Berkeley since 1972. He was one of the founders and developers of the generative semantics movement of the 1960s, and more recently has played the same role for cognitive linguistics. In collaborative work with Mark Johnson, he has focused on the role of metaphors in structuring our conceptual system; this collaboration resulted in *Metaphors We Live By* (1980). More recently he has published *Women, Fire, and Dangerous Things: What Categories Reveal about the Mind* (1987) and *Moral Politics: What Conservatives Know That Liberals Don't* (1996).

Langacker, Ronald (b. 1942, Fond du Lac, Wisconsin; Ph.D., Linguistics, University of Illinois, 1966). Langacker, who has spent his career in the linguistics department at the University of California at San Diego is a major contributor to the development of cognitive linguistics. He has rejected the idea of autonomous syntax and instead advocated the view that syntactic structures arise from cognition, especially spatial cognition. His major works include *Foundations of Cognitive Grammar* (vol. 1, 1987; vol. 2, 1991) and *Concept, Image, and Symbol: The Cognitive Basis of Grammar* (1991).

Lashley, Karl (b. 1890, Davis, West Virginia; d. 1958, Poitiers, France; Ph.D., Genetics, Johns Hopkins University, 1914). Lashley's career involved positions at the University of Minnesota, the University of Chicago, Harvard University, and Yerkes Laboratory of Primate Biology. Lashley was an ardent opponent of claims to cerebral localization of cognitive functions. In 1929 he published *Brain Mechanisms and Intelligence*, an account of his work on the effects of brain lesions on learning, memory, and discrimination in rats. He showed that the rate and accuracy of learning is proportionate to the amount of brain tissue available (the law of mass action), but that it is independent of the particular tissue that is available (the principle of equipotentiality).

Lenneberg, Eric H. (b. 1921, Düsseldorf; d. 1975, Ithaca, New York; Ph.D., Linguistics and Psychology, 1955). Working at Children's Hospital Medical Center in Boston and at Cornell University, Lenneberg was an early proponent of a biological approach to language. He argued that language is the manifestation of species-specific cognitive propensities and that there is a critical period for language acquisition. He used data from a variety of impaired populations to gain general insights into language processing. He is the author of the landmark volume *Biological Foundations of Language* (1967).

Levelt, Willem J. M. (b. 1938, Amsterdam; Ph.D., Psychology, Leiden, 1965). Levelt is chair of experimental psychology at Nijmegen University and the founding director of the Max Planck Institute for Psycholinguistics in Nijmegen, one of the foremost centers for psycholinguistic research. His early research focused on perception, both visual and auditory, while his subsequent research has focused on psycholinguistics, including language production. Among his books is *Speaking: From Intention to Articulation* (1989).

Luria, Alexander Romanovich (b. 1902, Kazan; d. 1977, Moscow; Dr of Pedagogical Sciences, Psychology, Moscow University, 1936, and M.D., Moscow University, 1943). Luria spent most of his career at Moscow University, where he headed the Department of Neuropsychology. He was strongly influenced by, and collaborated with, Vygotsky on studies of mental functioning in aphasia, and continued to work on brain organization, brain processes, and speech functions after Vygotsky's death. Luria postulated that the higher cortical functions are carried out through the interaction among cortical areas that work in a more general way but become associated through concrete activity and language. A founder of the field of neuropsychology, Luria developed a diagnostic system called "syndrome analysis." His earliest book on aphasia is *Traumatic Aphasia* (1947); later works include *Higher Cortical Functions in Man* (1962), *The Mind of a Mnemonist* (1968), *An Introduction to Neuropsychology* (1973), *The Neuropsychology of Memory* (1976), and *The Making of Mind* (1979).

Mandler, George (b. 1924, Vienna; Ph.D., Psychology, Yale, 1953). Mandler founded the Department of Psychology at the University of California at San Diego in 1965, and has spent the rest of his career there. In his research he emphasized organization in memory and later the distinction between activation and elaboration. He also developed a discrepancy–evaluation theory of emotion and has developed phenomenal and theoretical analyses of consciousness. Mandler edited *Psychological Review* from 1970 to 1976. His books include *The Language of Psychology* (1959), *Mind and Emotion* (1975), *Cognitive Psychology: An Essay in Cognitive Science* (1985), and *Human Nature Explored: Psychology, Evolution, Society* (1997).

Marr, David (b. 1945, Woodford, England; d. 1980). Trained in psychology, Marr held research fellowships at Trinity College and King's College, Cambridge, before moving to MIT, where he was a member of the Artificial Intelligence Lab from 1973 until his death. In his early work, Marr developed computational models of a number of neural systems, including the cerebellum and the hippocampus. His major contribution, though, concerned computational models of visual processes, including the extraction of lines from retinal images, the determination of depth and motion, and the representation of objects to facilitate recognition. His book *Vision: A Computational Investigation into the Human Representation and Processing of Visual Information* was published posthumously in 1982 and is still influential.

McCarthy, John (b. 1927, Boston; Ph.D., Mathematics, Princeton, 1951). McCarthy was a co-director of the Dartmouth conference in 1956 at which the term "artificial intelligence" was first employed; in 1958 he invented the programming language LISP, which has been one of the most used languages in AI. After co-founding the MIT AI Lab with Marvin Minsky, in 1962 McCarthy moved to Stanford University, where he founded and directed the Artificial Intelligence Lab at Stanford. His main research contributions are to commonsense knowledge and reasoning. Many of his papers are collected in *Formalizing Common Sense* (1991).

McCulloch, Warren S. (b. 1898, Orange, New Jersey; d. 1969; M.D., Columbia, 1927). McCulloch was director of the Neuropsychiatric Institute at the University of Illinois from 1941 to 1952, when he moved to the Research Laboratory of Electronics at MIT. While at the Neuropsychiatric Institute he began a collaboration with a young logician, Walter Pitts, that resulted in some of the first neural network models of mental processes. At MIT McCulloch and Pitts established a fruitful collaboration with Jerry Y. Lettvin and Humberto R. Maturna. Their investigation of feature detectors in the frog's retina resulted in the classic paper "What the frog's eye tells the frog's brain." McCulloch's interest in neural networks extended to questions of distributed control and reliable performance from unreliable parts; many of his influential papers were published in *Embodiments of Mind* (1965).

Miller, George A. (b. 1920, Charleston, West Virginia; Ph.D., Experimental Psychology, Harvard, 1946). Miller spent most of his career until 1958 at Harvard, where he played a pioneering role in the development of psycholinguistics and cognitive psychology and co-directed the Center for Cognitive Studies with Jerome Bruner. His 1960 book with Eugene Galanter and Karl Pribram, *Plans and the Structure of Behavior*, was a pioneering work in modeling information structures in the mind. After leaving Harvard, Miller went first to Rockefeller, then to Princeton, where he has remained since. Much of his recent research has been directed at the semantics of individual words, which he explored with Johnson-Laird in *Language and Perception* (1976) and more recently modeled in a large computer program called *WordNet*. Among his more recent books are *The Psychology of Communication* (1967), *Language and Speech* (1981), and *The Science of Words* (1991).

Milner, Brenda (b. 1915, Manchester; Ph.D., Psychology, McGill, 1952). Milner has spent her career at McGill University, where she is professor of psychology in the Department of Neurology and Neurosurgery and at the Montreal Neurological Institute. Her initial research focused on the effects of temporal lobe damage in humans, especially the consequences for memory. She performed the initial neuropsychological examinations of the memory losses in H.M., who had parts of both temporal lobes removed in 1953 and has since become one of the most widely studied amnesic patients. Milner's research has extended to the cognitive deficits following damage to other brain areas, especially frontal cortex.

Minsky, Marvin (b. 1927, New York City; Ph.D., Mathematics, Princeton, 1954). Minsky has spent his professional career at MIT, where he co-founded the Artificial Intelligence Lab with John McCarthy. He edited one of the first books on artificial intelligence, *Semantic Information Processing* (1968), in which most of the chapters were based on his students' Ph.D. dissertations. Minsky's research continued to generate both theoretical and practical advances in artificial intelligence, especially in the areas of knowledge representation, computational semantics, machine perception, and learning. In 1969 he and Seymour Papert published an influential book, *Perceptrons*, which contained a generally negative assessment of the potential for neural networks. Thereafter, Minsky's notion of a "frame" played a major role in the search for higher-level knowledge representations in AI. In 1985 he published *The Society of Mind*, proposing a theory of mind that he had been developing throughout the 1970s and 1980s. In his view, intelligence is not the product of any single mechanism, but comes from the managed interaction of a diverse variety of resourceful agents. Minsky was also one of the pioneers of intelligence-based mechanical robotics, designing and building some of the first mechanical hands with tactile sensors, visual scanners, and their software and computer interfaces.

Mishkin, Mortimer (b. 1926, Fitchburg, Massachusetts; Ph.D., Psychology, McGill, 1951). Mishkin has spent his career at the National Institutes of Mental Health, where he is currently chief of the Laboratory of Neuropsychology. Much of Mishkin's research has been devoted to studying neural pathways involved in perception and attention, recognition and recall, emotion and motivation, and volition and movement, as well as brain mechanisms of learning. In the early 1980s he, together with Leslie Ungerleider, developed an influential model of visual processing in which separate pathways were responsible for identifying and locating objects (the *what* and *where* pathways).

Neisser, Ulric (b. 1928, Kiel, Germany; Ph.D., Psychology, Harvard, 1956). Neisser's career has included appointments at Brandeis, the University of Pennsylvania, Cornell, and Emory University, where he founded the Emory Cognition Project and edited a number of volumes based on its conferences. In 1996 he returned to Cornell. Neisser is best known for three books: *Cognitive Psychology* (1967), which helped to establish that field; *Cognition and Reality* (1976), which attempted to reorient it with an infusion of ecological psychology; and *Memory Observed: Remembering in Natural Contexts* (1982), an edited book which introduced the ecological approach to the study of memory. Other research areas in which he has had an impact include divided attention and the effects of mental practice. Recently he has focused on characterizing several aspects of the self and self-knowledge (the ecological, interpersonal, remembered, private, and conceptual selves).

Newell, Allen (b. 1927, San Francisco; d. 1992, Pittsburgh; Ph.D., Industrial Administration, Carnegie Institute of Technology, 1957). Newell was a scientific staff member at the Rand Corporation when he met and began a collaboration with Herbert Simon; together they worked on computer simulations of cognitive processes such as chess playing and problem solving and developed the production system architecture that is widely used in AI. Newell joined Simon at the Carnegie Institute of Technology (now Carnegie–Mellon University), where he earned his Ph.D. and remained on the faculty for the rest of his career. Their work on problem solving and production systems culminated in *Human Problem Solving* (1972). With Simon, he also proposed the Physical Symbol System hypothesis, according to which the mind is defined as a system operating on physical symbols. Much of his later work centered on Soar, an architecture for intelligent problem solving and learning that emphasizes "chunking" as a strategy. In *Unified Theories of Cognition* (1990) Newell argued for the importance of developing architectures capable of performing all cognitive tasks, maintaining that Soar provided such an architecture.

Nisbett, Richard (b. 1941, Littlefield, Texas; Ph.D., Psychology, Columbia, 1966). Nisbett has been at the University of Michigan since 1971, where his research focuses on social psychology and on the relations between social and cognitive psychology. In 1980 he and Lee Ross wrote *Human Inference Strategies and Shortcomings of Social Judgment*, in which they attempted to demonstrate that errors in judgment are due to extending beyond their proper domain the same heuristics as account for successful human reasoning. In work with Timothy Wilson, Nisbett showed that people often confabulate their reasons for action. His collaboration with John Holland, Keith Holyoak, and Paul Thagard resulted in *Induction: Processes of Inference, Learning, and Discovery* (1986), which attempted to integrate a descriptive and normative analysis of human inductive inference.

Norman, Donald A. (b. 1935, New York City; Ph.D., Mathematical Psychology, University of Pennsylvania, 1962). After spending 1962–6 at the Center for Cognitive Studies at Harvard, Norman spent most of his academic career at the University of California at San Diego before joining Apple Corporation and then Hewlett Packard in the 1990s. While in the psychology department at UCSD he collaborated with Peter Lindsay on *Human Information Processing* (1972) and with David Rumelhart and the LNR research group on *Explorations in Cognition* (1975). He and Rumelhart established the Institute for Cognitive Science at UCSD, and when it became a department in 1988, Norman was its first chair. Initially known for his mathematical and computer simulation models of memory and cognition, Norman later focused on how cognition is distributed across people and the tools they construct and how tools and other artifacts can be better designed for use by humans. This work resulted in *The Design of Everyday Things* (1990) and *Things that Make us Smart* (1993).

Papert, Seymour (b. 1928, South Africa). Papert pursued mathematical research at Cambridge University from 1954 to 1958 and then worked with Jean Piaget at the University of Geneva from 1958 to 1963. Since 1963 he has been at MIT, where he became co-director of the Artificial Intelligence Lab and is now Lego Professor of Learning Research. At MIT he collaborated with Marvin Minsky on *Perceptrons* (1969), which temporarily helped reduce interest in neural networks and fostered increased pursuit of symbolic models. Much of his work has been directed to the use of computers and computer programming in education; he developed the Logo language, with which children can write programs to control the movements of mechanical turtles or perform other tasks. Among his books are *Mindstorms: Children, Computers and Powerful Ideas* (1980), *The Children's Machine: Rethinking School in the Age of the Computer* (1992), and *The Connected Family: Bridging the Digital Generation Gap* (1996).

Penfield, Wilder (b. 1891, Spokane, Washington; d. 1976, Montreal; M.D., Johns Hopkins University, 1918). Penfield taught at McGill from 1928 until 1960, where he founded the Montreal Neurological Institute, which he directed from 1934 to 1960. As a brain surgeon, Penfield specialized in the treatment of epileptics by removing portions of cortex. In order to determine which areas of cortex could not safely be removed, Penfield employed electrical stimulation, discovering that stimulation of certain parts of the cortex in human subjects can evoke vivid memories of past life experiences. Penfield presented the results of his research in a number of volumes: *Epilepsy and Cerebral Localization* (1941), *The Cerebral Cortex of Man* (1950), *The Excitable Cortex in Conscious Man* (1958), *Speech and Brain Mechanisms* (1959), and *The Mystery of the Mind* (1975).

Piaget, Jean (b. 1896, Neuchâtel, Switzerland; d. 1980, Geneva; Ph.D., Biology, University of Neuchâtel, 1918). Piaget spent most of his career at the University of Geneva, where he was professor of child psychology and the history of scientific thought from 1929 to 1971. He also founded and directed until his death the International Center of Genetic Epistemology in Geneva. Drawing upon detailed longitudinal observation of his own three children's early development, as well as analysis of older children's responses to ingenious tasks he devised, Piaget formulated a constructivist theory of development that emphasizes the interaction of nature and nurture via the child's own actions. He is best known for positing stages of development that the child traverses through processes of assimilation, accommodation, and equilibration, but these are just the most accessible aspects of an ambitious theory that combined mathematical, biological, and philosophical considerations. Among the most influential of his dozens of books are *The Language and Thought of the Child* (1926), *The Origins of Intelligence in Children* (1952), *The Growth of Logical Thinking from Childhood to Adolescence* (1958), and *The Principles of Genetic Epistemology* (1972).

Posner, Michael I. (b. 1936, Cincinnati, Ohio; Ph.D., Psychology, University of Michigan, 1962). Posner has been a professor of psychology at the University of Oregon since 1965. During the 1960s and 1970s, his work relied primarily on chronometric methods to identify component cognitive systems; the results are presented in *Chronometric Explorations of Mind* (1978). Subsequently, Posner began to draw more on neuroscience tools. From 1985 to 1988 he joined Marcus Raichle and Steven Petersen at Washington University, where they developed PET methods appropriate to imaging brain functions; many of these results are presented in *Images of Mind* (1994). Since 1988, Posner has been working on relating tools for identifying spatial locations involved in cognitive performance (PET and fMRI imaging) and tools for identifying the time of their involvement (ERP), with a particular focus on the plasticity of human attention and skill acquisition.

Pribram, Karl H. (b. 1919, Vienna; M.D., University of Chicago, 1941). Pribram spent much of his career at Yale and Stanford; he is now director of the Center for Brain Research at Radford University. His research efforts have been devoted to identifying nervous system substrates of primate behavior and determining the role of neocortex in emotion and cognition. Pribram collaborated with George Miller and Eugene Galanter on *Plans and the Structure of Behavior* (1960), which introduced TOTE units as an alternative to reflex arcs, so providing cognitive structure to cognition. He is best known for his proposed holographic modeling of memory. His books include *Languages of the Brain* (1971), *What Makes us Human* (1971), and *Brain and Perception: Holonomy and Structure in Figural Processing* (1991).

Putnam, Hilary (b. 1926, Chicago; Ph.D., Philosophy, University of California at Los Angeles, 1951). After teaching at Northwestern University, Princeton University, and MIT, Putnam moved to Harvard in 1976. Drawing on the theory of recursive functions and Turing machines, Putnam formulated a stance on the mind–body problem that he named "functionalism" in the 1950s, according to which it is organization, not material composition, that is relevant to the study of cognition. Putnam now believes that different computational (Turing machine) states can realize the same mental state, and thus that the mind cannot be identified with any particular computing machine. Putnam's work can be found in *Philosophical Papers* (3 vols, 1975–83), *Representation and Reality* (1988), and *Pragmatism: An Open Question* (1995).

Pylyshyn, Zenon W. (b. 1937, Montreal; Ph.D., Experimental Psychology, University of Saskatchewan, 1963). After spending much of his career at Western Ontario University, Pylyshyn moved to Rutgers University in 1994, where he is professor of psychology and director of the Center for Cognitive Science. One of Pylyshyn's major interests, developed in *Computation and Cognition: Toward a Foundation for Cognitive Science* (1984), has been in determining the nature of the human cognitive architecture, the level of organization at which the basis cognitive capacities are specified. Pylyshyn has also been a major critic of analog accounts of mental imagery and of connectionist models of cognition. In his research on visual attention, he has developed models of how the visual system tracks targets which it has already examined.

Quillian, M. Ross (b. 1931, Los Angeles; Ph.D., Psychology, Carnegie–Mellon, 1968). Quillian's career has been spent at the University of California at Irvine, in social science and then political science. His major contribution to cognitive science was the development of semantic networks, in which concepts are represented as nodes linked to other nodes.

Raichle, Marcus (b. 1937, Hoquiam, Washington; M.D., University of Washington, Seattle, 1964). Raichle has been at the Washington University School of Medicine since 1971, where he has been a leader in the development of brain imaging techniques and their application to cognitive function. He played a pivotal role in the development and application of positron emission tomography (PET), which was invented in Mike Ter-Pogossian's laboratory at Washington University in the early 1970s; he is now concentrating on fMRI. With Michael Posner, he published *Images of Mind* (1994), which provides an accessible overview of brain imaging research.

Rosch, Eleanor (Ph.D., Psychology, Harvard). Rosch, who has spent her career at the University of California at Berkeley, was a major instigator of contemporary research on concepts and categorization. She found that concepts such as *bird* exhibited a graded structure from more prototypical to less prototypical instances, and that a number of psychological measures, such as response time to evaluate whether an instance is a member of a category, correlated with the prototypicality of the instances. Much of this research is presented in *Cognition and Categorization* (1978). More recently, her interests have moved to Eastern psychologies and the psychology of religion, as manifested in *The Embodied Mind: Cognitive Science and Human Experience* (1991).

Rosenblatt, Frank (b. 1928, New Rochelle, New York; d. 1971, Easton, Maryland; Ph.D., Psychology, Cornell, 1956). Rosenblatt began his career at the Cornell Aeronautical Laboratory in Buffalo before returning to the main campus of Cornell in Ithaca as director of the Cognitive Systems Research Program. His main contribution was the development of a perceptron, a neural network system that learned to identify such objects as letters. He presented his work on perceptrons in his book *Principles of Neurodynamics* (1962). Rosenblatt also explored the possibility of transferring learning behaviour from trained to naive rats by injecting brain extracts from the former into the latter.

Rumelhart, David (b. 1942, Wessington Springs, South Dakota; Ph.D., Mathematical Psychology, Stanford, 1967). Rumelhart was in psychology at the University of California at San Diego from 1967 to 1987, when he moved to Stanford. While at UCSD he began a fruitful collaboration with Donald Norman and the LNR research group, resulting in *Explorations in Cognition* (1975). A subsequent collaboration with James McClelland and the PDP research group produced *Parallel Distributed Processing: Explorations in the Microstructure of Cognition* (2 vols, 1986), which played a major role in reintroducing connectionist models into cognitive science. With Hinton and Williams, Rumelhart discovered the technique of back-propagation for learning in multilayered nets. Since moving to Stanford, Rumelhart has concentrated on the development of neurally inspired computational architectures.

Schank, Roger (b. 1946, New York City; Ph.D., Linguistics, University of Texas, 1969). After appointments at Stanford and Yale, Schank is now director of the Institute for the Learning Sciences (ILS) at Northwestern University and president of the Learning Sciences Corporation. Beginning his career by building AI systems for natural language processing, Schank then collaborated with Robert Abelson on *Scripts, Plans, Goals, and Understanding* (1977), in which they developed the idea of scripts as higher-level knowledge structures. Schank extended this research into memory structures in *Dynamic Memory* (1982). His more recent work stresses the value of learning from experts, developing skills rather than perfecting routines, and applying the benefits of "just-in-time" training; this research is presented in *Engines for Education* (1995).

Searle, John R. (b. 1932, Denver; D.Phil., Philosophy, Oxford, 1959). Since 1959, Searle has been in the Department of Philosophy at the University of California at Berkeley. Much of Searle's early research was in the philosophy of language and focused on speech acts, culminating in *Speech Acts* (1969). More recently he has become a critic of what he terms "strong artificial intelligence," according to which all there is to having a mind is implementing the right computer program. His major argument, the Chinese room argument, was first presented in a 1980 *Behavioral and Brain Sciences* article, "Minds, Brains, and Programs." Searle's most recent books include *Intentionality: An Essay in the Philosophy of Mind* (1983) and *The Rediscovery of the Mind* (1992).

Selfridge, Oliver G. From 1951 to 1975 Selfridge was with the Lincoln Laboratory at MIT. He then moved to Bolt Beranek and Newman as senior scientist; in 1983 he moved again to GTE Laboratories as chief scientist in the Computer and Information Systems Laboratory, from which he retired in 1993. Selfridge was one of the earliest investigators to explore neural networks, and in the 1950s developed the Pandemonium model of perceptual recognition, in which a number of simple agents each tried to recognize its own target input and then competed to activate consistent decision units. Much of his subsequent research focused on machine learning.

Selz, Otto (b. 1881, Munich; d. 1943, Auschwitz; Ph.D., Psychology, University of Munich, 1909). Selz worked at the University of Bonn from 1912 to 1923 and at Handelshochschule in Mannheim from 1923 to 1933. A member of the Würzburg school of psychology, Selz studied the organized thinking process, stressing the notion that active processing takes place in the mind, and he called for a psychology of thinking that would be concerned with processes rather than content. His work was a major influence on the development of Herbert Simon's approach to studying human problem solving.

Shannon, Claude E. (b. 1916, Gaylord, Michigan; Ph.D., Mathematics, MIT, 1940). Shannon joined the staff of Bell Laboratories in 1941 to work on the problem of how to transmit information most efficiently; in the course of his research he developed a method for measuring information and became one of the pioneers of information theory, a full statement of which appeared in *The Mathematical Theory of Communication* (1949). From 1958 to 1980, Shannon taught at MIT.

Shepard, Roger N. (b. 1929, Palo Alto, California; Ph.D., Psychology, Yale, 1955). After stints with ATT Bell Laboratories and Harvard, Shepard moved to Stanford in 1968 and has remained there since. He is known both for his mathematical models of cognitive processes, especially his work on multidimensional scaling, and for his research on mental rotation of geometric figures. His books include *Mental Images and their Transformation* (1982) and *Mind Sights* (1990).

Sherington, Sir Charles Scott (b. 1857, London; d. 1952, Eastbourne, UK; M.B., Cambridge, 1885). Sherington held numerous positions in pathology and physiology, including at the University of Liverpool from 1895 to 1913 and Oxford from 1913 until 1935. He received the Nobel Prize in physiology and medicine in 1932. In the course of his research on reflex arcs he engaged in detailed neuroanatomy, which resulted in his discovery of gaps between neurons which he termed *synapses*. He studied reflexes as functional units operating not in isolation, but under control of higher levels of neural activity. His views on how reflexes were integrated into behavior were presented in *The Integrative Action of the Nervous System* (1906). His other books include *The Brain and its Mechanism* (1933) and *Man on his Nature* (1951).

Simon, Herbert A. (b. 1916, Milwaukee, Wisconsin; Ph.D., Political Science, University of Chicago, 1943). Simon's initial research focused on decision making in organizations, work for which he received the Nobel Prize in economics in 1978. Simon's interest gradually turned to the psychology of problem solving and the use of computers to model this process. In the 1950s he began a collaboration with Allen Newell, with whom he developed the production system architecture that is widely used in AI and produced some of the earliest computational models of human reasoning. This work culminated in *Human Problem Solving* (1972). With Newell, he also proposed the Physical Symbol System hypothesis. In his more recent work, Simon has often focused on modeling scientific discovery. Among his many books are *The Sciences of the Artificial* (1969), *Models of Discovery* (1977), *Models of Thought* (2 vols, 1979, 1989), and *Scientific Discovery* (1987).

Smith, Edward E. (b. 1940, Brooklyn, New York; Ph.D., Experimental Psychology, University of Michigan, 1966). After holding positions at the University of Wisconsin, Stanford, and Bolt Beranek and Newman, Smith moved to the University of Michigan in 1986, where he directs the Cognitive Science and the Cognitive Neuroscience Program. He has worked in the areas of perception, memory, and text processing; some of his more important contributions have been in the study of categorization and reasoning. In 1981 he and Douglas Medin published *Categories and Concepts*. His most recent work employs neuroimaging to study visual-spatial working memory, problem solving, and age-related changes in working memory.

Sperry, Roger W. (b. 1913, Hartford, Connecticut; d. 1994; Ph.D., Zoology, University of Chicago, 1941). Sperry, who spent most of his career at the California Institute of Technology, made extensive studies of patients in whom the two hemispheres of the cortex were separated in the course of treatment for epilepsy. He received the Nobel Prize in 1981 for this research on hemispheric interactions and specialization. His books include *Problems Outstanding in the Evolution of Brain Function* (1964) and *Science and Moral Priority* (1983).

Squire, Larry R. (b. 1941, Cherokee, Iowa; Ph.D., Psychology, MIT, 1968). Squire has spent his career at the University of California at San Diego, where his research has addressed the anatomy, physiology, and function of memory. His research examines amnesias as well as nonconscious learning and memory. Some of this research is reported in *Memory and Brain* (1987).

Sternberg, Saul H. (b. 1933, New York City; Ph.D., Social Relations, Harvard, 1960). After a short period at the University of Pennsylvania, Sternberg moved to Bell Laboratories, where he chaired the Human Information-Processing Research Department from 1970 to 1985. He then returned to the University of Pennsylvania. Among his accomplishments are the development of an additive factors approach to analyzing reaction time data and a model of exhaustive search for memory scanning tasks. His research has focused on such cognitive phenomena as visual processing, motor control, and perception of time.

Stevens, Stanley Smith (b. 1906, Ogden, Utah; d. 1973, Vail, Colorado; Ph.D., Psychology, Harvard, 1933). Stevens's career was spent at Harvard, where he established a psychophysics laboratory in which many of the founders of cognitive psychology began their careers. His major interest was in measurement and psychological scaling; among his contributions are the method of magnitude estimation and Stevens' law. In 1951 Stevens edited the *Handbook of Experimental Psychology*, which was rewritten in 1988 as *Stevens' Handbook of Experimental Psychology*.

Titchener, Edward Bradford (b. 1867, Chichester, England; d. 1927, Ithaca, New York; Ph.D., Psychology, University of Leipzig, 1892). A student of Wundt's who emphasized the introspectionist and structuralist aspects of his mentor's work, Titchener spent his career at Cornell University. His most important book was *Experimental Psychology* (4 vols, 1901–5).

Tolman, Edward (b. 1886, West Newton, Massachusetts; d. 1959, Berkeley; Ph.D., Psychology, Harvard, 1915). Tolman spent most of his career at the University of California at Berkeley, where he established a rat laboratory. To account for his rats' maze navigation and other abilities, Tolman posited such cognitive explanations as mental maps. Departing from standard behaviorism, he called his research program "purposive behaviorism." His major work is *Purposive Behavior in Animals and Men* (1932).

Tulving, Endel (b. 1927, Estonia; Ph.D., Experimental Psychology, Harvard, 1957). Currently, Tulving is Tanenbaum Chair in Cognitive Neuroscience at the Rotman

Research Institute, University of Toronto; since 1996 he has also held a permanent visiting position at Washington University in St Louis. Throughout his professional career, Tulving has studied human memory. Among the important concepts and distinctions he has introduced are cue-dependent forgetting, encoding specificity, episodic versus semantic memory, and availability versus accessibility of stored information. Most recently he has developed the HERA model of encoding/retrieval asymmetry of the frontal lobes. He is the author of *Elements of Episodic Memory* (1983).

Turing, Alan (b. 1912, London; d. 1953, Wilmslow, Cheshire, UK; Ph.D., Mathematics, Princeton, 1938). Turing's academic career was spent at Cambridge and Manchester; during World War II he played a major role in British attempts at Bletchley Park to break German cryptographic codes. This research played a major role in the development of his basic ideas of the computer. In addition to the development of physical machines, Turing's contributions include the Turing machine (a framework for computing any decidable function) and the Turing test (for evaluating whether machines are thinking).

Tversky, Amos (b. 1937, Haifa, Israel; d. 1996, Stanford, California; Ph.D., Psychology, University of Michigan, 1965). Tversky spent much of his career at Stanford University, where his research focused on similarity and scaling (including nearest neighbor analysis) and on human judgment and decision making. In research done collaboratively with Daniel Kahneman, Tversky marshaled evidence that human reasoning does not follow strict normative principles. Rather, people rely on heuristics that simplify the problem, but sometimes lead to error. Many of the major results of this research were published in *Judgment under Uncertainty: Heuristics and Biases* (1982).

von Helmholtz, Hermann (b. 1821, Potsdam, Germany; d. 1894, Berlin; M.D., Friedrich Wilhelm Medical Institute in Berlin, 1842). Helmholtz, who held positions at the universities of Königsberg, Bonn, and Heidelberg, was a major contributor to both physics and physiology. In *Die Lehre von den Tonempfindungen* (1863) and *Handbuch der physiologischen Optik* (1867), he defined the problems for the experimental psychology of visual and auditory perception for decades to follow. He articulated a conception of perception according to which it requires active, unconscious, automatic, logical inference on the part of the perceiver.

von Neumann, John (b. 1903, Budapest; d. 1957, Washington, DC; Ph.D., Mathematics, University of Budapest, 1926). After faculty appointments at the Universities of Berlin, Hamburg and Princeton, von Neumann spent most of his career at the Institute for Advanced Study at Princeton. He made many significant contributions to pure mathematics, quantum theory, the theory of electronic computing devices, and the theory of games. He played a major role in the development of computers and proposed the basic architecture of modern computers which bears his name. His research on the theory of games, conducted in collaboration with Oskar Morgenstern, focused on the study of risky decision making and resulted in *Theory of Games and Economic Behavior* (1944).

Vygotsky, Lev Semyonovich (b. 1896, Orsha, Belorus; d. 1934, Moscow). Much of Vygotsky's career was spent at Moscow University's Institute of Psychology. His approach to mind emphasized the role of social context in the development of individual minds. Well after his death, his work came to the attention of English speakers after the publication of *Thought and Language* (1962), and it has had increasing influence in the 1990s.

Wernicke, Carl (b. 1848, Tarnowitz, Upper Silesia; d. 1905, Dörreberg-im-Geratal; M.D., University of Breslau, 1870). Wernicke practiced medicine in Berlin from 1875 to 1885 and then taught psychiatry at Breslau from 1885 to 1904. He is most famous for discovering a second form of aphasia in addition to that studied by Broca. In Wernicke's aphasia, the ability to understand the meaning of spoken language is lost, while self-generated speech and the understanding of written language are preserved. Wernicke attributed the cause of this deficit to an area of the cerebral cortex that comprises parts of the first and second temporal gyri and the supramarginal angular gyrus, which has come to be known as "Wernicke's area."

Wertheimer, Max (b. 1880, Prague; d. 1943, New Rochelle, New York; Ph.D., Psychology, University of Würzburg, 1904). Wertheimer held positions in Frankfurt and Berlin before moving to the New School for Social Research in 1933, where he remained for ten years. In 1912, Wertheimer proposed an analysis of an apparent motion experiment according to which the perceived movement was a new phenomenon, a whole or *Gestalt*, that was not reducible to the elements that gave rise to it. This marked the beginning of the Gestalt school of psychology. He argued that it was the prior perception of the whole that determined how the parts would be perceived, not the other way around. Together with Koffka and Köhler, Wertheimer established a new journal, *Psychologische Forschung*, devoted to Gestalt ideas. Wertheimer's later attempts to extend the idea of *Gestalten* to creative thinking were published posthumously in *Productive Thinking* (1945).

Wiener, Norbert (b. 1894, Columbia, Missouri; d. 1964, Stockholm; Ph.D., Philosophy, Harvard, 1913). Wiener was at MIT from 1919 to 1960. He was a major contributor to the cybernetics movement, even naming it in his 1948 book *Cybernetics*. He defined "cybernetics" as a discipline concerned with the comparative study of control mechanisms in the nervous system and computers. Among Wiener's other books are *The Human Use of Human Beings* (1950) and *Cybernetics of the Nervous System* (1965).

Wiesel, Torsten (b. 1924, Uppsala; M.D., Karolinska Institute, 1954). Wiesel was at Harvard, where he often collaborated with David Hubel, from 1959 until 1983. Together they showed that individual neurons in visual cortex responded optimally to specific stimuli: a line or edge at a particular orientation and, for the lowest-level neurons, particular locations. They also found that cells are organized in ocular dominance columns; series of columns alternate between responsiveness to the right and left eye. Other research focused on binocular vision and the importance of early visual stimulation for normal development. They shared the Nobel Prize for medicine and physiology in 1981. In 1983 Wiesel moved to Rockefeller University.

Winograd, Terry (b. 1946, Maryland; Ph.D., Applied Mathematics, MIT, 1970). Winograd has held positions in computer science and linguistics at Stanford University since 1974. In his early research he developed SHRDLU, a program that manipulates simulated blocks in a simulated space, and answers questions about them. Later Winograd repudiated his approach. Following in the tradition of Heidegger and Maturana, Winograd rejects the attempt to understand human intelligence in terms of the manipulation of representations. Among his most important publications are

Understanding Natural Language (1972), *Language as a Cognitive Process*, Vol. 1: *Syntax* (1983), and *Understanding Computers and Cognition: A New Foundation for Design* (1987).

Winston, Patrick H. (b. 1943, Peoria, Illinois; Ph.D., Computer Science, MIT, 1970). Winston's career has been spent at the Artificial Intelligence Lab at MIT, where he is currently professor and director. His research focuses on how vision, language, and motor faculties interact to account for intelligence. He is the author of *Artificial Intelligence* (3 vols, 1977, 1984, 1992), *LISP* (1984), and *On to C++* (1994).

Wittgenstein, Ludwig (b. 1889, Vienna; d. 1951, Cambridge). First interested in mechanical engineering, Wittgenstein turned to mathematics and the foundations of mathematics. In the *Tractatus Logico-Philosophicus* (1922) he developed an analysis of linguistic meaning built on logic and an analysis of picturing. Wittgenstein then left philosophy to become a schoolteacher in Austria; he later returned to teach at Cambridge from 1929 to 1947. In *Philosophical Investigations* (1953) Wittgenstein challenged his earlier account of language and traditional approaches to philosophy generally. He rejected the view that to possess a concept, or understand a word, is to possess explicit knowledge of necessary and sufficient conditions. According to Wittgenstein, what is involved is an implicit, inarticulable knowledge of the "family resemblances" between situations and objects.

Wundt, Wilhelm (b. 1832, near Mannheim, Germany; d. 1920; M.D., University of Heidelberg, 1855). Wundt spent the early part of his career at Heidelberg before being appointed professor of scientific philosophy at the University of Leipzig in 1875, where he established the first institute for experimental psychology and created a journal devoted to experimental psychology, *Philosophische Studien*. Wundt was a major proponent of a scientific approach to psychology. In his *Principles of Physiological Psychology* (1874) he argued that conscious states could be scientifically studied through the systematic manipulation of antecedent variables, and analyzed by carefully controlled techniques of introspection. Wundt also emphasized chronometric studies of mental processes; with respect to "higher" cognitive phenomena like linguistic behavior, Wundt favored a nonexperimental, ethnographic methodology as expounded in his *Volkerpsychologie* ("ethnic psychology," or "group psychology"), a ten-volume work published in 1920.

Zadeh, Lotfi A. (b. 1921, Baku, Azerbaijan; Ph.D., Electrical Engineering, Columbia, 1949). Zadeh taught at Columbia University before moving in 1959 to the Department of Electrical Engineering and Computer Science at the University of California at Berkeley. Dissatisfied with classical logic as a tool for modeling human reasoning, Zadeh developed the formalization of "fuzzy logic," starting with his 1965 paper "Fuzzy Sets." He is the author of *Fuzzy Logic for the Management of Uncertainty* (1992).

Zurif, Edgar (b. 1940, Montreal; Ph.D., Psychology, University of Waterloo, 1967). Zurif has spent his career at the School of Medicine of Boston University and in linguistics and psychology at Brandeis University. His work has focused on neurolinguistics; he is particularly known for reinterpreting the deficits in Broca's versus Wernicke's aphasia in terms of syntax versus semantics rather than production versus comprehension.

Author index

Subject index